Chilton Book Company
REPAIR MANUAL

CHEVROLET FULL SIZE 1968-92

Covers all U.S. and Canadian models

President, Chilton Enterprises	David S. Loewith
Senior Vice President	Ronald A. Hoxter
Publisher and Editor-In-Chief	Kerry A. Freeman, S.A.E.
Managing Editors	Peter M. Conti, Jr. □ W. Calvin Settle, Jr., S.A.E.
Assistant Managing Editor	Nick D'Andrea
Senior Editors	Debra Gaffney □ Ken Grabowski, A.S.E., S.A.E.
	Michael L. Grady □ Richard J. Rivele, S.A.E.
	Richard T. Smith □ Jim Taylor
	Ron Webb
Director of Manufacturing	Mike D'Imperio
Editor	Michael W. Parks, S.A.E.

ONE OF THE *DIVERSIFIED PUBLISHING COMPANIES,*
A PART OF *CAPITAL CITIES/ABC, INC.*

SAFETY NOTICE

Proper service and repair procedures are vital to the safe, reliable operation of all motor vehicles, as well as the safety of those performing repairs. This book outlines procedures for servicing and repairing vehicles using safe effective methods. The procedures contain many NOTES, CAUTIONS and WARNINGS which should be followed along with standard safety procedures to eliminate the possibility of personal injury or improper service which could damage the vehicle or compromise its safety.

It is important to note that repair procedures and techniques, tools and parts for servicing motor vehicles, as well as the skill and experience of the individual performing the work vary widely. It is not possible to anticipate all of the conceivable ways or conditions under which vehicles may be serviced, or to provide cautions as to all of the possible hazards that may result. Standard and accepted safety precautions and equipment should be used during cutting, grinding, chiseling, prying, or any other process that can cause material removal or projectiles.

Some procedures require the use of tools specially designed for a specific purpose. Before substituting another tool or procedure, you must be completely satisfied that neither your personal safety, nor the performance of the vehicle will be endangered.

Although the information in this guide is based on industry sources and is as complete as possible at the time of publication, the possibility exists that the manufacturer made later changes which could not be included here. While striving for total accuracy, Chilton Book Company cannot assume responsibilty for any errors, changes, or omissions that may occur in the compilation of this data.

PART NUMBERS

Part numbers listed in the reference are not recommendations by Chilton for any product by brand name. They are references that can be used with interchange manuals and aftermarket supplier catalogs to locate each brand supplier's discrete part number.

SPECIAL TOOLS

Special tools are recommended by the vehicle manufacturer to perform their specific job. Use has been kept to a minimum, but where absolutely necessary, they are referred to in the text by the part number of the tool manufacturer. These tools can be purchased, under the appropiate part number, from the Service Tool Division, Kent-Moore Corporation, 1501 South Jackson Street, Jackson, MI 49203 or an equivalent tool can be purchased locally from a tool supplier or parts outlet. Before substituting any tool for the one recommended, read the SAFETY NOTICE at the top of this page.

ACKNOWLEDGEMENTS

Chilton Book Company expresses appreciation to Chevrolet Motor Division, General Motors Corporation their generous assistance

Copyright© 1992 by Chilton Book Company
All Rights Reserved
Published in Radnor, Pennsylvania 19089 by Chilton Book Company
ONE OF THE **DIVERSIFIED PUBLISHING COMPANIES**, A PART OF **CAPITAL CITIES/ABC, INC.**

Manufactured in the United States of America
34567890   1098765

Chilton's Repair Manual: Chevrolet Full Size 1968–92
ISBN 0–8019–8250–2 pbk.
Library of Congress Catalog Card No. 91–058813

CONTENTS

GENERAL INFORMATION and MAINTENANCE

- **1** How to use this book
- **2** Tools and Equipment
- **15** Routine Maintenance

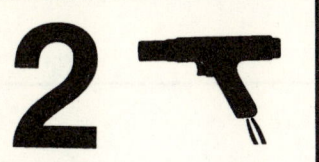

ENGINE PERFORMANCE and TUNE-UP

- **58** Tune-Up Performance
- **59** Tune-Up Specifications

ENGINE and ENGINE OVERHAUL

- **93** Engine Electrical System
- **113** Engine Service
- **118** Engine Specifications

EMISSION CONTROLS

- **192** Gasoline Engine Emissions
- **229** Diesel Engine Emission

FUEL SYSTEM

- **248** Carbureted Fuel System
- **286** Gasoline Fuel Injection System
- **292** Dielsel Fuel System

CHASSIS ELECTRICAL

- **312** Heating and Air Conditioning
- **320** Radio
- **322** Windshield Wipers
- **326** Instruments and Switches
- **334** Lighting
- **337** Circuit Protection

7 DRIVE TRAIN

340 Manual Transmission
344 Clutch
347 Automatic Transmission
364 Driveline

8 SUSPENSION and STEERING

374 Front Suspension
384 Rear Suspension
389 Steering

9 BRAKES

398 Brake Systems
407 Front Drum Brakes
408 Front Disc Brakes
414 Rear Drum Brakes
418 Anti-Lock Brake System

10 BODY

426 Exterior
438 Interior

11 MECHANIC'S DATA

446 Mechanic's Data
448 Glossary
454 Index

301 Chilton's Fuel Economy and Tune-Up Tips

429 Chilton's Body Repair Tips

General Information and Maintenance

HOW TO USE THIS BOOK

Chilton's Repair Manual for Chevrolet cars is intended to help you learn more about the inner workings of your car and save you money on its upkeep and operation.

The first two chapters will be the most used, since they contain maintenance and tune-up information and procedures. Studies have shown that a properly tuned and maintained car can get at least 10% better gas mileage than an out-of-tune car. The other chapters deal with the more complex systems of your car. Operating systems from engine through brakes are covered to the extent that the average do-it-yourselfer becomes mechanically involved. This book will not explain such things as rebuilding the differential for the simple reason that the expertise required and the investment in special tools make this task uneconomical. It will give you detailed instructions to help you change your own brake pads and shoes, replace spark plugs, and do many more jobs that will save you money, give you personal satisfaction, and help you avoid expensive problems.

A secondary purpose of this book is a reference for owners who want to understand their car and/or their mechanics better. In this case, no tools at all are required.

Before removing any bolts, read through the entire procedure. This will give you the overall view of what tools and supplies will be required. There is nothing more frustrating than having to walk to the bus stop on Monday morning because you were short one bolt on Sunday afternoon. So read ahead and plan ahead. Each operation should be approached logically and all procedures thoroughly understood before attempting any work.

All chapters contain adjustments, maintenance, removal and installation procedures, and repair or overhaul procedures. When repair is not considered practical, we tell you how to remove the part and then how to install the new or rebuilt replacement. In this way, you at least save the labor costs. Backyard repair of such components as the alternator is just not practical.

Two basic mechanic's rules should be mentioned here. One, whenever the left side of the car or engine is referred to, it is meant to specify the driver's side of the car. Conversely, the right side of the car means the passenger's side. Secondly, most screws and bolts are removed by turning counterclockwise, and tightened by turning clockwise.

Safety is always the most important rule. Constantly be aware of the dangers involved in working on an automobile and take the proper precautions. See the section in this chapter Servicing Your Vehicle Safely and the SAFETY NOTICE on the acknowledgement page.

Pay attention to the instructions provided. There are 3 common mistakes in mechanical work:

1. Incorrect order of assembly, disassembly or adjustment. When taking something apart or putting it together, doing things in the wrong order usually justs cost you extra time; however, it CAN break something. Read the entire procedure before beginning disassembly. Do everything in the order in which the instructions say you should do it, even if you can't immediately see a reason for it. When you're taking apart something that is very intricate (for example, a carburetor), you might want to draw a picture of how it looks when assembled at one point in order to make sure you get everything back in its proper position. (We will supply exploded view whenever possible). When making adjustments, especially tune-up adjustments, do them in order; often, one adjustment affects another, and you cannot expect even satisfactory results unless each adjust-

ns made only when it cannot be changed by any order.

2. Overtorquing (or undertorquing). While it is more common for over-torquing to cause damage, undertorquing can cause a fastener to vibrate loose causing serious damage. Especially when dealing with aluminum parts, pay attention to torque specifications and utilize a torque wrench in assembly. If a torque figure is not available, remember that if you are using the right tool to do the job, you will probably not have to strain yourself to get a fastener tight enough. The pitch of most threads is so slight that the tension you put on the wrench will be multiplied many, many times in actual force on what you are tightening. A good example of how critical torque is can be seen in the case of spark plug installation, especially where you are putting the plug into an aluminum cylinder head. Too little torque can fail to crush the gasket, causing leakage of combustion gases and consequent overheating of the plug and engine parts. Too much torque can damage the threads, or distort the plug which changes the spark gap.

There are many commercial products available for ensuring that fasteners won't come loose, even if they are not torqued just right (a very common brand is Loctite®). If you're worried about getting something together tight enough to hold, but loose enough to avoid mechanical damage during assembly, one of these products might offer substantial insurance. Read the label on the package and make sure the products is compatible with the materials, fluids, etc. involved before choosing one.

3. Crossthreading. This occurs when a part such as a bolt is screwed into a nut or casting at the wrong angle and forced. Crossthreading is more likely to occur if access is difficult. It helps to clean and lubricate fasteners, and to start threading with the part to be installed going straight in. Then, start the bolt, spark plug, etc. with your fingers. If you encounter resistance, unscrew the part and start over again at a different angle until it can be inserted and turned several turns without much effort. Keep in mind that many parts, especially spark plugs, used tapered threads so that gentle turning will automatically bring the part you're treading to the proper angle if you don't force it or resist a change in angle. Don't put a wrench on the part until its's been turned a couple of turns by hand. If you suddenly encounter resistance, and the part has not seated fully, don't force it. Pull it back out and make sure it's clean and threading properly.

Always take your time and be patient; once you have some experience, working on your car will become an enjoyable hobby.

TOOLS AND EQUIPMENT

Naturally, without the proper tools and equipment it is impossible to properly service your car. It would be impossible to catalog each tool that you would need to perform each or any operation in this book. It would also be unwise for the amateur to rush out and buy an expensive set of tool on the theory that he may need on or more of them at sometime.

The best approach is to proceed slowly gathering together a good quality set of those tools that are used most frequently. Don't be misled by the low cost of bargain tools. It is far better to spend a little more for better quality. Forged wrenches, 10 or 12 point sockets and fine tooth ratchets are by far preferable to their less expensive counterparts. As any good mechanic can tell you, there are few worse experiences than trying to work on a vehicle with bad tools. Your monetary savings will be far outweighed by frustration and mangled knuckles.

Begin accumulating those tools that are used most frequently; those associated with routine maintenance and tune-up.

In addition to the normal assortment of screwdrivers and pliers you should have the following tools for routine maintenance jobs:

1. SAE (or Metric) or SAE/Metric wrenches-sockets and combination open end-box end wrenches in sizes from $1/8''$ (3 mm) to $3/4''$ (19 mm) and a spark plug socket 13/16" or $5/8''$ depending on plug type).

If possible, buy various length socket drive extensions. One break in this department is that the metric sockets available in the U.S. will all fit the ratchet handles and extensions you may already have ($1/4''$, $3/8''$, and $1/2''$ drive).

2. Jackstands for support.
3. Oil filter wrench.
4. Oil filler spout for pouring oil.
5. Grease gun for chassis lubrication.
6. Hydrometer for checking the battery.
7. A container for draining oil.
8. Many rags for wiping up the inevitable mess.

In addition to the above items there are several others that are not absolutely necessary, but handy to have around. these include oil dry, a transmission funnel and the usual supply of lubricants, antifreeze and fluids, although these can be purchased as needed. This is a basic list for routine maintenance, but only your personal needs and desire can accurately determine you list of tools.

The second list of tools is for tune-ups. While the tools involved here are slightly more sophisticated, they need not be outrageously expensive. There are several inexpensive tach/dwell

GENERAL INFORMATION AND MAINTENANCE

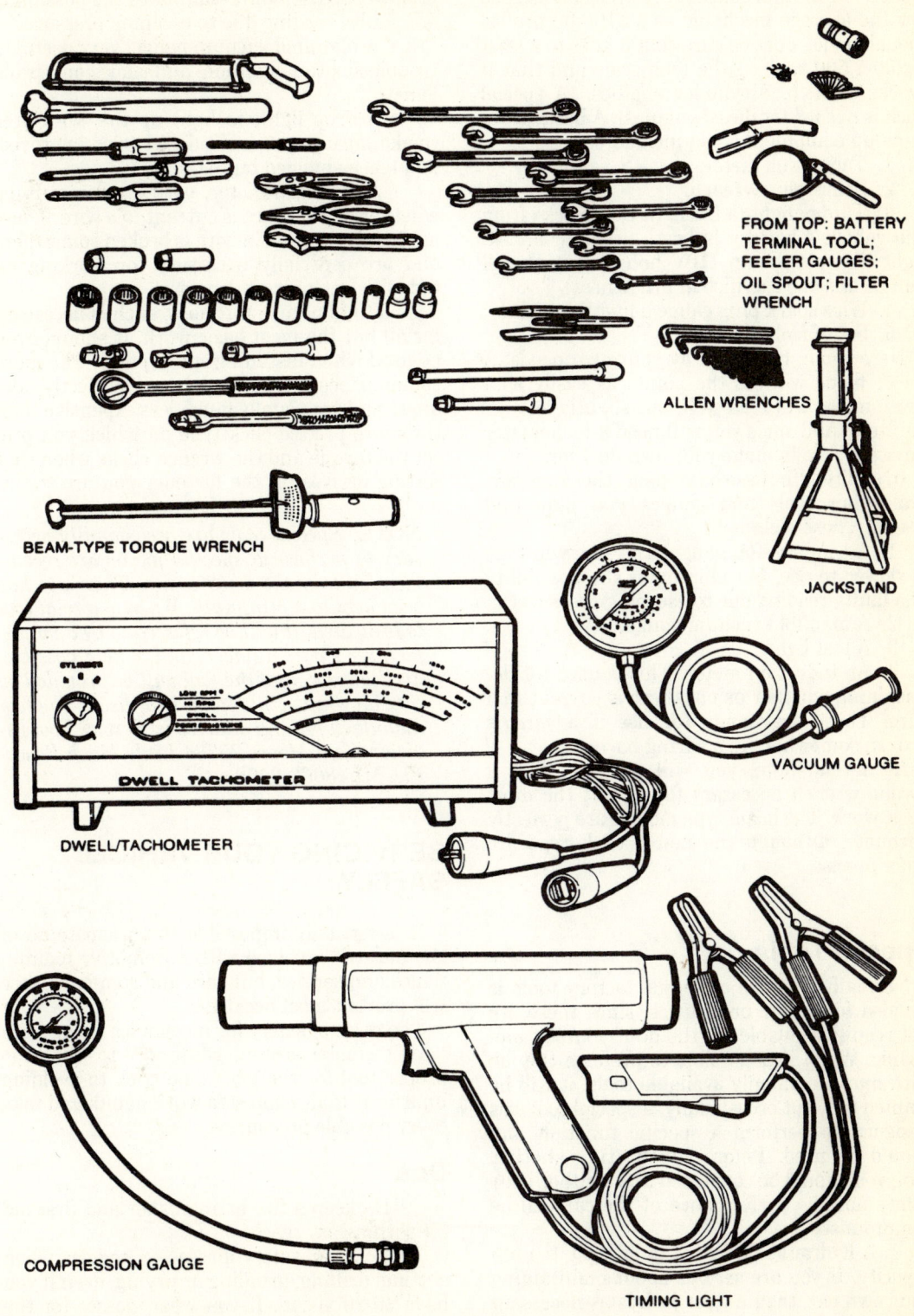

You need only a basic assortment of hand tools and test instruments for most maintenance and repair jobs

GENERAL INFORMATION AND MAINTENANCE

meters on the market that are every bit as good for the average mechanic as a $100.00 professional model. Just be sure that it goes to a least 1,200-1,500 rpm on the tach scale and that it works on 4, 6, 8 cylinder engines. (A special tach is needed for diesel engines). A basic list of tune-up equipment could include:

1. Tach/dwell meter.
2. Spark plug wrench.
3. Timing light (a DC light that works from the vehicle's battery is best, although an AC light that plugs into 110V house current will suffice at some sacrifice in brightness).
4. Wire spark plug gauge/adjusting tools.
5. Set of feeler blades.

Here again, be guided by your own needs. A feeler blade will set the points as easily as a dwell meter will read well, but slightly less accurately. And since you will need a tachometer anyway... well, make your own decision.

In addition to these basic tools, there are several other tools and gauges you may find useful. These include:

1. A compression gauge. The screw-in type is slower to use, but eliminates the possibility of a faulty reading due to escaping pressure.
2. A manifold vacuum gauge.
3. A test light.
4. An induction meter. This is used for determining whether or not there is current in a wire. These are handy for use if a wire is broken somewhere in a wiring harness.

As a final note, you will probably find a torque wrench necessary for all but the most basic work. The beam type models are perfectly adequate, although the newer click type are more precise.

Special Tools

Normally, the use of special factory tools is avoided for repair procedures, since these are not readily available for the do-it-yourself mechanic. When it is possible to preform the job with more commonly available tools, it will be pointed out, but occasionally, a special tool was designed to perform a specific function and should be used. Before substituting another tool, you should be convinced that neither your safety nor the performance of the car will be compromised.

• A hydraulic floor jack of at least 1 1/2 ton capacity. If you are serious about maintaining your own car, then a floor jack is as necessary as a spark plug socket. The greatly increased utility, strength, and safety of a hydraulic floor jack makes it pay for itself many times over through the years.

• A compression gauge. The screw-in type is slower to use but it eliminates the possibility of a faulty reading due to escaping pressure.

• A manifold vacuum gauge, very useful in troubleshooting ignition and emissions problems.

• A drop light, to light up the work area (make sure yours is Underwriter's approved, and has a shielded bulb).

• A volt/ohm meter, used for determining whether or not there is current in a wire. These are handy for use if a wire is broken somewhere and are especially necessary for working on today's electronics-laden cars.

As a final note, a torque wrench is necessary for all but the most basic work. It should even be used when installing spark plugs. The more common beam-type models are perfectly adequate and are usually much less expensive than the more precise click type on which you preset the torque and the wrench clicks when that setting arrives on the fastener you are torquing).

NOTE: *Special tools are occasionally necessary to perform a specific job or are recommended to make a job easier. Their use has been kept to a minimum. When a special tool is indicated, it will be referred to by a manufacturer's part number. and, where possible, an illustration of the tool will be provided so that an equivalent tool may be used. The tool manufacturer and address is: Service Tool Division Kent-Moore 29784 Little Mack Roseville, MI 48066-2298*

SERVICING YOUR VEHICLE SAFELY

It is virtually impossible to anticipate all of the hazards involved with automotive maintenance and service, but care and common sense will prevent most accidents.

The rules of safety for mechanics range from "don't smoke around gasoline," to "use the proper tool for the job." The trick to avoiding injuries is to develop safe work habits and take every possible precaution.

Dos

• Do keep a fire extinguisher and first aid kit within easy reach.

• Do wear safety glasses or goggles when cutting, drilling, grinding or prying, even if you have 20-20 vision. If you wear glasses for the sake of vision, they should be made of hardened glass that can serve also as safety glasses, or wear safety goggles over your regular glasses.

• Do shield your eyes whenever you work around the battery. Batteries contain sulphuric

acid. In case of contact with the eyes or skin, flush the area with water or a mixture of water and baking soda and get medical attention immediately.
• Do use safety stands for any under car service. Jacks are for raising cars; safety stands are for making sure the car stays raised until you want it to come down. Whenever the car is raised, block the wheels remaining on the ground and set the parking brake.
• Do use adequate ventilation when working with any chemicals or hazardous materials. Like carbon monoxide, the asbestos dust resulting from brake lining wear can be poisonous in sufficient quantities.
• Do disconnect the negative battery cable when working on the electrical system. The secondary ignition system can contain up to 40,000 volts.
• Do follow manufacturer's directions whenever working with potentially hazardous materials. Both brake fluid and antifreeze are poisonous if taken internally.
• Do properly maintain your tools. Loose hammerheads, mushroomed punches and chisels, frayed or poorly grounded electrical cords, excessively worn screwdrivers, spread wrenches (open end), cracked sockets, slipping ratchets, or faulty droplight sockets can cause accidents.
• Likewise, keep your tools clean; a greasy wrench can slip off a bolt head, ruining the bolt and often ruining your knuckles in the process.
• Do use the proper size and type of tool for the job being done.
• Do when possible, pull on a wrench handle rather than push on it, and adjust your stance to prevent a fall.
• Do be sure that adjustable wrenches are tightly closed on the nut or bolt and pulled so that the face is on the side of the fixed jaw.
• Do select a wrench or socket that fits the nut or bolt. The wrench or socket should sit straight, not cocked.

Always use jackstands when supporting the car, never use cylinder blocks or tire changing jacks

• Do strike squarely with a hammer; avoid glancing blows.
• Do set the parking brake and block the drive wheels if the work requires the engine running.

Don'ts

• Don't run the engine in a garage or anywhere else without proper ventilation-EVER! Carbon monoxide is poisonous; it takes a long time to leave the human body and you can build up a deadly supply of it in your system by simply breathing in a little every day. You may not realize you are slowly poisoning yourself. Always use power vents, windows, fans or open the garage doors.
• Don't work around moving parts while wearing a necktie or other loose clothing. Short sleeves are much safer than long, loose sleeves; hard-toed shoes with neoprene soles protect your toes and give a better grip on slippery surfaces. Jewelry such as watches, fancy belt buckles, beads or body adornment of any kind is not safe working around a car. Long hair should be tied back under a hat or cap.
• Don't use pockets for toolboxes. A fall or bump can drive a screwdriver deep into your body. Even a wiping cloth hanging from the back pocket can wrap around a spinning shaft or fan.
• Don't smoke when working around gasoline, cleaning solvent or other flammable material.
• Don't smoke when working around the battery. When the battery is being charged, it gives off explosive hydrogen gas.
• Don't use gasoline to wash your hands; there are excellent soaps available. Gasoline may contain lead, and lead can enter the body through a cut, accumulating in the body until you are very ill. Gasoline also removes all the natural oils from the skin so that bone dry hands will soak up oil and grease.
• Don't service the air conditioning system unless you are equipped with the necessary tools and training. The refrigerant, R-12, is extremely cold when compressed, and when released into the air will instantly freeze any surface it contacts, including your eyes. Although the refrigerant is normally non-toxic, R-12 becomes a deadly poisonous gas in the presence of an open flame. One good whiff of the vapors from burning refrigerant can be fatal.
• Don't use screwdrivers for anything other than driving screws! A screwdriver used as an prying tool can snap when you least expect it, causing injuries. At the very least, you'll ruin a good screwdriver.
• Don't use a bumper jack (that little

GENERAL INFORMATION AND MAINTENANCE

ratchet, scissors, or pantograph jack supplied with the car) for anything other than changing a flat! These jacks are only intended for emergency use out on the road; they are NOT designed as a maintenance tool. If you are serious about maintaining your car yourself, invest in a hydraulic floor jack of a least 1 1/2 ton capacity, and at least two sturdy jackstands.

SERIAL NUMBER IDENTIFICATION

Vehicle Identification Number

VIN Plate

The Vehicle Identification Number (VIN) is stamped on a plate located on the top left hand side of the instrument panel, so it can be seen by looking through the windshield.

The VIN is a thirteen digit (1968-80) or seventeen digit (1981-92) sequence of numbers and letters important for ordering parts and for servicing.

NOTE: *Model years appear in the VIN as the last digit of each particular year (6 is 1976, 8 is 1978, etc.) until 1980 (which is A). This is the final year under the thirteen digit*

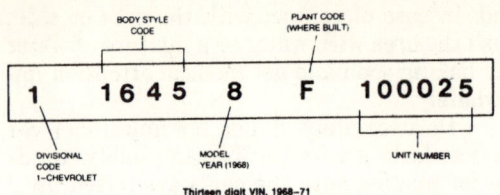

Thirteen digit VIN, 1968-71

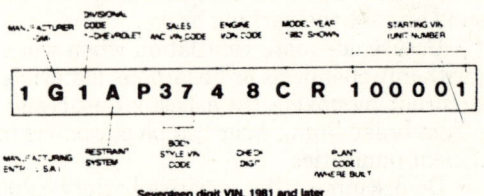

Seventeen digit VIN, 1981 and later

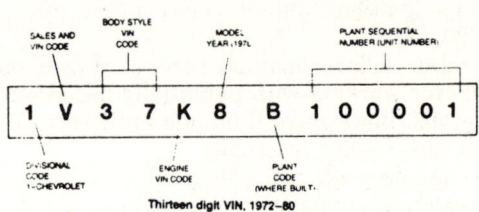

Thirteen digit VIN, 1972-80

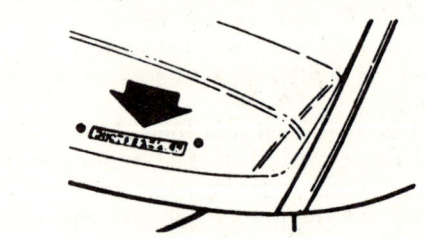

VIN location on 1968 and later cars

code. The seventeen digit VIN begins with 1981 (B) and continues 1982 (C), 1983 (D), etc. The letter I is not used in model year identification, 1987 (H) to 1988 (J).

Service Parts Identification Label

The service parts identification label has been developed and placed on the vehicle to aid in identifying parts and options originally installed on the vehicle. This is extremely helpful when purchasing a used vehicle, or restoring a used vehicle to its original state. The label is located on the underside of the luggage compartment lid on sedans, and on the inner right rear quarter panel of wagons.

Engine Serial Number

The engine serial number shows the manufacturing plant signified by a letter (F for Flint, T for Tonawanda, etc.), the month of manufacture, the day of manufacture, and the transmission and engine type represented by a two or

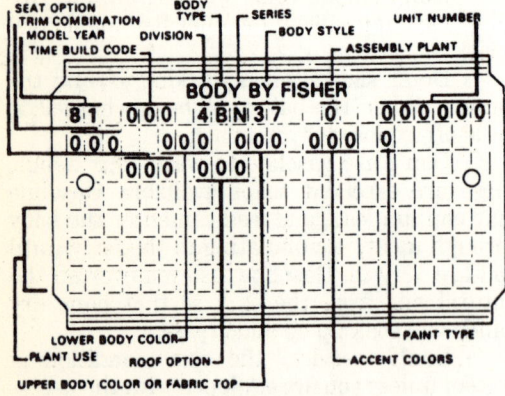

Body number plate — U.S. models

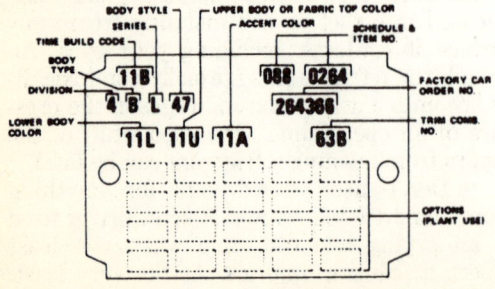

Body number plate — Canadian models

GENERAL INFORMATION AND MAINTENANCE

Engine Identification
1968–70

Engine identification code letter follows immediately after engine serial number.
6 Cyl.—pad at front right-hand side of cylinder block at rear of distributor.
V8—Pad at front right-hand side of cylinder block.

No. Cyls.	Cu. In. Displ.	Type	Year and Code		
			1968	1969	1970
6	230	HDC	BC	BC	
6	230	HDC, AC	BB	BB	
6	230	PG	BF	BF	
6	230	PG, w/ex. EM		AN	
6	230	Hyd., AC		AR	
6	230	PG, w/ex. EM, AC			
6	230	Hyd.		AD	
6	230	w/ex. EM			
6	230	w/ex. EM, AC			
6	230	PG, PCV, AC			
6	230	M.T.	BA	AM	
6	230	M.T.			
6	230	3 Spd. AC		AP	
6	230	PG			
6	230	PG, AC	BH	AQ	
6	250	3 Spd. or OD	CM	BE	CCL
6	250	3 Spd. AC	CN	BF	
6	250	3 Spd. or OD w/ex. EM			
6	250	3 Spd. AC w/ex. EM			
6	250	PG	CQ	BB	CCM
6	250	PG, AC	CR	BC	
6	250	Hyd.		BD	CCK
6	250	PG, w/ex. EM			
6	250	Hyd., AC		BH	
6	250	PG, AC w/ex. EM			
8	283	3 Spd.			
8	283	4 Spd.			
8	283	PG			
8	283	3 Spd., 4 BBL.			
8	283	PG, 3 BBL.			
8	283	w/ex. EM			
8	283	PG, w/ex. EM			
8	283	4 Spd., w/ex. EM			
8	283	4 BBL.; w/ex. EM			
8	283	PG, 4 BBL., w/ex. EM			
8	283	HDC			
8	307	Hyd.		DD	CNF
8	307	M.T.	DA	DA	CNC
8	307	4 Spd.	DE	DE	CND
8	307	PG	DB	DC	CNE
8	307	HDC	DN		
8	327	M.T.	EA		
8	327	HP			
8	327	w/ex. EM			
8	327	SHP	ES		
8	327	PG, w/ex. EM			
8	327	w/T. Ign.			
8	327	3 or 4 Spd. (325 H.P.)			
8	327	HDC (325 H.P.)	ES		
8	327	HDC (275 H.P.)	ED		
8	327	PG	EE		
8	327	PG, HP			
8	350	M.T.		HA	
8	350	Hyd.		HB	
8	350	2-BBL.		HC	
8	350	2-BBL., Hyd.		HD	
8	350	PG		HE	CNM(250)
8	350	PG, 2-BBL.		HF	

GENERAL INFORMATION AND MAINTENANCE

Engine Identification
1971–75

Engine identification code letter follows immediately after engine serial number.
6 Cyl.—pad at front right-hand side of cylinder block at rear of distributor.
V8—Pad at front right-hand side of cylinder block.

No. Cyls.	Cu. In. Displ.	Type	1971	1972	1973	1974	1975
6	250	PG		CBJ			
6	250	T.H.			CCA	CCX	
6	250	M.T.	CAA	CBG	CCC	CCR	D
6	250	M.T., w/NB2			CCD		
6	250	T.H., w/NB2			CCB	CCW	
8	307	T.H.		CTK	CMA		
8	307	M.T.			CHB		
8	307	PG		CKH			
8	307	M.T.	CCA	CKG			
8	300	T.H., w/NB2			CHC		
8	350	M.T.		CKK, CKA	CKA, CKB		
8	350	2-BBL., M.T.				CMC	H
8	350	2-BBL., T.H.				CMA	
8	350	PG		CKB, CDB			
8	350	T.H.		CT, CKD	CKL, CKJ		
8	350	M.T.	CGA(245)				
8	350	PG	CGB(245)				
8	350	M.T.	CGK(270)				
8	350	T.H. 350	CGL(270)				
			CJD(270)				
8	350	M.T.	CJJ(270)				
8	350	M.T., w/NB2		CKC, CKH			
8	350	T.H., w/NB2		CKD, CKK		CKD	
8	350	3-spd., 4-BBL.				CKH	J, T
8	400	T.H., 4-BBL.				CTC	U
8	400	T.H., 4-BBL., California				CTA	
8	402	M.T., HDC (330 hp)		CLA, CLS			
8	402	T.H. 400 (Mk. IV)	CLB	CLB			
8	402	4-spd. (Mk. IV)	CLL				
8	402	M.T. Police (Mk. IV)	CLR				
8	402	M.T. (Mk. IV)	CLS				
8	402	M.T. (Mk. IV)	CPR				
8	400	M.T. (MK. IV)	CPA	CPA			
			CPG				
			CPD				
			CPP				
8	454	T.H. 400 (450 hp)		CPD			
8	454	M.T.			CWA	CWA	
8	454	T.H.			CWB	CWX	
8	454	M.T., w/NB2			CWC		
8	454	T.H., w/NB2			CWD	CWD	

AC—air conditioned
HDC—heavy duty clutch
HP—high performance
M.T.—manual transmission
OD—overdrive
PG—powerglide transmission
PCV—positive crankcase ventilaton
w/ex. EM—with exhaust emission
w/T. Ign.—transistor ignition
4 BBL.—four barrel carburetor
T.H.—Turbo Hydra-Matic
#—Aluminum heads
NB2—Calif. only

GENERAL INFORMATION AND MAINTENANCE

1976–80

Model Year Code		Engine Code					
			Displacement				
Code	Year	Code	Cu. In.	Liters	Cyl.	Carb.	Eng. Mfg.
6	'76	D	250	4.1	6	1	Chev.
		Q	305	5.0	8	2	Chev.
		V	350	5.7	8	2	Chev.
		L	350	5.7	8	4	Chev.
		U	400	6.6	8	4	Chev.
		S	454	7.4	8	4	Chev.
7	'77	D	250	4.1	6	1	Chev.
		U	305	5.0	8	2	Chev.
		L	350	5.7	8	4	Chev.
8	'78	D	250	4.1	6	1	Chev.
		U	305	5.0	8	2	Chev.
		L	350	5.7	8	4	Chev.
9	'79	D	250	4.1	6	1	Chev.
		J	267	4.4	8	2	Chev.
		G	305	5.0	8	2	Chev.
		H	305	5.0	8	4	Chev.
		L	350	5.7	8	4	Chev.
A	'80	K	229	3.8	6	2	Chev.
		A	231	3.8	6	2	Buick
		J	267	4.4	8	2	Chev.
		H	305	5.0	8	4	Chev.
		N	350	5.7	8	Diesel	Olds.
		L	350	5.7	8	4	Chev.

The thirteen digit Vehicle Identification Number can be used to determine engine application and model year. The sixth digit indicates the model year, and the fifth digit identifies the factory installed engine.

1981–88

Engine Code						Model Year Code	
Code	Cu. In.	Liters	Cyl.	Carb.	Eng. Mfg.	Code	Year
K	229	3.8	6	2	Chev.	B	81
9	229	3.8	6	2	Chev.	C	82
A	231	3.8	6	2	Buick	D	83
Z	262	4.3	6	TBI	Chev.	E	84
J	267	4.4	8	2	Chev.	F	85
G	305	5.0	8	4	Chev.	G	86
H	305	5.0	8	4	Chev.	H	87
7	305	5.0	8	4	Chev.	J	88
6	350	5.7	8	4	Chev.		
N	350	5.7	8	Diesel	Chev.		
Y	307	5.0	8	4	Olds.		

The seventeen digit Vehicle Identification Number can be used to determine engine application and model year. The tenth digit indicates the model year, and the eighth digit identifies the factory installed engine.

10 GENERAL INFORMATION AND MAINTENANCE

1989–92

Code	Engine Code					Model Year Code	
	Cu. In.	Liters	Cyl.	Carb.	Eng. Mfg.	Code	Year
Z	262	4.3	6	FI	Chev.	K	89
2①	262	4.3	6	FI	Chev.	L	90
Y	307	5.0	8	4	Olds.	M	91
E	305	5.0	8	FI	Chev.	N	92
7	350	5.7	8	FI	Chev.		

The seventeen digit Vehicle Identification Number can be used to determine engine application and model year. The tenth digit indicates the model year, and the eighth digit identifies the factory installed engine.
① 1989 model year only

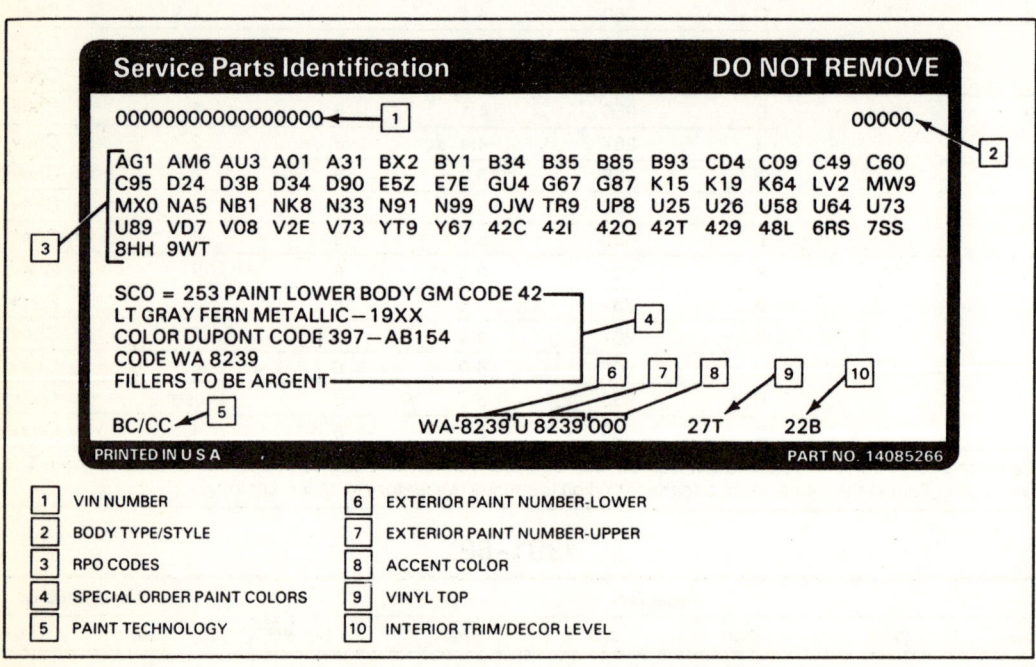

Service parts identification label, 1989 shown

three letter code. A typical engine serial number would be F1005FA. The F represents the manufacturing plant (Flint), the 10 signifies the month of manufacture (October), 05 signifies the day of manufacture and FA signifies the engine and transmission type. Beginning with 1968 cars, a VIN is stamped on the cylinder block next to the engine serial number. The VIN (up to 1971) is the same as the car serial number stamped on the instrument panel except that it does not include the four numbers representing body style. In 1972, the VIN changed somewhat. A typical VIN for the 1972 Chevrolet might be: 1Q87F2F000001 identifying this particular car as the first (000001)

1972 Chevrolet to roll off the Flint assembly line. It includes the manufacturer's identity number (number 1 representing Chevrolet products), a series code letter (B representing Impala), a two digit body style number (87 representing 2 dr. Sport Coupe), an engine code letter (letters listed below), a one digit vehicle year number (2 representing 1972), an assembly plant letter (F signifying Flint), and a unit number signifying order of production. According to the VIN, this car had a 307 cu. in. engine with 2-barrel carburetor represented by the engine code letter F.

This basic format is utilized through the 1981 vehicle year, although specific letter/

GENERAL INFORMATION AND MAINTENANCE

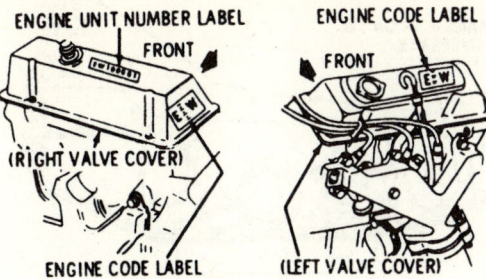

VIN label (decal) locations, 231 V6

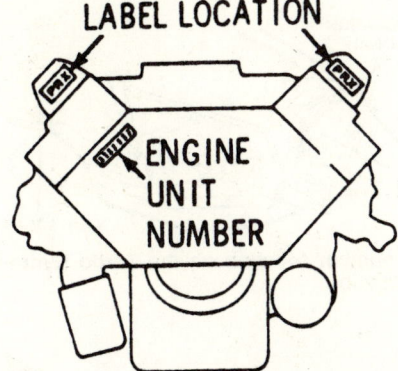

Some 305 and 350 V8s have VIN codes here

Engine serial numbers location — inline six cylinder engine

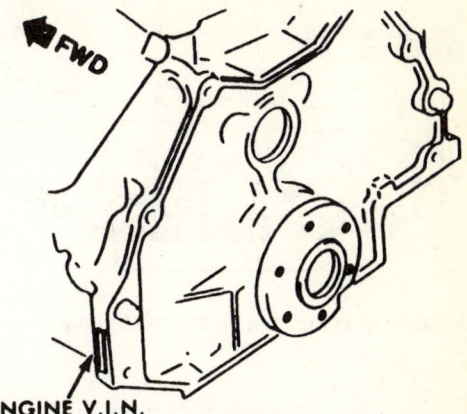

231 V6 engine serial number — at extreme left rear of cylinder block

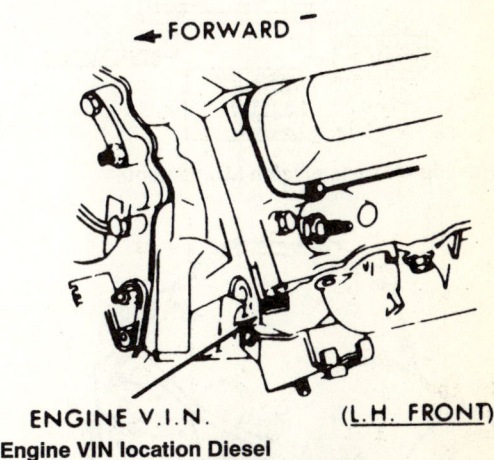

Engine VIN location Diesel

The serial number location on the 229, 262 V6 and all V8 engines is just below the right hand valve cover, stamped on the engine block

number designations may change from year to year.

Inline 6-Cylinder Engines

On 6-cylinder engines, the serial number is found on a pad at the front right hand side of the cylinder block, just to the rear of the distributor.

V8 and V6 Engines

On the 262 V6 (1985-92) and all V8 engines, except the 307 Oldsmobile produced engine, the serial number is found on a pad at the front right hand side of the cylinder block, just below the cylinder head. On the 231 V6, the number can be found on a pad on the left side of the cylinder block, where it meets the transmission. On the 307 Oldsmobile produced engine, the number can be found on a tag located on the oil fill tube assembly.

Transmission Serial Number

A transmission serial number is stamped on each transmission. Beginning with 1968 cars, a VIN is stamped on each cylinder block and on every transmission, in addition to the serial number. The VIN is the same as the vehicle serial number stamped on the instrument panel except that it does not include the four numbers representing body style. The location

12 GENERAL INFORMATION AND MAINTENANCE

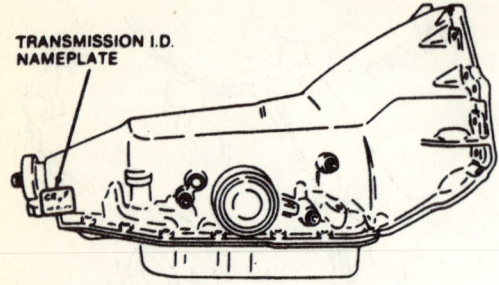

Serial number location on the THM-200-4R

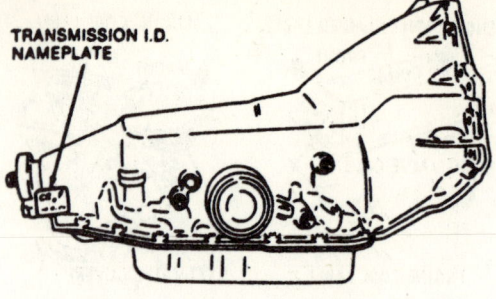

Serial number location — Turbo Hydra-Matic

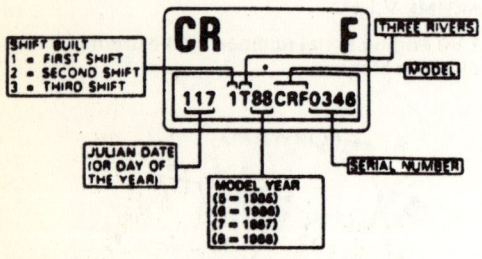

THM-200-4R transmission I.D. nameplate

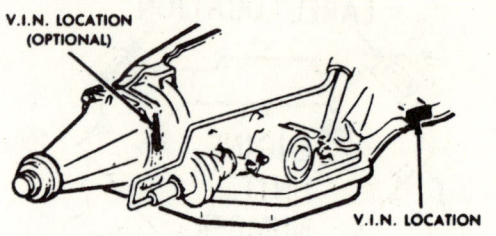

Serial number location on the Turbo Hydra-Matic 200 and 250

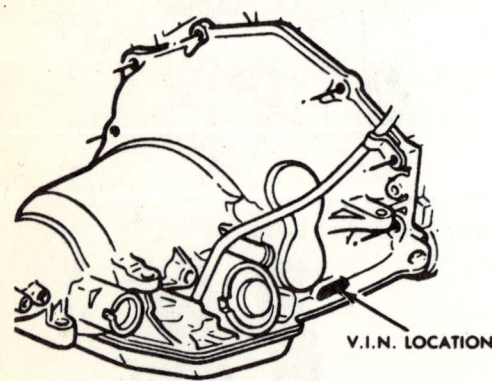

Serial number location on the Turbo Hydra-Matic 350 and 350C

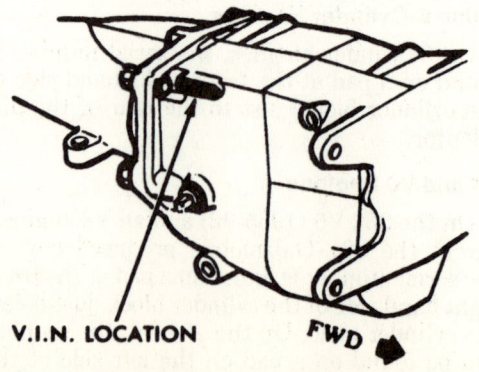

Serial number location on the powerglide

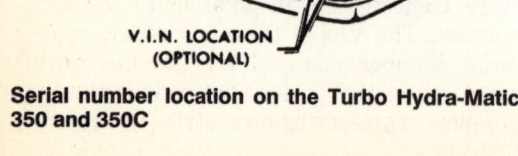

Manual transmission serial number location — 1979–81

GENERAL INFORMATION AND MAINTENANCE

Manual Transmissions,

	Years
Muncie 3-speed, Fully Synchronized	1969–74
Saginaw 3-speed, Fully Synchronized	1968–76
Muncie 4-speed	1968–74
Saginaw 4-speed	1968–76
Warner T-16, 3-speed	1968

Automatic Transmissions,

	Years
Power Glide	1968–80
Turbo Hydramatic 350	1969–81
Turbo Hydramatic 400	1968–72
Turbo Hydramatic 200	1978–81
Turbo Hydramatic 250	1974–78
Turbo Hydramatic 200C	1982–86
Turbo Hydramatic 250C	1982–84
Turbo Hydramatic 350C 4-speed	1982–84
Turbo Hydramatic 200-4R overdrive 4-speed	1978–90
Turbo Hydramatic 350C 3-speed	1982–83
Turbo Hydramatic 700-4R 4-speed	1982–89
Turbo Hydramatic 4L60 4-speed	1990–92

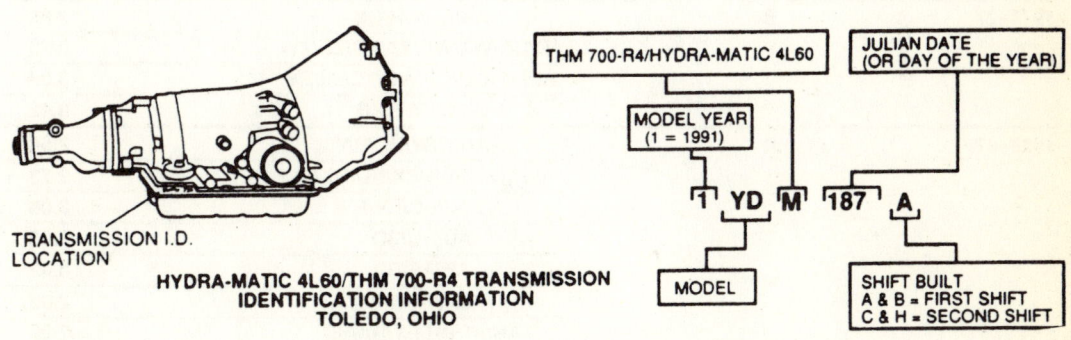

HYDRA-MATIC 4L60/THM 700-R4 TRANSMISSION IDENTIFICATION INFORMATION TOLEDO, OHIO

Serial number location — Saginaw 3 and 4 speed

of the transmission serial number on each transmission is as follows:

Rear Axle Identification

The axle identification number is located either on a metal tag (attached to the rear of the differential) or stamped onto the front right side of the axle tube, about three inches outboard from the differential cover.

For servicing the rear axle, the third letter must be known, which is the manufacturers identity. The codes are: B for Buick, C for Buf-

14 GENERAL INFORMATION AND MAINTENANCE

Drive Axle Application Chart

Year	Model	Axle Identification	Axle Type
1968–69	B	CA/CE/KG/KH	3.08
		CB/CG/KI/KL	3.36
		CC/CU/CI/K4/K5	3.73
		CD/CX/KX/CY	3.07
		CF/KY/KZ/CW	3.31
		CT/CH/CP/CZ/KC/KD/KE/KW	2.73
		CJ/CK/CL/CM/CN	2.56
		CQ/CX/CO	2.56
		CR/CV	3.70
		KA/KB/KF/KJ/KN	3.55
		KP/K2/K3	3.55
		KK/K6	4.10
		KM/K7	4.56
		KO/K8	4.88
1970–72	B	CF/CW/CCF/CCW/RU/RV	3.31
		CRU/CRV	3.31
		CH/CCH/CCP/CGC/KD	2.73
		CGD/CKC/CKD/GC/GD/GH	2.73
		CCA/CCE/GN/GF	3.08
		CCD/CCX	3.07
		CCN/CCD/CGA/CGB/CRJ	2.56
		CRK/GA/GB	2.56
		GI/CGG/CGI/GG	3.36
		CKF/CKJ	3.55
		CKK/CRW/RW	4.10
1973–77	B	AB/AH/CH/CB	2.56
		AC/CC/WA/WB/XA/XB/ZE/ZW	2.73
		AD/AF/CD/CF/WC/XC/ZJ/ZY	3.08
		AJ/CJ/WE/XE	3.42
1978–87	B	AA/AY/BY/RC/BA	2.56
		AB/AL/AX/BB/BX/RD/RX/BL	2.73
		AC/AV/BC/BV/RV	3.08
		AD/BU/BD	3.23
		AE/AS/BS	3.42
		AG/AW/BG/BW	2.93
		AH/AT/BH/BT/RA/SC	2.29
		AJ/AZ/BJ/BZ/PC/RB/SC	2.41
1988–92	B	GA/GS/GM/ZH/ZL	2.56
		GB/GT/GN/YL	2.73
		LG/LJ/ZE/ZV/YY/ZB/ZA	3.08
		DA/NC/LL/LM/DB/NA/DG	3.08
		LH/YE/YG/YM	3.23
		YC/YA/YH	2.93
		ZC/ZD/ZK/ZJ	3.42

GENERAL INFORMATION AND MAINTENANCE 15

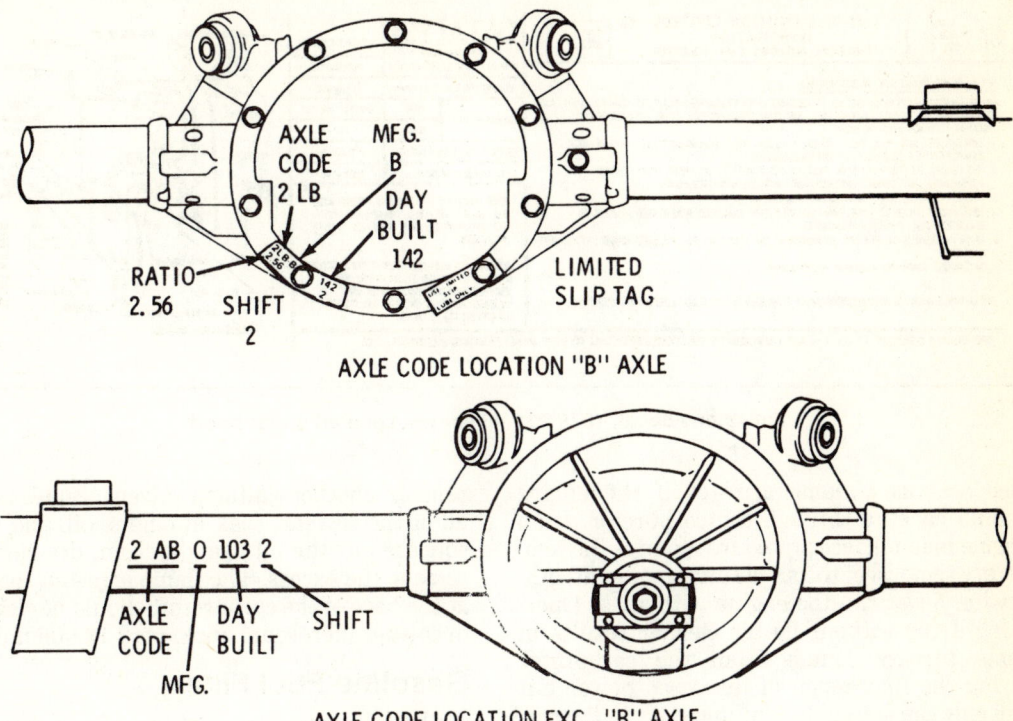

View of the rear axle identification numbers

falo, G for Chevrolet-Gear and Axle, K or M for GM of Canada, O for Oldsmobile, P for Pontiac and W for Warren. The rear axle identification number must known to order the exact replacement parts.

On all late model axle assemblies, the axle ratios are designed to meet emission standards for areas of operation. Due to the wide variety of axle ratios available since 1968 it is not possible to list each gear code and ratio number. Depending on the performance options on your vehicle the ratios start at, 2.29 and go as high as 4.88 (early production models) and 2.41–3.42 (later production models). If the identification tag is missing, or it does not have the gear ratio stamped on it, a quick way to determine the axle ratio is to:

1. Raise and safely support the rear of the vehicle on jack stands.
2. Mark the inside of the wheel with chalk or other suitable marker.
3. Mark the pinion yoke at the housing.
4. With transmission in neutral, turn the driveshaft (or pinion) and count the number of turns it takes to turn the wheel one complete revolution.
5. Lower vehicle.

You now have an estimated axle ratio, examples being; $2^1/_2$ turns of the pinion to 1 turn of the wheel equals about a 2.50 ratio. $3^1/_4$ turns of the pinion to 1 turn of the wheel equals about a 3.25 etc. Remember this is just an estimated ratio and the actual ratio may be slightly different upon ordering replacement parts.

Vehicle Emission Control Information Label

The Vehicle Emission Control Information Label is located in the engine compartment (fan shroud, radiator support, hood underside, etc) of every car produced by General Motors. The label contains important emission specifications and setting procedures, as well as a vacuum hose schematic with various emissions components identified.

When servicing your Chevrolet, this label should always be checked for up-to-date information pertaining specifically to your car.

NOTE: *Always follow the timing procedures on this label when adjusting ignition timing.*

ROUTINE MAINTENANCE

Air Cleaner

The air cleaner has a dual purpose. It not only filters the air going to the carburetor, but

16 GENERAL INFORMATION AND MAINTENANCE

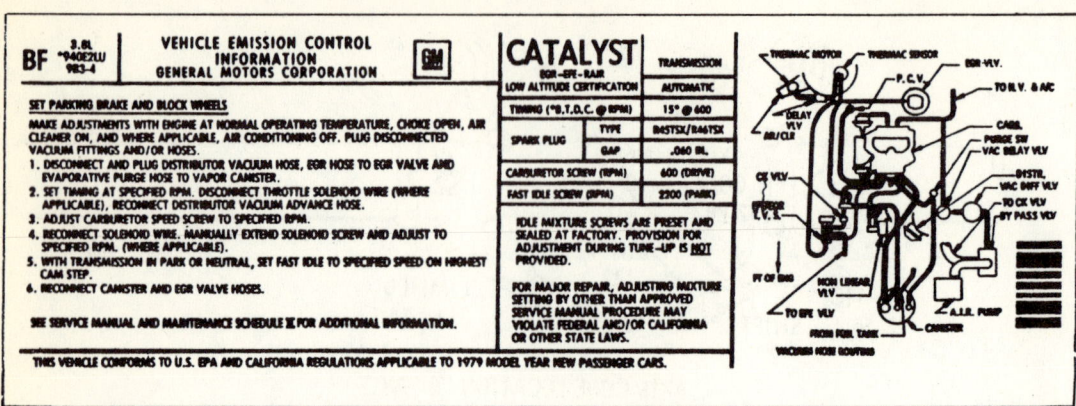

Emissions decal, 1979 231 V6 shown. Located under hood

also acts as a flame arrester if the engine should backfire through the carburetor. If an engine maintenance procedure requires the temporary removal of the air cleaner, remove it; otherwise, never run the engine without it. Operating a car without its air cleaner results in some throaty sounds from the carburetor giving the impression of increased power but will only cause trouble. Unfiltered air to the carburetor will eventually result in a dirty, inefficient carburetor and engine. A dirty carburetor increases the chances of carburetor backfire and, without the protection of an air cleaner, fire becomes a probable danger. The air cleaner assembly consists of the air cleaner itself, which is the large metal container that fits over the carburetor, the element (paper or polyurethane) contained within, and the flame arrester located in the base of the air cleaner. If your car is equipped with the paper element, it should be inspected at its first 12,000 miles, rechecked every 6,000 miles thereafter, and replaced after 24,000 miles. The 1975 and later air cleaners should be replaced at 30,000 mile intervals if the paper type (V6 and V8), and 15,000 miles if the oil wetted type (inline six). Inspections and replacements should be more frequent if the car is operated in a dirty, dusty environment. When inspecting the element, look for dust leaks, holes or an overly dirty appearance. If the element is excessively dirty, it may cause a reduction in clean air intake. If air has trouble getting through a dirty element, the carburetor fuel mixture will become richer (more gas, less air), the idle will be rougher, and the exhaust smoke will be noticeably black. To check the effectiveness of your paper element, remove the air cleaner assembly and, if the idle increases, then the element is restricting air flow and should be replaced. If a polyurethane element is installed, clean or replace it every 12,000 miles. If you choose to clean it, do so with kerosene or another suitable solvent. Squeeze out all of the solvent, soak in engine oil, and then squeeze out the oil using a clean, dry cloth to remove the excess. The flame arrester, located at the base of the carburetor, should be cleaned in solvent (kerosene) once every 12,000 miles.

Gasoline Fuel Filter

There are three types of fuel filters; internal (in the carburetor fitting), inline (in the fuel line) and in-tank (the sock on the fuel pickup tube).

CAUTION: *Before removing any component of the fuel system, refer to the Fuel Pressure Release procedures in this section and release the fuel pressure.*

REMOVAL AND INSTALLATION

Internal Filter

The carburetor inlet fuel filter should be replaced every 12,000 miles (15,000 miles for 1975-90 cars) or more often if necessary.

1. Disconnect the negative battery cable. Dis-

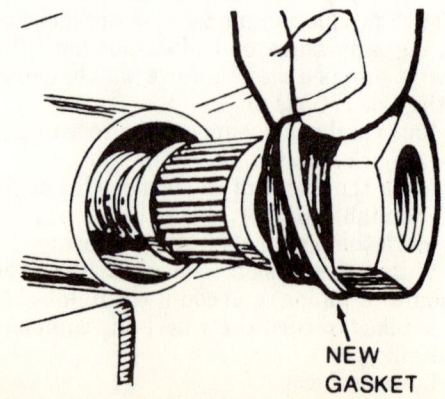

Install the new filter and spring. Certain early models use a bronze filter element, but most are made of paper

GENERAL INFORMATION AND MAINTENANCE

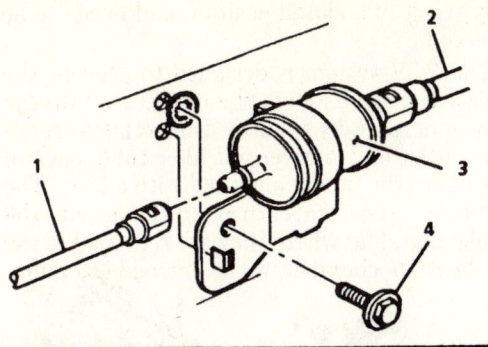

1	FUEL FEED LINE	3	IN-LINE FUEL FILTER
2	FUEL FEED LINE	4	IN-LINE FUEL FILTER ATTACHING BOLT

Inline fuel filter replacement — 1985 and later gasoline engines

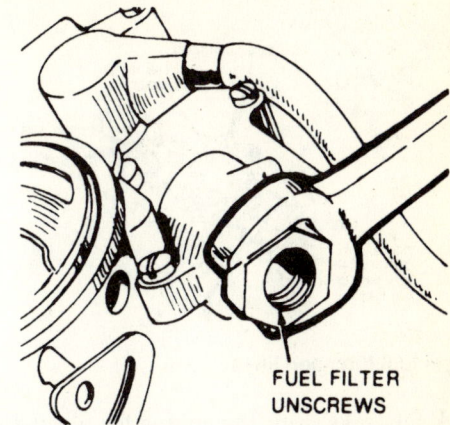

Remove the retaining nut and the filter will pop out under spring pressure

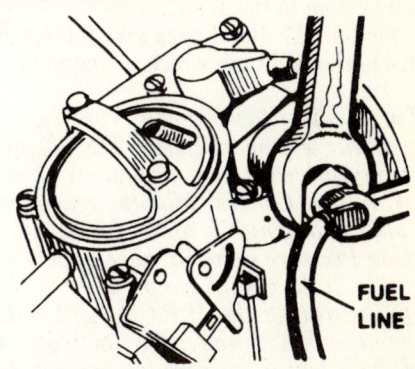

The fuel filter is located behind the large fuel line inlet nut on the carburetor

connect the fuel line connection at the fuel inlet filter nut on the carburetor.

2. Remove the fuel inlet filter from the carburetor.

3. Remove the filter and the spring.

NOTE: *If a check valve is not present with the filter, one must be installed when the filter is replaced.*

4. Install the spring, filter and check valve (must face the fuel line), then reverse the removal procedures.

5. Start the engine and check for leaks.

Inline Filter

1. Disconnect the negative battery cable. Disconnect the fuel lines.

2. Remove the fuel filter from the retainer or mounting bolt.

3. To install, reverse the removal procedures. Start the engine and check for leaks.

NOTE: *The filter has an arrow (fuel flow direction) on the side of the case, be sure to install it correctly in the system, the with arrow facing away from the fuel tank.*

In-Tank Filter

To service the in-tank fuel filter, refer to the Electric Fuel Pump Removal and Installation procedure in Chapter 5.

FUEL PRESSURE RELEASE

Carbureted Engine

To release the fuel pressure on the carbureted system, remove and replace the fuel tank cap.

Fuel Injected Engine

The TBI unit used on the fuel injected engines contains a constant bleed feature in the pressure regulator that relieves pressure any time the engine is turned off. Therefore, no special relieve procedure is required, however, a small amount of fuel may be released when the fuel line is disconnected.

CAUTION: *To reduce the chance of personal injury, cover the fuel line with cloth to collect the fuel and then place the cloth in an approved container.*

Diesel Fuel Filter

The diesel fuel filter is mounted on the rear of the intake manifold, and is larger than that on a gasoline engine because diesel fuel generally is dirtier (has more suspended particles than gasoline).

The diesel fuel filter should be changed every 30,000 miles or two years.

REMOVAL AND INSTALLATION

1. Disconnect the negative battery cable. With the engine cool, place absorbent rags underneath the fuel line fittings at the filter.

2. Disconnect the fuel lines from the filter.

3. Unbolt the filter from its bracket.

4. Install a new filter. Start the engine and

18 GENERAL INFORMATION AND MAINTENANCE

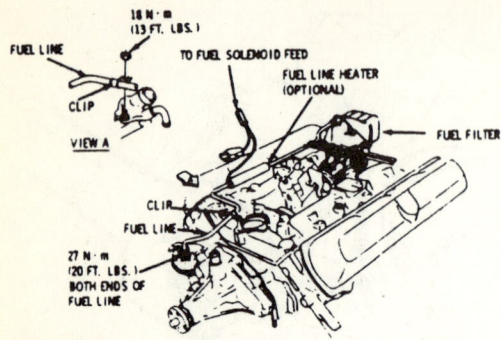

Diesel fuel filter and lines

check for leaks. Run the engine for about two minutes, then shut the engine off for the same amount of time to allow any trapped air in the injection system to bleed off.

Late model GM diesel cars also have a fuel filter inside the fuel tank which is maintenance-free.

NOTE: *If the filter element ever becomes clogged, the engine will stop. This stoppage is usually preceded by a hesitation or sluggish running. General Motors recommends that after changing the diesel fuel filter, the Housing Pressure Cold Advance be activated manually, if the engine temperature is about 125°F. Activating the H.P.C.A. will reduce engine cranking time. To activate the H.P.C.A. solenoid, disconnect the two lead connector at the engine temperature switch and bridge the connector with a jumper. After the engine is running, remove the jumper and reconnect the connector to the engine temperature switch. When the new filter element is installed, start the engine and check for leaks.*

Positive Crankcase Ventilation Valve (PCV)

The crankcase ventilation system (PCV) must be operating properly in order to allow evaporation of fuel vapors and water from the crankcase. This system should be checked at every oil change and serviced after one year or 12,000 miles. The PCV valve is replaced after 2 years or 24,000 miles. For 1975 and later cars, the service interval has been upgraded to one year or 15,000 miles, with PCV valve replacement scheduled for two years or 30,000 miles. Normal service entails cleaning the passages of the system hoses with solvent, inspecting them for cracks and breaks, and replacing them as necessary. The PCV valve contains a check valve and, when working properly, this valve will make a rattling sound when the outside case is tapped. If it fails to rattle, then it is prob-

ably stuck in a closed position and needs to be replaced.

The PCV system is designed to prevent the emission of gases from the crankcase into the atmosphere. It does this by connecting a crankcase outlet (valve cover, oil filler tube, back of engine) to the intake manifold with a hose. The crankcase gases travel through the hose to the intake manifold where they are returned to the combustion chamber to be burned. If main-

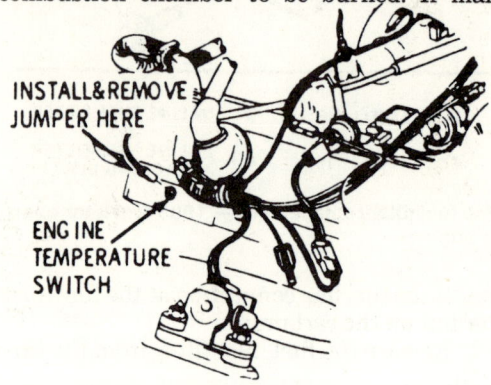

HDCA solenoid connection, diesels

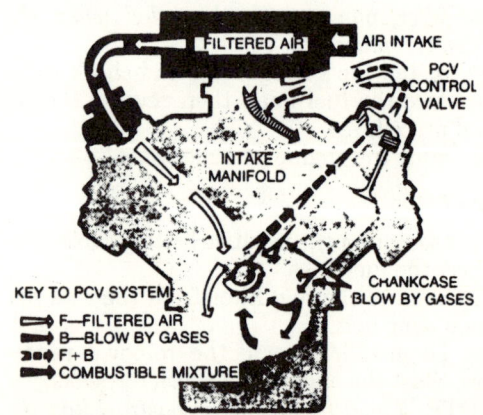

PCV system V8 engine shown

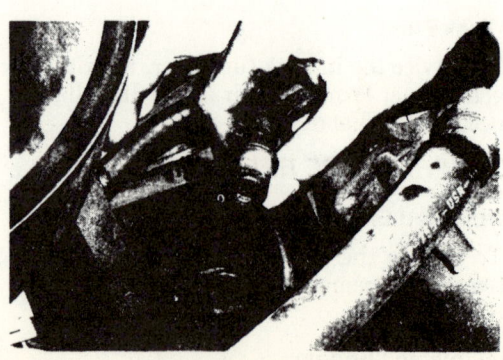

Pulling out the PCV valve from the rocker cover

GENERAL INFORMATION AND MAINTENANCE

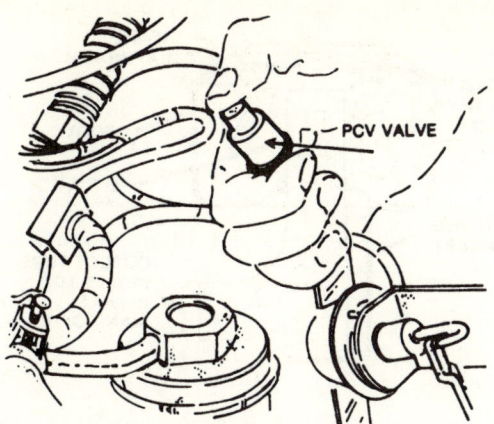

Checking PCV valve for vacuum

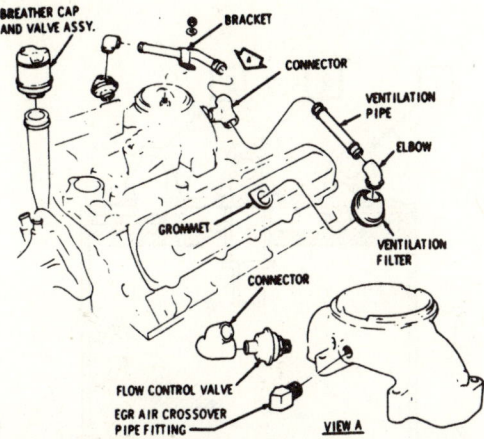

V8 diesel crankcase ventilation system, 1978–80

tained properly, this system reduces condensation in the crankcase and the resultant formation of harmful acids and oil dilution. A clogged PCV valve will often cause a slow or rough idle due to a richer fuel mixture. A car equipped with a PCV system has air going through a hose to the intake manifold from an outlet at the valve cover, oil filler tube, or rear of the engine. To compensate for this extra air going to the manifold, carburetor specifications require a richer (more gas) mixture at the carburetor. If the PCV valve or hose is clogged, this air doesn't go to the intake manifold and the fuel mixture is too rich. A rough, slow idle results. The valve should be checked before making any carburetor adjustments. Disconnect the valve from the engine or merely clamp the hose shut. If the engine speed decreases less than 50 rpm, the valve is clogged and should be replaced. If the engine speed decreases much more than 50 rpm, then the valve is good. The PCV valve is an inexpensive item and it is suggested that it be replaced. If the new valve doesn't noticeably improve engine idle, the problem might be a restriction in the PCV hose. For further details on PCV valve operation see Chapter 4.

Crankcase Depression Regulator and Flow Control Valve

Diesel Engine

The Crankcase Depression Regulator (CDR) found on 1981-85 diesel engine, and the flow control valve, used until 1980 are designed to scavenge crankcase vapors in basically the same manner as the PCV valve on gasoline engines. The valves are located either on the left rear corner of the intake manifold (CDR), or on the rear of the intake crossover pipe (flow control valve). On each system there are two ventilation filters, one per valve cover.

The filter assemblies should be cleaned every 15,000 miles by simply prying them carefully from the valve covers (be aware of the grommets underneath), and washing them out in solvent. The ventilation pipes and tubes should also be cleaned. Both the CDR and flow control valves should also be cleaned every 30,000 miles (the cover can be removed from the CDR; the flow control valve can simply be flushed with solvent). Dry each valve, filter, and hose with compressed air before installation.

NOTE: *Do not attempt to test the crankcase controls on these diesels. Instead, clean the valve cover filter assembly and vent pipes and check the vent pipes.*

Replace the breather cap assembly every 30,000 miles. Replace all rubber fittings as required every 15,000 miles.

Evaporative Emissions Control System

This system, standard since 1970, eliminates

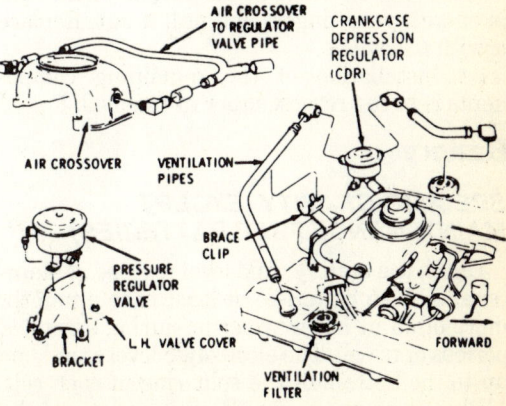

V8 diesel crankcase ventilation system, 1981–85

GENERAL INFORMATION AND MAINTENANCE

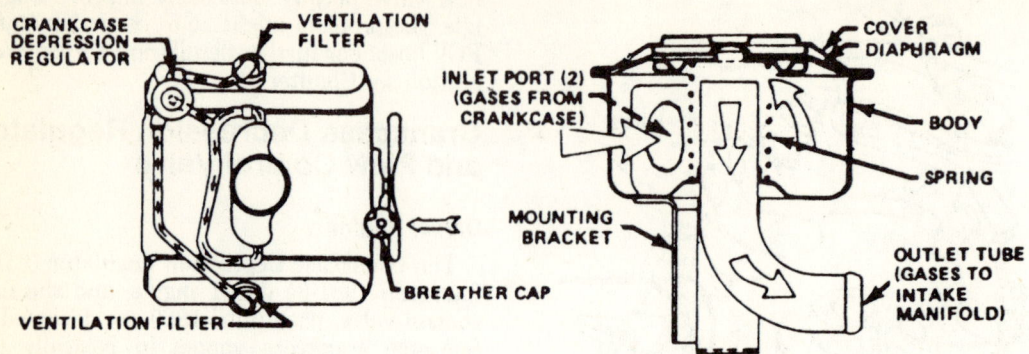

Diesel crankcase ventilation and crankcase depression regulator cutaway, 1980–85

the release of unburned fuel vapors into the atmosphere. The only periodic maintenance required is an occasional check of the connecting lines of the system for kinks or other damage and deterioration. Lines should only be replaced with quality fuel line or special hose marked **evap**. On the 1970-71 cars, every 12,000 miles or 12 months, the filter in the bottom of the carbon canister which is located in the engine compartment should be removed and replaced. On 1972-1976 cars, this service interval is 24,000 miles or 24 months. For 1977-92 cars, the mileage interval has been increased to 30,000 miles, while the time interval remains the same. For further details on the Evaporative Control System please refer to Chapter 4.

FILTER REPLACEMENT

1. Tag and disconnect all hoses connected to the charcoal canister.
2. Loosen the retaining clamps and then lift out the canister.
3. Grasp the filter in the bottom of the canister with your fingers and pull it out. Replace it with a new one.
4. Installation of the remaining components is in the reverse order of removal.

Battery

SPECIFIC GRAVITY (EXCEPT MAINTENANCE FREE BATTERIES)

Check the battery fluid level (except in Maintenance Free batteries) at least once a month, more often in hot weather or during extended periods of travel. The electrolyte level should be up to the bottom of the split ring in each cell. All batteries are equipped with an eye in the cap on one cell. If the eye glows or has an amber color to it, this means that the level is low and

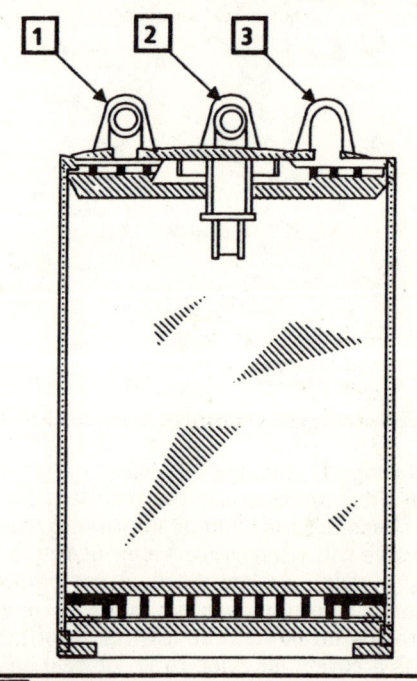

Evaporation canister

only distilled water should be added. Do not add anything else to the battery. If the eye has a dark appearance the battery electrolyte level is high enough. It is also wise to check each cell individually.

At least once a year, check the specific gravity of the battery. It should be between 1.20-1.26". Clean and tighten the clamps and apply

GENERAL INFORMATION AND MAINTENANCE

a thin coat of petroleum jelly to the terminals. This will help to retard corrosion. The terminals can be cleaned with a stiff wire brush or with an inexpensive terminal cleaner designed for this purpose.

If water is added during freezing weather, the car should be driven several miles to allow the electrolyte and water to mix. Otherwise the battery could freeze.

If the battery becomes corroded, a solution of baking soda and water will neutralize the corrosion. This should be washed off after making sure that the caps are securely in place. Rinse the solution off with cold water.

Some batteries were equipped with a felt terminal washer. This should be saturated with engine oil approximately every 6,000 miles. This will also help to retard corrosion.

If a fast charger is used while the battery is in the car, disconnect the battery before connecting the charger.

NOTE: *Keep flame or sparks away from the battery; it gives off explosive hydrogen gas.*

TESTING THE MAINTENANCE-FREE BATTERY

All later cars are equipped with maintenance-free batteries, which do not require normal attention as far as fluid level checks are concerned. However, the terminals require periodic cleaning, which should be performed at least once a year.

Battery State of Charge at Room Temperature

Specific Gravity Reading	Charged Condition
1.260–1.280	Fully Charged
1.230–1.250	¾ Charged
1.200–1.220	½ Charged
1.170–1.190	¼ Charged
1.140–1.160	Almost no Charge
1.110–1.130	No Charge

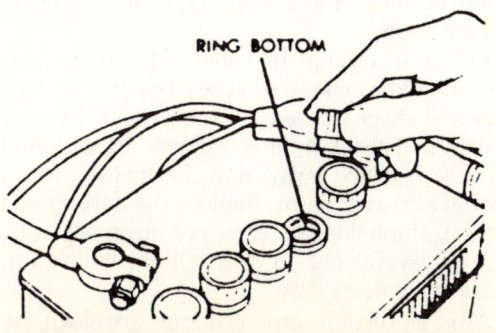

Fill each battery cell to the bottom of the split ring with distilled water

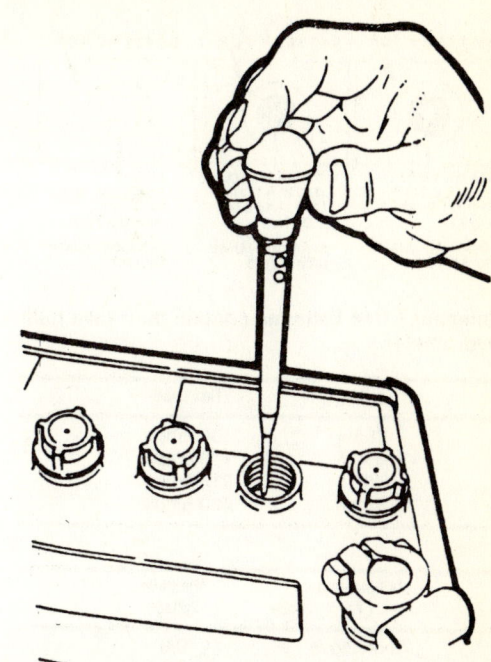

The specific gravity of the battery can be checked with a simple float-type hydrometer

The sealed top battery cannot be checked for charge in the normal manner, since there is no provision for access to the electrolyte. To check the condition of the battery:

1. If the indicator eye on top of the battery is dark, the battery has enough fluid. If the eye is light, the electrolyte fluid is too low and the battery must be replaced.

2. If a green dot appears in the middle of the eye, the battery is sufficiently charged. Proceed to Step 4. If no green dot is visible, charge the battery as in Step 3.

3. Charge the battery at this rate:

CAUTION: *Do not charge the battery for more than 50 amp/hours. If the green dot appears, or if electrolyte squirts out of the vent hole, stop the charge and proceed to Step 4.*

It may be necessary to tip the battery from side to side to get the green dot to appear after charging.

4. Connect a battery load tester and a voltmeter across the battery terminals (the battery cables should be disconnected from the battery). Apply a 300 amp load to the battery for 15 seconds to remove the surface charge. Remove the load.

5. Wait 15 seconds to allow the battery to recover. Apply the appropriate test load, as specified in the following chart: Apply the load for 15 seconds while reading the voltage. Disconnect the load.

6. Check the results against the following chart. If the battery voltage is at or above the

22 GENERAL INFORMATION AND MAINTENANCE

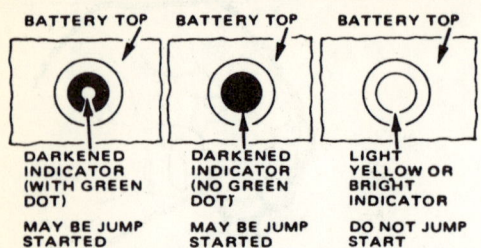

Charging Rate Amps	Time
75	40 min
50	1 hr
25	2 hr
10	5 hr

Maintenance free batteries contain their own built-in hydrometer

Battery	Test Load
Y85-4	130 amps
R85-5	170 amps
R87-5	210 amps
R89-5	230 amps

Temperature (°F)	Minimum Voltage
70 or above	9.6
60	9.5
50	9.4
40	9.3
30	9.1
20	8.9
10	8.7
0	8.5

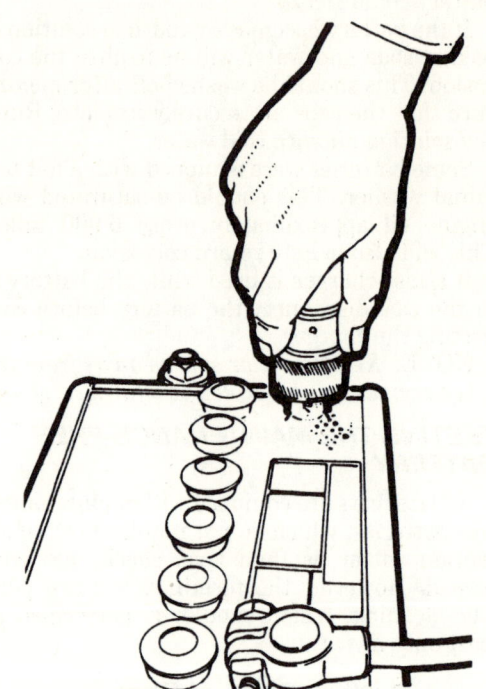

Clean the posts with a wire brush, or a terminal cleaner made for the purpose (shown)

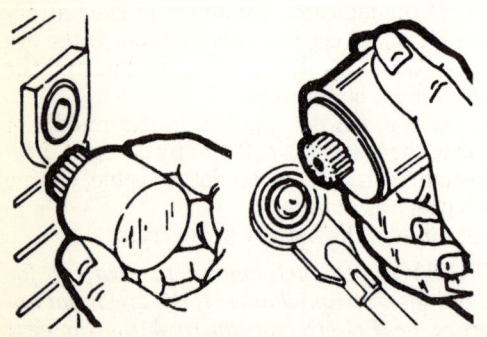

Special tools are available for cleaning the posts and clamps on side terminal batteries

specified voltage for the temperature listed, the battery is good. If the voltage falls below what's listed, the battery should be replaced.

CABLES AND CLAMPS

Once a year, the battery terminals and the cable clamps should be cleaned. Loosen the clamps and remove the cables, negative cable first. On batteries with posts on top, the use of a puller specially made for the purpose is recommended. These are inexpensive, and available in auto parts stores. Side terminal battery cables are secured with a bolt.

Clean the cable clamps and the battery terminal with a wire brush, until all corrosion, grease, etc. is removed and the metal is shiny. It is especially important to clean the inside of the clamp thoroughly, since a small deposit of foreign material or oxidation will prevent a sound electrical connection and inhibit either starting or charging. Special tools are available for cleaning these parts, one type for conventional batteries and another type for side terminal batteries.

Before installing the cables, loosen the battery holddown clamp or strap, remove the battery and check the battery tray. Clear it of any debris, and check it for soundness. Rust should be wire brushed away, and the metal given a coat of anti-rust paint. Replace the battery and tighten the holddown clamp or strap securely, but be careful not to overtighten, which will crack the battery case.

After the clamps and terminals are clean, reinstall the cables, negative cable last; do not hammer on the clamps to install. Tighten the

GENERAL INFORMATION AND MAINTENANCE

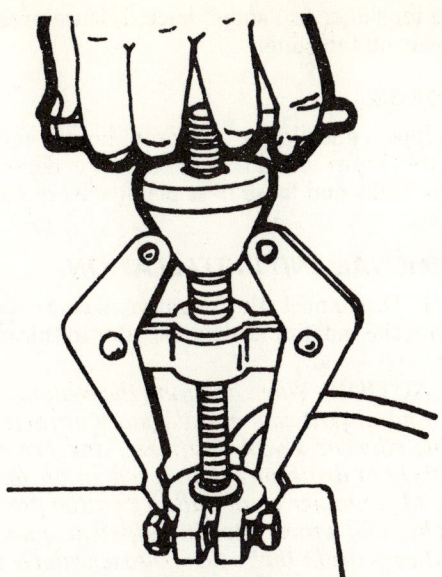

Pullers make clamp removal easier

clamps securely, but do not distort them. Give the clamps and terminals a thin external coat of grease after installation, to retard corrosion.

Check the cables at the same time that the terminals are cleaned. If the cable insulation is cracked or broken, or if the ends are frayed, the cable should be replaced with a new cable of the same length and gauge.

NOTE: *Keep flame or sparks away from the battery; it gives off explosive hydrogen gas. Battery electrolyte contains sulphuric acid. If you should splash any on your skin or in your eyes, flush the affected area with plenty of clear water; if it lands in your eyes, get medical help immediately.*

REPLACEMENT

When it becomes necessary to replace the battery, select a battery with a rating equal to or greater than the battery originally installed. Deterioration, embrittlement and just plain aging of the battery cables, starter motor, and associated wires makes the battery's job harder in successive years. The slow increase in electrical resistance over time makes it prudent to install a new battery with a greater capacity then the old. Details on battery removal and installation are covered in Chapter 3.

Manifold Heat Control Valve (Heat Riser) 1968-74

This valve is located in the exhaust manifold under the carburetor on in-line engines, and in either the right or left side exhaust manifold on V engines. It can be identified by looking for an external thermostatic spring and weight, and

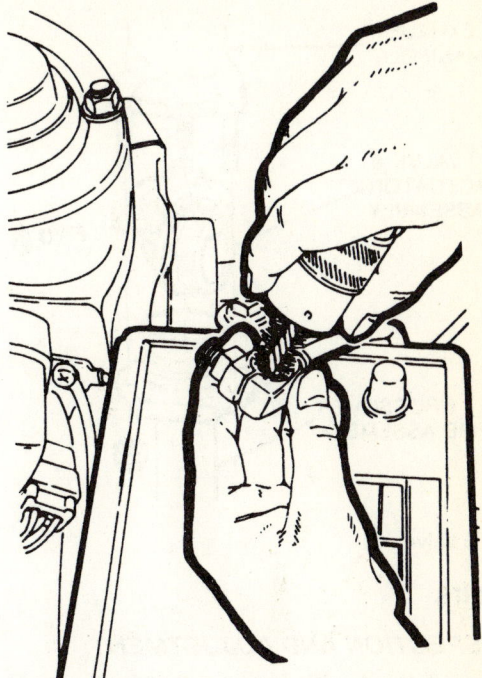

Clean the inside of the clamps with a wire brush, or the special tool

hinge pins that run through the walls of the manifold. Check the valve for free operation every 6,000 miles and, if it binds or is frozen, free it up with a solvent.

NOTE: *Certain early engines has the heat riser built into the manifold and could only be repaired by replacing the entire manifold.*

Early Fuel Evaporation (EFE) System 1975-92

This is a more effective form of heat riser which is vacuum actuated. It is used on all 1975-92 cars. It heats incoming mixture during the engine warm-up process, utilizing a ribbed heat exchanger of thin metal that is located in the intake manifold. This pre-heating allows the choke to open more rapidly, thus reducing emissions. Problem is this system might be indicted by poor engine operation during warm-up.

This valve should be checked initially at 6 months/7,500 miles, and, thereafter, at 18 months/22,500 mile intervals.

To check, move the valve through its full stroke by hand, making sure that the linkage does not bind and is properly connected. If the valve sticks, free it with a solvent. Also check that all vacuum hoses are properly connected and free of cracks or breaks. Replace hoses or broken or bent linkage parts as necessary.

24 GENERAL INFORMATION AND MAINTENANCE

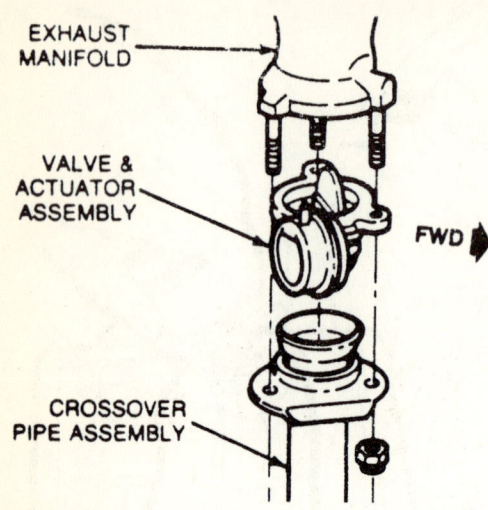

EFE valve

Belts

INSPECTION AND ADJUSTMENT

Check the drive belts every 7,500 miles or six months for evidence of wear such as cracking, fraying, and incorrect tension. Determine belt tension at a point halfway between the pulleys by pressing on the belt with moderate thumb pressure. If the distance between the pulleys (measured from the center of each pulley) is 13-16″, the belt should deflect $1/2$″ at the halfway point or 1.4″ if the distance is 7-10″. If the deflection is found to be too much or too little, loosen the mounting bolts and make the adjustments.

NOTE: *The replacement of the inner belt on multi-belted engines may require the removal of the outer belts.*

Before you attempt to adjust any of your engine's belts, you should take an old rag soaked in solvent and clean the mounting bolts of any road grime which has accumulated there. On some of the harder-to-reach belts, especially on late model V8's with air conditioning and power steering, it would be especially helpful to have a variety of socket extensions and universals to get to those hard-to-reach bolts.

NOTE: *When adjusting the air pump belt, if you are using a pry bar, make sure that you pry against the cast iron end cover and not against the aluminum housing. Excessive force on the housing itself will damage it.*

When replacing a serpentine belt, insert a $1/2$ in. breaker bar into the slot provided on the self adjusting belt tensioner and pry up enough to slip the belt out from under the pulley. Installation is the reverse of removal. Do not allow the tensioner to "snap" back, release pressure slowly on tensioner.

Hoses

Upper and lower radiator hoses and all heater hoses should be checked for deterioration, leaks and loose hose clamps every 15,000 miles.

REMOVAL AND INSTALLATION

1. Disconnect the negative battery cable. Drain the radiator as detailed later in this chapter.

CAUTION: *When draining the coolant, keep in mind that cats and dogs are attracted by the ethylene glycol antifreeze, and are quite likely to drink any that is left in an uncovered container or in puddles on the ground. This will prove fatal in sufficient quantity. Always drain the coolant into a sealable container. Coolant should be reused unless it is contaminated or several years old.*

2. Loosen the hose clamps at each end of the hose to be removed.
3. Working the hose back and forth, slide it off its connection and then install a new hose if necessary.
4. Position the hose clamps at least $1/4$″ from the end of the hose and tighten them.

NOTE: *Always make sure that the hose clamps are beyond the bead and place in the center of the clamping surface before tightening them.*

Cooling System

CAUTION: *Never remove the radiator cap under any conditions while the engine is running! Failure to follow these instructions could result in damage to the cooling system*

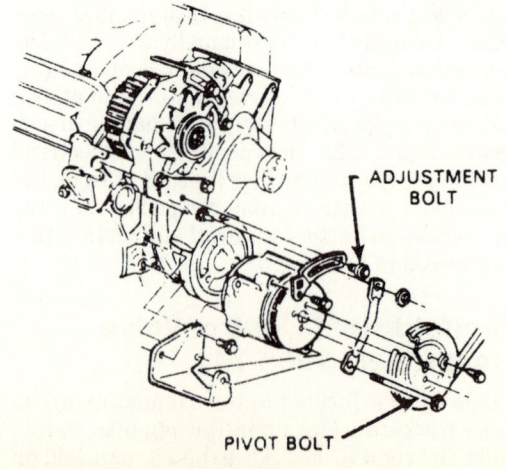

Air adjustment, 1980–88 V8

GENERAL INFORMATION AND MAINTENANCE

HOW TO SPOT WORN V-BELTS

V-Belts are vital to efficient engine operation—they drive the fan, water pump and other accessories. They require little maintenance (occasional tightening) but they will not last forever. Slipping or failure of the V-belt will lead to overheating. If your V-belt looks like any of these, it should be replaced.

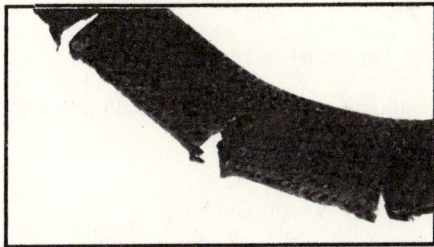
Cracking or weathering

This belt has deep cracks, which cause it to flex. Too much flexing leads to heat build-up and premature failure. These cracks can be caused by using the belt on a pulley that is too small. Notched belts are available for small diameter pulleys.

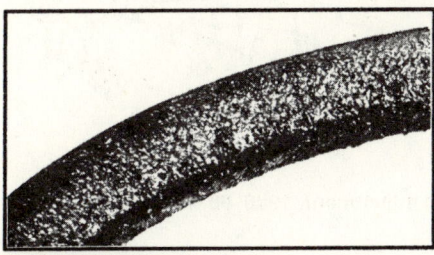
Softening (grease and oil)

Oil and grease on a belt can cause the belt's rubber compounds to soften and separate from the reinforcing cords that hold the belt together. The belt will first slip, then finally fail altogether.

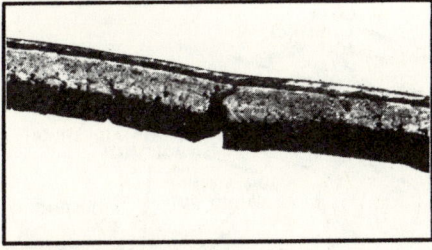
Glazing

Glazing is caused by a belt that is slipping. A slipping belt can cause a run-down battery, erratic power steering, overheating or poor accessory performance. The more the belt slips, the more glazing will be built up on the surface of the belt. The more the belt is glazed, the more it will slip. If the glazing is light, tighten the belt.

Worn cover

The cover of this belt is worn off and is peeling away. The reinforcing cords will begin to wear and the belt will shortly break. When the belt cover wears in spots or has a rough jagged appearance, check the pulley grooves for roughness.

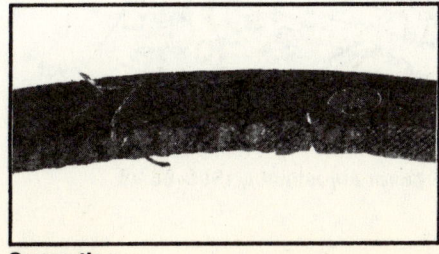
Separation

This belt is on the verge of breaking and leaving you stranded. The layers of the belt are separating and the reinforcing cords are exposed. It's just a matter of time before it breaks completely.

26 GENERAL INFORMATION AND MAINTENANCE

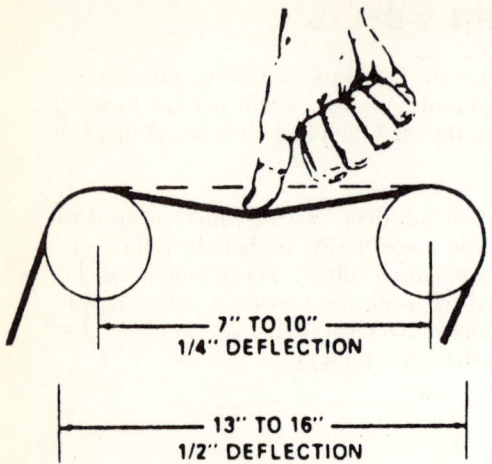

A gauge is recommended, but you can check the belt tension with thumb pressure

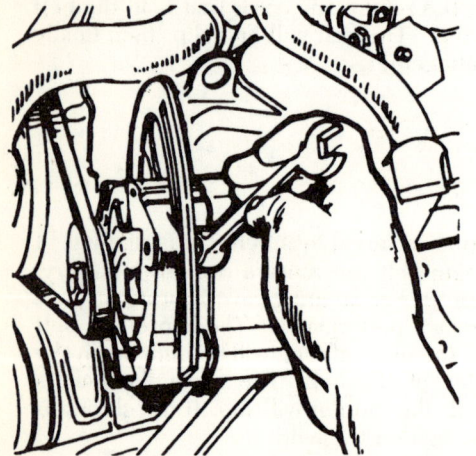

To adjust belt tension, or to replace belts, first loosen the component's mounting and adjusting bolts slightly

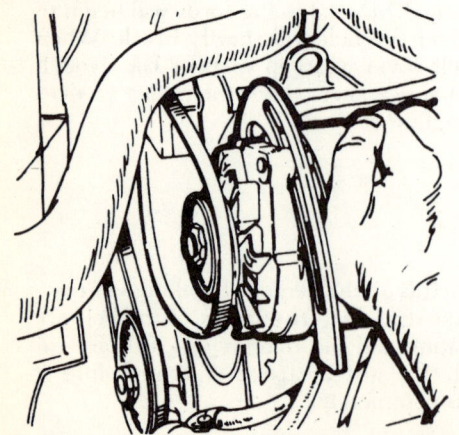

Push the component toward the engine and slip off the belt

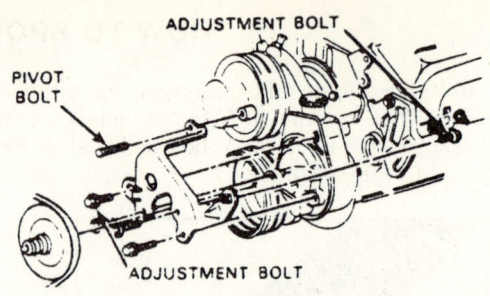

P/S pump adjustment, all except 1985–88 with A/C

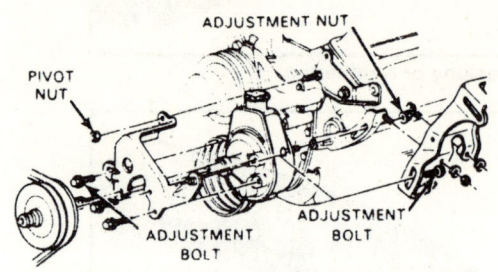

P/S pump adjustment, 1985–88 with A/C

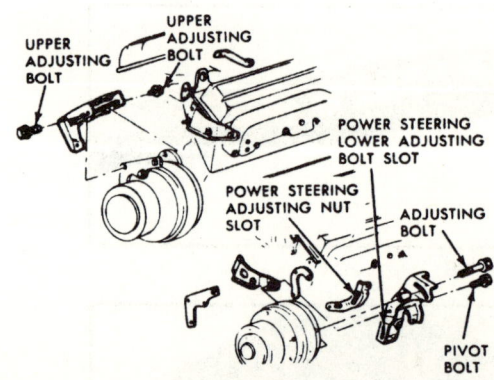

A/C compressor adjustment, 1980–84 V8

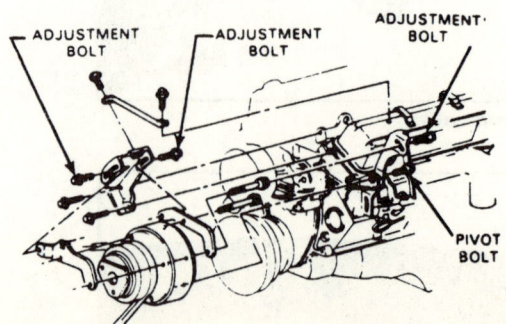

A/C compressor adjustment, 1985–88 V8

GENERAL INFORMATION AND MAINTENANCE

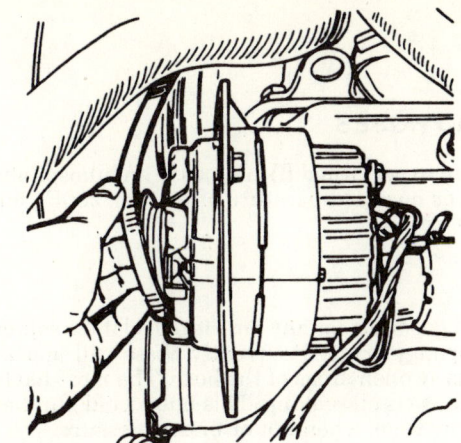

Slip the new belt over the pulley

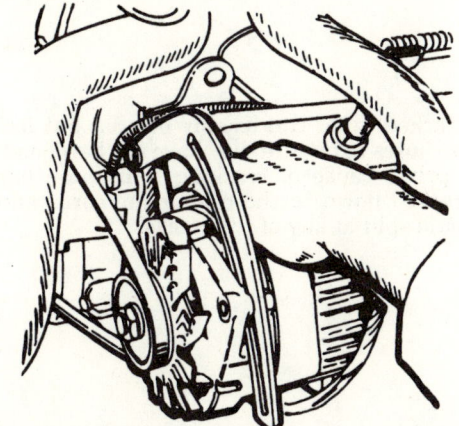

Pull outward on the component and tighten the mounting bolts

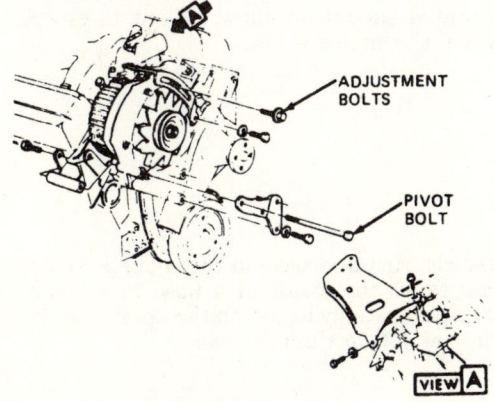

Generator adjustment, 1980–88 V8

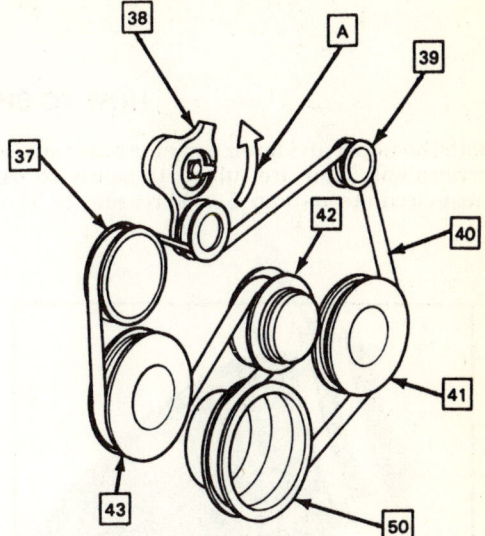

A	ROTATE TENSIONER IN DIRECTION SHOWN TO INSTALL OR REMOVE BELT
37	A/C COMPRESSOR
38	BELT TENSIONER
39	GENERATOR ASSEMBLY PULLEY
40	SERPENTINE BELT
41	P/S PUMP PULLEY
42	WATER PUMP PULLEY
43	AIR PUMP PULLEY
50	CRANKSHAFT PULLEY

Serpentine belt adjustment, 1989–92

or engine and/or personal injury. To avoid having scalding hot coolant or steam blow out of the radiator, use extreme care when removing the radiator cap from a hot radiator. Wait until the engine has cooled, then wrap a thick cloth around the radiator cap and turn it slowly to the first stop. Step back while the pressure is released from the cooling system. When you are sure the pressure has been released, press down on the radiator cap (still have the cloth in position) turn and remove the radiator cap.

Dealing with the cooling system can be dangerous matter unless the proper precautions are observed. It is best to check the coolant level in the radiator when the engine is cold. On early cars this is accomplished by carefully removing the radiator cap and checking that the coolant is within 2 2 inches of the bottom of the filler neck. On later cars, the cooling system has, as one of its components, a coolant recovery tank. If the coolant level is at or near the FULL COLD line (engine cold) or the FULL HOT line (engine hot), the level is satisfactory. Always be certain that the filler caps on both the radiator and the recovery tank are closed tightly.

In the event that the coolant level must be checked when the engine is hot on engines with-

HOW TO SPOT BAD HOSES

Both the upper and lower radiator hoses are called upon to perform difficult jobs in an inhospitable enviorment. They are subject to nearly 18 psi at under hood temperature often over 280F., and must circulate an hour-3 good reasons to have good hoses.

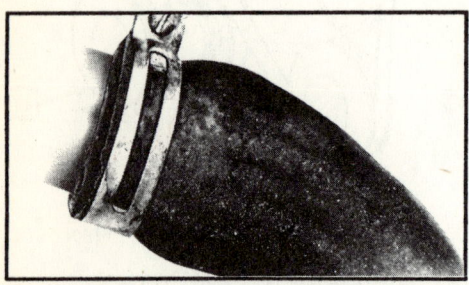
Swollen hose

A good test for any hose is to feel it for soft or spongy spots. Frequently these will appear as swollen areas of the hose. The most likely cause is oil soaking. This hose could burst at any time, when hot or under pressure.

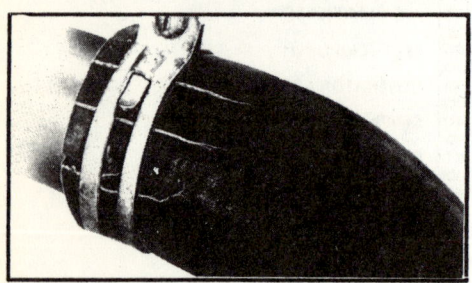
Cracked hose

Cracked hoses can usually be seen but feel the hoses to be sure they have not hardened; a prime cause of cracking. This hose has cracked down to the reinforcing cords and could split at any of the cracks.

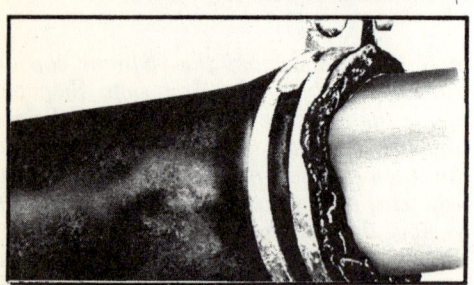
Frayed hose end (due to weak clamp)

Weakened clamps frequently are the cause of hose and cooling system failure. The connection between the pipe and hose has deteriorated enough to allow coolant to escape when the engine is hot.

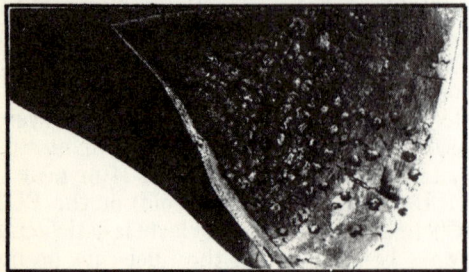
Debris in cooling system

Debris, rust and scale in the cooling system can cause the inside of a hose to weaken. This can usually be felt on the outside of the hose as soft or thinner areas.

GENERAL INFORMATION AND MAINTENANCE

out a coolant recovery tank, place a thick rag over the radiator cap and slowly turn the cap counterclockwise until it reaches the first detent. Allow all hot steam to escape. This will allow the pressure in the system to drop gradually, preventing an explosion of hot coolant. When the hissing noise stops, remove the cap the rest of the way.

If the coolant level is found to be low, add a 50/50 mixture of ethylene glycol-based antifreeze and clean water. On older cars, coolant must be added through the radiator filler neck. On newer cars with the recovery tank, coolant may be added either through the filler neck on the radiator or directly into the recovery tank.

CAUTION: *Never add coolant to a hot engine unless it is running. If it is not running you run the risk of cracking the engine block.*

If the coolant level is chronically low or rusty, refer to Cooling System Troubleshooting Chart at the end of this chapter.

At least once every 2 years, the engine cooling system should be inspected, flushed, and refilled with fresh coolant. If the coolant is left in the system too long, it loses its ability to pre-

Some radiator caps have pressure release levers

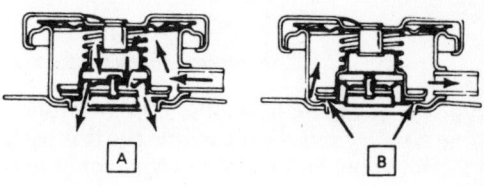

A VACUUM RELIEF
B PRESSURE RELIEF

Coolant recovery, closed system pressure cap

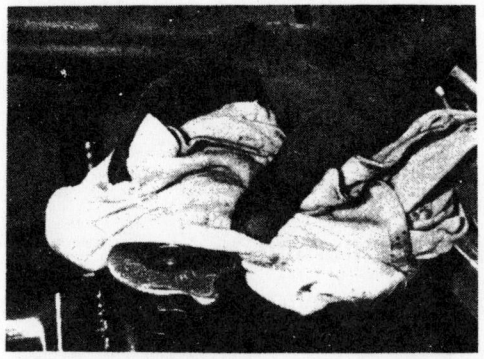
If the engine is hot, cover the radiator cap with a rag

On models without coolant recovery tank, the coolant level should be about 2 inches below the filler neck (engine cold)

vent rust and corrosion. If the coolant has too much water, it won't protect against freezing.

The pressure cap should be looked at for signs of age or deterioration. Fan belt and other drive belts should be inspected and adjusted to the proper tension. (See checking belt tension).

Hose clamps should be tightened, and soft or cracked hoses replaced. Damp spots, or accumulations of rust or dye near hoses, water pump or other areas, indicate possible leakage, which must be corrected before filling the system with fresh coolant.

CHECK THE RADIATOR CAP

While you are checking the coolant level, check the radiator cap for a worn or cracked gasket. It the cap doesn't seal properly, fluid will be lost and the engine will overheat.

Worn caps should be replaced with a new one.

CLEAN RADIATOR OF DEBRIS

Periodically clean any debris — leaves, paper, insects, etc. — from the radiator fins. Pick the large pieces off by hand. The smaller pieces can be washed away with water pressure from a hose.

Carefully straighten any bent radiator fins

GENERAL INFORMATION AND MAINTENANCE

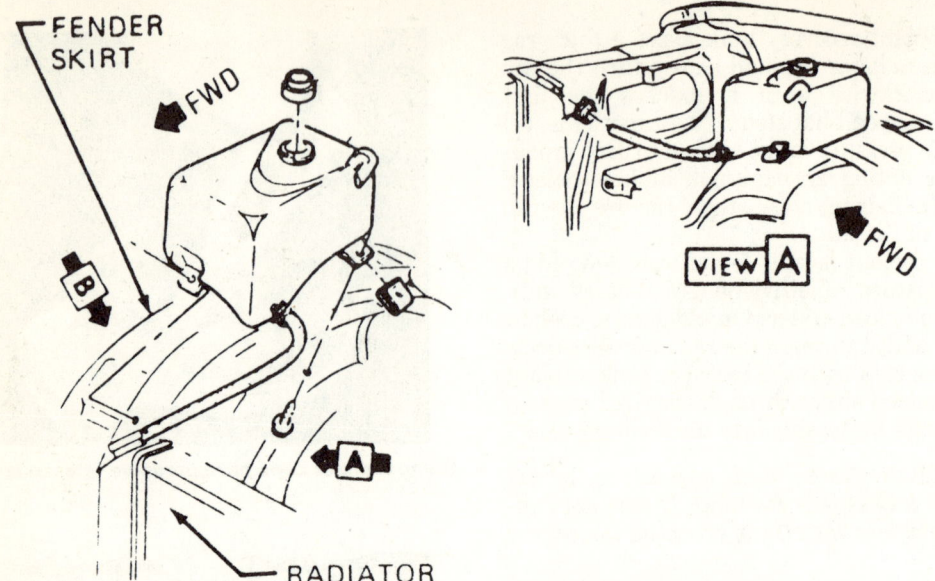

The coolant recovery tank is mounted on the right fender skirt

with a pair of needle nose pliers. Be careful — the fins are very soft. Don't wiggle the fins back and forth too much. Straighten them once and try not to move them again.

DRAIN AND REFILL THE COOLING SYSTEM

Completely draining and refilling the cooling system every two years at least will remove accumulated rust, scale and other deposits. Coolant in late model cars is a 50/50 mixture of ethylene glycol and water for year round use. Use a good quality antifreeze with water pump lubricants, rust inhibitors and other corrosion inhibitors along with acid neutralizers.

1. Drain the existing antifreeze and coolant. Open the radiator and engine drain petcocks, or disconnect the bottom radiator hose, at the radiator outlet.

CAUTION: *When draining the coolant, keep in mind that cats and dogs are attracted by the ethylene glycol antifreeze, and are quite likely to drink any that is left in an uncovered container or in puddles on the ground. This will prove fatal in sufficient quantity. Always drain the coolant into a sealable container. Coolant should be reused unless it is contaminated or several years old.*

2. Close the petcock or reconnect the lower hose and fill the system with water.
3. Add a can of quality radiator flush.
4. Idle the engine until the upper radiator hose gets hot.
5. Drain the system again.
6. Repeat this process until the drained water is clear and free of scale.
7. Close all petcocks and connect all the hoses.
8. If equipped with a coolant recovery system, flush the reservoir with water and leave empty.
9. Determine the capacity of your coolant system (see capacities specifications). Add a 50/50 mix of quality antifreeze (ethylene glycol) and water to provide the desired protection.
10. Run the engine to operating temperature.
11. Stop the engine and check the coolant level.
12. Check the level of protection with an antifreeze tester, replace the cap and check for leaks.

Air Conditioning

GENERAL SERVICING PROCEDURES

NOTE: *It is recommended that a qualified technician perform the following services.*

The most important aspect of air conditioning service is the maintenance of a pure and adequate charge of refrigerant in the system. A refrigeration system cannot function properly if a significant percentage of the charge is lost. Leaks are common because the severe vibration encountered in an automobile can easily cause a sufficient cracking or loosening of the air conditioning fittings; as a result, the extreme operating pressures of the system force refrigerant out.

The problem can be understood by considering what happens to the system as it is operated with a continuous leak. Because the expansion valve regulates the flow of refrigerant to the evaporator, the level of refrigerant there is fairly constant. The receiver/drier stores any excess of refrigerant, and so a loss will first

GENERAL INFORMATION AND MAINTENANCE

appear there as a reduction in the level of liquid. As this level nears the bottom of the vessel, some refrigerant vapor bubbles will begin to appear in the stream of liquid supplied to the expansion valve. This vapor decreases the capacity of the expansion valve very little as the valve opens to compensate for its presence. As the quantity of liquid in the condenser decreases, the operating pressure will drop there and throughout the high side of the system. As the R-12 continues to be expelled, the pressure available to force the liquid through the expansion valve will continue to decrease, and, eventually, the valve's orifice will prove to be too much of a restriction for adequate flow even with the needle fully withdrawn.

At this point, low side pressure will start to drop, and severe reduction in cooling capacity, marked by freeze-up of the evaporator coil, will result. Eventually, the operating pressure of the evaporator will be lower than the pressure of the atmosphere surrounding it, and air will be drawn into the system wherever there are leaks in the low side.

Because all atmospheric air contains at least some moisture, water will enter the system and mix with the R-12 and the oil. Trace amounts of moisture will cause sludging of the oil, and corrosion of the system. Saturation and clogging of the filter/drier, and freezing of the expansion valve orifice will eventually result. As air fills the system to a greater and greater extent, it will interfere more and more with the normal flows of refrigerant and heat.

From this description, it should be obvious that much of the repairman's time will be spent detecting leaks, repairing them, and then restoring the purity and quantity of the refrigerant charge. A list of general precautions that should be observed while doing this follows:

1. Keep all tools as clean and dry as possible.
2. Thoroughly purge the service gauges and hoses of air and moisture before connecting them to the system. Keep them capped when not in use.
3. Thoroughly clean any refrigerant fitting before disconnecting it, in order to minimize the entrance of dirt into the system.
4. Plan any operation that requires opening the system beforehand, in order to minimize the length of time it will be exposed to open air. Cap or seal the open ends to minimize the entrance of foreign material.
5. When adding oil, pour it through an extremely clean and dry tube or funnel. Keep the oil capped whenever possible. Do not use oil that has not been kept tightly sealed.
6. Use only refrigerant 12. Purchase refrigerant intended for use in only automatic air conditioning systems. Avoid the use of refrigerant 12 that may be packaged for another use, such as cleaning, or powering a horn, as it is impure.
7. Completely evacuate any system that has been opened to replace a component, or that has leaked sufficiently to draw in moisture and air. This requires evacuating air and moisture with a good vacuum pump for at least one hour. If a system has been open for a considerable length of time it may be advisable to evacuate the system for up to 12 hours (overnight).
8. Use a wrench on both halves of a fitting that is to be disconnected, so as to avoid placing torque on any of the refrigerant lines.
9. When overhauling a compressor, pour some of the oil into a clean glass and inspect it. If there is evidence of dirt or metal particles, or both, flush all refrigerant components with clean refrigerant before evacuating and recharging the system. In addition, if metal particles are present, the compressor should be replaced.
10. Schrader valves may leak only when under full operating pressure. Therefore, if leakage is suspected but cannot be located, operate the system with a full charge of refrigerant and look for leaks from all Schrader valves. Replace any faulty valves.

Additional Preventive Maintenance Checks

ANTIFREEZE

In order to prevent heater core freeze-up during A/C operation, it is necessary to maintain permanent type antifreeze protection of +15°F, or lower. A reading of −15°F is ideal since this protection also supplies sufficient corrosion inhibitors for the protection of the engine cooling system.

NOTE: *The same antifreeze should not be used longer than the manufacturer specifies.*

RADIATOR CAP

For efficient operation of an air conditioned car's cooling system, the radiator cap should have a holding pressure which meets manufacturer's specifications. A cap which fails to hold these pressures should be replaced.

CONDENSER

Any obstruction of or damage to the condenser configuration will restrict the air flow which is essential to its efficient operation. It is therefore a good rule to keep this unit clean and in proper physical shape.

NOTE: *Bug screens are regarded as obstructions.*

GENERAL INFORMATION AND MAINTENANCE

CONDENSATION DRAIN TUBE

This single molded drain tube expels the condensation, which accumulates on the bottom of the evaporator housing, into the engine compartment. If this tube is obstructed, the air conditioning performance can be restricted and condensation buildup can spill over onto the vehicle's floor.

SAFETY PRECAUTIONS

Because of the importance of the necessary safety precautions that must be exercised when working with air conditioning systems and R-12 refrigerant, a recap of the safety precautions are outlined.

1. Avoid contact with a charged refrigeration system, even when working on another part of the air conditioning system or car. If a heavy tool comes into contact with a section of copper tubing or a heat exchanger, it can easily cause the relatively soft material to rupture.

2. When it is necessary to apply force to a fitting which contains refrigerant, as when checking that all system couplings are securely tightened, use a wrench on both parts of the fitting involved, if possible. This will avoid putting torque on refrigerant tubing. (It is advisable, when possible, to use tube or line wrenches when tightening these flare nut fittings.)

3. Do not attempt to discharge the system by merely loosening a fitting, or removing the service valve caps and cracking these valves. Precise control is possible only when using the service gauges. Place a rag under the open end of the center charging hose while discharging the system to catch any drops of liquid that might escape. Wear protective gloves when connecting or disconnecting service gauge hoses.

4. Discharge the system only in a well ventilated area, as high concentrations of the gas can exclude oxygen and act as an anesthetic. When leak testing or soldering, this is particularly important, as toxic gas is formed when R-12 contacts any flame.

5. Never start a system without first verifying that both service valves are back-seated, if equipped, and that all fittings throughout the system are snugly connected.

6. Avoid applying heat to any refrigerant line or storage vessel. Charging may be aided by using water heated to less than 125° to warm the refrigerant container. Never allow a refrigerant storage container to sit out in the sun, or near any other source of heat, such as a radiator.

7. Always wear goggles when working on a system to protect the eyes. If refrigerant contacts the eyes, it is advisable in all cases to see a physician as soon as possible.

8. Frostbite from liquid refrigerant should be treated by first gradually warming the area with cool water, and then gently applying petroleum jelly. A physician should be consulted.

9. Always keep refrigerant drum fittings capped when not in use. Avoid sudden shock to the drum, which might occur from dropping it, or from banging a heavy tool against it. Never carry a drum in the passenger compartment of a car.

10. Always completely discharge the system before painting the car (if the paint is to be baked on), or before welding anywhere near refrigerant lines.

Air Conditioning Tools and Gauges

Test Gauges

Most of the service work performed in air conditioning requires the use of a set of two gauges, one for the high (head) pressure side of the system, the other for the low (suction) side.

The low side gauge records both pressure and vacuum. Vacuum readings are calibrated from 0 to 30 inches and the pressure graduations read from 0 to no less than 60 psi.

The high side gauge measures pressure from 0 to at least 600 psi.

Both gauges are threaded into a manifold that contains two hand shut-off valves. Proper manipulation of these valves and the use of the attached test hoses allow the user to perform the following services:

1. Test high and low side pressures.
2. Remove air, moisture, and contaminated refrigerant.
3. Purge the system (of refrigerant).
4. Charge the system (with refrigerant).

The manifold valves are designed so they have no direct effect on gauge readings, but serve only to provide for, or cut off, flow of refrigerant through the manifold. During all testing and hook-up operations, the valves are kept in a closed position to avoid disturbing the refrigeration system. The valves are opened only to purge the system of refrigerant or to charge it.

When purging the system, the center hose is uncapped at the lower end, and both valves are cracked open slightly. This allows refrigerant pressure to force the entire contents of the system out through the center hose. During charging, the valve on the high side of the manifold is closed, and the valve on the low side is cracked open. Under these conditions, the low pressure in the evaporator will draw refrigerant from the relatively warm refrigerant storage container into the system.

GENERAL INFORMATION AND MAINTENANCE

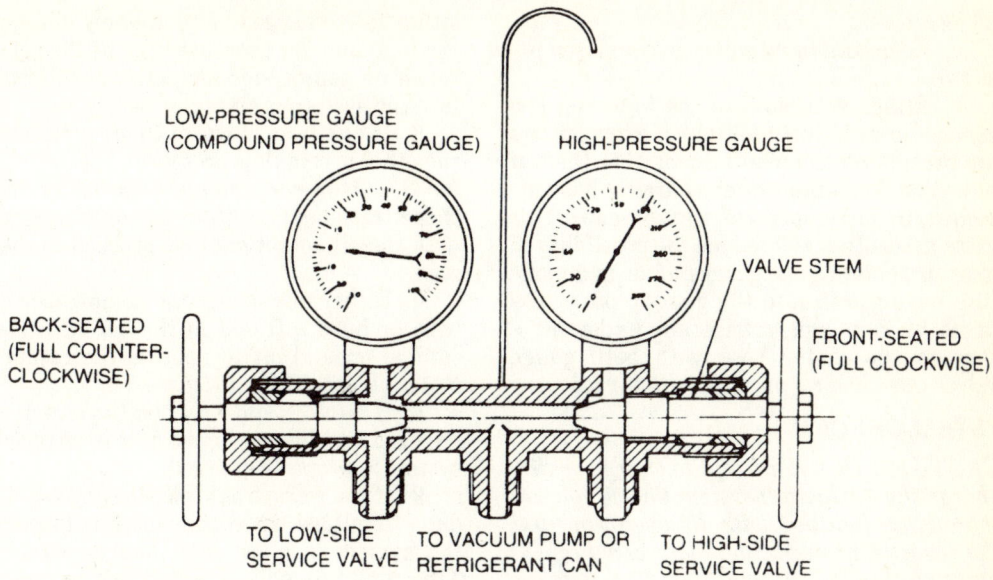

Manifold gauge set

Service Valves

For the user to diagnose an air conditioning system he or she must gain "entrance" to the system in order to observe the pressures. There are two types of terminals for this purpose, the hand shut off type and the familiar Schrader valve.

The Schrader valve is similar to a tire valve stem and the process of connecting the test hoses is the same as threading a hand pump outlet hose to a bicycle tire. As the test hose is threaded to the service port the valve core is depressed, allowing the refrigerant to enter the test hose outlet. Removal of the test hose automatically closes the system.

Extreme caution must be observed when removing test hoses from the Schrader valves as some refrigerant will normally escape, usually under high pressure. (Observe safety precautions.)

Some systems have hand shut-off valves (the stem can be rotated with a special ratcheting box wrench) that can be positioned in the following three ways:

1. FRONT SEATED – Rotated to full clockwise position.

    a. Refrigerant will not flow to compressor, but will reach test gauge port. COMPRESSOR WILL BE DAMAGED IF SYSTEM IS TURNED ON IN THIS POSITION.

    b. The compressor is now isolated and ready for service. However, care must be exercised when removing service valves from the compressor as a residue of refrigerant may still be present within the compressor. Therefore, remove service valves slowly observing all safety precautions.

2. BACK SEATED – Rotated to full counter clockwise position. Normal position for system while in operation. Refrigerant flows to compressor but not to test gauge.

3. MID-POSITION (CRACKED) – Refrigerant flows to entire system. Gauge port (with hose connected) open for testing.

USING THE MANIFOLD GAUGES

The following are step-by-step procedures to guide the user to correct gauge usage.

1. WEAR GOGGLES OR FACE SHIELD DURING ALL TESTING OPERATIONS. BACKSEAT HAND SHUT-OFF TYPE SERVICE VALVES.

2. Remove caps from high and low side service ports. Make sure both gauge valves are closed.

3. Connect low side test hose to service valve that leads to the evaporator (located between the evaporator outlet and the compressor).

4. Attach high side test hose to service valve that leads to the condenser.

5. Mid-position hand shutoff type service valves.

6. Start engine and allow for warm-up. All testing and charging of the system should be done after engine and system have reached normal operation temperatures (except when using certain charging stations).

7. Adjust air conditioner controls to maximum cold.

8. Observe gauge readings.

When the gauges are not being used it is a

good idea to:

a. Keep both hand valves in the closed position.

b. Attach both ends of the high and low service hoses to the manifold, if extra outlets are present on the manifold, or plug them if not. Also, keep the center charging hose attached to an empty refrigerant can. This extra precaution will reduce the possibility of moisture entering the gauges. If air and moisture have gotten into the gauges, purge the hoses by supplying refrigerant under pressure to the center hose with both gauge valves open and all openings unplugged.

SYSTEM CHECKS

CAUTION: *Do not attempt to charge or discharge the refrigerant system unless you are thoroughly familiar with its operation and the hazards involved. The compressed refrigerant used in the air conditioning system expands and evaporates (boils) into the atmosphere at a temperature of –21.7°F (–29.8°C) or less. This will freeze any surface that it comes in contact with, including your eyes. In addition, the refrigerant decomposes into a poisonous gas in the presence of flame.*

1968-76 cars equipped with factory installed air conditioners have a sight glass for checking the refrigerant charge. The sight glass is on top of the VIR (valves-in-receiver) which is located in the front of the engine compartment, usually on the left side of the radiator.

1977-92 cars utilize C.C.O.T system which does not include a sight glass. The (C.C.O.T) Cycling Clutch Orfice Tube refrigeration system is designed to cycle the compressor on and off to maintain desired cooling and to prevent evaporator freeze.

NOTE: *If your car is equipped with an aftermarket air conditioner, the following system checks may not apply. Contact the manufacturer of the unit for instructions on system checks.*

1968-76

This test works best if the outside air temperature is warm (above 70°F).

1. Place the automatic transmission in Park or the manual in Neutral. Set the parking brake.
2. With the help of the friend, run the engine at a fast idle (about 1500 rpm).
3. Set the control for maximum cold with the blower on high.
4. Look at the sight glass on top of the VIR. If a steady stream of bubbles is present in the sight glass, the system is low on charge. Very lightly there is a leak in the system.
5. If no bubbles are present, the system is either fully charged or completely empty. Feel the high and low pressure lines at the compressor, if no appreciable temperature difference is felt, the system is empty or nearly so.
6. If one hose is warm (high pressure) and the other is cold (low pressure), the system may be OK. However, you are probably making these tests because there is something wrong with the air conditioner, so proceed to the next step.
7. Either disconnect the compressor clutch wire or have a friend in the car turn the fan control on and off to operate the compressor clutch. Watch the sight glass.
8. If bubbles appear when the clutch is disengaged and disappear when it is engaged, the system is properly charged.
9. If the refrigerant takes more than 45 seconds to bubble when the clutch is disengaged, the system is more than likely overcharged. This condition will usually result in poor cooling at low speeds.

NOTE: *If it is determined that the system has a leak, it should be repaired as soon as possible. Leaks may allow moisture to enter the system, causing an expensive rust problem.*

1977-92

The air conditioning system on these cars does not incorporate a sight glass.

1. Run the engine until it reaches normal operating temperature.
2. Open the hook and all doors.
3. Turn the air conditioning on, move the temperature selector to the first detent to the right of COLD (outside air) and then turn the blower on HI.
4. Idle the engine at 1,000 rpm.
5. Feel the temperature of the evaporator inlet and the accumulator outlet with the compressor clutch engaged.
6. Both lines should be cold. If the inlet pipe is colder than the outlet pipe, the system is low on charge. Do not attempt to charge the system yourself.

CHARGING THE SYSTEM

CAUTION: *Never attempt to charge the system by opening the high pressure gauge control while the compressor is operating. The compressor accumulating pressure can burst the refrigerant container, causing sever personal injuries.*

Basic System

In this procedure the refrigerant enters the suction side of the system as a vapor while the compressor is running. Before proceeding, the system should be in a partial vacuum after ade-

GENERAL INFORMATION AND MAINTENANCE

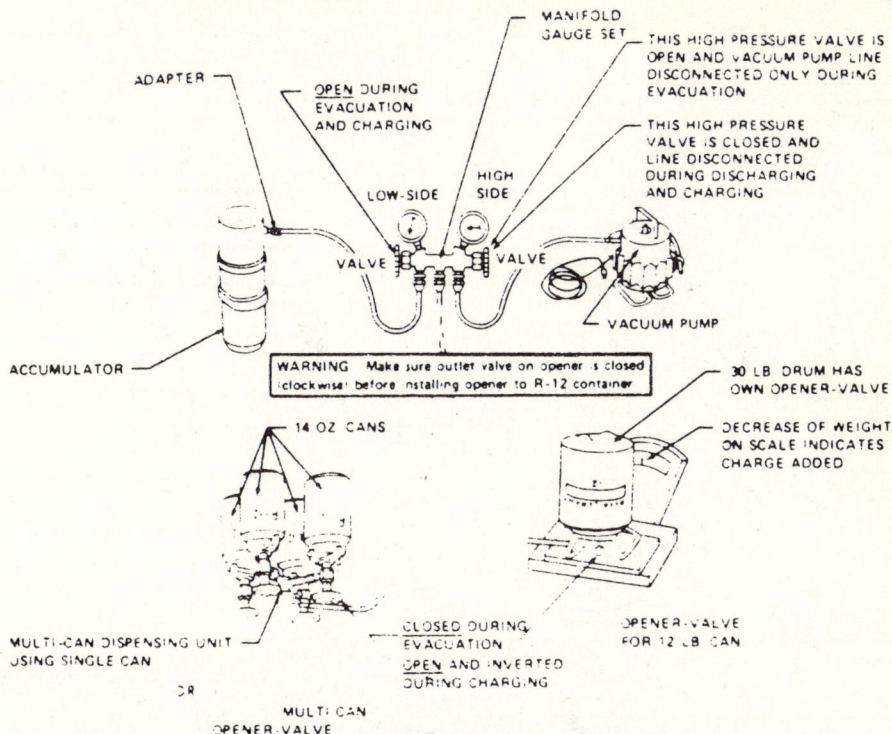

Gauge connections for discharge, evacuation and charging the system

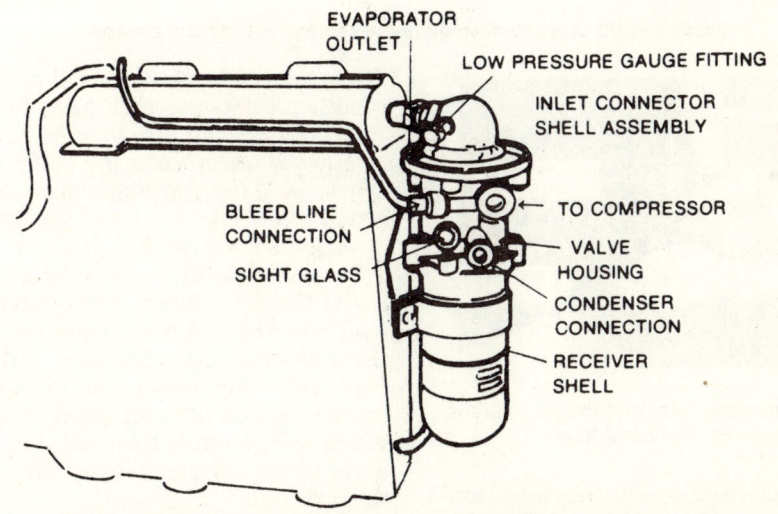

Air conditioning VIR assembly showing sight glass and connections

quate evacuation. Both hand valves on the gauge manifold should be closed.

1. Attach both test hoses to their respective service valve ports. Mid-position manually operated service valves, if present.
2. Install dispensing valve (closed position) on the refrigerant container. (Single and multiple refrigerant manifolds are available to accommodate one to four 15 oz. cans.)
3. Attach center charging hose to the refrigerant container valve.
4. Open dispensing valve on the refrigerant can.
5. Loosen the center charging hose coupler where it connects to the gauge manifold to allow the escaping refrigerant to purge the hose of contaminants.
6. Tighten center charging hose connection.
7. Purge the low pressure test hose at the gauge manifold.
8. Start the engine, roll down the windows

36 GENERAL INFORMATION AND MAINTENANCE

Check item \ Amount of refrigerant	Almost no refrigerant	Insufficient	Suitable	Too much refrigerant
Temperature of high pressure and low pressure lines.	Almost no difference between high pressure and low pressure side temperature	High pressure side is warm and low pressure side is fairly cold.	High pressure side is hot and low pressure side is cold.	High pressure side is abnormally hot.
State in sight glass.	Bubbles flow continuously. Bubbles will disappear and something like mist will flow when refrigerant is nearly gone.	The bubbles are seen at intervals of 1-2 seconds	Almost transparent. Bubbles may appear when engine speed is raised and lowered. No clear difference exists between these two conditions.	No bubbles can be seen.
Pressure of system.	High pressure side is abnormally low.	Both pressure on high and low pressure sides are slightly low.	Both pressures on high and low pressure sides are normal.	Both pressures on high and low pressure sides are abnormally high.
Repair	Stop compressor immediately and conduct an overall check.	Check for gas leakage, repair as required, replenish and charge system.		Discharge refrigerant from service valve of low pressure side

Using a sight glass to determine the relative refrigerant charge

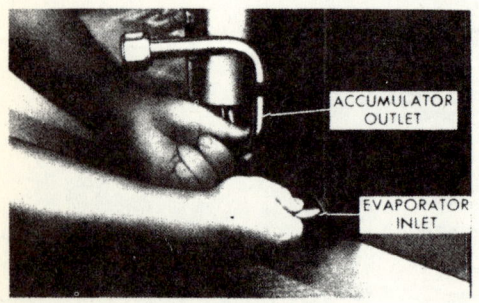

Checking the evaporator inlet and the accumulator outlet line temperatures, 1978 and later

and adjust the air conditioner to maximum cooling. The engine should be at normal operating temperature before proceeding. The heated environment helps the liquid vaporize more efficiently.

9. Crack open the low side hand valve on the manifold. Manipulate the valve so that the refrigerant that enters the system does not cause the low side pressure to exceed 40 psi. Too sudden a surge may permit the entrance of unwanted liquid to the compressor. Since liquids cannot be compressed, the compressor will suffer damage if compelled to attempt it. If the suction side of the system remains in a vacuum the system is blocked. Locate and correct the condition before proceeding any further.

NOTE: *Placing the refrigerant can in a container of warm water (no hotter than 125°F) will speed the charging process. Slight agitation of the can is helpful too, but be careful not to turn the can upside down.*

Some manufacturers allow for a partial charging of the A/C system in the form of a liquid (can inverted and compressor off) by opening the high side gauge valve only, and putting the high side compressor service valve in the middle position (if so equipped). The remainder of the refrigerant is then added in the form of a gas in the normal manner, through the suction side only.

Systems With A Sight Glass

The air conditioning systems that use a sight glass as a means to check the refrigerant level, should be carefully checked to avoid under or over charging. The gauge set should be attached to the system for verification of pressures.

To check the system with the sight glass, clean the glass and start the engine. Operate the air conditioning controls on maximum for approximately five minutes to stabilize the system. The room temperature should be above

GENERAL INFORMATION AND MAINTENANCE

70°F. Check the sight glass for one of the following conditions:

1. If the sight glass is clear, the compressor clutch is engaged, the compressor discharge line is warm and the compressor inlet line is cool, the system has a full charge of refrigerant.

2. If the sight glass is clear, the compressor clutch is engaged and there is no significant temperature difference between the compressor inlet and discharge lines, the system is empty or nearly empty. By having the gauge set attached to the system a measurement can be taken. If the gauge reads less than 25 psi, the low pressure cutoff protection switch has failed.

3. If the sight glass is clear and the compressor clutch is disengaged, the clutch is defective, or the clutch circuit is open, or the system is out of refrigerant. By-pass the low pressure cutoff switch momentarily to determine the cause.

4. If the sight glass shows foam or bubbles, the system can be low on refrigerant. Occasional foam or bubbles is normal when the room temperature is above 110°F or below 70°F. To verify, increase the engine speed to approximately 1,500 rpm and block the airflow through the condenser to increase the compressor discharge pressure to 225-250 psi. If the sight glass still shows bubbles or foam, the refrigerant level is low.

CAUTION: *Do not operate the engine any longer than necessary with the condenser airflow blocked. This blocking action also blocks the cooling system radiator and will cause the system to overheat rapidly.*

When the system is low on refrigerant, a leak is present or the system was not properly charged. Use a leak detector and locate the problem area and repair. If no leakage is found, charge the system to its capacity.

CAUTION: *It is not advisable to add refrigerant to a system utilizing the suction throttling valve and a sight glass, because the amount of refrigerant required to remove the foam or bubbles will result in an overcharge and potentially damaged system components.*

Systems Without A Sight Glass

When charging the CCOT system, attach only the low pressure line to the low pressure gauge port, located on the accumulator. Do not attach the high pressure line to any service port or allow it to remain attached to the vacuum pump after evacuation. Be sure both the high and the low pressure control valves are closed on the gauge set. To complete the charging of the system, follow the outline supplied.

1. Start the engine and allow to run at idle, with the cooling system at normal operating temperature.

2. Attach the center gauge hose to a single or multi-can dispenser.

3. With the multi-can dispenser inverted, allow one pound or the contents of one or two 14 oz. cans to enter the system through the low pressure side by opening the gauge low pressure control valve.

4. Close the low pressure gauge control valve and turn the A/C system on to engage the compressor. Place the blower motor in its high mode.

5. Open the low pressure gauge control valve and draw the remaining charge into the system. Refer to the capacity chart at the end of this section for the individual car or system capacity.

6. Close the low pressure gauge control valve and the refrigerant source valve, on the multi-can dispenser. Remove the low pressure hose from the accumulator quickly to avoid loss of refrigerant through the Schrader valve.

7. Install the protective cap on the gauge port and check the system for leakage.

8. Test the system for proper operation.

REFRIGERANT CAPACITIES:

- 1968–72: 3 lbs.
- 1973–80: $3^{3}/_{4}$ lbs.
- 1981–88: $3^{1}/_{4}$ lbs.
- 1989–90: $3^{1}/_{8}$ lbs.
- 1991–92: $3^{1}/_{2}$ lbs.

Leak Testing the System

There are several methods of detecting leaks in an air conditioning system; among them, the two most popular are (1) halide leak-detection or the "open flame method," and (2) electronic leak-detection.

The halide leak detection is a torch like device which produces a yellow-green color when refrigerant is introduced into the flame at the burner. A purple or violet color indicates the presence of large amounts of refrigerant at the burner.

An electronic leak detector is a small portable electronic device with an extended probe. With the unit activated the probe is passed along those components of the system which contain refrigerant. If a leak is detected, the unit will sound an alarm signal or activate a display signal depending on the manufacturer's design. It is advisable to follow the manufacturer's instructions as the design and function of the detection may vary significantly.

CAUTION: *Care should be taken to operate either type of detector in well ventilated areas, so as to reduce the chance of personal injury, which may result from coming in con-*

GENERAL INFORMATION AND MAINTENANCE

tact with poisonous gases produced when R-12 is exposed to flame or electric spark.

Windshield Wipers

REFILL REPLACEMENT

For maximum effectiveness and longest element life, the windshield and wiper blades should be kept clean. Dirt, tree sap, road tar and so on will cause streaking, smearing and blade deterioration if left on the glass. It is advisable to wash the windshield carefully with a commercial glass cleaner at least once a month. Wipe off the rubber blades with the wet rag afterwards. do not attempt to move the wipers back and forth by hand; damage to the motor and drive mechanism will result.

If the blades are found to be cracked, broken or torn, they should be replaced immediately. Replacement intervals will vary with usage, although ozone deterioration usually limits blade life to about one year. If the wiper pattern is smeared or streaked, or if the blade chatters across the glass, the blades should be replaced. It is easiest and most sensible to replace them in pairs.

There are basically three different types of wiper blade refills, which differ in their method of replacement. One type has two release buttons, approximately $1/3$ of the way up from the ends of the blade frame. Pushing the buttons down releases a lock and allows the rubber blade to be removed from the frame. The new blade slides back into the frame and locks in place.

The second type of refill has two metal tabs which are unlocked by squeezing them together. The rubber blade can then be withdrawn from the frame jaws. A new one is installed by inserting it into the front frame jaws and sliding it rearward to engage the remaining frame jaws. There are usually four jaws; be certain when installing that the refill is engaged in all of them. At the end of its travel, the tabs will lock into place on the front jaws of the wiper blade frame.

The third type is a refill made from polycarbonate. The refill has a simple locking device at one end which flexes downward out of the groove into which the jaws of the holder fit, allowing easy release. By sliding the new refill through all the jaws and pushing through the slight resistance when it reaches the end of its travel, the refill will lock into position.

Regardless of the type of refill used, make sure that all the frame jaws are engaged as the refill is pushed into place and locked. The metal blade holder and frame will scratch the glass if allowed to touch it.

Tires and Wheels

INFLATION

Tires should be checked weekly for proper air pressure. A chart, located either in the glove compartment or on the driver's or passenger's door, gives the recommended inflation pressures. Maximum fuel economy and tire life will result if the pressure is maintained at the highest figure given on the chart. Pressures should be checked before driving since pressure can increase as much as six pounds per square inch (psi) due to heat buildup. It is a good idea to have you own accurate pressure gauge, because not all gauges on service station air pumps can be trusted. When checking pressures, do not neglect the spare tire. Note that some spare tires require pressures considerably higher than those used in the other tires.

While you are about the task of checking air pressure, inspect the tire treads for cuts, bruises and other damage. Check the air valves to be sure that they are tight. Replace any missing valve caps.

Check the tires for uneven wear that might indicate the need for front end alignment or tire rotation. Tires should be replaced when a tread wear indicator appears as a solid band across the tread.

TIRE DESIGN

When buying new tires, give some thought to the following points, especially if you are considering a switch to larger tires or a different profile series:

1. All four tires must be of the same construction type. This rule cannot be violated, Radial, bias, and bias-belted tires must not be mixed.

2. The wheels should be the correct width for the tire. Tire dealers have charts of tire and rim compatibility. A mis-match will cause sloppy handling and rapid tire wear. The tread width should match the rim width (inside bead to inside bead) within an inch. For radial tires, the rim should be 80% or less of the tire (not tread) width.

3. The height (mounted diameter) of the new tires can change speedometer accuracy, engine speed at a given road speed, fuel mileage, acceleration, and ground clearance. Tire manufacturers furnish full measurement specifications.

4. The spare tire should be usable, at least for short distance and low speed operation, with the new tires.

5. There shouldn't be any body interference when loaded, on bumps, or in turns.

GENERAL INFORMATION AND MAINTENANCE

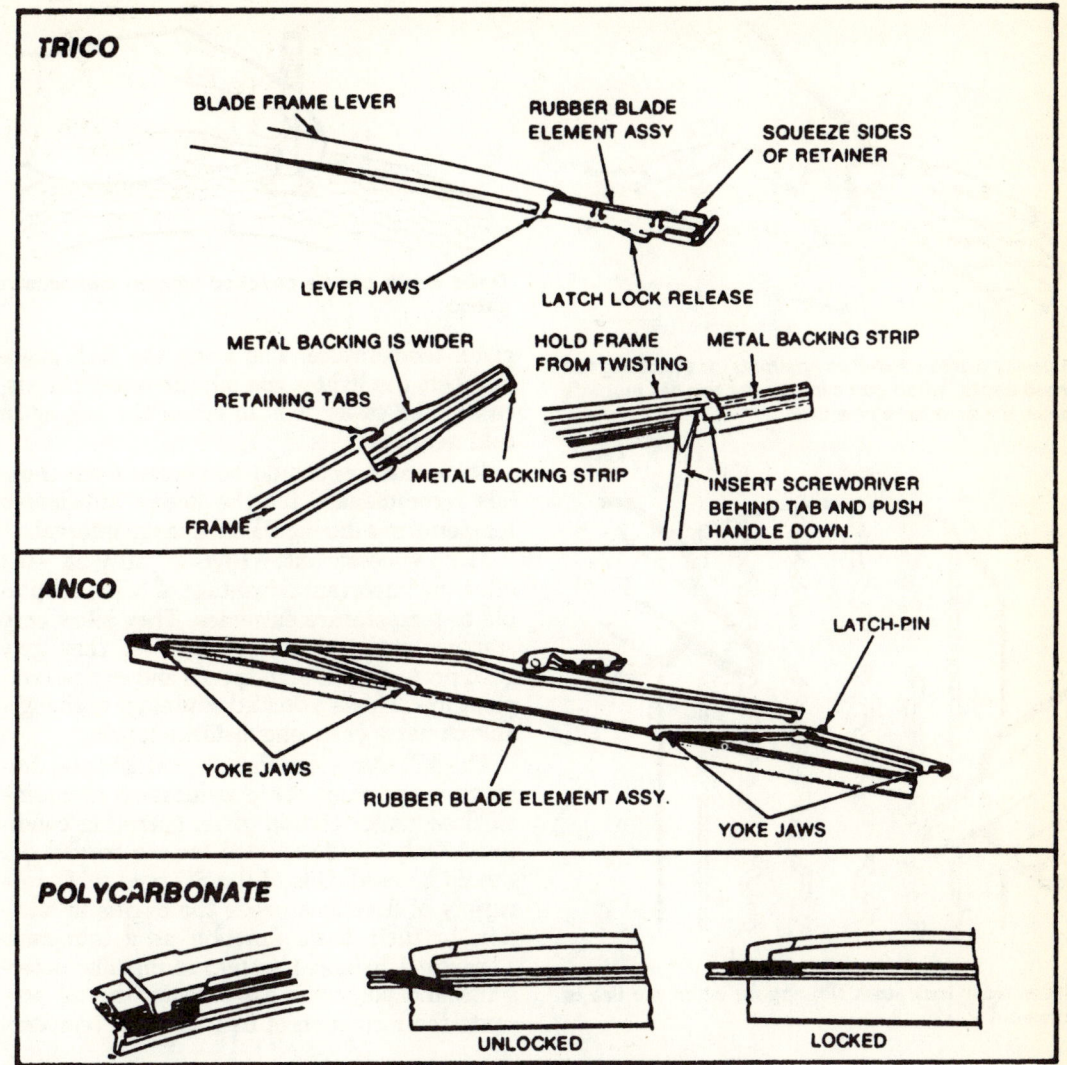

The three types of wiper blade retention

TIRE ROTATION

Tire rotation is recommended every 6,000 miles or so, to obtain maximum tire wear. The pattern you use depends on whether or not your car has a usable spare. Radial tires should not be cross-switched (from one side of the car to the other); they last longer if their direction of rotation is not changed. Snow tires sometimes have directional arrows molded into the side of the carcass; the arrow shows the direction of rotation. They will wear very rapidly if the rotation is reversed. Studded tires will lose their studs if their rotational direction is reversed.

NOTE: *Mark the wheel position or direction or rotation on radial tires or studded snow tires before removing them.*

STORAGE

Store the tires at the proper inflation pressure if they are mounted on wheels. Keep them in a cool dry place, laid on their sides. If the tires are stored in the garage or basement, do not let them stand on a concrete floor; set them on strips of wood.

FLUIDS AND LUBRICANTS

Oil and Fuel Recommendations

Oil

The SAE (Society of Automotive Engineers) grade number indicates the viscosity of the engine oil and thus its ability to lubricate a

GENERAL INFORMATION AND MAINTENANCE

A penny works as well as anything for checking tire tread depth; when you can see the top of Lincoln's head, it's time for a new tire

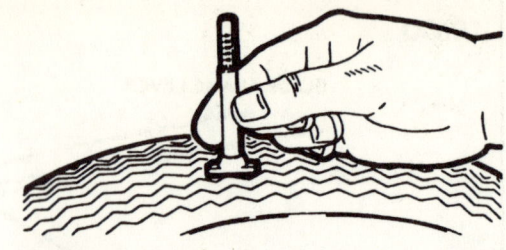

Tread depth can be checked with an inexpensive gauge

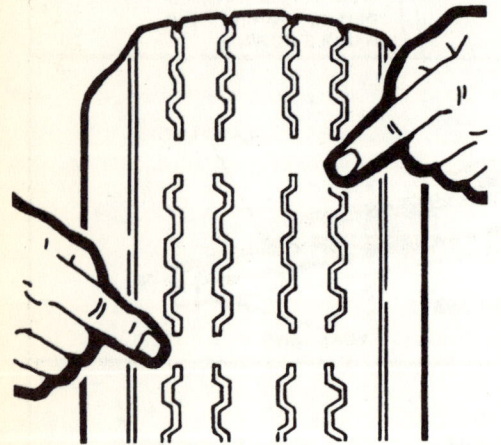

Tread wear indicators will appear when the tire is worn out

given temperature. The lower the SAE grade number, the lighter the oil; the lower the viscosity, the easier it is to crank the engine in cold weather.

Oil viscosities should be chosen from those oils recommended for the lowest anticipated temperatures during the oil change interval.

Multi-viscosity oils (10W-30, 20W-50 etc.) offer the important advantage of being adaptable to temperature extremes. They allow easy starting at low temperatures, yet they give good protection at high speeds and engine temperatures. This is a decided advantage in changeable climates or in long distance touring.

The API (American Petroleum Institute) designation indicates the classification of engine oil used under certain given operating conditions. Only oils designated for use Service SG should be used. Oils of the SG type perform a variety of functions inside the engine in addition to their basic function as a lubricant. Through a balanced system of metallic detergents and polymeric dispersants, the oil prevents the formation of high and low tempera-

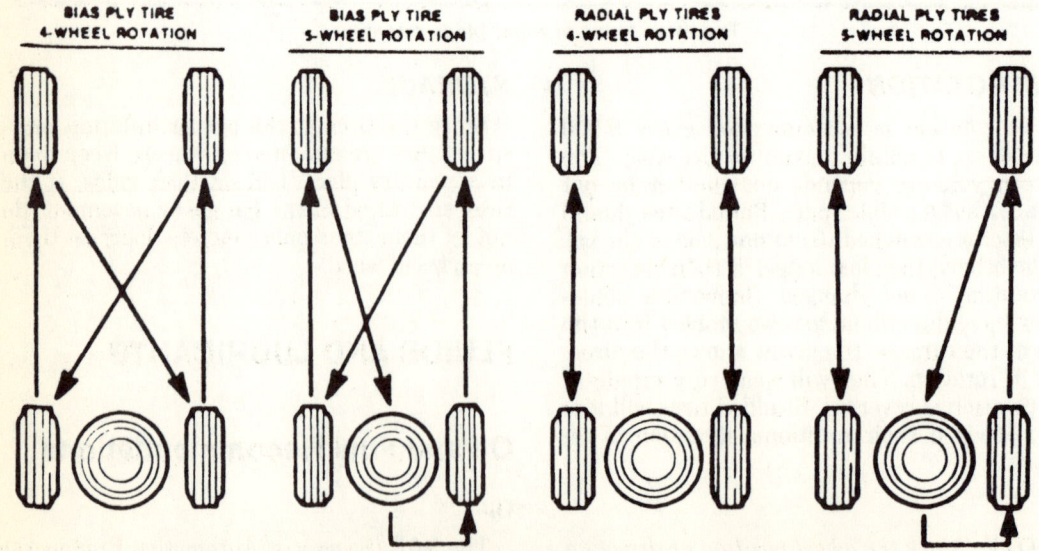

Tire rotation diagrams: note that radials should not be cross-switched

GENERAL INFORMATION AND MAINTENANCE

ture deposits and also keeps sludge and particles of dirt in suspension. Acids, particularly sulfuric acid, as well as other byproducts of combustion, are neutralized. Both the SAE grade number and the API designation can be found on top of the oil can.

For recommended oil viscosities, refer to the chart.

Synthetic Oil

There are excellent synthetic and fuel-efficient oils available that, under the right circumstances, can help provide better fuel mileage and better engine protection. However, these advantages come at a price, which can be three or four times the price per quart of conventional motor oils.

Before pouring any synthetic oils into your car's engine, you should consider the condition of the engine and the type of driving you do. Also, check the car's warranty conditions regarding the use of synthetics.

Generally, it is best to avoid the use of synthetic oil in both brand new and older, high mileage engines. New engines require a proper break-in, and the synthetics are so slippery that they can prevent this; most manufacturers recommend that you wait at least 5,000 miles before switching to a synthetic oil. Conversely, older engines are looser and tend to use more oil; synthetics will slip past worn parts more readily than regular oil, and will be used up faster. If you car already leaks and/or uses oil (due to worn parts or bad seals or gaskets), it will leak and use more with a slippery synthetic inside.

Consider your type of driving. If most of your accumulated mileage is on the highway at higher, steadier speed, a synthetic oil will reduce friction and probably help deliver better

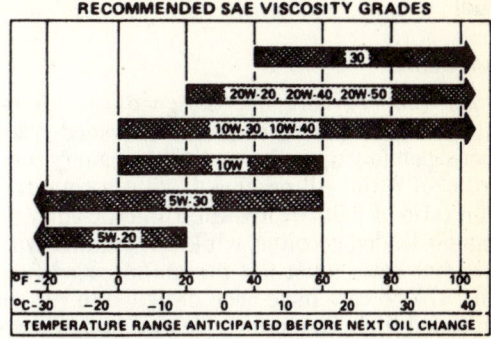

NOTICE: Do not use SAE 5W-20 oils for continuous high-speed driving.

Oil viscosity chart

fuel mileage. Under such ideal highway conditions, the oil change interval can be extended, as long as the oil filter will operate effectively for the extended life of the oil. If the filter can't do its job for this extended period, dirt and sludge will build up in your engine's crankcase, sump, oil pump and lines, no matter what type of oil is used. If using synthetic oil in this manner, you should continue to change the oil filter at the recommended intervals.

Vehicles used under harder, stop-and-go, short hop circumstances should always be serviced more frequently, and for these vehicles synthetic oil may not be a wise investment. Because of the necessary shorter change interval needed for this type of driving, you cannot take advantage of the long recommended change interval of most synthetic oils.

Finally, most synthetic oils are not compatible with conventional oils and cannot be added to them. This means you should always carry a couple of quarts of synthetic oil with you while on a long trip, as not all service stations carry this oil.

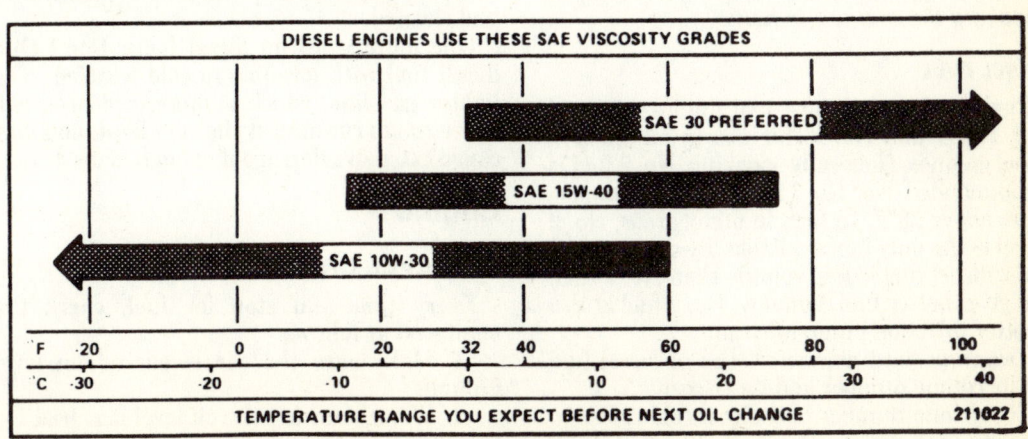

Diesel oil viscosity chart

Fuel

GASOLINE

All 1968-74 cars are designed to run on either regular leaded or premium leaded grade fuel depending upon the particular engine's compression ratio. All engines having a compression ratio of 9.0:1 or less can run efficiently on regular leaded gasoline, while any engines with a higher ratio must use premium leaded fuel. All 1975-88 cars have been designed to run on unleaded fuel. The use of a leaded fuel in a car requiring unleaded fuel will plug the catalytic converter and render it inoperative. It will also increase exhaust backpressure to the point where engine output will be severely reduced. In all cases, the minimum octane rating of the unleaded fuel being used must be at least 91 RON (87 CLC). All unleaded fuels sold in the U.S. are required to meet this minimum rating.

The use of a fuel too low in octane (a measurement of anti-knock quality) will result in spark knock. Since many factors such as altitude, terrain, air temperature and humidity affect operating efficiency, knocking may result even though the recommended fuel is being used. If persistent knocking occurs, it may be necessary to switch to a higher grade of fuel. Continuous or heavy knocking may result in engine damage.

NOTE: *Your engine's fuel requirement can change with time, mainly due to carbon buildup, which will in turn change the compression ratio. If your engine pings, knocks, or diesels (runs with the ignition off) switch to a higher grade of fuel. Sometimes just changing brands will cure the problem. If it becomes necessary to retard the timing from the specifications, don't change it more than a few degrees. Retarded timing will reduce power output and fuel mileage, in addition to making the engine run hotter.*

DIESEL FUEL

Fuel markers produce two grades of diesel fuel, No. 1 and No. 2, for use in automotive diesel engines. Generally speaking, No. 2 fuel is recommended over No. 1 for driving in temperature above 20°F. In fact, in many areas, No. 2 diesel is the only fuel available. By comparison, No. 2 diesel fuel is less volatile than No. 1 fuel, and gives better fuel economy. No. 2 fuel is also a better injection pump lubricant.

Two important characteristics of diesel fuel are its cetane number and its viscosity.

The cetane number of a diesel fuel refers to the ease with which a diesel fuel ignites. High cetane numbers mean that the fuel will ignite with relative ease or that it ignites well at low temperatures. Naturally, the lower the cetane number, the higher the temperature must be to ignite the fuel. Most commercial fuels have cetane numbers that range from 35 to 65. No. 1 diesel fuel generally has a higher cetane rating than No. 2 fuel.

Viscosity is the ability of a liquid, in this case diesel fuel, to flow. Using straight No. 2 diesel fuel below 20°F can cause problems, because this fuel tends to become cloudy, meaning wax crystals begin forming in the fuel (20°F is often call the cloud point for No. 2 fuel). In extreme cold weather, No. 2 fuel can stop flowing altogether. In either case, fuel flow is restricted, which can result in a no start condition or poor engine performance. Fuel manufacturers often winterize No. 2 diesel fuel by using various fuel additives and blends (NO. 1 diesel fuel, kerosene, etc.) to lower its winter time viscosity. Generally speaking, though, No. 1 diesel fuel is more satisfactory in extremely cold weather.

NOTE: *No. 1 and No. 2 diesel fuels will mix and burn with no ill effects, although the engine manufacturers will undoubtedly recommend one or the other. Consult the owner's manual for information.*

Depending on local climate, most fuel manufacturers make winterized No. 2 fuel available seasonally.

Many automobile manufacturers publish pamphlets giving the locations of diesel fuel stations nationwide. Contact the local dealer for information.

NOTE: *Do not substitute home heating oil for automotive diesel fuel.*

While in some cases, home heating oil refinement levels equal those of diesel fuel, many times they are far below diesel engine requirements. The result of using dirty home heating oil will be a clogged fuel system, in which case the entire system may have to be dismantled and cleaned.

One more word on diesel fuels. Don't thin diesel fuel with gasoline in cold weather. The lighter gasoline, which is more explosive, will cause rough running at the very least, and may cause extensive damage if enough is used.

Engine

OIL LEVEL CHECK

Every time you stop for fuel, check the engine oil as follows:

1. Make sure the car is parked on level ground.
2. When checking the oil level it is best for the engine to be a normal operating temperature, although checking the oil immediately after stopping will lead to a false reading. Wait

GENERAL INFORMATION AND MAINTENANCE

a few minutes after turning off the engine to allow the oil to drain back into the crankcase.

3. Open the hood and locate the dipstick which will be on either the right or left side depending upon your particular engine. Pull the dipstick from its tube, wipe it clean and then reinsert it.

4. Pull the dipstick out again and, holding it horizontally, read the oil level. The oil should be between the FULL and ADD marks on the dipstick. If the oil is below the ADD mark, add oil of the proper viscosity through the capped opening in the top of the cylinder head cover. See the Oil and Fuel Recommendations chart in this chapter for the proper viscosity and rating of oil to use.

5. Replace the dipstick and check the oil level again after adding any oil. Be careful not to over fill the crankcase. Approximately one quart of oil will raise the level from the ADD mark to the FULL mark. Excess oil will generally be consumed at an accelerated rate.

OIL AND FILTER CHANGE

The oil should be changed every four months or 6,000 miles on all 1968-74 cars. On 1975-78 cars, the interval is six months or 7,500 miles. 1979-92 cars increased the time interval to 12 months while keeping the mileage (7,500) the same (Diesel interval is 5,000 miles). Make sure that you change the oil based on whichever interval comes first.

NOTE: *Chevrolet recomends changing the filter at every other oil change. WE recommend a new filter with EVERY oil change, you could only benefit from this practice.*

The oil drain plug is located on the bottom of the oil pan (bottom of the engine, underneath the car). the oil filter is located on the right side of the inline 6-cylinder engine and on the left side of all other engines.

The mileage figures given are the Chevrolet recommended intervals assuming normal driving and conditions. If your car is used under dusty, polluted or off-road conditions, change the oil and filter more often than specified. The

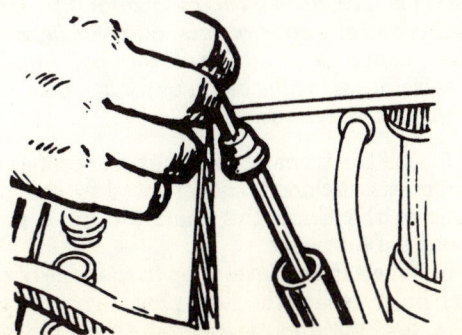

The oil level is checked with the dipstick

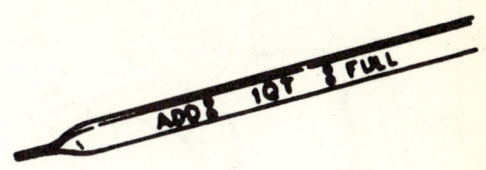

The oil level should be between the ADD and FULL marks on the dipstick

same goes for cars driven in stop-and-go traffic or only for short distances at a time. Always drain the engine oil after the engine has been running long enough to bring it up to normal operating temperature. Hot oil will flow easier and more contaminants will be removed along with the oil than if it were drained cold. To change the oil and filter:

1. Run the engine until it reaches normal operating temperature.
2. Raise and support the car safely.
3. Slide a drain pan of a least 6 quarts capacity under the oil pan.
4. Loosen the drain plug. Turn the plug out by hand. By keeping an inward pressure on the plug as you unscrew it, oil won't escape past the threads and you can remove it without being burned by hot oil.
5. Allow the oil to drain completely and then install the drain plug. don't overtighten the plug, or you'll be buying a new pan or a trick replacement plug for stripped threads.
6. Using a strap wrench, remove the oil filter. Keep in mind that it's holding about one quart of dirty, hot oil.

NOTE: *A Chilton environmental tip. Used engine oil contains heavy metals, and has been determined to be hazardous to the environment. Recycling engine oil is the best way for disposal. Recycle whenever possible.*

7. Empty the old filter into the drain pan and dispose of the filter.
8. Using a clean rag, wipe off the filter adapter on the engine block. Be sure that the rag doesn't leave any lint which could clog an oil passage.
9. Coat the rubber gasket on the filter with fresh oil. Spin it onto the engine by hand; when the gasket touches the adapter surface, give it another $1/2$-$1/3$ turn. No more, or you'll squash the gasket and it will leak.
10. Refill the engine with the correct amount of fresh oil. See the Capacities chart.
11. Check the oil level on the dipstick. It is normal for the level to be a bit above the full mark. Start the engine and allow it to idle for a few minutes.

CAUTION: *Do not run the engine above idle speed until it has built up oil pressure, indicated when the oil light goes out.*

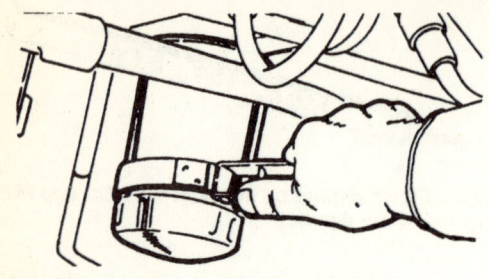

Remove the oil filter with a strap wrench

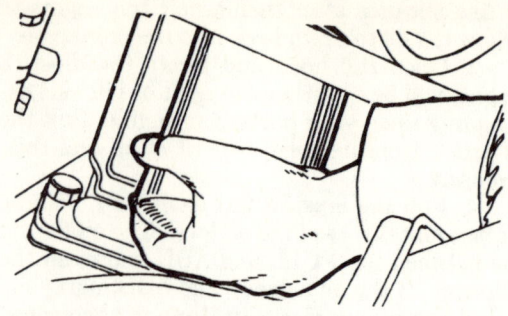

Install the new oil filter by hand

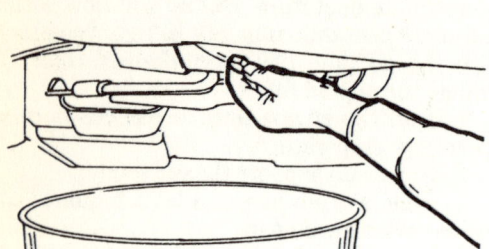

By keeping an inward pressure on the plug as you unscrew it, oil won't escape past the threads

Coat the new oil filter gasket with clean oil

12. Shut off the engine, allow the oil to drain for a minute, and check the oil level. Check around the filter and drain plug for any leaks, and correct as necessary.

Manual Transmission

FLUID RECOMMENDATIONS AND LEVEL CHECK

The oil in the manual transmission should be checked at least every 6,000 miles for 1968-74 cars or every 7,500 miles for all 1975 and later cars.

1. With the car parked on a level surface, remove the filler plug from the side of the transmission housing.

2. If the lubricant begins to trickle out of the hole, there is enough and you need not go any further. Otherwise, carefully insert your finger (watch out for sharp threads) and check to see if the oil is up the edge of the hole.

3. If not, add oil through the hole until the level is at the edge of the hole. Most gear lubricants come in a plastic squeeze bottle with a nozzle; making additions simple. You can also use a common kitchen baster. Use only standard GL-5 hypoid-type gear oil - SAE 80W or SAE 80W/90.

4. Replace the filler plug, run the engine and check for leaks.

DRAIN AND REFILL

There is no recommended interval for the manual transmission but it is always a good idea to change the fluid if you have purchased the car used or if it has been driven in water high enough to reach the axles.

1. The oil must be hot before it is drained. Drive the car until the engine reaches normal operating temperature.

2. Remove the filter plug to provide a vent.

3. Place a large container underneath the transmission and then remove the drain plug.

4. Allow the oil to drain completely. Clean off the drain plug and replace it; tighten it until it is just snug.

NOTE: *A Chilton environmental tip. Used oil contains heavy metals, and has been determined to be hazardous to the environment. Recycling oil is the best way for disposal. Recycle whenever possible.*

5. Fill the transmission with the proper lubricant as detained earlier in this chapter. Refer to the Capacities chart for the correct amount of lubricant.

6. When the oil level is up to the edge of the filler hole, replace the filler plug. Drive the car for a few minutes, stop, and check for any leaks.

GENERAL INFORMATION AND MAINTENANCE

Automatic Transmission

FLUID RECOMMENDATIONS AND LEVEL CHECK

Check the automatic transmission fluid level at least every 6,000 miles (7,500 miles for 1975-92 cars). The dipstick can be found in the rear of the engine compartment. The fluid level should be checked only when the transmission is hot (normal operating temperature). The transmission is considered hot after about 20 miles of highway driving.

1. Park the car on a level surface with the engine idling. Shift the transmission into Neutral and set the parking brake.
2. Remove the dipstick, wipe it clean and then reinsert it firmly. Be sure that it has been pushed all the way in. Remove the dipstick again and check the fluid level while holding it horizontally. With the engine running, the fluid level should be between the second notch and the FULL HOT line. If the fluid must be checked when it is cool, the level should be between the first and second notches.
3. If the fluid level is below the second notch (engine hot) or the first notch (engine cold), add DEXRON® (1968-75) or DEXRON®II (1976-92) automatic transmission fluid through the dipstick tube. this is easily done with the aid of a funnel. Check the level often as you are filling the transmission, be extremely careful not to overfill it. Overfilling will cause slippage, seal damage and overheating. Approximately one pint of ATF will raise the fluid level from one notch/line to the other.

NOTE: *Always use DEXRON® or DEXRON®II AFT. The use of AFT Type F or any other fluid will cause severe damage to the transmission.*

The fluid on the dipstick should always be a bright red color. If it is discolored (brown or black), or smells burnt, serious transmission troubles, probably due to overheating, should be suspected. The transmission should be inspected by a qualified technician to locate the cause of the burnt fluid.

DRAIN AND REFILL PAN AND FILTER SERVICE

The procedures for automatic transmission fluid drain and refill, filter change and band adjustment are all detailed in Chapter 7.

Rear Axle

FLUID RECOMMENDATIONS AND LEVEL CHECK

The oil in the differential should be checked

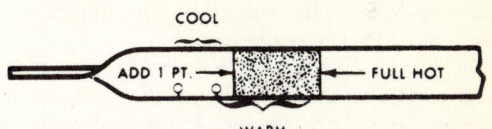

Automatic transmission dipstick marks; the proper level is within the shaded area

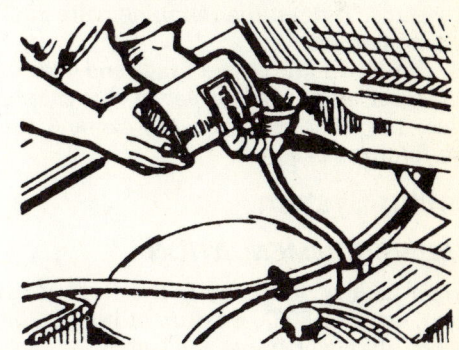

Add automatic transmission fluid through the dipstick tube

at least every 6,000 miles (7,500 miles for 1975-92 cars.

1. With the car on a level surface, remove the filler plug from the front side of the differential.
2. If the oil begins to trickle out of the hole, there is enough. Otherwise, carefully insert you finger (watch out for sharp threads) into the hold and check that the oil is up the bottom edge of the filler hole.
3. If not, add oil through the hole until the level is at the edge of the hole. Most gear oils come in a plastic squeeze bottle with a nozzle; making additions is simple. You can also use a common kitchen baster. Use only standard GL-5 hypoid-type gear oil - SAE 80W or SAE 80W/90.

NOTE: *On all cars equipped with the positraction/limited slip rear axle, GM recommends that you use only the special lubricant which is available at your local Chevrolet parts department.*

DRAIN AND REFILL

There is no recommended change interval for the rear axle but it is always a good idea to change the fluid if you have purchased the car used or if it has been driven in water high enough to reach the axle.

1. Park the car on a level surface and set the parking brake.
2. Remove the filler plug.
3. Place a large container underneath the rear axle.
4. Unscrew the retaining bolts and remove

46 GENERAL INFORMATION AND MAINTENANCE

the rear cover. This will allow the lubricant to drain out into the container.

NOTE: *A Chilton environmental tip. Used oil contains heavy metals, and has been determined to be hazardous to the environment. Recycling oil is the best way for disposal. Recycle whenever possible.*

5. Install the rear cover using a new gasket and sealant. Tighten the retaining bolts in the crosswise pattern.

6. Refill with the proper grade and quantity of lubricant as detailed earlier in this chapter. Replace the filler plug, run the car and then check for any leaks.

Cooling System

FLUID RECOMMENDATION

When adding or changing the fluid in the system, create a 50/50 mixture of high quality ethylene glycol antifreeze and water.

LEVEL CHECK

The fluid level may be checked by observing the fluid level marks of the recovery tank. The level should be below the ADD mark when the system is cold. At normal operating temperatures, the level should be between the ADD and the FULL marks. Only add coolant to bring the level to the FULL mark.

CAUTION: *Should it be necessary to remove the radiator cap, make sure that the system has had time to cool, reducing the internal pressure.*

DRAIN, FLUSH AND REFILL

The cooling system should be drained, thoroughly flushed and refilled at least every 30,000 miles or 24 months. These operations should be done with the engine cold.

1. Remove the radiator and recovery tank caps. Run the engine until the upper radiator hose gets hot. This means that the thermostat is open and the coolant is flowing through the system.

2. Turn the engine OFF and place a large container under the radiator. Open the drain valve at the bottom of the radiator. Open the block drain plugs to speed up the draining process.

CAUTION: *When draining the coolant, keep in mind that cats and dogs are attracted by the ethylene glycol antifreeze, and are quite likely to drink any that is left in an uncovered container or in puddles on the ground. This will prove fatal in sufficient quantity. Always drain the coolant into a sealable container. Coolant should be reused unless it is contaminated or several years old.*

3. Close the drain valves and add water until the system is full. Repeat the draining and filling process several times, until the liquid is nearly colorless.

4. After the last draining, fill the system with a 50/50 mixture of ethylene glycol and water. Run the engine until the system is hot and add coolant, if necessary. Replace the caps and check for any leaks.

Brake Master Cylinder

FLUID RECOMMENDATIONS AND LEVEL CHECK

The brake master cylinder is located under the hood, in the left rear section of the engine compartment. It is divided into two sections (reservoirs) and the fluid must be kept within 1/4" of the top edge of both reservoirs. The level should be checked at least every 6,000 miles (7,500 miles for 1975-92 cars).

NOTE: *Any sudden decrease in the level of fluid indicates a possible leak in the system*

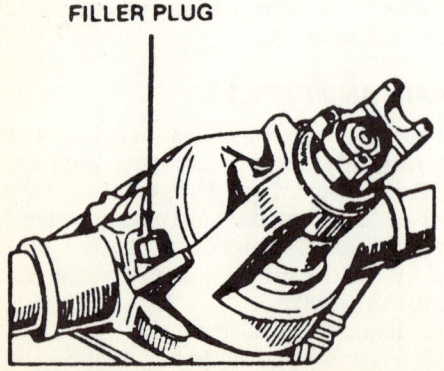

Remove the filter plug to check the lubricant level in the rear axle

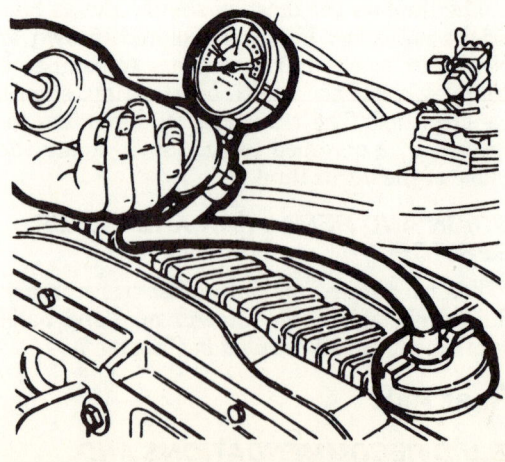

The system should be pressure tested at least once a year

GENERAL INFORMATION AND MAINTENANCE

Coolant protection can be checked with a simple float-type tester

wheels are turned straight ahead. If the level is low, fill the pump reservoir with DEXRON® Automatic Transmission Fluid on 1976 and earlier cars. 1977-92 cars require GM power steering fluid, until the fluid level measures full on the reservoir dipstick. Low fluid level usually produces a moaning sound as the wheels are turned (especially when standing still or parking) and increases steering wheel effort.

Chassis Greasing

Chassis greasing can be performed with a pressurized grease gun or it can be performed at home by using a hand-operated grease gun. Wipe the grease fittings clean before greasing in order to prevent the possibility of forcing any dirt into the component.

Body Lubrication

Transmission Shift Linkage

Lubricate the manual transmission shift linkage contact points with the EP grease used for chassis greasing, which should meet GM specification 6031M. The automatic transmission

and should be checked out immediately.

To check the fluid level, simply pry off the retaining bar and then lift off the top cover of the master cylinder. When making additions of brake fluid, use only fresh, uncontaminated brake fluid which meets or exceeds DOT 3 standards. Be careful not to spill any brake fluid on painted surfaces, as it eats paint. Do not allow the brake fluid container or the master cylinder reservoir to remain open any longer than necessary; brake fluid absorbs moisture from the air, reducing its effectiveness and causing corrosion in the lines. Although it is not mandatory We recommend you perododdicly flush the brake lines with new brake fluid, this will help stop the corrosion buildup in the brake system. Use the brake bleeding method, described in chapter 9, to accomplish this procedure.

CAUTION: *When flushing the brake system DO NOT use any solvents or cleaners! Use only new brake fluid which meets or exceeds DOT 3 standards.*

NOTE: *The reservoir cover on some later models (1978-92) may be without a retaining bail. If so, simply pry the cover off with your fingers.*

Power Steering Pump

FLUID RECOMMENDATIONS AND LEVEL CHECKS

Power steering fluid level should be checked at least once every 6,000 miles (7,500 miles for 1975-92 cars). To prevent possible overfilling, check the fluid level only when the fluid has warmed to operating temperatures and the

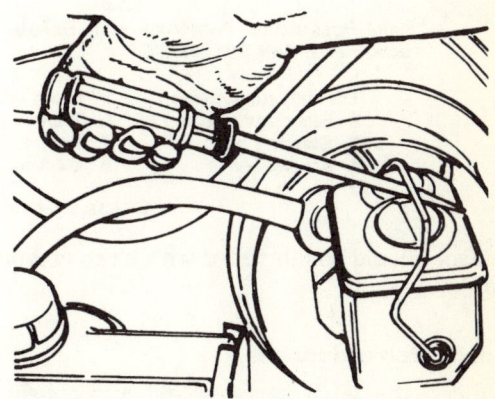

Pry the retaining bail from the master cylinder reservoir cap to check the fluid level

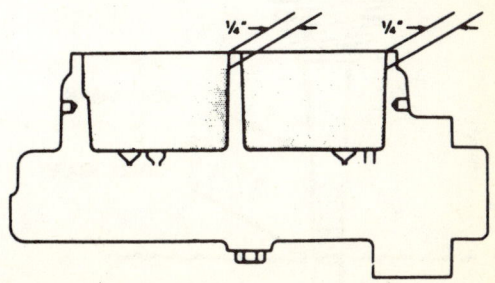

The fluid level in the master cylinder reservoir should be within 1/4 in. of the top edge.

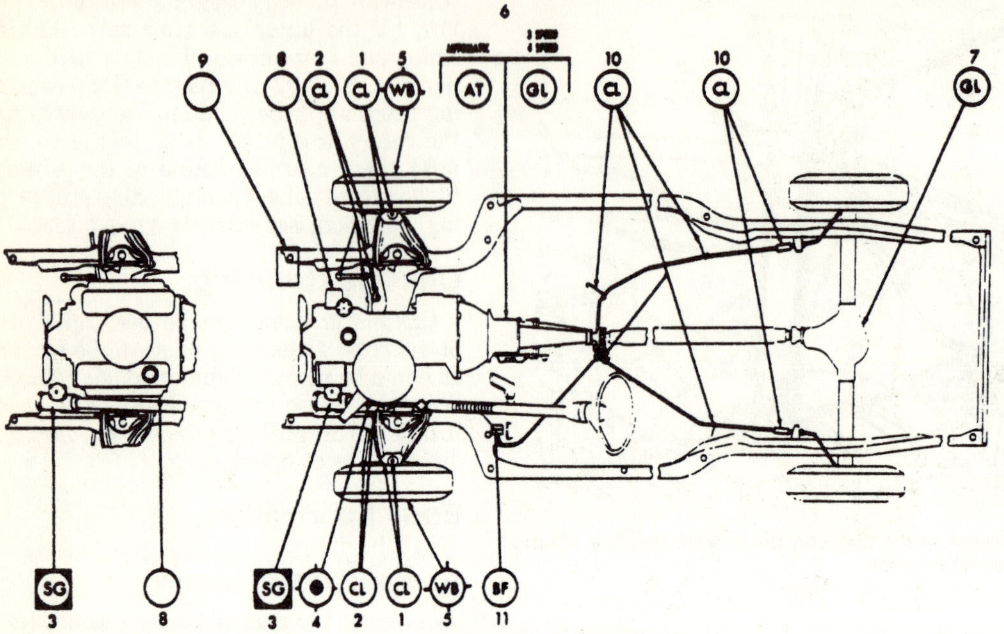

48 GENERAL INFORMATION AND MAINTENANCE

- ⊙ Lubricate every 6000 miles
- ⊙ Replace every 12,000 miles
- ⊙ Replace every 24,000 miles
- ○ Check for grease leakage every 36,000 miles

GL —Multi-purpose Gear Lubricant*
WB —Wheel Bearing Lubricant
CL —Chassis Lubricant
AT —DEXRON® Automatic Transmission Fluid
BF —Hydraulic Brake Fluid
SG —Steering Gear Lubricant

*Refill Positraction Rear Axle with Special Positraction Rear Axle Lubricant Only.

1. Front suspension
2. Steering linkage
3. Steering gear
4. Air cleaner
5. Front wheel bearings
6. Transmission
*7. Rear axle
8. Oil filter
9. Battery
10. Parking brake
11. Brake master cylinder

Chassis lubrication, 1975 and later

linkage should be lubricated with clean engine oil.

Hood Latch and Hinges

Clean the latch surfaces and apply clean engine oil to the latch pilot bolts and the spring anchor. Use the engine oil to lubricate the hood hinges as well. Use a chassis grease to lubricate all the pivot points in the latch release mechanism.

Door Hinges

The gas tank filler door, car door, and rear hatch or trunk lid hinges should be wiped clean and lubricated with clean engine oil. Silicone spray also works well on these parts, but must be applied more often. Use engine oil to lubricate the trunk or hatch lock mechanism and the lock bolt and striker. The door lock cylinders can be lubricated easily with a shot of silicone spray or one of the may dry penetrating lubricants commercially available.

Parking Brake Linkage

Use chassis grease on the parking brake cable where it contacts the guides, links, levers, and pulleys. The grease should be water resistant one for durability under the car.

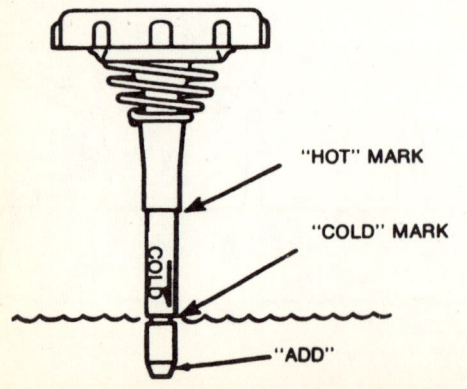

Use the dipstick to check the power steering fluid

GENERAL INFORMATION AND MAINTENANCE

Accelerator Linkage

Lubricate the carburetor stud, carburetor lever, and the accelerator pedal lever at the support inside the car with clean engine oil.

Wheel Bearings

Once every 12 months or 12,000 miles, clean and repack wheel bearings with a wheel bearing grease. Use only enough grease to completely coat the rollers. Remove any excess grease from the exposed surface of the hub and seal.

It is important that wheel bearings be properly adjusted after installation. Improperly adjusted wheel bearings can cause steering instability, front-end shimmy and wander, and increased tire wear. For complete lubrication and adjustment procedures, see the Wheel Bearing section in Chapter 8.

TRAILER TOWING

General Recommendations

Your car was primarily designed to carry passengers and cargo. It is important to remember that towing a trailer will place additional loads on your car's engine, drive train, steering, braking and other systems. However, if you find it necessary to tow a trailer, using the proper equipment is a must.

Local laws may require specific equipment such as trailer brakes or fender mounted mirrors. Check your local laws.

Trailer Weight

The weight of the trailer is the most important factor. A good weight-to-horsepower ratio is about 35:1, 35 lbs. of GCW (Gross Combined Weight) for every horsepower your engine develops. Multiply the engine's rated horsepower by 35 and subtract the weight of the car passengers and luggage. The result is the approximate ideal maximum weight you should tow, although a numerically higher axle ratio can help compensate for heavier weight.

Hitch Weight

Figure the hitch weight to select a proper hitch. Hitch weight is usually 9-11% of the trailer gross weight and should be measured with the trailer loaded. Hitches fall into three types: those that mount on the frame and rear bumper or the bolt-on or weld-on distribution type used for larger trailers. Axle mounted or clamp-on bumper hitches should never be used.

Check the gross weight rating of your trailer. Tongue weight is usually figured as 10% of gross trailer weight. Therefore, a trailer with a maximum gross weight of 2,000 lb. will have a maximum tongue weight of 200 lb. Class I trailers fall into this category. Class II trailers are those with a gross weight rating of 2,000-3,500 lb., while Class III trailers fall into the 3,500-6,000 lb. category. Class IV trailers are those over 6,000 lb. and are for use with fifth wheel trucks, only.

When you've determined the hitch that you'll need, follow the manufacturer's installation instructions, exactly, especially when it comes to fastener torques. The hitch will be subjected to a lot of stress and good hitches come with hardened bolts. Never substitute an inferior bolt for a hardened bolt.

Cooling

ENGINE

One of the most common, if not THE most common, problems associated with trailer towing is engine overheating.

If you have a standard cooling system, without an expansion tank, you'll definitely need to get an aftermarket expansion tank kit, preferably one with at least a 2 quart capacity. These kits are easily installed on the radiator's overflow hose, and come with a pressure cap designed for expansion tanks.

Another helpful accessory is a Flex Fan. These fans are large diameter units and are designed to provide more airflow at low speeds, with blades that have deeply cupped surfaces. The blades then flex, or flatten out, at high speed, when less cooling air is needed. These fans are far lighter in weight than stock fans, requiring less horsepower to drive them. Also, they are far quieter than stock fans.

If you do decide to replace your stock fan with a flex fan, note that if your car has a fan clutch, a spacer between the flex fan and water pump hub will be needed.

Aftermarket engine oil coolers are helpful for prolonging engine oil life and reducing overall engine temperatures. Both of these factors increase engine life.

While not absolutely necessary in towing Class I and some Class II trailers, they are recommended for heavier Class II and all Class III towing.

Engine oil cooler systems consist of an adapter, screwed on in place of the oil filter, a remote filter mounting and a multi-tube, finned heat exchanger, which is mounted in front of the radiator or air conditioning condenser.

TRANSMISSION

An automatic transmission is usually recommended for trailer towing. Modern automatics have proven reliable and, of course, easy to operate, in trailer towing.

The increased load of a trailer, however, causes an increase in the temperature of the automatic transmission fluid. Heat is the worst enemy of an automatic transmission. As the temperature of the fluid increases, the life of the fluid decreases.

It is essential, therefore, that you install an automatic transmission cooler.

The cooler, which consists of a multi-tube, finned heat exchanger, is usually installed in front of the radiator or air conditioning compressor, and hooked inline with the transmission cooler tank inlet line. Follow the cooler manufacturer's installation instructions.

Select a cooler of at least adequate capacity, based upon the combined gross weights of the car and trailer.

Cooler manufacturers recommend that you use an aftermarket cooler in addition to, and not instead of, the present cooling tank in your radiator. If you do want to use it in place of the radiator cooling tank, get a cooler at least two sizes larger than normally necessary.

NOTE: *A transmission cooler can, sometimes, cause slow or harsh shifting in the transmission during cold weather, until the fluid has a chance to come up to normal operating temperature. Some coolers can be purchased with or retrofitted with a temperature bypass valve which will allow fluid flow through the cooler only when the fluid has reached operating temperature, or above.*

Handling A Trailer

Towing a trailer with ease and safety requires a certain amount of experience. It's a good idea to learn the feel of a trailer by practicing turning, stopping and backing in an open area such as an empty parking lot.

PUSHING AND TOWING

Push Starting

This is the last recommended method of starting a car and should be used only in an extreme case. Chances of body damage are high, so be sure that the pushcar's bumper does not override your bumper. If your Chevrolet has an automatic transmission it cannot be push started. In an emergency, you can start a manual transmission equipped car by pushing. With the bumpers evenly matched, get in your car, switch on the ignition, and place the gearshift in Second or Third gear. Do not engage the clutch. Start off slowly. When the speed of the car reaches about 15-20 mph, release the clutch.

Towing

The car can be towed safely (with the transmission in Neutral) from the front at speeds of 35 mph or less. The car must either be towed with the rear wheels off the ground or the driveshaft disconnected if: towing speeds are to be over 35 mph, or towing distance is over 50 miles, or transmission or rear axle problems exist.

When towing the car on its front wheels, the steering wheel must be secured in a straight-ahead position and the steering column unlocked. Tire-to-ground clearance should not exceed 6″ during towing.

JUMP STARTING A DUAL-BATTERY DIESEL

Vehicles equipped with a diesel engine utilize two 12 volt batteries, one on either side of the engine compartment. The batteries are connected in a parallel circuit (positive terminal to positive terminal, negative terminal to negative terminal). Hooking the batteries up in parallel circuit increases battery cranking power without increasing total battery voltage output. Output remains at 12 volts. On the other hand, hooking two 12 volt batteries up in a series circuit (positive terminal to negative terminal, positive terminal to negative terminal) increases total battery output to 24 volts (12 volts plus 12 volts).

CAUTION: *NEVER hook the batteries up in a series circuit or the entire electrical system will go up in smoke, especially the starter!*

In the event that a diesel needs to be jump started, use the following procedure.

1. Turn all lights off.
2. Turn on the heater blower motor to remove transient voltage.
3. Connect one jumper cable to the passenger side battery positive (+) terminal and the other cable clamp to the positive (+) terminal to the booster (good) battery.
4. Connect one end of the other jumper cable to the negative (−) terminal of the booster (good) battery and the other cable clamp to an engine bolt head, alternator bracket or other solid, metallic point on the diesel engine. DO NOT connect this clamp to the negative (−) terminal of the bad battery.

GENERAL INFORMATION AND MAINTENANCE

CAUTION: *Be very careful to keep the jumper cables away from moving parts (cooling fan, belts, etc.) on both engines.*

5. Start the engine of the donor car and run it at moderate speed.
6. Start the engine of the diesel.
7. When the diesel starts, remove the cable from the engine block before disconnecting the positive terminal.

JACKING

The standard jack utilizes slots in the bumper to raise the car. the jack supplied with the car should never be used for any service operation other than tire changing. Never get under the car while it is supported by only a jack. Always block the wheels when changing tires.

There service operations in this book often require that one end or the other, or both, of the car be raised and safely supported. The ideal method, of course, would be a hydraulic hoist. Since this is beyond both the resource and requirement of the do-it-yourselfer, a small hydraulic, screw or scissors jack will suffice for the procedures in this guide. Two sturdy jackstands should be acquired if you intend to work under the car at any time. An alternate method of raising the car would be drive-on ramps. These are available commercially or can be fabricated from heavy boards or steel. Be sure to block the wheels when using ramps. Never use concrete blocks to support the car. They may break if the load is not evenly distributed.

Regardless of the method of jacking or hoisting the car, there are only certain areas of the undercarriage and suspension you can safely use to support it. See the illustration below, and make sure that only the shaded areas are used. In addition, be especially careful on cars built after 1974 that you do not damage the catalytic converter. Remember that various cross braces and supports on a lift can sometimes contact low hanging parts of the car.

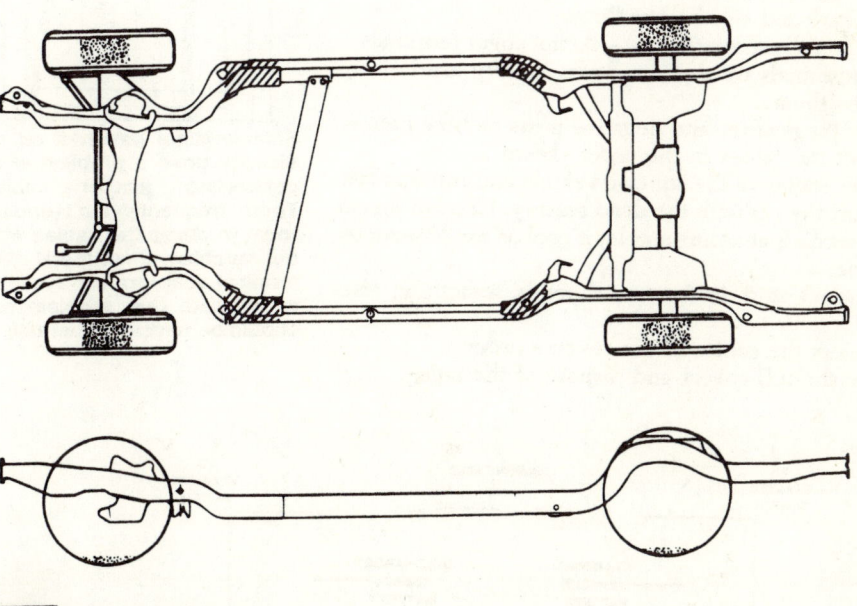

DRIVE ON HOIST

FLOOR JACK OR JOIST LIFT: DO NOT LIFT AT REAR AXLE WHEN EQUIPPED WITH REAR STABILIZER.

Vehicle hoisting and jacking points

GENERAL INFORMATION AND MAINTENANCE

JUMP STARTING A DEAD BATTERY

The chemical reaction in a battery produces explosive hydrogen gas. This is the safe way to jump start a dead battery, reducing the chances of an accidental spark that could cause an explosion.

Jump Starting Precautions

1. Be sure both batteries are of the same voltage.
2. Be sure both batteries are of the same polarity (have the same grounded terminal).
3. Be sure the vehicles are not touching.
4. Be sure the vent cap holes are not obstructed.
5. Do not smoke or allow sparks around the battery.
6. In cold weather, check for frozen electrolyte in the battery. Do not jump start a frozen battery.
7. Do not allow electrolyte on your skin or clothing.
8. Be sure the electrolyte is not frozen.

CAUTION: *Make certain that the ignition key, in the vehicle with the dead battery, is in the OFF position. Connecting cables to vehicles with on-board computers will result in computer destruction if the key is not in the OFF position.*

Jump Starting Procedure

1. Determine voltages of the two batteries; they must be the same.
2. Bring the starting vehicle close (they must not touch) so that the batteries can be reached easily.
3. Turn off all accessories and both engines. Put both cars in Neutral or Park and set the handbrake.
4. Cover the cell caps with a rag—do not cover terminals.
5. If the terminals on the run-down battery are heavily corroded, clean them.
6. Identify the positive and negative posts on both batteries and connect the cables in the order shown.
7. Start the engine of the starting vehicle and run it at fast idle. Try to start the car with the dead battery. Crank it for no more than 10 seconds at a time and let it cool off for 20 seconds in between tries.
8. If it doesn't start in 3 tries, there is something else wrong.
9. Disconnect the cables in the reverse order.
10. Replace the cell covers and dispose of the rags.

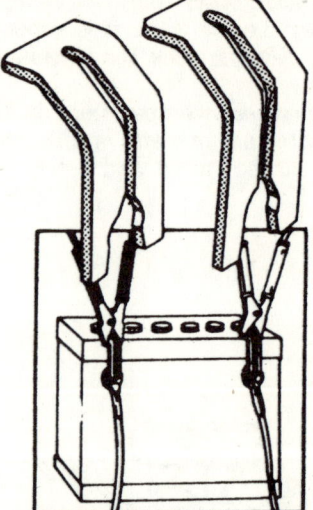

Side terminal batteries occasionally pose a problem when connecting jumper cables. There frequently isn't enough room to clamp the cables without touching sheet metal. Side terminal adaptors are available to alleviate this problem and should be removed after use.

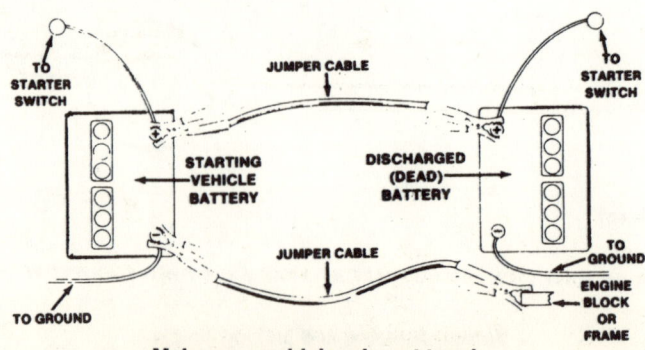

Make sure vehicles do not touch

This hook–up for negative ground cars only

GENERAL INFORMATION AND MAINTENANCE

Maintenance Intervals
Diesel Engined Cars

When to Perform Services (Months or Miles, Whichever Occurs First)	Services
LUBRICATION AND GENERAL MAINTENANCE	
Every 5,000 miles (8,000 km)	*ENGINE OIL—Change *OIL FILTER—Change ●CHASSIS—Lubricate ●FLUID LEVELS—Check
See Explanation	TIRES—Rotation REAR AXLE OR FINAL DRIVE—Check lube
Every 12 months or 15,000 miles (24,000 km)	*COOLING SYSTEM—Check *CRANKCASE VENTILATION—Service
Every 30,000 miles (48,000 km)	WHEEL BEARINGS—Repack
See Explanation	AUTOMATIC TRANSMISSION—Change fluid and filter
SAFETY MAINTENANCE	
At first 5,000 miles (8,000 km) Then at 15,000/30,000/45,000 miles	*EXHAUST SYSTEM—Check condition
Every 12 months or 10,000 miles (16,000 km)	TIRES, WHEEL AND DISC BRAKE—Check SUSPENSION AND STEERING—Check BRAKES AND POWER STEERING—Check
Every 5,000 miles (8,000 km)	*DRIVE BELTS—Check condition and adjustment
Every 12 months or 15,000 miles (24,000 km)	DRUM BRAKES AND PARKING BRAKE—Check THROTTLE LINKAGE—Check operation BUMPERS—Check condition
EMISSION CONTROL MAINTENANCE	
At first 5,000 miles (8,000 km) Then at 15,000/30,000/45,000 miles	EXHAUST PRESSURE REGULATOR VALVE
At first 5,000 miles (8,000 km) Then at 30,000 miles (48,000 km)	ENGINE IDLE SPEED—Adjust
Every 30,000 miles (48,000 km)	AIR CLEANER—Replace FUEL FILTER—Replace

● Also a Safety Service
* Also on Emission Control Service

Recommended Lubricants

Item	Lubricant
Engine Oil (Gasoline)	API "SE", "SF/CC" or "SF/CD"
Engine Oil (Diesel)	API "SF/CC", "SF/CD" or "SE/CC"
Manual Transmission	SAE 80W GL-5 or SAE 80W/90 GL-5
Automatic Transmission	DEXRON® or DEXRON® II ATF
Rear Axle—Standard	SAE 80W GL-5 or SAE 80W/90 GL-5
Positraction/Limited Slip	GM Part #1052271 or 1052272
Power Steering Reservoir	DEXRON® ATF—1964-76 Power Steering Fluid—1977 and later
Brake Fluid	DOT 3
Antifreeze	Ethylene Glycol
Front Wheel Bearings	GM Wheel Bearing Grease
Clutch Linkage	Engine Oil
Hood and Door Hinges	Engine Oil
Chassis Lubrication	NLGI #1 or NLGI #2
Lock Cylinders	WD-40 or Powdered Graphite

GENERAL INFORMATION AND MAINTENANCE

Capacities

Year	Engine No. Cyl. Displacement (cu. in.)	Engine Crankcase Add 1 qt. for New Filter	Transmission Pts. to Refill After Draining - Manual 3-Speed	Transmission Pts. to Refill After Draining - Manual 4-Speed	Transmission Pts. to Refill After Draining - Automatic ●	Drive Axle (pts.)	Gasoline Tank (gals.)	Cooling System (qts.) With Heater	Cooling System (qts.) With A/C
1968	6-230	4	3④	3④	6③	3.5	20	12	12
	6-250	4	3④	3④	6③	3.5	20	12	12
	8-307	4	3④	3④	6③	3.5	20	17	17
	8-327	4	2.5	3④	6③	3.5	20	16	16
	8-350	4	2.5	3④	6③	3.5	20	16	16
	8-396	4	2.5	3④	6⑦③	3.5	20	23	23
1969	6-230	4	3④	—	6⑧③	3.5③	20	13	13
	6-250	4	3④	—	6⑧③	3.5③	20	13	13
	8-307	4	3④	3④	6⑧③	3.5③	20	17	18
	8-350	4	3④	3④	6⑧③	3.5③	20	16	17
	8-396	4	3④	3④	8③	3.5③	20	23	24
1970	6-230	4	3	—	6⑧③	3.75⑨	20⑫	12	13
	6-250	4	3	—	6⑧③	3.75⑨	20⑫	12	13
	8-307	4	3	—	6⑧③	3.75⑨	20⑫	15	16
	8-350	4	3	3	6.5⑧③	3.75⑨	20⑫	16	16
	8-400	4	3	3	8③	3.75⑨	20⑫	16	16
	8-396	4	3	3	8③	3.75⑨	20⑫	23	24
	8-402	4	3	3	8③	3.75⑨	20⑫	23	24
	8-454	4	3	3	8③	3.75⑨	20⑫	22	23
1971	6-250	4	3	—	6③	3.75	19⑫	12	—
	8-307	4	3	—	6⑧③	3.75	19⑫	15	16
	8-350	4	3	3	6.5⑧③	3.75	19⑫	16	16
	8-402	4	3	3	8③	3.75	19⑫	23	23
	8-454	4	—	3	8③	3.75	19⑫	22	23
1972	6-250	4	3	—	6⑧③	4.25	19⑫	12	—
	8-307	4	3	—	6⑧③	4.25	19⑫	15	16
	8-350	4	3	3	6.5⑧③	4.25	19⑫	16	16
	8-402	4	—	3	8③	4.25⑤	19⑫	24	24
	8-454	4	—	3	8③	4.25⑤	19⑫	23	24
1973	6-250	4	3	—	6⑧③	4.25	22	12.5	—
	8-307	4	3	—	5③	4.25	22	16	17
	8-350	4	3	3	5③	4.25	22	16	17
	8-454	4	—	3	8③	4.25⑤	22	23	24
1974	6-250	4	3	—	8③	4.25	22	12.5	—
	8-350	4	3	3	8③	4.25	22	16	17
	8-400	4	—	—	8③	4.25⑤	22	16	17
	8-454	4	—	3	9③	4.9	22	23	24
1975	6-250	4	3	—	8③	4.25	22	14	16
	8-350	4	3	3	8③	4.25	22	17	18
	8-400	4	—	—	8③	4.25⑤	22	17	18
	8-454	4	—	3	9③	4.9	22	23	23
1976–77	6-250	4	3	—	8③	4.25	22	15	17
	8-305	4	3	—	8③	4.25	22	17	18
	8-350	4	—	3	8③	4.25	22	17	18
	8-400	4	—	—	8③	4.25	22	17	18

GENERAL INFORMATION AND MAINTENANCE

Capacities

Year	Engine No. Cyl. Displacement (cu. in.)	Engine Crankcase Add 1 qt. for New Filter	Transmission Pts. to Refill After Draining			Drive Axle (pts.)	Gasoline Tank (gals.)	Cooling System (qts.)	
			Manual					With Heater	With A/C
			3-Speed	4-Speed	Automatic ●				
1978	V6-200	4	3	—	6③	3.5	25⑫	16.8	18.8
	V6-231	4⑩	3	3	6③	3.5	25⑫	14.7	14.7
	6-250	4	3	—	6③	3.5	25⑫	14.6	14.6
	8-305	4	—	3	6③	3.5⑨	25⑫	19.2	19.2
	8-350	4	—	3	6③	3.5⑨	25⑫	19.2	19.2
1979	V6-200	4	3	—	8③	3.25	25⑫	18.5	18.5
	V6-231	4⑩	3	—	8③	3.25	25⑫	15.5	15.5
	8-267	4	—	3.4	8③	3.25	25⑫	19.2	19.2
	8-305	4	—	3.4	8③	3.5⑨	25⑫	19.2	19.2
	8-350	4	—	3.4	8③	3.5⑨	25⑫	19.2	19.2
1980–82	V6-229	4⑩	3	—	8③	3.25	25⑫	18.5	18.5
	V6-231	4⑩	3	—	8③	3.25	25⑫	15.5	15.5
	8-267	4	—	—	8③	3.25	25⑫	21	21
	8-305	4	—	3.4	8③	3.5⑨	25⑫	19	19
	8-350	4	3.0	—	6	3.5	25⑫	15.0	15.0
	8-350	6	—	—	6	3.5	25⑫	18.0	18.0
1983–84	V6-229	4	—	—	6	3.5	25⑫	15.0	15.0
	V6-231	4	—	—	6	3.5	25⑫	15.0	15.0
	8-305	4	—	—	6	3.5	25⑫	15.0	15.0
	8-350	4	—	—	6	3.5	25⑫	15.0	15.0
	8-350	6	—	—	6	3.5	25⑫	18.0	18.0
1985	V6-231	4	—	—	6	3.5	25⑫	15.0	15.0
	V6-262	4	—	—	6	3.5	25⑫	15.0	15.0
	8-305	4	—	—	6	3.5	25⑫	15.0	15.0
	8-350	4	—	—	6	3.5	25⑫	15.0	15.0
	8-305	6	—	—	6	3.5	25⑫	15.0	15.0
1986–88	V6-262	4	—	—	6	3.5	25⑫	15.0	15.0
	8-305	4	—	—	6	3.5	25⑫	15.0	15.0
	8-307⑪	4	—	—	6	3.5	25⑫	15.0	15.0
	8-350	4	—	—	6	3.5	25⑫	15.0	15.0
1989–92	V6-262	4	—	—	7⑬	⑭	25⑫	16.7	⑨
	8-305	4	—	—	7⑬	⑭	25⑫	16.7	⑨
	8-307⑪	4	—	—	7⑬	⑭	25⑫	16.7	⑨
	8-350	4	—	—	7⑬	⑭	25⑫	16.7	⑨

● Specifications do not include torque converter
① Figure given is for dry refill
② 18 qts.—300 hp eng., 19 qts.—350 hp eng.
③ Figure given is for drain and refill
④ 3.5 pts. with heavy duty trans.
⑤ 4 pts. with 8.875 in. ring gear
⑥ 16 qts.—350 hp eng.
⑦ 8 pts.—THM 400
⑧ 5 pts.—THM 350
⑨ With HD radiator add 0.6 qts.
⑩ Figure is the same with or without a filter change
⑪ Oldsmobile Produced Engine
⑫ 22 gals. Station Wagon—23 gals. 1990–91 All Models
⑬ 700R4 10 pts.
⑭ 7½ ring gear—3.5 pts.
8½ ring gear—4.25 pts.
8¾ ring gear—5.4 pts.

GENERAL INFORMATION AND MAINTENANCE

Maintenance Intervals
Gasoline Engined Cars—1977 and Later

When to Perform Services (Months or Miles, Whichever Occurs First)	Services
LUBRICATION AND GENERAL MAINTENANCE	
Every 12 months or 7,500 miles (12,000 km)	● CHASSIS—Lubricate ● FLUID LEVELS—Check CLUTCH PEDAL TREE TRAVEL—Check Adjust
See Explanation of Maintenance Schedule	*ENGINE OIL—Change *ENGINE OIL FILTER—Replace TIRES—Rotation (Radial Tires) REAR AXLE OR FINAL DRIVE—Check Lube
Every 12 months or 15,000 miles (24,000 km)	*COOLING SYSTEM—See Explanation of Maintenance Schedule
Every 30,000 miles (48,000 km)	WHEEL BEARINGS—Repack CLUTCH CROSS SHAFT—Lubricate
See Explanation	AUTOMATIC TRANSMISSION—Change fluid and service filter
SAFETY MAINTENANCE	
Every 12 months or 7,500 miles (12,000 km)	TIRES, WHEELS AND DISC BRAKES—Check condition *EXHAUST SYSTEM—Check condition SUSPENSION & STEERING SYSTEM—Check condition BRAKES AND POWER STEERING—Check all lines and hoses
Every 12 months or 15,000 miles (24,00 km)	*DRIVE BELTS—Check condition and adjustment ① DRUM BRAKES AND PARKING BRAKE—Check condition of linings; adjust parking brake THROTTLE LINKAGE—Check operation and condition BUMPERS—Check condition *FUEL CAP, TANK AND LINES—Check
EMISSION CONTROL MAINTENANCE	
At first 6 months or 7,500 miles (12,000 km)—Then at 24-month 30,000 mile (48,000 km) intervals as indicated in Log. Except Choke Which Requires Service at 45,000 miles (72,000 km)	CARBURETOR CHOKE & HOSES—Check ② ENGINE IDLE SPEED—Check adjustment ② EFE SYSTEM—Check operation (if so equipped) CARBURETOR—Torque attaching bolts or nuts to manifold ②
Every 30,000 miles (48,000 km)	THERMOSTATICALLY CONTROLLED AIR CLEANER—Check operation VACUUM ADVANCE SYSTEM AND HOSES—Check ③ SPARK PLUG WIRES—Check IDLE STOP SOLENOID AND OR DASH POT OR ISC—Check operation SPARK PLUGS—Replace ② ENGINE TIMING ADJUSTMENT AND DISTRIBUTOR—Check AIR CLEANER AND PCV FILTER ELEMENT—Replace ② PCV VALVE—Replace EGR VALVE—Service

● Also a Safety Service
* Also an Emission Control Service
① In California, a separately driven air pump belt check is recommended but not required at 15,000 miles (24,000 km) and 45,000 miles (72,000 km).
② Only these emission control maintenance items are considered to be required maintenance as defined by the California Air Resources Board (ARB) regulation and are, according to such regulation, the minimum maintenance an owner in California must perform to fulfill the minimum requirements of the emission warranty. All other emission maintenance items are recommended maintenance as defined by such regulation. General Motors urges that all emission control maintenance items be performed.
③ Not applicable on vehicle equipped with electronic spark timing (EST).

GENERAL INFORMATION AND MAINTENANCE

Maintenance Intervals
Gasoline Engined Cars—1970-76

Interval At Which Services Are To Be Performed	Service
LUBRICATION AND GENERAL MAINTENANCE	
Every 6 months or 7,500 miles	*CHASSIS—Lubricate ●*FLUID LEVELS—Check *ENGINE OIL—Change
At first oil change—then every 2nd	*ENGINE OIL FILTER—Replace (V-6 Replace each oil change)
See Explanation of Maintenance Schedule	TIRES—Rotate DIFFERENTIAL
Every 12 months	AIR CONDITIONING SYSTEM—Check charge & hose condition. TEMPMATIC AIR FILTER—Replace every other year.
Every 12 months or 15,000 miles	*COOLING SYSTEM—See Explanation of Maintenance Schedule
Every 30,000 miles	WHEEL BEARINGS—Clean and repack *AUTOMATIC TRANS.—Change fluid and service filter MANUAL STEERING GEAR—Check seals CLUTCH CROSS SHAFT—Lubricate
SAFETY MAINTENANCE	
Every 6 months or 7,500 miles	TIRES AND WHEELS—Check condition *EXHAUST SYSTEM—Check condition of system *DRIVE BELTS—Ck. cond. & adjustment. Replace every 30,000 miles FRONT AND REAR SUSPENSION & STEERING SYSTEM— Ck. cond. BRAKES AND POWER STEERING—Check all lines and hoses
Every 12 months or 15,000 miles	DRUM BRAKES AND PARKING BRAKE—Check condition of linings; adjust parking brake THROTTLE LINKAGE—Check operation and condition UNDERBODY—Flush and check condition BUMPERS—Check condition
EMISSION CONTROL MAINTENANCE	
At 1st 6 months or 7,500 miles—then at 18 month/22,500 mile Intervals Therafter	THERMOSTATICALLY CONTROLLED AIR CLEANER—Check operation CARBURETOR CHOKE—Check operation ENGINE IDLE SPEED ADJUSTMENT EFE VALVE—Check operation CARBURETOR—Torque attaching bolts or nuts to manifold
Every 12 months or 15,000 miles	CARBRETOR FUEL INLET FILTER—Replace VACUUM ADVANCE SYSTEM AND HOSES—Check oper. PCV SYSTEMS—See Explanation of Maintenance Schedule
Every 18 months or 22,500 miles	IDLE STOP SOLENOID OR DASHPOT—Check operation SPARK PLUG AND IGNITION COIL WIRES—Inspect and clean
Every 22,500 miles	SPARK PLUGS—Replace ENGINE TIMING ADJUSTMENT & DISTRIBUTOR CHECK
Every 24 months or 30,000 miles	ECS SYSTEM—See Explanation of Maintenance Schedule FUEL CAP, TANK AND LINES—Check condition
Every 30,000 miles	AIR CLEANER ELEMENT—Replace

*Also Required Emission Control Maintenance
● Also a Safety Service

Engine Performance and Tune-Up

TUNE-UP PROCEDURES

In order to extract the full measure of performance and economy from your engine it is essential that it is properly tuned at regular intervals. A regular tune-up will keep your cars engine running smoothly and well prevent the annoying breakdowns and poor performance associated with an untuned engine.
NOTE: *All 1968-74 cars use a conventional breaker point ignition system. In 1975, Chevrolet switched to a full electronic ignition system known as HEI.*
A complete tune-up should be performed at least every 15,000 miles (12,000 miles for early cars) or twelve months, whichever comes first.
NOTE: *1981 and later cars have increased their interval to 30,000 miles.*
This interval should be halved if the car is operated under severe conditions such as trailer towing, prolonged idling, start-and-stop driving, or if starting or running problems are noticed. It is assumed that the routine maintenance described in Chapter 1 has been kept up, as this will have a decided effect on the results of a tune-up. All of the applicable steps of a tune-up should be followed in order, as the result is a cumulative one.

If the specifications on the underhood tune-up sticker in the engine compartment of your car disagree with the Tune-Up Specifications chart in this chapter, the figures on the sticker must be used. The sticker often reflects changes made during the production run.

Spark Plugs

A typical spark plug consists of a metal shell surrounding a ceramic insulator. A metal electrode extends downward through the center of the insulator and protrudes a short distance.
Located at the end of the plug and attached these electrodes (measured in thousandths of an inch) is called spark plug gap. The spark plug in no way produces a spark but merely provides a gap across which the current can arc. The coil produces 20,000-25,000 Volts (the HEI transistorized ignition produces considerably more voltage than the standard type, approximately 50,000 volts), which travels to the distributor where it is distributed through the spark plug wires to the plugs. The current passes along the center electrode and jumps the gap to the side electrode and, in so doing, ignites the air/fuel mixture in the combustion chamber. All plugs used since 1969 have a resistor built into the center electrode to reduce interference to any nearby radio and television receivers. The resistor also cuts down on erosion of plug electrodes caused by excessively long sparking. Resistor spark plug wiring is original equipment on all cars.

Spark plug life and efficiency depend upon condition of the engine and the temperatures to which the plug is exposed. Combustion chamber temperatures are affected by many factors such as compression ratio of the engine, fuel/

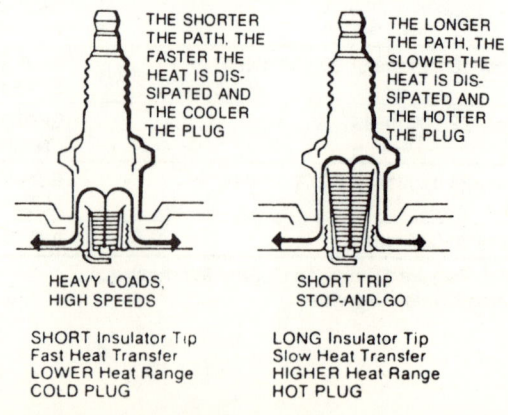

Spark plug heat range

ENGINE PERFORMANCE AND TUNE-UP

Gasoline Engine Tune-Up Specifications

When analyzing the compression test results, look for uniformity among cylinders rather than specific pressures.

Year	Engine No. Cyl. Displacement (cu. in.)	hp	Spark Plugs Type	Gap (in.)	Distributor Point Dwell (deg.)	Point Gap (in.)	Ignition Timing ▲ (deg.)■● Man. Trans.	Auto. Trans.	●Valves■ Intake Opens (deg.)	Fuel Pump Pressure (psi)	●Idle Speed▲ Man. Trans.	Auto. Trans.
1968	6-230	140	46N	0.035	31–34	0.019	TDC	4B	16	3½–4½	700	500②/400
	6-250	155	46N	0.035	31–34	0.019	TDC	4B	16	3½–4½	700	500②/400
	8-307	200	45S	0.035	28–32	0.019	2B	2B	28	5–6½	700	600
	8-327	275	44	0.035	28–32	0.019	TDC	4B	28	5–6½	700	600
	8-327	325	44	0.035	28–32	0.019	4B	—	40	5–6½	750②	—
	8-396	325	43N	0.035	28–32	0.019	4B	4B	28	5–6½	700②	600②
	8-396	350	43N	0.035	28–32	0.019	TDC	4B	40	7¼–8½	700	600
	8-396	375	R-43N	0.035	28–32	0.019	4B	—	44	5–8½	750	—
1969	6-250	155	R-46N	0.035	31–34	0.019	TDC	4B	16	3½–4½	700	550/400②
	8-307	200	R-45S	0.035	28–32	0.019	2B	2B	28	5–6½	700	600
	8-350	250	R-44	0.035	28–32	0.019	TDC	4B	28	5–6½	700	600
	8-350	300	R-44	0.035	28–32	0.019	TDC	4B	28	5–6½	700	600
	8-396	325	R-44N	0.035	28–32	0.019	4B	4B	28	5–8½	800	600
	8-396	350	R-43N	0.035	28–32	0.019	TDC	4B	56	5–8½	800	600
	8-396	375	R-43N	0.035	28–32	0.019	4B	4B	44	5–8½	750	750/400
1970	6-250	155	R-46T	0.035	31–34	0.019	TDC	4B	16	3½–4½	750	600/400
	8-307	200	R-43	0.035	28–32	0.019	2B	8B	28	5–6½	700	600/450
	8-350	250	R-44	0.035	28–32	0.019	TDC	4B	28	5–6½	750	600/450
	8-350	300	R-44	0.035	28–32	0.019	TDC	4B	28	5–6½	700	600
	8-396	350	R-44T	0.035	28–32	0.019	TDC	4B	56	5–8½	700	600
	8-396	375	R-43T	0.035	28–32	0.019	4B	4B	NA	5–8½	750	700
	8-400	265	R-44	0.035	28–32	0.019	4B	8B	28	5–8½	700	600/450
	8-402	330	R-44T	0.035	28–32	0.019	4B	4B	28	5–8½	700	600
	8-454	360	R-43T	0.035	28–32	0.019	6B	6B	56	5–8½	700	600
	8-454	390	R-43T	0.035	28–32	0.019	6B	6B	NA	5–8½	700	600
	8-454	450	R-43T	0.035	28–32	0.019	4B	4B	56	5–8½	700	700
1971	6-250	145	R-46TS	0.035	31–34	0.019	4B	4B	16	3½–4½	550	500
	8-307	200	R-45TS	0.035	29–31	0.019	4B	8B	28	5–6½	600	550
	8-350	245	R-45TS	0.035	29–31	0.019	2B	6B	28	7–8½	600	550
	8-350	270	R-44TS	0.035	29–31	0.019	4B	8B	28	7–8½	600	550
	8-400	255	R-44TS	0.035	29–31	0.019	4B	8B	28	7–8½	600	550
	8-402	300	R-44TS	0.035	29–31	0.019	8B	8B	28	7–8½	600	600
	8-454	365	R-42TS	0.035	29–31	0.019	8B	8B	56	7–8½	600	600
	8-454	425	R-42TS	0.035	29–31	0.019	8B	12B	44	7–8½	700	700
1972	6-250	110	R-46TS	0.035	31–34	0.019	4B	4B	16	3½–4½	700	600
	8-307	130	R-44T	0.035	29–31	0.019	4B	8B	28	5–6½	900	600

Gasoline Engine Tune-Up Specifications

When analyzing the compression test results, look for uniformity among cylinders rather than specific pressures.

Year	Engine No. Cyl. Displacement (cu. in.)	hp	Spark Plugs Type	Gap (in.)	Distributor Point Dwell (deg.)	Distributor Point Gap (in.)	Ignition Timing ▲ (deg.)■ ● Man. Trans.	Ignition Timing ▲ (deg.)■ ● Auto. Trans.	● Valves■ Intake Opens (deg.)	Fuel Pump Pressure (psi)	● Idle Speed ▲ Man. Trans.	● Idle Speed ▲ Auto. Trans.
	8-350	165	R-44T	0.035	29–31	0.019	6B	6B	28	7–8½	900	600
	8-350	175	R-44T	0.035	29–31	0.019	4B	8B	28	7–8½	800	600
	8-402	240	R-44T	0.035	29–31	0.019	8B	8B	30	7–8½	750	600
	8-454	270	R-44T	0.035	29–31	0.019	8B	8B	56	7–8½	750	600
1973	6-250	100	R-46T	0.035	31–34	0.019	6B	6B	16	3½–4½	700/450	600/450
	8-307	115	R-44T	0.035	29–31	0.019	4B	8B	28	5–6½	900/450	600/450
	8-350	145	R-44T	0.035	29–31	0.019	8B	8B	28	7–8½	900/450	600/450
	8-350	175	R-44T	0.035	29–31	0.019	8B	12B	28	7–8½	900/450	600/450
	8-454	245	R-44T	0.035	29–31	0.019	10B	10B	55	7–8½	900/450	600/450
1974	6-250	100	R-46T	0.035	31–34	0.019	6B	6B	16	4–5	800/450	600/450
	8-350	145	R-44T	0.035	29–31	0.019	4B	8B	28	7½–9	900/450	600/450
	8-350	160	R-44T	0.035	29–31	0.019	4B	8B	44	7½–9	900/450	600/450
	8-400	150	R-44T	0.035	29–31	0.019	—	8B	28	7½–9	—	600/450
	8-400	180	R-44T	0.035	29–31	0.019	—	8B	44	7½–9	—	600/450
	8-454	235	R-44T	0.035	29–31	0.019	10B	10B	55	7½–9	800/450	600/450
1975	6-250	105	R-46TX	0.060	Electronic		10B	10B	16	4–5	850/425	550/425 (600/425)
	8-350	145	R-44TX	0.060	Electronic		6B	6B	28	7½–9	800	600
	8-350	155	R-44TX	0.060	Electronic		—	6B	28	7½–9	—	600
	8-400	175	R-44TX	0.060	Electronic		—	8B	28	7½–9	—	600
	8-454	215	R-44TX	0.060	Electronic		—	16B	55	7½–9	—	600/500
1976	6-250	105	R-46TS	0.035	Electronic		6B	6B	16	3½–4½	850	550 (600)
	8-305	140	R-45TS	0.045	Electronic		—	8B (TDC)	28	7–8½	—	600
	8-350	145	R-45TS	0.045	Electronic		—	6B	28	7–8½	—	600
	8-350	165	R-45TS	0.045	Electronic		—	8B (6B)	28	7–8½	—	600
	8-400	175	R-45TS	0.045	Electronic		—	8B	28	7–8½	—	600
1977	6-250	110	R-46TS	0.035	Electronic		6B @ 850	8B @ 600	16	4–5	750	550
	8-305	145	R-45TS	0.045	Electronic		—	8B @ 500	28	7½–9	—	500
	8-350	170	R-45TS	0.045	Electronic		—	8B @ 500	28	7½–9	—	500

ENGINE PERFORMANCE AND TUNE-UP

Gasoline Engine Tune-Up Specifications

When analyzing the compression test results, look for uniformity among cylinders rather than specific pressures.

Year	Engine No. Cyl. Displacement (cu. in.)	hp	Spark Plugs Type	Gap (in.)	Distributor Point Dwell (deg.)	Distributor Point Gap (in.)	Ignition Timing ▲ (deg.)■ ● Man. Trans.	Ignition Timing ▲ (deg.)■ ● Auto. Trans.	● Valves■ Intake Opens (deg.)	Fuel Pump Pressure (psi)	● Idle Speed ▲ Man. Trans.	● Idle Speed ▲ Auto. Trans.
1978	6-231	105	R-46TSX	0.060	Electronic		15B	15B	17	6–7	600	500
	8-305	145	R-45TS	0.045	Electronic		4B	④	28	7.5–9	600	500
	8-350	170	R-45TS	0.045	Electronic		—	8B	2B	7.5–9	—	500
1979	6-231	all	R-46TSX	0.060	Electronic		15B	15B	16	4.2–5.7	600	500
	8-267	all	R-45TS	0.045	Electronic		4B	10B	2B	7.5–9.0	600	500
	8-305	all	R-43TS	0.045	Electronic		4B	4B	28	7.5–9.0	600	500
	8-350	all	R-43TS	0.045	Electronic		—	8B	2B	7.5–9.0	—	500
1980	6-229	115	R-T5TS⑤	0.045	Electronic		8B	12B	42	4.6–6.0	700	600
	6-231	110	R-45TSX	0.060	Electronic		—	15B	16	4.25–5.75	—	560 (600)
	8-267	120	R-45TS	0.045	Electronic		—	4B	28	7.5–9.0	—	500
	8-305	155	R-45TS	0.045	Electronic		4B	4B	28	7.5–9.0	700	500 (550)
1981	6-229	110	R-45TS	0.045	Electronic		6B	6B	42	4.5–6.0	⑥	⑥
	6-231	110	R-45TS8	0.080	Electronic		—	15B	16	4.25–5.75	—	⑥
	8-267	115	R-45TS	0.045	Electronic		—	6B	28	7.5–9.0	—	500
	8-305	150	R-45TS	0.045	Electronic		6B	6B	28	7.5–9.0	700	500
1982–84	6-229	110	R-45TS	0.045	Electronic		—	6B⑥	42	4.5–6.0	—	⑥
	6-231	110	R-45TS	0.045	Electronic		—	15B⑥	16	4.5–6.0	—	⑥
	8-267	115	R-45TS	0.045	Electronic		—	6B⑥	44	7.5–9.0	—	⑥
	8-305	145⑦	R-45TS	0.045	Electronic		—	6B⑥	44	7.5–9.0	—	⑥
	8-350	NA	R-45TS	0.045	Electronic		—	⑥	44	7.5–9.0	—	⑥
1985–88	6-262	140	R-43TS	0.035	Electronic		—	⑥	44	9.0–13.0	—	⑥
	8-305	145	R-45TS	0.045	Electronic		—	⑥	44	7.5–9.0	—	⑥
	8-307①	NA	R-46SX	0.040	Electronic		—	⑥	16	7.5–9.0	—	⑥
	8-350	155	R-45TS	0.045	Electronic		—	⑥	44	7.5–9.0	—	⑥
1989–92	6-262	140	R-45TS	⑥	Electronic		—	⑥	45	9.0–13.0	—	⑥
	8-305	170	R-45TS	⑥	Electronic		—	⑥	45	11	—	⑥
	8-307①	148	FR3L56	⑥	Electronic		—	⑥	16	6.0–7.5	—	⑥
	8-350	195	R-43TS⑥	⑥	Electronic		—	⑥	45	9.0–13	—	⑥

NOTE: The underhood specification sticker often reflects tune-up specification changes made in production. Sticker figures must be used if they disagree with those in this chart.

▲ See text for procedure
● Figures in parentheses indicate California
■ All figures Before Top Dead Center
*When two idle speed figures are separated by a slash, the lower figure is with the idle speed solenoid disconnected
NA—Not available
① Oldsmobile Produced Engine
② A/C on
③ Lower figure is with idle solenoid disconnected
④ 49 states—4B
   Calif.—6B
   High alt.—8B
⑤ With A/T—R-45TS
⑥ See underhood specifications sticker
⑦ 150 hp, 1983–84

ENGINE PERFORMANCE AND TUNE-UP

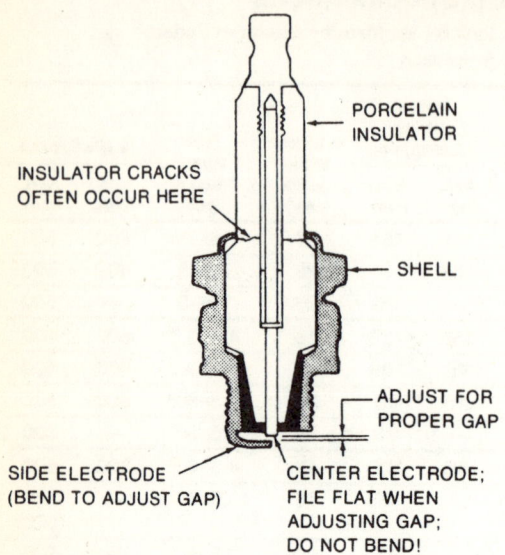

Cross section of a spark plug

air mixtures, exhaust emission equipment, and the type of driving you do. Spark plugs are designed and classified by number according to the heat range at which they will operate most efficiently. The amount of heat that the plug absorbs is determined by the length of the lower insulator. The longer the insulator (it extends farther into the engine), the hotter the plug will operate; the shorter it is, the cooler it will operate. A plug that has a short path for heat transfer and remains too cool will quickly accumulate deposits of oil and carbon since it is not hot enough to burn them off. This leads to plug fouling and consequently to misfiring. A plug that has a long path of heat transfer will have no deposits but, due to the excessive heat, the electrodes will burn away quickly and, in some instances, pre-ignition may result. Pre-ignition takes place when plug tips get so hot that they glow sufficiently to ignite the fuel/air mixture before the spark does. This early ignition will usually cause a pinging (sounding much like castanets) during low speeds and heavy loads. In severe cases, the heat may become enough to start the fuel/air mixture burning throughout the combustion chamber rather than just to the front of the plug as in normal operation. At this time, the piston is rising in the cylinder making its compression stroke. The burning mass is compressed and an explosion results producing tremendous pressure. Something has to give, and it does; pistons are often damaged. Obviously, this detonation (explosion) is a destructive condition that can be avoided by installing a spark plug designed and specified for your particular engine.

A set of spark plugs usually requires replacement after 10,000-12,000 miles depending on the type of driving (this interval has been increased to 22,500 miles for all 1975-79 cars and

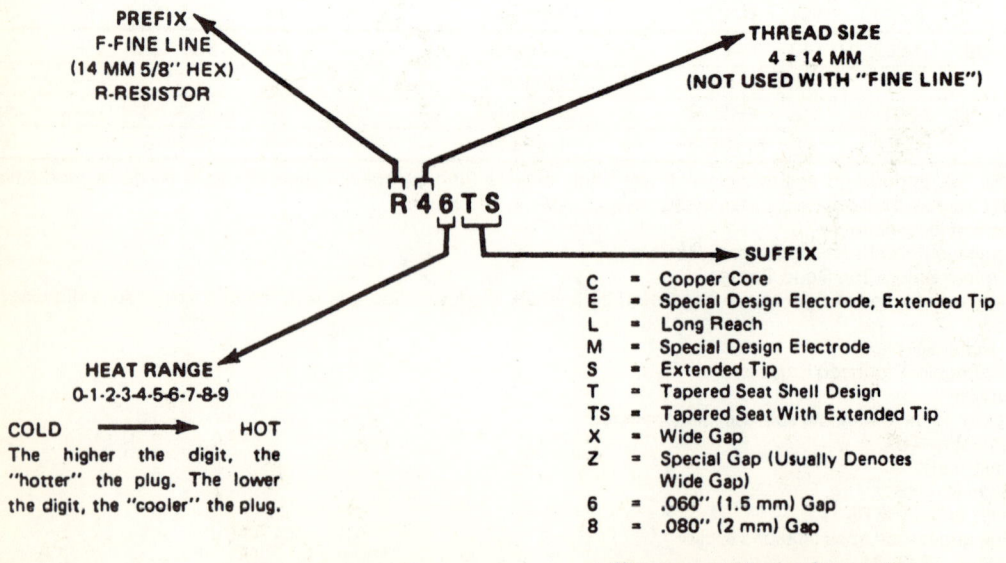

Ac Delco spark plug identification

ENGINE PERFORMANCE AND TUNE-UP

30,000 miles for all 1980 and later cars). The electrode on a new spark plug has a sharp edge but, with use, this edge becomes rounded by erosion causing the plug gap to increase. In normal operation, plug gap increases about 0.001" in every 1,000-2,000 miles. As the gap increases, the plug's voltage requirement also increases. It requires a greater voltage to jump the wider gap and about two to three times as much voltage to fire a plug a high speed and acceleration than at idle.

The higher voltage produced by the HEI ignition coil is one of the primary reasons for the prolonged replacement interval for spark plugs in 1975 and later cars. A consistently hotter spark prevents the fouling of plugs for much longer than could normally be expected; this spark is also able to jump across a larger gap more efficiently than a spark from a conventional system. However, even plugs used with the HEI system wear after time in the engine.

Worn plugs become obvious during acceleration. Voltage requirement is greatest during acceleration and a plug with an enlarged gap may require more voltage than the coil is able to produce. As a result, the engine misses and sputters until acceleration is reduced. Reducing acceleration reduces the plug's voltage requirement and the engine runs smoother. Slow, city driving is hard on plugs. The long periods of idle experienced in traffic creates an overly rich gas mixture. The engine is not running fast enough to completely burn the gas and, consequently, the plugs are fouled with gas deposits and engine idle becomes rough. In many cases, driving under the right conditions can effectively clean these fouled plugs.

NOTE: *There are several reasons why a spark plug will foul and you can usually learn which is at fault by just looking at the plug. A few of the most common reasons for plug fouling, and a description of the fouled plug's appearance, can be found in the color insert in this book.*

Accelerate your car to the speed where the engine begins to miss and then slow down to the point where the engine smooths out. Run at this speed for a few minutes and then accelerate again to the point of engine miss. With each repetition this engine miss should occur at increasingly higher speeds and then disappear altogether. Do not attempt to shortcut this procedure by hard acceleration. This approach will compound problems by fusing deposits into a hard permanent glaze. Dirty, fouled plugs may be cleaned by sandblasting. Many shops have a spark plug sandblaster. After sandblasting, the electrode should be filed to a sharp, square shape and then gapped to specifications. Gapping a plug too close will produce a rough idle while gapping it too wide will increase its voltage requirement and cause missing at high speed and during acceleration.

The type of driving you do may require a change in spark plug heat range. If the majority of you driving is done in the city and rarely at high speeds, plug fouling may necessitate changing to a plug with a heat range one number higher than that specified by the vehicle manufacturer. For example, a 1970 Chevrolet with a 350 cu. in. (300 hp) engine requires an R44 plug. Frequent city driving may foul these plugs making engine operation rough. An R45 is the next hottest plug in the AC heat range (the higher the AC number, the hotter the plug) and its insulator is longer than the R44 so that it can absorb and retain more heat than the shorter R44. This hotter R45 burns off deposits even at low city speeds but would be too hot for prolonged turnpike driving. Using this plug at high speed would create dangerous pre-ignition. On the other hand, if the aforementioned Chevrolet were used almost exclusively for long distance high speed driving, the specified R44 might be too hot resulting in rapid electrode wear and dangerous pre-ignition. In this case, it might be wise to change to a colder R43. If the car is used for abnormal driving (as in the examples above), or the engine has been modified for higher performance, then change to a plug of a different heat range may be necessary. For a modified car it is always wise to go to a colder plug as a protection against pre-ignition. It will require more frequent plug cleaning, but destructive detonation during acceleration will be avoided.

REMOVAL

When you're removing spark plugs, you should work on one at a time. Don't start by removing the plug wires all at once because unless you number them, they're going to get mixed up. On some cars though, it will be more convenient for you to remove all the wires before you start to work on the plugs. If this is necessary, take a minute before you begin and number the wires with tape before you take them off. The time you spend here will pay off later on.

1. Twist the spark plug boot and remove the boot from the plug. You may also use a plug wire removal tool designed especially for this purpose. Do not pull on the wire itself. When the wire has been removed, take a wire brush and clean the area around the plug. Make sure that all the grime is removed so that none will enter the cylinder after the plug has been removed.

2. Remove the plug using the proper size socket, extensions, and universals as necessary.

64 ENGINE PERFORMANCE AND TUNE-UP

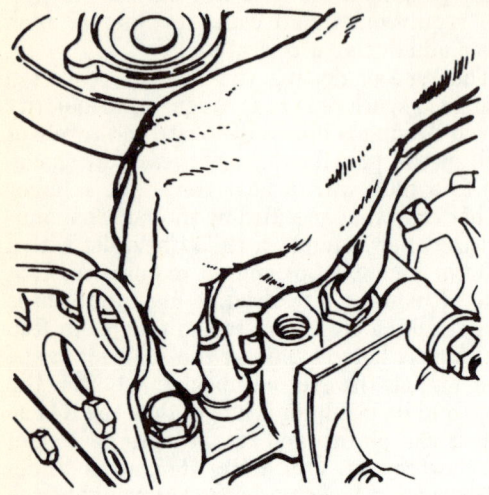

Twist and pull on the rubber boot to remove the spark plug wires; never pull on the wire itself

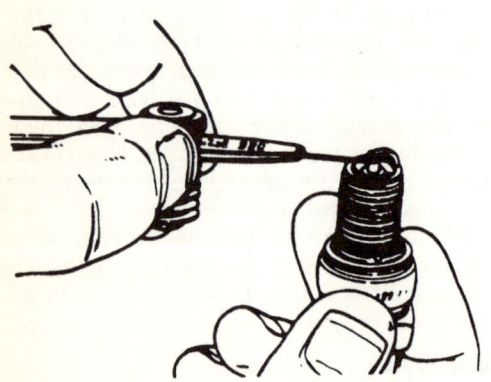

Always use a wire gauge to check the electrode gap

For 1968-69 engines, and all V8s 1968-71, use a 13/16″ spark plug socket. 6-cylinder engines for 1970 and V8s from 1972 are equipped with tapered seat plugs which require a 5/8″ socket.

3. If removing the plug is difficult, drip some penetrating oil on the plug threads, allow it to work, then remove the plug. Also, be sure that the socket is straight on the plug, especially on those hard to reach plugs.

INSPECTION

Check the plugs for deposits and wear. If they are not going to be replaced, clean the plugs thoroughly. Remember that any kind of deposit will decrease the efficiency of the plug. Plugs can be cleaned on a spark plug cleaning machine, which can sometimes be found in service stations, or you can do an acceptable job of cleaning with a stiff brush. If the plugs are cleaned, the electrodes must be filed flat. use an ignition points file, not an emery board or the like, which will leave deposits. The electrodes must be filed perfectly flat with sharp edges; rounded edges reduce the spark plug voltage by as much as 50%.

Check spark plug gap before installation. The ground electrode (the L-shaped one connected to the body of the plug) must be parallel to the center electrode and the specified size wire gauge (see Tune-Up Specifications) should pass through the gap with a slight drag. Always check the gap on new plugs, too; they are not always set correctly at the factory. Do not use a flat feeler gauge when measuring the gap, because the reading will be inaccurate. Wire gap-

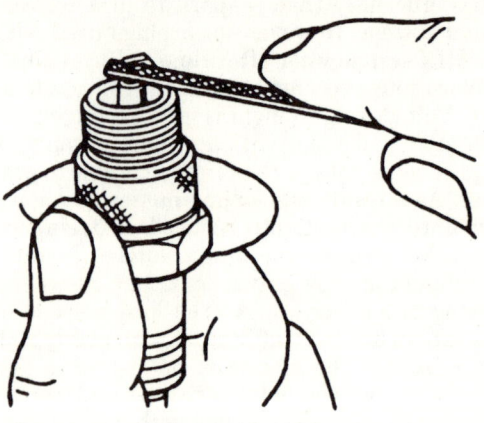

Plugs that are in good condition can be filed and re-used

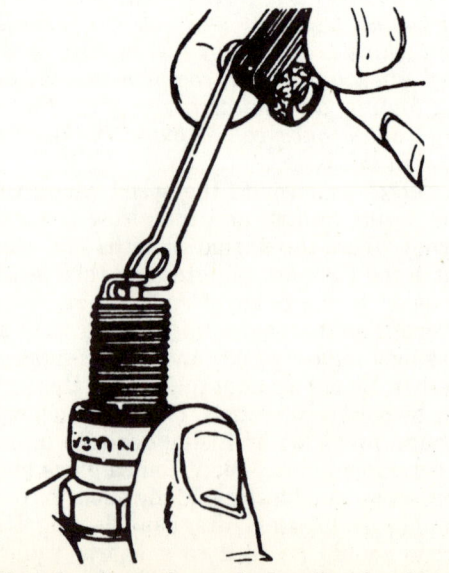

Adjust the electrode gap by bending the side electrode

ENGINE PERFORMANCE AND TUNE-UP

ping tools usually have a bending tool attached. Use that to adjust the side electrode until the proper distance is obtained. Absolutely never bend the center electrode. Also, be careful not to bend the side electrode too far or too often; it may weaken and break off within the engine, requiring removal of the cylinder head to retrieve it.

INSTALLATION

1. Lubricate the threads of the spark plugs with a drop of oil. Install the plugs and tighten them hand-tight. Take care not to cross-thread them.
2. Tighten the spark plugs with a socket. Do not apply the same amount of force you would use for a bolt; just snug them in. If a torque wrench is available, tighten to 11-15 ft. lbs.
3. Install the wires on their respective plugs. Make sure the wires are firmly connected. You will be able to feel them click into place.

CHECKING AND REPLACING SPARK PLUG WIRES

Every 15,000 miles, inspect the spark plug wires for burns, cuts, or breaks in the insulation. Check the boots and the nipples on the distributor cap. Replace any damaged wiring.

Every 45,000 miles or so, the resistance of the wires should be checked with an ohmmeter. Wires with excessive resistance will cause misfiring, and may make the engine difficult to start in damp weather. Generally, the useful life of the cables is 45,000-60,000 miles.

To check resistance, remove the distributor cap, leaving the wires in place. Connect one lead of an ohmmeter to an electrode within the cap; connect the other lead to the corresponding spark plug terminal (remove it from the spark plug for this test). Replace any wire which shows a resistance over 30,000Ω. Generally speaking, resistance should not be over 25,000Ω, and 30,000Ω must be considered the outer limit of acceptability.

It should be remembered that resistance is also a function of length; the longer the wire, the greater the resistance. Thus, if the wires on your car are longer than the factory originals, resistance will be higher, quite possibly outside these limits.

When installing new wires, replace them one at a time to avoid mixups. Start by replacing the longest one first. Install the boot firmly over the spark plug. Route the wire over the same path as the original. Insert the nipple firmly onto the tower on the distributor cap, then install the cap cover and latches to secure the wires.

NOTE: *For further information on spark plug wires, refer to the color insert on Spark Plug Analysis.*

FIRING ORDERS

NOTE: *To avoid confusion, remove and tag the wires one at a time, for replacement.*

Breaker Points and Condenser

The points function as a circuit breaker for the primary circuit of the ignition system. The ignition coil must boost the 12 volts of electrical pressure supplied by the battery to as much as 25,000 volts in order to fire the plugs. To do this, the coil depends on the points and the con-

1986–90 307 Oldsmobile produced V8 engine. Engine firing order: 1–8–4–3–6–5–7–2 distributor rotation: counter clockwise

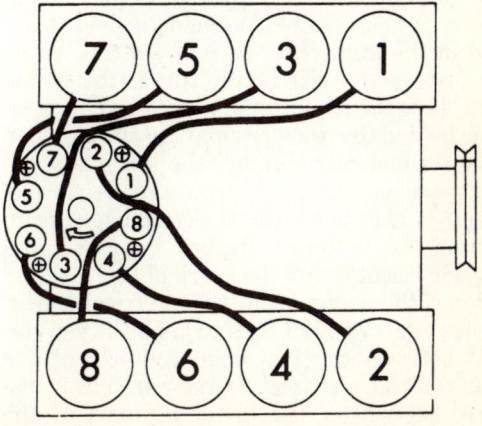

1975–92 V8 engine firing order, except 1986–90 307 Oldsmobile produced engine

66 ENGINE PERFORMANCE AND TUNE-UP

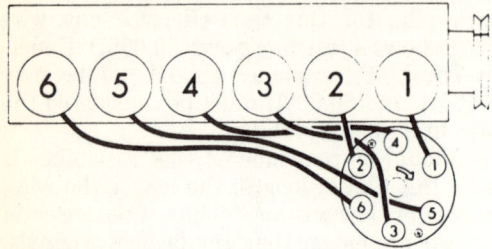

Inline six cylinder (1968–79) firing order

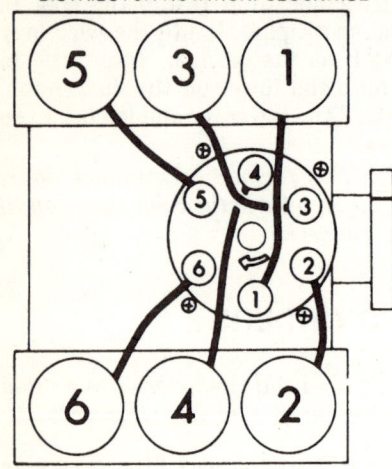

the 231 (1980–83) V6 firing order

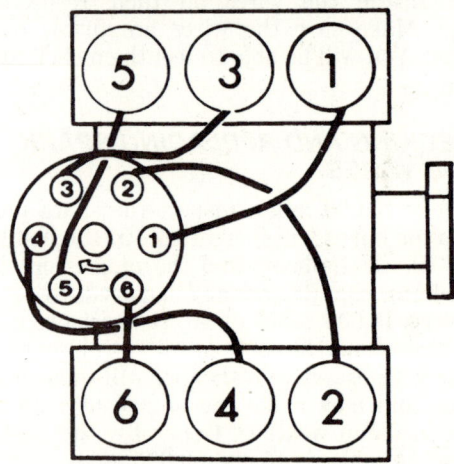

The 200 (1978–79), 229 (1980–83) and 262 (1984 and later) V6 firing order

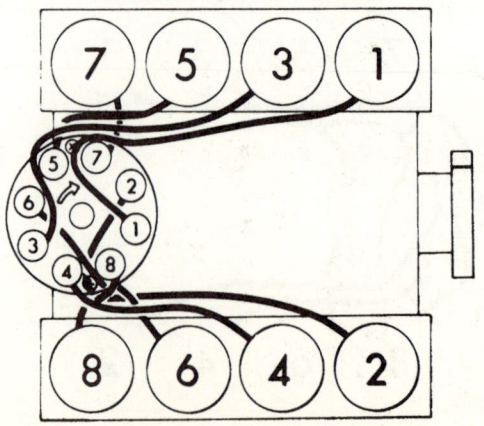

1968–74 V8 firing order

and thus prevent the points from becoming pitted or burned.

If you have ever wondered why it is necessary to tune-up your engine occasionally, consider the fact that the ignition system must complete the above cycle each time a spark plug fires. On a 4-cylinder, 4-cycle engine, two of the four plugs must fire once for every engine revolution. If the idle speed of you engine is 800 revolutions per minute (800 rpm), the breaker points open and close two times for each revolution. For every minute your engine idles, your points open and close 1,600 times (2 x 800 = 1,600). And that is just at idle. What about at 60 mph?

There are two ways to check breaker point gap: with a feeler gauge or with a dwell meter. Either way you set the points, you are adjusting denser to make a clean break in the primary circuit.

The coil has both primary and secondary circuits. When the ignition is turned on, the battery supplies voltage through the coil and onto the points. The points are connected to ground, completing the primary circuit. As the current passes through the coil, a magnetic field is created in the iron center core of the coil. When the cam in the distributor turns, the points open, breaking the primary circuit. The magnetic field in the primary circuit of the coil then collapses and cuts through the secondary circuit windings around the iron core. Because of the physical principle called electromagnetic induction, the battery voltage is increased to a level sufficient to fire the spark plugs.

When the points open, the electrical charge in the primary circuit tries to jump the gap created between the two open contacts of the points. If this electrical charge were not transferred elsewhere, the metal contacts of the points would start to change rapidly.

The function of the condenser is to absorb excessive voltage from the points when they open

ENGINE PERFORMANCE AND TUNE-UP

ing the amount of time (in degrees of distributor rotation) that the points will remain open. If you adjust the points with a feeler gauge, you are setting the maximum amount of points will open when the rubbing block on the points is on a high point of the distributor cam. when you adjust the points with a dwell meter, you are measuring the number of degrees (of distributor cam rotation) that the points will remain closed before they start to open as a high point of the distributor cam approaches the rubbing block of the points.

If you still do not understand how the points function, take a friend, go outside, and remove the distributor cap from your engine. Have you friend operate the starter (make sure that the transmission is not in gear) as you look at the exposed parts of the distributor.

There are two rules that should always be followed when adjusting or replacing points. The points and condenser are a matched set; never replace one without replacing the other. If you change the point gap or dwell of the engine, you also change the ignition timing. Therefore, if you adjust the points, you must also adjust the timing.

REMOVAL AND INSTALLATION

1968-74

The usual procedure is to replace the condenser each time that point set is replaced. Although this is not always necessary, it is easy to do at this time and the cost is negligible. Every time you adjust or replace the breaker points, the ignition timing must be checked and, if necessary, adjusted. no special equipment other than a feeler gauge is required for point replacement or adjustment, but a dwell meter is strongly advised. A magnetic tool is handy to prevent the small points and condenser screws from falling down into the distributor.

Point sets using the push-in type wiring terminal should be used on those distributors equipped with an R.F.I. (Radio Frequency Interference) shield (1970-74). Points using a lockscrew-type terminal may short out due to contact between the shield and the screw.

1. Push down on the spring-loaded V8 distributor cap retaining screws and give them a half turn to release. Unscrew the captive 6-cylinder cap retaining screws. Remove the cap. You might have to unclip or detach some or all of the plug wires to remove the cap. If so, number the wires and the cap for removal.

2. Clean the cap inside and out with a clean rag. Check for cracks and carbon paths. A carbon path shows up as a dark line, usually from one of the cap sockets or inside terminals to a ground. Check the condition of the carbon button inside the center of the cap and inside terminals. Replace the cap as necessary. Carbon paths cannot usually be successfully scraped off. It is better to replace the cap.

3. Pull the 6-cylinder rotor up and off the shaft. Remove the two screws and lift the round V8 rotor off. There is less danger of losing the screws if you just back them out all the way and lift them off with the rotor. Clean off the metal outer tip if it is burned or corroded. don't file it. Replace the rotor as necessary or if one cam with your tune-up kit.

4. Remove the radio frequency interference shield if your distributor has one. Watch out for those little screws! The factory says that the points don't need to be replace if they are only slightly rough or pitted. However, sad experience shows that it is more economical and reliable in the long run to replace the point set while the distributor is open, than to have to do this at a later (and possibly more inconvenient) time.

5. Pull one of the two wire terminals from the point assembly. One wire comes from the condenser and the other comes from within the distributor. The terminals are usually held in place by spring tension only. There might be a clamp screw securing the terminals on some older versions. There is also available a one-

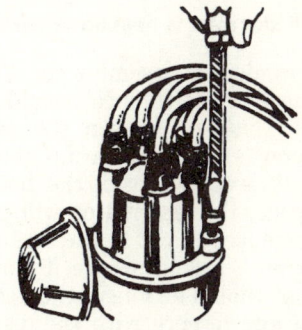

Six cylinder distributor caps are retained by screws

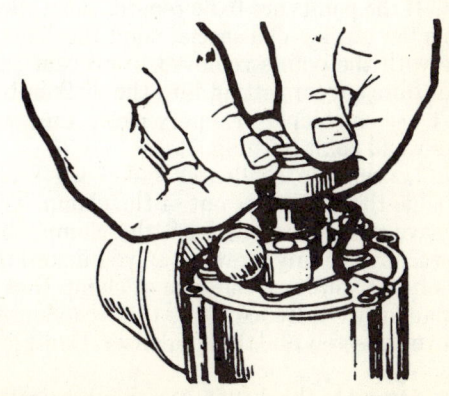

Pull the six cylinder rotor straight up to remove

68 ENGINE PERFORMANCE AND TUNE-UP

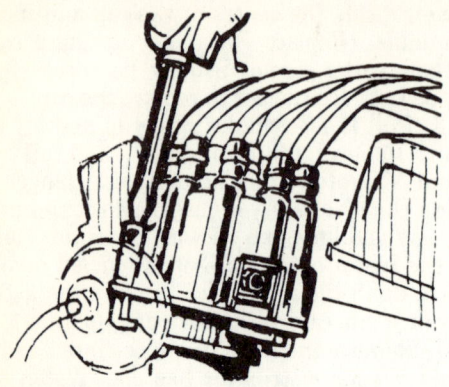

Eight cylinder distributor caps are retained by push and turn latches

Install the point set on the breaker plate, then attach wires

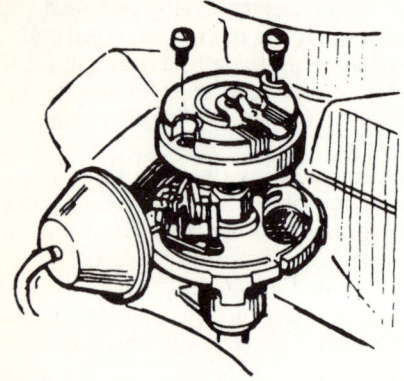

The eight cylinder rotor is held on by two screws

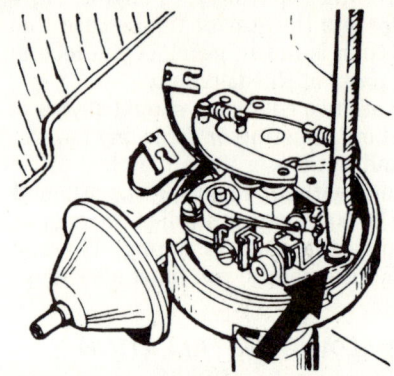

The points on all models are retained by screws; use a magnetic screwdriver to avoid dropping them

piece point/condenser assembly for V8s. The radio frequency interference shield is not needed with this set. Loosen the point set holddown screw(s). Be very careful not to drop any of these little screws inside the distributor. If this happens, the distributor will probably have to be removed to get at the screw. If the holddown screw is lost elsewhere, it must be replace with one that is no longer than the original to avoid interference with the distributor workings. Remove the point set, even if it is to be reused.

6. If the points are to be reused, clean them with a few strokes of a special point file. This is done with the points removed to prevent tiny metal filings from getting into the distributor. Don't use sandpaper or emery cloth; they will cause rapid point burning.

7. Loosen the condenser holddown screw and slide the condenser out of the clamp. This will save you a struggle with the clamp, condenser, and the tiny screw when you install the new one. If you have the type of clamp that is permanently fastened to the condenser, remove the screw and the condenser. Don't lose the screw.

8. Attend to the distributor cam lubricator. If you have the round kind, turn it around on its shaft at the first tune-up and replace it at the second. If you have the long kind, switch ends at the first tune-up and replace it at the second.

NOTE: *Don't oil or grease the lubricator. The foam is impregnated with a special lubricant.*

If you didn't get any lubricator at all, or if it looks like someone took it off, don't worry. You don't really need it. just rub a matchhead size dab of grease on the cam lobes.

9. Install the new condenser. If you left the clamp in place, just slide the new condenser into the clamp.

10. Replace the point set and tighten the screw on a V8. Leave the screw slightly loose on a 6-cylinder. Replace the two wire terminals, making sure that the wires don't interfere with anything. Some V8 distributors have a ground wire that must go under one of the screws.

11. Check that the contacts meet squarely. If they don't, bend the tab supporting the fixed contact.

NOTE: *If you are installing preset points on a V8, go ahead to Step 16. If they are preset, it will say so on the package. It would be a good idea to make a quick check on point*

ENGINE PERFORMANCE AND TUNE-UP

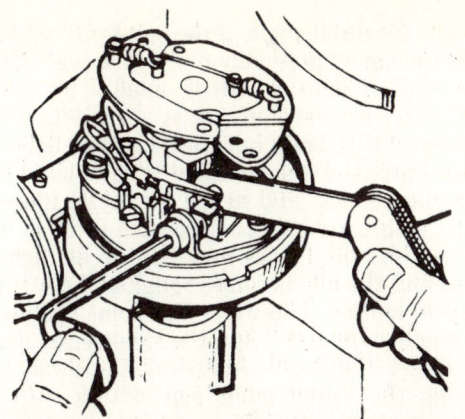

You will need an allen wrench to adjust the eight cylinder point gap

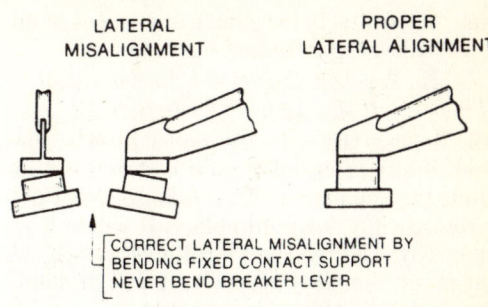

Check the points for proper alignment after installation

gap, anyway. Sometimes those preset points aren't.

12. Turn the engine until a high point on the cam that opens the points contacts the rubbing block on the point arm. You can turn the engine by hand if you can get a wrench on the crankshaft pulley nut, or you can grasp the fan belt and turn the engine with the spark plugs removed.

CAUTION: *If you try turning the engine by hand, be very careful not to get your fingers pinched in the pulleys.*

On a manual transmission you can push it forward in High gear. Another alternative is to bump the starter switch or use a remote starter switch.

13. On the 6-cylinder engine, there is a screwdriver slot near the contacts. Insert a suitable tool and lever the points open or closed until they appear to be at about the gap specified in the Tune-Up Specifications. On the V8 engine, simply insert a 1/8" allen wrench into the adjustment screw and turn. The wrench sometimes comes with a tune-up kit.

14. Insert the correct size feeler gauge and adjust the gap until you can push the gauge in and out between the contacts with a slight drag, but without disturbing the point arm. This operation takes a bit of experience to obtain the correct feel. Check by trying the gauges 0.001-0.002" larger and smaller than the setting size. The larger one should disturb the point arm, while the smaller one should not drag at all. Tighten the 6-cylinder point set holddown screw. Recheck the gap, because it often changes when the screw is tightened.

15. After all the point adjustments are complete, pull a white index card through (be-

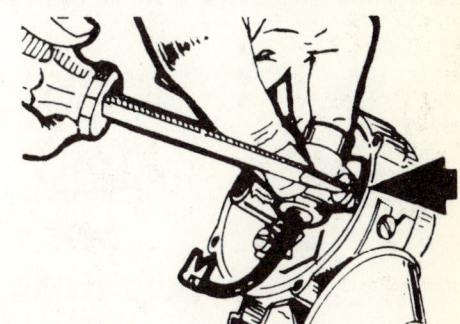

The condenser is held in place by a screw and a clamp

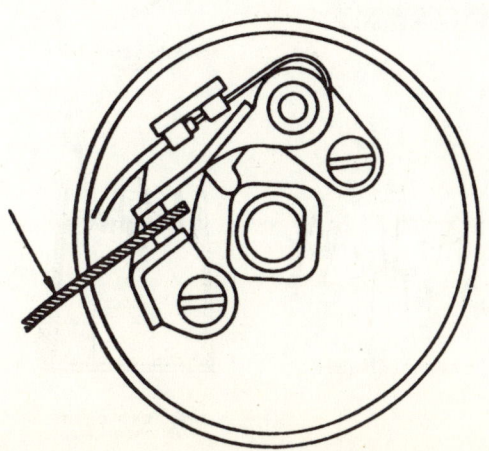

The arrow indicates the feeler gauge used to check the point gap

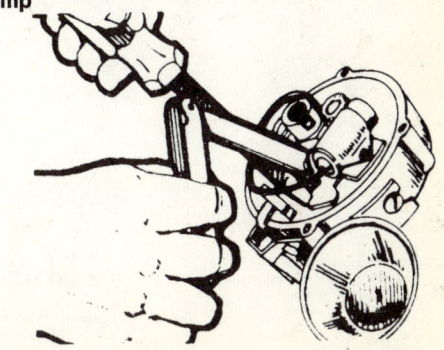

Use a screwdriver to lever the points closer together or further apart on six cylinder models

ENGINE PERFORMANCE AND TUNE-UP

tween) the contacts to remove any traces of oil. Oil will cause rapid contact burning.

NOTE: *You can adjust 6-cylinder dwell at this point, if you wish. Refer to Step 18.*

16. Replace the radio frequency interference shield, if any. You don't need it if you are installing the one-piece point/condenser set. Push the rotor firmly down into place. It will only go on one way. Tighten the V8 rotor screws. If the rotor is not installed properly, it will probably break when the starter is operated.

17. Replace the distributor cap.

18. Check the dwell meter if it is available.

NOTE: *This hookup may not apply to electronic, capacitive discharge, or other special ignition systems. Some dwell meters won't work at all with such systems.*

1975 and Later

These engines use the breakerless HEI (High Energy Ignition) system. Since there is no mechanical contact, there is no wear or need for periodic service. There is an item in the distributor that resembles a condenser; it is a radio interference suppression capacitor which requires no service.

Dwell Angle

Dwell angle is the amount of time (measured in degrees of distributor cam rotation) that the contact points remain closed. Initial point gap determines dwell angle. If the points are set too wide they open gradually and dwell angle (the time they remain closed) is small. This wide gap causes excessive arcing at the points and, because of this, point burning. This small dwell doesn't give the coil sufficient time to build up maximum energy and so coil output decreases. If the points are set too close, the dwell is increased but the points may bounce at higher speed and the idle becomes rough and starting is made harder. The wider the point opening, the smaller the dwell and the smaller the gap, the larger the dwell. Adjusting the dwell by making the initial point gap setting with a feeler gauge is sufficient to get the car started but a finer adjustment should be made. A dwell meter is needed to check the adjustment.

Connect the red lead (positive) wire of the meter to the distributor primary wire connection on the Positive (+) side of the coil, and the black ground (negative) wire of the meter to a good ground on the engine. The dwell angle may be checked either with the engine cranking or running, although the reading will be more accurate if the engine is running. With the engine cranking, the reading will fluctuate between 0° dwell and the maximum figure of that angle. While cranking, the maximum figure is the correct one.

NOTE: *Dwell angle is permanently set electronically on HEI distributors, requiring no adjustment or checking.*

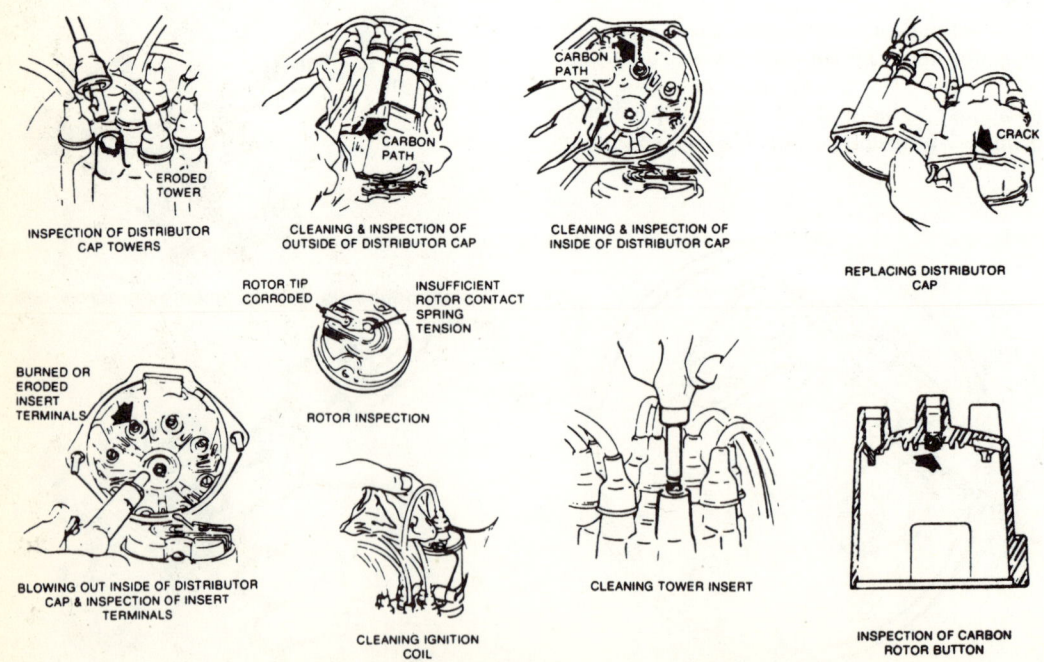

Distributor cap and rotor checkpoints

ENGINE PERFORMANCE AND TUNE-UP

ADJUSTMENT

1968-74

Dwell can be checked with the engine running or cranking. Decrease dwell by increasing the point gap; increase by decreasing the gap. Dwell angle is simply the number of degrees of distributor shaft rotation during which the points stay closed. Theoretically, if the point gap is correct, the dwell should also be correct or nearly so. Adjustment with a dwell meter produces more exact, consistent results since it is a dynamic adjustment. If dwell varies more than 3° from idle speed to 1,750 engine rpm, the distributor is worn.

1. To adjust dwell on a 6-cylinder engine, trial and error point adjustments are required. On a V8 engine, simply open the metal window on the distributor and insert a 1/8" allen wrench. Turn until the meter shows the correct reading. Be sure to snap the window closed.

2. An approximate dwell adjustment can be made without a meter on a V8 engine. Turn the adjusting screw clockwise until the engine begins to misfire, then turn it out 1/2 turn.

3. If the engine won't start, check:

   a. That all the spark plug wires are in place.

   b. That the rotor has been installed.

   c. That the two (or three) wires inside the distributor are connected.

   d. That the points open and close when the engine turns.

   e. That the gap is correct and the holddown screw (on a 6-cylinder engine) is tight.

4. After the first 200 miles or so on a new set of points, the point gap often closes up due to initial rubbing block wear. For best performance, recheck the dwell (or gap) at this time. This quick initial wear is the reason the factory recommends 0.003" more gap on new points.

5. Since changing the gap affects the ignition timing, the timing should be checked and adjusted as necessary after each point replacement or adjustment.

1974-88

The dwell angle on these cars is preset at the factory and not adjustable

High Energy Ignition (HEI) System

NOTE: *This book contains simple testing procedures for your car's electronic ignition. More comprehensive testing in this system and other electronic control systems on your car can be found in CHILTON'S GUIDE TO ELECTRONIC ENGINE CONTROLS, book part number 7535 for early HEI, or number 8173 for the most updated HEI, available at most book stores and auto parts stores or available directly from Chilton Co.*

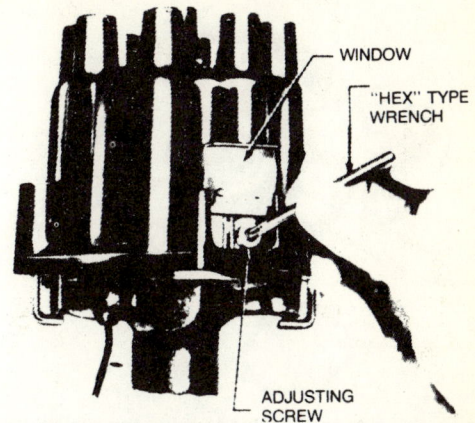

To set the dwell on a V8, lift the window and then turn the adjusting screw

The General Motors HEI system is a pulse-triggered, transistor-controlled, inductive discharge ignition system. Except on inline 6-cylinder engines through 1977, the entire HEI system is contained within the distributor cap. Inline 6-cylinder engines through 1977 have an external coil. Otherwise, the systems are the same.

The distributor, in addition to housing the mechanical and vacuum advance mechanisms, contains the ignition coil (except on some inline 6-cylinder), the electronic control module, and the magnetic pick-up assembly contains a permanent magnet, a pole piece with internal teeth, and a pick-up coil (not to be confused with the ignition coil).

For 1981 and later an HEI distributor with Electronic Spark Timing is used (for more information on EST, refer to Chapter 4). This system uses a one piece distributor cap, similar to 1980.

All spark timing changes in the 1981 and later distributors are done electronically by the Electronic Control Module (ECM) which monitors information from various engine sensors, computes the desired spark timing and then signals the distributor to change the timing accordingly. No vacuum or mechanical advance systems are used whatsoever.

In the HEI system, as in other electronic ignition systems, the breaker points have been replaced with an electronic switch, a transistor, which is located within the control module. This switching transistor performs the same function the points did in a conventional ignition system; it simply turns coil primary current on and off at the correct time. Essentially

ENGINE PERFORMANCE AND TUNE-UP

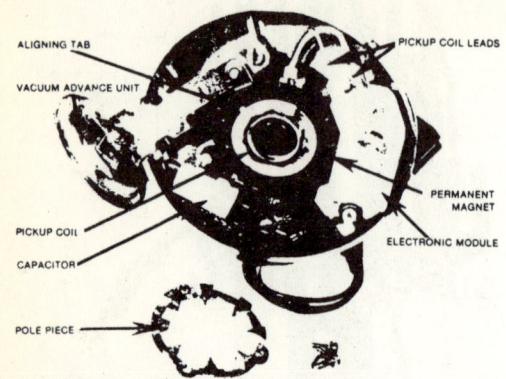

All HEI circuitry is contained within the distributor body (1980 shown)

then, electronic and conventional ignition systems operate on the same principle.

The module which houses the switching transistor is controlled (turned on and off) by a magnetically generated impulse induced in the pick-up coil. When the teeth of the rotating timer align with the teeth of the pole piece, the induced voltage in the pick-up coil signals the electronic module to open the coil primary circuit. The primary current then decreases, and a high voltage is induced in the ignition coil secondary windings which is then directed through the rotor and high voltage leads (spark plug wires) to fire the spark plugs.

In essence then, the pick-up coil module system simply replaces the conventional breaker points and condenser. The condenser found within the distributor is for radio suppression purposes only and has nothing to do with the ignition process. The module automat-

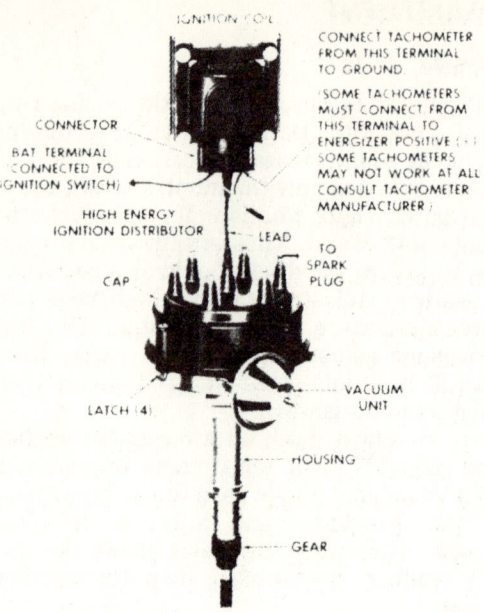

Early six cylinder engines with the HEI distributor had an external coil

ically controls the dwell period, increasing it with increasing engine speed. Since dwell is automatically controlled, it cannot be adjusted. The module itself is non-adjustable and non-repairable and must be replaced if found defective.

HEI SYSTEM PRECAUTIONS

Before going on to troubleshooting, it might be a good idea to take note of the following precautions:

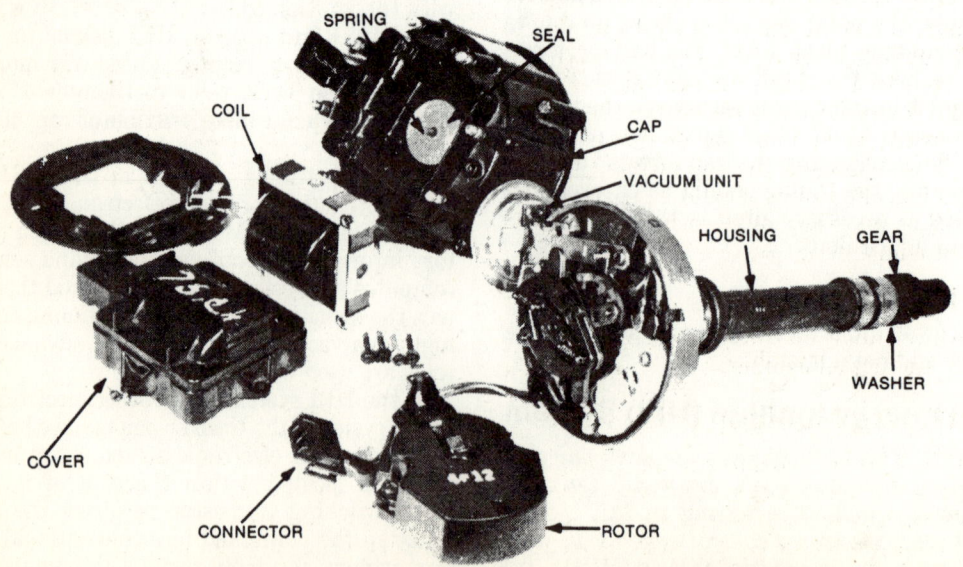

V8 HEI distributor components, 1975–80 (1978–80 6 cyl. similar)

ENGINE PERFORMANCE AND TUNE-UP 73

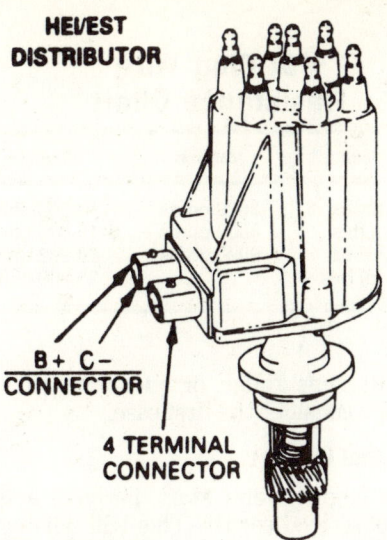

1989–92 HEI distributor used with a seperate coil

Timing Light Use

Inductive pick-up timing lights are the best kind to use with HEI. Timing lights which connect between the spark plug and the spark plug wire occasionally (not always) give false readings.

Spark Plug Wires

The plug wires used with HEI systems are of a different construction than conventional wires. When replacing them, make sure you get the correct wires, since conventional wires won't carry the voltage. Also, handle them care-

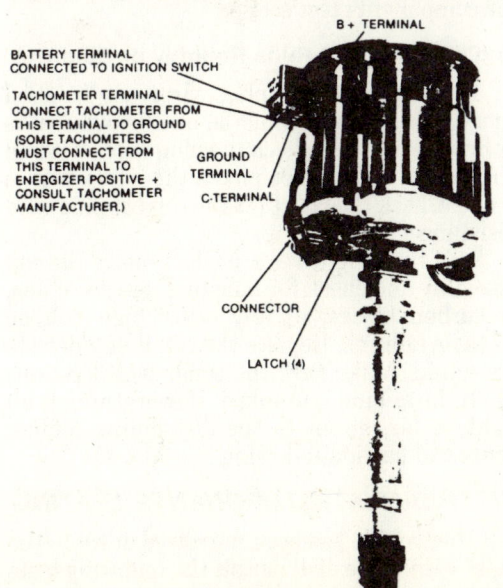

Typical HEI distributor connections

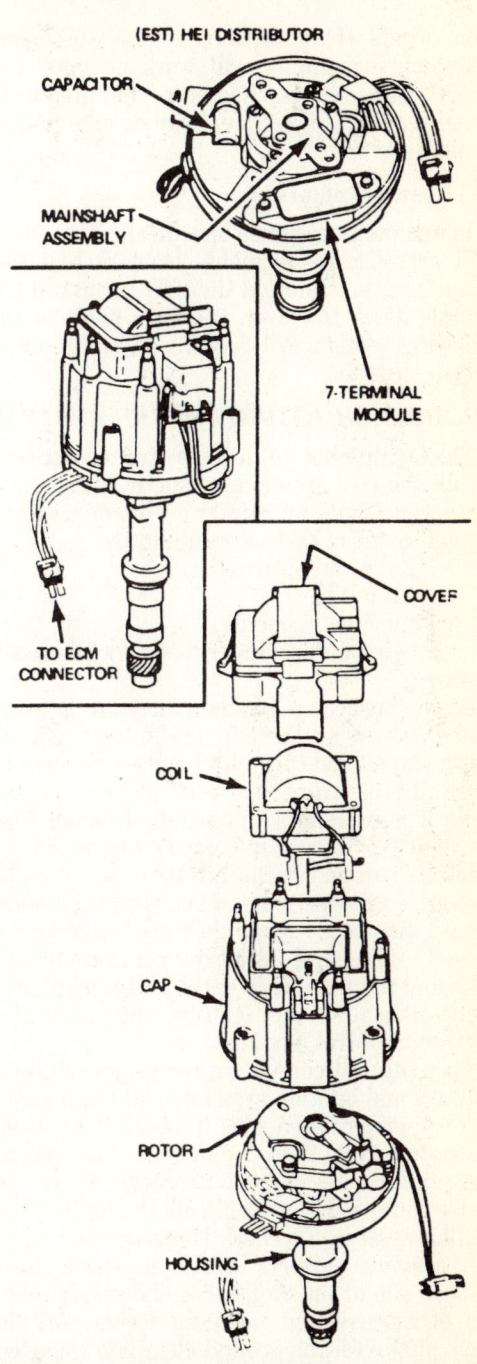

HEI EST distributor components, 1981 (note absence of vacuum advance unit)

fully to avoid cracking or splitting them and never pierce them.

Tachometer Use

Not all tachometers will operate or indicate correctly when used on a HEI system. While some tachometers may give a reading, this does not necessarily mean the reading is correct. In addition, some tachometers hook up differently from others. If you can't figure out whether or not your tachometer will work on your car, check with the tachometer manufacturer. Dwell readings, or course, have no significance at all.

HEI System Testers

Instruments designed specifically for testing HEI systems are available from several tool manufacturers. Some of these will even test the module itself. However, the tests given in the following section will require only ohmmeter and a voltmeter.

TROUBLESHOOTING THE HEI SYSTEM

The symptoms of a defective component within the HEI system are exactly the same as those you would encounter in a conventional system. Some of these symptoms are:
- Hard or no Starting
- Rough Idle
- Poor Fuel Economy
- Engine misses under load or while accelerating

If you suspect a problem in your ignition system, there are certain preliminary checks which you should carry out before you begin to check the electronic portions of the system. First, it is extremely important to make sure the vehicle battery is in a good state of charge. A defective or poorly charged battery will cause the various components of the ignition system to read incorrectly when they are being tested. Second, Make sure all wiring connections are clean and tight, not only at the battery, but also at the distributor cap, ignition coil, and at the electronic control module.

Since the only change between electronic and conventional ignition systems is in the distributor component area, it is imperative to check the secondary ignition circuit first. If the secondary circuit checks out properly, then the engine condition is probably not the fault of the ignition system. To check the secondary ignition system, perform a simple spark test. Remove one of the plug wires and insert some sort of extension in the plug socket. An old spark plug with the ground electrode removed makes a good extension. Hold the wire and extension about 1/4" away from the block and crank the engine. If a normal spark occurs, then the problem is most likely not in the ignition system. Check for fuel system problems, or fouled spark plugs.

If, however, there is no spark or a weak spark, then further ignition system testing will have to be done. Troubleshooting techniques fall into two categories, depending on the nature of the problem. The categories are (1) Engine cranks, but won't start or (2) Engine runs, but runs rough or cuts out. To begin with, let's consider the first case.

HEI Plug Wire Resistance Chart

Wire Length	Minimum	Maximum
0–15 inches	3000 ohms	10,000 ohms
15–25 inches	4000 ohms	15,000 ohms
25–35 inches	6000 ohms	20,000 ohms
Over 35 inches		25,000 ohms

Engine Fail to Start

If the engine won't start, perform a spark test as described earlier. This will narrow the problem area down considerably. If no spark occurs, check for the presence of normal battery voltage of the battery (BAT) terminal in the distributor cap. The ignition switch must be in the ON position for this test. Either a voltmeter or a test light may be used for this test. Connect the test light wire to ground and probe end to the BAT terminal at the distributor. If the light comes on, you have voltage on the distributor. If the light fails to come on, this indicates an open circuit in the ignition primary wiring leading to the distributor. In this case, you will have to check wiring continuity back to the ignition switch using a test light. If there is battery voltage at the BAT terminal, but no spark at the plugs, then the problem lies within the distributor assembly. Go on to the distributor components test section.

Engine Runs, But Runs Roughly or Cuts Out

1. Make sure the plug wires are in good shape first. There should be no obvious cracks or breaks. You can check the plug wires with an ohmmeter, but do not pierce the wires with a probe. Check the chart for the correct plug wire resistance.

2. If the plug wires are OK, remove the cap assembly and check for moisture, cracks, chips, or carbon tracks, or any other high voltage leaks or failures. Replace the cap if any defects are found. Make sure the timer wheel rotates when the engine is cranked. If everything is all right so far, go on to the distributor components test section following.

DISTRIBUTOR COMPONENTS TESTING

If the trouble has been narrowed down to the units within the distributor, the following tests can help pinpoint the defective component. An ohmmeter with both high and low ranges should be used. These tests are made with the cap assembly removed and the battery wire dis-

ENGINE PERFORMANCE AND TUNE-UP

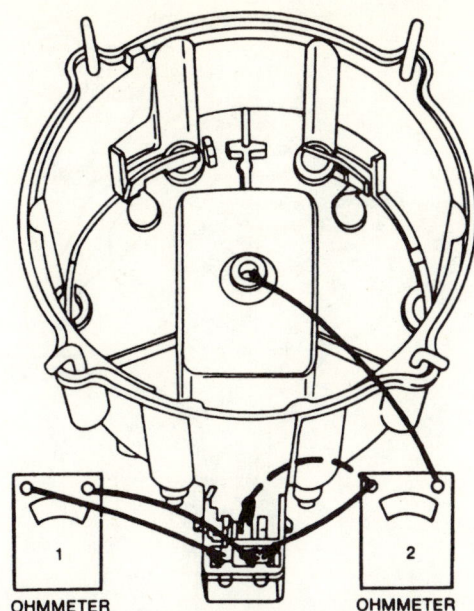

Ohmmeter 1 shows the primary coil resistance connection. Ohmmeter 2 shows the secondary resistance connection (1980 shown, most models similar)

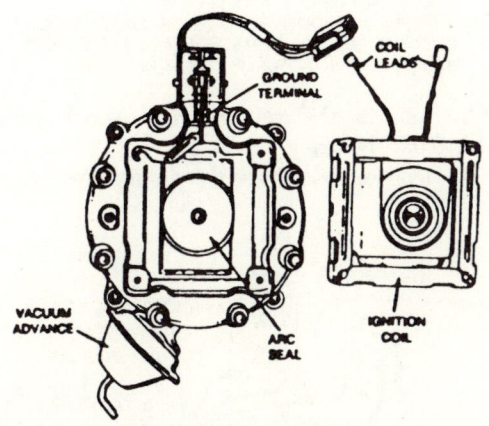

Check the condition of the arc seal under the coil

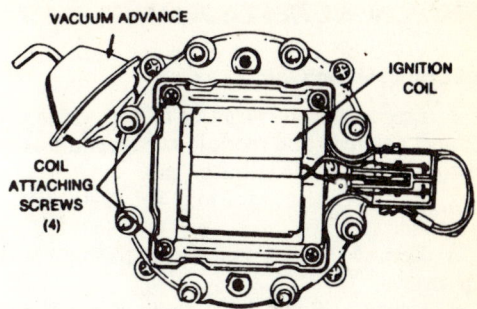

The coil on all but the 1975–77 six cylinders engines is accessible by removing the four attaching screws

connected. If a tachometer is connected to the TACH terminal, disconnect it before making these tests.

1. Connect an ohmmeter between the TACH and BAT terminals in the distributor cap. The primary coil resistance should be less than 1Ω.

2. To check the coil secondary resistance, connect an ohmmeter between the rotor button and BAT terminal. Note the reading. Connect the ohmmeter between the rotor button and the TACH terminal. Note the reading. The resistance in both cases should be between 6,000 and 30,000Ω. Be sure to test between the rotor button and both the BAT and TACH terminals.

3. Replace the coil only if the readings in Step 1 and Step 2 are infinite.

NOTE: *These resistance checks will not disclose shorted coil windings. This condition can only be detected with scope analysis or a suitably designed coil tester. If these instruments are unavailable, replace the coil with a known good coil as a final coil test.*

4. To test the pick-up coil, first disconnect the white and green module leads. Set the ohmmeter on the high scale and connect it between a ground and either the white or green lead. Any resistance measurement less than infinity requires replacement of the pick-up coil.

5. Pick-up coil continuity is tested by connecting the ohmmeter (on low range) between the white and green leads. Normal resistance is between 650 and 850Ω, or 500 and 1500Ω on 1977 and later cars. Move the vacuum advance arm while performing this test. This will detect any break in coil continuity. Such a condition can cause intermittent misfiring. Replace the pick-up if the reading is outside the specified limits.

6. If no defects have been found at this time, you still have a problem, then the module will have to be checked. If you do not have access to a module tester, the only possible alternative is a substitution test. If the module fails the substitution test, replace it.

HEI SYSTEM MAINTENANCE

Except for periodic checks of the spark plug wires, and an occasional check of the distributor cap for cracks (see Steps 1 and 2 under Engine Runs, But Runs Rough or Cuts Out for details), no maintenance is required on the HEI System. No periodic lubrication is necessary; engine oil lubricates the lower bushing, and an oil-filled reservoir lubricates the upper bushing.

76 ENGINE PERFORMANCE AND TUNE-UP

COMPONENT REPLACEMENT

Integral Ignition Coil

1. Disconnect the negative battery cable. Disconnect the feed and module wire terminal connectors from the distributor cap.
2. Remove the ignition set retainer, if equipped.
3. Remove the 4 coil cover-to-distributor cap screws.
4. Using a blunt drift, press the coil wire spade terminals up out of the distributor cap.
5. Lift the coil up out of the distributor cap.
6. Remove and clean the coil spring, rubber seal washer and coil cavity of the distributor cap.
7. Coat the rubber seal with a dielectric lubricant furnished in the replacement ignition coil package.
8. Install the coil assembly in the distributor cap, along with the coil spring.
9. Connect the coil wire spade terminals to their respective places in the distributor housing.
10. Install the coil cap and retaining screws. Install the ignition set retainer, if equipped.
11. Connect the distributor cap electrical connections. Connect the negative battery cable.

Distributor Cap

1. Disconnect the negative battery cable. Remove the feed and module wire terminal connectors from the distributor cap.

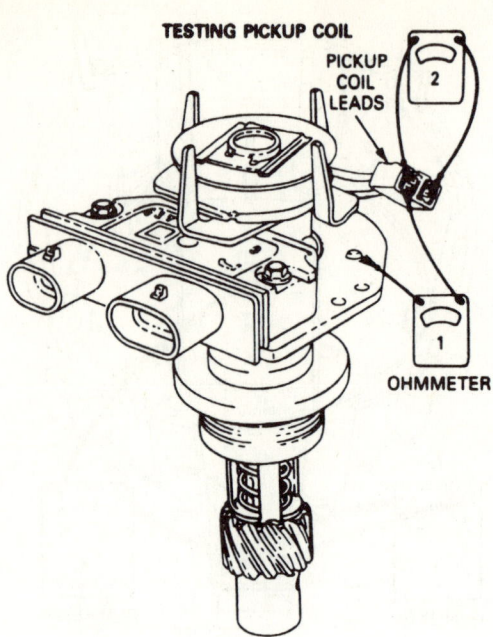

3. REMOVE ROTOR AND PICKUP COIL LEADS FROM MODULE.
4. CONNECT OHMMETER PART 1 AND PART 2.
5. OBSERVE OHMMETER. FLEX LEADS BY HAND TO CHECK FOR INTERMITTENT OPENS.
   STEP 1 — SHOULD READ INFINITE AT ALL TIMES. IF NOT, PICKUP COIL IS DEFECTIVE.
   STEP 2 — SHOULD READ ONE STEADY VALUE BETWEEN 500-1500 OHMS AS LEADS ARE FLEXED BY HAND. IF NOT, PICKUP COIL IS DEFECTIVE.

2. Remove the retainer and spark plug wires from the cap.
3. Depress and release the 4 distributor cap-

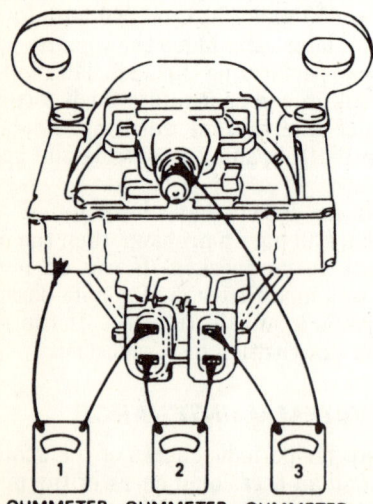

2. CHECK IGNITION COIL WITH OHMMETER FOR OPENS AND GROUNDS:
   STEP 1. — USE HIGH SCALE. SHOULD READ VERY HIGH (INFINITE). IF NOT, REPLACE COIL.
   STEP 2. — USE LOW SCALE. SHOULD READ VERY LOW OR ZERO. IF NOT, REPLACE COIL.
   STEP 3. — USE HIGH SCALE. SHOULD NOT READ INFINITE. IF IT DOES, REPLACE COIL.

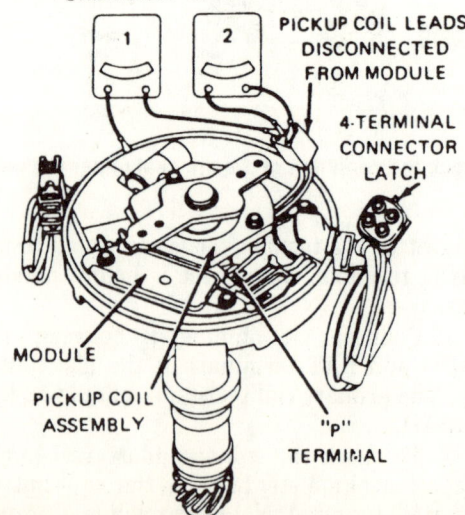

Ohmmeter 1 shows the connections for testing the pickup coil. Ohmmeter 2 shows the connections for testing the pickup coil continuity (1980 shown, most models similar)

ENGINE PERFORMANCE AND TUNE-UP

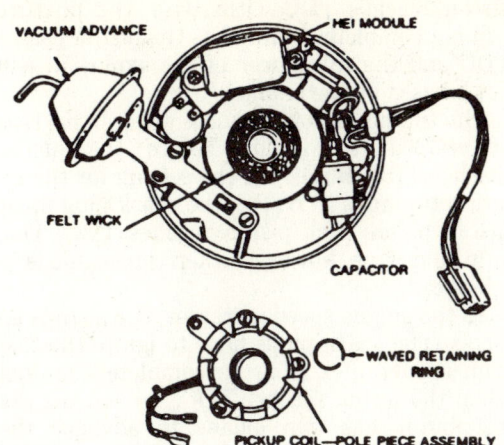

Pickup coil removal (1981 and later models don't have the vacuum advance unit)

to-housing retainers and lift off the cap assembly.

4. Remove the 4 coil cover screws and cover.

5. Using a finger or a blunt drift, push the spade terminals up out of the distributor cap.

6. Remove all 4 coil screws and lift the coil, coil spring and rubber seal washer out of the cap coil cavity.

7. Using a new distributor cap, reverse the above procedure to assemble being sure to clean and lubricate the rubber seal washer with dielectric lubricant.

Rotor

1. Disconnect the negative battery cable. Disconnect the feed and module wire connector from the distributor.

2. Depress and release the 4 distributor cap-to-housing retainers and lift off the cap assembly.

3. Remove the two rotor attaching screws and rotor.

4. Reverse the above procedure to install.

Vacuum Advance (1975-80)

1. Disconnect the negative battery cable. Remove the distributor cap and rotor as previously described.

2. Disconnect the vacuum hose from the vacuum advance unit.

3. Remove the two vacuum advance retaining screws, pull the advance unit outward, rotate and disengage the operating rod from its tang.

4. Reverse the above procedure to install.

Module

1. Disconnect the negative battery cable. Remove the distributor cap and rotor as previously described.

2. Disconnect the harness connector and pick-up coil spade connectors from the module. Be careful not to damage the wires when removing the connector.

3. Remove the two screws and module from the distributor housing.

4. Coat the bottom of the new module with dielectric lubricant supplied with the new module. Reverse the above procedure to install.

HEI SYSTEM TACHOMETER HOOKUP

There is a terminal marked TACH on the distributor cap. Connect one tachometer lead to this terminal and the other lead to a ground. On some tachometers, the leads must be con-

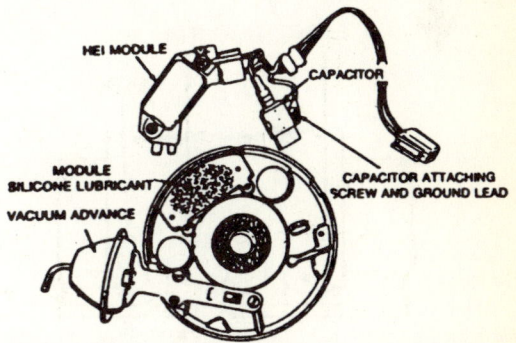

Module replacement; be sure to coat the mating surfaces with silicone lubricant

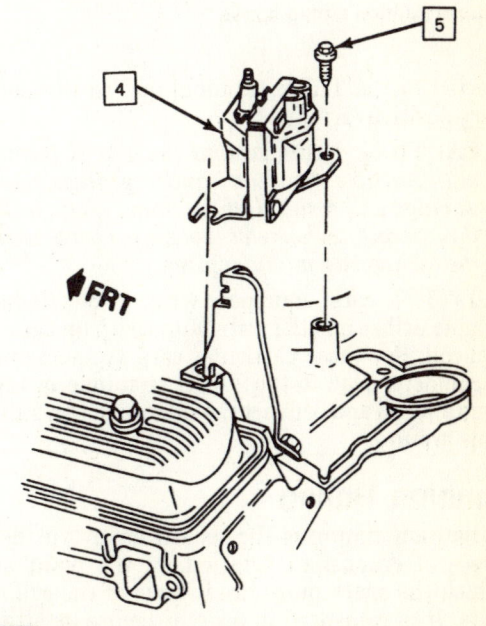

| 4 | COIL |
| 5 | BOLT |

Coil removal and installation; non integral type 1989–92

ENGINE PERFORMANCE AND TUNE-UP

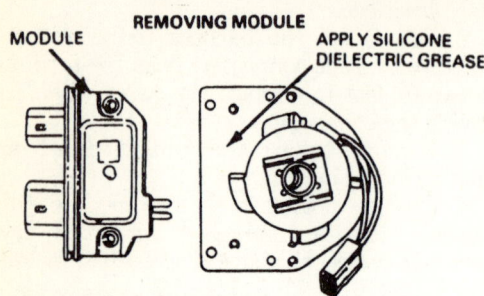

1989–92 module removal installation

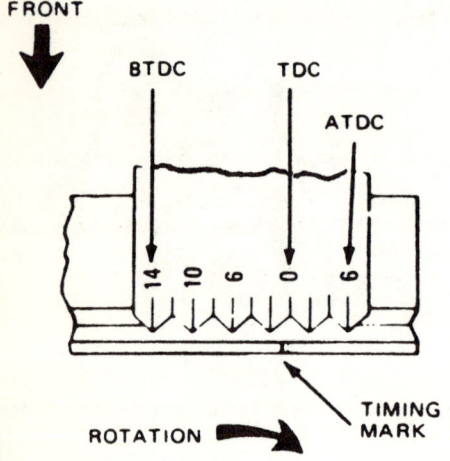

typical ignition timing marks

nected to the TACH terminal and to the battery positive terminal.

CAUTION: *Never ground the TACH terminal; serious module and ignition coil damage will result. If there is any doubt as to the correct tachometer hookup, check with the tachometer manufacturer.*

1975-77 cars equipped with a 6-cylinder engine utilize an HEI distributor with an external coil. For these particular cars, connect one tachometer lead to the TACH terminal on the ignition coil and connect the other one to a suitable ground.

Ignition Timing

Ignition timing is the measurement, in degrees of crankshaft rotation, of the point at which the spark plugs fire in each of the cylinders. It is measured in degrees before or after Top Dead Center (TDC) of the compression stroke.

Because it takes a fraction of a second for the spark plug to ignite the mixture in the cylinder, the spark plug must fire a little before the piston reaches TDC. Otherwise, the mixture will not completely ignited as the piston passes TDC and the full power of the explosion will not be used by the engine.

The timing measurement is given in degrees of crankshaft rotation before the piston reaches TDC (BTDC). If the setting for the ignition timing is 5° BTDC, the spark plug must fire 5° before each piston reaches TDC. This only holds true, however, when the engine is at idle speed.

As the engine speed increases, the pistons go faster. The spark plugs have to ignite the fuel even sooner if it is to be completely ignited when the piston reaches TDC. To do this, the distributor has two means to advance the timing of the spark as the engine speed increases. This is accomplished by centrifugal weights within the distributor, and a vacuum diaphragm mounted on the side of the distributor (non HEI EST distributors only).

If the ignition is set too far advanced (BTDC), the ignition and expansion of the fuel in the cylinder will occur too soon and tend to force the piston down while it is still traveling up. This causes engine ping. If the ignition spark is set too far retarded, after TDC (ATDC), the piston will have already passed TDC and started on its way down when the fuel is ignited. This will cause the piston to be forced down for only portion of its travel. This will result in poor engine performance and lack of power.

Timing marks consist of a notch on the rim of the crankshaft pulley and a scale of degrees attached to the front of the engine. The notch corresponds to the position of the piston in the number 1 cylinder. A stroboscopic (dynamic) timing light is used, which is hooked into the circuit of the No. 1 cylinder spark plug. Every time the spark plug fires, the timing light flashes. By aiming the timing light at the timing marks, the exact position of the piston within the cylinder can be read, since the stroboscopic flash makes the mark on the pulley appear to be standing still. Proper timing is indicated when the notch is aligned with the correct number on the scale.

There are three basic types of timing lights available. The first is a simple neon bulb with two wire connections (one for the spark plug and one for the plug wire, connecting the light in series). This type of light is quite dim, and must be held closely to the marks to be seen, but it is quite inexpensive. The second type of light operated from the vehicle's battery. Two alligator clips connect to the battery terminals, while a third wire connects to the spark plug with an adapter. This type of light is more expensive, but the xenon bulb provides a nice

ENGINE PERFORMANCE AND TUNE-UP

bright flash which can even be seen in sunlight. The third type replace the battery source with 110 volt house current. Some timing lights have other functions built into them, such as dwell meters, tachometers, or remote starting switches. These are convenient, in that they reduce the tangle of wires under the hood, but may duplicate the functions of tools you already have.

If your car has electronic ignition, you should use a timing light with an inductive pickup. This pickup simply clamps onto the No. 1 spark plug wire, eliminating the adapter. It is not susceptible to cross firing or false triggering, which may occur with a conventional light, due to the greater voltages produced by electronic ignition.

CHECKING AND ADJUSTMENT

1. Warm the engine to normal operating temperature. Shut off the engine and connect the timing light to the No. 1 spark plug (left front on V8 and V6 or front on an inline 6-cylinder). Do not, under any circumstances, pierce a wire to hook up a light.

2. Clean off the timing marks and mark the pulley or damper notch and the timing scale with while chalk or paint. The timing notch on the damper or pulley can be elusive. Bump the engine around with the starter or turn the crankshaft with a wrench on the front pulley bolt to get it to an accessible position.

3. Disconnect and plug the vacuum advance hose (if equipped) at the distributor, to prevent any distributor advance. The vacuum line is the rubber hose connected to the metal cone-shaped canister on the side of the distributor. A short screw, pencil, or a golf tee can be used to plug the hose.

NOTE: *1981 cars equipped with Electronic Spark Timing have no vacuum advance, therefore you may skip the previous step, but you must disconnect the four terminal EST connector before going on.*

4. Start the engine and adjust the idle speed to the specified rpm in the Tune-Up Specifications chart. Some cars require that the timing be set with the transmission in Neutral. You can disconnect the idle solenoid, if any, to get the speed down. Otherwise, adjust the idle speed screw. This is to prevent any centrifugal advance of timing in the distributor.

The tachometer hookup for 1968-74 cars is the same as that shown for the dwell meter in the Tune-Up section. On 1975-77 HEI systems, the tachometer connects to the TACH terminal on the distributor for V8s, or on the coil for 6-cylinder, and to a ground. For 1978 and later cars, all tachometer connections are to the TACH terminal. Some tachometers must connect to the TACH terminal and to the positive battery terminal. Some tachometers won't work at all with HEI. Consult the tachometer manufacturer if the instructions supplied with the unit do not give the proper connection.

WARNING: *Never ground the HEI TACH terminal; serious system damage will result, including module burnout.*

5. Aim the timing light at the timing marks. Be careful not to touch the fan, which may appear to be standing still. Keep your clothes and hair, and the light's wires clear of the fan, belts and pulleys. If the pulley or damper notch isn't aligned with the proper timing mark (see the Tune-Up Specifications chart), the timing will have to be adjusted.

NOTE: *TDC or Top Dead Center corresponds to 0°; B, or BTDC, or Before Top Dead Center, may be shown as BEFORE; A or ATDC, or After Top Dead Center, may be shown as AFTER.*

6. Loosen the distributor base clamp locknut. You can buy special wrenches which make this task a lot easier on V8s and V6s. Turn the distributor slowly to adjust the timing, holding it by the body and not the cap. Turn the distributor in the direction of rotor rotation (found in the Firing Order illustration in this Chapter) to retard, and against the direction to advance.

NOTE: *The 231 V6 engine has two timing marks on the crankshaft pulley. One timing mark is 1/8" wide and the other, four inches away, is ‹1/16" wide. The smaller mark is used for setting the timing with a hand-held timing light. The larger mark is used with the magnetic probe and is only of use to a dealer or garage. Make sure you set the timing using the smaller mark.*

7. Tighten the locknut. Check the timing, in case the distributor moved as you tightened it.

8. Replace the distributor vacuum hose, if removed. Correct the idle speed.

9. Shut off the engine, reconnect the EST wire (if so equipped), disconnect the timing light and tachometer.

Valve Lash

Hydraulic valve lifters rarely require adjustment, and are not adjusted as part of a normal tune-up. All adjustment procedures concerning them will be found in Chapter 3.

Solid Lifters (1968-71)

Before adjusting solid lifters, thoroughly warm the engine. The solid lifters are generally found on certain early model high-performance engines.

ENGINE PERFORMANCE AND TUNE-UP

ENGINE RUNNING

1. Run the engine to reach normal operating temperature.

2. Remove the valve covers and gaskets by tapping the end of the cover rearward. Do not attempt to pry the cover off.

3. To avoid being splashed with hot oil, use oil deflector clips. Place one each oil hole in the rocker arm.

4. Measure between the rocker arm and the valve stem with a flat feeler gauge, then adjust the rocker arm stud nut until clearance agrees with the specification in the chart.

5. After adjusting all the valves, stop the engine, clean the gasket surfaces, and install the valve covers with new gaskets.

ENGINE NOT RUNNING

These are initial adjustments usually required after assembling an engine or doing a valve job. They should be followed up by an adjustment with the engine running as described above.

1. Set the engine to the No. 1 firing position.

2. Adjust the clearance between the valve stems and the rocker arms with a feeler gauge. Check the chart for the proper clearance. Adjust the following valves in the No. 1 firing position: Intake No. 2, 7, Exhaust No. 4, 8.

3. Turn the crankshaft ½ revolution clockwise. Adjust the following valves: Intake no. 1,8, Exhaust No. 3, 6.

4. Turn the crankshaft ½ revolution clockwise to No. 6 firing position. Adjust the following valves in the No.6 firing position: Intake No. 3, 4, Exhaust No. 5, 7.

5. Turn the crankshaft ½ revolution clockwise. Adjust the following valves: Intake No. 5, 6, Exhaust No. 1, 2.

6. Run the engine until the normal operating temperature is reached. Reset all clearances, using the procedures listed above under Engine Running.

Hydraulic Lifters — 1972-88

All cars, with the exception of those few already discussed, use a hydraulic tappet system with adjustable rocker mounting nuts to obtain zero lash. No periodic adjustment is necessary.

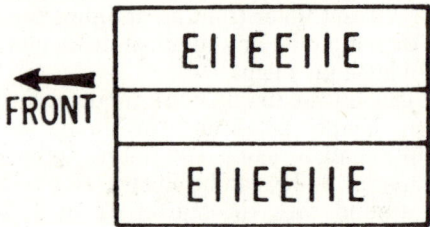

Valve arrangement, Chevrolet small block V8

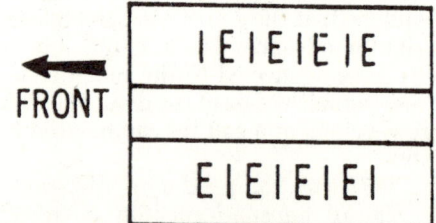

Big block V8 valve arrangement

Inline six cylinder valve arrangement

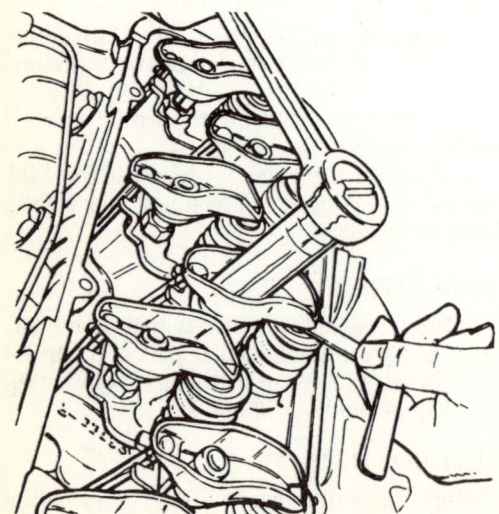

Adjusting the solid valve lifters

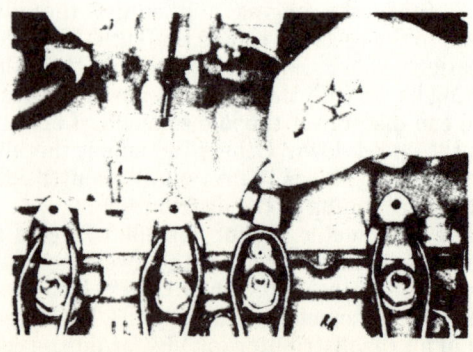

Oil deflector clips will prevent splatter when adjusting the valves with the engine running

ENGINE PERFORMANCE AND TUNE-UP

Carburetor

Idle mixture and speed adjustments are critical aspects of exhaust emission control. It is important that all tune-up instructions be carefully followed to ensure satisfactory engine performance and minimum exhaust pollution. The different combinations of emission systems application on the different engines have resulted in a great variety of tune-up specifications. See the Tune-Up Specifications at the beginning of this section. Beginning in 1968, all cars have a decal conspicuously placed in the engine compartment giving tune-up specifications.

When adjusting a carburetor with two idle mixture screws, adjust them alternately and evenly, unless otherwise stated.

IDLE SPEED AND MIXTURE ADJUSTMENT

See Chapter 5 for illustrations and adjustment specifications of Carter and Rochester carburetors. In the following adjustment procedures the term lean roll means turning the mixture adjusting screws in (clockwise) from optimum setting to obtain an obvious drop in engine speed (usually 20 rpm).

1968-69

Adjust with the air cleaner installed.

NOTE: *Turn off the air conditioner (if applicable) unless your car is a 1968 car with the 6-cylinder engine and automatic transmission. These cars should have the air conditioner turned on when setting the idle.*

1. Turn in the idle mixture screws until they seat gently, then back them out three turns.
2. Start the engine and allow it to reach operating temperature. Make sure the choke is fully open and the preheater valve is open, then adjust the idle speed screw to obtain the specified idle speed (automatic in Drive, manual in Neutral).
3. Adjust the idle mixture screws to obtain the highest steady idle speed, then readjust the idle speed screw to obtain the specified speed. On cars with an idle stop solenoid adjust as follows:

   a. Adjust the idle speed to 500 RPM (6-cylinder engine) or 600 rpm (1968 V8 engine) by turning the hex on the solenoid plunger. Refer to the tune-up decal in the engine compartment for 1969 idle speeds.

   b. Disconnect the wire at the solenoid. This allows the throttle lever to seat against the idle screw.

   c. Adjust the idle screw to obtain 400 rpm (1968 cars), then reconnect the wire. On 1969 cars, adjust the idle to that specified on the tune-up decal in the engine compartment.

4. Adjust one mixture screw to obtain a 20 rpm drop in idle speed, and back out the screw 1/4 turn from this point.
5. Repeat Steps three and four for the second mixture screw (if so equipped).
6. Readjust the idle speed to obtain the specified idle speed.

1970 Initial Adjustments

Adjust with the air cleaner installed.

1. Disconnect the fuel tank line from the vapor canister (EEC).
2. Connect a tachometer to the engine, start the engine and allow it to reach operating temperature. Make sure the choke and preheater valves are fully open.
3. Turn off the air conditioner and set the parking brake. Disconnect and plug the distributor vacuum line.
4. Make the following adjustments:

6-250

1. Turn in the mixture screw until it gently seats, then back out the screw four turns.
2. Adjust the solenoid screw to obtain 830 rpm for manual transmissions (in Neutral) or 630 rpm for automatic transmission (in Drive).
3. Adjust the mixture screw to obtain 750 rpm for manual transmissions (in Neutral) or 600 rpm for automatic transmissions (in Drive).
4. Disconnect the solenoid wire and set the idle speed to 400 rpm, then reconnect it.
5. Reconnect the distributor vacuum line.

V8-307, 400

1. Turn in the mixture screws until they seat gently, then back them out four turns.
2. Adjust the carburetor idle speed screw to obtain 800 rpm for manual transmissions (in Neutral), or adjust the solenoid screw to obtain 630 rpm for automatic transmissions (in Drive).
3. Adjust both mixture screws equally inward to obtain 700 rpm for manual transmissions, 600 rpm for automatic transmissions (in Drive).
4. On cars with automatic transmissions, disconnect the solenoid wire, set the carburetor idle screw to obtain 450 rpm and reconnect the solenoid.
5. Reconnect the distributor vacuum line.

V8-350 (250 HP)

1. Turn in the mixture screws until they gently seat, then back them out four turns.
2. Adjust the solenoid screw to obtain 830 rpm for manual transmissions (in neutral), or

ENGINE PERFORMANCE AND TUNE-UP

630 rpm for automatic transmissions (in Drive).

3. Adjust both mixture screw equally inward to obtain 750 rpm for manual transmissions or 600 rpm for automatics (in Drive).

4. Disconnect the solenoid wire, set the carburetor idle screw to obtain 450 rpm, and reconnect the solenoid.

5. Reconnect the distributor vacuum line.

V8-350 (300 HP), 8-402 (330 HP)

1. Turn in both mixture screws until they gently seat, then back them out four turns.

2. Adjust the carburetor idle screw to obtain 775 rpm for manual transmission, 630 rpm for automatics (in Drive).

3. Adjust the mixture screw equally to obtain 700 rpm for manual transmission, 600 rpm for automatics (in Drive).

4. Reconnect the distributor vacuum line.

V8-402 (350 HP), 8-454

1. Turn in both mixture screws until they gently seat, then back them out four turns.

2. Adjust the carburetor idle screw to obtain 700 rpm for manual transmission or 630 rpm for automatics (in Drive).

3. On cars equipped with automatic transmission: adjust the mixture screws equally to obtain 600 rpm with the transmission in Drive.

4. On cars equipped with manual transmissions: Turn in one mixture screw until the speed drops to 400 rpm, then adjust the carburetor idle screw to obtain 700 rpm. Turn in the other mixture screw until the speed drops 400 rpm, then regain 700 rpm by adjusting the carburetor idle screw.

5. Reconnect the distributor vacuum line.

1971-72 Initial Adjustments

Adjust with air cleaner installed. The following initial idle adjustments are part of the normal engine tune-up. There is a tune-up decal placed conspicuously in the engine compartment out lining the specific procedure and settings for each engine application. Follow all of the instructions when adjusting the idle. These tuning procedures are necessary to obtain the delicate balance of variables for the maintenance of both reliable engine performance and efficient exhaust emission control.

NOTE: *All engines have limiter caps on the mixture adjusting screws. The idle mixture is preset and the limiter caps installed at the factory in order to meet emission control standards. Do not remove these limiter caps unless all other possible causes of poor idle condition have been thoroughly checked out. The solenoid used on 1971 carburetors is different from the one used on earlier cars. The Combination Emission Control System (C.E.C.) solenoid valve regulates distributor vacuum as a function of transmission gear position. The C.E.C solenoid is adjusted only after: 1) replacement of the solenoid, 2) major carburetor overhaul, or 3) after the throttle body is removed or replaced.*

All initial adjustments described below are made:

1. With the engine warmed up and running.

2. With the choke fully open.

3. With the fuel tank line disconnected from the Evaporative Emission canister on all cars.

4. With the vacuum hose disconnected at the distributor and plugged.

Be sure to reconnect the distributor vacuum hose and to connect the fuel tank-to-evaporative emission canister line or install the gas cap when idle adjustments are complete.

6-250

1. Adjust the carburetor idle speed screw to obtain 550 rpm (700 rpm of 1972) for manual transmissions (in Neutral) or 550 rpm (600 rpm for 1972) for automatics (in Drive). Do not adjust the solenoid screw. Using the solenoid screw to set idle or incorrectly adjusting it may result in a decrease in engine braking.

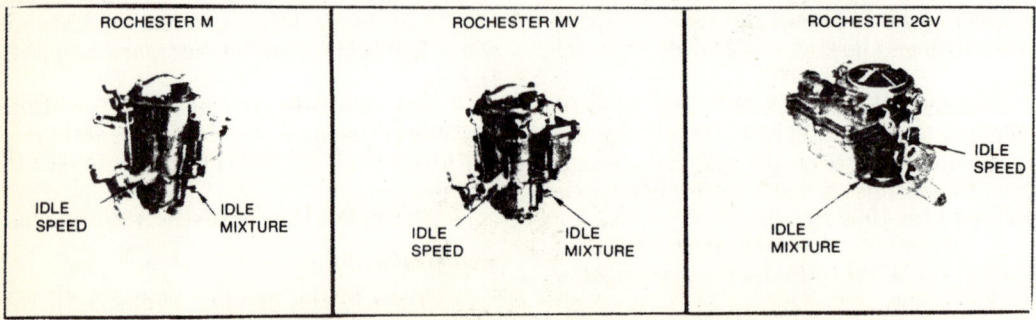

1968–70 one and two barrel carburetor adjustment screw locations

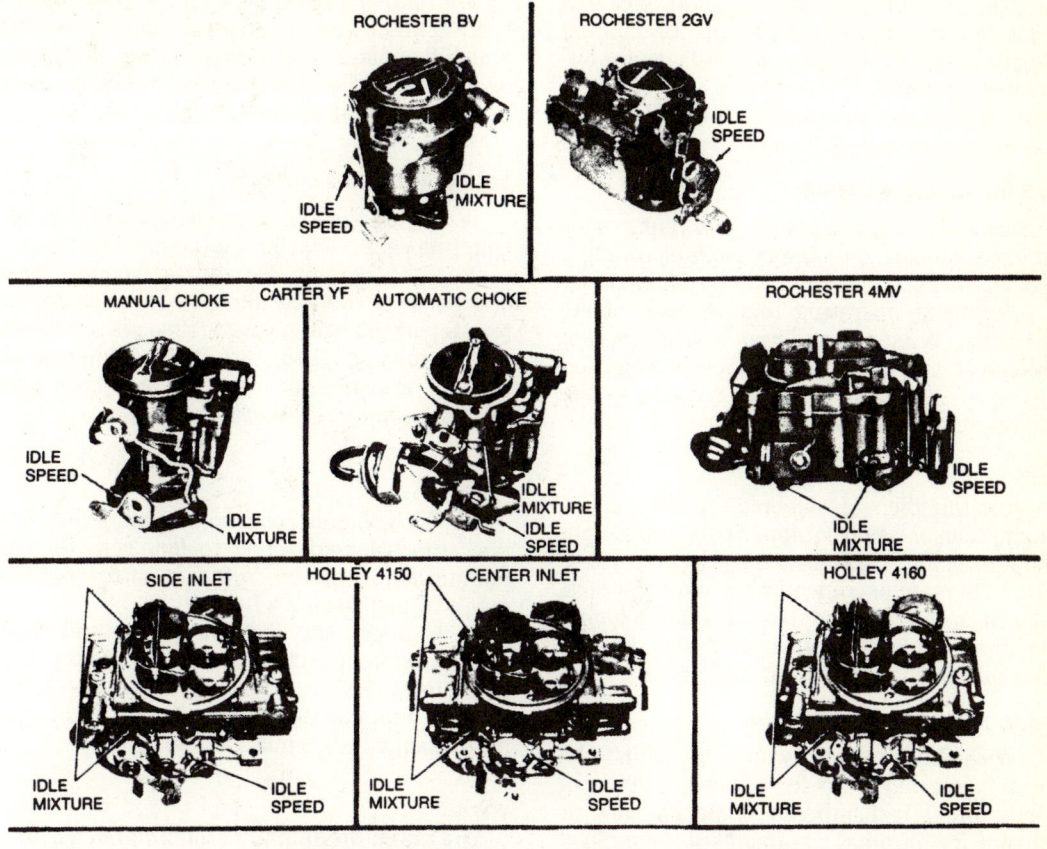

Idle speed and mixture screw location, 1964–72

2. Reconnect the vapor line and distributor vacuum advance line.

V8-307 AND 8-350 (2-BBL)

1. On 1971 cars, adjust the carburetor idle speed screw to obtain 600 rpm for manual transmission (in Neutral) with the air conditioning turned off, or 550 rpm for automatic transmissions (in Drive) with the air conditioning turned on. Do not adjust the solenoid screw. On 1972 cars, turn the air conditioning off and adjust the idle stop solenoid screw to obtain 900 rpm for manual transmissions (in Neutral) or 600 rpm for automatics (in Drive). Place the transmission in Park or Neutral and adjust the fast idle cam screw to get 1,850 rpm on 307 engines and 2,200 rpm on 350 engines.

2. Reconnect the vapor line and distributor vacuum advance line.

V8-350 (4-BBL)

1. On 1971 cars, adjust the carburetor idle speed screw to obtain 600 rpm for manual transmissions (in Neutral) with the air conditioning turned off, or 550 rpm for automatics (in Drive) with the air conditioning turned on. do not adjust the solenoid screw. On 1972 cars, turn the air conditioning off and adjust the idle stop solenoid screw to get 800 rpm for manual transmissions (in Neutral) or 600 rpm for automatic transmissions (in Drive).

2. For both 1971 and 1972 cars, place the fast idle cam follower on the second step of the fast idle cam, turn the air conditioning off and adjust the fast idle to 1,350 rpm for manual transmissions (in Neutral) or 1,500 rpm for automatics (in Park).

3. Reconnect the vapor line and the distributor vacuum advance line on all cars.

V8-402, 8-454

1. On 1971 cars, turn off the air conditioner and adjust the carburetor idle speed screw to obtain 600 rpm with manual transmissions in Neutral and automatics in Drive. Do not adjust the solenoid screw. On 1972 cars, turn off the air conditioning and adjust the idle stop solenoid screw to 800 rpm (in Neutral) for manual transmissions and 600 rpm (in Drive) for automatics.

ENGINE PERFORMANCE AND TUNE-UP

2. On both 1971 and 1972 cars, place the fast idle cam follower on the second stop of the fast idle cam, turn off the air conditioner and adjust the fast idle to 1,350 rpm for manual transmissions (in Neutral) or 1,500 rpm for automatics (in Park).

3. Reconnect the vapor line and the distributor vacuum line on 1971 and 1972 cars.

1973 Initial Adjustments

All cars are equipped with idle limiter caps and idle solenoids. disconnect the fuel tank line from the evaporative canister. The engine must be running at operating temperature, choke off, parking brake on, and rear wheels blocked. Disconnect the distributor vacuum hose and plug it. After adjustment, reconnect the vacuum and evaporative hoses.

6-250

Adjust the idle stop solenoid for 700 rpm on manual transmission equipped cars or 600 rpm on automatic transmission equipped cars (in Drive). On manual transmission equipped cars, make no attempt to adjust the CEC solenoid (the larger of the two carburetor solenoids) or a decrease in engine braking could result.

V8-307, 8-350 AND 8-400 (2-BBL)

1. With the air conditioning Off, adjust the idle stop solenoid screw for a speed of 900 rpm on manual transmission equipped cars or 600 rpm for automatic transmission equipped cars (in Drive).

2. Disconnect the idle stop solenoid electrical connector and adjust the idle speed screw (screw resting on lower stop of the cam) for 450 rpm on all 307 cu. in. engines, 400 rpm on 350 and 400 engines with automatic transmissions, or 500 rpm on 350 and 400 cu. in. engine with manual transmissions.

V8-350 AND 8-400 (4-BBL)

1. Adjust the idle stop solenoid screw to 900 rpm (manual), 600 rpm (automatic in Drive).

2. Connect the distributor vacuum hose and position the fast idle cam follower on the top step of the fast idle cam (turn air conditioning off) and adjust the fast idle to 1,300 rpm on manual transmission 350 engines; 1,600 rpm for all automatics in Park.

V8-454

1. With the air conditioning off, adjust the idle stop solenoid screw to 900 rpm for the manual transmission; 700 rpm with the automatic transmission in Drive.

2. Connect the distributor vacuum hose and place fast idle cam follower on the top step of the fast idle cam. Adjust the fast idle to 1,300 rpm for manual transmission; and 1,600 rpm for automatic transmissions (in Park).

1974

The same preliminary adjustments as for 1973 apply.

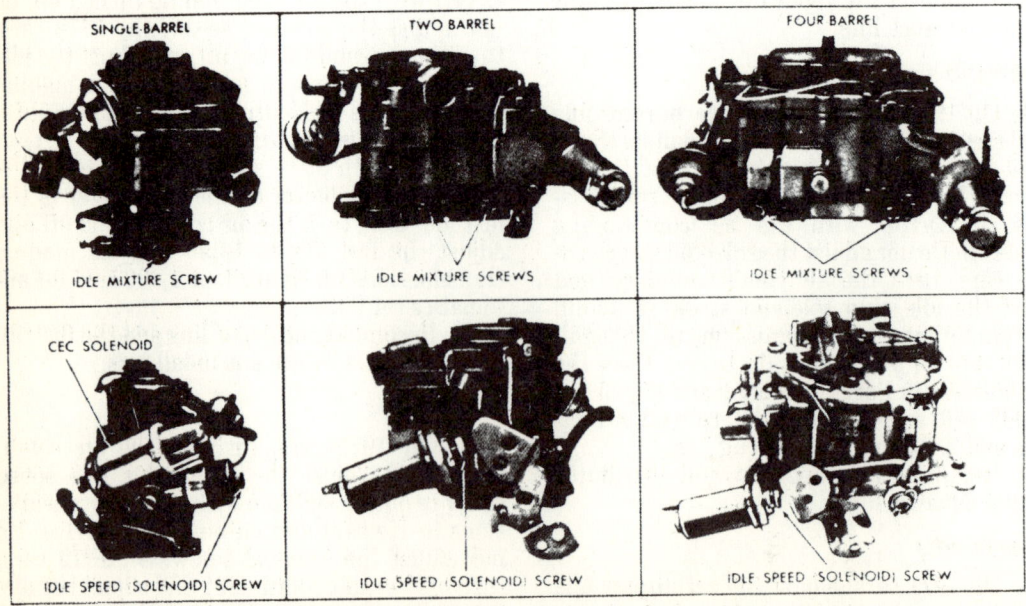

Idle speed and mixture screw location, 1973–75

ENGINE PERFORMANCE AND TUNE-UP

6-250

Using the hex nut on the end of the solenoid body, turn the entire solenoid to get 850 rpm for the manual transmission, 600 rpm for automatic transmissions in Drive.

V8-350 AND 8-400 (2-BBL)

1. Turn the air conditioning off. Adjust the idle stop solenoid screw for 900 rpm on manual;
600 rpm on automatic (in Drive).
2. De-energize the solenoid and adjust the carburetor idle cam screw (on low step of cam) for 400 rpm on automatic transmission equipped cars (in Drive); 500 rpm on 350 engines with manual transmission.

V8-350 AND 8-400 (4-BBL)

1. Turn the air conditioning off. Adjust the idle stop solenoid screw for 900 rpm on manual transmission equipped cars or 600 rpm on automatic transmission equipped cars (in Drive).
2. Connect the distributor vacuum hose. Position the fast idle cam follower on the top step of the fast idle cam and adjust the fast idle speed to 1,300 rpm on manual; 1,600 on automatic (in Park).

V8-454

1. With the air conditioning off, adjust the idle stop solenoid screw for 800 rpm with the manual transmission; 600 with the automatic transmission in Drive.
2. Reconnect the distributor vacuum advance hose and place the fast idle cam follower on the top step of the fast idle cam. With the air conditioning off, adjust the fast idle to 1,600

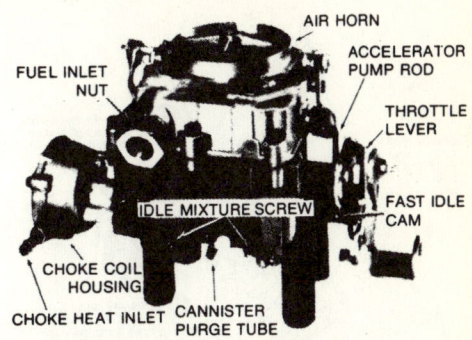

Idle mixture screw location, 1976–77 2GC

rpm for manual transmissions; 1,500 rpm for all automatics in Park.

1975-76

6-250 (1-BBL)

1. Idle speed is adjusted with the engine at normal operating temperature, air cleaner on, choke open, and air conditioning off (air conditioning equipped cars). Hook up a tachometer to the engine.
2. Block the rear wheels and apply the parking brake.
3. Disconnect the fuel tank hose from the evaporative canister.
4. Disconnect and plug the distributor vacuum advance hose.
5. Start the engine and check the ignition timing. Adjust if necessary.

Reconnect the vacuum hose. to adjust the idle speed, turn the solenoid in or out to obtain the higher of the two specifications listed on the decal. With the automatic transmission in

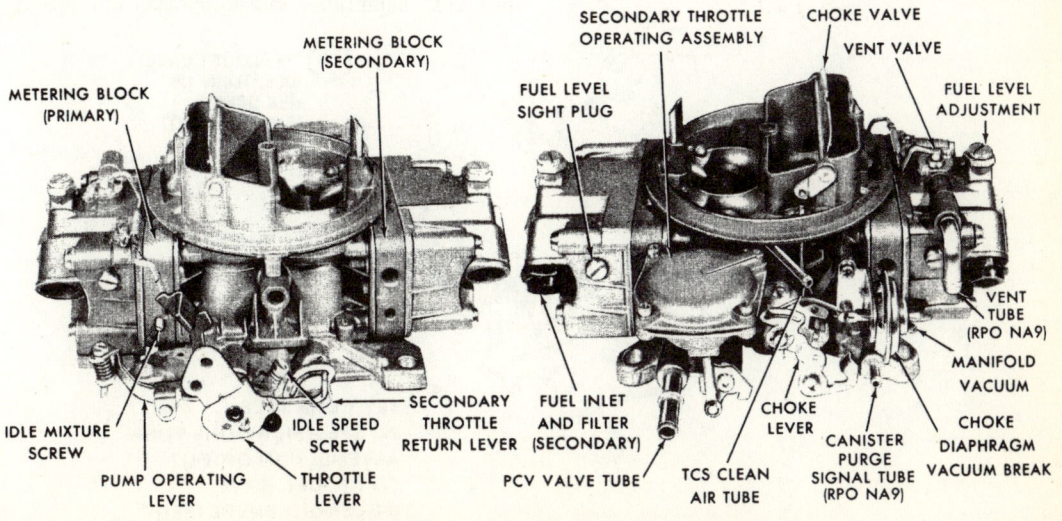

Idle speed and mixture screw location, 1972 Holley 4150

ENGINE PERFORMANCE AND TUNE-UP

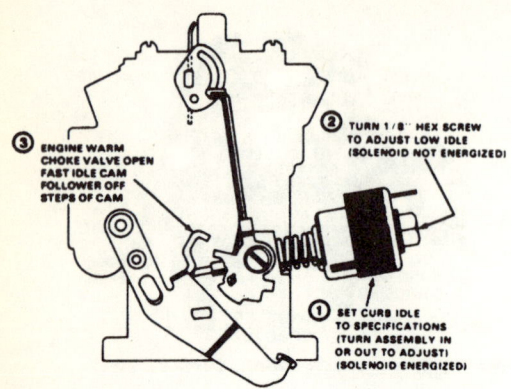

Idle speed adjustment screw location, 1976–79 1 BBL

Drive or the manual transmission in Neutral, disconnect the solenoid electrical connector and turn the 1/8″ Allen screw in the end of the solenoid body to the lower idle speed.

6. Place the shift lever in Drive on cars equipped with automatic transmissions and have an assistant apply the brakes. On cars equipped with manual transmissions, put it into neutral.

7. Cut the tab off the mixture limiter cap, but don't remove the cap. Turn the screw counterclockwise until the highest idle speed is reached.

8. Set the idle speed to the higher of the two listed idle speed by turning the solenoid in or out.

9. Check the tachometer and turn the mixture screw clockwise until the idle speed is at the lower of the two listed idle speeds.

10. Shut off the engine, remove the tachometer, and reconnect the carbon canister hose.

V8-350 1975 (2-BBL)
V8-350 1976 (2-BBL)

1. Idle speed is adjusted with the engine at normal operating temperature, air cleaner on, choke open, and air conditioning off. Hook up a tachometer to the engine.

2. Block the rear wheels and apply the parking brake.

3. Disconnect the fuel tank hose from the evaporative canister.

4. Disconnect and plug the distributor vacuum advance hose.

5. Start the engine and check the ignition timing. Adjust if necessary. Reconnect the vacuum hose.

6. Adjust the idle speed screw to the specified rpm. If the figures given in the Tune-Up Specifications chart differ from those on the tune-up decal, those on the decal take precedence. Automatic transmissions should be in Drive, manual transmissions should be in Neutral.

CAUTION: *Make doubly sure that the rear wheels are blocked and the parking brake applied.*

7. Adjust the idle speed to the higher of the two figures on the tune-up decal. Back out the two mixture screws equally until the highest idle is reached. Reset the speed if necessary to the higher one on the tune-up decal. Next, turn the screws in equally until the lower of the two figures on the decal is obtained.

8. Shut off the engine, reconnect hose to evaporative canister, and remove blocks from wheels.

V8-350, 8-400 AND 8-454

1975 4-barrel carburetors are equipped with idle stop solenoids. There are two idle speeds,

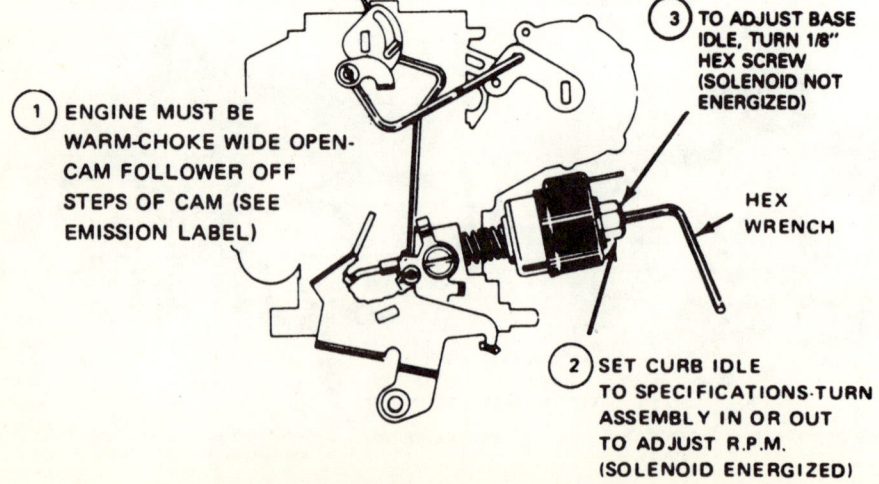

Idle speed adjustment for the six cylinder engine, 1976–78

ENGINE PERFORMANCE AND TUNE-UP

one with the solenoid energized and second with the solenoid de-energized. Both are set with the solenoid. The slower speed (solenoid de-energized) is necessary to prevent dieseling by allowing the throttle plate to close further than at a normal idle speed.

1. Idle speed is set with the engine at normal operating temperature, air cleaner on, choke open, and air conditioning off. Hook up a tachometer to the engine.

2. Block the rear wheels and apply the parking brake.

3. Disconnect the fuel tank hose from the evaporative canister.

4. Disconnect and plug the distributor vacuum advance hose.

5. Start the engine and check the ignition timing. Adjust if necessary. Reconnect the vacuum hose.

6. Disconnect the electrical connector at the idle solenoid.

7. Set the transmission in Drive. Adjust the low idle speed screw for the lower of the two figures given for idle speed.

CAUTION: *Make sure that the drive wheels are blocked and the parking brake is applied.*

8. Reconnect the idle solenoid and open the throttle slightly to extend the solenoid plunger.

9. Turn the solenoid plunger screw in or out to obtain the higher of the two idle speed figures (this is normal curb-idle)

10. To adjust the mixture, break off the limiter caps. Make sure that the idle is at the higher of the two speeds listed on the decal. Turn the mixture screws out equally to obtain the highest idle. Reset the idle speed with the plunger screw if necessary. Turn the mixture screw in until the lower of the two figures on the decal is obtained.

11. Shut off the engine, remove blocks from drive wheels, and reconnect hose to evaporative canister.

NOTE: *For 1976, the idle solenoid has been dropped. To adjust the idle, follow the preceding Steps 1-5. Skip Steps 6, 7, and 8, then follow Steps 9 and 10, adjusting the idle speed screw.*

1977

1. First satisfy all the following requirements:

   a. Set parking brake and block drive wheels.

   b. Bring the engine to operating temperature.

   c. Remove the air cleaner for access, but make sure all hoses stay connected.

   d. Consult the Emission Control Information label under the hood, and disconnect and plug hoses as required by the instructions there.

   e. Connect an accurate tach to the engine.

2. Set ignition timing as described previously.

3. Remove the cap(s) from the idle mixture screw(s). Remove caps carefully, to prevent bending these screws.

4. Turn in the screw(s) till they seat very lightly, then back screw(s) out just far enough to permit the engine to run.

5. Put automatic transmission in Drive.

6. Back out screw(s) $1/8$ turn at a time, going alternately from screw to screw after each $1/8$ turn where there are two screws, until the highest possible idle speed is achieved. Then, set the idle speed as follows: 250 CID engine with manual transmission - 950; with automatic - 575; with automatic in California - 640; with

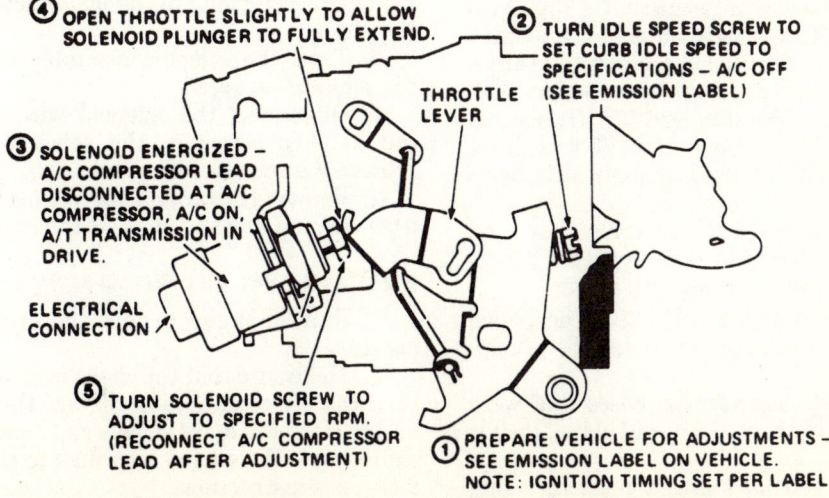

Idle speed adjustment for the V8, 4 BBL with solenoid, 1977

ENGINE PERFORMANCE AND TUNE-UP

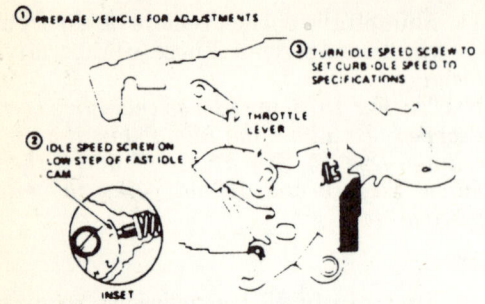

Idle speed adjustment for the V8, 4 BBL without solenoid, 1977

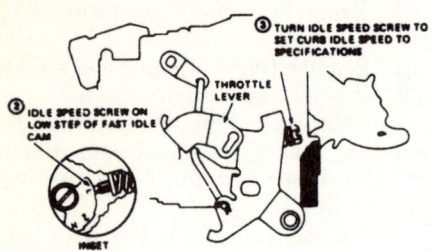

Idle speed adjustment for the 2 BBL without solenoid, 1978–79

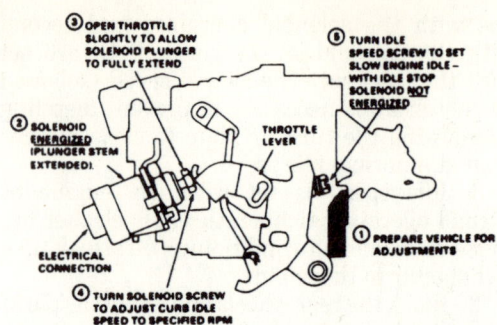

Idle speed adjustment for the 2 BBL with solenoid, 1978–79

automatic used in higher altitude area - 650; 305 V8 with manual transmission - 650; with automatic - 550; standard 350 V8 with automatic - 550; 350 V8 used at high altitudes - 650.

7. After setting the idle speed, repeat the mixture adjustment to ensure that mixture is at the point where highest idle speed is obtained. Then, if idle speed has increased, repeat idle speed adjustment of Step 6.

8. Now, turn screw(s) in, going evenly in $1/8$ turn increments where there are two, until the following idle speed are obtained: 250 CID engine with manual transmission - 750; with automatic - 550; with automatic in California - 600; with automatic used in high altitude areas - 600; 305 V8 with manual transmission - 600; with automatic - 500; standard 350 V8, manual transmission - 700; standard 350 V8 with automatic - 500; 350 V8 used at high altitudes - 600.

9. Reset idle speed to the value shown on the engine compartment sticker, if that differs from the final setting in the step above.

10. Check and adjust fast idle as described on the engine compartment sticker. See Chapter 5.

11. Reconnect any vacuum hoses that were disconnect for the procedure, and install the air cleaner.

12. If idle speed has changed, reset according to the engine compartment sticker. Disconnect tach.

1978-80

These cars have sealed idle mixture screws; in most cases these are concealed under staked-in plugs. Idle mixture is adjustable only during carburetor overhaul, and requires the addition of propane as an artificial mixture enrichener.

See the emission control label in the engine compartment for procedures and specifications not supplied here. Prepare the car for adjustment (engine warm, choke open, fast idle screw off the fast idle cam) as per the label instructions.

1-BBL

1. Run the engine to normal operating temperature.
2. Make sure that the choke is fully opened.
3. Turn the air conditioning off and disconnect the vacuum line at the vapor canister. Plug the line.
4. Set the parking brake, block the drive wheels and place the transmission in Drive (AT) or Neutral (MT). Connect a tachometer to the engine according to the manufacturer's instructions.
5. Turn the solenoid assembly to achieve the solenoid-on speed.
6. Disconnect the solenoid wire and turn the $1/8''$ hex screw in the solenoid end, to achieve the solenoid-off speed.
7. Remove the tachometer, connect the canister vacuum line and shut off the engine.

2-BBL AND 4-BBL (ALL BUT V8-350)

1. Run the engine to normal operating temperature.
2. Make sure that the choke is fully opened, turn the air conditioning off, set the parking brake, block the drive wheels and connect a tachometer to the engine according to the manufacturer's instructions.
3. Disconnect and plug the vacuum hoses at the EGR valve and the vapor canister.

ENGINE PERFORMANCE AND TUNE-UP

4. Place the transmission in Park (AT) or Neutral (MT).

5. Disconnect and plug the vacuum advance hose at the distributor. Check and adjust the timing.

6. Connect the distributor vacuum line.

7. On manual transmission equipped cars without air conditioning and without solenoid: place the idle speed screw on the low step of the fast idle cam and turn the screw to achieve the specified idle speed.

8. If equipped with air conditioning: set the idle speed screw to the specified rpm. Disconnect the compressor clutch wire and turn the air conditioning on. Open the throttle momentarily to extend the solenoid plunger. Turn the solenoid screw to obtain the specified rpm.

9. On automatic transmission equipped cars without air conditioning: manual transmission cars without air conditioning, solenoid-equipped carburetor: momentarily open the throttle to extend the solenoid plunger. Turn the solenoid screw to obtain the specified rpm. Disconnect the solenoid wire and turn the idle speed screw to obtain the slow engine idle speed.

V8-350

1. Run the engine to normal operating temperature.

2. Set the parking brake and block the drive wheels.

3. Connect a tachometer to the engine according to the manufacturer's instructions.

4. Disconnect and plug the purge hose at the vapor canister. disconnect and plug the EGR vacuum hose at the EGR valve.

5. Turn the air conditioning off.

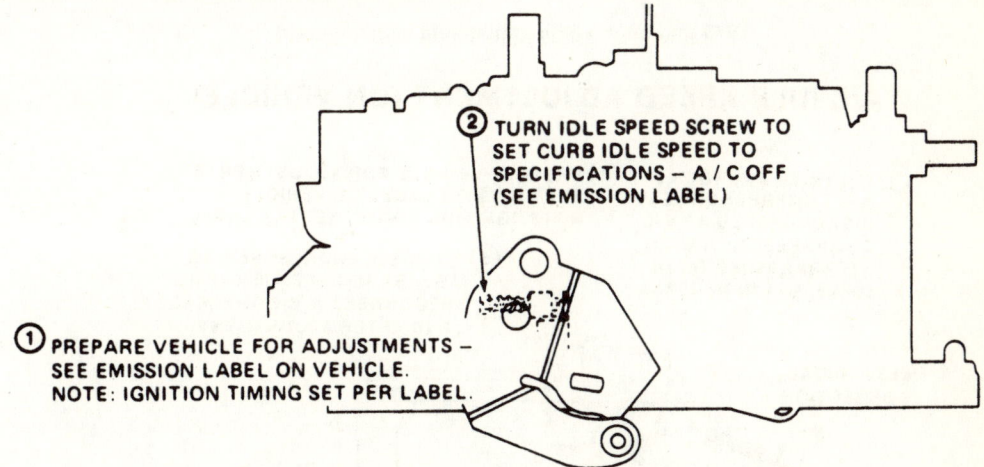

1979 and later 4 BBL adjustment without solenoid

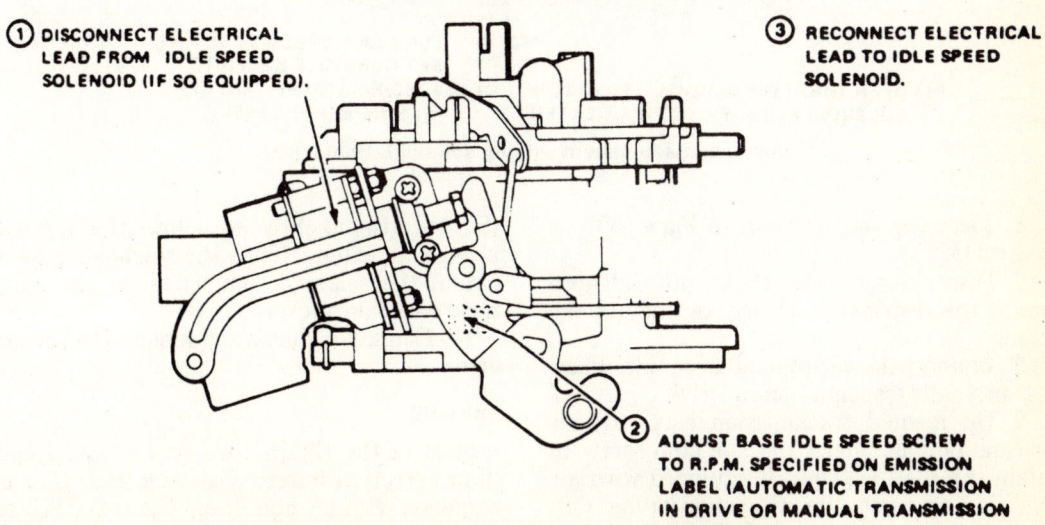

Idle speed adjustment with solenoid, 1980 2 BBL

ENGINE PERFORMANCE AND TUNE-UP

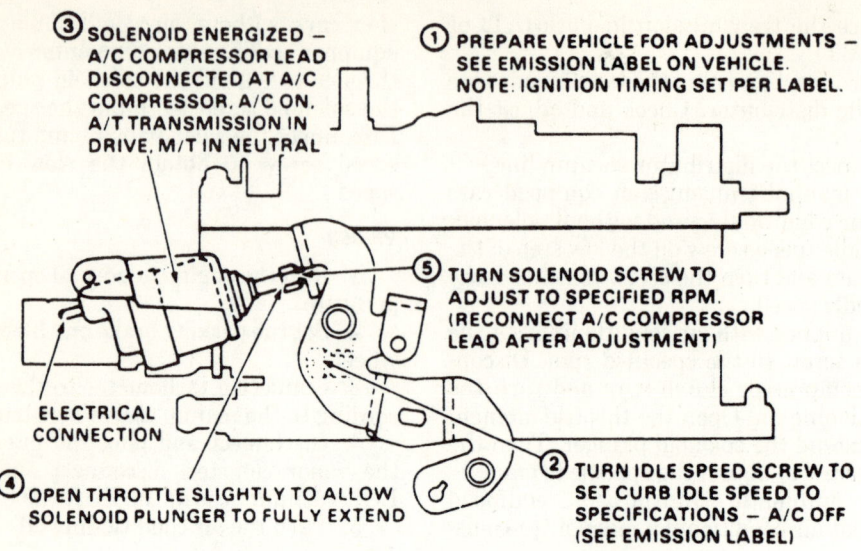

1979 and later 4 BBL adjustment with solenoid

A/C IDLE SPEED ADJUSTMENT (ON VEHICLE)

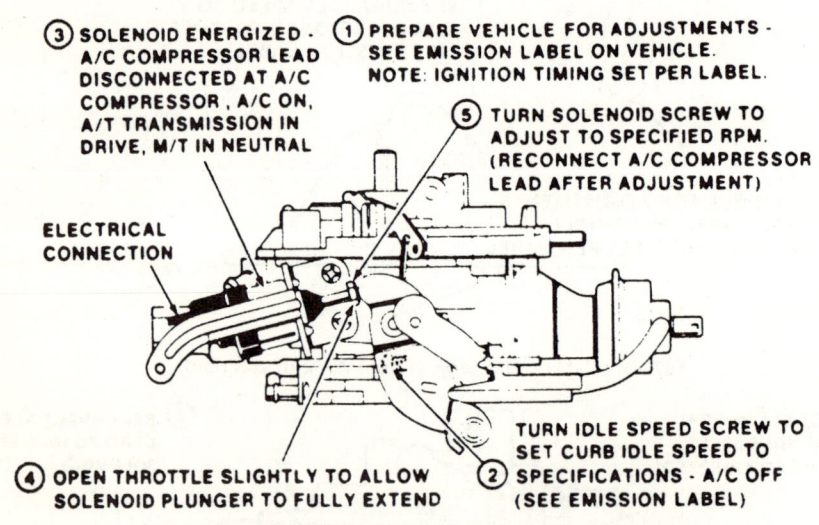

Idle speed adjustment without solenoid, 1980 2 BBL

6. Place the transmission in Park (AT) or Neutral (MT).

7. Disconnect and plug the vacuum advance line at the distributor. Check and adjust the timing.

8. Connect the vacuum advance line. Place the automatic transmission in Drive.

9. On manual transmission cars without air conditioning: adjust the idle stop screw to obtain the specified rpm. If equipped with air conditioning: with the air conditioning Off, adjust the idle stop screw to obtain the specified rpm. Disconnect the compressor clutch wire and turn the air conditioning On. Open the throttle slightly to allow the solenoid plunger to extend. Turn the solenoid screw to obtain the solenoid rpm listed on the under hood emission sticker.

10. Connect all hoses and remove the tachometer.

1981-82

Most of the E2ME (two barrel) and E4ME (four barrel) carburetors used on these cars are equipped with an Idle Speed Control (ISC) assembly, monitored by the ECM, which controls engine idle speed. The curb idle is programmed into the ECM and is not adjustable. Some cars

ENGINE PERFORMANCE AND TUNE-UP

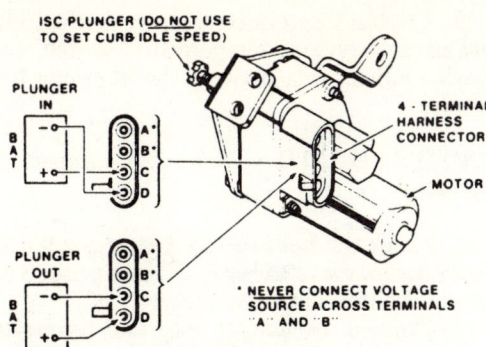

Typical idle speed control assembly (ISC)

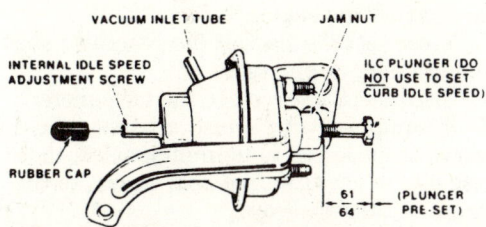

Typical idle load compensator (ILC)

with air conditioning may be equipped with an Idle Speed Solenoid (ISS), if so, refer to the 1978-80 procedure.

1983-88

The E2ME (two barrel) and the E4ME (four barrel) carburetors used on these cars all are equipped with an Idle Speed Control (ISC) assembly, monitored by the ECM, which controls engine idle speed. The curb idle is programmed into the ECM and is not adjustable.

On the E4MC carburetors, an Idle Load Compensator (ILC) mounted on the float bowl is used to control curb idle speeds. No attempt should be made to adjust this since it is controlled by the ECM.

On cars that do not include an ISC or ILC but are equipped with air conditioning, an Idle Speed Solenoid (ISS) is used to maintain curb idle speed any time the air conditioner compressor clutch is engaged. If so refer to the 1978-80 procedures.

Throttle Body Injection (TBI)

1985-92

The throttle body injected cars are controlled by a computer which supplies the correct amount of fuel during all engine operating conditions; no adjustment is necessary.

Diesel Fuel Injection

NOTE: *GM diesel engines are equipped with Roosa-Master, CAV Lucas, or Stanadyne injection pumps. The Roosa-Master and Stanadyne pumps are nearly identical.*

IDLE SPEED ADJUSTMENT

A special tachometer with an RPM counter suitable for diesel engine rpm readings is necessary for this adjustment; a standard tach suitable for gasoline engines will not work.

1. Place the transmission in Park, block the rear wheels and firmly set the parking brake.

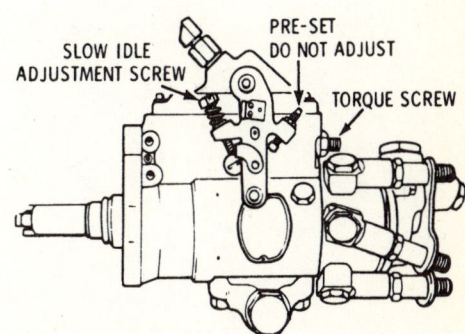

CAV-Lucas injection pump slow idle screw

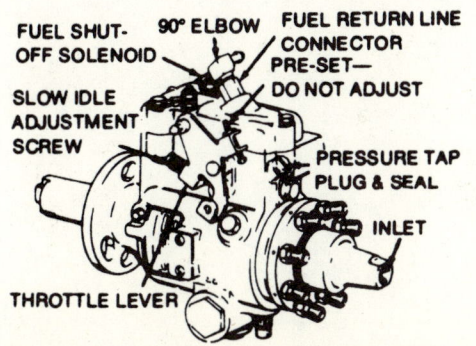

Roosa-Master injection pump slow idle screw

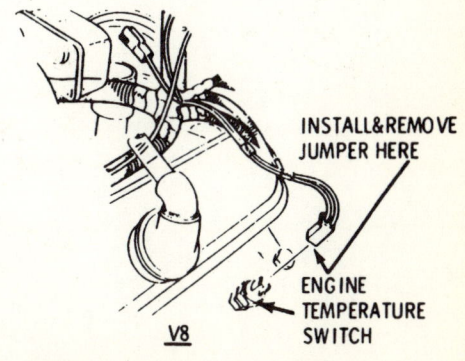

Fast idle temperature switch (V8 diesel)

ENGINE PERFORMANCE AND TUNE-UP

2. If necessary, adjust the throttle linkage as described in Chapter 7.

3. Start the engine and allow it to warm up for 10-15 minutes.

4. Shut off the engine and remove the air cleaner assembly.

5. Clean off any grime from the timing probe holder on the front cover; also clean off the crankshaft balancer rim.

6. Install the magnetic probed end of the tachometer fully into the timing probe holder. Complete the remaining tachometer connections according to the tach manufacturer's instructions.

7. Make sure all electrical accessories are OFF.

NOTE: *At no time should either the steering wheel or brake pedal be touched.*

8. Start the engine and place the transmission in Drive (after first making sure the parking brake is firmly applied).

9. Check the slow idle speed reading against the one printed on the underhood emissions sticker. Reset if necessary.

10. Unplug the connector from the fast idle cold advance (engine temperature) switch, and install a jumper wire between the connector terminals.

NOTE: *DO NOT allow the jumper to ground.*

11. Check the fast idle speed and reset if necessary according to the specification printed on the underhood emissions sticker.

12. Remove the jumper wire and reconnect it to the temperature switch.

13. Recheck the slow idle speed and reset if necessary.

14. Shut off the engine.

15. Reconnect the leads at the generator and air conditioning compressor.

16. Disconnect and remove the tachometer.

17. If equipped with cruise control, adjust the servo throttle rod to minimum slack, then put the clip in the first free hole closest to the bellcrank or throttle lever.

18. Install the air cleaner assembly.

Engine and Engine Overhaul

3

ENGINE ELECTRICAL

Understanding the Engine Electrical System

The engine electrical system can be broken down into three separate and distinct systems:
1. The starting system
2. The charging system
3. The ignition system.

BATTERY AND STARTING SYSTEM

Basic Operating Principals

The battery is the first link in the chain of mechanisms which work together to provide cranking of the automobile engine. In most modern vehicles, the battery is a lead-acid electrochemical device consisting of six 2 volt (2V) subsections connected in series so the unit is capable of producing approximately 12V of electrical pressure. Each subsection, or cell, consists of a series of positive and negative plates held a short distance apart in a solution of sulfuric acid and water. The two types of plates are of dissimilar metals. This causes a chemical reaction to be set up, and it is this reaction which produces current flow from the battery when its positive and negative terminals are connected to an electrical appliance such as a lamp or motor. The continued transfer of electrons would eventually convert the sulfuric acid in the electrolyte to water, and make the two plates identical in chemical composition. As electrical energy is removed from the battery, its voltage output tends to drop. Thus, measuring battery voltage and battery electrolyte composition are two ways of checking the ability of the unit to supply power. During the starting of the engine, electrical energy is removed from the battery. However, if the charging circuit is in good condition and the operating conditions are normal, the power removed from the battery will be replaced by the generator (or alternator) which will force electrons back through the battery, reversing the normal flow, and restoring the battery to its original chemical state.

The battery and starting motor are linked by very heavy electrical cables designed to minimize resistance to the flow of current. Generally, the major power supply cable that leaves the battery goes directly to the starter, while other electrical system needs are supplied by a smaller cable. During the starter operation, power flows from the battery to the starter and is grounded through the vehicles frame and the battery's negative ground strap.

The starting motor is a specially designed, direct current electric motor capable of producing a very great amount of power for its size. One thing that allows the motor to produce a great deal of power is its tremendous rotating speed. It drives the engine through a tiny pinion gear (attached to the starter's armature), which drives the very large flywheel ring gear at a greatly reduced speed. Another factor allowing it to produce so much power is that only intermittent operation is required of it. Thus, little allowance for air circulation is required, and the windings can be built into a very small space.

The starter solenoid is a magnetic device which employs the small current supplied by the starting switch circuit of the ignition switch. This magnetic action moves a plunger which mechanically engages the starter and electrically closes the heavy switch which connects it to the battery. The starting switch circuit consists of the starting switch contained within the ignition switch, a transmission neutral safety switch or clutch pedal switch, and the wiring necessary to connect these with the starter solenoid or relay.

ENGINE AND ENGINE OVERHAUL

A pinion, which is a small gear, is mounted to a one-way drive clutch. this clutch is splined to the starter armature shaft. When the ignition switch is moved to the start position, the solenoid plunger slides the pinion toward the flywheel ring gear via a collar and spring. If the teeth on the pinion and flywheel match properly, the pinion will engage the flywheel immediately. IF the gear teeth butt one another, the spring will be compressed and will force the gears to mesh as soon as the starter turns far enough to allow them to do so. As the solenoid plunger reaches the end of its travel, it closes the contacts that connect the battery and starter and then the engine is cranked.

As soon as the engine starts, the flywheel ring gear begins turning fast enough to drive the pinion at an extremely high rate of speed. At this point, the one-way clutch begins allowing the pinion to spin faster that the starter shaft so that the starter will not operate at excessive speed. When the ignition switch is released from the starter position, the solenoid is de-energized, and a spring contained within the solenoid assembly pulls the gear out of mesh and interrupts the current flow to the starter.

Some starters employ a separate relay, mounted away from the starter, to switch the motor and solenoid current on and off. The relay thus replaces the solenoid electrical switch, but does not eliminate the need for a solenoid mounted on the starter used to mechanically engage the starter drive gears. The relay is used to reduce the amount of current the starting switch must carry.

THE CHARGING SYSTEM

Basic Operating Principals

The automobile charging system provides electrical power for operation of the vehicle's ignition and starting systems and all the electrical accessories. The battery serves as an electrical surge of storage tank, storing (in chemical form) the energy originally produced by the engine driven generator. The system also provides a means of regulating generator output to protect the battery from being overcharged and to avoid excessive voltage to the accessories.

The storage battery is a chemical device incorporating parallel lead plates in a tank containing a sulfuric acid-water solution. Adjacent plates are slightly dissimilar, and the chemical reaction of the two dissimilar plates produces electrical energy when the battery is connected to a load such as the starter motor. The chemical reaction is reversible, so that when the generator is producing a voltage (electrical pressure) greater then that produced by the battery, electricity is forced into the battery, and the battery is returned to its fully charged state.

The vehicle's generator is driven mechanically, through V belts, by the engine crankshaft. It consists of two coils of fine wire, one stationary (the stator), and one movable (the rotor). The rotor may also be known as the armature and consists of fine wire wrapped around an iron core which is mounted on a shaft. The electricity which flows through the two coils of wire (provided initially by the battery in some cases) creates an intense magnetic field around both rotor and stator, and the interaction between the two fields creates voltage, allowing the generator to power the accessories and charge the battery.

There are two types of generators; the earlier is the direct current (DC) type. The current produced by the DC generator is generated in the armature and carried off the spinning armature by stationary brushes contacting the commutator. The commutator is a series of smooth metal contact plates on the end of the armature. The commutator plates, which are separated from one another by a very short gap, are connected to the armature circuits so that current will flow in one direction only in wires carrying the generator output. The generator stator consists of two stationary coils of wire which draw some of the output current of the generator to form a powerful magnetic field and create the interaction of fields which generates the voltage. The generator field is wired in series with the regulator.

Newer automobiles use alternating current generators or alternators because they are more efficient, can be rotated at higher speeds, and have fewer brush problems, In an alternator, the field rotates while all the current produced passes only through the stator windings. The brushes bear against continuous slip rings rather than a commutator. This causes the current produced to periodically reverse the direction of its flow. diodes (electrical one-way switches) block the flow of current from traveling in the wrong direction. A series of diodes is wired together to permit the alternating flow of the stator to be converted to a pulsating, but unidirectional flow of current from traveling in the wrong direction. A series of diodes is wires together to permit the alternating flow of the stator to be converted to a pulsating, but unidirectional flow at the alternator output. The alternator's field is wires in series with the voltage regulator.

The regulator consist of several circuits. Each circuit has a core, or magnetic coil of wire, which operates a switch. Each switch is connected to ground through on or more resistors. The coil of wire responds directly to system voltage. When the voltage reaches the required level, the magnetic field created by the winding of wire closes the switch and inserts a resistance into the generator field circuit, thus reducing the output. The contacts of the switch cycle open and close many times each second to precisely control voltage.

While alternators are self-limiting as far as maximum current is concerned. DC generators employ a current regulating circuit which responds directly to the total amount of current flowing through the generator circuit rather than to the output voltage. The current regulator is similar to the voltage regulator except all system current must flow through the energizing coil on its way to the various accessories.

SAFETY PRECAUTIONS

Observing these precautions will ensure safe handling of the electrical system components, and will avoid damage to the vehicle's electrical system:

  a. Be absolutely sure of the polarity of a booster battery before making connections. Connect the cables positive to positive, and negative to negative. Connect positive cables first and then make the last connection to ground on the body of the booster vehicle so that arcing cannot ignite hydrogen gas that may have accumulated near the battery. Even momentary connection of a booster battery with the polarity reversed will damage alternator diodes.

  b. Disconnect both vehicle battery cables before attempting to charge a battery.

  c. Never ground the alternator or generator output or battery terminal. Be cautious when using metal tools around a battery to avoid creating a short circuit between the terminals.

  d. Never ground the field circuit between the alternator and regulator.

  e. Never run an alternator or generator without load unless the field circuit is disconnected.

  f. Never attempt to polarize an alternator.

  g. Keep the regulator cover in place when taking voltage and current limiter readings.

  h. Use insulated tools when adjusting the regulator.

  i. Whenever DC generator-to-regulator wires have been disconnected, the generator must be repolarized. To do this with an externally grounded, light duty generator, momentarily place a jumper wire between the battery terminal and the generator terminal of the regulator. With an internally grounded heavy duty unit, disconnect the wire to the regulator field terminal and touch the regulator battery terminal with it.

ENGINE ELECTRICAL

High Energy Ignition (HEI) Distributor

The Delco-Remy High Energy Ignition (HEI) System is breakerless, pulse triggered, transistor controlled, inductive discharge ignition system available as an option in 1974 and standard in 1975.

There are only nine external electrical connections; the ignition switch feed wire, and the eight spark plug leads. On eight cylinder vehicles through 1977, and all 1978–90 vehicles, the ignition coil is located with the distributor cap, connecting directly to the rotor. Some models in 1989–90, and all models from 1991–92, utilize a separate coil mounted away from the distributor.

The magnetic pick-up assembly located inside the distributor contains a permanent magnet, a pole piece with internal teeth, and a pick-up coil. When the teeth of the rotating timer core and pole piece align, an induced voltage in the pick-up coil signals the electronic module to open the coil primary circuit. As the primary current decreases, a high voltage is induced in the secondary windings of the ignition coil, directing a spark through the rotor and high voltage leads to fire the spark plugs. The dwell period is automatically controlled by the electronic module and is increased with increasing engine rpm. The HEI System features a longer spark duration which is instrumental in firing lean and EGR diluted fuel/air mixtures. The condenser (capacitor) located within the HEI distributor is provided for noise (static) suppression purposes only and is not a regularly replaced ignition system component.

As already noted in Chapter 2, 1981 models continue to use the HEI distributor although it now incorporates an Electronic Spark Timing System. With the new EST system, all spark timing changes are performed electronically by the Electronic Control Module (ECM) which monitors information from various engine sensors, computes the desired spark timing accordingly. Because all timing changes are controlled electronically, no vacuum or mechanical advance systems are used whatsoever.

ENGINE AND ENGINE OVERHAUL

Ignition Coil

TESTING

1968-74

PRIMARY CIRCUIT WITH VOLTMETER

A quick, tentative check of the 12 volt ignition primary circuit (including ballast resistor) can be made with a simple voltmeter, as follows:

1. With engine at operating temperature, but stopped, and the distributor side of the ignition coil grounded with a jumper wires, hook up a voltmeter between the ignition coil (switch side) and a good ground.
2. Jiggle the ignition switch (switch on) and watch the meter. An unstable needle will indicate a defective ignition switch.
3. With the ignition switch on (engine stopped) the voltmeter should read 5.5 to 7 volts for 12 volt systems.
4. Crank the engine. Voltmeter should read at least 9 volts during cranking period.
5. Now remove the jumper wire from the coil. Start the engine. voltmeter should read from 9.0 volts to 11.5 volts (depending upon generator output) while running.

PRIMARY CIRCUIT WITH OHMMETER

To check ignition coil resistance, primary side, switch ohmmeter to low scale. Connect the ohmmeter leads across the primary terminals of the coil and read the low ohms scale.

Coils requiring ballast resistors should read about 1.0 ohm resistance. 12 volt coils, not requiring external ballast resistors, should read about 4.0 ohms resistance.

SECONDARY CIRCUIT WITH OHMMETER

To check ignition coil resistance, secondary side, switch ohmmeter to high scale. Connect one test lead to the distributor cap end of the coil secondary cable. Connect the other test lead to the distributor terminal of the coil. A coil in satisfactory condition should show between 4K and 8K on the scale. Some special coils (Mallory, etc.) may show a resistance as high as 13K. If the reading is much lower than 4K, the coil probably has shorted secondary turns. If the reading is extremely high (40K or more) the secondary winding is either open, there is a bad connection at the coil terminal, or resistance is high in the cable.

If both primary and secondary windings of the coil test good, but the ignition system is still unsatisfactory, check the system further.

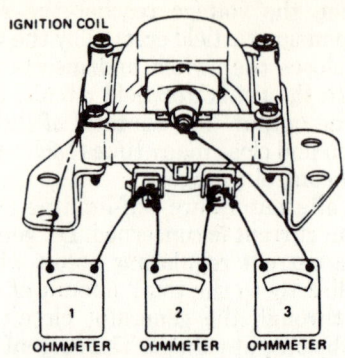

To test HEI ignition coil on external coil models, attach an ohmmeter as shown: test 1, use high scale. Reading should be very high or infinite. Test 2, use low scale. Reading should be very low or zero. Test 3, use high scale. Reading should not be infinite. If any test proved otherwise, replace coil

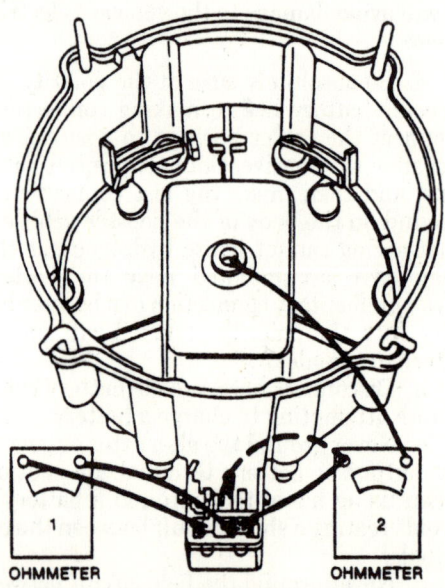

Ohmmeter 1 shows the primary coil resistance connection. Ohmmeter 2 shows the secondary resistance connection (1980 shown, most models similar)

1975-92

An ohmmeter with both high and low ranges should be used. These tests are made with the cap assembly removed and the battery wire disconnected. If a tachometer is connected to the TACH terminal, disconnect it before making these tests.

1. Connect an ohmmeter between the TACH and BAT terminals in the distributor cap. The primary coil resistance should be less than one ohm.
2. To check the coil secondary resistance,

ENGINE AND ENGINE OVERHAUL

connect an ohmmeter between the rotor button and the BAT terminal. Note the reading. Connect the ohmmeter between the rotor button and the TACH terminal. Note the reading. The resistance in both cases should be between 6,000 and 30,000 ohms. Be sure to test between the rotor button and both the BAT and TACH terminals.

3. Replace the coil only if the readings in Step 1 and Step 2 are infinite.

NOTE: *These resistance checks will not disclose shorted coil windings. This condition can only be detected with scope analysis or a suitable designed coil tester. If these instruments are unavailable, replace the coil with a known good coil as a final coil test.*

REMOVAL AND INSTALLATION

1968-74, 1989–92 Non Integral Coil

1. Disconnect the negative battery cable. Remove the ignition switch-to-coil lead from the coil.
2. Unfasten the distributor leads from the coil.
3. Remove the screws which secure the coil to the engine and lift it off.
4. Installation is the reverse of removal.

1975-90 Integral Coil

1. Disconnect the negative battery cable. Disconnect the feed and module wire terminal connectors from the distributor cap.
2. Remove the ignition wire set retainer, if equipped.
3. Remove the 4 coil cover-to-distributor cap screws and the coil cover.
4. Remove the 4 coil-to-distributor cap screws.
5. Using a blunt drift, press the coil wire spade terminals up out of distributor cap.
6. Lift the coil up out of the distributor cap.
7. Remove and clean the coil spring, rubber seal washer and coil cavity of the distributor cap.
8. Install the coil spring and rubber seal washer. Install the coil in the distributor cavity.
9. Install the coil connections in the cap. Install the coil retaining screws.
10. Install the coil assembly cover. Install the ignition wires. Connect the feed and module wire terminal. Connect the negative battery cable.

Ignition Module

REMOVAL AND INSTALLATION

1. Disconnect the negative battery cable. Remove the distributor cap and rotor as previously described.
2. Disconnect the harness connector and pick-up coil spade connectors from the module (note their positions).
3. Remove the two screws and module from the distributor housing.
4. Coat the bottom of the new module with silicone dielectric compound.

NOTE: *If a five terminal or seven terminal module is replaced, the ignition timing must be checked and reset as necessary.*

5. Installation is the reverse of the removal procedure.

Distributor

REMOVAL AND INSTALLATION

Point-Type Ignition

1968-74

1. Disconnect the negative battery cable. Remove the distributor cap and position it out of the way.
2. Disconnect the primary coil wire and the vacuum advance hose.
3. Scribe a mark on the distributor body and the engine block showing their relationship. Mark the distributor housing to show the direction in which the rotor is pointing. Note the positioning of the vacuum advance unit.
4. Remove the holddown bolt and clamp and remove the distributor.
5. To install the distributor with the engine undisturbed:
   a. Reinsert the distributor into its opening, aligning the previously made marks on the housing and the engine block.
   b. The rotor may have to be turned either way a slight amount to align the rotor-to-housing marks.
   c. Install the retaining clamp and bolt. Install the distributor cap, primary wire or electrical connector, and the vacuum hose.
   d. Start the engine and check the ignition timing.
6. To install the distributor with the engine disturbed:
   a. Turn the engine to bring the No. 1 piston to the top of its compression stroke. This may be determined by inserting a rag into the No. 1 spark plug hole and slowly turning the engine over. When the timing mark on the crankshaft pulley aligns with the 0 on the timing scale and the rag is blown out by compression, the No. 1 piston is at top dead center (TDC).

NOTE: *On Mark IV (big block) V8 engines there is a punch mark on the distributor drive gear which indicates the rotor position.*

ENGINE AND ENGINE OVERHAUL

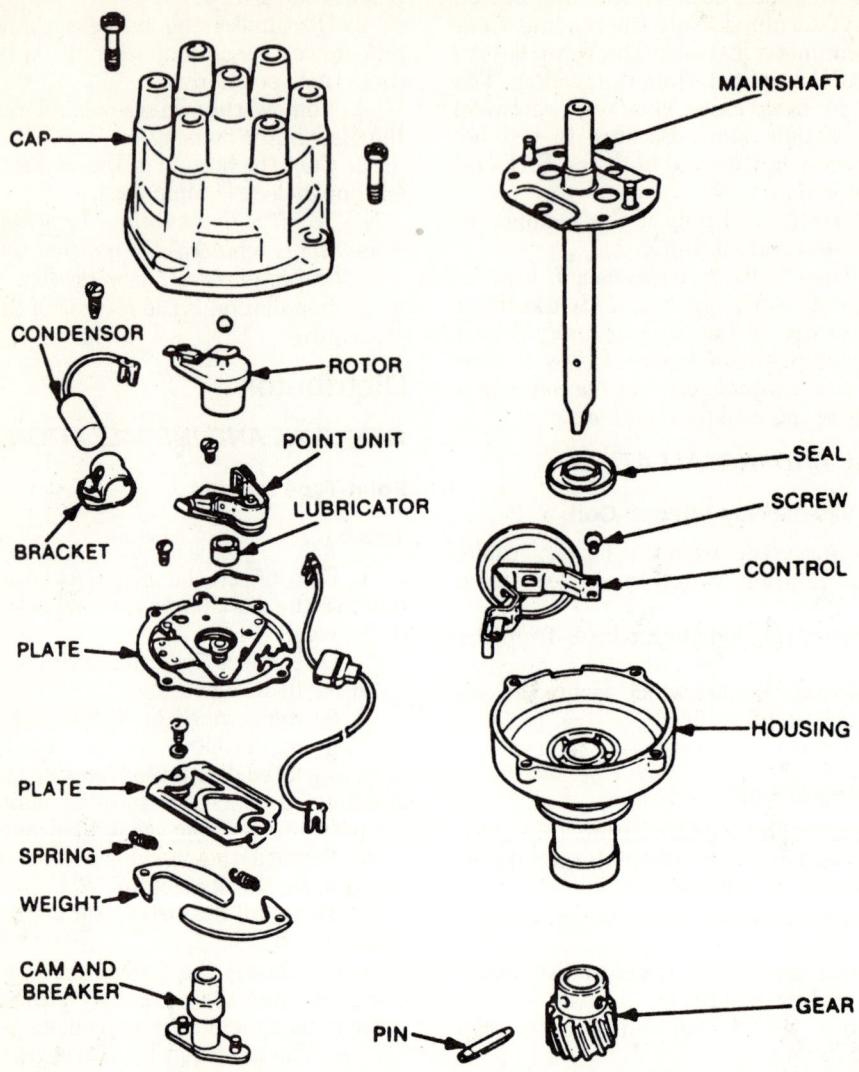

Exploded view of the six cylinder points-type distributor

Thus, the distributor may be installed with the cap in place. Align the punch mark 2 degrees clockwise from the No. 1 cap terminal, then rotate the distributor body 1/8 turn counterclockwise and push the distributor down into the block.

b. Install the distributor to the engine block so that the vacuum advance unit points in the correct direction.

c. Turn the rotor so that it will point to the No. 1 terminal in the cap.

d. Install the distributor into the engine block. It may be necessary to turn the rotor a little in either direction in order to engage the gears.

e. Tap the starter a few times to ensure that the oil pump shaft is mated to the distributor shaft.

f. Bring the engine to No. 1 TDC again and check to see that the rotor is indeed pointing toward the No. 1 terminal of the cap.

g. After correct positioning is assured, turn the distributor housing so that the points are just opening. Tighten the retaining clamp.

h. Install the cap and primary wire. Check the ignition timing. Install the vacuum hose.

ENGINE AND ENGINE OVERHAUL

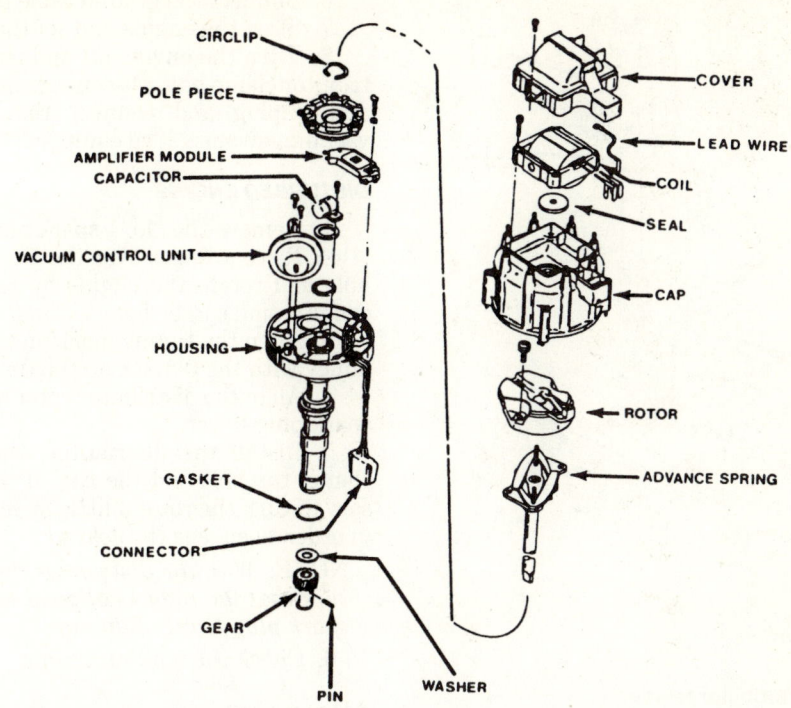

Exploded view of the HEI distributor (1981 models don't have the vacuum advance unit

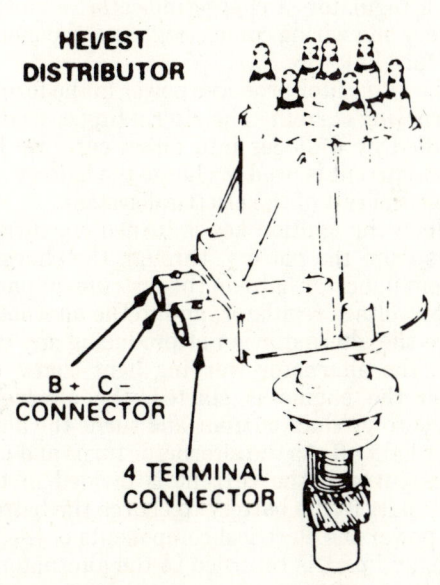

HEI distributor used with a separately mounted coil, 1990–92 models

HEI Distributor

1975-92

1. Disconnect the negative battery cable.
2. Tag and disconnect the feed and module terminal connectors from the distributor cap.
3. Disconnect the hose at the vacuum advance, if equipped.
4. Depress and release the 4 distributor cap-to-housing retainers (2 retainers on the non integral coil distributor) and lift off the cap assembly.
5. Using crayon or chalk, make locating marks on the rotor and module and on the distributor housing and engine for installation purposes.
6. Loosen and remove the distributor clamp bolt and clamp, and lift the distributor out of the engine. Noting the relative position of the rotor and module alignment marks, make a second mark on the rotor to align it with the mark on the module.

UNDISTURBED ENGINE

1. With a new O-ring on the distributor housing and the second mark on the rotor aligned with the mark on the module, install the distributor, taking care to align the mark on the housing with the one on the engine. It may be necessary to lift the distributor and turn the rotor slightly to align the gears and the oil pump driveshaft.
2. With the respective marks aligned, install the clamp and bolt finger-tight.
3. Install and secure the distributor cap.

ENGINE AND ENGINE OVERHAUL

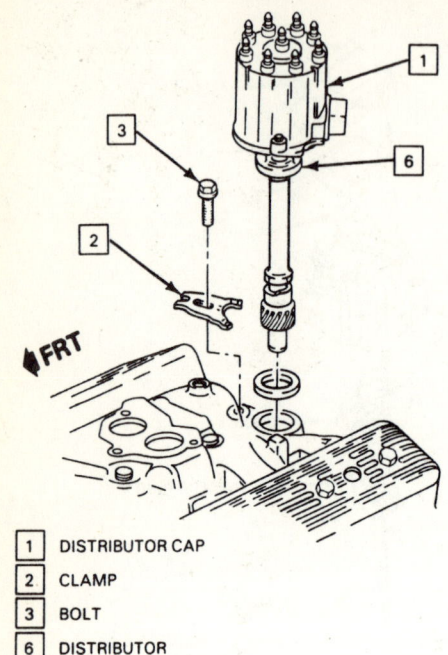

1	DISTRIBUTOR CAP
2	CLAMP
3	BOLT
6	DISTRIBUTOR

Non integral distributor removal

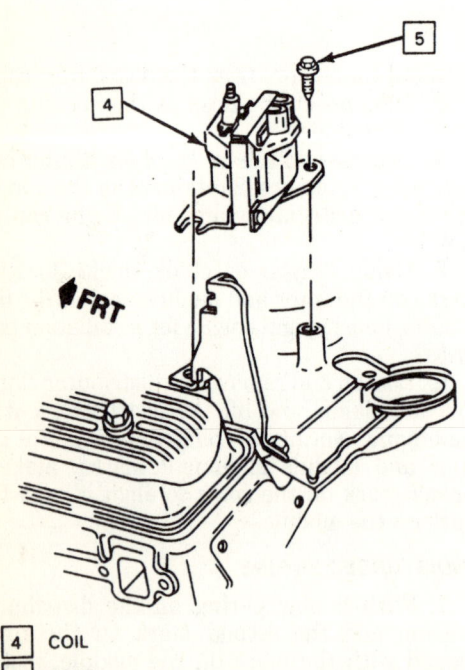

4	COIL
5	BOLT

Coil installation, non integral

4. Connect the feed and module connectors to the distributor or distributor cap.
5. Connect a timing light to the engine and plug the vacuum hose, if so equipped.
6. Connect the ground cable to the battery.
7. Start the engine and set the timing.
8. Turn the engine off and tighten the distributor clamp bolt. Disconnect the timing light and unplug and connect the hose to the vacuum advance, if so equipped.

DISTURBED ENGINE

1. Remove the No. 1 spark plug.
2. Place a finger over the No. 1 spark plug hole and rotate the engine by hand until the compression can be felt.
3. Align the timing mark on the crankshaft pulley with the **0** mark on the timing plate.
4. Align the distributor rotor near the No. 1 spark plug tower.
5. Install the distributor, the hold down clamp, the bolt and the cap. It may be necessary to turn the rotor a little in either direction in order to engage the gears.

NOTE: *With the distributor installed, make sure that the rotor is aligned with the No. 1 spark plug tower of the cap.*

6. Check the ignition timing.

Alternator

The alternator charging system is a negative (−) ground system which consists of an alternator, a regulator, a charge indicator, a storage battery and wiring connecting the components, and fuse link wire.

The alternator produces power in the form of alternating current. The alternating current is rectified by 6 diodes into direct current. The direct current is used to charge the battery and power the rest of the electrical system.

When the ignition key is turned on, current flows from the battery, through the charging system indicator light on the instrument panel, to the voltage regulator, and to the alternator. Since the alternator is not producing any current, the alternator warning light comes on. When the engine is started, the alternator begins to produce current and turns the alternator light off. As the alternator turns and produces current, the current is divided in two ways: part to the battery to charge the battery and power the electrical components of the vehicle, and part is returned to the alternator to enable it to increase its output. In this situation, the alternator is receiving current from the battery and from itself. A voltage regulator is wired into the current supply to the alternator to prevent it from receiving too much current which would cause it to put out too much current. Conversely, if the voltage regulator does not allow the alternator to receive enough current, the battery will not be fully charged and will eventually go dead.

The battery is connected to the alternator at all times, whether the ignition key is turned on or not. If the battery were shorted to ground, the alternator would also be shorted. This would damage the alternator. To prevent this, a fuse link is installed in the wiring between the battery and the alternator. If the battery is shorted, the fuse link is melted, protecting the alternator.

The alternating current generator (alternator) supplies a continuous output of electrical energy at all engine speeds. The alternator generates electrical energy and recharges the battery by supplying it with electrical current. This unit consists of four main assemblies: two end frame assemblies, a rotor assembly, and a stator assembly. The rotor assembly is supported in the drive end frame by a ball bearing and at the other end by a roller bearing. These bearings are lubricated during assembly and re-

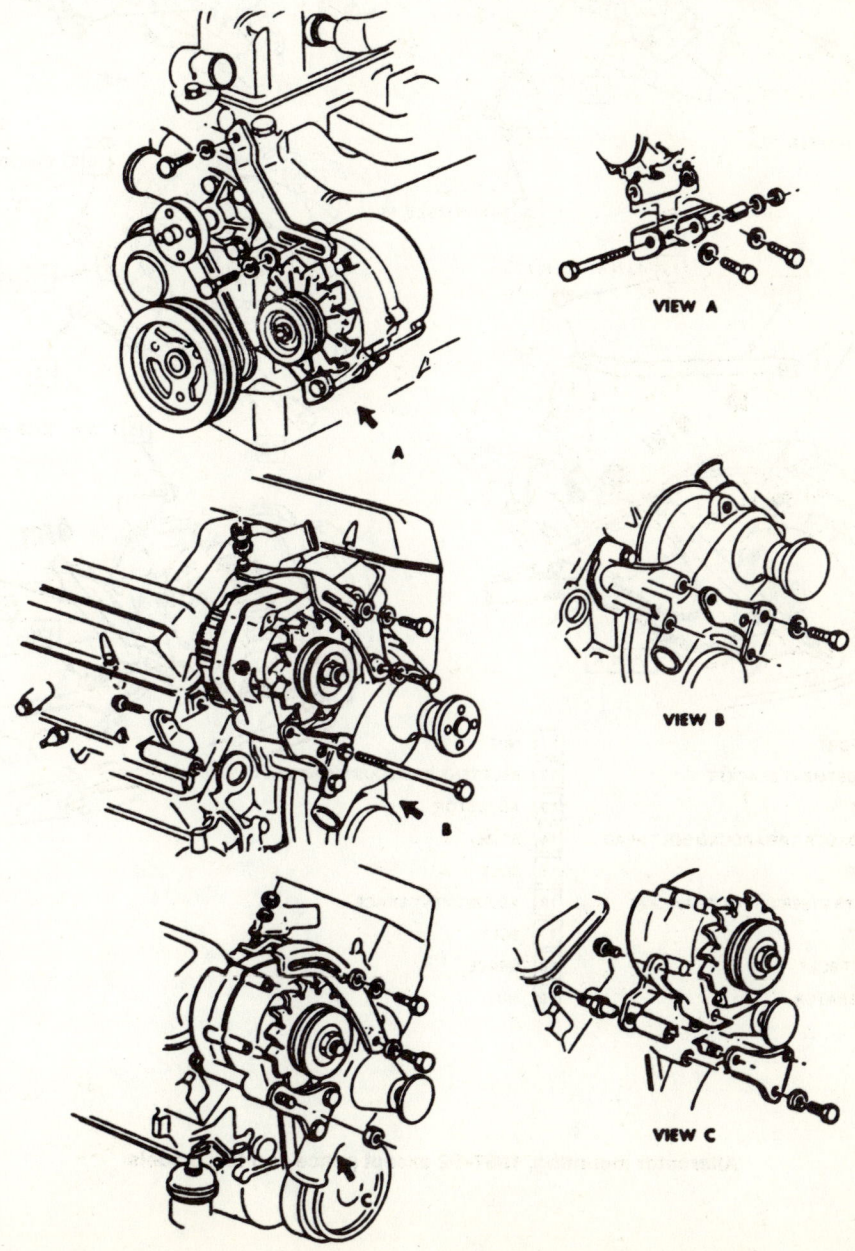

1968–86 alternator mounting; inline six cylinder (top), V6 and small block V8 (center) and big block V8 (bottom)

102 ENGINE AND ENGINE OVERHAUL

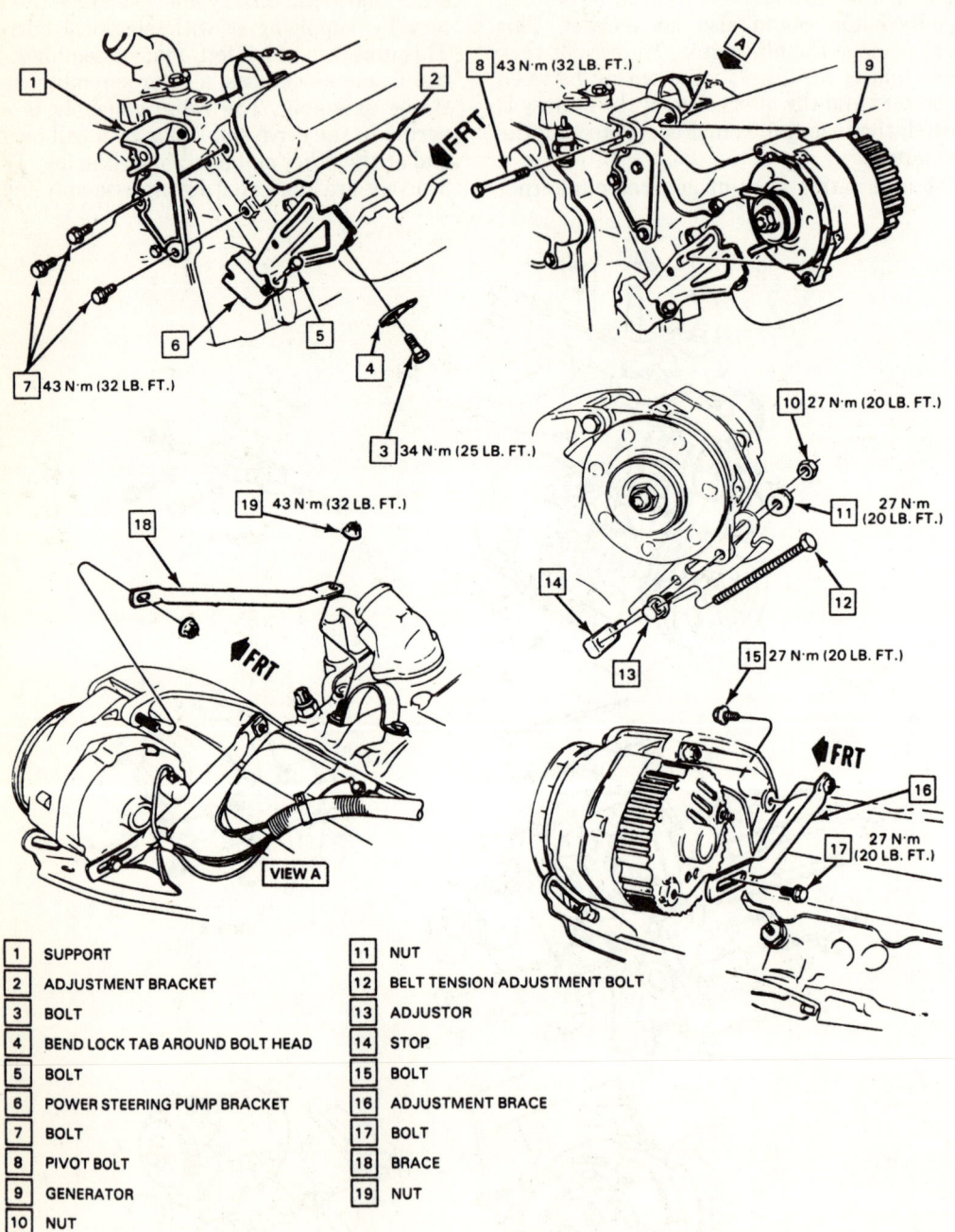

1 SUPPORT	11 NUT
2 ADJUSTMENT BRACKET	12 BELT TENSION ADJUSTMENT BOLT
3 BOLT	13 ADJUSTOR
4 BEND LOCK TAB AROUND BOLT HEAD	14 STOP
5 BOLT	15 BOLT
6 POWER STEERING PUMP BRACKET	16 ADJUSTMENT BRACE
7 BOLT	17 BOLT
8 PIVOT BOLT	18 BRACE
9 GENERATOR	19 NUT
10 NUT	

Alternator mounting, 1987–92 except police and taxi models

ENGINE AND ENGINE OVERHAUL

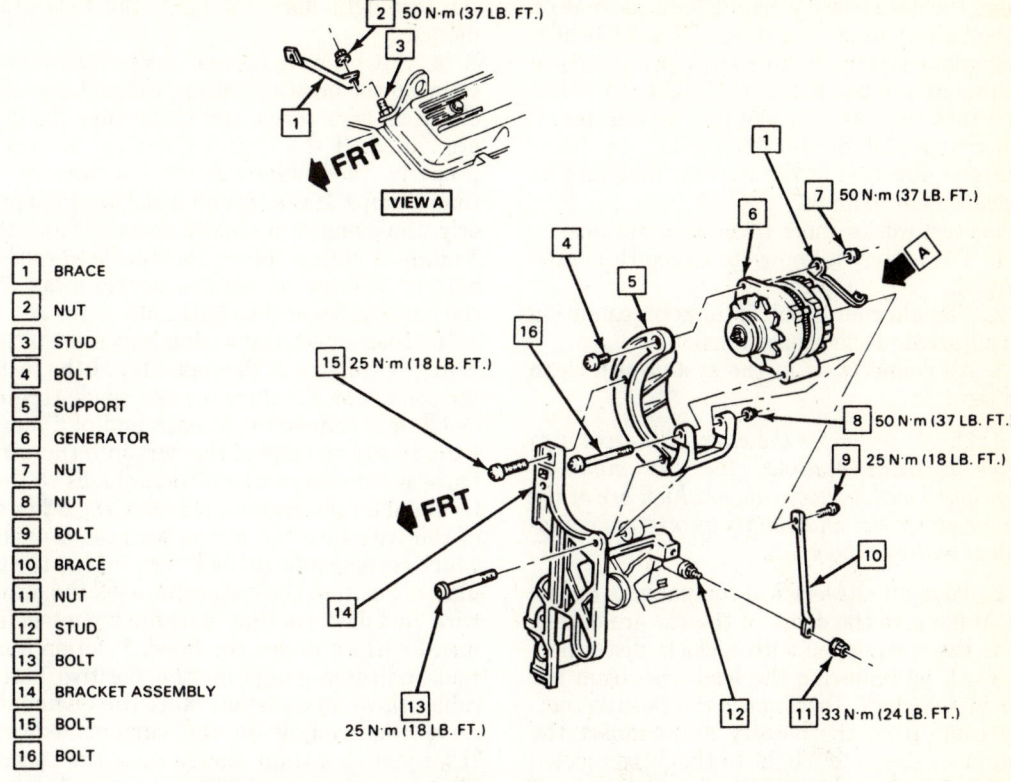

1	BRACE
2	NUT
3	STUD
4	BOLT
5	SUPPORT
6	GENERATOR
7	NUT
8	NUT
9	BOLT
10	BRACE
11	NUT
12	STUD
13	BOLT
14	BRACKET ASSEMBLY
15	BOLT
16	BOLT

Heavy duty alternator mounting, 1987–92 police and taxi models

quired no maintenance. There are six diodes in the end frame assembly. These diodes are electrical check valves that also change the alternating current developed within the stator windings to a direct (DC) current at the output (BAT) terminal. Three of these diodes are negative and are mounted flush with the end frame while the other three are positive and are mounted into a strip called a heat sink. The positive diodes are easily identified as the ones within small cavities or depressions.

ALTERNATOR PRECAUTIONS

To prevent damage to the alternator and regulator, the following precautions should be taken when working with the electrical system.
1. Never reverse the battery connections.
2. Booster batteries for starting must be connected properly: positive-to-positive and negative-to-ground.
3. Disconnect the battery cables before using a fast charger; the charger has a tendency to force current through the diodes in the opposite direction for which they were designed. This burns out the diodes.
4. Never use a fast charger as a booster for starting the vehicle.
5. Never disconnect the voltage regulator while the engine is running.
6. Avoid long soldering times when replacing diodes or transistors. Prolonged heat is damaging to AC generators.
7. Do not use test lamps of more than 12 volts (V) for checking diode continuity.
8. Do not short across or ground any of the terminals on the AC generator.
9. The polarity of the battery, generator, and regulator must be matched and considered before making any electrical connections within the system.
10. Never operate the alternator on an open circuit. make sure that all connections within the circuit are clean and tight.
11. Disconnect the battery terminals when performing any service on the electrical system. This will eliminate the possibility of accidental reversal of polarity.
12. Disconnect the battery ground cable if arc welding is to be done on any part of the car.

CHARGING SYSTEM TROUBLESHOOTING

There are many possible ways in which the charging system can malfunction. Often the source of a problem is difficult to diagnose, requiring special equipment and a good deal of experience. This is usually not the case, however, where the charging system fails completely and

104 ENGINE AND ENGINE OVERHAUL

causes the dash board warning light to come on or the battery to become dead. To troubleshoot a complete system failure only two pieces of equipment are needed: a test light, to determine that current is reaching a certain point; and a current indicator (ammeter), to determine the direction of the current flow and its measurement in amps.

This test works under three assumptions:

1. The battery is known to be good and fully charged.
2. The alternator belt is in good condition and adjusted to the proper tension.
3. All connections in the system are clean and tight.

NOTE: *In order for the current indicator to give a valid reading, the car must be equipped with battery cables which are of the same gauge size and quality as original equipment battery cables.*

1. Turn off all electrical components on the car. Make sure the doors of the car are closed. If the car is equipped with a clock, disconnect the clock by removing the lead wire from the rear of the clock. Disconnect the positive battery cable from the battery and connect the ground wire on a test light to the disconnected positive battery cable. Touch the probe end of the test light to the positive battery post. The test light should not light. If the test light does light, there is a short or open circuit on the car.

2. Disconnect the voltage regulator wiring harness connector at the voltage regulator. Turn on the ignition key. Connect the wire on a test light to a good ground (engine bolt). Touch the probe end of a test light to the ignition wire connector into the voltage regulator wiring connector. This wire corresponds to the **I** terminal on the regulator. If the test light goes on, the charging system warning light circuit is complete. If the test light does not come on and the warning light on the instrument panel is on, either the resistor wire, which is parallel with the warning light, or the wiring to the voltage regulator, is defective. If the test light does not come on and the warning light is not on, either the bulb is defective or the power supply wire from the battery through the ignition switch to the bulb has an open circuit. Connect the wiring harness to the regulator.

3. Examine the fuse link wire in the wiring harness from the starter relay to the alternator. If the insulation on the wire is cracked or split, the fuse link may be melted. Connect a test light to the fuse link by attaching the ground wire on the test light to an engine bolt and touching the probe end of the light to the bottom of the fuse link wire where it splices into the alternator output wire. If the bulb in the test light does not light, the fuse link is melted.

4. Start the engine and place a current indicator on the positive battery cable. Turn off all electrical accessories and make sure the doors are closed. If the charging system is working properly, the gauge will show a draw of less than 5 amps. If the system is not working properly, the gauge will show a draw of more than 5 amps. A charge moves the needle toward the battery, a draw moves the needle away from the battery. Turn the engine off.

5. Disconnect the wiring harness from the voltage regulator at the regulator at the regulator connector. Connect a male spade terminal (solderless connector) to each end of a jumper wire. Insert one end of the wire into the wiring harness connector which corresponds to the **A** terminal on the regulator. Insert the other end of the wire into the wiring harness connector which corresponds to the **F** terminal on the regulator. Position the connector with the jumper wire installed so that it cannot contact any metal surface under the hood. Position a current indicator gauge on the positive battery cable. Have an assistant start the engine. Observe the reading on the current indicator. Have your assistant slowly raise the speed of the engine to about 2,000 rpm or until the current indicator needle stops moving, whichever comes first. Do not run the engine for more than a short period of time in this condition. If the wiring harness connector or jumper wire becomes excessively hot during this test, turn off the engine and check for a grounded wire in the regulator wiring harness. If the current indicator shows a charge of about three amps less than the output of the alternator, the alternator is working properly. If the previous tests showed a draw, the voltage regulator is defective. If the gauge does not show the proper charging rate, the alternator is defective.

REMOVAL AND INSTALLATION

1. Disconnect the battery negative battery cable.
2. Tag and disconnect the alternator wiring.
3. Remove the alternator brace bolt. As required, if the vehicle is equipped with power steering loosen the pump brace and mount nuts. Detach the drive belt.
4. Support the alternator and remove the mount bolts. Remove the unit from the vehicle.
5. Installation is the reverse of the removal procedure.

BELT TENSION ADJUSTMENT

NOTE: *On some models it may be necessary to remove the lower splash shield to gain clear-*

ance when installing a new drive belt.

Belt tension should be checked with a gauge made for the purpose. If a gauge is not available, tension can be checked with moderate thumb pressure applied to the belt at its longest span midway between pulleys. If the belt has a free span less than 12″, it should deflect approximately $1/4″$. If the span is longer than 12″, deflection can range between $1/4″$ and $3/8″$.

To adjust or replace belts:

If equipped with a serpentine belt, these models have an automatic belt tensioner and no further adjustment is nessecary, replace the belt if worn or looseness is discovered.

1. Loosen the driven accessory's pivot and mounting bolts.
2. Move the accessory toward or away from the engine until the tension is correct. You can use a wood hammer handle or broomstick as a lever, but do not use anything metallic.
3. Tighten the bolts and re-check the tension. If new belts have been installed, run the engine for a few minutes, then re-check and re-adjust as necessary.

NOTE: *If the driven component has two drive belts, the belts should be replaced in pairs to maintain proper tension.*

It is better to have belts too loose than too tight, because overtight belts will lead to bearing failure, particularly in the water pump and alternator. However, loose belts place an extremely high impact load on the driven components due to the whipping action of the belt.

Regulator

The voltage regulator combines with the battery and alternator to comprise the charging system. Just as the name implies, the voltage regulator regulates the alternator voltage output to a safe amount. A properly working

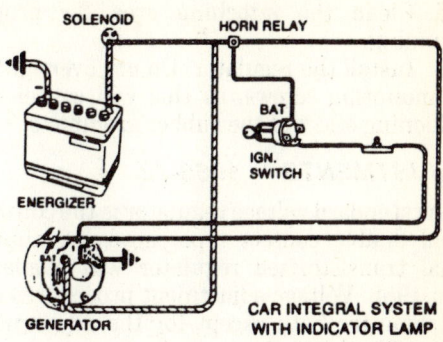

Integral voltage regulator charging system schematic

regulator prevents excessive voltage from burning out wiring, bulbs, or contact points, and prevents overcharging of the battery. Mechanical adjustments (air gap, point opening) must be followed by electrical adjustments and not vice versa.

Since 1973 all GM vehicles are equipped with alternators which have built-in solid state voltage regulators. The regulator is in the end frame (inside) of the alternator and requires no adjustment. The following adjustments apply to 1968-73 vehicles.

NOTE: *Although standard since 1973, this integral alternator/regulator has been available as an option since 1969.*

REMOVAL AND INSTALLATION

1. Disconnect the negative battery cable.
2. Disconnect the wiring harness from the regulator.
3. Remove the mounting screws and remove the regulator.
4. Make sure that the regulator base gasket is in place before installation.

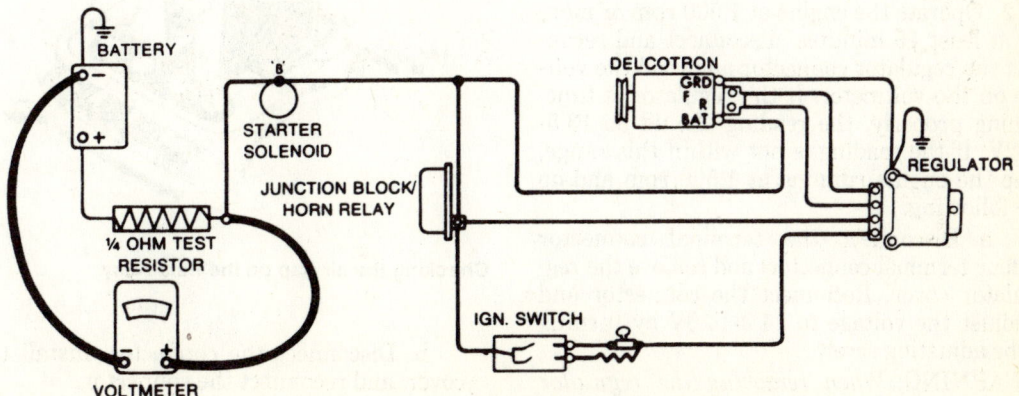

Schematic for testing the regulator voltage setting

ENGINE AND ENGINE OVERHAUL

5. Clean the attaching area for proper grounding.

6. Install the regulator. Do not overtighten the mounting screws, as this will cancel the cushioning effect of the rubber grommets.

ADJUSTMENTS — 1968-72

The standard voltage regulator is the conventional double contact type; however, an optional transistorized regulator was available from 1968. Voltage adjustment procedures are the same for both except for the adjustment points. The double contact adjustment screw is under the regulator cover, the 1968 transistorized regulator is adjusted externally after removing an Allen screw from the adjustment hole.

Field Relay Adjustments (Mechanical)

As explained earlier, mechanical adjustments must be made first and then follow by electrical adjustments.

Point Opening

Using a feeler gauge, check the point opening as illustrated. To change the opening, carefully bend the armature stop. The point opening for all regulators should be 0.014″.

Air Gap

Check the air gap with the points just touching. The gap should be 0.067″. If the point opening setting is correct, then the relay will operate OK even if the air gap is off. To adjust air gap, bend the flat contact spring.

Voltage Adjustment (Electrical)

1. Connect a $1/4\Omega$, 25 watt fixed resistor (a knife blade switch using a $1/4\Omega$ resistor) into the charging circuit (as illustrated) at the battery positive terminal. One end of the resistor connects to the battery positive terminal while the other connects to the voltmeter.

2. Operate the engine at 1,500 rpm or more for at least 15 minutes. disconnect and reconnect the regulator connector and read the voltage on the voltmeter. If the regulator is functioning properly, the reading should be 13.5-15.2V. If the reading is not within this range, keep the engine running at 1,500 rpm and do the following:

   a. Disconnect the terminal connector (four terminal connector) and remove the regulator cover. Reconnect the connector and adjust the voltage to 14.2-14.6V by turning the adjusting screw.

   WARNING: *When removing the regulator cover ALWAYS disconnect the connector first to prevent regulator damage by short circuits.*

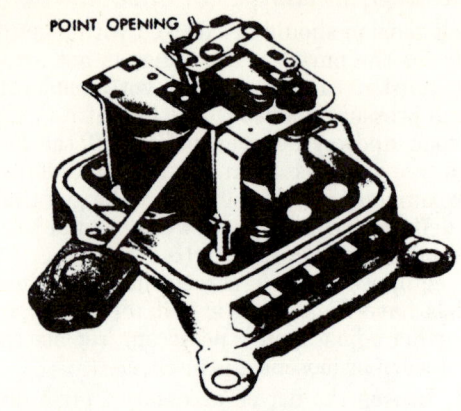

Checking the field relay point opening

Voltage setting adjustment

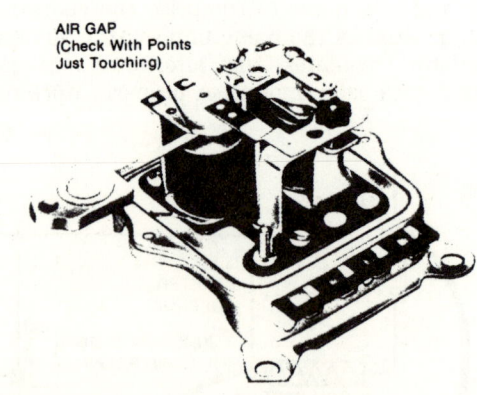

Checking the air gap on the field relay

   b. Disconnect the connector, install the cover, and reconnect the connector.

   c. Increase the regulator temperature by running the engine at 1,500 rpm for 10 more minutes.

Alternator and Regulator Specifications

Year	Alternator Part No. or Manufacturer	Alternator Field Current @ 12 V	Alternator Output (amps)	Regulator Field Relay Air Gap (in.)	Regulator Field Relay Point Gap (in.)	Regulator Field Relay Volts to Close	Regulator Air Gap (in.)	Regulator Point Gap (in.)	Regulator Volts @ 75°
1968	1100813	2.2–2.6	37	0.15	0.30	2.3–2.7	0.067	0.014	13.5–14.4
	1100693	2.2–2.6	37	0.15	0.30	2.3–2.7	0.067	0.014	13.5–14.4
1969	1100834	2.2–2.6	37	0.15	0.30	2.3–2.7	0.067	0.014	13.5–14.4
	1100836	2.2–2.6	37	0.15	0.30	2.3–2.7	0.067	0.014	13.5–14.4
1970	1100834	2.2–2.6	37	0.15	0.30	2.3–2.7	0.067	0.014	13.5–14.4
	1100837	2.2–2.6	37	0.15	0.30	2.3–2.7	0.067	0.014	13.5–14.4
1971	1100838	2.2–2.6	37	0.15	0.30	2.3–2.7	0.067	0.014	13.5–14.4
	1100839	2.2–2.6	37	0.15	0.30	2.3–2.7	0.067	0.014	13.5–14.4
1972	1100566	2.2–2.6	35	0.15	0.30	1.5–3.2	0.067	0.014	13.8–14.8
	1100917	2.8–3.2	59	0.30	0.30	1.5–3.2	0.067	0.014	13.8–14.8
	1100843	2.8–3.2	58	Integrated with Alternator					13.8–14.8
1973	1100497	2.8–3.2	36	Integrated with Alternator					13.8–14.8
	1100934	2.8–3.2	37	Integrated with Alternator					13.8–14.8
1974	1100934	4–4.5	37	Integrated with Alternator					13.8–14.8
	1102347	4–4.5	61	Integrated with Alternator					13.8–14.8
	1100497	4–4.5	37	Integrated with Alternator					13.8–14.8
	1100573	4–4.5	42	Integrated with Alternator					13.8–14.8
	1100597	4–4.5	61	Integrated with Alternator					13.8–14.8
	1100560	4–4.5	55	Integrated with Alternator					13.8–14.8
	1100575	4–4.5	55	Integrated with Alternator					13.8–14.8
1975	1100497	4–4.5	37	Integrated with Alternator					13.8–14.8
	1102397	4–4.5	37	Integrated with Alternator					13.8–14.8
	1102483	4–4.5	37	Integrated with Alternator					13.8–14.8
	1100560	4–4.5	55	Integrated with Alternator					13.8–14.8
	1100575	4–4.5	55	Integrated with Alternator					13.8–14.8
	1100597	4–4.5	61	Integrated with Alternator					13.8–14.8
	1102347	4–4.5	61	Integrated with Alternator					13.8–14.8
1976–79	1102491	4–4.5	37	Integrated with Alternator					13.8–14.8
	1102480	4–4.5	61	Integrated with Alternator					13.8–14.8
	1102486	4–4.5	61	Integrated with Alternator					13.8–14.8
	1102394	4–4.5	37	Integrated with Alternator					13.8–14.8
1980–81	1103161	4–4.5	37	Integrated with Alternator					13.8–14.8
	1103118	4–4.5	37	Integrated with Alternator					13.8–14.8
	1103043	4–4.5	42	Integrated with Alternator					13.8–14.8
	1103162	4–4.5	37	Integrated with Alternator					13.8–14.8
	1103092	4–4.5	55	Integrated with Alternator					13.8–14.8
	1103088	4–4.5	55	Integrated with Alternator					13.8–14.8
	1103100	4–4.5	55	Integrated with Alternator					13.8–14.8
	1103085	4–4.5	55	Integrated with Alternator					13.8–14.8
	1103044	4–4.5	63	Integrated with Alternator					13.8–14.8
	1103091	4–4.5	63	Integrated with Alternator					13.8–14.8
	1103169	4–4.5	63	Integrated with Alternator					13.8–14.8
	1103102	4–4.5	63	Integrated with Alternator					13.8–14.8
	1103122	4–4.5	63	Integrated with Alternator					13.8–14.8

Alternator and Regulator Specifications

Year	Part No. or Manufacturer	Alternator Field Current @ 12 V	Output (amps)	Field Relay Air Gap (in.)	Field Relay Point Gap (in.)	Field Relay Volts to Close	Regulator Air Gap (in.)	Regulator Point Gap (in.)	Volts @ 75°
	1101044	4–4.5	70		Integrated with Alternator				13.8–14.8
	1101071	4–4.5	70		Integrated with Alternator				13.8–14.8
1982–84	1103161	4–4.5	37		Integrated with Alternator				13.8–14.8
	1103118	4–4.5	37		Integrated with Alternator				13.8–14.8
	1103043	4–4.5	42		Integrated with Alternator				13.8–14.8
	1103162	4–4.5	37		Integrated with Alternator				13.8–14.8
	1103092	4–4.5	55		Integrated with Alternator				13.8–14.8
	1103088	4–4.5	55		Integrated with Alternator				13.8–14.8
	1103100	4–4.5	55		Integrated with Alternator				13.8–14.8
	1103085	4–4.5	55		Integrated with Alternator				13.8–14.8
	1103044	4–4.5	63		Integrated with Alternator				13.8–14.8
	1103091	4–4.5	63		Integrated with Alternator				13.8–14.8
	1103169	4–4.5	63		Integrated with Alternator				13.8–14.8
	1101044	4–4.5	70		Integrated with Alternator				13.8–14.8
	1101066	4–4.5	70		Integrated with Alternator				13.8–14.8
	1101071	4–4.5	70		Integrated with Alternator				13.8–14.8
	1100226	4–4.5	37		Integrated with Alternator				13.8–14.8
	1100246	4–4.5	63		Integrated with Alternator				13.8–14.8
	1100270	4–4.5	78		Integrated with Alternator				13.8–14.8
	1100239	4–4.5	55		Integrated with Alternator				13.8–14.8
	1100247	4–4.5	63		Integrated with Alternator				13.8–14.8
	1100200	4–4.5	78		Integrated with Alternator				13.8–14.8
	1100230	4–4.5	42		Integrated with Alternator				13.8–14.8
	1100260	4–4.5	78		Integrated with Alternator				13.8–14.8
	1100263	4–4.5	78		Integrated with Alternator				13.8–14.8
	1105022	4–4.5	78		Integrated with Alternator				13.8–14.8
	1100237	4–4.5	55		Integrated with Alternator				13.8–14.8
	1100228	4–4.5	37		Integrated with Alternator				13.8–14.8
	1100300	4–4.5	63		Integrated with Alternator				13.8–14.8
	1105041	4–4.5	78		Integrated with Alternator				13.8–14.8
1985	1100246	4–4.5	66		Integrated with Alternator				13.8–14.8
	1100237	4–4.5	56		Integrated with Alternator				13.8–14.8
	1105652	4–4.5	78		Integrated with Alternator				13.8–14.8
	1105521	4–4.5	78		Integrated with Alternator				13.8–14.8
	1105520	4–4.5	56		Integrated with Alternator				13.8–14.8
	1105652	4–4.5	78		Integrated with Alternator				13.8–14.8
	1105523	4–4.5	56		Integrated with Alternator				13.8–14.8
1986–88	1100200	—	78		Integrated with Alternator				13.5–16.0
	1100239	—	56		Integrated with Alternator				13.5–16.0
	1105197	—	70		Integrated with Alternator				13.5–16.0
	1105651	—	94		Integrated with Alternator				13.5–16.0
	1105673	—	56		Integrated with Alternator				13.5–16.0
	1105652	—	78		Integrated with Alternator				13.5–16.0

Alternator and Regulator Specifications

Year	Alternator			Regulator					
				Field Relay			Regulator		
	Part No. or Manufacturer	Field Current @ 12 V	Output (amps)	Air Gap (in.)	Point Gap (in.)	Volts to Close	Air Gap (in.)	Point Gap (in.)	Volts @ 75°
	1105676	—	56	Integrated with Alternator					13.5–16.0
	1105674	—	66	Integrated with Alternator					13.5–16.0
1989–90	1101229	—	85	Integrated with Alternator					13.0–16.0
	1101253	—	85	Integrated with Alternator					13.0–16.0
	1101254	—	100	Integrated with Alternator					13.0–16.0
	1101454	—	120	Integrated with Alternator					13.0–16.0
1991	1101599	—	100	Integrated with Alternator					13.0–16.0
	1101454	—	120	Integrated with Alternator					13.0–16.0
1992	10479906	—	100	Integrated with Alternator					13.0–16.0
	10479907	—	124	Integrated with Alternator					13.0–16.0

d. Disconnect and reconnect the connector and read the voltmeter. A reading of 13.5-15.2V indicates a good regulator.

Battery

REMOVAL AND INSTALLATION

1. Disconnect the negative cable from the terminal, and then the positive cable. On non side terminal batteries special pullers are available to remove the cable clamps, if they seem stuck.

NOTE: *To avoid sparks, always disconnect the ground cable first, and connect it last.*

2. Remove the battery holddown clamp.
3. Remove the battery, being careful not to spill the acid.

NOTE: *Spilled acid can be neutralized with a baking soda/water solution. If you somehow get acid into your eyes, flush it out with lots of water and get to a doctor.*

4. Clean the battery posts thoroughly before reinstalling, or when installing a new battery.
5. Clean the cable clamps, using a wire brush, both inside and out.
6. Install the battery and holddown clamp or strap. Connect the positive, and the negative cable (see Note above). Do not hammer the cables onto the terminal posts. The complete terminals should be coated lightly (externally) with petroleum jelly or grease to help prevent corrosion. There are also felt washers impregnated with an anti-corrosion substance which are slipped over the battery posts before installing the cables; these are available in most auto parts stores.

WARNING: *Make absolutely sure that the battery is connected properly (positive to positive, negative to negative) before you turn the ignition key. reversed polarity can burn out your alternator and regulator in a matter of seconds.*

Starter

REMOVAL AND INSTALLATION

NOTE: *The starters on some engines require the addition of shims to provide proper clearance between the starter pinion gear and the flywheel. These shims are available in 0.015" sizes from Chevrolet dealers. Flat washers can be used if shims are unavailable.*

1. Disconnect the negative battery cable.
2. Raise and support the vehicle safely.
3. Disconnect all wiring from the starter. Replace each nut as the connector is removed, as thread sizes differ from connector to connector. Note or tag the wiring positions for installation.
4. Remove the front bracket from the starter and the two mounting bolts. On engines with a solenoid heat shield, remove the front bracket upper bolt and detach the bracket from the starter.
5. Remove the front bracket bolt or nut. Lower the starter front end first, and then remove the unit from the vehicle.
6. Reverse the removal procedures to install the starter. Torque the two mounting bolts to 25-35 ft. lbs.

SHIMMING THE STARTER

Starter noise during cranking and after the engine fires is often a result of too much or too little distance between the starter pinion gear and the flywheel. A high pitched whine during cranking (before the engine fires) can be caused by the pinion and flywheel being too far apart. Likewise, a whine after the engine starts (as the key is released) is often a result of the pinion-flywheel relationship being too close. In both cases flywheel damage can occur. Shims are available in 0.015" sizes to properly adjust the starter on its mount. You will also need a flywheel turning tool, available at most auto parts stores or from any auto tool store or salesperson.

If your vehicles starter emits the above noises, follow the shimming procedure below:

1. Disconnect the negative battery cable.
2. Remove the flywheel inspection cover on the bottom of the bellhousing.
3. Using the flywheel turning tool, turn the flywheel and examine the flywheel teeth. If damage is evident, the flywheel should be replaced.
4. Insert a suitable tool into the small hole in the bottom of the starter, then move the starter pinion and clutch assembly so the pinion and flywheel teeth mesh. If necessary, rotate the flywheel so that a pinion tooth is directly in the center of the two flywheel teeth and on the centerline of the two gears, as shown in the accompanying illustration.
5. Check the pinion-to-flywheel clearance

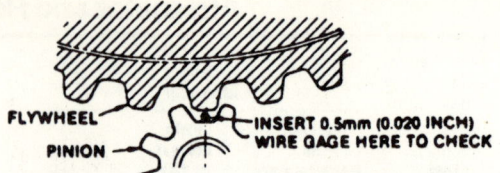

Checking the gap between starter pinion and flywheel

by using a 0.020" wire gauge (a spark plug wire gauge may work here, or you can make your own). Make sure you center the pinion tooth between the flywheel teeth and the gauge - NOT in the corners, as you may get a false reading. If the clearance is under this minimum, shim the starter away from the flywheel by adding shim(s) one at a time to the starter mount. Check clearance after adding each shim.

6. If the clearance is a good deal over 0.020" (in the vicinity of a 0.050" plug), shim the starter towards the flywheel. Broken or severely mangled flywheel teeth are also a good indicator that the clearance here is too great. Shimming the starter towards the flywheel is done by adding shims to the outboard starter mounting pad only. Check the clearance after each shim is added. A shim of 0.015" at this location will decrease the clearance about 0.010".

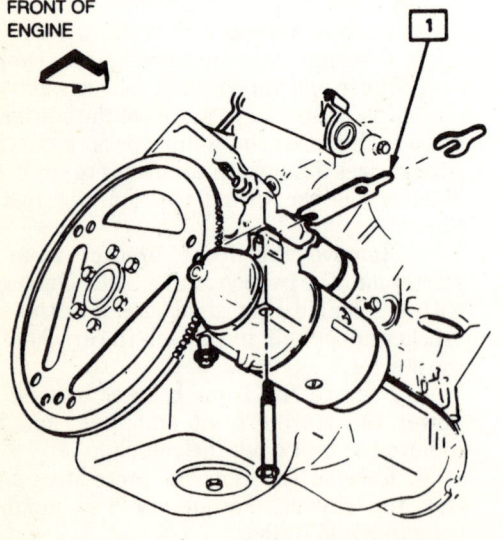

1. Use shims as required
2. Shield

Starter noise diagnostic procedure
1. Starter noise during cranking: remove 1–.015" double shim or add single .015" shim to *outer* bolt only.
2. High pitched whine after engine fires: add .015" double shims until noise disappears.

See text for complete procedure.

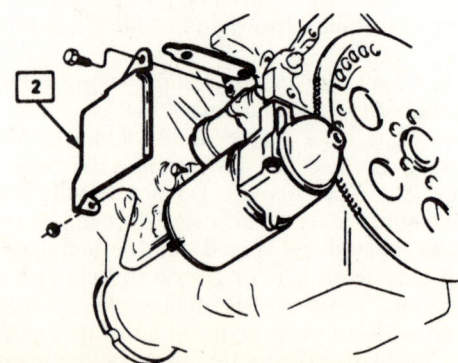

Starter motor mounting; V6 at left, diesel at right. Others similar

Battery and Starter Specifications

Year	Engine No. Cyl. Displacement (cu. in.)	Battery Ampere Hour Capacity	Battery Volts	Terminal Grounded	Starter Lock test Amps	Starter Lock test Volts	Starter Torque (ft. lbs.)	Starter No-Load Test Amps	Starter No-Load Test Volts	Starter No-Load Test RPM	Brush Spring Tension (oz.)
1968–69	6, 8-307	45	12	Neg			Not Recommended	—	10.6	—	35
	8-302, 327, 350, 396	61	12	Neg			Not Recommended	—	9	—	35
1970–71	6, 8-307	45	12	Neg			Not Recommended	50–80	9	5,500–10,500	35
	8-350	61	12	Neg			Not Recommended	55–80	9	3,500–6,000	35
	8-402 (396)	61	12	Neg			Not Recommended	65–95	9	7,500–10,500	35
	8-454	62	12	Neg			Not Recommended	65–95	9	7,500–10,500	35
1972	6-250	45	12	Neg			Not Recommended	50–80	9	5,500–10,500	35
	8-307, 350, 402	61	12	Neg			Not Recommended	55–80①	9	5,500–10,500	35
	8-454	76	12	Neg			Not Recommended	65–95	9	7,500–10,500	35
1973	6-250	45	12	Neg			Not Recommended	50–80	9	5,500–10,500	35
	8-307	61	12	Neg			Not Recommended	50–80	9	5,500–10,500	35
	8-350, 454	76	12	Neg			Not Recommended	65–95	9	7,500–10,500	35
1974	6-250	2300②	12	Neg			Not Recommended	50–80	9	5,500–10,500	35
	8-350, 400	2900②	12	Neg			Not Recommended	65–95	9	7,500–10,500	35
	8-454	3750②	12	Neg			Not Recommended	65–95	9	7,500–10,500	35
1975–76	6-250	2500②	12	Neg			Not Recommended	50–80	9	5,500–10,500	35
	8-350, 400	3200②	12	Neg			Not Recommended	65–95	9	7,500–10,500	35
	8-454	4000②	12	Neg			Not Recommended	65–95	9	7,500–10,500	35
1977–78	6-250	275③	12	Neg			Not Recommended	50–80	9	5,500–10,500	35
	8-305	350③	12	Neg			Not Recommended	50–80	9	5,500–10,500	35
	8-350	350③	12	Neg			Not Recommended	65–95	9	7,500–10,500	35
1979	6-250	275③	12	Neg			Not Recommended	60–88	10.6	6,500–10,100	35
	6-231	350③	12	Neg			Not Recommended	50–80	10.6	7,500–11,400	35
	6-200	350③	12	Neg			Not Recommended	50–80	10.6	7,500–11,400	35
	8-305	350③	12	Neg			Not Recommended	50–80	10.6	7,500–11,400	35
	8-350	350③	12	Neg			Not Recommended	65–95	10.6	7,500–10,500	35

Battery and Starter Specifications

Year	Engine No. Cyl. Displacement (cu. in.)	Battery Ampere Hour Capacity	Volts	Terminal Grounded	Starter Lock test Amps	Volts	Torque (ft. lbs.)	No-Load Test Amps	Volts	RPM	Brush Spring Tension (oz.)
1980–81	6-229	350③	12	Neg			Not Recommended	50–80	10.6	7,500–11,400	35
	6-231	350③	12	Neg			Not Recommended	50–80	10.6	7,500–11,400	35
	8-267	250③	12	Neg			Not Recommended	50–80	10.6	7,500–11,400	35
	8-305	350③	12	Neg			Not Recommended	50–80	10.6	7,500–11,400	35
1982–84	6-229	350③④	12	Neg			Not Recommended	60–80	10.6	7,500–11,400	35
	6-231	350③④	12	Neg			Not Recommended	60–80	10.6	7,500–11,400	35
	6-263	465③⑥	12	Neg			Not Recommended	160–220	10.6	4,000–5,500	35
	6-267	350③	12	Neg			Not Recommended	45–70	10.6	7,500–11,400	35
	8-305	350③	12	Neg			Not Recommended	44–70	10.6	7,500–11,400	35
	8-350	465③⑥	12	Neg			Not Recommended	160–220	10.6	4,000–5,500	35
1985	6-262	630③	12	Neg			Not Recommended	50–75	10	6,000–11,900	35
	8-305	500③	12	Neg			Not Recommended	50–75	10	6,000–11,900	35
1986–88	6-262	630③	12	Neg			Not Recommended	70–120	10	5,500–10,700	—
	8-305	525③	12	Neg			Not Recommended	70–120	10	5,500–10,700	—
1989–90	6-262	525③	12	Neg			Not Recommended	50–75	10	6,000–11,900	—
	8-305	570③	12	Neg			Not Recommended	70–110	10	6,500–10,700	—
	8-350	570③⑦	12	Neg			Not Recommended	70–110	10	6,500–10,700	—
1991–92	6-305	525③⑦	12	Neg			Not Recommended	45–75	10	6,500–11,000	—
	8-350	525③⑦	12	Neg			Not Recommended	70–110	10	6,500–10,700	—

① 350 & 402 use 454 starter
② Cranking power in watts @ 0°F
③ Cranking power in amps @ 0°F
④ minute reserve capacity
⑤ 75 minute reserve capacity
⑥ 115 minute reserve capacity, 2 batteries used
⑦ Police, Taxi 730 cranking amps

ENGINE AND ENGINE OVERHAUL

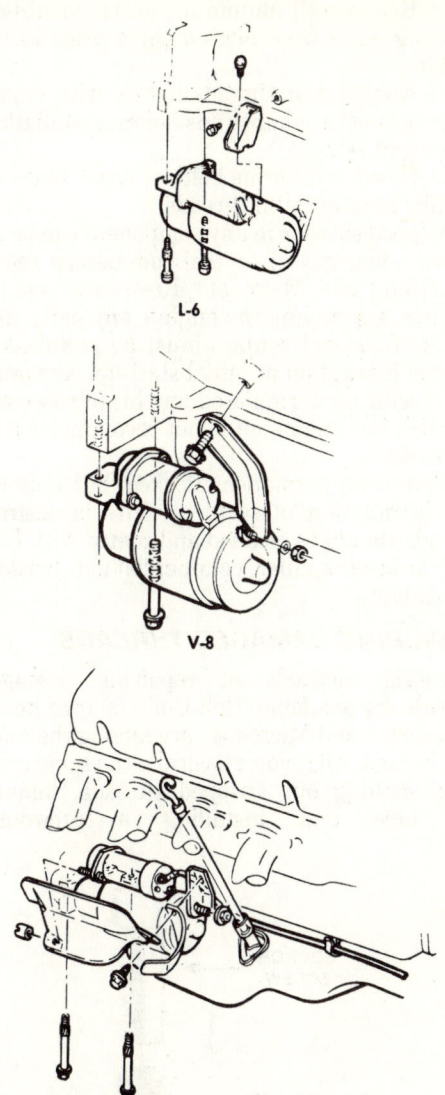

V8 with solenoid heat shield

The small block family of V8 engines, which has included the 267, 283, 305, 307 (except the 1986-88 Oldsmobile produced engine), 327, 350 and 400 cu. in. blocks, have all evolved from the design of the 1955 265 cu. in. V8. It was this engine that introduced the ball joint type rocker arm design which is now used by many vehicle makers. The Chevrolet built V6s are also similar.

This line of engines features a great deal of interchangeability, and later parts may be utilized on earlier engines for increased reliability and/or performance.

The 396, 402 and 454 engines are known as the big blocks, or less frequently, the Mark IV engines. They are available in the high performance SS versions of the Chevrolet, and feature many tuning modifications such as high lift camshafts, solid lifters (in some cases), high compression ratios and large carburetors. These big block engines are similar to their small block little brothers in basic design, but parts cannot be interchanged between the small and big blocks.

The 350 V8 diesel is derived from the 350 gasoline engine, except that the cylinder block, crankshaft, main bearings, connecting rods and wrist pins are heavier duty in the diesel (due to the much higher compression ratio). The diesel cylinder heads, intake manifold, ignition and fuel systems are also different from their gasoline engine counterparts. Aircraft-type hydraulic roller valve lifters are used in later production diesel engines.

The Buick built 231 V6 is the only engine used which is substantially different. This engine follows Buick V8 practice in that its valve gear incorporates rocker shafts instead of the ball joint style rockers on the other engines.

The Oldsmobile built 307 V8 engine is used in the Chevrolet station wagon from 1986-88. This engine is a smaller version of the Oldsmobile 350 engine which has been used in Oldsmobiles since the late 1960's.

ENGINE MECHANICAL

Design

All engines, whether inline sixes (L6) V6 or V8, are water cooled, overhead valve powerplants. Most engines use cast iron blocks and heads, with the exception of some high performance 8-454s, which use aluminum heads.

The crankshaft in the L6-230 and L6-250 cu. in. engines is supported in seven main bearings, with the thrust being taken by the No. 7 bearing. The camshaft is low in the block and is gear driven. Relatively long pushrods actuate the valve through ball jointed rocker arms.

Engine Overhaul Tips

Most engine overhaul procedures are fairly standard. In addition to specific parts replacement procedures and complete specifications for your individual engine, this chapter also is a guide to accept rebuilding procedures. Examples of standard rebuilding practice are shown and should be used along with specific details concerning your particular engine.

Competent and accurate machine shop services will ensure maximum performance, reliability and engine life.

In most instances it is more profitable for the do-it-yourself mechanic to remove, clean and in-

114 ENGINE AND ENGINE OVERHAUL

spect the component, buy the necessary parts and deliver these to a shop for actual machine work.

On the other hand, much of the rebuilding work (crankshaft, block, bearings, piston rods, and other components) is well within the scope of the do-it-yourself mechanic.

TOOLS

The tools required for an engine overhaul or parts replacement will depend on the depth of your involvement. With a few exceptions, they will be the tools found in a mechanic's tool kit (see Chapter 1). More in-depth work will require any or all of the following:

- a dial indicator (reading in thousandths) mounted on a universal base
  - micrometers and telescope gauges
  - jaw and screw-type pullers
  - scraper
  - valve spring compressor
  - ring groove cleaner
  - piston ring expander and compressor
  - ridge reamer
  - cylinder hone or glaze breaker
  - Plastigage®
  - engine crane and engine stand

The use of most of these tools is illustrated in this chapter. Many can be rented for a one-time use from a local parts jobber or tool supply house specializing in automotive work.

Occasionally, the use of special tools is called for. See the information on Special Tools and Safety Notice in the front of this book before substituting another tool.

INSPECTION TECHNIQUES

Procedures and specifications are given in this chapter for inspecting, cleaning and assessing the wear limits of most major components. Other procedures such as Magnaflux® and Zyglo® can be used to locate material flaws and stress cracks. Magnaflux® is a magnetic process applicable only to ferrous materials. The Zyglo® process coats the material with a fluorescent dye penetrant and can be used on any material Check for suspected surface cracks can be more readily made using spot check dye. The dye is sprayed onto the suspected area, wiped off and the area sprayed with a developer. Cracks will show up brightly.

OVERHAUL TIPS

Aluminum has become extremely popular for use in engines, due to its low weight. Observe the following precautions when handling aluminum parts:

- Never hot tank aluminum parts (the caustic hot tank solution will eat the aluminum.
- Remove all aluminum parts (identification tag, etc.) from engine parts prior to the tanking.
- Always coat threads lightly with engine oil or anti-seize compounds before installation, to prevent seizure.
- Never overtorque bolts or spark plugs especially in aluminum threads.

Stripped threads in any component can be repaired using any of several commercial repair kits (Heli-Coil®, Microdot, Keenserts®, etc.).

When assembling the engine, any parts that will be frictional contact must be prelubed to provide lubrication at initial start-up. Any product specifically formulated for this purpose can be used, but engine oil is not recommended as a prelube.

When semi-permanent (locked, but removable) installation of bolts or nuts is desired, threads should be cleaned and coated with Loctite® or other similar, commercial non-hardening sealant.

REPAIRING DAMAGED THREADS

Several methods of repairing damaged threads are available. Heli-Coil® (shown here), Keenserts® and Microdot® are among the most widely used. All involve basically the same principle — drilling out stripped threads, tapping the hole and installing a prewound

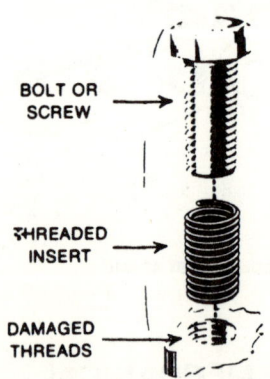

Damaged bolt holes can be repaired with thread repair inserts

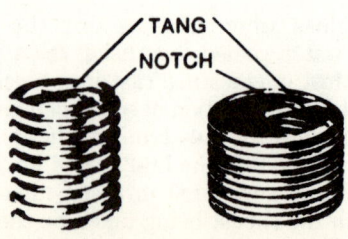

Standard thread repair insert (left) and spark plug thread insert (right)

ENGINE AND ENGINE OVERHAUL

Standard Torque Specifications and Fastener Markings

In the absence of specific torques, the following chart can be used as a guide to the maximum safe torque of a particular size/grade of fastener.

- There is no torque difference for fine or coarse threads.
- Torque values are based on clean, dry threads. Reduce the value by 10% if threads are oiled prior to assembly.
- The torque required for aluminum components or fasteners is considerably less.

U.S. Bolts

SAE Grade Number Number of lines always 2 less than the grade number. Bolt Size (Inches)—(Thread)	1 or 2			5			6 or 7		
	Maximum Torque			Maximum Torque			Maximum Torque		
	Ft./Lbs.	Kgm	Nm	Ft./Lbs.	Kgm	Nm	Ft./Lbs.	Kgm	Nm
¼ — 20	5	0.7	6.8	8	1.1	10.8	10	1.4	13.5
— 28	6	0.8	8.1	10	1.4	13.6			
⁵⁄₁₆ — 18	11	1.5	14.9	17	2.3	23.0	19	2.6	25.8
— 24	13	1.8	17.6	19	2.6	25.7			
⅜ — 16	18	2.5	24.4	31	4.3	42.0	34	4.7	46.0
— 24	20	2.75	27.1	35	4.8	47.5			
⁷⁄₁₆ — 14	28	3.8	37.0	49	6.8	66.4	55	7.6	74.5
— 20	30	4.2	40.7	55	7.6	74.5			
½ — 13	39	5.4	52.8	75	10.4	101.7	85	11.75	115.2
— 20	41	5.7	55.6	85	11.7	115.2			
⁹⁄₁₆ — 12	51	7.0	69.2	110	15.2	149.1	120	16.6	162.7
— 18	55	7.6	74.5	120	16.6	162.7			
⅝ — 11	83	11.5	112.5	150	20.7	203.3	167	23.0	226.5
— 18	95	13.1	128.8	170	23.5	230.5			
¾ — 10	105	14.5	142.3	270	37.3	366.0	280	38.7	379.6
— 16	115	15.9	155.9	295	40.8	400.0			
⅞ — 9	160	22.1	216.9	395	54.6	535.5	440	60.9	596.5
— 14	175	24.2	237.2	435	60.1	589.7			
1 — 8	236	32.5	318.6	590	81.6	799.9	660	91.3	894.8
— 14	250	34.6	338.9	660	91.3	849.8			

Metric Bolts

Relative Strength Marking Bolt Markings Bolt Size Thread Size x Pitch (mm)	4.6, 4.8			8.8		
	Maximum Torque			Maximum Torque		
	Ft. Lbs.	Kgm	Nm	Ft. Lbs.	Kgm	Nm
6 x 1.0	2–3	.2–.4	3–4	3–6	.4–.8	5–8
8 x 1.25	6–8	.8–1	8–12	9–14	1.2–1.9	13–19
10 x 1.25	12–17	1.5–2.3	16–23	20–29	2.7–4.0	27–39
12 x 1.25	21–32	2.9–4.4	29–43	35–53	4.8–7.3	47–72
14 x 1.5	35–52	4.8–7.1	48–70	57–85	7.8–11.7	77–110
16 x 1.5	51–77	7.0–10.6	67–100	90–120	12.4–16.5	130–160
18 x 1.5	74–110	10.2–15.1	100–150	130–170	17.9–23.4	180–230
20 x 1.5	110–140	15.1–19.3	150–190	190–240	26.2–46.9	160–320
22 x 1.5	150–190	22.0–26.2	200–260	250–320	34.5–44.1	340–430
24 x 1.5	190–240	26.2–46.9	260–320	310–410	42.7–56.5	420–550

ENGINE AND ENGINE OVERHAUL

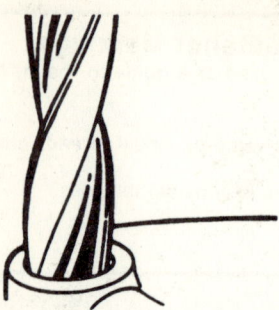

Drill out the damaged threads with specified drill. Drill completely through the hole or to the bottom of a blind hole

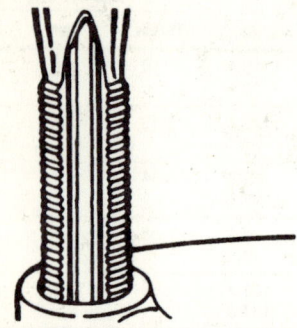

With the tap supplied, tap the hole to receive the thread insert. Keep the tap well oiled and back it out frequently to avoid clogging the threads

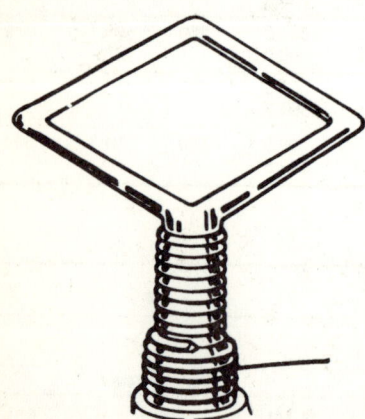

Screw the threaded insert onto the installation tool until the tang engages the slot. Screw the insert into the tapped hole until it's 1/4–1/2 turns below the top surface. After installation, break off the tang with a hammer and punch

insert—making welding, plugging and oversize fasteners unnecessary.

Two types of thread repair inserts are usually supplied: a standard type for most Inch Coarse, Inch Fine, Metric Course and Metric Fine thread sizes and a spark lug type to fit most spark plug port sizes. Consult the individual manufacturer's catalog to determine exact applications. Typical thread repair kits will contain a selection of prewound threaded inserts, a tap (corresponding to the outside diameter threads of the insert) and an installation tool. Spark plug inserts usually differ because they require a tap equipped with pilot threads and a combined reamer/tap section. Most manufacturers also supply blister-packed thread repair inserts separately in addition to a master kit containing a variety of taps and inserts plus installation tools.

Before effecting a repair to a threaded hole, remove any snapped, broken or damaged bolts or studs. Penetrating oil can be used to free frozen threads. The offending item can be removed with locking pliers or with a screw or stud extractor. After the hole is clear, the thread can be repaired, as shown in the series of accompanying illustrations.

Checking Engine Compression

A noticeable lack of engine power, excessive oil consumption and/or poor fuel mileage measured over an extended period are all indicators of internal engine wear. Worn piston rings, scored or worn cylinder bores, blown head gaskets, sticking or burnt valves and worn valve seats are all possible culprits here. A check of each cylinder's compression will help you locate the problems.

As mentioned in the Tools and Equipment section of Chapter 1, a screw-in type compression gauge is more accurate that the type you simply hold against the spark plug hole, although it takes slightly longer to use. It's worth it to obtain a more accurate reading. Follow the procedures below.

Gasoline Engine

1. Warm up the engine to normal operating temperature.
2. Stop the engine. Remove all the spark plugs.
3. Disconnect the high tension lead from the ignition coil.
4. On fully open the throttle either by operating the carburetor throttle linkage by hand or by having an assistant floor the accelerator pedal.
5. Screw the compression gauge into the no.1 spark plug hole until the fitting is snug.

WARNING: *Be careful not to crossthread the plug hole. On aluminum cylinder heads use extra care, as the threads in these heads are easily ruined.*

6. Ask an assistant to depress the accelerator pedal fully on both carbureted and fuel injected vehicles. Then, while you read the compression gauge, ask the assistant to crank the

ENGINE AND ENGINE OVERHAUL

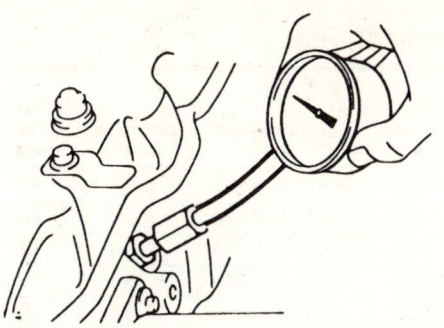

The screw-in type compression gauge is more accurate

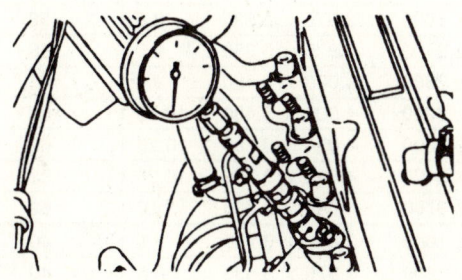

Diesel engines require a special compression gauge adaptor

engine two or three times in short bursts using the ignition switch.

7. Read the compression gauge at the end of each series of cranks, and record the highest of these readings. Repeat this procedure for each of the engine's cylinders. Compare the highest reading of each cylinder to the compression pressure specification in the Tune-Up Specifications chart in Chapter 2. The specs in this chart are maximum values.

A cylinder's compression pressure is usually acceptable if it is not less than 80% of maximum. The difference between any two cylinders should be no more than 12-14 pounds.

8. If a cylinder is unusually low, pour a tablespoon of clean engine oil into the cylinder through the spark plug hole and repeat the compression test. If the compression comes up after adding the oil, it appears that the cylinder's piston rings or bore are damaged or worn. If the pressure remains low, the valves may not be seating properly (a valve job is needed), or the head gasket may be blown near that cylinder. If compression in any two adjacent cylinders is low, and if the addition of oil doesn't help the compression, there is leakage past the head gasket. Oil and coolant water in the combustion chamber can result from this problem.

There may be evidence of water droplets on the engine dipstick when a head gasket has blown.

Diesel Engine

Checking cylinder compression on diesel engines is basically the same procedure as on a gas engine except for the following:

1. A special compression gauge adaptor suitable for diesel engines (because these engines have much greater compression pressures) must be used.

2. Remove the injector tubes and remove the injectors from each cylinder.

WARNING: *Don't forget to remove the washer underneath each injector. Otherwise, it may get lost when the engine is cranked.*

3. When fitting the compression gauge adaptor to the cylinder head, make sure the bleeder of the gauge (if equipped) is closed.

4. When reinstalling the injector assemblies, install new washers underneath each injector.

Engine

REMOVAL AND INSTALLATION

CAUTION: *Please refer to Chapter 1 before discharging the compressor or disconnecting air conditioning lines. Damage to the air conditioning system or personal injury could result.*

Inline 6-Cylinder Gasoline Engine

1. Scribe alignment marks around the hood hinges and remove the hood.
2. Disconnect the negative battery cable. Remove the air cleaner.
3. Drain the cooling system.

CAUTION: *When draining the coolant, keep in mind that cats and dogs are attracted by the ethylene glycol antifreeze, and are quite likely to drink any that is left in an uncovered container or in puddles on the ground. This will prove fatal in sufficient quantity. Always drain the coolant into a sealable container. Coolant should be reused unless it is contaminated or several years old.*

4. Disconnect the radiator and the heater hoses, then remove the radiator and the fan shroud.

NOTE: *If equipped with an automatic transmission, disconnect and plug the oil cooler lines at the radiator.*

5. Disconnect and label the wires at the ignition coil, the starter, the alternator, the temperature switch and the oil pressure switch.
6. Disconnect accelerator control cable at the inlet manifold, the fuel inlet line from the fuel pump, the hoses from the vapor canister

General Engine Specifications

Year	Engine No. Cyl. Displacement (cu. in.)	Type Carburetor	Horsepower @ rpm■	Torque @ rpm (ft. lbs.)■	Bore and Stroke (in.)	Compression Ratio	Oil Pressure @ 2000 rpm (psi)
1968	6-230	1-bbl	145 @ 4400	220 @ 1600	3.875 × 3.250	8.5:1	58
	6-250	1-bbl	155 @ 4200	235 @ 1600	3.875 × 3.530	8.5:1	58
	8-307	2-bbl	200 @ 4600	300 @ 2400	3.875 × 3.250	9.0:1	58
	8-327	2-bbl	210 @ 4600	320 @ 2400	4.001 × 3.250	8.75:1	58
	8-327	4-bbl	275 @ 4800	355 @ 3200	4.001 × 3.250	10.0:1	58
	8-396	4-bbl	325 @ 4800	410 @ 3200	4.094 × 3.760	10.25:1	62
	8-396	4-bbl	350 @ 5200	415 @ 3400	4.094 × 3.750	10.25:1	62
	8-396	4-bbl	375 @ 5600	415 @ 3600	4.094 × 3.760	11.0:1	62
1969	6-230	1-bbl	140 @ 4400	220 @ 1600	3.875 × 3.250	8.5:1	58
	6-250	1-bbl	155 @ 4200	235 @ 1600	3.875 × 3.530	8.5:1	58
	8-307	2-bbl	200 @ 4600	300 @ 2400	3.875 × 3.250	9.0:1	58
	8-350	2-bbl	250 @ 4800	345 @ 2800	4.000 × 3.480	9.0:1	62
	8-350	4-bbl	300 @ 4800	380 @ 3200	4.000 × 3.480	10.25:1	62
	8-396	4-bbl	325 @ 4800	410 @ 3200	4.094 × 3.760	10.25:1	62
	8-396	4-bbl	350 @ 5200	415 @ 3400	4.094 × 3.760	10.25:1	62
	8-396	4-bbl	375 @ 5200	415 @ 3600	4.094 × 3.760	11.0:1	62
1970	6-230	1-bbl	140 @ 4400	220 @ 1600	3.875 × 3.250	8.5:1	40
	6-250	1-bbl	155 @ 4200	235 @ 1600	3.875 × 3.530	8.5:1	40
	8-307	2-bbl	200 @ 4600	300 @ 2400	3.875 × 3.250	9.0:1	40
	8-350	2-bbl	250 @ 4800	345 @ 2800	4.000 × 3.480	9.0:1	40
	8-350	4-bbl	300 @ 4800	380 @ 3200	4.000 × 3.480	10.25:1	40
	8-400	2-bbl	265 @ 4400	400 @ 2400	4.125 × 3.760	9.0:1	40
	8-402	4-bbl	330 @ 4800	410 @ 3200	4.126 × 3.760	10.25:1	40
	8-402	4-bbl	350 @ 5200	415 @ 3400	4.126 × 3.760	10.25:1	40
	8-454	4-bbl	360 @ 4400	500 @ 3200	4.251 × 4.000	10.25:1	40
1971	6-250	1-bbl	145 @ 4200	230 @ 1600	3.875 × 3.250	8.5:1	40
	8-307	2-bbl	200 @ 4600	300 @ 2400	3.875 × 3.250	8.5:1	40
	8-350	2-bbl	245 @ 4800	350 @ 2800	4.000 × 3.480	8.5:1	40
	8-350	4-bbl	270 @ 4800	360 @ 3200	4.000 × 3.480	8.5:1	40
	8-350	4-bbl	330 @ 5000	275 @ 5600	4.000 × 3.480	9.0:1	40
	8-402	4-bbl	300 @ 4800	400 @ 3200	4.126 × 3.760	8.5:1	40
	8-454	4-bbl	365 @ 4800	465 @ 3200	4.251 × 4.000	8.5:1	40
	8-454	4-bbl	425 @ 5600	475 @ 4000	4.251 × 4.000	9.0:1	40
1972	6-250	1-bbl	110 @ 3800	185 @ 1600	3.875 × 3.530	8.5:1	40
	8-307	2-bbl	130 @ 4000	230 @ 2400	3.875 × 3.250	8.5:1	40
	8-350	2-bbl	165 @ 4000	280 @ 2400	4.000 × 3.480	8.5:1	40
	8-350	4-bbl	200 @ 4400	300 @ 2800	4.000 × 3.480	8.5:1	40
	8-350	4-bbl	225 @ 5600	280 @ 4000	4.000 × 3.480	9.0:1	40
	8-402	4-bbl	240 @ 4400	345 @ 3200	4.126 × 3.760	8.5:1	40
	8-454	4-bbl	270 @ 4000	390 @ 3200	4.251 × 4.000	8.5:1	40
1973	6-250	1-bbl	100 @ 3800	175 @ 1600	3.875 × 3.530	8.25:1	40
	8-307	2-bbl	115 @ 4000	205 @ 2000	3.875 × 3.250	8.5:1	40
	8-350	2-bbl	145 @ 4000	255 @ 2400	4.000 × 3.480	8.5:1	40
	8-350	4-bbl	175 @ 4400	270 @ 2400	4.000 × 3.480	8.5:1	40
	8-350	4-bbl	245 @ 5200	280 @ 4000	4.000 × 3.480	9.0:1	40
	8-454	4-bbl	245 @ 4000	375 @ 2800	4.251 × 4.000	8.5:1	40

ENGINE AND ENGINE OVERHAUL

General Engine Specifications

Year	Engine No. Cyl. Displacement (cu. in.)	Type Carburetor	Horsepower @ rpm■	Torque @ rpm (ft. lbs.)■	Bore and Stroke (in.)	Compression Ratio	Oil Pressure @ 2000 rpm (psi)
1974	6-250	1-bbl	100 @ 3600	175 @ 1800	3.875 × 3.530	8.25:1	40
	8-350②	2-bbl	145 @ 3600	250 @ 2200	4.000 × 3.480	8.5:1	40
	8-350③	4-bbl	160 @ 3800	245 @ 2400	4.000 × 3.480	8.5:1	40
	8-350	4-bbl	185 @ 4000	270 @ 2600	4.000 × 3.480	8.5:1	40
	8-400②	2-bbl	150 @ 3200	295 @ 2600	4.126 × 3.750	8.5:1	40
	8-400③	4-bbl	180 @ 3800	290 @ 2400	4.126 × 3.750	8.5:1	40
	8-454	4-bbl	235 @ 4000	360 @ 2800	4.251 × 4.000	8.25:1	40
1975	6-250	1-bbl	105 @ 3800	185 @ 1200	3.875 × 3.530	8.25:1	40
	8-350②	2-bbl	145 @ 3800	250 @ 2200	4.000 × 3.480	8.5:1	40
	8-350③	4-bbl	155 @ 3800	250 @ 2400	4.000 × 3.480	8.5:1	40
	8-400	4-bbl	175 @ 3600	305 @ 2000	4.126 × 4.000	8.5:1	40
	8-454②	4-bbl	215 @ 4000	350 @ 2400	4.251 × 4.000	8.15:1	40
1976–77	6-250	1-bbl	105 @ 3800	185 @ 1200	3.875 × 3.530	8.25:1	40
	8-305	2-bbl	140 @ 3800	245 @ 2000	3.736 × 3.480	8.5:1	40
	8-350	2-bbl	145 @ 3800	250 @ 2200	4.000 × 3.480	8.5:1	40
	8-350	4-bbl	165 @ 3800	260 @ 2400	4.000 × 3.480	8.5:1	40
	8-400	4-bbl	175 @ 3600	305 @ 2000	4.126 × 4.000	8.5:1	40
1978–79	6-250	1-bbl	110 @ 3800	190 @ 1600	3.875 × 3.530	8.1:1	40
	8-305	4-bbl	155 @ 3800	260 @ 2800	3.736 × 3.480	8.4:1	45
	8-305	2-bbl	145 @ 3800	245 @ 2400	3.736 × 3.480	8.5:1	40
	8-267	2-bbl	125 @ 3800	215 @ 2400	3.500 × 3.480	8.2:1	45
	8-350	4-bbl	170 @ 3800	270 @ 2400	4.000 × 3.480	8.5:1	40
1980	6-231	2-bbl	110 @ 3800	190 @ 1600	3.800 × 3.400	8.0:1	45
	8-267	2-bbl	120 @ 3600	215 @ 2000	3.500 × 3.480	8.3:1	45
	8-305	4-bbl	155 @ 4000	240 @ 1600	3.736 × 3.480	8.6:1	45
	8-305 (Calif.)	4-bbl	155 @ 4000	230 @ 2400	3.736 × 3.480	8.6:1	45
	8-350	4-bbl	170 @ 3800	270 @ 2400	4.000 × 3.480	8.5:1	40
	8-350	Diesel	105 @ 3200	200 @ 1600	4.057 × 3.385	22.5:1	45①
1981	6-229	2-bbl	110 @ 4200	170 @ 2000	3.736 × 3.480	8.6:1	45
	6-231	2-bbl	110 @ 3800	190 @ 1600	3.800 × 3.400	8.0:1	45
	8-267	2-bbl	115 @ 4000	200 @ 2400	3.500 × 3.480	8.3:1	45
	8-305	4-bbl	150 @ 3800	240 @ 2400	3.736 × 3.480	8.6:1	45
	8-350	4-bbl	170 @ 3800	270 @ 2400	4.000 × 3.480	8.5:1	40
	8-350	Diesel	105 @ 3200	200 @ 1600	4.057 × 3.385	22.5:1	45①
1982–84	6-229	2-bbl	110 @ 4200	170 @ 2000	3.736 × 3.480	8.6:1	45
	6-231	2-bbl	110 @ 3800	190 @ 1600	3.800 × 3.400	8.0:1	45
	8-267	2-bbl	115 @ 4000	205 @ 2400	3.500 × 3.480	8.3:1	45
	8-305	4-bbl	145 @ 4000	240 @ 1600	3.736 × 3.480	8.6:1	45
	8-350	Diesel	105 @ 3200	200 @ 1600	4.057 × 3.385	22.5:1	45①
1985	6-231	2-bbl	110 @ 3800	190 @ 1600	3.800 × 3.400	8.0:1	45
	6-262	TBI	130 @ 3600	218 @ 2000	4.000 × 3.480	9.3:1	45
	8-305	4-bbl	134 @ 4800	319 @ 3200	3.736 × 3.480	9.5:1	45
	8-305	4-bbl	150 @ 3800	240 @ 2400	3.736 × 3.480	8.6:1	45
	8-350	4-bbl	170 @ 3800	270 @ 2400	4.000 × 3.480	8.5:1	40
	8-350	Diesel	105 @ 3200	200 @ 1600	4.057 × 3.385	22.5:1	45①

ENGINE AND ENGINE OVERHAUL

General Engine Specifications

Year	Engine No. Cyl. Displacement (cu. in.)	Type Carburetor	Horsepower @ rpm∎	Torque @ rpm (ft. lbs.)∎	Bore and Stroke (in.)	Compression Ratio	Oil Pressure @ 2000 rpm (psi)
1986–88	6-262	TBI	130 @ 3600	218 @ 2000	4.000 × 3.480	9.3:1	45
	8-305	4-bbl	134 @ 4800	319 @ 3200	3.736 × 3.480	9.5:1	45
	8-305	4-bbl	150 @ 3800	240 @ 2400	3.736 × 3.480	8.6:1	45
	8-307	4-bbl	148 @ 3800	250 @ 2400	3.800 × 3.385	8.0:1	40①
	8-350	4-bbl	170 @ 3800	270 @ 2400	4.000 × 3.480	8.5:1	40
1989–90	6-262	TBI	140 @ 4000	225 @ 2000	4.000 × 3.480	9.3:1	18
	8-305	TBI	170 @ 4400	255 @ 2400	3.740 × 3.480	9.3:1	18
	8-307	4-bbl	140 @ 3200	255 @ 2000	3.800 × 3.385	8.0:1	30①
	8-350	TBI	195 @ 4200	295 @ 2400	4.000 × 3.480	9.3:1	18
1991–92	8-305	TBI	170 @ 4200	255 @ 2400	3.740 × 3.480	9.3:1	18
	8-350	TBI	195 @ 4200	295 @ 2400④	4.000 × 3.480	9.8:1	18

TBI—Throttle Body Injection
∎Starting 1972, horsepower and torque are SAE net figures. They are measured at the rear of the transmission with all accessories installed and operating. Since the figures vary when a given engine is installed in different models, some are representative rather than exact.
① Oil pressure at 1500 rpm
② Not available—Calif.
③ Calif. only
④ 1992 Police, 205 hp, 300 ft. lbs.

and (if equipped) the power brake vacuum line from the intake manifold.

7. If equipped with power steering, remove the power steering pump and move it aside.

8. Raise and support the vehicle safely. Drain the engine oil

CAUTION: *The EPA warns that prolonged contact with used engine oil may cause a number of skin disorders, including cancer! You should make every effort to minimize your exposure to used engine oil. Protective gloves should be worn when changing the oil. Wash your hands and any other exposed skin areas as soon as possible after exposure to used engine oil. Soap and water, or waterless hand cleaner should be used.*

NOTE: *A Chilton environmental tip. Used oil contains heavy metals, and has been determined to be hazardous to the environment. Recycling oil is the best way for disposal. Check local laws and recycle whenever possible.*

9. Disconnect the exhaust pipe from the exhaust manifold and (if equipped) the converter bracket from the rear transmission mount.

10. Remove the starter and the flywheel splash shield or the converter housing cover.

11. If equipped with an automatic transmission, remove the converter-to-flexplate bolts.

12. Attach a vertical hoist to the engine and support the transmission with a floor jack, then raise the engine slightly.

13. Remove the engine mount through bolts and the bell housing-to-engine bolts.

14. Carefully remove the engine from the vehicle.

To Install:

15. Attach the engine hoist and lower the engine into position.

NOTE: *Make sure the torque converter bolts are aligned with the mounting holes on the flexplate, before attaching bell housing to engine.*

16. Support the transmission with a floor jack and install the bell housing-to-engine bolts and the engine mount through bolts.

NOTE: *Make sure the torque converter bolts are aligned with the mounting holes on the flexplate, before attaching bell housing to engine.*

17. If equipped with an automatic transmission, install the converter-to-flexplate bolts.

18. Install the starter and the flywheel splash shield or the converter housing cover.

19. Connect the exhaust pipe to the exhaust manifold and (if equipped) the converter bracket to the rear transmission.

20. Lower the vehicle.

21. If equipped with power steering, install the power steering pump.

ENGINE AND ENGINE OVERHAUL

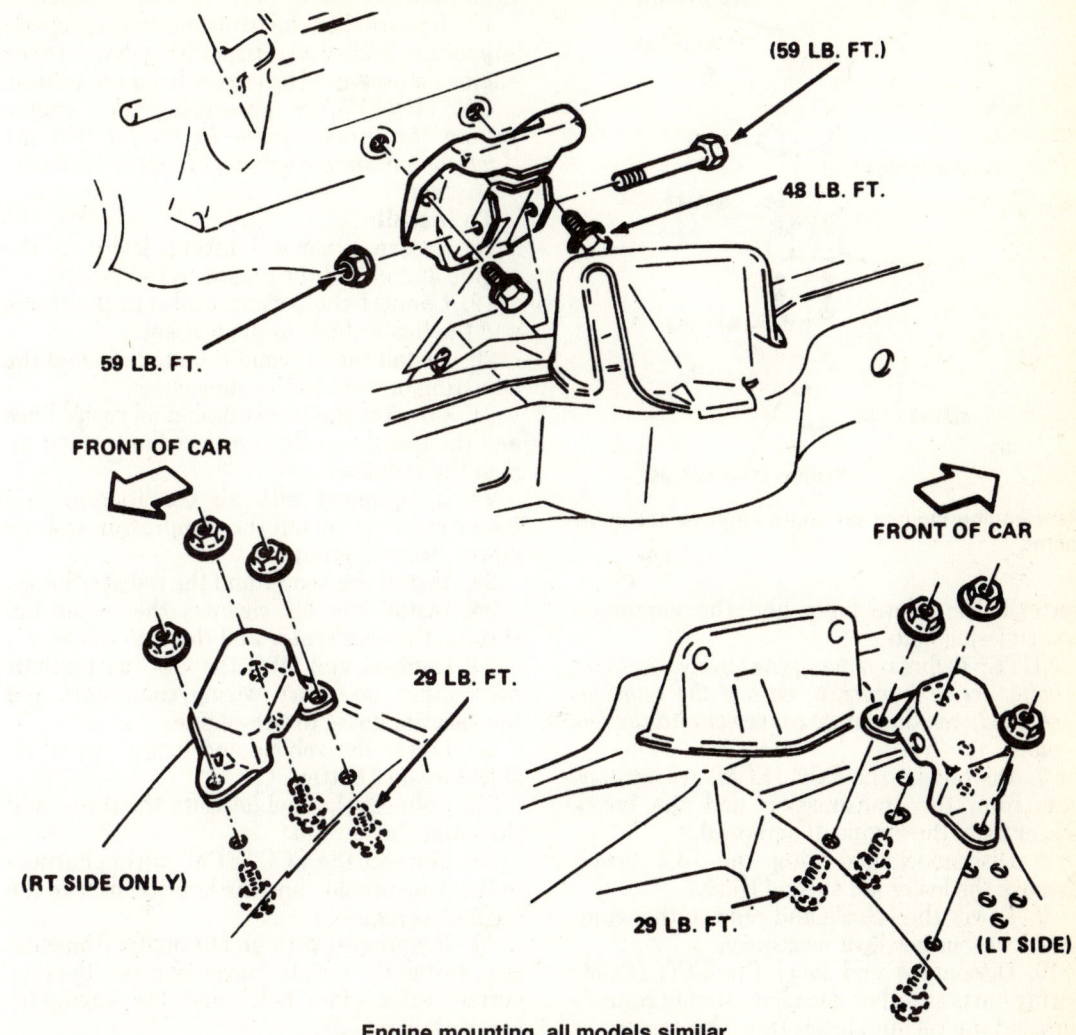

Engine mounting, all models similar

22. Connect accelerator control cable at the inlet manifold, the fuel inlet line to the fuel pump, the hoses to the vapor canister and (if equipped) the power brake vacuum line to the intake manifold.

23. Connect and labeled wires at the ignition coil, the starter, the alternator, the temperature switch and the oil pressure switch.

NOTE: *If equipped with an automatic transmission, connect the oil cooler lines at the radiator.*

24. Install the radiator and fan shroud, then connect the heater and radiator hoses.

25. Fill the cooling system and the crankcase.

26. Connect the negative battery cable and remove the air cleaner.

27. Install the hood.

WARNING: *Please refer to Chapter 1 before charging the compressor or damage to the air conditioning system or personal injury could result.*

V6 Gasoline Engine

1. Scribe alignment marks at the hood hinges and remove the hood.
2. Disconnect the negative battery cable.
3. Disconnect the exhaust pipe from the exhaust manifold.
4. Remove the bell housing cover and drain the transmission oil cooler lines at the oil pan.
5. Remove the left engine mount through bolt and loosen the right engine mount through bolt.
6. If equipped with an automatic transmission, remove the torque converter cover, the con-

122 ENGINE AND ENGINE OVERHAUL

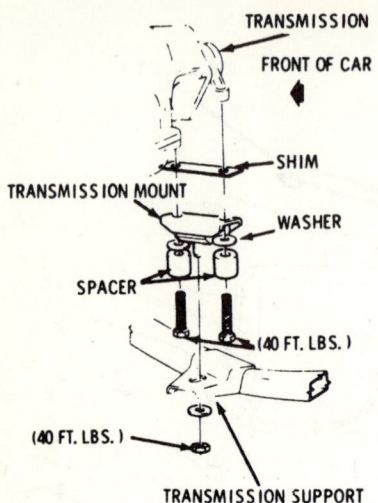

Rear engine mounts on some engines are shimmered

verter-to-flex plate bolts and the engine-to-transmission bolts.

NOTE: *Before removing the torque converter bolts, scribe a mark to ensure the relationship between the torque converter and the flex plate.*

7. Disconnect the CCC (ECM) wiring harness from the transmission and the knock sensor from the engine, if equipped.
8. Disconnect and plug the fuel hoses. Remove the lower fan shroud bolts.
9. Lower the vehicle and remove the windshield washer bottle, if necessary.
10. Disconnect and label the CCC (ECM) wiring harness, other necessary wiring connectors and the vacuum hoses from the engine.
11. Remove the air cleaner, the upper fan shroud, the accelerator and the T.V. cables.
12. Drain the cooling system, then remove the heater and the radiator hoses.

CAUTION: *When draining the coolant, keep in mind that cats and dogs are attracted by the ethylene glycol antifreeze, and are quite likely to drink any that is left in an uncovered container or in puddles on the ground. This will prove fatal in sufficient quantity. Always drain the coolant into a sealable container. Coolant should be reused unless it is contaminated or several years old.*

13. If equipped with air conditioning and power steering, remove the compressor and the power steering pump, then move them aside.
14. Disconnect the transmission oil cooler lines and the overflow tube from the radiator, then remove the radiator.
15. Remove the air conditioning hose and the adjusting bracket from the alternator.
16. Disconnect the battery cables from the frame and the heater hose from the bracket.
17. Be sure that the transmission is properly supported. Secure a vertical lifting device to the engine and remove the engine from the vehicle.

CAUTION: *When separating the engine from the transmission, be careful that the torque converter does not pull out of the transmission.*

To install:
18. Secure a vertical lifting device to the engine and install the engine in the vehicle.
19. Connect the battery cables to the frame and the heater hose to the bracket.
20. Install the air conditioning hose and the adjusting bracket to the alternator.
21. Connect the transmission oil cooler lines and the overflow tube to the radiator, then install the radiator.
22. If equipped with air conditioning and power steering, install the compressor and the power steering pump.
23. Install the heater and the radiator hoses.
24. Install the air cleaner, the upper fan shroud, the accelerator and the T.V. cables.
25. Connect and label the CCC wiring harness, other necessary wiring connectors and the vacuum hoses to the engine.
26. Lower the vehicle and install the windshield washer bottle.
27. Connect the fuel hoses to the frame and the lower fan shroud.
28. Connect the CCC (ECM) wiring harness to the transmission and the knock sensor to the engine, as required.
29. If equipped with an automatic transmission, install the torque converter cover, the converter-to-flex plate bolts and the engine-to-transmission bolts.
30. Install the left engine mount through bolt and tighten the right engine mount through bolt.
31. Install the bell housing cover and drain the transmission oil cooler lines at the oil pan.
32. Connect the exhaust pipe to the exhaust manifold.
33. Connect the negative battery cable.
34. Refill the radiator with the proper coolant mixture.
35. Install the hood.

V8 Gasoline Engine

1. Scribe alignment marks on hood and remove hood from hinges.
2. Disconnect the negative battery cable.
3. Drain cooling system, then remove the heater hoses and the radiator hoses from the engine.

CAUTION: *When draining the coolant, keep in mind that cats and dogs are attracted by the ethylene glycol antifreeze, and are quite*

Valve Specifications

Year	Engine No. Cyl. Displacement (cu. in.)	Seat Angle (deg.)	Face Angle (deg.)	Spring Test Pressure (lbs. @ in.)	Spring Installed Height (in.)	Stem to Guide Clearance (in.) Intake	Stem to Guide Clearance (in.) Exhaust	Stem Diameter (in.) Intake	Stem Diameter (in.) Exhaust
1968	6-230	46①	45	59 @ 1.66	1²¹/₃₂	0.0010–0.0037	0.0015–0.0052	0.3414	0.3414
	6-250	46①	45	59 @ 1.66	1²¹/₃₂	0.0010–0.0037	0.0015–0.0052	0.3414	0.3414
	8-307	46①	45	80 @ 1.70	1⁵/₃₂	0.0010–0.0037	0.0010–0.0047	0.3414	0.3414
	8-327	46①	45	80 @ 1.70	1⁵/₃₂	0.0010–0.0037	0.0010–0.0047	0.3414	0.3414
	8-350	46①	45	80 @ 1.70	1⁵/₃₂	0.0010–0.0037	0.0010–0.0047	0.3414	0.3414
	8-396	46①	45	90 @ 1.88	1⁷/₈	0.0010–0.0035	0.0012–0.0047	0.3719	0.3717
1969	6-230	46①	45	59 @ 1.66	1²¹/₃₂	0.0010–0.0037	0.0015–0.0052	0.3414	0.3414
	6-250	46①	45	59 @ 1.66	1²¹/₃₂	0.0010–0.0037	0.0015–0.0052	0.3414	0.3414
	8-307	46①	45	80 @ 1.70	1²³/₃₂	0.0010–0.0037	0.0012–0.0049	0.3414	0.3414
	8-350	46①	45	80 @ 1.70	1⁵/₃₂	0.0010–0.0037	0.0010–0.0047	0.3414	0.3414
	8-396	46①	45	90 @ 1.88	1⁷/₈	0.0010–0.0035	0.0012–0.0047	0.3719	0.3719
	8-396②	46①	45	100 @ 1.88	1⁷/₈	0.0010–0.0035	0.0012–0.0047	0.3719	0.3719
1970	6-230	46①	45	59 @ 1.66	1²¹/₃₂	0.0010–0.0037	0.0015–0.0052	0.3414	0.3414
	6-250	46①	45	59 @ 1.66	1²¹/₃₂	0.0010–0.0037	0.0015–0.0052	0.3414	0.3414
	8-307	46①	45	80 @ 1.70	1²³/₃₂	0.0010–0.0037	0.0012–0.0049	0.3414	0.3414
	8-350	46①	45	80 @ 1.70	1²³/₃₂	0.0010–0.0037	0.0012–0.0049	0.3414	0.3414
	8-400	46①	45	80 @ 1.70	1⁷/₈	0.0010–0.0035	0.0012–0.0047	0.3414	0.3414
	8-402	46①	45	75 @ 1.88③	1⁷/₈	0.0010–0.0035	0.0012–0.0047	0.3719	0.3717
	8-454	46①	45	75 @ 1.88③	1⁷/₈	0.0010–0.0035	0.0012–0.0047	0.3717	0.3719
1971	6-250	46	45	60 @ 1.66	1²¹/₃₂	0.0010–0.0037	0.0015–0.0052	0.3414	0.3414
	8-307	46	45	80 @ 1.70	1²³/₃₂	0.0010–0.0037	0.0012–0.0049	0.3414	0.3414
	8-350	46	45	80 @ 1.70	1²³/₃₂	0.0010–0.0037	0.0012–0.0049	0.3414	0.3414
	8-402	46	45	75 @ 1.88③	1⁷/₈	0.0010–0.0037	0.0012–0.0047	0.3719	0.3717
	8-454	46	45	75 @ 1.88③	1⁷/₈	0.0010–0.0037	0.0012–0.0047	0.3719	0.3717
1972	6-250	46	45	60 @ 1.66	1²¹/₃₂	0.0010–0.0037	0.0015–0.0052	0.3414	0.3414
	8-307	46	45	80 @ 1.70	1²³/₃₂	0.0010–0.0037	0.0012–0.0049	0.3414	0.3414
	8-350	46	45	80 @ 1.70	1²³/₃₂	0.0010–0.0037	0.0012–0.0049	0.3414	0.3414
	8-402	46	45	75 @ 1.88③	1⁷/₈	0.0010–0.0037	0.0012–0.0047	0.3719	0.3717
	8-454	46	45	75 @ 1.88③	1⁷/₈	0.0010–0.0037	0.0012–0.0047	0.3719	0.3717
1973	6-250	46	45	60 @ 1.66	1²¹/₃₂	0.0010–0.0027	0.0015–0.0032	0.3414	0.3414
	8-307	46	45	80 @ 1.61	1⁵/₈	0.0010–0.0037	0.0012–0.0047	0.3414	0.3414
	8-350	46	45	80 @ 1.70	1²³/₃₂	0.0012–0.0027	0.0012–0.0029	0.3414	0.3414
	8-454	46	45	80 @ 1.88	1⁷/₈	0.0010–0.0027	0.0012–0.0027	0.3719	0.3717
1974	6-250	46	45	60 @ 1.66	1²¹/₃₂	0.0010–0.0027	0.0012–0.0027	0.3414	0.3414
	8-350	46	45	80 @ 1.70	1²³/₃₂	0.0010–0.0027	0.0010–0.0027	0.3414	0.3414
	8-400	46	45	80 @ 1.70	1²³/₃₂	0.0010–0.0027	0.0010–0.0027	0.3414	0.3414
	8-454	46	45	80 @ 1.88	1⁷/₈	0.0010–0.0027	0.0010–0.0027	0.3719	0.3719
1975–77	6-250	46	45	60 @ 1.66	1²¹/₃₂	0.0010–0.0027	0.0010–0.0027④	0.3414	0.3414
	8-305	46	45	80 @ 1.70	1²³/₃₂	0.0010–0.0027	0.0010–0.0027	0.3414	0.3414
	8-350	46	45	80 @ 1.70⑤	1²³/₃₂	0.0010–0.0027	0.0010–0.0027	0.3414	0.3414
	8-400	46	45	80 @ 1.70⑤	1²³/₃₂	0.0010–0.0027	0.0010–0.0027	0.3414	0.3414
	8-454	46	45	90 @ 1.88	1⁷/₈	0.0010–0.0027	0.0010–0.0027	0.3719	0.3717
1978–79	6-250	46	45	175 @ 1.26	1²¹/₃₂	0.0010–0.0027	0.0015–0.0032	0.3414	0.3414
	8-267	46	45	200 @ 1.25	1²³/₃₂	0.0010–0.0027	0.0010–0.0027	0.3414	0.3414
	8-305	46	45	200 @ 1.25	1²³/₃₂⑥	0.0010–0.0027	0.0010–0.0027	0.3414	0.3414
	8-350	46	45	200 @ 1.25	1²³/₃₂⑥	0.0010–0.0027	0.0010–0.0027	0.3414	0.3414

ENGINE AND ENGINE OVERHAUL

Valve Specifications

Year	Engine No. Cyl. Displacement (cu. in.)	Seat Angle (deg.)	Face Angle (deg.)	Spring Test Pressure (lbs. @ in.)	Spring Installed Height (in.)	Stem to Guide Clearance (in.)		Stem Diameter (in.)	
						Intake	Exhaust	Intake	Exhaust
1980–81	6-231	45	45	168 @ 1.33	1 47/64	0.0015–0.0032	0.0015–0.0032	0.3407	0.3409
	8-267	46	45	200 @ 1.25	1 23/32	0.0010–0.0027	0.0010–0.0027	0.3414	0.3414
	8-305	46	45	200 @ 1.25	1 23/32	0.0010–0.0027	0.0010–0.0027	0.3414	0.3414
	8-350	46	45	200 @ 1.25	1 23/32 ⑧	0.0010–0.0027	0.0010–0.0027	0.3414	0.3414
	8-350 ⑦	45 ⑨	44 ⑩	205 @ 1.30	1.67	0.0010–0.0027	0.0015–0.0032	0.3429	0.3424
1982–85	6-231	45	45	168 @ 1.33	1 47/64	0.0015–0.0032	0.0015–0.0032	0.3407	0.3409
	6-262	45	46	189 @ 1.30	1 43/64	0.0010–0.0027	0.0015–0.0032	0.3429	0.3423
	8-267	46	45	200 @ 1.25	1 23/32	0.0010–0.0027	0.0010–0.0027	0.3414	0.3414
	8-305	46	45	200 @ 1.25	1 23/32	0.0010–0.0027	0.0010–0.0027	0.3414	0.3414
	8-350	45	46	189 @ 1.30	1 43/64	0.0010–0.0027	0.0015–0.0032	0.3429	0.3423
	8-350 ⑦	45 ⑨	44 ⑩	205 @ 1.30	1.67	0.0010–0.0027	0.0015–0.0032	0.3429	0.3424
1986–88	6-262	46	45	200 @ 1.25	1 23/32	0.0010–0.0027	0.0010–0.0027	0.3414	0.3414
	8-305	46	45	200 @ 1.25	1 23/32	0.0010–0.0027	0.0010–0.0027	0.3414	0.3414
	8-307 ⑧	45	44	187 @ 1.27	1.67	0.0010–0.0027	0.0015–0.0032	0.3429	0.3429
	8-350	45	46	189 @ 1.30	1 43/64	0.0010–0.0027	0.0015–0.0032	0.3429	0.3423
1989–90	6-262	46	45	200 @ 1.25	1 23/32	0.0011–0.0027	0.0011–0.0027	0.3414	0.3414
	8-305	46	44	190 @ 1.27	1 23/32	0.0011–0.0027	0.0011–0.0027	0.3429	0.3425
	8-350	45	44	190 @ 1.27	1 23/32	0.0011–0.0027	0.0011–0.0027	0.3429	0.3425
1991–92	8-305	46	45	200 @ 1.25	1 23/32	0.0011–0.0027	0.0011–0.0027	0.3414	0.3414
	8-350	46	45	200 @ 1.25	1 23/32	0.0011–0.0027	0.0011–0.0027	0.3414	0.3414

① 45° on aluminum heads
② 350 hp
③ Inner spring—30 @ 1.78
④ 1976 and later: 0.0015–0.0032
⑤ 80 @ 1.61 for exhaust
⑥ Exhaust valve—1 19/32
⑦ Diesel engine
⑧ Oldsmobile produced engine
⑨ Exhaust 31°
⑩ Exhaust 30°

likely to drink any that is left in an uncovered container or in puddles on the ground. This will prove fatal in sufficient quantity. Always drain the coolant into a sealable container. Coolant should be reused unless it is contaminated or several years old.

4. Remove the upper fan shroud and the fan assembly.
5. If equipped with air conditioning and power steering, remove the compressor and the power steering pump, then move them aside.
6. Disconnect the accelerator and the T.V. cables.
7. Remove the transmission oil cooler lines, if equipped from the radiator and remove the radiator.
8. Disconnect and label the vacuum hoses and the CCC (ECM) wiring harness connectors from the engine, as required.
9. Remove the AIR pipe from the converter, as required.
10. Remove the windshield washer bottle, if necessary.
11. Disconnect and mark the wiring harness at the bulkhead and related engine wiring.
12. Remove the distributor cap and the cruise control cable, if equipped.
13. Disconnect the positive battery cable from the battery and the frame. Disconnect the negative battery cable from the air conditioning hose/alternator bracket.
14. Raise and support the vehicle safely.
15. Remove the crossover pipe and the catalytic converter as an assembly, if equipped.
16. If equipped with an automatic transmis-

ENGINE AND ENGINE OVERHAUL

Camshaft Specifications

All measurements given in inches.

Year	Engine ID/VIN	Engine cu. in. (liter)	Journal Diameter 1	2	3	4	5	Elevation In.	Ex.	Bearing Clearance	Camshaft End Play
1968–70	—	L6 230①	1.8682–1.8692	1.8682–1.8692	1.8682–1.8692	1.8682–1.8692	1.8682–1.8692	0.1896	0.1896	—	0.003–0.008
	—	L6 250①	1.8682–1.8692	1.8682–1.8692	1.8682–1.8692	1.8682–1.8692	1.8682–1.8692	0.2217	0.2217	—	0.003–0.008
	—	V8 307①	1.8682–1.8692	1.8682–1.8692	1.8682–1.8692	1.8682–1.8692	1.8682–1.8692	0.2600	0.2733	—	—
	—	V8 327①	1.8682–1.8692	1.8682–1.8692	1.8682–1.8692	1.8682–1.8692	1.8682–1.8692	0.3234	0.3234	—	—
	—	V8 350①	1.8682–1.8692	1.8682–1.8692	1.8682–1.8692	1.8682–1.8692	1.8682–1.8692	0.2600	0.2733	—	—
	—	V8 396①	1.9487–1.9497	1.9487–1.9497	1.9487–1.9497	1.9487–1.9497	1.9487–1.9497	②	②	—	—
	—	V8 400①	1.9487–1.9497	1.9487–1.9497	1.9487–1.9497	1.9487–1.9497	1.9487–1.9497	0.2365	0.2411	—	—
	—	V8 402①	1.9487–1.9497	1.9487–1.9497	1.9487–1.9497	1.9487–1.9497	1.9487–1.9497	0.2343	0.2343	—	—
	—	V8 454①	1.9487–1.9497	1.9487–1.9497	1.9487–1.9497	1.9487–1.9497	1.9487–1.9497	0.2434 ③	③ 0.2529	—	—
1971–72	—	L6 250①	1.8682–1.8692	1.8682–1.8692	1.8682–1.8692	1.8682–1.8692	1.8682–1.8692	0.2217	0.2217	—	0.001–0.005
	—	V8 307①	1.8682–1.8692	1.8682–1.8692	1.8682–1.8692	1.8682–1.8692	1.8682–1.8692	0.2600	0.2733	—	0.004–0.012
	—	V8 350①	1.8682–1.8692	1.8682–1.8692	1.8682–1.8692	1.8682–1.8692	1.8682–1.8692	0.2600 ④	④ 0.2733	—	0.004–0.012
	—	V8 400①	1.9482–1.9492	1.9482–1.9492	1.9482–1.9492	1.9482–1.9492	1.9482–1.9492	0.2235	0.2411	—	—
	—	V8 402①	1.9482–1.9492	1.9482–1.9492	1.9482–1.9492	1.9482–1.9492	1.9482–1.9492	0.2343	0.2343	—	—
	—	V8 454①	1.9482–1.9492	1.9482–1.9492	1.9482–1.9492	1.9482–1.9492	1.9482–1.9492	0.2714	0.2824	—	—
1973–75	D	L6 250①	1.8677–1.8697	1.8677–1.8697	1.8677–1.8697	1.8677–1.8697	1.8677–1.8697	0.2217	0.2217	—	0.001–0.005
	—	V8 307①	1.8682–1.8692	1.8682–1.8692	1.8682–1.8692	1.8682–1.8692	1.8682–1.8692	0.2600	0.2733	—	0.003–0.008
	Q	V8 350①	1.8682–1.8692	1.8682–1.8692	1.8682–1.8692	1.8682–1.8692	1.8682–1.8692	0.2600 ⑤	⑤ 0.2733	—	0.003–0.008
	U	V8 400①	1.9482–1.9492	1.9482–1.9492	1.9482–1.9492	1.9482–1.9492	1.9482–1.9492	0.2235	0.2411	—	—
	S	V8 454①	1.9482–1.9492	1.9482–1.9492	1.9482–1.9492	1.9482–1.9492	1.9482–1.9492	0.2590	0.2590	—	—
1976–79	D	L6 250①	1.8677–1.8697	1.8677–1.8697	1.8677–1.8697	1.8677–1.8697	1.8677–1.8697	0.2217	0.2315	—	0.003–0.008
	J	V8 267 (4.4)	1.8682–1.8692	1.8682–1.8692	1.8682–1.8692	1.8682–1.8692	1.8682–1.8692	0.2485	0.2600	—	0.004–0.012
	Q	V8 305 (5.0)	1.8682–1.8692	1.8682–1.8692	1.8682–1.8692	1.8682–1.8692	1.8682–1.8692	0.2485	0.2733	—	0.004–0.012
	L	V8 350 (5.7)	1.8682–1.8692	1.8682–1.8692	1.8682–1.8692	1.8682–1.8692	1.8682–1.8692	0.2600	0.2733	—	0.004–0.012
	U	V8 400 (6.6)	1.9482–1.9492	1.9482–1.9492	1.9482–1.9492	1.9482–1.9492	1.9482–1.9492	0.2600	0.2733	—	—
1980–84	A	V6 231 (3.8)	1.8682–1.8692	1.8682–1.8692	1.8682–1.8692	1.8682–1.8692	1.8682–1.8692	0.2340	0.2570	—	0.004–0.012
	9	V6 229 (3.8)	1.8682–1.8692	1.8682–1.8692	1.8682–1.8692	1.8682–1.8692	⑥	0.2340	0.2570	—	0.004–0.012

ENGINE AND ENGINE OVERHAUL

Camshaft Specifications
All measurements given in inches.

Year	Engine ID/VIN	Engine cu. in. (liter)	Journal Diameter 1	2	3	4	5	Elevation In.	Ex.	Bearing Clearance	Camshaft End Play
	J	V8 267 (4.4)	1.8682–1.8692	1.8682–1.8692	1.8682–1.8692	1.8682–1.8692	1.8682–1.8692	0.2485	0.2600	—	0.004–0.012
	H	V8 305 (5.0)	1.8682–1.8692	1.8682–1.8692	1.8682–1.8692	1.8682–1.8692	1.8682–1.8692	0.2690	0.2760	—	0.004–0.012
	L	V8 350 (5.7)	1.8682–1.8692	1.8682–1.8692	1.8682–1.8692	1.8682–1.8692	1.8682–1.8692	0.2570	0.2690	—	0.004–0.012
	N	V8 350 (5.7)	⑦	—	—	—	—	—	—	0.0020–0.0058	0.011–0.077
1985–88	A	V6 231 (3.8)	1.8682–1.8692	1.8682–1.8692	1.8682–1.8692	1.8682–1.8692	⑥	0.2340	0.2570	—	0.004–0.012
	Z	V6 262 (4.3)	1.8682–1.8692	1.8682–1.8692	1.8682–1.8692	1.8682–1.8692	⑥	0.2340	0.2570	—	0.004–0.012
	G	V8 305 (5.0)	1.8682–1.8692	1.8682–1.8692	1.8682–1.8692	1.8682–1.8692	1.8682–1.8692	0.2340	0.2570	—	0.004–0.012
	Y	V8 307 (5.0)⑧	⑨	—	—	—	—	0.2470	0.2510	—	0.006–0.022
	G	V8 350 (5.7)	1.8682–1.8692	1.8682–1.8692	1.8682–1.8692	1.8682–1.8692	1.8682–1.8692	0.2570	0.2690	—	0.004–0.012
	N	V8 350 (5.7)	⑦	—	—	—	—	—	—	0.0020–0.0058	0.011–0.077
1989–92	Z	V6 262 (4.3)	1.8682–1.8692	1.8682–1.8692	1.8682–1.8692	1.8682–1.8692	⑥	0.2340	0.2570	—	0.004–0.012
	E	V8 305 (5.0)	1.8682–1.8692	1.8682–1.8692	1.8682–1.8692	1.8682–1.8692	1.8682–1.8692	0.2340	0.2570	—	0.004–0.012
	Y	V8 307 (5.0)⑧	⑨	—	—	—	—	0.2470	0.2510	—	0.006–0.022
	7	V8 350 (5.7)	1.8682–1.8692	1.8682–1.8692	1.8682–1.8692	1.8682–1.8692	1.8682–1.8692	0.2330 ⑩	⑩ 0.2560	—	0.004–0.012

① Cubic Inch Displacement only
② STD—Intake: 0.2343, Exhaust: 0.2343
   350 hp—Intake: 0.2714, Exhaust: 0.2824
   375 hp—Intake: 0.3057, Exhaust: 0.3057
③ 360 hp—Intake: 0.2655, Exhaust: 0.2824
   390 hp—Intake: 0.2714, Exhaust: 0.2824
   450 hp—Intake: 0.3057, Exhaust: 0.3057
④ 330 hp—Intake: 0.3057, Exhaust: 0.3234
⑤ 245 hp—Intake: 0.3000, Exhaust: 0.3070
⑥ V6—4 cam bearing journals
⑦ Diesel—No. 1—2.0365–2.0357
   Engine—No. 2—2.0165–2.0157
          No. 3—1.9965–1.9957
          No. 4—1.9765–1.9757
          No. 5—1.9565–1.9557
⑧ 307 (5.0) Oldsmobile Produced Engine
⑨ No. 1—2.0365–2.0352
   No. 2—2.0166–2.0152
   No. 3—1.9965–1.9952
   No. 4—1.9765–1.9752
   No. 5—1.9565–1.9552
⑩ 1992 Police Engine
   Intake: 0.2570, Exhaust: 0.2690

Crankshaft and Connecting Rod Specifications

(All measurements are given in in.)

Year	Engine No. Cyl. Displacement (cu. in.)	Crankshaft				Connecting Rod		
		Main Brg. Journal Dia.	Main Brg. Oil Clearance	Shaft End-Play	Thrust on No.	Journal Diameter	Oil Clearance	Side Clearance
1968	6-230	2.2983–2.2993	0.0003–0.0029	0.002–0.006	7	1.999–2.000	0.0007–0.0027	0.009–0.013
	6-250	2.2983–2.2993	0.0003–0.0029	0.002–0.006	7	1.999–2.000	0.0007–0.0027	0.009–0.013
	8-307	2.4484–2.4493①	0.0008–0.002③	0.003–0.011	5	2.099–2.100	0.0007–0.0028	0.009–0.013
	8-327	2.4484–2.4493①	0.0008–0.002③	0.003–0.011	5	2.099–2.100	0.0007–0.0027	0.009–0.013
	8-350	2.4484–2.4493①	0.0008–0.002③	0.003–0.011	5	2.099–2.100	0.0007–0.0028	0.009–0.013
	8-396	④	⑥	0.006–0.010	5	2.199–2.200	0.0009–0.0025	0.015–0.021
	8-396 (375 HP)	⑤	0.0013–0.0025⑦	0.006–0.010	5	2.1985–2.1995	0.0014–0.0030	0.019–0.025
1969	6-230	2.2983–2.2993	0.0003–0.0029	0.002–0.006	7	1.999–2.000	0.0007–0.0027	0.009–0.013
	6-250	2.2983–2.2993	0.0003–0.0029	0.002–0.006	7	1.999–2.000	0.0007–0.0027	0.009–0.013
	8-307	2.4479–2.4488	0.0008–0.002⑦	0.003–0.011	5	2.099–2.100	0.0007–0.0027	0.009–0.013
	8-350	2.4479–2.4488	0.0008–0.002③	0.003–0.011	5	2.099–2.100	0.0007–0.0028	0.009–0.013
	8-396	④	⑥	0.006–0.010	5	2.199–2.200	0.0009–0.0025	0.015–0.021
	8-396 (375 HP)	⑤	0.0013–0.0025⑦	0.006–0.010	5	2.1985–2.1995	0.0014–0.0030	0.019–0.025
1970	6-230	2.2983–2.2993	0.0003–0.0029	0.002–0.006	7	1.999–2.000	0.0007–0.0027	0.009–0.013
	6-250	2.2983–2.2993	0.0003–0.0029	0.002–0.006	7	1.999–2.000	0.0007–0.0027	0.009–0.013
	8-307	2.4484–2.4493①	0.0003–0.0015⑧	0.002–0.006	5	2.099–2.100	0.0007–0.0028	0.008–0.014
	8-350	2.4484–2.4493①	0.0003–0.0015⑧	0.002–0.006	5	2.099–2.100	0.0007–0.0028	0.008–0.014
	8-400 (Monte Carlo)	2.6584–2.6493⑫	0.0008–0.0020⑮	0.002–0.006	5	2.099–2.100	0.0009–0.0025	0.008–0.014
	8-402	2.7487–2.7496⑨	0.0007–0.0019⑩	0.006–0.010	5	2.199–2.200	0.0009–0.0025	0.013–0.023
	8-454	2.7485–2.7494⑤	0.0013–0.0025⑪	0.006–0.010	5	2.199–2.200	0.0009–0.0025	0.015–0.021
1971	6-250	2.2983–2.2993	0.0003–0.0029	0.002–0.006	7	1.999–2.000	0.0007–0.0027	0.009–0.014
	8-307	2.4484–2.4493⑭	0.0008–0.0020⑮	0.002–0.006	5	2.099–2.100	0.0013–0.0035	0.008–0.014
	8-350	2.4484–2.4493⑭	0.0008–0.0020⑮	0.002–0.006	5	2.099–2.100	0.0013–0.0035	0.008–0.014
	8-402	2.7487–2.7496⑨	0.0007–0.0019⑩	0.006–0.010	5	2.199–2.200	0.0009–0.0025	0.013–0.023
	8-454 (365 HP)	2.7485–2.7494⑤	0.0013–0.0025⑪	0.006–0.010	5	2.199–2.200	0.0009–0.0025	0.015–0.021
	8-454 (425 HP)	2.7481–2.7490①	0.0013–0.0025⑬	0.006–0.010	5	2.1985–2.1995	0.0009–0.0025	0.019–0.025
1972	6-250	2.2983–2.2993	0.0003–0.0029	0.002–0.006	7	1.999–2.000	0.0007–0.0027	0.009–0.014

Crankshaft and Connecting Rod Specifications

(All measurements are given in in.)

Year	Engine No. Cyl. Displacement (cu. in.)	Crankshaft				Connecting Rod		
		Main Brg. Journal Dia.	Main Brg. Oil Clearance	Shaft End-Play	Thrust on No.	Journal Diameter	Oil Clearance	Side Clearance
	8-307	2.4484–2.4493⑭	0.0008–0.0020⑮	0.002–0.006	5	2.099–2.100	0.0013–0.0035	0.008–0.014
	8-350	2.4484–2.4493⑭	0.0008–0.0020⑮	0.002–0.006	5	2.099–2.100	0.0013–0.0035	0.008–0.014
	8-402	2.7487–2.7496⑨	0.0007–0.0019⑩	0.006–0.010	5	2.199–2.200	0.0009–0.0025	0.013–0.023
	8-454	2.7485–2.7494⑤	0.0013–0.0025⑪	0.006–0.010	5	2.199–2.200	0.0009–0.0025	0.015–0.021
1973	6-250	2.3004	0.0003–0.0029	0.002–0.006	7	1.9999–2.000	0.0007–0.0027	0.009–0.014
	8-307, 350	2.4502⑯	0.0008–0.0020⑮	0.002–0.006	5	2.099–2.100	0.0013–0.0035	0.008–0.014
	8-454	2.7492⑰	0.0007–0.0019⑱	0.006–0.010	5	2.199–2.200	0.0009–0.0025	0.015–0.023
1974–77	6-250	2.2988	0.0003–0.0029	0.002–0.006	7	1.9928–2.000	0.0007–0.0027	0.007–0.016
	8-305	2.4489⑳	⑲	0.002–0.006	5	2.099–2.100	0.0013–0.0035	0.008–0.014
	8-350	2.4489⑳	⑲	0.002–0.006	5	2.099–2.100	0.0013–0.0035	0.008–0.014
	8-400	2.6489㉑	0.0008–0.0002㉒	0.002–0.006	5	2.099–2.100	0.0015–0.0035	0.008–0.014
	8-454	2.7490	0.0013–0.0025㉓	0.006–0.010	5	2.099–2.200	0.0009–0.0025	0.015–0.021
1978–79	6-200	2.4489	0.0011–0.0023㉔	0.002–0.006	5	2.0988–2.0998	0.0013–0.0035	0.008–0.014
	6-231	2.4995	0.0004–0.0015	0.004–0.008	2	2.2487–2.2495	0.0005–0.0026	0.006–0.027
	6-250	2.2988	0.0010–0.0024	0.002–0.006	7	1.9980–2.0000	0.0010–0.0026	0.006–0.017
	8-267	2.4489㉕	0.0020–0.0035⑮	0.002–0.007	5	2.0978–2.0988	0.0013–0.0035	0.006–0.016
	8-305	2.4489㉕	0.0011–0.0023㉔	0.002–0.006	5	2.0988–2.0998	0.0013–0.0035	0.008–0.014
	8-350	2.4489㉕	0.0011–0.0023㉔	0.002–0.006	5	2.0988–2.0998	0.0013–0.0035	0.008–0.014
1980–81	6-231	2.4995	0.0004–0.0015	0.004–0.008	2	2.2485–2.2487	0.0005–0.0026	0.006–0.027
	8-267	㉖	㉗	0.002–0.006	5	2.0986–2.0998	0.0013–0.0035	0.006–0.014
	8-305	㉖	㉗	0.002–0.006	5	2.0986–2.0998	0.0013–0.0035	0.006–0.014
	8-350	2.4489㉕	0.0011–0.0023㉔	0.002–0.006	5	2.0988–2.0998	0.0013–0.0035	0.008–0.014
	8-350㉜	2.9993–3.0003	0.0005–0.0021	0.005–0.0021㉞	3	2.1238–2.1248	0.0005–0.0026	0.006–0.020
1982–85	6-231	2.4995	0.0004–0.0015	0.004–0.008	2	2.2485–2.2487	0.0008–0.0026	0.006–0.027
	6-262	2.4484–2.4493㉚	0.0008–0.0020㉛	0.002–0.006	5	2.0986–2.0998	0.0013–0.0035	0.006–0.014
	6-267	2.4489㉕	0.0020–0.0035⑮	0.002–0.007	5	2.0978–2.0988	0.0013–0.0035	0.006–0.016
	8-305	2.4489㉕	0.0011–0.0023㉔	0.002–0.007	5	2.0978–2.0988	0.0013–0.0035	0.006–0.016

Crankshaft and Connecting Rod Specifications

(All measurements are given in in.)

Year	Engine No. Cyl. Displacement (cu. in.)	Crankshaft				Connecting Rod		
		Main Brg. Journal Dia.	Main Brg. Oil Clearance	Shaft End-Play	Thrust on No.	Journal Diameter	Oil Clearance	Side Clearance
	8-350	2.4489㉕	0.0011–0.0023㉔	0.002–0.006	5	2.0988–2.0988	0.0013–0.0035	0.008–0.016
	8-350㉜	2.9993–3.0003	0.0005–0.0021	0.005–0.0021㉞	3	2.1238–2.1248	0.005–0.0026	0.006–0.020
1986–88	6-262	2.4484–2.4493㉚	0.0008–0.0020㉛	0.002–0.006	5	2.0986–2.0998	0.0013–0.0035	0.006–0.014
	8-305	2.4489㉕	0.0011–0.0023㉔	0.002–0.007	5	2.0978–2.0988	0.0013–0.0035	0.006–0.016
	8-307㉝	2.4990–2.4995㉟	0.0005–0.0021㉞	0.0035–0.0135	3	2.1238–2.1248	0.0004–0.0033	0.006–0.020
	8-350	2.4489㉕	0.0011–0.0023㉔	0.002–0.006	5	2.0988–2.0998	0.0013–0.0035	0.008–0.014
1989–90	6-262	㉖	0.0011–0.0020㊱	0.001–0.007	5	2.8487–2.2498	0.0013–0.0035	0.006–0.014
	8-305	㉖	0.0011–0.0020㊱	0.001–0.007	5	2.0986–2.0998	0.0013–0.0035	0.006–0.014
	8-307㉝	2.4985–2.4995㉟	0.0005–0.0021㉞	0.0035–0.0135	3	2.1238–2.1248	0.0004–0.0033	0.006–0.020
	8-350	㉖	0.0011–0.0020㊱	0.001–0.007	5	2.0986–2.0998	0.0013–0.0035	0.006–0.014
1991–92	8-305	㉖	0.0011–0.0020㊱㊲	0.002–0.007	5	2.0983–2.0998	0.0013–0.0035	0.006–0.014
	8-350	㉖	0.0011–0.0020㊱㊲	0.002–0.007	5	2.0983–2.0998	0.0013–0.0035	0.006–0.014

① No. 5—2.4478–2.4488
② No. 1—0.0008–0.002
   Nos. 2–4—0.0018–0.002
   No. 5—0.0010–0.0036
③ No. 5—0.0018–0.0034
④ Nos. 1-2—2.7484–2.7493
   Nos. 3-4—2.7481–2.7490
   No. 5—2.7478–2.7488
⑤ No. 1—2.7484–2.7493
   Nos. 2-4—2.7481–2.7490
   No. 5—2.7478–2.7488
⑥ Nos. 1-2—0.0010–0.0022
   Nos. 3-4—0.0013–0.0025
   No. 5—0.0015–0.0031
⑦ No. 5—0.0015–0.0031
⑧ Nos. 2-4—0.0006–0.0018
   No. 5—0.0008–0.0023
⑨ Nos. 3-4—2.7481–2.7490
   No. 5—2.7473–2.7483
⑩ Nos. 2-4—0.0013–0.0025
   No. 5—0.0019–0.0025
⑪ No. 5—0.0024–0.0040
⑫ No. 5—2.6479–2.6488
⑬ No. 5—0.0029–0.0045
⑭ Nos. 2-4—2.4481–2.4490
   No. 5—2.4479–2.4488
⑮ Nos. 2-4—0.0011–0.0023
   No. 5—0.0017–0.0033
⑯ No. 5—2.4508
⑰ Nos. 2-4—2.7504
   No. 5—2.7499
⑱ Nos. 2-4—0.0013–0.0028
   No. 5—0.0019–0.0035
⑲ w/Auto trans.—No. 1—0.0019–0.0031
   Nos. 2-4—0.0013–0.0025
   No. 5—0.0023–0.0033
⑳ Nos. 2-4—2.4486
   No. 5—2.4485
㉑ No. 5—2.6485
㉒ Nos. 2-4—0.001–0.0023
   No. 5—0.0017–0.0033
㉓ No. 5—0.0024–0.0070
㉔ No. 1—0.0008–0.0020
㉕ Nos. 2-4—2.4486
   No. 5—2.4485
㉖ No. 1—2.4484-2.4493
   Nos. 2,3,4—2.4481–2.4490
   No. 5—2.4481–2.4488
㉗ No. 1—0.0008–0.0020
   Nos. 2-4—.0011–.0023
   No. 5—.0017–.0032
㉘ Nos. 1,2,3—.0005–.0021
   No. 4—.0020–.0034
㉙ Nos. 1,2,3—.0005–.0021
   No. 4—.0015–.0031
㉚ Intermediate: 2.4481–2.4490
   Rear: 2.4479–2.4488
㉛ Intermediate: 0.0011–0.0023
   Rear: 0.0017–0.0032
㉜ Diesel Engine
㉝ Oldsmobile Produced Engine
㉞ No. 5—0.0015–0.0031
㉟ No. 1—2.4993–2.4998
㊱ No. 5—0.0020–0.0032
㊲ No. 1—0.008–0.0020

ENGINE AND ENGINE OVERHAUL

Piston and Ring Specifications

(All measurements are given in inches. To convert inches to metric units, refer to the Metric Information section.)

Year	Engine Type Disp. cu. in.	Piston-to Bore Clearance	Ring Gap			Ring Side Clearance		
			Top Compression	Bottom Compression	Oil Control	Top Compression	Bottom Compression	Oil Control
1968–79	6-250	0.0005–0.0015	0.010–0.020	0.010–0.020	0.015–0.055	0.0012–0.0027	0.0012–0.0032	0.0000–0.0050
1968–73	8-307	0.0005–0.0011	0.010–0.020	0.010–0.020	0.015–0.055	0.0012–0.0032	0.0012–0.0027	0.0000–0.0050
1968–70	8-396	0.0010–0.0018①	0.010–0.020	0.010–0.020	0.010–0.030	0.0017–0.0032	0.0017–0.0032	0.0005–0.0065
1969	8-350	0.0005–0.0011	0.013–0.025	0.013–0.025	0.0015–0.0055	0.0012–0.0032	0.0012–0.0027	0.0000–0.0050
1970–79	8-350	0.0007–0.0013	0.010–0.020②	0.013–0.025②	0.015–0.055	0.0012–0.0032④	0.0012–0.0027⑤	0.0000–0.0050③⑧
1970–76	8-400	0.0034	0.010–0.020	0.010–0.020	0.015–0.055	0.0012–0.0027⑥	0.0012–0.0032⑥	0.0000–0.0050
1970–76	8-454	0.0049⑦	0.010–0.020	0.010–0.020	0.015–0.055	0.0017–0.0032	0.0017–0.0032	0.0005–0.0065
1971–72	8-402	0.0018–0.0026	0.010–0.020	0.010–0.020	0.015–0.055	0.0017–0.0032	0.0017–0.0032	0.0005–0.0065
1976–77	8-305	0.0007–0.0017	0.010–0.020	0.010–0.025	0.015–0.055	0.0012–0.0032	0.0012–0.0032	0.0000–0.0050
1978–79	6-200	0.0007–0.0017	0.010–0.020	0.010–0.025	0.010–0.030	0.0012–0.0032	0.0012–0.0032	0.002–0.007
1978–84	6-231	0.0008–0.0020	0.010–0.020	0.010–0.020	0.015–0.035	0.003–0.005	0.003–0.005	0.0035 max.
1978–92	8-305	0.0007–0.0017	0.010–0.020	0.010–0.025	0.015–0.055	0.0012–0.0032	0.0012–0.0032	0.002–0.007
1980–84	6-229	0.0007–0.0017	0.010–0.020	0.010–0.025	0.015–0.055	0.0012–0.0032	0.0012–0.0032	0.002–0.007
1979–84	8-267	0.0007–0.0017	0.010–0.020	0.010–0.025	0.015–0.055	0.0012–0.0032	0.0012–0.0032	0.002–0.007
1980–84	8-350⑨	0.005–0.006	0.015–0.025	0.015–0.025	0.015–0.055	0.005–0.007	0.0018–0.0038	0.001–0.005
1986–90⑩	8-307	0.0008–0.0018	0.009–0.019	0.009–0.019	0.015–0.055	0.0020–0.0040	0.0020–0.0040	0.000–0.0035
1982–84	6-263⑨	0.0035–0.0045	0.015–0.025	0.015–0.025	0.015–0.055	0.005–0.007	0.003–0.005	0.001–0.005
1985–90	6-262	0.0007–0.0017	0.010–0.020	0.010–0.025	0.015–0.055	0.0012–0.0032	0.0012–0.0032	0.002–0.007
1980–92	8-350	0.0005–0.0022	0.010–0.020	0.018–0.026	0.015–0.055	0.0012–0.0032	0.0012–0.0032	0.002–0.007

① 0.0036–0.0044; 11:1 compression
② 325, 350 hp: Top 0.010–0.020
　　　　　　　 2nd 0.013–0.023
③ 1978–82: 0.002–0.007
④ 0.0012–0.0027 on 1975 2-bbl
⑤ 165, 245, 250 hp: 0.0012–0.0032
⑥ 300 hp: Top 0.0017–0.0032
　　　　　2nd 0.0017–0.0032
⑦ 425 hp—1971: 0.0065
　　1973–75: 0.0035
⑧ 1978–80: 0.015–0.050
⑨ Diesel
⑩ Oldsmobile Produced Engine

Torque Specifications
All readings in ft. lbs.

Year	Engine ID/VIN	Engine Displacement liter (cc)	Cylinder Head Bolts	Main Bearing Bolts	Rod Bearing Bolts	Crankshaft Damper Bolts	Flywheel Bolts	Manifold Intake	Manifold Exhaust	Spark Plugs	Plug Nut
1968–75	In-Line	6-230	95	65	36	—	60	30⑧	27⑦	20	100
	In-Line	6-250	95	65	36	—	60	30⑧	27⑦	20	100
1968–77	V8	305	60–70	75②⑨	45	60⑥	60	30	⑤	20	100
	V8	307	60–70	75②⑨	45	60⑥	60	30	⑤	20	100
	V8	327	60–70	75②⑨	45	60⑥	60	30	⑤	20	100
	V8	350	60–70	75②⑨	45	60⑥	60	30	⑤	20	100
	V8	400	60–70	75②⑨	45	60⑥	60	30	⑤	20	100
1968–76	V8	396	80①	105③	50	85⑥	65	30	30	20	100
	V8	402	80①	105③	50	85⑥	65	30	30	20	100
	V8	427	80①	105③	50④	85	65	30	30	20	100
	V8	454	80①	105③	50④	85	65	30	30	20	100
1978–84	6	200	65	70	45	60	60	30	20⑩	20	100
	V8	267	65	70	45	60	60	30	20⑩	20	100
	V8	305 (5.0)	65	70	45	60	60	30	20⑩	20	100
	V8	350 (5.7)	65	70	45	60	60	30	20⑩	20	100
1980–84	V6	231 (3.8)	80	100	40	175	60	45	25	20	100
1982–85	Diesel	350 (5.7)	130⑬	120	42	⑮	60	40	25	12⑱	100
1985–88	V6	262 (4.3)	60–75	70–85	45	65–75	70	25–45	20	22	100
	V8	305 (5.0)	60–75	70–85	45	65–75	60	25–45	20⑯	22	100
	V8	307 (5.0)⑫	125⑪	⑰	42	200–310	46	40⑪	25	22	100
	V8	350 (5.7)	60–75	70–85	45	65–75	60	25–45	20⑯	22	100
1989–90	V6	262 (4.3)	65	65	44	43	74	35	26⑲	22	100
	V8	305 (5.0)	68	77	44	43	74	35	26⑲	22	100
	V8	350 (5.7)	68	77	44	43	74	35	26⑲	22	100
1991–92	V8	305 (5.0)	68	77	44	43	74	35	26⑲	22	100
	V8	350 (5.7)	68	77	44	43	74	35	26⑲	22	100

① Aluminum Heads—Short bolts 65, Long bolts 75
② Engines with 4-bolt mains—Outer bolts 65
③ 1966–68 2-bolt mains 95
   1966–67 4 bolt mains 115
④ 7/16 Rod bolts—70
⑤ Center bolts—25–30, end bolts 15–20
⑥ Where applicable
⑦ Exhaust-to-intake
⑧ Manifold-to-head
⑨ 70 starting 1976
⑩ Inside bolts on 350—30 ft. lbs.
⑪ Dip bolt in oil
⑫ Oldsmobile Produced Engine
⑬ Dip bolts in engine oil before torquing
⑮ Pulley-to-balancer bolts, 27 ft. lbs.
   Balancer-to-crankshaft bolt, 271–420 ft. lbs.
⑯ Inside bolts: 25 ft. lbs.
⑰ No. 1–No. 4—80
   No. 5—120
⑱ Glow Plug, Diesel
⑲ Nuts, 20 ft. lbs.

sion, remove the torque converter cover and the torque converter bolts.

NOTE: *Before removing the torque converter bolts, scribe a mark to ensure the relationship between the torque converter and the flex plate.*

17. Remove the engine-to-mount bolts.
18. Disconnect and plug the fuel line at the fuel pump.
19. If equipped with an automatic transmission, disconnect the torque converter clutch wiring from the transmission. Disconnect the transmission oil cooler lines from the clip at the engine oil pan.
20. Remove engine-to-transmission bolts.
21. Lower the vehicle. Properly support the transmission.
22. Secure a vertical lifting device to the engine and remove the engine from the vehicle.

CAUTION: *When removing the engine from the transmission, be careful that the torque converter does not pull out of the transmission.*

To Install:

23. Secure a vertical lifting device to the engine and install the engine to the vehicle.
24. Support the transmission.
25. Install the engine-to-transmission bolts.

NOTE: *Make sure the torque converter bolts are aligned with the mounting holes on the flexplate, before attaching bell housing to engine.*

26. If equipped with an automatic transmission, connect the torque converter clutch wiring to the transmission. Connect the transmission oil cooler lines to the clip at the engine oil pan.
27. Connect the fuel line to the fuel pump.
28. Install the engine-to-mount bolts.
29. If equipped with an automatic transmission, install the torque converter cover and the torque converter bolts.
30. As required, install the crossover pipe and the catalytic converter as an assembly.
31. Lower the vehicle.
32. Connect the positive battery cable to the battery and the frame. Connect the negative battery cable to the air conditioning hose/alternator bracket.
33. Install the distributor cap and the cruise control cable (if equipped).
34. Connect the wiring harness at the bulkhead and related engine wiring.
35. As required, install the windshield washer bottle.
36. As required, install the AIR pipe to the converter.
37. Connect the vacuum hoses and the CCC (ECM) wiring harness connectors to the engine, if equipped.
38. Install the transmission oil cooler lines, if equipped to the radiator and install the radiator.
39. Connect the accelerator and the T.V. cables.
40. If equipped with air conditioning and power steering, install the compressor and the power steering pump.
41. Install the upper fan shroud and the fan assembly.
42. Install the heater hoses and the radiator hoses to the engine.
43. Fill the cooling system.
44. Connect the negative battery cable.
45. Install the hood.

Diesel Engine

1. Drain the cooling system.

CAUTION: *When draining the coolant, keep in mind that cats and dogs are attracted by the ethylene glycol antifreeze, and are quite likely to drink any that is left in an uncovered container or in puddles on the ground. This will prove fatal in sufficient quantity. Always drain the coolant into a sealable container. Coolant should be reused unless it is contaminated or several years old.*

2. Remove the air cleaner.
3. Mark the hood to hinge position and remove the hood.
4. Disconnect the ground cables from the batteries.
5. Disconnect the ground wires at the fender panels and the ground strap at the cowl.
6. Disconnect the radiator hoses, cooler lines, heater hoses, vacuum hoses, power steering pump hoses, air conditioning compressor (hoses attached), fuel inlet hose and all attached wiring.
7. Remove the bellcrank clip.
8. Disconnect the throttle and transmission cables.
9. Remove the upper radiator support and the radiator.
10. Raise and support the vehicle.
11. Disconnect the exhaust pipes at the manifold.
12. Remove the torque converter cover and the three bolts holding the converter to the flywheel.
13. Remove the engine mount bolts or nuts.
14. Remove the transmission to engine retaining bolts. Remove the starter.
15. Lower the vehicle and attach a hoist to the engine.
16. Properly support the transmission.
17. Using the proper lifting equipment, remove the engine from the vehicle.

To Install:

18. Using the proper equipment, lower the

ENGINE AND ENGINE OVERHAUL

engine into position and support the transmission with a transmission jack.

19. Install the transmission-to-engine bolts.
20. Install the starter.
21. Install the engine mount bolts or nuts.
22. Install the torque converter cover and the three bolts holding the converter to the flywheel. Torque to 40 ft. lbs.
23. Connect the exhaust pipes at the manifold.
24. Lower the vehicle.
25. Install the upper radiator support and the radiator.
26. Connect the throttle and transmission cables.
27. Install the bellcrank clip.
28. Connect the radiator hoses, cooler lines, heater hoses, vacuum hoses, power steering pump hoses, air conditioning compressor (hoses attached), fuel inlet hose and all attached wiring.
29. Connect the ground wires at the fender panels and the ground strap at the cowl.
30. Connect the ground cables to the batteries.
31. Install the hood.
32. Install the air cleaner.
33. Fill the cooling system.

Rocker Arm Cover

REMOVAL AND INSTALLATION

Inline 6-Cylinder Gasoline Engine

1. Disconnect the negative battery cable. At the rocker arm cover, disconnect the ventilation hoses. Remove the air cleaner.
2. Disconnect the wires, the fuel and the

Installing RTV sealant to the rocker arm cover

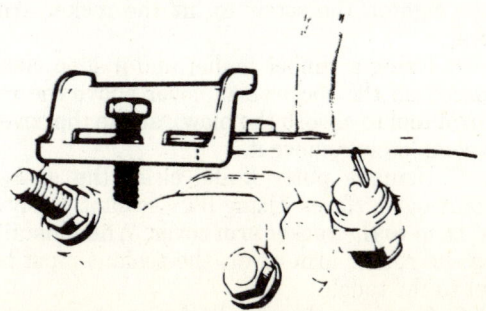

Removing the rocker arm cover on diesel engines

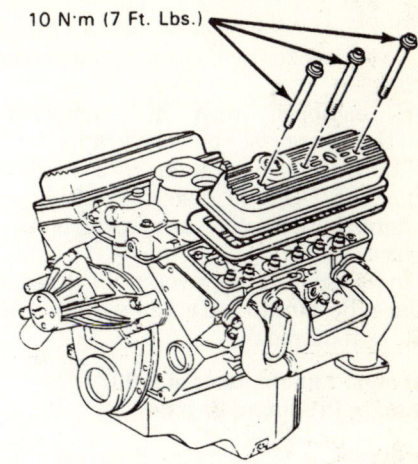

Rocker arm cover installation, V6

vacuum tubes from the rocker arm cover clips.
3. If equipped, disconnect the air injection hose from the check valve of the AIR pipe.
4. Remove the cover-to-cylinder head screws and the cover by rotating it from under the air pipe, if equipped.

NOTE: *DO NOT pry on the cover to remove it. If it sticks, use your hand palm or a rubber mallet to bump it rearwards, from the front.*

5. Using a gasket scraper, clean the gasket mounting surfaces.
6. To install, reverse the removal procedures. Start the engine and check for leaks.

NOTE: *When installing the cover, use a new gasket (1968-76) or an $1/8''$ bead of RTV sealant (1977-79).*

V6 Gasoline Engine

RIGHT SIDE

1. Disconnect the negative battery cable. Remove the air cleaner.
2. Disconnect the dipstick tube bracket from the alternator bracket. Then the dipstick tube head.
3. At the engine, disconnect the heater hoses from the evaporator case and bracket.
4. At the intake manifold, disconnect the heater hose and the wiring harness bracket. Remove the rocker arm cover bolts and the breather pipe.
5. Disconnect and plug the fuel lines. Disconnect the clips and the spark plug wires at the distributor.
6. Remove the rocker arm cover retaining bolts. Remove the rocker arm cover.
7. Installation is the reverse of the removal procedure. Be sure to use a new gasket or RTV sealant, as required.

ENGINE AND ENGINE OVERHAUL

LEFT SIDE

1. Disconnect the negative battery cable.
2. At the rocker arm cover, disconnect the PCV valve.
3. If equipped with air conditioning, remove the compressor and the brace.

NOTE: *When removing the air conditioning compressor, DO NOT disconnect the hoses.*

4. Remove the rocker arm cover bolts and the cover.
5. Installation is the reverse of the removal procedure. Be sure to use a new gasket or RTV sealant, as required.

V8 Gasoline Engine, Except 8-307 Oldsmobile Produced Engine

RIGHT SIDE

1. Disconnect the negative battery cable. Remove the air cleaner.
2. On the 1985-92 vehicles, disconnect the ECM wire harness from the intake manifold and the oxygen sensor.
3. At the exhaust manifold, disconnect the AIR hose.
4. Disconnect the wires from the alternator, the choke and the spark plugs, then the harness from the rocker cover, position it to the side.
5. Remove the EGR valve. Remove the rocker arm cover bolts and the cover.
6. Installation is the reverse of the removal procedure. be sure to use a new gasket or RTV sealant, as required.

LEFT SIDE

1. Disconnect the negative battery cable. Remove the air cleaner.
2. Disconnect the power brake pipe from the carburetor and the booster.
3. Disconnect the AIR hose from the exhaust manifold. Remove the PCV valve.
4. Disconnect the wire from the oxygen sensor. Remove the rocker arm cover bolts and the cover.
5. Installation is the reverse of the removal procedure. Be sure to use a new gasket or RTV sealant, as required.

307 Olsmobile Produced Engine

RIGHT SIDE

1. Disconnect the negative battery cable. Remove the air cleaner.
2. Disconnect the spark plug wires. Remove the air conditioning compressor bracket, as required.
3. Disconnect the required electrical wires. Remove the PCV valve.
4. Remove the valve cover retaining bolts.

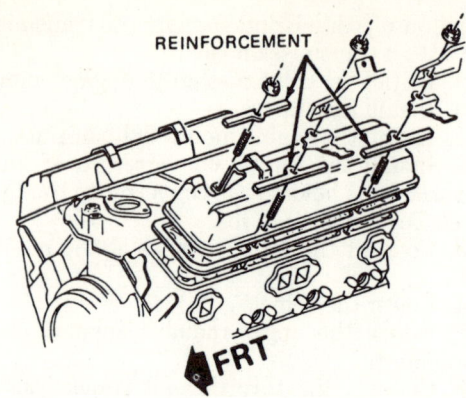

Rocker arm cover installation, V8

Remove the valve cover.

5. Installation is the reverse of the removal procedure. Be sure to use a new gasket or RTV sealant, as required.

LEFT SIDE

1. Disconnect the negative battery cable. Remove the air cleaner.
2. Disconnect the spark plug wires. Remove the EGR valve assembly to gain clearance.
3. Disconnect the required electrical wires. Remove the side air cleaner assembly.
4. Remove the valve cover retaining bolts. Remove the valve cover.
5. Installation is the reverse of the removal procedure. Be sure to use a new gasket or RTV sealant, as required.

V8 Diesel Engine

1. Refer to the Injection Lines, Removal and Installation procedures, in the Diesel Fuel System of Chapter 5 and remove the fuel injection lines.
2. Remove the rocker arm cover-to-cylinder head screws and any accessory mounting brackets, if necessary.
3. Using the Valve Cover Removal tool No. J-34144 or BT-8315, place it midway between the ends of the valve cover (on the upper side), then tighten the screw to lift the rocker arm cover.
4. Using a rubber mallet and a shop cloth (placed on the rocker arm cover above the removal tool to absorb the blow), strike the cover to completely remove it.
5. Using a putty knife, clean the gasket mounting surfaces. Using RTV sealant, apply a 1/4" bead to the rocker arm cover. When installing the rocker arm cover, the sealant must be wet to the touch.
6. Continue the installation in the reverse order of the removal procedure.

ENGINE AND ENGINE OVERHAUL

Rocker Arms

REMOVAL AND INSTALLATION

All Gasoline Engines, Except 231 V6 Engine and 307 Oldsmobile Produced Engine

Rocker arms are removed by removing the adjusting nut. Be sure to adjust the valve lash after replacing the rocker arms. Coat the replacement rocker arm and ball with SAE 90 gear oil before installation. Make sure the valves are closed on the cylinder you are working on before installation.

NOTE: *When replacing an exhaust rocker, move an old intake rocker to the exhaust rocker arm stud and install the new rocker arm on the intake stud. This will prevent burning of the new rocker arm on the exhaust position.*

Rocker arms studs that have damaged threads or are loose in the cylinder heads may be replaced by reaming the bore and installing oversize studs. Oversizes available are 0.003" and 0.013". The bore may also be tapped and screw-in studs installed. Several aftermarket companies produce complete rocker arm stud kits with installation tools. Mark IV and late model high performance small block engines use screw-in studs and pushrod guide plates.

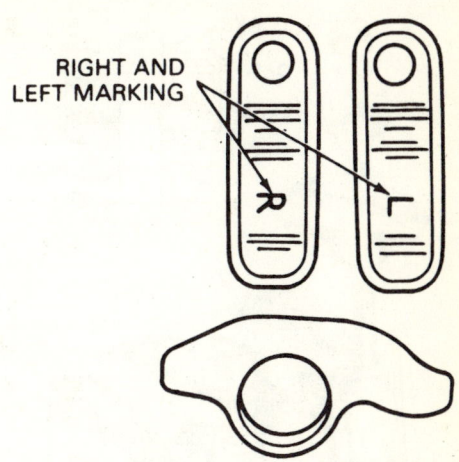

231 V6 replacement rocker arm identification

V6-231 Gasoline Engine

1. Disconnect the negative battery cable. Remove the rocker arm covers.
2. Remove the rocker arm shaft assembly bolts.
3. Remove the rocker arm shaft assembly.
4. To remove the rocker arms from the shaft, the nylon arm retainers must be removed. They can be removed with a pair of

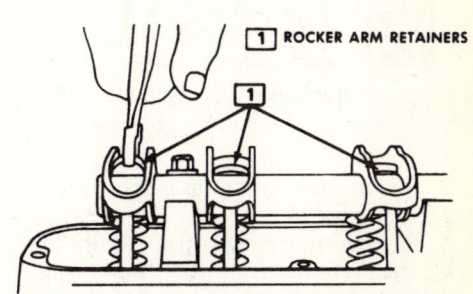

Removing nylon retainers, 231 V6 rocker shafts

water pump pliers, or they can be broken by hitting them below the head with a chisel.

5. Remove the rocker arms from the shaft. Make sure you keep them in order. Also note that the external rib on each arm points away from the rocker arm shaft bolt located between each pair of rocker arms.
6. If you are installing new rocker arms, note that the replacement rocker arms are marked **R** and **L** for right and left side installation. Do not interchange them.
7. Install the rocker arms on the shaft and lubricate them with oil.
8. Center each arm on the 1/4" hole in the shaft. Install new nylon rocker arm retainers in the holes using a 1/2" drift.
9. Locate the push rods in the rocker arms and insert the shaft bolts. Tighten the bolts a little at a time until they are tight.
10. Install the rocker covers using new gaskets or RTV sealant, as required..

8-307 Oldsmobile Produced Engine

1. Disconnect the negative battery cable. Remove the valve cover.

Position of rocker arms on shaft, 231 V6

ENGINE AND ENGINE OVERHAUL

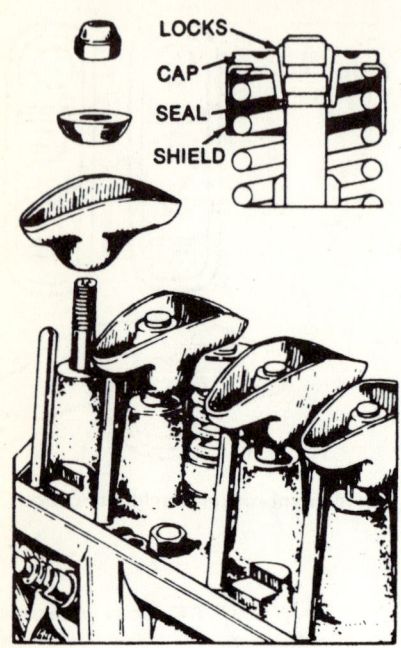

Inline six cylinder rocker arm components (V8s similar)

2. Remove the retaining bolts from a set of rocker arm assemblies.
3. Remove the rocker arms, bolts and the rocker arm retainer from their mounting.
4. Installation is the reverse of the removal procedure.
5. Be sure to use new gaskets or RTV sealant, as required.

V8 Diesel Engine

NOTE: *When the diesel engine rocker arms are removed or loosened, the lifters must be bled down to prevent oil pressure buildup inside each lifter, which could cause it to raise up higher than normal and bring the valves within striking distance of the pistons.*

1. Disconnect the negative battery cable. Remove the valve cover.
2. Remove the rocker arm pivot bolts, the bridged pivot and rocker arms.
3. Remove each rocker set as a unit.
4. To install, lubricate the pivot wear points and position each set of rocker arms in its proper location. Do not tighten the pivot bolts for fear of bending the pushrods when the engine is turned. Bleed down the valve lifters.

Diesel Engine Lifter Bleed Down

1. Before installing any removed rocker arms, rotate the crankshaft so that No. 1 cylinder is 32°BTDC. This is 50mm (2") counterclockwise from the 0 degree pointer on the timing indicator on the front of the engine. If only the right valve cover was removed, remove the glow plug from No. 1 cylinder to determine if the piston is in the correct position. If the left valve cover was removed, rotate the crankshaft until the No. 5 cylinder intake valve pushrod ball is 7.0mm (0.28") above the No. 5 cylinder exhaust valve pushrod ball.

WARNING: *Use only hand wrenches to torque the rocker arm pivot bolts to avoid pushrod damage.*

2. If the No. 5 cylinder pivot and rocker arms were removed, install them. Torque the bolts alternately between the intake and exhaust valves until the intake valve begins to open, then stop.
3. Install the remaining rocker arms except No. 3 and No. 8 intake valves, if these rockers were removed.
4. If removed, install but do not torque the No. 3 valve pivots beyond the point that the valve would be fully open. This is indicated by a strong resistance while still turning the pivot retaining bolts. Going beyond this will bend the pushrod. Torque the bolts slowly, allowing the lifter to bleed down.
5. Finish torquing the No. 5 cylinder rocker arm pivot bolt slowly. Do not go beyond the point that the valve would be fully open. this is indicated by a strong resistance while still turning the pivot retaining bolts. Going beyond this will bend the pushrod.
6. Do not turn the engine crankshaft for at least 45 minutes, allowing the lifters to bleed down. This is important.
7. Finish reassembling the engine as the lifters are being bled.

Valve Lash Adjustment

Most engines described in this book use hydraulic lifters, which require no periodic adjustment. In the event of cylinder head removal or any operation that requires disturbing the

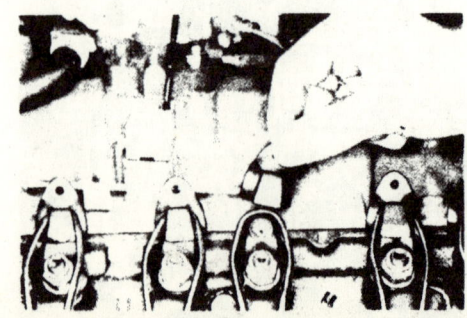

Oil splash stopper clips will prevent splatter when adjusting the valves with the engine running

ENGINE AND ENGINE OVERHAUL

rocker arms, the valves will have to be adjusted, except on the 307 Oldsmobile produced engine.

Solid Lifters (1968-71)

ENGINE RUNNING

NOTE: *Before adjusting solid lifters, thoroughly warm the engine. The solid lifters are generally found on certain high performance engines.*

1. Run the engine to reach normal operating temperature.
2. Remove the valve covers and gaskets by tapping the end of the cover rearward. Do not attempt to pry the cover off.
3. To avoid being splashed with hot oil, use oil deflector clips. Place one on each oil hole in the rocker arm.
4. Measure between the rocker arm and the valve stem with a flat feeler gauge, then adjust the rocker arm stud nut until clearance agrees with the specifications in the chart.
5. After adjusting all the valves, stop the engine, clean the gasket surfaces, and install the valve covers with new gaskets.

ENGINE NOT RUNNING

These are initial adjustments usually required after assembling an engine or doing a valve job. They should be followed up by an adjustment with the engine running as described above. If the camshaft was replaced with an aftermarket high performance one, the adjustment procedure is the same except you will need the camshaft manufacturer's clearance specifications.

1. Set the engine to the No. 1 firing position.
2. Adjust the clearance between the valve stems and the rocker arms with a feeler gauge. Check the Chart for the proper clearance. Adjust the following valves in the No. 1 firing position: Intake No. 2, 7, Exhaust No. 4, 8.
3. Turn the crankshaft $1/2$ revolution clockwise. Adjust the following valves: Intake No. 1, 8, Exhaust No. 3, 6.
4. Turn the crankshaft $1/2$ revolution clockwise to No. 6 firing position. Adjust the following valves in the NO. 6 firing position: Intake No. 3, 4, Exhaust No. 5, 7.
5. Turn the crankshaft $1/2$ revolution clockwise, Adjust the following valves: Intake No. 5, 6, Exhaust No. 1, 2.
6. Run the engine until the normal operating temperature is reached. Reset all clearances, using the procedure listed above under Engine Running.

Hydraulic Lifters

1972-88

NOTE: *The 307 Oldsmobile produced engine does not require that the valves be manually adjusted.*

1. Remove the rocker covers and gaskets.
2. Adjust the valves on inline 6-cylinder engines as follows:

a. Mark the distributor housing with a piece of chalk at the No. 1 and 6 plug wire positions. Remove the distributor cap with the plug wires attached.

b. Crank the engine until the distributor rotor points to the NO. 1 cylinder and the No. 1 piston is at TDC (both No. 1 cylinder valves closed). At this point, adjust the following valves:

- No. 1 - Exhaust and Intake
- No. 2 - Intake
- No. 3 - Exhaust

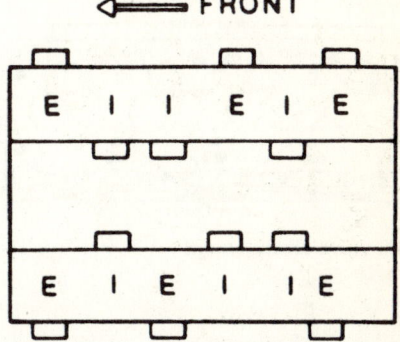

Valve arrangement of the Chevrolet built V6 engines (E-exhaust; I-intake)

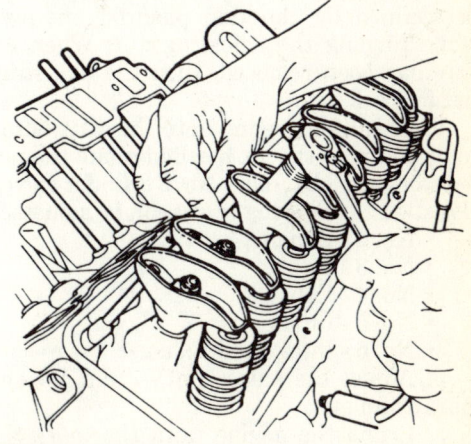

V8 valve adjustment

ENGINE AND ENGINE OVERHAUL

Inline six cylinder valve adjustment

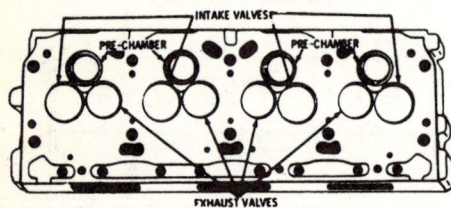

Diesel V8 valve arrangement

- No. 4 - Intake
- No. 5 - Exhaust

c. Back out the adjusting nut until lash is felt at the pushrod, then turn the adjusting nut in until all lash is removed. This can be determined by checking pushrod end-play while turning the adjusting nut. When all play has been removed, turn the adjusting nut in 1 full turn.

d. Crank the engine until the distributor rotor points to the no. 6 cylinder and the no. 6 piston is at TDC (both No. 6 cylinder valves closed). The following valves can be adjusted:
- No. 2 - Exhaust
- No. 3 - Intake
- No. 4 - Exhaust
- No. 5 - Intake
- No. 6 - Intake and Exhaust

3. Adjust the valves on V6 and V8 engines as follows:

a. Crank the engine until the mark on the damper aligns with the TDC or 0 degree mark on the timing tab and the engine is in the No. 1 firing position. This can be determined by placing the fingers on the No. 1 cylinder valves as the marks align. If the valves do not move, it is in the No. 1 firing position. If the valves move, it is in the No. 6 firing position (no. 4 on V6) and the crankshaft should be rotated one more revolution to the No. 1 firing position.

b. With the engine in the No. 1 firing position, the following valves can be adjusted:
- V8 - Exhaust: 1, 3, 4, 8
- V8 - Intake: 1, 2, 5, 7
- V6 - Exhaust: 1, 5, 6
- V6 - Intake: 1, 2, 3

c. Back out the adjusting nut until lash is felt at the pushrod, then turn the adjusting nut in until all lash is removed. This can be determined by checking pushrod end-play while turning the adjusting nut. When all play has been removed, turn the adjusting nut in:
- $1/2$-$1 1/4$ additional turn (flat lifter)
- $3/4$-$1 1/4$ additional turn (roller lifter V8)
- $3/4$ additional turn (roller lifter V6)

d. Crank the engine 1 full revolution until the marks are again in alignment. This is the No. 6 firing position (no. 4 on V6). The following valves can now be adjusted:
- V8 - Exhaust: 2, 5, 6, 7
- V8 - Intake: 3, 4, 6, 8
- V6 - Exhaust: 2, 3, 4
- V6 - Intake: 4, 5, 6

4. Reinstall the rocker arm covers using new gaskets or sealer.

5. Install the distributor cap and wire assembly.

6. Adjust the carburetor idle speed.

Thermostat

REMOVAL AND INSTALLATION

1. Disconnect the negative battery cable. Drain the radiator until the level is below the thermostat level (below the level of the intake manifold).

CAUTION: *When draining the coolant, keep in mind that cats and dogs are attracted by the ethylene glycol antifreeze, and are quite likely to drink any that is left in an uncov-*

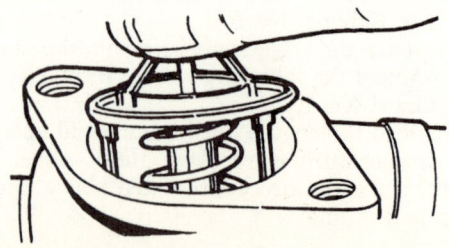

Removing thermostat from water outlet elbow

ENGINE AND ENGINE OVERHAUL 139

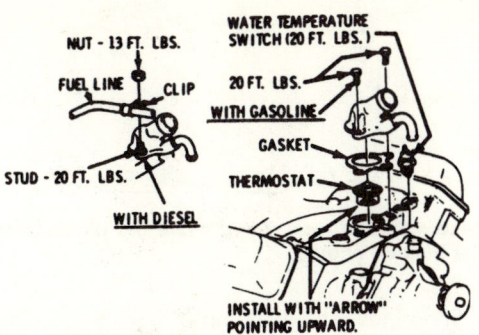

Diesel (left) and gasoline engine thermostat location and outlet elbow mounting

ered container or in puddles on the ground. This will prove fatal in sufficient quantity. Always drain the coolant into a sealable container. Coolant should be reused unless it is contaminated or several years old.

2. Remove the water outlet elbow assembly from the engine. Remove the thermostat from inside the elbow.

3. Install new thermostat in the reverse order of removal, making sure the spring side is inserted into the elbow. Clean the gasket surfaces on the water outlet elbow and the intake manifold. Use a new gasket when installing the elbow to the manifold. On later models the thermostat housing may have been sealed with RTV sealant. If so, place a $1/8''$ bead of RTV sealer all around the thermostat housing sealing surface on the intake manifold and install the housing while it is still wet. Refill the radiator to approximately $2 1/2''$ below the filler neck.

NOTE: *If the thermostat is equipped with a pin hole, be sure to install pin side facing upwards.*

Intake Manifold

REMOVAL AND INSTALLATION

Inline 6-Cylinder Gasoline Engine

NOTE: *The 1975-79 engines are equipped with an intake manifold which is integral with the cylinder head and must be removed along with the cylinder head..*

1. Disconnect the negative battery cable. Remove the air cleaner.
2. Disconnect the throttle rods at the bellcrank and remove the throttle return spring.
3. Disconnect and plug the fuel line at the carburetor. Disconnect the vacuum lines at the carburetor.
4. Disconnect the crankcase ventilation hose from the valve cover and the evaporation control hose from the carbon canister.

5. Disconnect the exhaust pipe from the manifold and throw away the packing.
6. Remove the manifold assembly and scrape off the gaskets.
7. Check the condition of the manifold. If a manifold is cracked or distorted 0.030" or more, it should be replaced to prevent exhaust leakage. To detect distortion, lay a straightedge along the length of the exhaust port faces. If, at any point, a gap of 0.030" or more exists between the straightedge and the manifold, distortion of that amount is present.
8. Be removing one bolt and two nuts, the manifold can be separated.

To Install:

9. Position the manifold assembly to the cylinder block. Be sure to use new gaskets. Install the retaining bolts. Install the exhaust pipe, as required.

NOTE: *The center and end bolts torque different.*

10. Install the PCV valve hose and the evaporation control hose. Connect the fuel line. Connect the vacuum lines.
11. Connect the throttle rods at the bellcrank. Connect the throttle return spring.
12. Install the air cleaner. Connect the negative battery cable. Start the engine and check for leaks, correct as required.

All Gasoline Engines, Except V6-231 and 8-307 Oldsmobile Produced Engine

1. Remove the air cleaner.
2. Drain the radiator.

CAUTION: *When draining the coolant, keep in mind that cats and dogs are attracted by the ethylene glycol antifreeze, and are quite likely to drink any that is left in an uncovered container or in puddles on the ground. This will prove fatal in sufficient quantity. Always drain the coolant into a sealable con-*

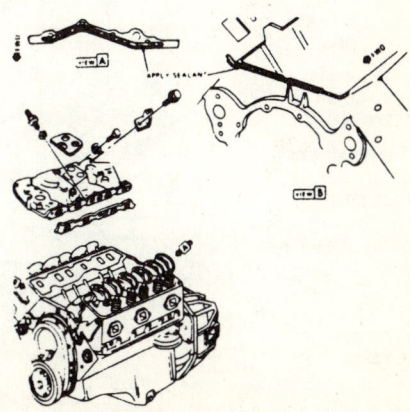

V6 intake manifold seal location except 231 V6 engine

ENGINE AND ENGINE OVERHAUL

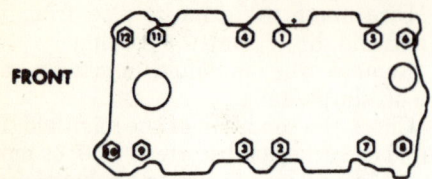

Intake manifold torque sequence small block Chevrolet built V8 engine

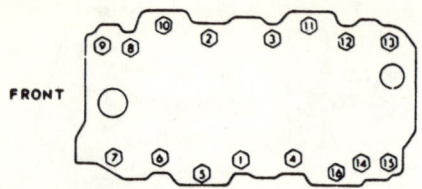

Intake manifold torque sequence big block Chevrolet built V8 engine

tainer. Coolant should be reused unless it is contaminated or several years old.

3. Disconnect:
 a. Negative battery cable.
 b. Upper radiator and heater hoses at the manifold.
 c. Crankcase ventilation hoses as required.
 d. Fuel line at the carburetor, or the fuel line clips and lines at the throttle body, if equipped with fuel injection.
 e. Accelerator linkage and TV cables, if equipped.
 f. Vacuum hose at the distributor, if equipped.
 g. Power brake hose at the carburetor base or manifold, if applicable.
 h. Ignition coil, the spark plug wires and temperature sending switch wires.
 i. Water pump bypass at the water pump (Mark IV engine only).
 j. If equipped with CCC (ECM), disconnect the electrical harness and lay it aside.

4. Remove the distributor cap and scribe the rotor position relative to the distributor body.

5. Remove the distributor.

6. If applicable, remove the alternator upper bracket. As required, remove the oil filler bracket, air cleaner bracket, air conditioning compressor and bracket, and accelerator bellcrank.

7. Remove the manifold-to-head attaching bolts, then remove the manifold and carburetor or throttle body as an assembly.

8. If the manifold is to be replaced, transfer the carburetor (and mounting studs), water outlet and thermostat (use a new gasket) heater hose adapter, EGR valve (use new gasket) and, if applicable, TVS switch and the choke coil. 1975-79 engines use a new carburetor heat choke tube which must be transferred to a new manifold.

9. Using a gasket scraper clean the gasket and seal surfaces of the cylinder heads and manifold.

To Install:

10. Install the manifold end seals, folding the tabs if applicable, and the manifold/head gas-

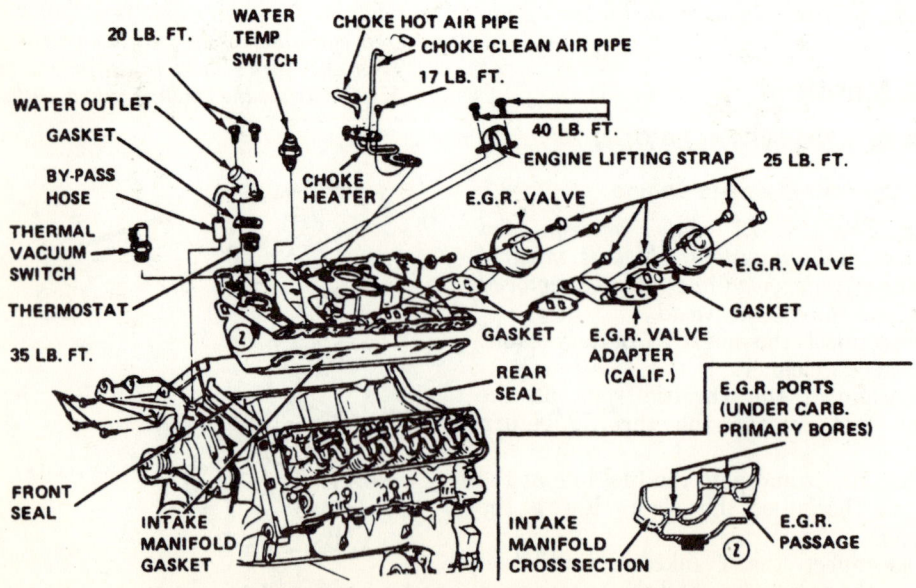

Typical gasoline V6 and V8 intake manifold installation

ENGINE AND ENGINE OVERHAUL

kets, using a sealing compound around the water passages.

NOTE: *1974-75 V8 engines require a new intake manifold side gasket on 4-bbl engines. The new gasket has restricted crossover ports. The 350 2-bbl uses a restricted crossover gasket on the right hand side and an open gasket on the left. The 350 4-bbl uses restricted crossover gaskets on both sides. The correct gaskets are essential.*

11. When installing the manifold, care should be taken not to dislocate the end seals. It is helpful to use a pilot in the distributor opening. Tighten the manifold bolts to 30-35 ft. lbs. in the sequence illustrated.

12. If applicable, install the alternator upper bracket. As required, install the oil filler bracket, air cleaner bracket, air conditioning compressor and bracket, and accelerator bellcrank.

13. Install the distributor.

14. Install the distributor cap and scribe the rotor position relative to the distributor body.

15. Connect:
   a. Negative battery cable.
   b. Upper radiator and heater hoses at the manifold.
   c. Crankcase ventilation hoses as required.
   d. Fuel line at the carburetor, or the fuel line clips and lines at the throttle body, if equipped with fuel injection.
   e. Accelerator linkage and TV cables, if equipped.
   f. Vacuum hose at the distributor, if equipped.
   g. Power brake hose at the carburetor base or manifold, if applicable.
   h. Ignition coil, the spark plug wires and temperature sending switch wires.
   i. Water pump bypass at the water pump (Mark IV only).
   j. If equipped with CCC (ECM), connect the electrical harness to its connectors.

16. Install the air cleaner.

17. Refill the cooling system. Start the engine, adjust the timing (if necessary) and check for leaks.

V6-231 Gasoline Engine

NOTE: *A special wrench adapter, available from Snap-On and several other tool manufacturers, is necessary to remove the left front intake manifold bolt.*

1. Disconnect the negative battery cable. Drain the cooling system. Remove the upper radiator hose, and the coolant bypass hose from the manifold.

CAUTION: *When draining the coolant, keep in mind that cats and dogs are attracted by the ethylene glycol antifreeze, and are quite likely to drink any that is left in an uncovered container or in puddles on the ground. This will prove fatal in sufficient quantity. Always drain the coolant into a sealable container. Coolant should be reused unless it is contaminated or several years old.*

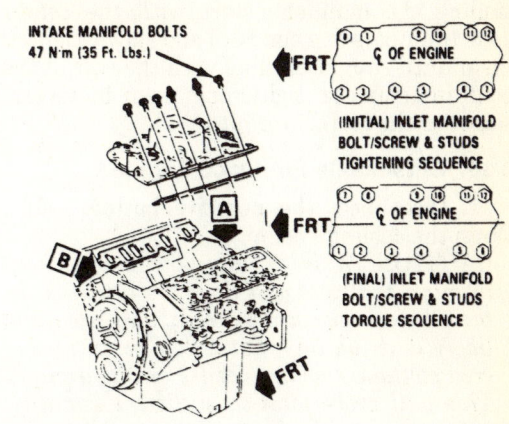

Intake manifold torque sequence, 262 V6

2. Remove the air cleaner. Disconnect the throttle linkage from the carburetor. Remove the linkage bracket from the manifold. If the vehicle is equipped with an automatic transmission, remove the downshift linkage.

3. Disconnect the fuel line from the carburetor. If equipped with power brakes, disconnect the power brake line from the manifold. Disconnect the choke pipe and all vacuum lines. Disconnect the anti-dieseling solenoid wire.

4. Remove the manifold bolts. It will be necessary to remove the distributor cap and rotor to gain access to the front left manifold bolt. This is the special bolt, known as a Torx® bolt. Remove the plug wires from the plugs.

5. Remove the manifold.

6. Installation is in the reverse order of removal. Use a new gasket and seals. Coat the ends of the seals with a non-hardening silicone sealer. The pointed end of the seal should be a snug it against the block and head. When in-

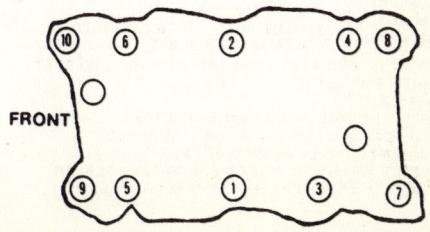

Intake manifold torque sequence, 231 V6

ENGINE AND ENGINE OVERHAUL

stalling the manifold, start with the center bolts (numbers one and two) and slowly tighten them until snug. Continue with the rest of the bolts in sequence, tightening them in several stages to the correct torque.

8-307 Oldsmobile Produced Engine

1. Disconnect the negative battery cable. Drain the cooling system.

CAUTION: *When draining the coolant, keep in mind that cats and dogs are attracted by the ethylene glycol antifreeze, and are quite likely to drink any that is left in an uncovered container or in puddles on the ground. This will prove fatal in sufficient quantity. Always drain the coolant into a sealable container. Coolant should be reused unless it is contaminated or several years old.*

2. Remove the air cleaner assembly. As needed, remove the carburetor assembly from the intake manifold. If equipped with cruise control remove the servo assembly.

3. Remove the top radiator hose. Disconnect the fuel line and remove it for accessibility. As necessary, remove the alternator and the air conditioning compressor mounting bolts.

4. Disconnect all required electrical wires, vacuum lines and throttle cables.

5. Remove the intake manifold retaining bolts. Remove the intake manifold from the engine.

To Install:

6. Coat both sides of the gasket surface that seal the intake manifold to the cylinder heads with G.M. sealer # 1050026 or the equivalent. Use an RTV sealer on the end seals. Position the intake manifold gasket to the cylinder heads. Install the end seals, making sure that the ends are positioned under the cylinder heads.

7. Install the manifold to the engine. Install the retaining bolts. Torque them to specification.

8. If removed, install the carburetor. Connect all the required electrical connections, vacuum lines and throttle linkages.

9. Connect the top radiator hose. Connect the fuel line. Fill the cooling system with the proper solution.

10. Install the air cleaner. Connect the battery cable. Start the engine and check for leaks.

Diesel Engine

1. Disconnect the negative battery cable. Remove the air cleaner.

2. Drain the radiator and loosen the upper bypass hose clamp. Remove the thermostat housing bolts, the housing and the thermostat from the intake manifold.

CAUTION: *When draining the coolant, keep in mind that cats and dogs are attracted by the ethylene glycol antifreeze, and are quite likely to drink any that is left in an uncovered container or in puddles on the ground. This will prove fatal in sufficient quantity. Always drain the coolant into a sealable container. Coolant should be reused unless it is contaminated or several years old.*

3. Remove the breather pipes from the rocker covers and the air crossover. Remove the air crossover.

4. Disconnect the throttle rod and the return spring. If equipped with cruise control, remove the servo.

5. Remove the hairpin clip at the bellcrank

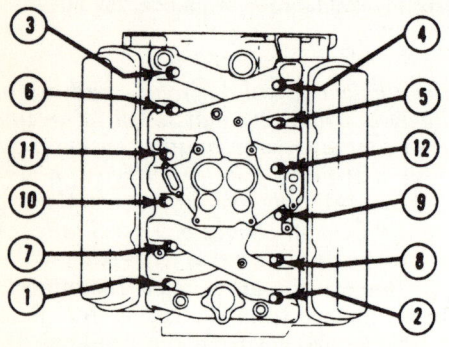

1. LUBRICATE ENTIRE BOLT IN ENGINE OIL
2. TIGHTEN ALL BOLTS IN SEQUENCE SHOWN TO 20 N·m (15 LBS. FT.)
3. RETIGHTEN IN SEQUENCE SHOWN TO 54 N·m (40 LBS. FT.)

Intake manifold bolt torque sequence 307 Oldsmobile produced engine

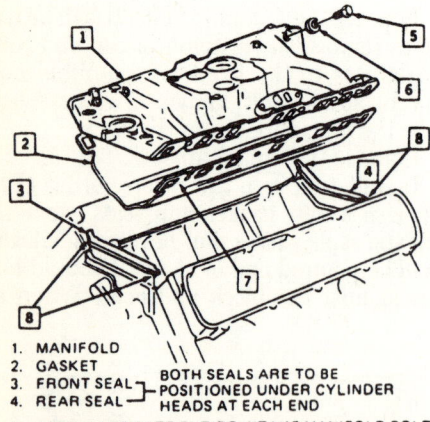

1. MANIFOLD
2. GASKET
3. FRONT SEAL — BOTH SEALS ARE TO BE POSITIONED UNDER CYLINDER
4. REAR SEAL — HEADS AT EACH END
5. BOLT (LUBRICATE ENTIRE INTAKE MANIFOLD BOLT WITH ENGINE OIL)
6. WASHER
7. APPLY 1050026 OR EQUIVALENT SEALER TO BOTH SIDES OF THE INTAKE MANIFOLD GASKET AT ALL THE PORT AREAS.
8. APPLY 1052915, GE 1673 OR EQUIVALENT SEALER TO BOTH ENDS OF INTAKE MANIFOLD SEALS.

Intake manifold installation 307 Oldsmobile produced engine

ENGINE AND ENGINE OVERHAUL

and disconnect the cables. Remove the throttle cable from the bracket on the manifold; position the cable away from the engine. Disconnect and label any wiring as necessary.

6. Remove the alternator bracket if necessary. If equipped with air conditioning, remove the compressor mounting bolts and move the compressor aside, without disconnecting any of the hoses. Remove the compressor mounting bracket from the intake manifold.

7. Disconnect the fuel line from the pump and the fuel filter. Remove the fuel filter and bracket.

8. Remove the fuel injection pump and lines. See Chapter 5, Fuel System, for procedures.

9. Disconnect and remove the vacuum pump or oil pump drive assembly from the rear of the engine.

10. Remove the intake manifold drain tube.

11. Remove the intake manifold bolts and the manifold. Remove the adapter seal and the injection pump adapter.

12. Clean the mating surfaces of the cylinder heads and the intake manifold using a gasket scraper.

To Install:

13. Coat both sides of the gasket surface that seal the intake manifold to the cylinder heads with G.M. sealer # 1050026 or the equivalent. Use an RTV sealer only on the end seals. Position the intake manifold gaskets on the cylinder heads. Install the end seals, making sure that the ends are positioned under the cylinder heads.

14. Carefully lower the intake manifold into place on the engine.

15. Clean the intake manifold bolts thoroughly, then dip them in clean engine oil. Install the bolts and tighten to 15 ft. lbs. in the sequence shown. Next, tighten all the bolts to 30 ft. lbs., in sequence, and finally tighten to 40 ft. lbs. in sequence.

16. Install the intake manifold drain tube and clamp.

17. Install injection pump adapter. See Chapter 5. If a new adapter is not being used, skip steps 4 and 9. 18. Install the intake manifold drain tube.

19. Connect and install the vacuum pump or oil pump drive assembly at the rear of the engine.

20. Install the fuel injection pump and lines. See Chapter 5, Fuel System, for procedures.

21. Connect the fuel line at the pump and the fuel filter. Install the fuel filter and bracket.

22. Install the alternator bracket if necessary. If equipped with air conditioning, install the compressor mounting bracket and bolts.

23. Install the throttle cable at the bracket

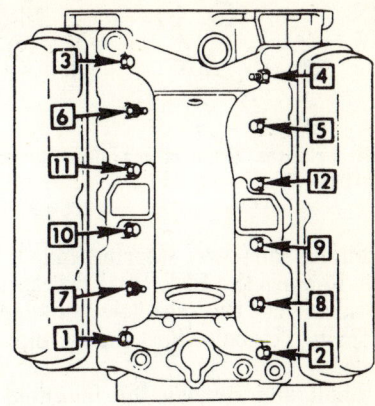

350 diesel intake manifold torque sequence

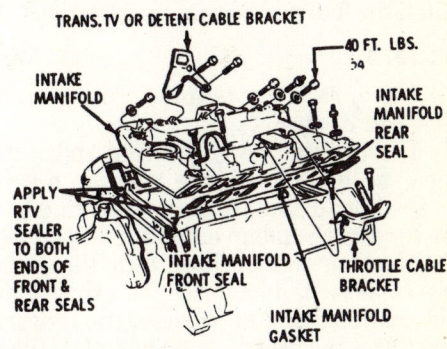

Diesel intake manifold and gaskets

on the manifold. Connect any wiring as necessary.

24. Connect the throttle rod and the return spring. If equipped with cruise control, install the servo.

25. Install the air crossover. Install the breather pipes on the rocker covers and the air crossover.

26. Install the thermostat.

27. Fill the cooling system.

28. Connect the negative battery cable.

29. Install the air cleaner.

Exhaust Manifold

REMOVAL AND INSTALLATION

Inline 6-Cylinder Gasoline Engine

1968-74

Exhaust manifold removal for these model years is covered in the Intake Manifold, Removal and Installation section earlier in this chapter.

1975-76

1. Disconnect the negative battery cable. Remove the air cleaner.

ENGINE AND ENGINE OVERHAUL

Inline six cylinder exhaust manifold torque sequence, 1975-79

2. Remove the power steering and air pump brackets. Remove the EFE valve bracket.
3. Disconnect the throttle linkage and return spring. Unbolt the exhaust pipe from the flange.
4. Unbolt and remove the manifold from the engine.
5. Reverse the procedure for installation. Tighten the four end bolts to specifications last.

1975-1979

1. Disconnect the negative battery cable. Remove air cleaner.
2. Remove the power steering and/or AIR pumps and brackets, where they are present. (You need not disconnect power steering hoses - just support the pump out of the way).
3. Working from below, with the vehicle safely supported, disconnect the exhaust pipe at the manifold and at the catalytic converter bracket near the transmission mount. If the vehicle uses an exhaust manifold mounted converter, disconnect the pipe at the bottom of the converter and remove the converter.
4. Working from above, remove the rear heat shield and accelerator cable bracket.
5. Remove the exhaust manifold bolts, and pull off the manifold.

To Install:

6. If the manifold is to be replaced, transfer the EFE valve, actuator, and rod assembly to the new part.
7. Clean and then inspect the manifold carefully for cracks, and for free operation of the EFE valve. Repair or replace parts, and free up the EFE valve with solvent, if necessary.
8. Make sure the gasket surface is clean and free of deep scratches. Position a new gasket on the manifold, and then put the manifold in position on the block, and install the bolts hand tight.
9. Torque all bolts in the proper order to the specified torque (see illustration).
10. Install the rear heat shield and the accelerator cable bracket.
11. Working from underneath, connect the exhaust pipe at the manifold flange and connect the converter bracket at the transmission mount. If the vehicle has an exhaust manifold converter, first install the converter to the manifold loosely; then attach the exhaust pipe to the converter and align the exhaust system; finally, torque the converter mounting bolts to 15 ft. lbs., in an X pattern, and torquing in several stages.
12. Working from above, install the power steering and A.I.R. pumps as necessary. Then, install the air cleaner and connect the battery ground cable.

V6 and V8 Gasoline Engines

1968-84

1. Disconnect the negative battery cable.
2. If equipped with AIR (Air Injection Reaction), remove the air injector manifold assembly. The $1/4''$ pipe threads in the manifold are straight threads. Do not use a $1/4''$ tapered pipe tap to clean the threads.
3. If applicable, remove the air cleaner preheater shroud.
4. Remove the spark plug wire heat shields. On Mark IV engine, remove spark plugs.
5. On the left exhaust manifold, disconnect and remove the alternator.

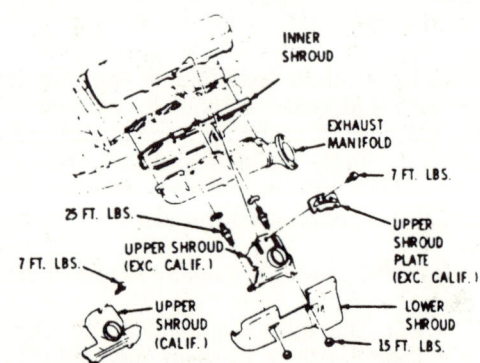

Typical exhaust manifold and hot air shrouds

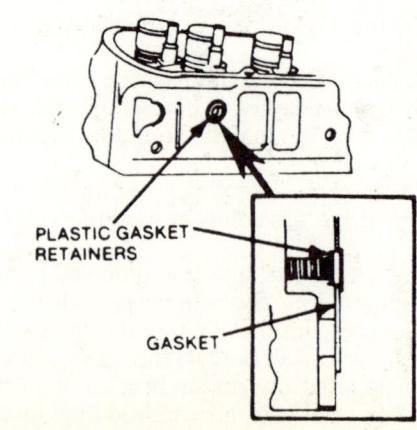

Plastic manifold gasket retainers

ENGINE AND ENGINE OVERHAUL   145

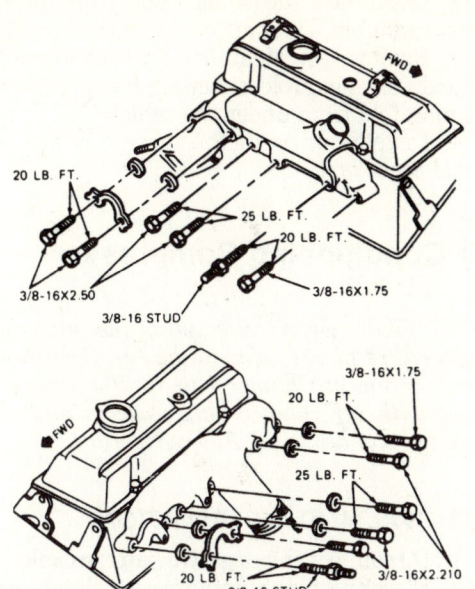

Exhaust manifold installation, V6, V8 (1980 shown, others similar)

6. Disconnect the exhaust pipe from the manifold and hang it from the frame out of the way.
7. Bend the locktabs and remove the end bolts, then the center bolts. Remove the manifold.

NOTE: *A ‹9/16" thin wall 6-point socket, sharpened at the leading edge and tapped onto the head of the bolt, simplified bending the locktabs. When installing a new manifold on the right side on 1978-84 V8s you must transfer the heat stove from the old manifold to the new one.*

8. Position the manifold and install the end bolts, then the center bolts. Bend the locktabs.
9. Connect the exhaust pipe at the manifold.
10. On the left exhaust manifold, install the alternator.
11. Install the spark plug wire heat shields. On Mark IV engine, install spark plugs.
12. If applicable, install the air cleaner preheater shroud.
13. If equipped with AIR (Air Injection Reaction), install the air injector manifold assembly.
14. Connect the negative battery cable.

V6 Gasoline Engine

RIGHT SIDE (1985-91)

1. Disconnect the negative battery cable. Remove the air cleaner. Raise and support the vehicle safely. Disconnect the spark plug wires.
2. Remove the exhaust pipe bolts, then lower the vehicle. If equipped with air management valve, disconnect the bracket.
3. If equipped with AIR, disconnect the hoses, then the pipes to the converter, the cylinder head and the exhaust manifold.
4. Remove the exhaust manifold bolts and the manifold.
5. Installation is the reverse of the removal procedure. Be sure to use new gaskets, as required.

LEFT SIDE (1985-91)

1. Disconnect the negative battery cable. Raise and support the vehicle safely. Remove the exhaust pipe bolts, then lower the vehicle.
2. If equipped with air conditioning, remove the compressor and the rear adjusting brace.
3. If equipped with power steering, remove the power steering pump and the rear lower power steering adjusting brace.
4. Disconnect the spark plug wires from the spark plugs. Remove the exhaust manifold bolts and the manifold.
5. Installation is the reverse of the removal procedure. be sure to use new gaskets, as required.

V8 Engine, Except 8-307 Oldsmobile produced Engine

RIGHT SIDE (1985-92)

1. Disconnect the negative battery cable. Raise and support the vehicle safely. Remove the exhaust pipe bolts, then lower the vehicle.
2. Remove the air cleaner. Disconnect the spark plug wires from the spark plugs, the vacuum hoses from the carbon canister and the AIR hoses.
3. Loosen the alternator belt, then remove the lower alternator bracket and the AIR valve. Disconnect the converter's AIR pipe from the back of the manifold.
4. Remove the exhaust manifold bolts and the manifold.
5. Installation is the reverse of the removal procedure. be sure to use new gaskets, as required.

LEFT SIDE (1985-92)

1. Disconnect the negative battery cable. Raise and support the vehicle safely. Remove the exhaust pipe from the manifold, and lower the vehicle.
2. Disconnect the AIR hose. If equipped with air conditioning, loosen the bracket at the front of the head, then remove the rear bracket and the compressor.
3. If equipped with power steering, remove

146 ENGINE AND ENGINE OVERHAUL

the power steering pump and the lower adjusting bracket.
4. Remove the exhaust manifold bolts, the wire loom holder at the valve cover and the manifold.
5. Installation is the reverse of the removal procedure.

307 Oldsmobile Produced Engine

RIGHT SIDE

1. Disconnect the negative battery cable. As required, remove the air manifold pipes for accessibility.
2. Raise and support the vehicle safely. Disconnect the exhaust pipe at the exhaust manifold. Disconnect the crossover pipe. Lower the vehicle.
3. Remove the exhaust manifold retaining bolts. Remove the exhaust manifold from the engine.
4. Installation is the reverse of the removal procedure.

LEFT SIDE

1. Disconnect the negative battery cable. As required, remove the air manifold pipes for accessibility.
2. Raise and support the vehicle safely. Disconnect the exhaust pipe at the exhaust manifold. Lower the vehicle.
3. Remove the lower alternator bracket, as needed. Remove the heat shield assembly retaining bolts. Remove the heat shield.
4. Remove the exhaust manifold retaining bolts. Remove the exhaust manifold from the engine.
5. Installation is the reverse of the removal procedure.

Diesel Engine

LEFT SIDE

1. Disconnect the negative battery cable. Remove the air cleaner.
2. Remove the alternator lower bracket. Raise and support the vehicle safely.
3. Disconnect the exhaust pipe at the exhaust manifold. Lower the vehicle.
4. Remove the exhaust manifold retaining bolts. remove the exhaust manifold from the engine.
5. Installation is in the reverse of the removal procedure.

RIGHT SIDE

1. Disconnect the negative battery cable. Raise and support the vehicle safely.

2. Disconnect the exhaust pipe from the exhaust manifold. Disconnect the crossover pipe.
3. Remove the right front wheel. Remove the exhaust manifold retaining bolts. Remove the manifold from under the vehicle.
4. Installation is in the reverse of the removal procedure.

Air Conditioning Compressor

CAUTION: *Before attempting the following procedure please refer to the Air Conditioning section in Chapter 1 to familiarize yourself with air conditioning systems and the cautions of handling A/C refrigerant.*

REMOVAL AND INSTALLATION

1. Disconnect the negative battery cable.
2. Disconnect the compressor clutch coil wire.
3. Remove the engine fan and fan shroud.
4. Remove the drive belt from the compressor.
5. Remove the compressor bracket.
6. Refer to Chapter 1 and discharge the air conditioning system.
7. Remove the screw attaching the muffler to the compressor support.
8. Remove the hose and muffler assembly from the compressor.
9. Remove the vacuum pump hose from the metal vacuum line.
10. Remove the four nuts and one bolt and spacer from the compressor support.
11. Remove the compressor support by pulling forward.
12. Pull the compressor forward to remove.
To Install:
13. Position the compressor to its mounting.
14. Install the compressor support by pulling forward.
15. Install the four nuts and one bolt and spacer from the compressor support.
16. Install the vacuum pump hose to the metal vacuum line.
17. Install the hose and muffler assembly to the compressor.
18. Install the screw attaching the muffler to the compressor support.
19. Install the compressor bracket.
20. Install the drive belt to the compressor and adjust.
21. Install the engine fan and fan shroud.
22. Connect the compressor clutch coil wire.
23. Refer to Chapter 1 and charge the air conditioning system.
24. Connect the negative battery cable.

ENGINE AND ENGINE OVERHAUL

Radiator

REMOVAL AND INSTALLATION

1. Disconnect the negative battery cable. Drain cooling system.

CAUTION: *When draining the coolant, keep in mind that cats and dogs are attracted by the ethylene glycol antifreeze, and are quite likely to drink any that is left in an uncovered container or in puddles on the ground. This will prove fatal in sufficient quantity. Always drain the coolant into a sealable container. Coolant should be reused unless it is contaminated or several years old.*

2. If necessary, remove the fan assembly. Remove the upper fan shroud or the upper support.
3. Disconnect upper and lower radiator hoses.
4. Disconnect and plug the oil cooler lines, if equipped with an automatic transmission.
5. Remove the radiator from the vehicle.

NOTE: *If necessary to remove the fan assembly keep it in an upright position to prevent the fluid from leaking, if the vehicle is equipped with air conditioning.*

6. Installation is the reverse of the removal procedure.
7. Fill the radiator with the proper solution. Start the engine and check for leaks.

Condenser

REMOVAL AND INSTALLATION

1. Disconnect the negative battery cable.
2. Discharge the air conditioning system.

NOTE: *Refer to Chapter 1 for Discharging, Evacuating and Charging of the air conditioning System.*

3. As required, remove the upper radiator shroud. Disconnect the air conditioning lines at the condenser.
4. Push the radiator forward and pull the condenser out from the top.
5. Installation is the reverse of the removal procedure.
6. Be sure to properly recharge the air conditioning system.

Water Pump

The water pump is a die cast, centrifugal-type with sealed bearings. Since it is pressed together, it must be serviced as a unit.

REMOVAL AND INSTALLATION

Except Diesel Engine and 8-307 Oldsmobile Produced Engine

1. Disconnect the negative battery cable.

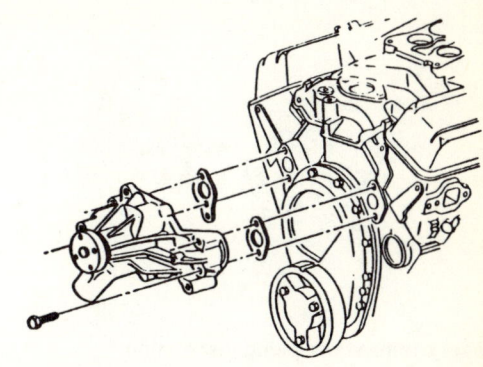

V6, V8 water pump installation

Drain the cooling system.

CAUTION: *When draining the coolant, keep in mind that cats and dogs are attracted by the ethylene glycol antifreeze, and are quite likely to drink any that is left in an uncovered container or in puddles on the ground. This will prove fatal in sufficient quantity. Always drain the coolant into a sealable container. Coolant should be reused unless it is contaminated or several years old.*

2. If necessary, remove the fan shroud or the upper radiator support.
3. Remove the necessary drive belts.
4. Remove the fan and water pump pulley.
5. Remove the alternator and the power steering pump (if equipped) brackets, then move the units aside.
6. Remove the heater hose and the lower radiator hose from the pump.
7. Remove the water pump retaining bolts and the pump. Clean the gasket mounting surfaces.

NOTE: *Use an anti-seize compound on the water pump bolt threads.*

8. To install, use new gaskets and reverse the removal procedures. Torque the water pump, the alternator and the power steering (if equipped) mounting bolts to 30 ft. lbs. Adjust the drive belts and fill the cooling system.

NOTE: *If a belt tensioning gauge is available, adjust the belts to 100-130 lbs. of tension on new belts or to 70 lbs. on used belts. If the gauge is not available, adjust the belts so that a $1/4$-$1/2$" deflection can be made on the longest span of the belt under moderate thumb pressure.*

Diesel Engine and 8-307 Oldsmobile Produced Engine

1. Disconnect the negative battery cable. Drain the cooling system.

ENGINE AND ENGINE OVERHAUL

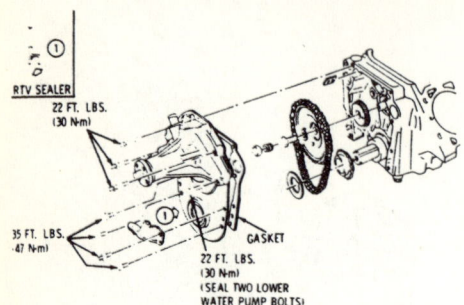

Diesel engine water pump installation

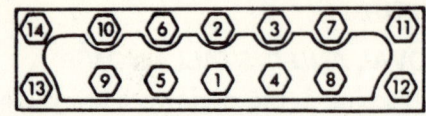

Cylinder head torque sequence — inline six cylinder

CAUTION: *When draining the coolant, keep in mind that cats and dogs are attracted by the ethylene glycol antifreeze, and are quite likely to drink any that is left in an uncovered container or in puddles on the ground. This will prove fatal in sufficient quantity. Always drain the coolant into a sealable container. Coolant should be reused unless it is contaminated or several years old.*

2. Disconnect the lower radiator hose, the heater hose and the by-pass hose from the water pump.
3. Remove the fan assembly, the drive belts and the water pump pulley.
4. Remove the alternator, the power steering pump and the air conditioning compressor brackets, then move the units aside.
5. Remove the water pump mounting bolts and the pump. Clean the gasket mounting surfaces.
6. To install, use new gaskets, sealant and reverse the removal procedures. Torque the water pump bolts to 22 ft. lbs. Adjust the belts and refill the cooling system.

NOTE: *Apply sealer to the lower water pump bolts.*

Cylinder Head

REMOVAL AND INSTALLATION

NOTE: *The engine should be overnight cold before the cylinder head is removed to prevent warpage.*

Inline 6-Cylinder Gasoline Engine

1. Disconnect the negative battery cable.
2. Drain cooling system. Remove air cleaner and disconnect the PCV hose.

CAUTION: *When draining the coolant, keep in mind that cats and dogs are attracted by the ethylene glycol antifreeze, and are quite likely to drink any that is left in an uncovered container or in puddles on the ground. This will prove fatal in sufficient quantity. Always drain the coolant into a sealable container. Coolant should be reused unless it is contaminated or several years old.*

3. Disconnect accelerator pedal rod at the bellcrank on the manifold, the fuel and the vacuum lines at the carburetor.
4. Disconnect the exhaust pipe at the manifold flange, then remove the manifold bolts, the clamps, the manifolds and the carburetor as an assembly.
5. Remove the fuel and the vacuum line retaining clip from the water outlet. Disconnect the wiring harness from the temperature sending unit and the coil, leaving the harness clear of the clips on the rocker arm cover.
6. Disconnect the radiator hose at the water outlet housing and the negative battery cable from the cylinder head.
7. Disconnect the wires and remove the spark plugs.
8. Remove the rocker arm cover, then back off the rocker arm nuts. Pivot the rocker arms to clear the push rods and remove the push rods.
9. Remove the cylinder head bolts, the cylinder head and the gasket.
10. Using a gasket scraper, clean the gasket mounting surfaces.

To Install:

11. Using a new gasket, position the head on the block. Oil the cylinder head bolts and install them. Torque the cylinder head bolts a little at a time using the torquing sequence.
12. Install the push rods. Pivot the rocker arms over the push rods, then tighten the rocker arm nuts.
13. Adjust the valves.
14. Install the rocker arm cover.
15. Install the spark plugs and wires.
16. Connect the radiator hose at the water outlet housing and the negative battery cable at the cylinder head.
17. Install the fuel and vacuum line retaining clip at the water outlet. Connect the wiring harness at the temperature sending unit and the coil.
18. Install the manifolds and the carburetor as an assembly. Connect the exhaust pipe at the manifold flange.
19. Connect accelerator pedal rod at the bellcrank on the manifold, the fuel and the vacuum lines at the carburetor.
20. Fill cooling system.

ENGINE AND ENGINE OVERHAUL

21. Install air cleaner.
22. Connect the PCV hose.
23. Connect the negative battery cable.

V6 and V8 Gasoline engine Except V6-231

1. Remove the intake manifold.
2. Remove the alternator's lower mounting bolt and move the unit aside.
3. Remove the exhaust manifolds, the rocker arm covers and the rocker arm assemblies. Remove the pushrods.
4. As required, remove the air injection manifold assemblies. Drain the cooling system.

CAUTION: *When draining the coolant, keep in mind that cats and dogs are attracted by the ethylene glycol antifreeze, and are quite likely to drink any that is left in an uncovered container or in puddles on the ground. This will prove fatal in sufficient quantity. Always drain the coolant into a sealable container. Coolant should be reused unless it is contaminated or several years old.*

5. Remove the diverter valve, if equipped. Remove the cylinder head bolts and the cylinder heads from the engine.
6. Installation is the reverse of the removal procedure.
7. On the 307 Oldsmobile produced engine,

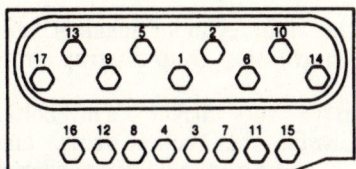

Cylinder head torque sequence small block Chevrolet built V8 engine

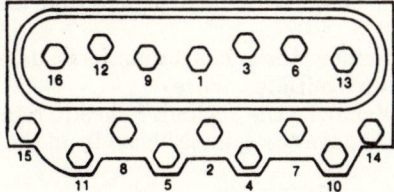

Cylinder head bolt torque sequence big block Chevrolet built V8 engine

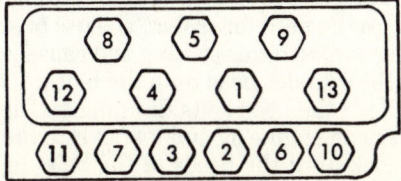

Cylinder head bolt torque sequence V6 engine except 231 V6 engine

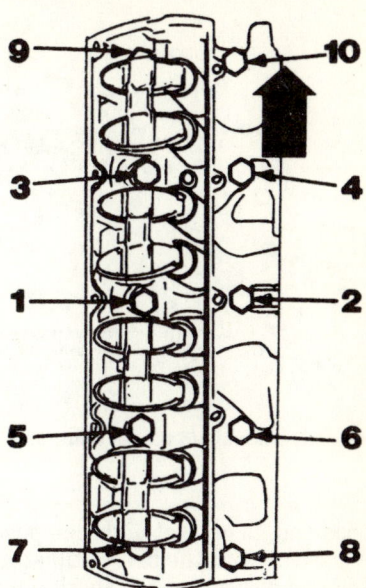

Cylinder head bolt torque sequence 307 Oldsmobile produced engine

be sure to coat the cylinder head retaining bolts with clean engine oil prior to installation.

8. Be sure to use new gaskets or RTV sealant, as required. Torque the cylinder head bolts to specification and in the proper sequence.

V6-231 Engine

NOTE: *On vehicles equipped with AIR, disconnect the rubber hose at the injection tubing check valve. This way the tubing will not have to be removed from the exhaust manifold.*

1. Remove the intake manifold.
   a. Loosen and remove all drive belts.
   b. Tag and disconnect the wires leading from the rear of the alternator.
   c. Remove the air conditioning compressor, if equipped and position it out of the way with all the hoses still connected.
   d. Remove the alternator and its mounting bracket.
3. When removing the left cylinder head:
   a. Remove the oil gauge rod.
   b. Remove the power steering pump, if

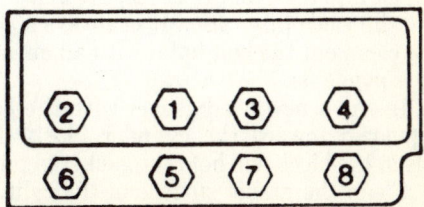

Cylinder head torque sequence, 231 V6

ENGINE AND ENGINE OVERHAUL

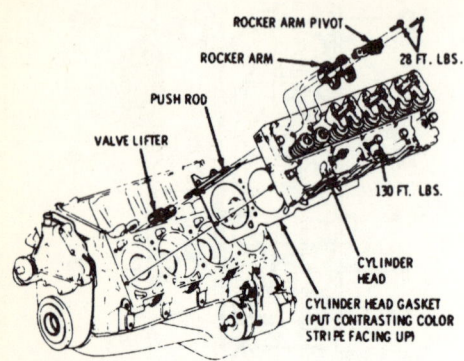

Cylinder head removal, V8 diesel engines

350 V8 diesel cylinder head torque sequence

equipped and its bracket and then position it out of the way with the hoses still attached.

4. Tag and disconnect the spark plug wires and then remove the spark plug wires clips from the cylinder head cover studs.

5. Remove the exhaust manifold mounting bolts from the head which is being removed, and then pull the manifold away from the head.

6. Use an air hose if available, or a bunch of clean rags and clean the dirt off the head and surrounding areas thoroughly. It is extremely important to avoid getting dirt into the hydraulic valve lifters.

7. Remove the cylinder head cover from the top of the head that you wish to remove.

8. Remove the rocker arm and shaft assembly from the cylinder head and then remove the pushrods.

NOTE: *If the valve lifters are to serviced, remove them at this time. Otherwise, protect the lifters and the camshaft from dust and dirt by covering the entire area with a clean cloth. Whenever the lifters or the pushrods are removed from the head it is always a good idea to place them in a wooden block with numbered holes to keep them identified as to their position in the engine.*

9. Loosen and remove all the cylinder head bolts and then lift off the cylinder head.

To install:

10. Clean the engine block gasket surface thoroughly. Make sure that no foreign material has fallen into the cylinder bores, the bolt holes or into the valve lifter area. It is always a good idea to clean out the bolt holes with an air hose if one is available.

11. Install a new head gasket with the bead facing down toward the cylinder block. The dowels in the block will hold the gasket in place.

12. Clean the gasket surface of the cylinder head and carefully set it into place on the dowels in the cylinder block.

13. Use a heavy body thread sealer on all of the head bolts since the bolt holes go all the way through into the coolant.

14. Install the head bolts. Tighten the bolts a little at a time about three times around in the sequence shown in the illustration. Tighten the bolts to a final torque equal to that given in the Torque Specifications chart.

15. Installation of the remaining components is in the reverse order of removal.

Diesel Engine

1. Remove the intake manifold, using the procedure outlined above.

2. Remove the rocker arm covers, after removing any accessory brackets with interfere with cover removal.

3. Disconnect and label the glow plug wiring.

4. If the right cylinder head is being removed, remove the ground strap from the head.

5. Remove the rocker arm bolts, the bridged pivots, the rocker arms, and the pushrods, keeping all the parts in order so that they can be returned to their original locations. It is a good practice to number or mark the parts to avoid interchanging them.

6. Remove the fuel return lines from the nozzles.

7. Remove the exhaust manifolds, using the procedures outlined above.

8. Remove the cylinder head retaining bolts. Remove the cylinder head from the vehicle.

To Install:

9. First clean the mating surfaces thoroughly. Install new head gaskets on the engine block. DO NOT coat the gaskets with any sealer. The gaskets have a special coating that eliminates the need for sealer. The use of sealer will interfere with this coating and cause leaks. Install the cylinder head onto the block.

10. Clean the head bolts thoroughly. Dip the bolts in clean engine oil and install into the cylinder block until the heads of the bolts lightly contact the cylinder head.

11. Tighten the bolts, in the sequence illustrated, to 100 ft. lbs. When all bolts have been

ENGINE AND ENGINE OVERHAUL 151

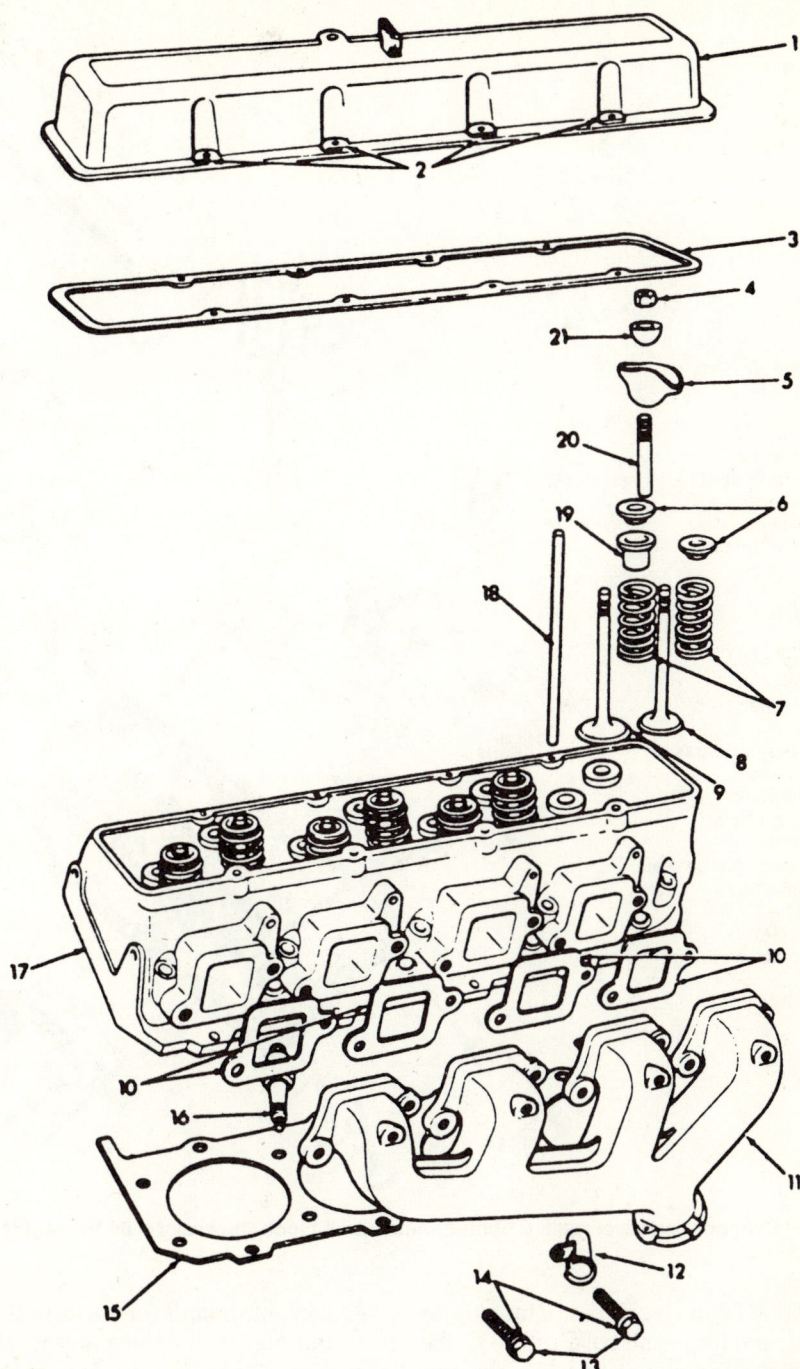

1. Valve cover
2. Screw reinforcements
3. Gasket
4. Adjusting nut
5. Rocker arm
6. Valve spring retainer
7. Valve spring
8. Exhaust valve
9. Intake valve
10. Gasket
11. Exhaust manifold
12. Spark plug shield
13. Bolt
14. Washer
15. Head gasket
16. Spark plug
17. Cylinder head
18. Pushrod
19. Spring shield
20. Rocker arm stud
21. Rocker arm ball

Exploded view of the big block cylinder head

152 ENGINE AND ENGINE OVERHAUL

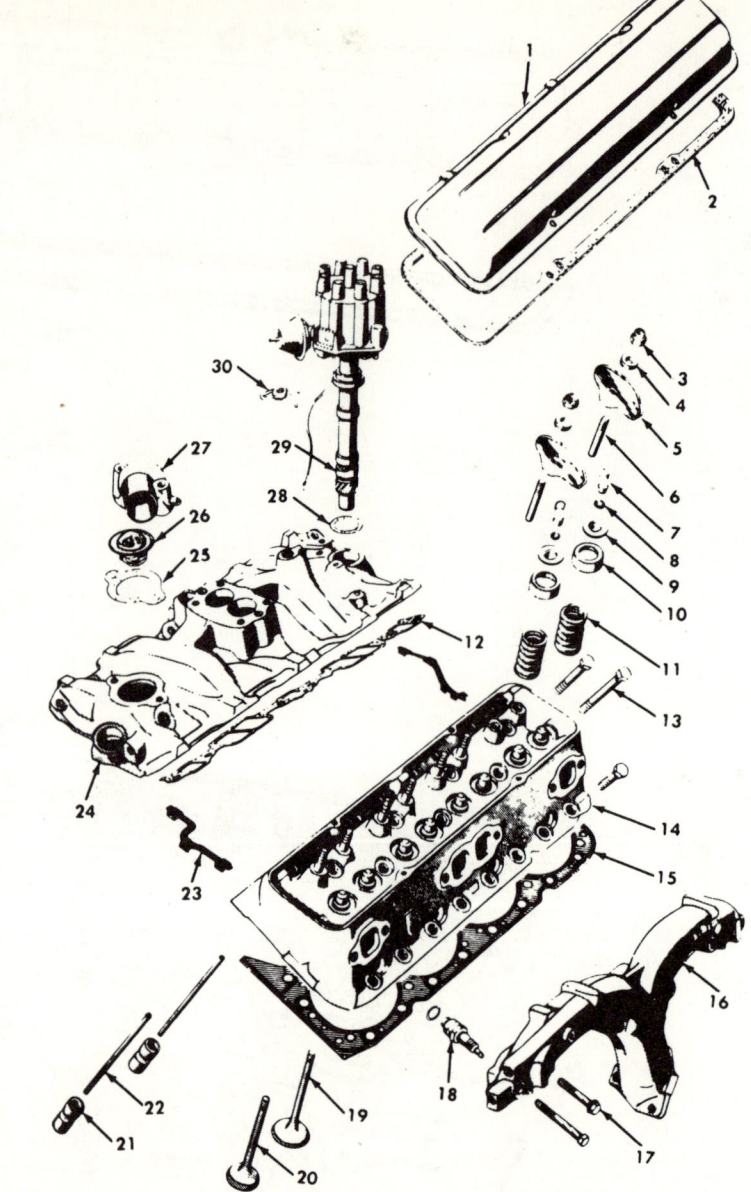

1. Rocker arm cover
2. Gasket
3. Nut
4. Ball
5. Rocker arms
6. Rocker arm studs
7. Valve keeper locks
8. O-ring seals
9. Valve spring cap
10. Shield
11. Spring
12. Gasket
13. Bolts
14. Cylinder head
15. Head gasket
16. Exhaust manifold
17. Bolts
18. Spark plug and gasket
19. Intake valve
20. Exhaust valve
21. Hydraulic lifters
22. Push rods
23. Intake manifold gaskets
24. Intake manifold
25. Gasket
26. Thermostat
27. Thermostat housing
28. Gasket
29. Distributor
30. Clamp

Cylinder head and related components small block Chevrolet built V8 engine

tightened to this figure, begin the tightening sequence again, and torque all bolts to 130 ft. lbs.

12. Install the exhaust manifolds, the fuel return lines, the glow plug wiring, and the ground strap for the right cylinder head.

13. Install the valve train assembly. Refer to Diesel Engine, Rocker Arm Replacement, above, for valve lifter bleeding procedures.

14. Install the intake manifold.

15. Install the rocker covers. The valve covers are sealed with RTV silicone sealer instead of a gasket. Use GM No. 1052434 or its equivalent. Install the cover to the head within 10 minutes (while the sealer is still wet).

CLEANING AND INSPECTION

Chip carbon away from the valve heads, combustion chambers, and ports, using a chisel made of hardwood. Remove the remaining deposits with a stiff wire brush.

NOTE: *Be sure that the deposits are actually removed, rather than burnished. Have the cylinder head hot-tanked to remove grease, corrosion, and scale from the water passages.*

ENGINE AND ENGINE OVERHAUL 153

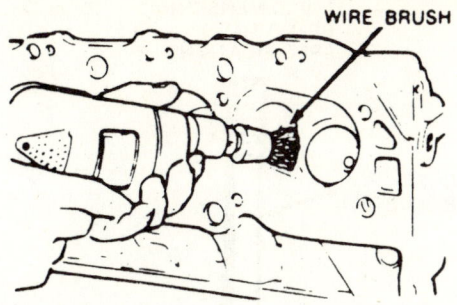

Removing the carbon from the cylinder head with a wire brush and an electric drill

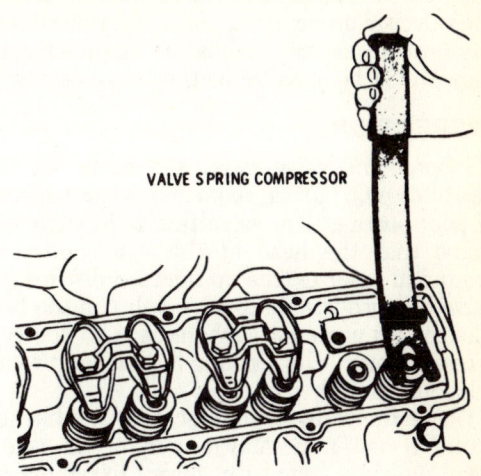

Removing the valve springs

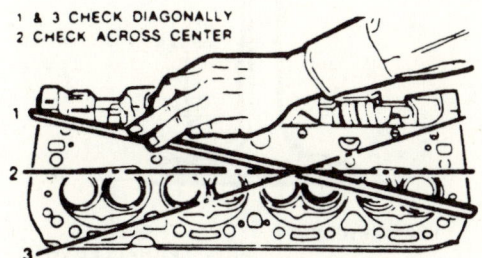

Checking the cylinder head for warpage

Clean the remaining cylinder head parts in an engine cleaning solvent. Do not remove the protective coating from the spring.

Place a straightedge across the gasket surface of the cylinder head. Using feeler gauges, determine the clearance at the center of the straightedge. If warpage exceeds 0.003″ in a 6″ span, or 0.006″ over the total length, the cylinder head must be resurfaced.

NOTE: *If warpage exceeds the manufacturer's maximum tolerance for material removal, the cylinder head must be replaced. when milling the cylinder heads of V-type engines, the intake manifold mounting position is altered, and must be corrected by milling the manifold flange a proportionate amount.*

RESURFACING

NOTE: *This procedure should only be performed by a machine shop.*

Valves and Springs

REMOVAL AND INSTALLATION

1. Remove the heads, and place on a clean surface.
2. Using a suitable spring compressor (for pushrod-type overhead valve engines), compress the valve spring and remove the valve spring cap key. Release the spring compressor and remove the valve spring and cap (and valve rotator on some engines).

NOTE: *Use care in removing the keys; they are easily lost.*

3. Remove the valve seals from the intake valve guides. Throw these old seals away, as you'll be installing new seals during reassembly.
4. Slide the valves out of the head from the combustion chamber side.
5. Make a holder for the valves out of a piece of wood or cardboard, as outlined for the pushrod in Cylinder Head Removal. Make sure

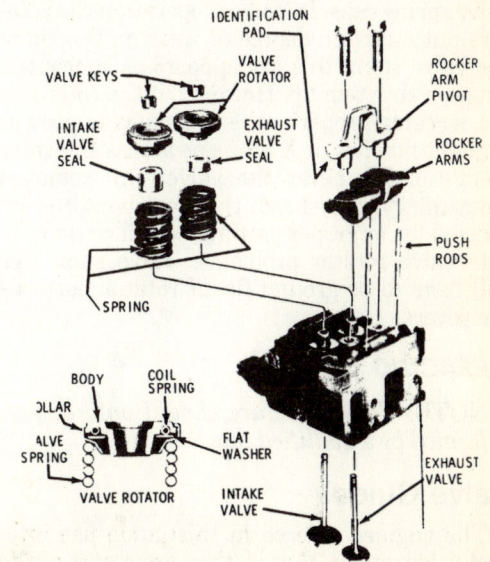

Valve train exploded view, all engine similar except 231 V6 (with rocker shafts)

154   ENGINE AND ENGINE OVERHAUL

you number each hole in the cardboard to keep the valves in proper order. Slide the valve out of the head from the combustion chamber side; they MUST be installed as they were removed.

INSPECTION

Inspect the valve faces and seats (in the head) for pits, burned spots and other evidence of poor seating. If a valve face is in such bad shape that the head of the valve must be ground in order to true up the face, discard the valve because the sharp edge will run too hot. The correct angle for valve faces is 45 degrees. We recommend the refacing be done at a reputable machine shop.

Check the valve stem for scoring and burned spots. If not noticeably scored or damaged, clean the valve stem with solvent to remove all gum and varnish. Clean the valve guides using solvent and an expanding wire type valve guide cleaner. If you have access to a dial indicator for measuring valve stem-to-guide clearance, mount it so that the stem of the indicator is at 90 degrees to the valve stem,, and as close to the valve guide as possible. Move the valve off its seat, and measure the valve guide-to-stem clearance by rocking the stem back and forth to actuate the dial indicator. Measure the valve stems using a micrometer, and compare to specifications to determine whether stem or guide wear is responsible for the excess clearance. If a dial indicator and micrometer are not available to you, take your cylinder head and valves to a reputable machine shop of inspection.

Some of the engines covered in this guide are equipped with valve rotators, which double as valve spring caps. In normal operation the rotators put a certain degree of wear on the tip of the valve stem; this ear appears as concentric rings on the stem tip. However, if the rotator is not working properly, the wear may appear as straight notches or **X** patterns across the valve stem tip. Whenever the valves are removed from the cylinder head, the tips should be inspected for improper pattern, which could indicate valve rotator problems. Valve stem tips will have to be ground flat if rotator patterns are severe.

REFACING

NOTE: *This procedure should only be performed by a qualified machine shop.*

Valve Guides

The engines covered in this guide use integral valve guides; that is, they are a part of the cylinder head and cannot be replaced. The guides can, however, be reamed oversize if they are found to be worn past an acceptable limit.

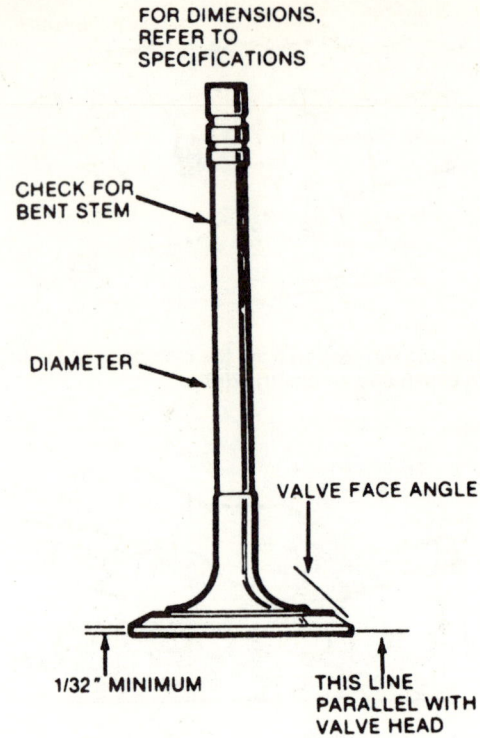

Critical valve dimensions

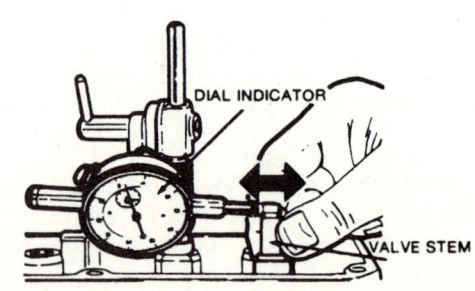

Checking the valve stem to guide clearance

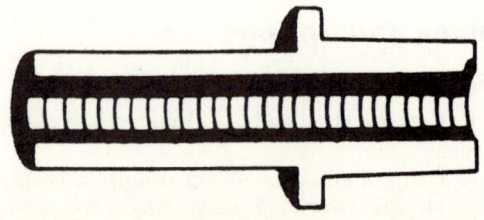

Cutaway of a knurled valve guide

ENGINE AND ENGINE OVERHAUL

Have the valve seat concentricity checked at a machine shop

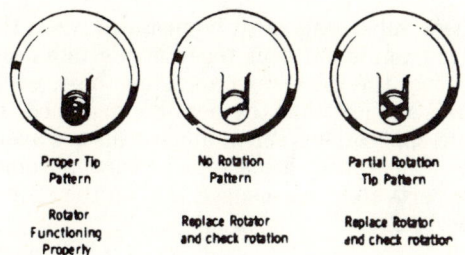

Valve stem wear

Occasionally, a valve guide bore will be oversize as manufactured. These are marked on the inboard side of the cylinder heads on the machined surface just above the intake manifold.

If the guides must be reamed (this service is available at most machine shops) then valves with oversize stems must be fitted. Valves are usually available in 0.001", 0.003" and 0.005" stem oversizes. Valve guides which are not excessively worn or distorted may, in some cases, be knurled rather than reamed. Knurling is a process which the metal on the valve guide bore is displaced and raised, thereby reducing clearance. Knurling also provides excellent oil control. The option of knurling rather than reaming valve guides should be discussed with a reputable machinist or engine specialist.

CYLINDER HEAD CLEANING AND INSPECTION

NOTE: *Any diesel cylinder head work should be handled by a reputable machine shop familiar with diesel engines. Disassembly, valve lapping, and assembly can be completed by following the gasoline engine procedures.*

Gasoline Engine

Once the complete valve train has been removed from the cylinder head(s), the head itself can be inspected, cleaned and machined (if necessary). Set the head(s) on a clean work space, so the combustion chambers are facing up. Begin cleaning the chambers and ports with a hardwood chisel or other non-metallic tool (to avoid nicking or gouging the chamber, ports, and especially the valve seats). Chip away the major carbon deposits, then remove the remainder of carbon with a wire brush fitted to an electric drill.

NOTE: *Be sure that the carbon is actually removed, rather than just burnished. After decarbonizing is completed, take the head(s) to a machine shop and have the head hot tanked. In this process, the head is lowered into a hot chemical bath that very effectively cleans all grease, corrosion, and scale from all internal and external head surfaces. Also have the machinist check the valve seats and re-cut them if necessary. When you bring the clean head(s) home, place them on a clean surface. Completely clean the entire valve train with solvent.*

CHECKING FOR HEAD WARPAGE

Lay the head down with the combustion chambers facing up. Place a straightedge across the gasket surface of the head, both diagonally and straight across the center. Using a flat feeler gauge, determine the clearance at the center of the straightedge. If warpage exceeds 0.003" in a 6" span, or 0.006" over the total length, the cylinder head must be resurfaced

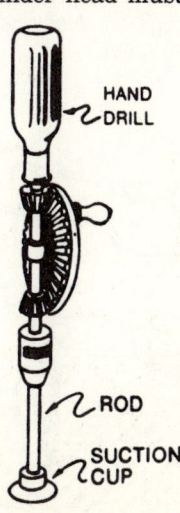

Home-made valve lapping tool

ENGINE AND ENGINE OVERHAUL

(which is akin to planing a piece of wood). Resurfacing can be performed at most machine shops.

NOTE: *When resurfacing the cylinder head(s) of V6 or V8 engines, the intake manifold mounting position is altered, and must be corrected by machining a proportionate amount from the intake manifold flange.*

LAPPING THE VALVES

When valve faces and seats have been refaced and recut, or if they are determined to be in good condition, the valves must be lapped in to ensure efficient sealing when the valve closes against the seal.

1. Invert the cylinder head so that the combustion chambers are facing up.
2. Lightly lubricate the valve stems with clean oil, and coat the valve seats with valve grinding compound. Install the valves in the head as numbered.
3. Attach the suction cup of a valve lapping tool to a valve head. You'll probably have to moisten the cup to securely attach the tool to the valve.
4. Rotate the tool between the palms, changing position and lifting the tool often to prevent grooving. Lap the valve until a smooth, polished seat is evident (you may have to add a bit more compound after some lapping is done).
5. Remove the valve and tool, and remove ALL traces of grinding compound with solvent-soaked rag, or rinse the head with solvent.

NOTE: *Valve lapping can also be done by fastening a suction cup to a piece of drill rod in a hand eggbeater type drill. Proceed as above, using the drill as a lapping tool. Due to the higher speeds involved when using the hand drill, care must be exercised to avoid grooving the seat. Lift the tool and change direction of rotation often.*

Valve Springs

HEIGHT AND PRESSURE CHECK

1. Place the valve spring on a flat, clean surface next to a square.
2. Measure the height of the spring, and rotate it against the edge of the square to measure distortion (out-of-roundness). If spring height varies between springs by more than 1/16," or if the distortion exceeds 1/16", replace the spring.

A valve spring tester is needed to test spring test pressure, so the valve springs must usually be taken to a professional machine shop for this test. Spring pressure at the installed and compressed heights is checked, and a tolerance of plus or minus 5 lbs. (plus or minus 1 lb. on the 231 V6) is permissible on the spring covered in this guide.

VALVE INSTALLATION

New valve seals must be installed when the valve train is put back together. Certain seals slip over the valve stem and guide boss, while others require that the boss be machined. In some applications Teflon guide seals are available. Check with a machinist and/or automotive parts store for a suggestion on the proper seals to use.

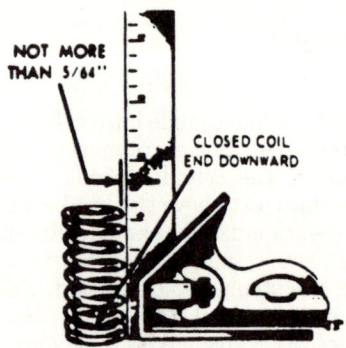

Check the valve spring free length and squareness

Lapping the valves by hand

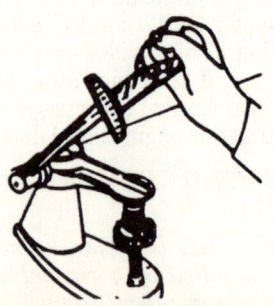

Have the valve spring test pressure checked professionally

ENGINE AND ENGINE OVERHAUL   157

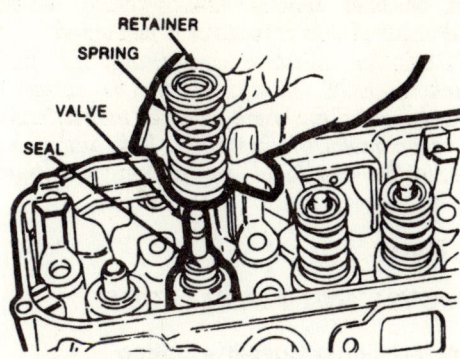

Installing valve stem seals

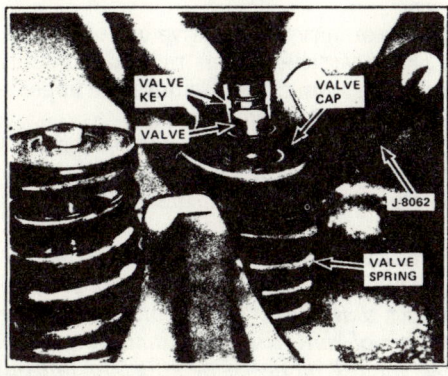

Removing the valve key and cap. A magnet is useful here in removing the keys

NOTE: *Remember that when installing valve seals, a small amount of oil is able to pass the seal to lubricate the valve guides; otherwise, excessive wear will result.*

To install the valve and rocker assembly:

1. Lubricate the valve stems with clean engine oil.
2. Install the valves in the cylinder head, one at a time, as numbered.
3. Lubricate and position the seals and valve springs, again a valve at a time.
4. Install the spring retainers, and compress the springs.
5. With the valve key groove exposed above the compressed valve spring, wipe some wheel bearing grease around the groove. This will retain the keys as you release the spring compressor.
6. Using needlenosed pliers (or your fingers), place the keys in the key grooves. The grease should hold the keys in place. Slowly release the spring compressor; the valve cap or rotator will raise up as the compressor is released, retaining the key.

7. Install the rocker assembly, and install the cylinder head(s).

VALVE ADJUSTMENT

In the event of cylinder head removal or any operation that requires disturbing or removing the rocker arms, the rocker arms must be adjusted. See the Valve Lash procedure earlier in this chapter.

Valve Lifters

REMOVAL AND INSTALLATION

Gasoline and Diesel Engines

NOTE: *Valve lifters and pushrods should be kept in order so they can be reinstalled in their original position. Some engines will have both standard size 0.010" oversize valve lifters as original equipment. The oversize lifters are etched with an O on their sides; the cylinder block will also be marked with an O if the oversize lifter is used.*

1. Remove the intake manifold and gasket.
2. Remove the valve covers.
3. On vehicles equipped with roller lifters, remove the valve lifter retainer and restrictor.
4. Remove the rocker arm assemblies and pushrods.
5. If the lifters are coated with varnish, apply carburetor cleaning solvent to the lifter body. The solvent should dissolve the varnish in about 10 minutes.
6. Remove the lifters. On diesel engines and the 307 Oldsmobile produced engine, remove the lifter retainer guide bolts, and remove the guides. A special tool for removing lifters is available, and is helpful for this procedure.

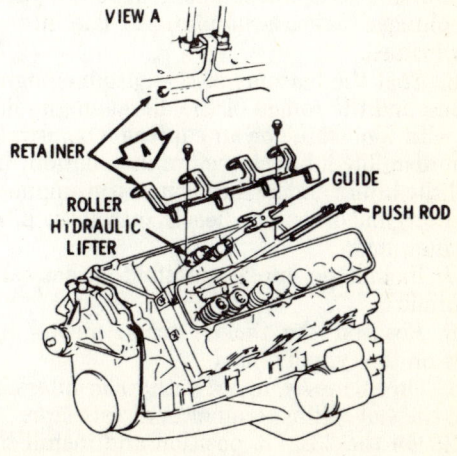

Diesel engine and 307 Oldsmobile produced engine valve lifter guide and retainer

ENGINE AND ENGINE OVERHAUL

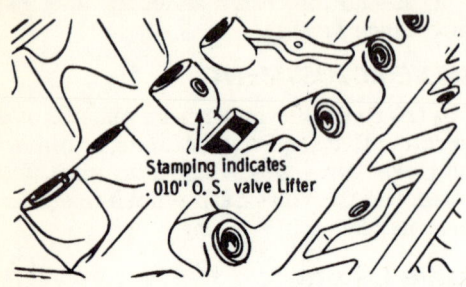

Oversize valve lifter bore marking

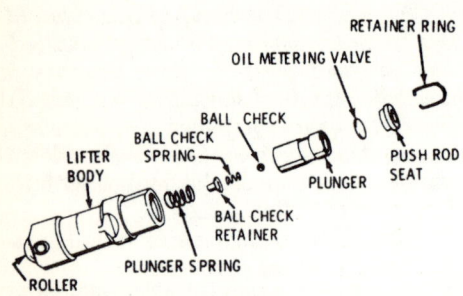

Hydraulic valve lifter, roller type

To Install:

7. New lifters must be primed before installation, as dry lifters will seize when the engine is started. On the diesel lifters, submerge the lifters in clean diesel fuel or kerosene and work the lifter plunger up and down to prime. On gasoline engine lifters, submerge the lifters in SAE 10 oil, which is very thin. Carefully insert the end of a $1/8''$ (3mm) drift into the lifter and push down on the plunger. Hold the plunger down while the lifter is still submerged; do not pump the plunger. Release the plunger. The lifter is now primed.

8. Coat the bottoms of the gasoline engine lifters, and the rollers of the diesel engine lifters with Molykote® or an equivalent molybdenum-disulfide lubricate before installation. Install the lifters and pushrods into the engine in their original order. On diesels, install the lifter retainer guide.

9. Install the intake manifold gaskets and manifold.

10. Position the rocker arms, pivots and bolts on the cylinder head.

11. On engines equipped with roller lifters install the valve lifter retainer and restrictor.

12. On the 231 V6, position and install the rockers and rocker shafts, Refer to the Rocker Arm Removal and Installation procedure in this chapter for lifter bleed-down. New lifters

must be bled down; valve-to-piston contact could occur if this procedure is neglected.

NOTE: *An additive containing EP lube, such as EOS, should always be added to crankcase oil for break-in when new lifters or a new camshaft is installed. This additive is generally available in automotive parts stores.*

Oil Pan

REMOVAL AND INSTALLATION

Inline 6-Cylinder Gasoline Engine

NOTE: *Pan removal for all engines may be easier if the engine is turned to No. 1 cylinder firing position. This positions the crankshaft in the path of least resistance for pan removal.*

1. Disconnect the negative battery cable.
2. Remove the upper radiator mounting bolts or side mount bolt. Remove the upper and lower hoses for the water pump.
3. Install a piece of heavy cardboard between the fan and the radiator.
4. Disconnect the fuel suction line from the fuel pump.
5. Raise the vehicle and drain the oil.

CAUTION: *The EPA warns that prolonged contact with used engine oil may cause a number of skin disorders, including cancer! You should make every effort to minimize your exposure to used engine oil. Protective gloves should be worn when changing the oil. Wash your hands and any other exposed skin areas as soon as possible after exposure to used engine oil. Soap and water, or waterless hand cleaner should be used.*

NOTE: *A Chilton environmental tip. Used oil contains heavy metals, and has been determined to be hazardous to the environment. Recycling oil is the best way for disposal. Check local laws and recycle whenever possible.*

6. Remove the starter.
7. Remove the flywheel lower pan or converter lower pan and splash shield.
8. Rotate the crankshaft until the timing mark on the damper is at the six o'clock position.
9. Remove the brake line retaining bolts from the crossmember and move the brake line out of the way.
10. Remove the through-bolts from the front motor mounts.
11. Remove the oil pan bolts.
12. Slowly raise the engine until the motor mounts can be removed from the frame brackets.
13. Remove the mounts and continue to

ENGINE AND ENGINE OVERHAUL

raise the engine until it has been raised three inches.

14. Remove the oil pan by pulling it down from the engine and then twisting it into the opening left by the removal of the left engine mount.

15. When the pan is clear of the engine, tilt the front up and remove it by pulling it down and to the rear.

To Install:

16. Install the oil pan gaskets to the engine block.

17. Install the oil pan and torque the side bolt to 6-8 ft. lbs. and the end bolts to 9-12 ft. lbs.

18. Install the motor mounts.

19. Install the brake line retaining bolts at the crossmember.

20. Install the flywheel lower pan or converter lower pan and splash shield.

21. Install the starter.

22. Connect the fuel suction line at the fuel pump.

23. Remove the cardboard from between the fan and the radiator.

24. Install the upper radiator mounting bolts or side mount bolt. Install the upper and lower hoses.

25. Connect the negative battery cable.

26. Fill the crankcase.

V6 Gasoline Engine

1. Disconnect the negative battery cable.
2. Remove the upper half of the radiator fan shroud.
3. Raise the front of the vehicle and drain the oil.

CAUTION: *The EPA warns that prolonged contact with used engine oil may cause a number of skin disorders, including cancer! You should make every effort to minimize your exposure to used engine oil. Protective gloves should be worn when changing the oil. Wash your hands and any other exposed skin areas as soon as possible after exposure to used engine oil. Soap and water, or waterless hand cleaner should be used.*

NOTE: *A Chilton environmental tip. Used oil contains heavy metals, and has been determined to be hazardous to the environment. Recycling oil is the best way for disposal. Check local laws and recycle whenever possible.*

4. Unscrew the exhaust pipe cross-over tube mounting nuts at the manifold. Lower the crossover tube.

5. On vehicles equipped with an automatic transmission, remove the torque converter cover, and the oil cooler lines at the oil pan.

6. Remove the upper bolt on the starter

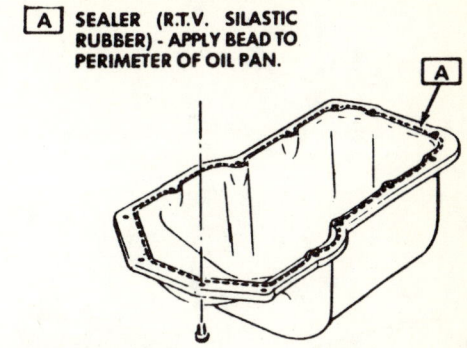

Oil pan, V6–231

brace and then remove the inboard starter bolt and swing the starter assembly aside.

NOTE: *If equipped with Air Injection Reaction (AIR) system, disconnect the AIR hose from the converter pipe and the AIR pipe from the exhaust manifold.*

7. Loosen and remove the left hand motor mount through-bolt and then loosen the through-bolt on the right hand mount.

8. Raise the engine and then reinstall the through-bolt in the left hand motor mount. Do not tighten the bolt.

9. Unscrew the attaching bolts and remove the oil pan from under the engine.

10. Position the oil pan to the engine. Be sure to use a new gasket or RTV sealant, as required.

11. Install the motor mount retaining bolts. As required, install the AIR injection assembly.

12. Install the starter bolt. Install the torque converter cover. Install the transmission oil cooler lines.

13. Lower the vehicle. If removed install the upper half of the radiator fan shroud. Install the exhaust pipe to manifold retaining bolts.

14. Fill the crankcase with the proper grade and type engine oil.

15. Connect the negative battery cable. Start the engine and check for leaks.

V8 Gasoline Engine

1968-78

1. Disconnect the negative battery cable.
2. As a precaution, remove the distributor cap to keep it from getting broken when the engine is raised.
3. Remove the fan shroud retaining bolts.
4. On some vehicles, it may be necessary to remove the radiator upper mounting panel.
5. Raise the vehicle and support it safely. Drain the engine oil.

ENGINE AND ENGINE OVERHAUL

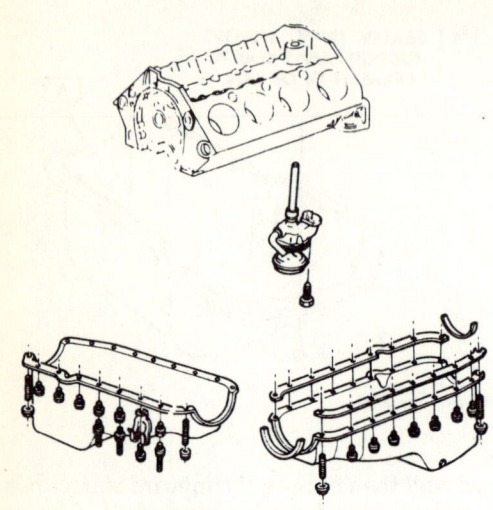

Oil pan, small block V8

CAUTION: *The EPA warns that prolonged contact with used engine oil may cause a number of skin disorders, including cancer! You should make every effort to minimize your exposure to used engine oil. Protective gloves should be worn when changing the oil. Wash your hands and any other exposed skin areas as soon as possible after exposure to used engine oil. Soap and water, or waterless hand cleaner should be used.*

NOTE: *A Chilton environmental tip. Used oil contains heavy metals, and has been determined to be hazardous to the environment. Recycling oil is the best way for disposal. Check local laws and recycle whenever possible.*

6. Disconnect the exhaust pipes or crossover pipes. As required, remove the starter.
7. On automatic transmission equipped vehicles, remove the converter housing underpan and splash shield.
8. Rotate the crankshaft until the timing mark on the torsional dampener is at the six o'clock position.
9. The starter can be swung out of the way by disconnecting the brace at the starter, removing the inboard starter bolt and loosening the outboard starter bolts. On some engines the fuel pump may have to be removed.
10. Remove the front engine mount throughbolts.
11. Raise the engine and insert blocks, at least 3" thick, under the engine mounts.
12. Remove the oil pan bolts and remove the oil pan.

To Install:

13. Position the oil pan to the engine. Be sure to use a new gasket or RTV sealant, as required.
14. Remove the wood blocks. Install the motor mount retaining bolts. As required, install the AIR injection assembly.
15. Install the starter. Install the torque converter cover. Install the transmission oil cooler lines.
16. Lower the vehicle. If removed install the upper half of the radiator fan shroud. Install the exhaust pipe to manifold retaining bolts.
17. Install the distributor cap assembly. Fill the crankcase with the proper grade and type engine oil.
18. Connect the negative battery cable. Start the engine and check for leaks.

1979-88

1. Disconnect the negative battery cable.
2. Remove the air cleaner, the upper radiator mounting panel and the fan shroud.
3. Remove the distributor cap. Remove the fan assembly.
4. Disconnect the AIR hose from the converter pipe and the AIR pipe from the exhaust manifold, as required.
5. Raise and support the vehicle safely. Drain the engine oil.

CAUTION: *The EPA warns that prolonged contact with used engine oil may cause a number of skin disorders, including cancer! You should make every effort to minimize your exposure to used engine oil. Protective gloves should be worn when changing the oil. Wash your hands and any other exposed skin areas as soon as possible after exposure to used engine oil. Soap and water, or waterless hand cleaner should be used.*

NOTE: *A Chilton environmental tip. Used oil contains heavy metals, and has been determined to be hazardous to the environment. Recycling oil is the best way for disposal. Check local laws and recycle whenever possible.*

6. Remove the exhaust crossover pipe from the manifold and the catalytic converter. As required, remove the starter.
7. If equipped with an automatic transmission, remove torque converter housing cover plate and disconnect the transmission oil cooler lines at the oil pan. If equipped with a manual transmission, remove the flywheel housing cover plate.
8. Rotate crankshaft until timing mark on torsional damper is at 6 o'clock position, this positions the crankshaft throw in the horizontal place.
9. Remove front engine mount through bolts.
10. Raise engine and insert blocks under

ENGINE AND ENGINE OVERHAUL

engine mounts. The block thickness should be 3"

11. Remove the oil pan bolts and lower the pan.

To Install:

12. Position the oil pan to the engine. Be sure to use a new gasket or RTV sealant, as required.

13. Remove the wood blocks. Install the motor mount retaining bolts. As required, install the AIR injection assembly.

14. Install the starter. Install the torque converter cover. Install the transmission oil cooler lines.

15. Lower the vehicle. If removed install the upper half of the radiator fan shroud. Install the exhaust pipe to manifold retaining bolts.

16. Install the distributor cap assembly. Fill the crankcase with the proper grade and type engine oil.

17. Connect the negative battery cable. Start the engine and check for leaks.

Diesel Engine

1. Disconnect the negative battery cables. Remove the engine oil dipstick.
2. Remove the vacuum pump assembly. Remove the upper radiator support and fan shroud.
3. Raise and support the vehicle safely. Drain the oil. Remove the starter.

CAUTION: *The EPA warns that prolonged contact with used engine oil may cause a number of skin disorders, including cancer! You should make every effort to minimize your exposure to used engine oil. Protective gloves should be worn when changing the oil. Wash your hands and any other exposed skin areas as soon as possible after exposure to used engine oil. Soap and water, or waterless hand cleaner should be used.*

NOTE: *A Chilton environmental tip. Used oil contains heavy metals, and has been determined to be hazardous to the environment. Recycling oil is the best way for disposal.*

Check local laws and recycle whenever possible.

4. Remove the flywheel cover. Disconnect the exhaust and crossover pipes.
5. Remove the oil cooler lines at the filter base. Properly support the engine with a jack.
6. Remove the engine mounts from the block. Raise the front of the engine high enough the remove the oil pan.
7. Remove the oil pan retaining bolts. Remove the oil pan from the vehicle.

To Install:

8. Position the oil pan to the engine. Be sure to use a new gasket or RTV sealant, as required. Install the attaching bolts.
9. Remove the wood blocks. Install the motor mount retaining bolts. As required, install the AIR injection assembly.
10. Install the starter. Install the torque converter cover. Install the transmission oil cooler lines.
11. Lower the vehicle. If removed install the upper half of the radiator fan shroud. Install the exhaust pipe to manifold retaining bolts.
12. Install the distributor cap assembly. Fill the crankcase with the proper grade and type engine oil.
13. Connect the negative battery cable. Start the engine and check for leaks.

Oil Pump

REMOVAL AND INSTALLATION

Inline 6-Cylinder Gasoline Engine

1. Remove the oil pan as previously described.
2. Remove the two flange mounting bolts, the pick-up bolt, then remove the pump and screen together.
3. To install, align the oil pump driveshafts to match with the distributor tang and position

Diesel engine oil cooler line location

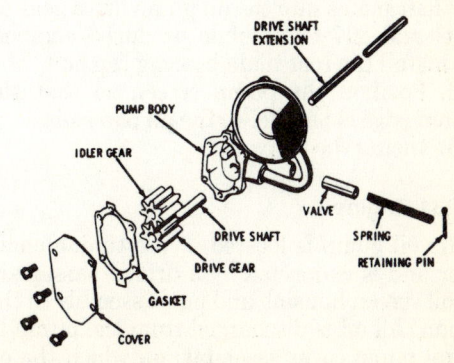

Oil pump, exploded view of early model

162 ENGINE AND ENGINE OVERHAUL

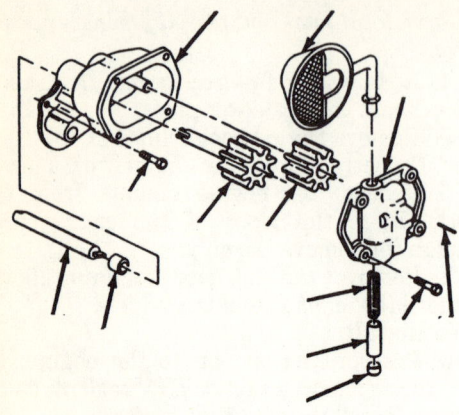

IDLER GEAR
PUMP BODY
PICKUP SCREEN AND PIPE ASSEMBLY
DRIVESHAFT
RETAINING PIN
SPRING
PRESSURE REGULATOR VALVE
PLUG
DRIVE GEAR AND SHAFT
RETAINER
COVER
COVER BOLT

Oil pump, exploded view of late model

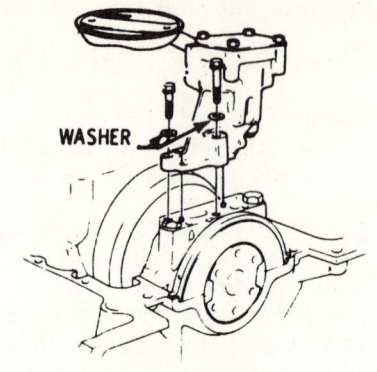

Diesel engine oil pump assembly

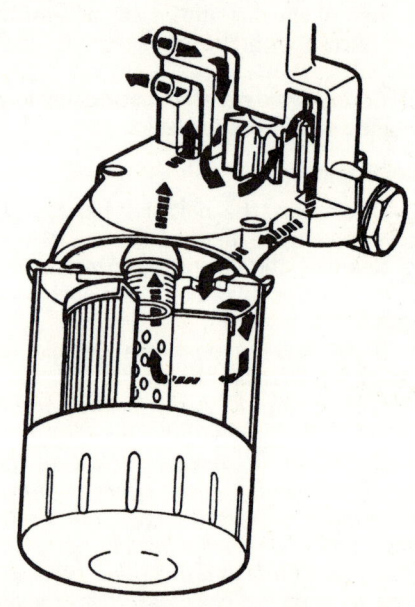

Oil pump and filter assembly, V6–231

the flange over the distributor lower bushing. Install the pump mounting bolts.

4. Install the oil pan.

Gasoline and Diesel Engines Except 231 V6

1. Remove the oil pan as previously described.
2. Remove the pump-to-rear main bearing cap bolts and remove the pump and extension shaft.
3. To install align the slot on the top end of the extension shaft with the drive tang on the lower end of the distributor driveshaft (or until the shaft mates into the oil pump drive gear on diesel and 307 Oldsmobile produced engines) and install the rear main bearing cap bolt.
4. Position the pump screen so that the bottom edge is parallel to the oil pan rails.
5. Install the oil pan.

V6-231 Engine

The oil pump is located in the timing chain cover and is connected by a drilled passage to the oil screen housing and pipe assembly in the oil pan. All oil is discharged from the pump to the oil pump cover assembly, on which the oil filter is mounted.

1. Disconnect the negative battery cable. To remove the oil pump cover and gears, first remove the oil filter.
2. Remove the screws which attach the oil pump cover assembly to the timing chain cover.
3. Remove the cover assembly and slide out the oil pump gears. Clean the gears and inspect them for any obvious defects such as chipping or scoring.
4. Remove the oil pressure relief valve cap, spring and valve. Clean them and inspect them for wear or scoring. Check the relief valve spring to see that it is not worn on its side or collapsed. Replace the spring if it seems questionable.
5. Check the relief valve for a correct fit in its bore. It should be an easy slip fit and no more. If any perceptible shake can be felt, the valve and/or cover should be replaced.

ENGINE AND ENGINE OVERHAUL

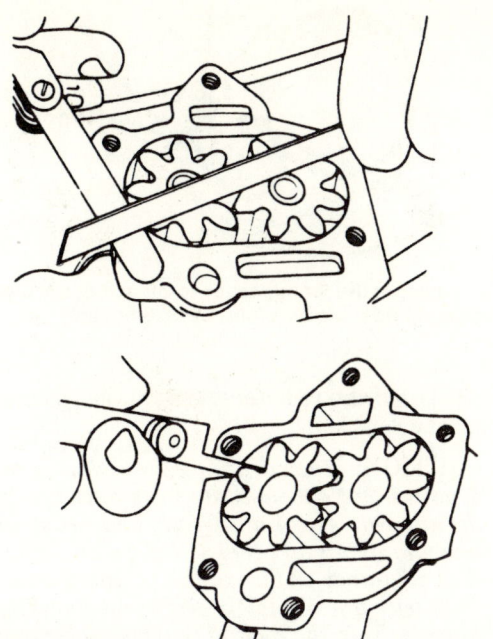

Use a feeler gauge to check oil pump side clearance

feeler gauge between the gear teeth and the side wall of the pump body. Clearance should be between 0.002″ and 0.005″.

4. Check the pump cover flatness by placing a straight edge across the cover face, with the feeler gauge between the straight edge and the cover. If clearance is 0.001″ or more, replace the cover.

5. Installation is the reverse of removal.

6. Pack the oil pump gear cavity full of petroleum jelly. Do not use gear lube. Reinstall the oil pump gears so that the petroleum jelly is forced into every cavity of the gear pocket, and between the gear teeth. There must be no air spaces. This step is very important.

WARNING: *Unless the pump is primed this way, it won't produce any oil pressure when the engine is started.*

Gasoline And Diesel Engines Except V6-231 Engine

1. Remove the oil pump and pump drive shaft extension, as previously described.

2. Remove the cotter pin, spring and the pressure regulator valve.

NOTE: *Place your thumb over the pressure regulator bore before removing the cotter pin, as the spring is under pressure.*

3. Remove the oil pump cover attaching screws and remove the oil pump cover and gasket. Clean the pump in solvent or kerosene, and wash out the pick-up screen.

4. Remove the drive gear and idler gear from the pump body.

5. Check the gears for scoring and other damage. Install the gears if in good condition, or replace them if damaged. Check gear end clearance by placing a straight edge over the gears and measure the clearance between the straight edge and the gasket surface with a feeler gauge.

6. Check gear side clearance by inserting the feeler gauge between the gear teeth and the side wall of the pump body. Clearance should be between 0.002″ and 0.005″.

7. Pack the inside of the pump completely with petroleum jelly. DO NOT use engine oil. The pump MUST be primed this way or it won't produce any oil pressure when the engine is started.

8. Install the cover screws and tighten alternately and evenly to 8 ft. lbs.

9. Position the pressure regulator valve into the pump cover, closed end first, then install the spring and retaining pin.

NOTE: *When assembling the driveshaft extension to the drive shaft, the end of the extension nearest the washers must be inserted into the drive shaft.*

To Install:

6. Lubricate the pressure relief valve and spring and place them in the cover. Install the cap and the gasket. Torque the cap to 35 ft. lbs.

7. Pack the oil pump gear cavity full of petroleum jelly. Do not use gear lube. Reinstall the oil pump gears so that the petroleum jelly is forced into every cavity of the gear pocket, and between the gear teeth. There must be no air spaces. This step is very important. Unless the pump is packed, it may not begin to pump oil as soon as the engine is started.

8. Install the cover assembly using a new gasket and sealer. Tighten the screws to 10 ft. lbs.

9. Install the oil filter.

OVERHAUL

V6-231 Engine

1. Remove the oil pump cover and gears, as described above.

2. Clean and inspect all parts as described.

3. Install the oil pump gears (if removed) and the shaft in the oil pump body section of the timing chain cover to check the gear end clearance and gear side clearance. Check gear end clearance by placing a straight edge over the gears and measure the clearance between the straight edge and the gasket surface. Clearance should be between 0.002″ and 0.006″. Check gear side clearance by inserting the

164 ENGINE AND ENGINE OVERHAUL

10. Insert the driveshaft extension through the opening in the main bearing cap and block until the shaft mates into the distributor drive gear.

11. Install the pump onto the rear main bearing cap and install the attaching bolts. Torque the bolts to 35 ft. lbs.

12. Install the oil pan.

Timing Gear Cover

REMOVAL AND INSTALLATION

Inline 6-Cylinder Gasoline engine

1. Disconnect the negative battery cable. Drain the engine oil. Remove the oil pan on 1968-72 vehicles.

CAUTION: *The EPA warns that prolonged contact with used engine oil may cause a number of skin disorders, including cancer! You should make every effort to minimize your exposure to used engine oil. Protective gloves should be worn when changing the oil. Wash your hands and any other exposed skin areas as soon as possible after exposure to used engine oil. Soap and water, or waterless hand cleaner should be used.*

NOTE: *A Chilton environmental tip. Used oil contains heavy metals, and has been determined to be hazardous to the environment. Recycling oil is the best way for disposal. Check local laws and recycle whenever possible.*

When the timing gear cover is replaced on most engines, the oil pan front seal must be modified

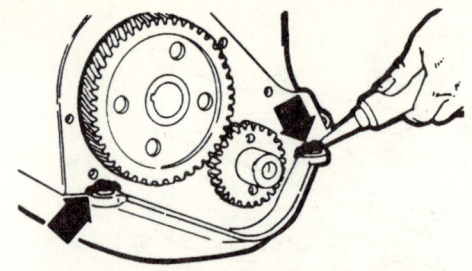

Use sealer at the timing cover-to-oil pan and the oil pan-to-cylinder block joints, L6–250 engine

2. Drain the radiator. Remove the radiator from the vehicle.

CAUTION: *When draining the coolant, keep in mind that cats and dogs are attracted by the ethylene glycol antifreeze, and are quite likely to drink any that is left in an uncovered container or in puddles on the ground. This will prove fatal in sufficient quantity. Always drain the coolant into a sealable container. Coolant should be reused unless it is contaminated or several years old.*

3. Remove the fan, pulley, and belt. Remove any power steering and/or AIR pump drive belts. Remove any braces for the above pumps which will interfere with cover removal and position the pumps out of the way.

4. Remove the crankshaft pulley and damper. Use the puller tool No. J-16516 to remove the damper. do not attempt to pry or hammer the damper off, or it will be damaged.

5. Remove the retaining bolts, and remove the cover on 1968-72 vehicles.

6. On 1973-79 vehicles, pull the cover forward slightly and cut the oil pan front seal off flush with the block. Remove the cover. On in-

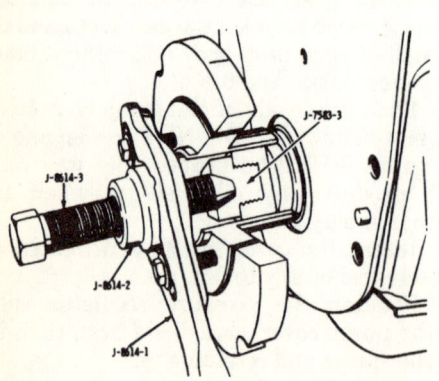

Removing harmonic balancer using a puller

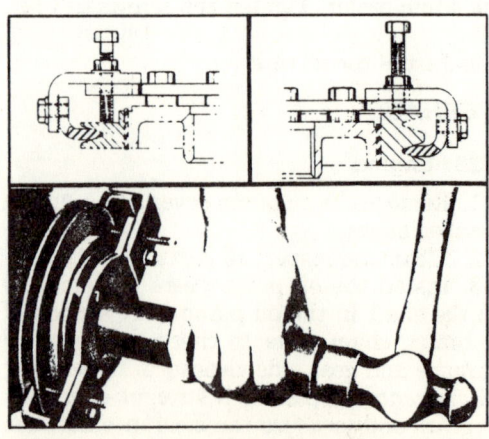

Installing torsional damper L6 engine

ENGINE AND ENGINE OVERHAUL

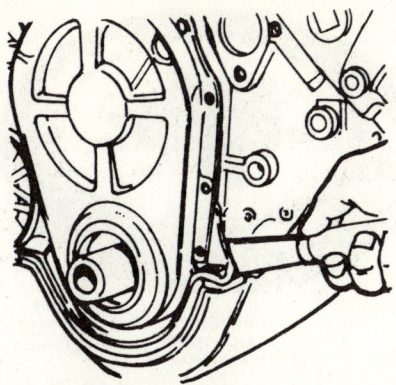

Timing cover installation Chevrolet built V8 engines, waterpump removed

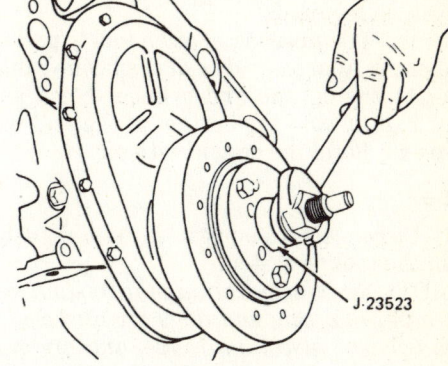

Installing the damper pulley

stallation, cut the tabs off a new oil pan front seal and install it on the cover.

7. On installation, coat the gasket with sealer and use a $1/8''$ bead of silicone sealer at the oil pan-to-cylinder block joint. Replace the damper before tightening the cover bolts down, so that the cover seal will align. The damper must be drawn into place using installation tool J-22197. Hammering it will destroy it.

NOTE: *When installing the timing cover, place the centering tool J-23042 inside the oil seal, slide the timing cover into position, install 2 screws and remove the tool.*

8. Replace the oil pan if it was removed, and fill the crankcase with oil.

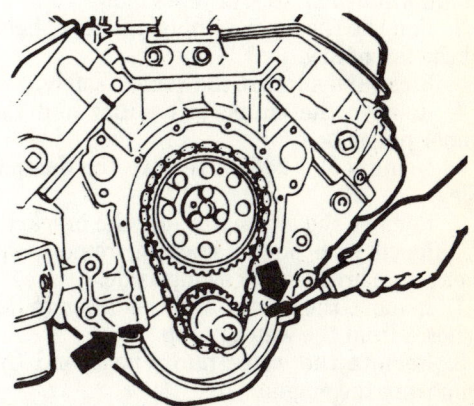

Apply sealer to the front pads at the area shown

V6 and V8 Gasoline Engines Except V6-231 Engine and 8-307 Oldsmobile Produced Engine

1968-74

1. Drain the oil and remove the oil pan. The pan need not be removed on 1974 and later small block engines.

CAUTION: *The EPA warns that prolonged contact with used engine oil may cause a number of skin disorders, including cancer! You should make every effort to minimize your exposure to used engine oil. Protective gloves should be worn when changing the oil. Wash your hands and any other exposed skin areas as soon as possible after exposure to used engine oil. Soap and water, or waterless hand cleaner should be used.*

NOTE: *A Chilton environmental tip. Used oil contains heavy metals, and has been determined to be hazardous to the environment. Recycling oil is the best way for disposal. Check local laws and recycle whenever possible.*

2. Drain and remove the radiator.

CAUTION: *When draining the coolant, keep in mind that cats and dogs are attracted by the ethylene glycol antifreeze, and are quite likely to drink any that is left in an uncovered container or in puddles on the ground. This will prove fatal in sufficient quantity. Always drain the coolant into a sealable container. Coolant should be reused unless it is contaminated or several years old.*

3. Remove the fan, pulley, and belt. Remove any power steering and/or AIR pump drive belts. Remove any braces for these pumps which will interfere with cover removal and position the pumps out of the way.

4. Remove the water pump. Remove the crankshaft pulley and damper. Use a puller on the damper. Do not attempt to pry or hammer the damper off.

5. Remove the retaining bolts, and remove the timing cover.

6. Installation is the reverse of the removal

ENGINE AND ENGINE OVERHAUL

procedure. Be sure to use new gaskets or RTV sealant, as required.

7. Use a damper installation tool to pull the damper on. Apply an $1/8''$ bead of silicone rubber sealer to the oil pan and cylinder block joint faces. Lightly coat the bottom of the seal with engine oil. Refill the engine with oil.

1975-92

1. Disconnect the negative battery cable. Drain the cooling system.

CAUTION: *When draining the coolant, keep in mind that cats and dogs are attracted by the ethylene glycol antifreeze, and are quite likely to drink any that is left in an uncovered container or in puddles on the ground. This will prove fatal in sufficient quantity. Always drain the coolant into a sealable container. Coolant should be reused unless it is contaminated or several years old.*

2. Remove the fan assembly, the drive belts and the fan pulley.
3. Raise and support the vehicle safely.
4. Remove the crankshaft pulley and the damper pulley bolt.
5. Using tool J-23523, remove the damper pulley.
6. Remove the alternator and the brackets. If equipped with power steering, remove the lower pump bracket and swing aside.
7. Remove the heater and the lower radiator hoses from the water pump.
8. Remove the water pump bolts and the pump from the engine.
9. Remove the timing cover bolts and the timing cover.
10. Using a gasket scraper, clean the gasket mounting surfaces.

To Install:

11. Install the timing cover. Be sure to use a new gasket or RTV sealant, as required.
12. Install the water pump. Install the heater and the lower radiator hoses to the water pump.
13. Install the alternator and the brackets. If equipped with power steering, install the lower pump bracket. Using the proper tool install the damper pulley.
14. Install the crankshaft pulley and the damper pulley bolt.
15. Install the fan assembly, the drive belts and the fan pulley.
16. Install the negative battery cable. Refill the cooling system. Start the engine and check for leaks.

V6-231 Engine

1. Disconnect the negative battery cable. Drain the cooling system. Remove the radiator hoses. Remove the radiator.

V6–231 harmonic balancer timing marks

CAUTION: *When draining the coolant, keep in mind that cats and dogs are attracted by the ethylene glycol antifreeze, and are quite likely to drink any that is left in an uncovered container or in puddles on the ground. This will prove fatal in sufficient quantity. Always drain the coolant into a sealable container. Coolant should be reused unless it is contaminated or several years old.*

2. Remove all the drive belts. Remove the fan and fan pulley. Remove the bypass hose.
3. Remove the crank pulley, the fuel pump and the distributor. Note the position of the distributor rotor before you remove the distributor.
4. Remove the alternator and its bracket.
5. Remove the harmonic balancer. You'll need a puller.
6. Remove the two bolts which attach the oil pan to the front cover. Remove the bolts which attach the cover to the block.
7. Remove the cover and remove the gasket material.

To Install:

NOTE: *Before installing the timing cover remove the oil pump cover and pack the space around the oil pump gears with petroleum jelly. Do not use gear lube. There must be no air space left inside the pump. Reinstall the pump cover using a new gasket. This step is very important since the oil pump may lose its prime any time the pump, pump cover or timing cover is disturbed. If the pump is not packed, it may not begin to pump oil as soon as the engine is started.*

8. Position the cover to the engine block. Be sure to use a new gasket or RTV sealant, as required.
9. Install the two bolts which attach the oil pan to the front cover. Install the bolts which attach the cover to the block.
10. Install the harmonic balancer. Install the alternator and its bracket.

ENGINE AND ENGINE OVERHAUL

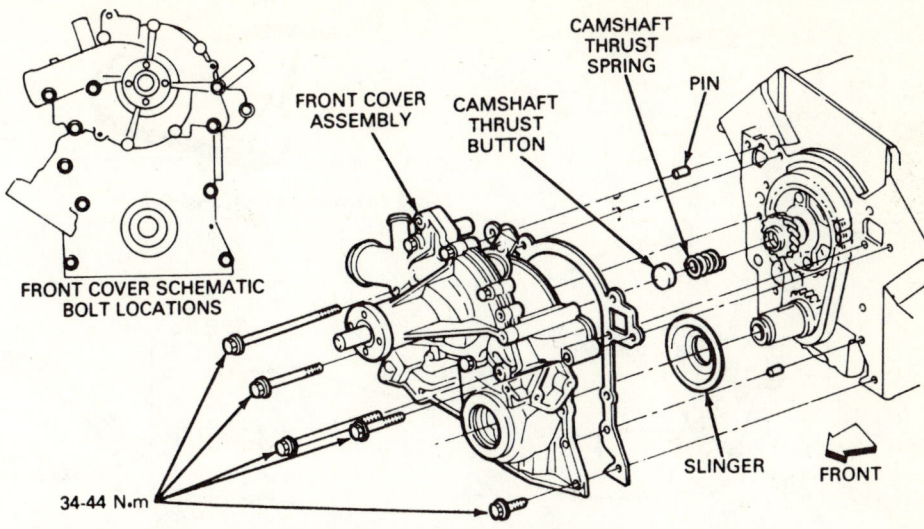

Timing chain cover, V6-231

11. Install the crank pulley, the fuel pump and the distributor. Note the position of the distributor rotor before you install the distributor.

12. Install all of the drive belts. Install the fan and fan pulley. Install the bypass hose. Install the radiator. Install the radiator hoses.

13. Connect the negative battery cable. Refill the cooling system. Start the engine and check for leaks.

Diesel Engine and 8-307 Oldsmobile Produced Engine

1. Disconnect the negative battery cable. Drain the cooling system and disconnect the radiator hoses.

CAUTION: *When draining the coolant, keep in mind that cats and dogs are attracted by the ethylene glycol antifreeze, and are quite likely to drink any that is left in an uncovered container or in puddles on the ground. This will prove fatal in sufficient quantity. Always drain the coolant into a sealable container. Coolant should be reused unless it is contaminated or several years old.*

2. Remove all belts, fan and pulley, crankshaft pulley and balancer, using a balancer puller.

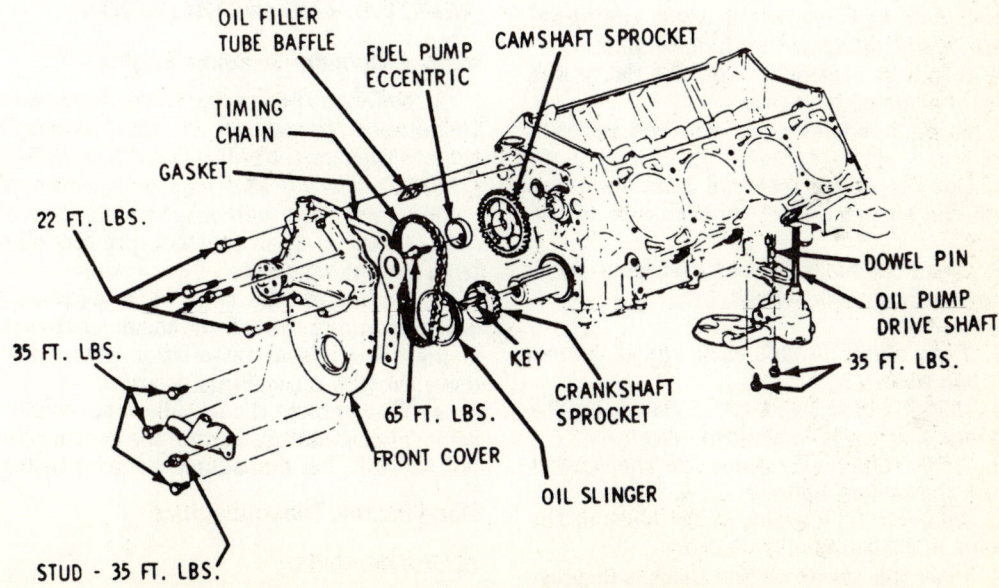

Front cover and timing assembly, V8-307 Oldsmobile produced engine and diesel

ENGINE AND ENGINE OVERHAUL

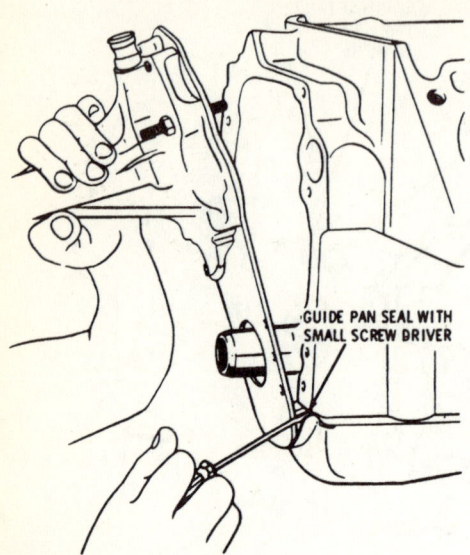

Timing cover installation diesel engine and 307 Oldsmobile produced engine

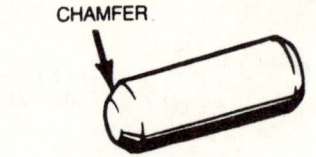

Front cover dowel pin chamfer

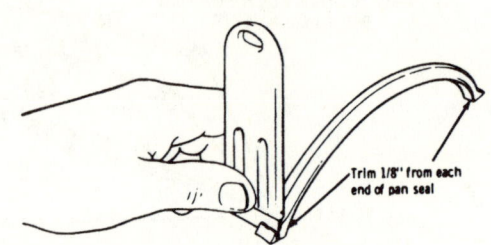

Trimming pan seal with razor blade

WARNING: *The use of any other type of puller, such as a universal claw type which pulls on the outside of the hub, can destroy the balancer. The outside ring of the balancer is bonded in rubber to the hub. Pulling on the outside will break the bond. The timing mark is on the outside ring. If it is suspected that the bond is broken, check that the center of the keyway is 16 degrees from the center of the timing slot. In addition, there are chiseled aligning marks between the weight and the hub.*

3. Unbolt and remove the cover, timing indicator, water pump and both dowel pins.
4. It may be necessary to grind a flat on the cover for gripping purposes.
5. Grind a chamfer on one end of each dowel pin.
6. Cut the excess material from the front end of the oil pan gasket on each side of the block.
7. Clean the block, oil pan and front cover mating surfaces with solvent.

To Install:

8. Trim about 1/8" off each end of a new front pan seal.
9. Install a new front cover gasket on the block and a new seal in the front cover.
10. Apply an RTV sealer to the gasket around the coolant holes.
11. Apply an RTV sealer to the block at the junction of the pan and front cover.
12. Place the cover on the block and press down to compress the seal. Rotate the cover left and right and guide the pan seal into the cavity using a small suitable tool. Oil the bolt threads and install two bolts to hold the cover in place. Install both dowel pins (chamfered end first), then install the remaining front cover bolts.
13. Apply a lubricant, compatible with rubber, on the balancer seal surface.
14. Install the balancer and bolt. Torque the bolt to 200-300 ft. lbs.
15. Install the other parts in the reverse order of removal.
16. Fill the cooling system using the proper solution. start the engine and check for leaks.

Timing Cover Oil Seal

REMOVAL AND INSTALLATION

Inline 6-Cylinder Gasoline Engine

1. Refer to the Timing Cover Removal and Installation procedures in this section and remove the damper pulley from the crankshaft.

NOTE: *The oil seal may be removed from the timing cover without removing the cover.*

2. Using a small pry ball, pry the oil seal from the timing cover.
3. To install the new oil seal, place the seal's open end toward the inside of the cover. Using the oil seal installation tool J-23042, drive the new oil seal into position.
4. To complete the installation, reverse the removal procedures. Torque the damper pulley bolt to 60 ft. lbs. and adjust the drive belts.

Gasoline and Diesel Engines

COVER REMOVED

1. Refer to the Timing Cover Removal and Installation procedures in this section and

ENGINE AND ENGINE OVERHAUL

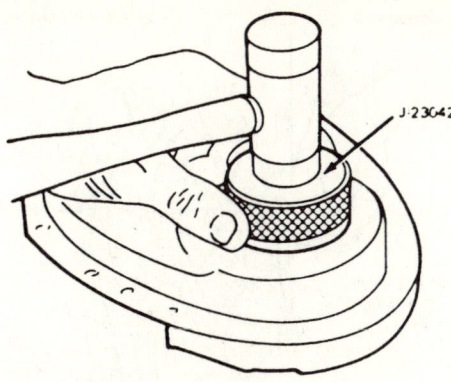

Installing a new front oil seal

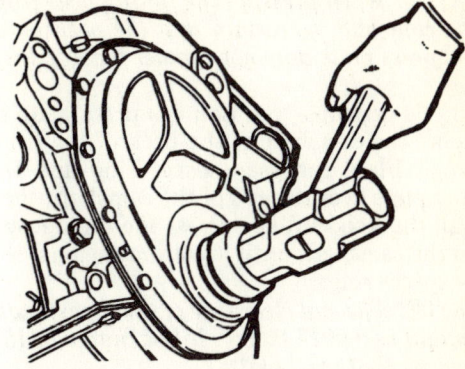

Oil seal installation with the cover installed

Oil seal installation with the cover removed

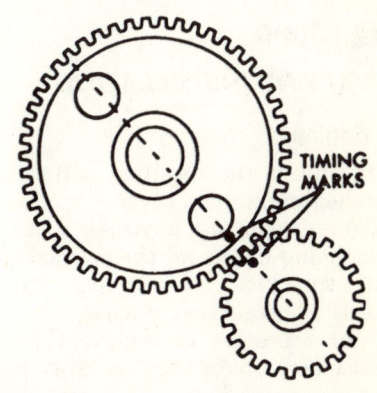

Timing gear alignment — inline six cylinder

The inline six cylinder crankshaft gear is removed with a gear puller

remove the timing cover.

2. Using a small pry bar, pry the oil seal from the timing cover.

3. Using tool J-23042, drive the new oil seal into the timing cover.

NOTE: *When installing the new oil seal, be sure to support the rear side of the timing cover.*

4. Install the timing cover to the engine.

COVER INSTALLED

1. Refer to the Timing Cover Removal and Installation procedures in this section and remove the balancer from the crankshaft.

2. Using a small pry bar, pry the oil seal from the timing cover.

3. Place the new seal (open end toward the engine) on the timing cover and drive it into the cover using tool J-23042.

4. To complete the installation, reverse the removal procedures. Torque the balancer bolt to 65-75 ft. lbs.

Timing Gear

REMOVAL AND INSTALLATION

Inline 6-Cylinder Gasoline Engine

NOTE: *The timing gear is pressed onto the camshaft. To remove or install the timing gear an arbor must be used.*

1. Disconnect the negative battery cable. Refer to the Camshaft Removal and Installation procedures in this section and remove the camshaft from the engine.

2. Using an arbor press, a press plate and a gear removal tool J-971, press the timing gear from the camshaft.

ENGINE AND ENGINE OVERHAUL

NOTE: *When pressing the timing gear from the camshaft, be certain that the position of the press plate does not contact the woodruff key.*

3. To assemble, position the press plate to support the camshaft at the back of the front journal. Place the gear spacer ring and the thrust plate over the end of the camshaft, then install the woodruff key. Press the timing gear onto the camshaft, until it bottoms against the gear spacer ring.

NOTE: *The end clearance of the thrust plate should be 0.0015-0.005". If less than 0.0015", replace the thrust plate.*

4. To complete the installation, align the marks on the timing gears and reverse the removal procedures.

Timing Chain

REMOVAL AND INSTALLATION

V6-231 Engine

1. Disconnect the negative battery cable. Remove the timing chain cover.
2. Before removing anything else, make sure the timing marks on the crankshaft and camshaft sprockets are aligned. This will greatly ease reinstallation of parts.

It is not necessary to remove the timing chain tensioners unless they are worn or damaged.

3. Remove the front crankshaft oil slinger.
4. Remove the bolt and the special washer that hold the camshaft distributor drive gear and fuel pump eccentric at the forward end of the camshaft.
5. Using a pair of prybars, alternately pry the camshaft sprocket, then the crankshaft sprocket forward until the camshaft sprocket is free.
6. After the camshaft sprocket and chain are removed, finish pulling the crankshaft sprocket off the crankshaft.

To Install:

7. If the engine has not been disturbed, proceed to Step Eleven.
8. If the engine has been disturbed, turn the crankshaft so that the number one piston is at top dead center.
9. Temporarily install the sprocket key and the camshaft sprocket on the camshaft. Turn the camshaft so that the index mark of the sprocket is pointing downward. Remove the key and the sprocket from the camshaft.
10. Assemble the time chain and sprockets. Install the keys, sprockets, and chain assembly on the crankshaft and camshaft so that the index marks of both the sprockets are aligned. It will be necessary to hold the chain tensioners

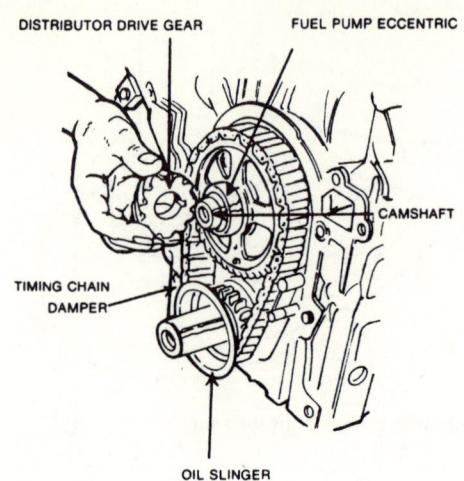

Timing chain and gears, V6-231

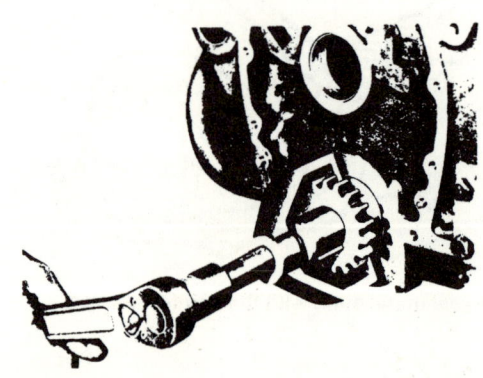

V6 and V8 crankshaft sprocket removal using special puller

out of the way while installing the timing chain and sprocket assembly.

11. Install the front oil slinger on the crankshaft with the concave side toward the front of the engine.
12. Install the fuel pump eccentric with the oil groove forward.
13. Install the distributor drive gear on the camshaft.
14. Install the timing chain cover.

V6 and V8 Gasoline Engines Except V6-231 Engine and 8-307 Oldsmobile Produced Engine

1. Disconnect the negative battery cable. Refer to the Timing Cover Removal and Installation procedures in this section and remove the water pump and the timing cover.
2. Turn the crankshaft until the mark on

ENGINE AND ENGINE OVERHAUL

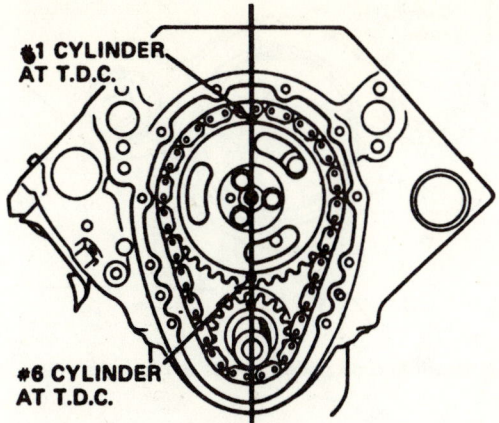

Timing sprocket alignment, V6 and V8 (1980–92) gasoline engines

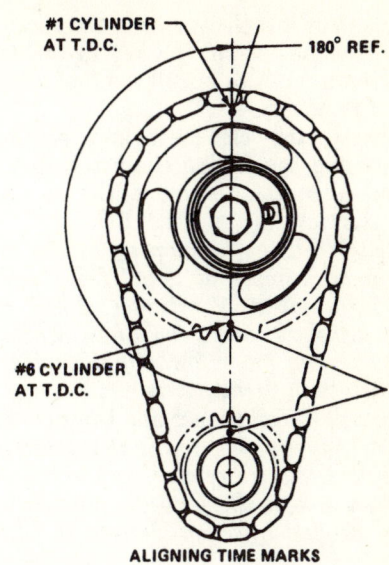

Timing sprocket alignment 307 Oldsmobile produced engine

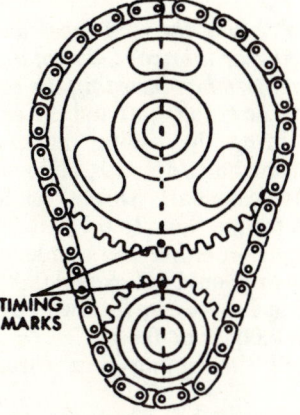

Timing sprocket alignment, V6 and V8 (1968–79) gasoline engines

the camshaft sprocket aligns with the mark on the crankshaft sprocket.

3. Remove the camshaft sprocket bolts, the camshaft sprocket, the timing chain and the crankshaft sprocket, if necessary.

NOTE: *When installing the timing chain, install the sprockets with the timing marks facing each other; this position is TDC of the No. 6 cyl. (V8) or No. 4 cyl. (V6). To locate the TDC of the No. 1 cyl., turn the crankshaft one full revolution, the camshaft timing mark will now be at the top of the sprocket.*

4. To install, use new gaskets, sealant and reverse the removal procedures. Torque the camshaft sprocket bolts to 13–23 ft. lbs. Check and/or adjust the engine timing.

8-307 Oldsmobile Produced Engine

1. Disconnect the negative battery cable. Remove the timing chain cover.
2. Remove the crankcase oil slinger. Remove the camshaft thrust button and spring.
3. Remove the fuel pump. Discard the fuel pump gasket. Remove the fuel pump eccentric.
4. Remove the spark plugs. Remove the camshaft sprocket. Remove the timing chain.

To Install:
NOTE: *Be sure that the timing marks are in alignment before removing the timing chain and the camshaft sprocket.*

5. Position the camshaft sprocket and the timing chain to its mounting on the engine.
6. Install the fuel pump eccentric, flat side toward the engine. Install the camshaft retaining bolt finger tight.
7. Rotate the crankshaft until the keyways are aligned.

NOTE: *When the timing marks are aligned, the NO. 6 piston is at TDC. When both timing marks are on top, No.1 cylinder is in the firing position.*

8. Tighten the camshaft sprocket to 65 ft. lbs. Install the camshaft thrust button and spring.
9. Install the crankshaft oil slinger. Install the spark plugs. Install the front cover.
10. Start the engine and check for proper operation. Check for leaks.

Diesel Engine

1. Disconnect the negative battery cable. Remove the front cover.
2. Remove the oil slinger and camshaft bolts, and remove the camshaft and crankshaft sprockets and timing chain as a unit.
3. Remove the fuel pump eccentric from

ENGINE AND ENGINE OVERHAUL

the crankshaft if replacement is necessary.

4. To install, install the key in the crankshaft if removed. Also install the fuel pump eccentric if removed.

5. Install the camshaft and crankshaft sprockets and the timing chain together as a unit and align the timing marks as shown. Torque the camshaft sprocket bolt to 65 ft. lbs.

NOTE: *When the two timing marks are in alignment, number six is at TDC. To obtain TDC for number one cylinder, slowly rotate the crankshaft one revolution. This will bring the cam mark to the top; number one will then be in firing position.*

6. Install the oil slinger and front cover.

NOTE: *Any time the timing chain and gears are replaced, it will be necessary to retime the injection system. Refer to the paragraph on Diesel Engine Injection Timing in Chapter 5.*

Camshaft

REMOVAL AND INSTALLATION

Inline 6-Cylinder Gasoline Engine

Due to the length of the inline 6-cylinder camshaft, a large amount of working room will be required in front of the engine to remove the camshaft. There are two ways to go about this task: either remove the engine from the vehicle, or remove the radiator, grille and all supports which are mounted directly in front of the engine. If the second alternative is chosen, you must also disconnect the motor mounts and raise the front of the engine enough to gain the clearance necessary to remove the cam from the engine.

1. In addition to removing the timing gear cover, remove the grille and radiator.
2. Remove the valve cover and gasket, loosen all the valve rocker arm nuts and pivot the arms clear of the pushrods.
3. Remove the distributor and the fuel pump.
4. Remove the coil, the side cover and its gasket. Remove the pushrods and valve lifters.
5. Remove the two camshaft thrust plate retaining screws by working through the holes in the camshaft gear.
6. Remove the camshaft and gear assembly by pulling it out through the front of the block.

To Install:

NOTE: *If renewing either the camshaft or the camshaft gear, the gear must be pressed off the camshaft. The replacement parts must be assembled in the same manner (under pressure). In placing the gear on the camshaft, press the gear onto the shaft until it bottoms against the gear spacer ring. The end clear-*

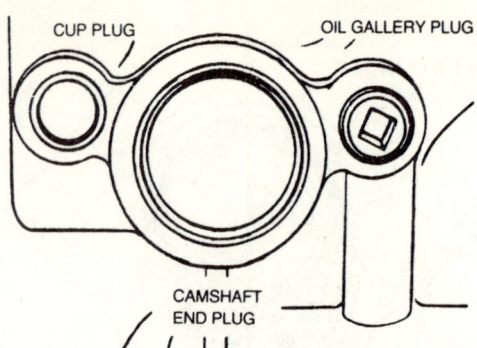

Camshaft and oil gallery plugs at rear of block

ance of the thrust plate should be 0.001-0.005".

7. Install the camshaft assembly in the engine.

NOTE: *Pre-lube the cam lobes with E.O.S. or SAE 90 gear lubricant. do not dislodge the cam bearings when inserting the camshaft.*

8. Turn the crankshaft and the camshaft to align and bring the timing marks together. Push the camshaft into this aligned position. Install the camshaft thrust plate-to-block screws and torque them to 6-7$\frac{1}{2}$ ft. lbs.
9. Runout on either the crankshaft or the camshaft gear should not exceed 0.003".
10. Backlash between the two gears should be between 0.004" and 0.006".
11. Install the timing gear cover and its gasket.
12. Install the oil pan and gaskets.
13. Install the harmonic balancer.
14. Line up the keyway in the balancer with the key on the crankshaft and the drive balancer onto the shaft until it bottoms against the crankshaft gear.
15. Install the valve lifters and pushrods. Install the side cover with new gasket. Attach the coil wires; install the fuel pump.
16. Install the distributor and set the timing as described under Distributor Installation at the beginning of this section.
17. Pivot the rocker arms over the pushrods and then adjust the valves.
18. Add oil to the engine. Install and adjust the fan belt.
19. Install the radiator or shroud.
20. Install the grille assembly.
21. Fill the cooling system, start the engine and check for leaks.
22. Check and adjust the timing.

V6 and V8 Gasoline engine except V6-231 Engine

1. Refer to the Timing Chain Removal and Installation procedures in this section and

ENGINE AND ENGINE OVERHAUL

remove the camshaft sprocket and the timing chain.

NOTE: *If the camshaft sprocket is tight on the camshaft, use a plastic hammer to bump it loose.*

2. On the V8, remove the oil cooler lines and the hoses from the radiator, then the radiator.

CAUTION: *When draining the coolant, keep in mind that cats and dogs are attracted by the ethylene glycol antifreeze, and are quite likely to drink any that is left in an uncovered container or in puddles on the ground. This will prove fatal in sufficient quantity. Always drain the coolant into a sealable container. Coolant should be reused unless it is contaminated or several years old.*

3. Remove the intake manifold and the rocker arm covers.

4. On the V8, remove the AIR pump bracket and disconnect the fuel lines at the fuel pump, then remove the fuel pump.

5. If equipped with air conditioning on the V8, remove the compressor and the condenser, then move them aside.

6. Remove the rocker arm assemblies, the push rods and the valve lifters.

7. Install two ‹5/16›"-18 x 4" bolts in the camshaft and carefully pull it from the front of the engine.

NOTE: *When removing or replacing the camshaft, be careful not to damage the camshaft bearings.*

8. To install, reverse the removal procedures. Torque the camshaft mounting bolts to 13-23 ft. lbs. Check and/or adjust the engine timing. Refill the cooling system.

V6-231 Engine

1. Disconnect the negative battery cable. Drain the engine coolant and remove the radiator and radiator hoses.

CAUTION: *When draining the coolant, keep in mind that cats and dogs are attracted by the ethylene glycol antifreeze, and are quite likely to drink any that is left in an uncovered container or in puddles on the ground. This will prove fatal in sufficient quantity. Always drain the coolant into a sealable container. Coolant should be reused unless it is contaminated or several years old.*

2. Remove the water pump and all the drive belts. Remove the alternator.

3. Remove the crankshaft pulley and the vibration damper.

4. Remove the intake manifold. Mark the location of the distributor and remove the distributor.

5. Remove the fuel pump. Remove the timing chain cover and the oil pump.

6. Remove the timing chain and the camshaft sprocket, along with the distributor drive gear and the fuel pump eccentric.

7. Remove the rocker arm covers and the rocker arm assemblies. Mark the pushrods and remove them. Remove the lifters. mark them so they can be returned to their original position.

8. Carefully remove the camshaft from the engine. Make sure you don't damage the bearings.

9. Clean all gasket and seal surfaces, use new gaskets and reverse the order of removal to install. Remember to pack the oil pump with petroleum jelly.

Diesel Engine

NOTE: *If equipped with air conditioning, the system must be discharged by an air conditioning specialist before the camshaft is removed. The condenser must also be removed from the vehicle. Removal of the camshaft also requires removal of the injection pump drive and driven gears, removal of the intake manifold, disassembly of the valve lifters, and re-timing of the injection pump.*

1. Disconnect the negative battery cables. Drain the coolant. Remove the radiator.

CAUTION: *When draining the coolant, keep in mind that cats and dogs are attracted by the ethylene glycol antifreeze, and are quite likely to drink any that is left in an uncovered container or in puddles on the ground. This will prove fatal in sufficient quantity. Always drain the coolant into a sealable container. Coolant should be reused unless it is contaminated or several years old.*

2. Remove the intake manifold and gasket and the front and rear intake manifold seals. Refer to the Intake Manifold Removal and Installation procedures.

3. Remove the balancer pulley and balancer. See Caution under Diesel Engine Front Cover Removal and Installation, above. Remove the engine front cover using the appropriate procedure.

4. Remove the valve covers. Remove the rocker arms, pushrods and valve lifters; see the procedure earlier in this section. Be sure to keep the parts in order to that they may be returned to their original position.

5. Remove the camshaft sprocket retaining bolt, and remove the timing chain and sprockets, using the procedure outlined earlier.

6. Position the camshaft dowel pin at the 3 o'clock position.

7. Push the camshaft rearward and hold it there, being careful not to dislodge the oil gallery plug at the rear of the engine. Remove the fuel injection pump drive gear by sliding it from

ENGINE AND ENGINE OVERHAUL

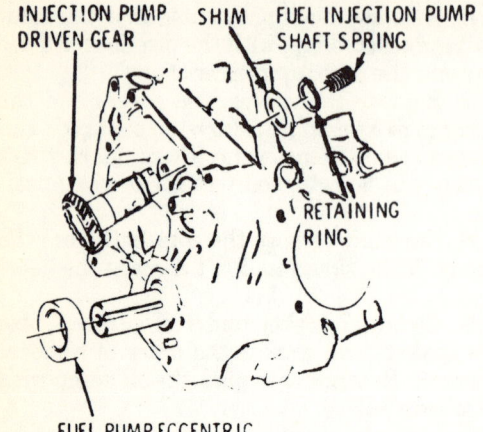

Diesel injection pump driven gear and shim

the camshaft while rocking the pump driven gear.

8. To remove the fuel injection pump driven gear, remove the pump adapter, the snapring, and remove the selective washer. Remove the driven gear and spring.

9. Remove the camshaft by sliding it out the front of the engine. On the V6, install a longer bolt into the front hole on the camshaft, to act as a handle. Be extremely careful not to allow the cam lobes to contact any of the bearings, or the journals to dislodge the bearings during camshaft removal. do not force the camshaft, or bearing damage will result.

10. If either the injection pump drive or driven gears are to be replaced, replace both gears.

To Install:

11. Coat the camshaft and the cam bearings with a heavy-weight engine oil, GM lubricant #1052365 or the equivalent.

12. Carefully slide the camshaft into position in the engine.

13. Fit the crankshaft and camshaft sprockets, aligning the timing marks as shown in the timing chain removal and installation procedures, above. Remove the sprockets without disturbing the timing.

14. Install the injection pump driven gear, spring, shim, and snapring. Check the gear end play. If the endpay is not within 0.002-0.006" on engines through 1979 and 0.002-0.015" on 1980-85 engines, replace the shim to obtain the specified clearance. Shims are available is 0.003" increments, from 0.008" to 0.0115".

15. Position the camshaft dowel pin at the 3 o'clock position. Align the zero marks on the pump drive gear and pump driven gear. Hold the camshaft in the rearward position and slide the pump drive gear onto the camshaft. Install the camshaft bearing retainer.

16. Install the timing chain and sprockets, making sure the timing marks are aligned.

17. Install the lifters, pushrods, and rocker arms. See Rocker Arm Replacement, Diesel Engine for lifter bleed down procedures. Failure to bleed down the lifters could bend valves when the engine is turned over.

18. Install the injection pump adapter and injection pump. See the appropriate sections under Fuel System above for procedures.

19. Install the remaining components in the reverse order of removal.

CAMSHAFT INSPECTION

Completely clean the camshaft with solvent, paying special attention to cleaning the oil holes. Visually inspect the cam lobes and bearing journals for excessive wear. If the lobe is questionable, have the cam checked at a reputable machine shop; if a journal or lobe is worn, the camshaft must be reground or replaced. Also have the camshaft checked for straightness on a dial indicator.

NOTE: *If a cam journal is worn, there is a good chance that the bearings are worn.*

Camshaft Bearings

REMOVAL AND INSTALLATION

If excessive camshaft wear is found, or if the engine is being completely rebuilt, the camshaft bearings should be replaced.

NOTE: *The front and rear bearings should be removed last, and installed first. Those bearings act as guides for the other bearings and pilot.*

1. Drive the camshaft rear plug from the block.

2. Assemble the removal puller with is shoulder on the bearing to be removed. Gradually tighten the puller nut until the bearing is removed.

3. Remove the remaining bearings, leaving the front and rear for last. To remove these, reverse the position of the puller, so as to pull the bearings towards the center of the block. Leave

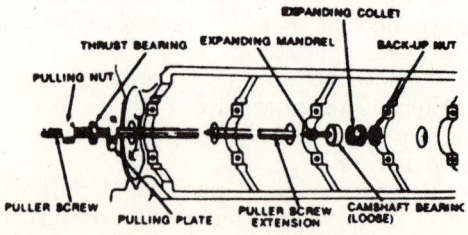

Remove camshaft bearings with a puller

ENGINE AND ENGINE OVERHAUL

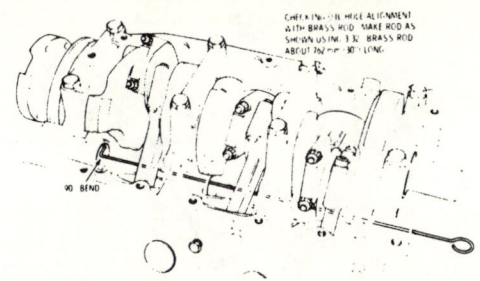

Check cam bearing alignment with this home-made tool

the tool in this position, pilot the new front and rear bearings on the installer, and pull them into position.

4. Return the puller to its original position and pull the remaining bearings into position.

NOTE: *Ensure that the oil holes align when installing the bearings. This is very important! You can make a simple tool out of the piece of ‹3/32›" brass rod to check alignment. See the illustration.*

5. Replace the camshaft rear plug, and stake it into position.

Pistons and Connecting Rods

REMOVAL AND INSTALLATION

Before removing the pistons, the top of the cylinder bore must be examined for a ridge. A ridge at the top of the bore is the result of normal cylinder wear, caused by the piston rings only traveling so far up the bore in the course of the piston stroke. The ridge can be felt by hand; it must be removed before the pistons are removed.

A ridge reamer is necessary for this operation. Place the piston at the bottom of its stroke, and cover it with a rag. Cut the ridge away with the ridge reamer, using extreme care to avoid cutting too deeply. Remove the rag, and remove the cuttings that remain on the piston with a magnet and a rag soaked in clean oil. Make sure the piston top and cylinder bore are absolutely clean before moving the piston.

1. Remove intake manifold and cylinder heads.
2. Remove oil pan.
3. Remove oil pump assembly if necessary.
4. Matchmark the connecting rod cap to the connecting rod with a scribe; each cap must be reinstalled on its proper rod in the proper direction. Remove the connecting rod bearing cap and the rod bearing. Number the top of each piston with silver paint or a felt-tip pen for later assembly.
5. Cut lengths of $3/8$" diameter hose to use as rod bolt guides. Install the hose over the threads of the rod bolts, to prevent the bolt threads from damaging the crankshaft journals and cylinder walls when the piston is removed.

6. Squirt some clean engine oil onto the cylinder wall from above, until the wall is coated. Carefully push the piston and rod assembly up and out of the cylinder by tapping on the bottom of the connecting rod with a wooden hammer handle.

7. Place the rod bearing cap back on the connecting rod, and install the nut temporarily. Using a number stamp or punch, stamp the cyl-

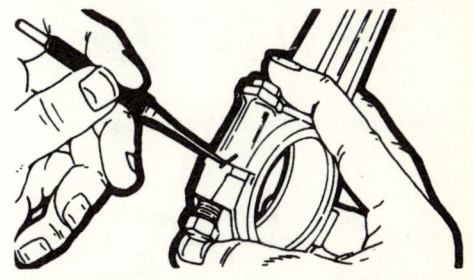

Match connecting rods to their caps with a scribe mark

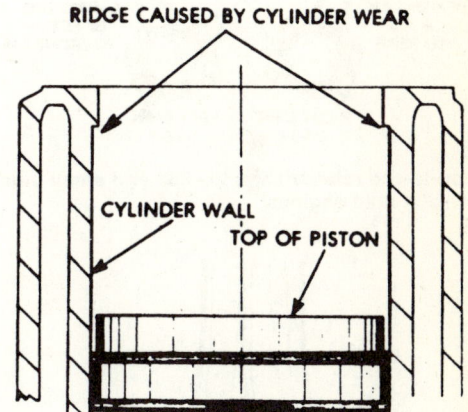

This ridge must be removed before pistons are removed

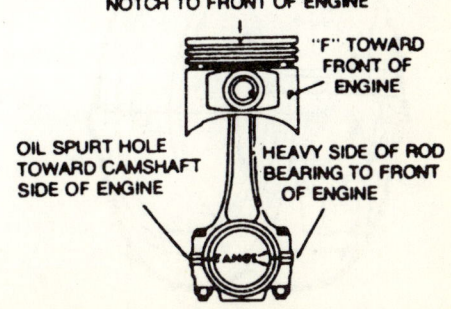

inline six cylinder piston-to-rod relationship

176 ENGINE AND ENGINE OVERHAUL

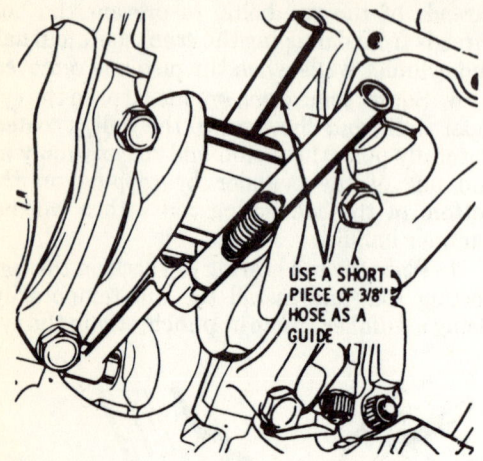

Cut rubber hose for con-rod bolt guides

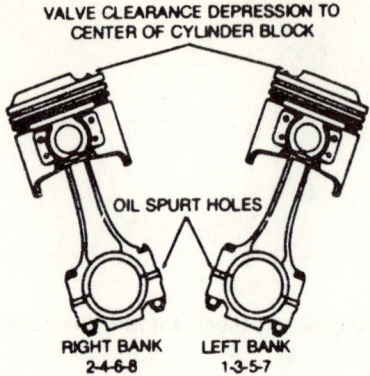

Big block (Mark IV) piston-to-rod relationship

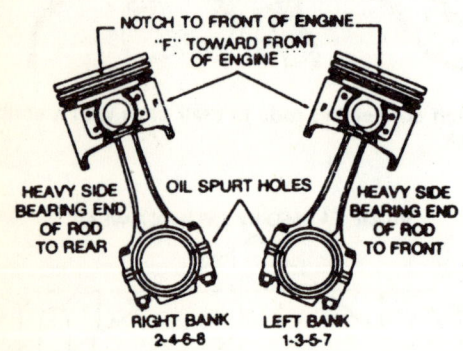

Piston-to-rod relationship, V6–262 and small block Chevrolet built engines

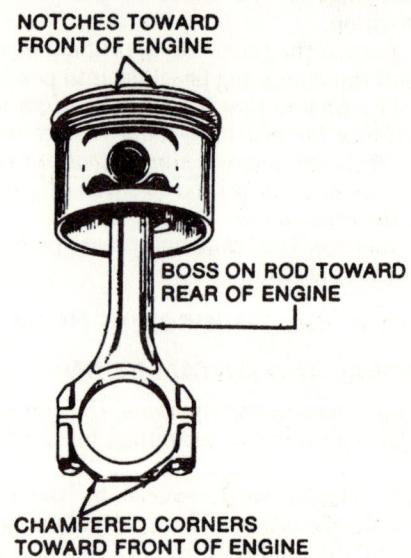

Left bank piston and rod assembly, V6–231

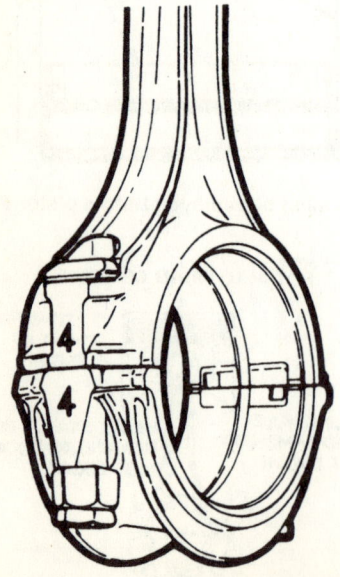

Match the connecting rods to their cylinders with a number stamp

inder number on the side of the connecting rod and cap; this will help keep the proper piston and rod assembly on the proper cylinder.

NOTE: *On V6 engines, starting at the front the cylinders are numbered 2-4-6 on the right bank and 1-3-5 on the left. On all V8s, starting at the front the right bank cylinders are 2-4-6-8 and the left bank 1-3-5-7.*

8. Remove remaining pistons in similar manner.

On all engines, the notch on the piston will face the front of the engine for assembly. The chamfered corners of the bearing caps should face toward the front of the left bank and toward the rear of the right bank, and the boss on the connecting rod should face toward the front of the engine for the right bank and to the rear of the engine on the left.

On various engines, the piston compression

rings are marked with a dimple, a letter **T**, a letter **O**, **GM** or the word **TOP** to identify the side of the ring which must face toward the top of the piston.

CLEANING AND INSPECTING

A piston ring expander is necessary for removing piston rings without damaging them; any other method (screwdriver blades, pliers, etc.) usually results in the ring being bent, scratched or distorted, or the piston itself being damaged. When the rings are removed, clean the ring grooves using an appropriate ring groove cleaning too, using care not to cut too deeply. Thoroughly clean all carbon and varnish from the piston with solvent.

WARNING: *Do not use a wire brush or caustic solvent (acids, etc.) on piston. Inspect the pistons for scuffing, scoring, cranks, pitting, or excessive ring groove wear. If these are evident, the piston must be replaced.*

The piston should also be checked in relation to the cylinder diameter. Using a telescoping gauge and micrometer, or a dial gauge, measure the cylinder bore diameter perpendicular (90%) to the piston pin, $2^{1}/_{2}''$ below the cylinder block deck (surface where the block mates with the heads). Then, with the micrometer, measure the piston perpendicular to its wrist pin on the skirt. The difference between the two measurements is the piston clearance.

If the clearance is within specifications or slightly below (after the cylinders have been bored or honed), finish honing is all that is necessary. If the clearance is excessive, try to obtain a slightly larger piston to bring clearance to within specifications. If this is not possible obtain the first oversize piston and hone (or if necessary, bore) the cylinder to size. Generally, if the cylinder bore is tapered 0.005" or more or is out-of-round 0.003" or more, it is advisable to rebore for the smallest possible oversize piston and rings. After measuring, mark pistons with a felt-tip pen for reference and for assembly.

NOTE: *Cylinder block boring should be performed by a reputable machine shop with the proper equipment. In some cases, cleanup honing can be done with the cylinder block in the vehicle, but most excessive honing and all cylinder boring must be done with the block stripped and removed from the vehicle.*

Piston Ring and Wrist Pin

REMOVAL

Some of the engines covered in this guide utilize pistons with pressed-in wrist pins; these must be removed by a special press designed for this purpose. Other pistons have their wrist pins secured by snaprings, which are easily removed with snapring pliers. Separate the piston from the connecting rod.

A piston ring expander is necessary for removing piston rings without damaging them; any other method (screwdriver blades, pliers, etc.) usually results in the rings being bent, scratched or distorted, or the piston itself being damaged. When the rings are removed, clean the ring grooves using an appropriate ring groove cleaning tool, using care not to cut too deeply. Thoroughly clean all carbon and varnish from the piston with solvent.

WARNING: *Do not use a wire brush or caustic solvent (Acids, etc.) on pistons. Inspect the pistons for scuffing, scoring, cracks, pitting, or excessive ring groove wear. If these are evident, the piston must be replaced.*

The piston should also be checked in relation to the cylinder diameter. Using a telescoping gauge and micrometer, or a dial gauge, measure the cylinder bore diameter perpendicular (90%) to the piston pin, $2^{1}/_{2}''$ below the cylinder block deck (surface where the block mates with the heads). Then, with the micrometer, measure the piston perpendicular to its wrist pin on the skirt. The difference between the two measurements is the piston clearance. If the clearance is within specifications or slightly below

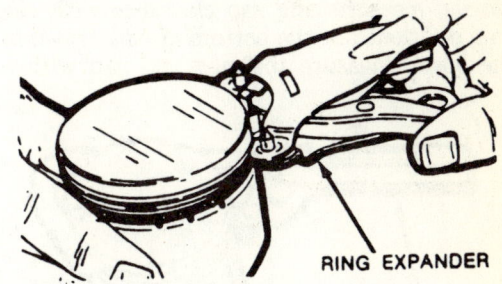

Removing the piston rings

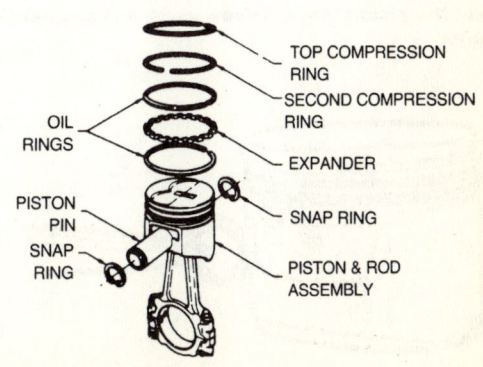

Piston rings and wrist pin

178 ENGINE AND ENGINE OVERHAUL

(after the cylinders have been bored or honed), finish honing is all that is necessary. If the clearance is excessive, try to obtain a slightly larger piston to bring clearance to within specifications. If this is not possible obtain the first oversize piston and hone (or if necessary, bore) the cylinder to size, Generally, if the cylinder bore is tapered 0.005″ or more or is out-of-round 0.003″ or more, it is advisable to rebore for the smallest possible oversize piston and rings.

After measuring, mark pistons with a felt-tip pen for reference and for assembly.

NOTE: *Cylinder honing and/or boring should be performed by a reputable, professional mechanic with the proper equipment. In some cases, clean-up honing can be done with the cylinder block in the vehicle, but most excessive honing and all cylinder boring must be done with the block stripped and removed from the vehicle.*

PISTON RING END GAP

Piston ring end gap should be checked while the rings are removed from the pistons. Incorrect end gap indicates the wrong size rings are being used; ring breakage could occur.

Compress the piston rings to be used in a cylinder, one at a time, into that cylinder. Squirt clean oil into the cylinder, so that the rings approximately 1″ below the deck of the block (on diesels, measure ring gap clearance with the ring positioned at the bottom of ring travel in the bore). Measure the ring end gap with a

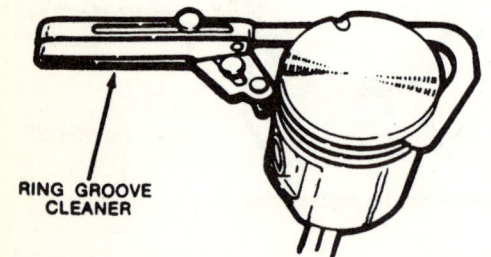

Clean the piston ring grooves using a ring groove cleaner

Removing wrist pin clips

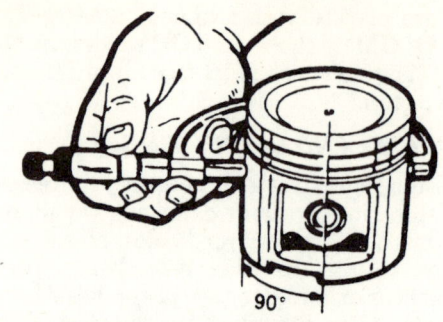

Measuring the piston prior to fitting

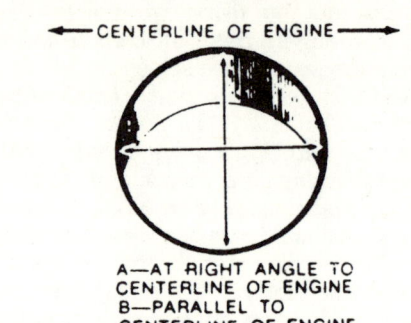

A—AT RIGHT ANGLE TO CENTERLINE OF ENGINE
B—PARALLEL TO CENTERLINE OF ENGINE

Cylinder bore measuring points

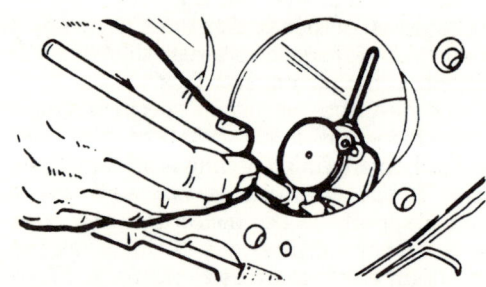

Measuring cylinder bore with a dial gauge

feeler gauge, and compare to the Ring Gap chart in this chapter. Carefully pull the ring out of the cylinder and file the ends squarely with a fine file to obtain the proper clearance.

PISTON RING SIDE CLEARANCE CHECK AND INSTALLATION

Check the pistons to see that the ring grooves and oil return holes have been properly cleaned. Slide a piston ring into its groove, and check the side clearance with a feeler gauge. On gasoline engines, make sure you insert the gauge between the ring and its lower land (lower edge of the groove), because any wear that occurs forms a step at the inner portion of the lower land. On diesels, insert the gauge between the ring and the upper land. If the piston grooves have worn to the extent that relatively

ENGINE AND ENGINE OVERHAUL

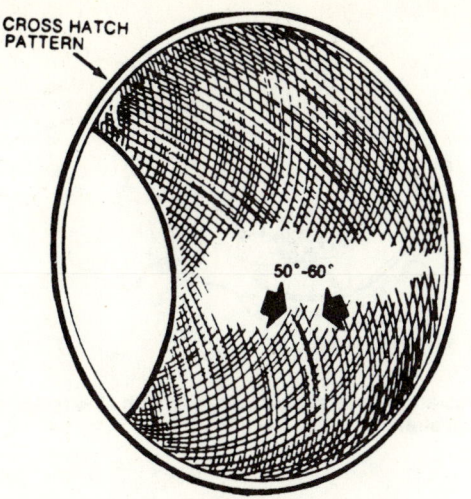

Cylinder bore cross-hatching after honing

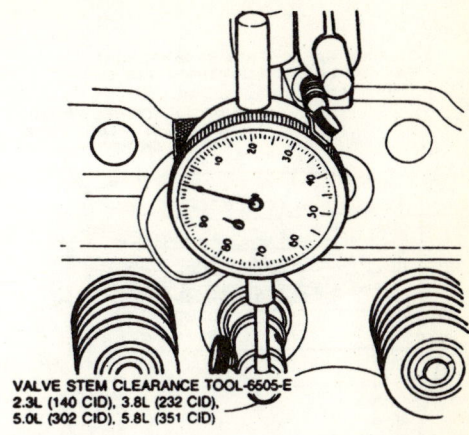

Checking piston ring end gap with a feeler gauge

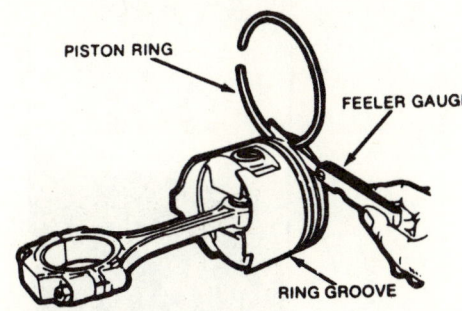

Checking ring side clearance

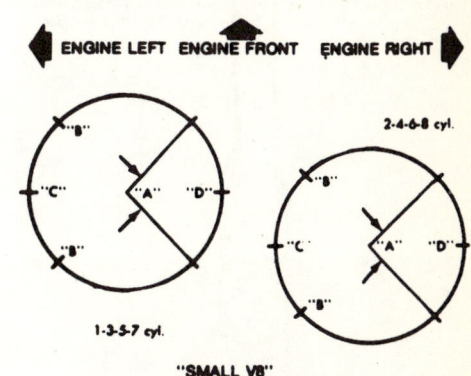

"A" OIL RING SPACER GAP
(Tang in Hole or Slot within Arc)
"B" OIL RING RAIL GAPS
"C" 2ND COMPRESSION RING CAP
"D" TOP COMPRESSION RING GAP

Ring gap locations, all gas engines except V6–231

high steps exist on the lower land, the piston should be replaced, because these will interfere with the operation of the new rings and ring clearance will be excessive. Piston rings are not furnished in oversize widths to compensate for ring groove wear.

Install the rings on the piston, lowest ring first, using a piston ring expander. There is a high risk of breaking or distorting the rings, or scratching the piston, if the rings are installed by hand or other means.

Position the rings on the piston as illustrated; spacing of the various piston ring gaps is crucial to proper oil retention and even cylinder wear. When installing new rings, refer to the installation diagram furnished with the new parts.

Connecting Rod Bearings

Connecting rod bearings for the engines covered in this guide consist of two halves or shells which are interchangeable in the rod and cap.

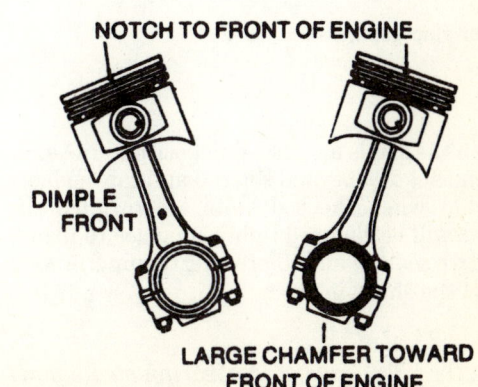

Piston and connecting rod

180 ENGINE AND ENGINE OVERHAUL

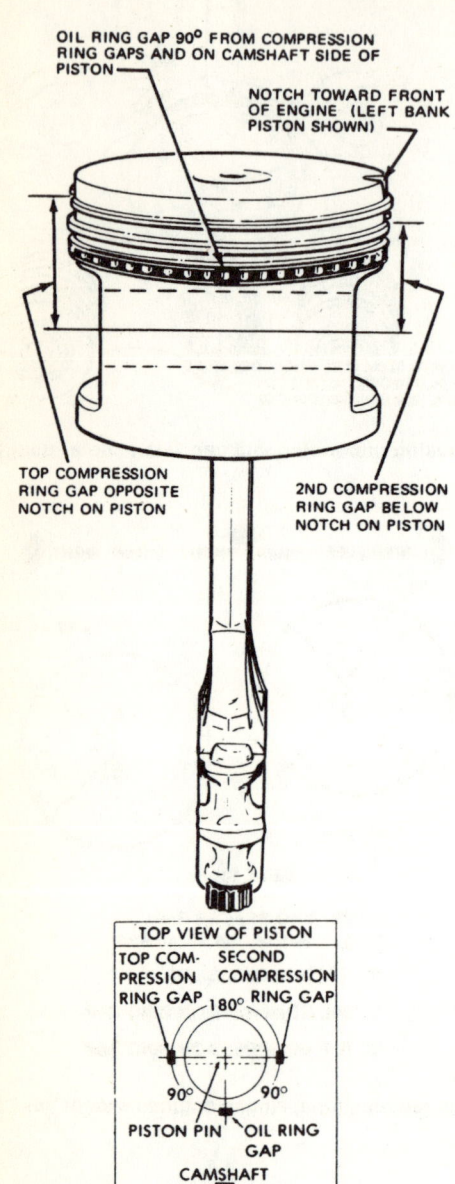

Piston ring gap location, V6–231

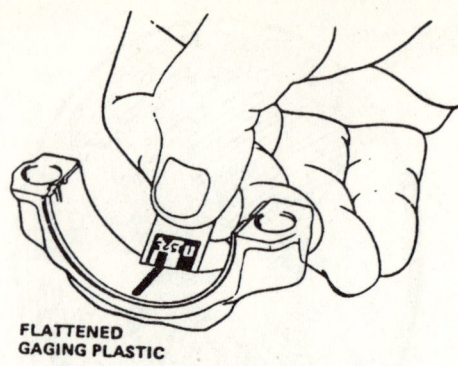

Checking rod bearing clearance with Plastigage or equivalent

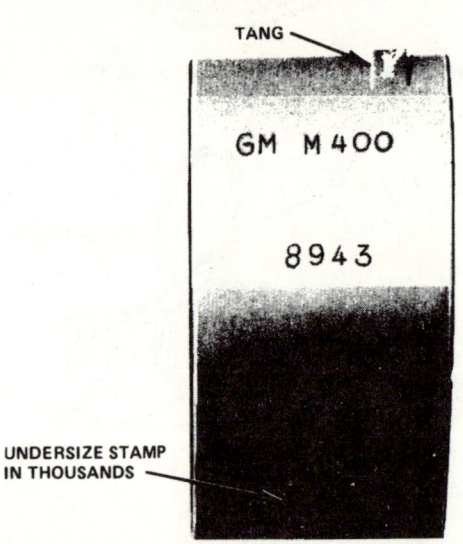

Undersize marks are stamped on the bearing shells. Tang fits in the notches in the rod cap

If a rod bearing becomes noisy or is worn so that its clearance on the crank journal is sloppy, a new bearing of the correct undersize must be selected and installed since there is no provision for adjustment.

WARNING: *Under no circumstances should the rod end or cap be filed to adjust the bearing clearance, nor should shims of any kink be used.*

Inspect the rod bearings while the rod assemblies are out of the engine. If the shells are scored or show flaking, they should be replaced. If they are in good shape check for proper clearance on the crank journal (see below). Any scoring or ridges on the crank journal means the crankshaft must be replaced, or reground and fitted with undersized bearings.

When the shells are placed in position, the ends extend slightly beyond the rod and cap surfaces so that when the rod bolts are torqued the shells will be clamped tightly in place to insure positive seating and to prevent turning. A tang holds the shells in place.

NOTE: *The ends of the bearing shells must never be filed flush with the mating surface of the rod and cap.*

ENGINE AND ENGINE OVERHAUL

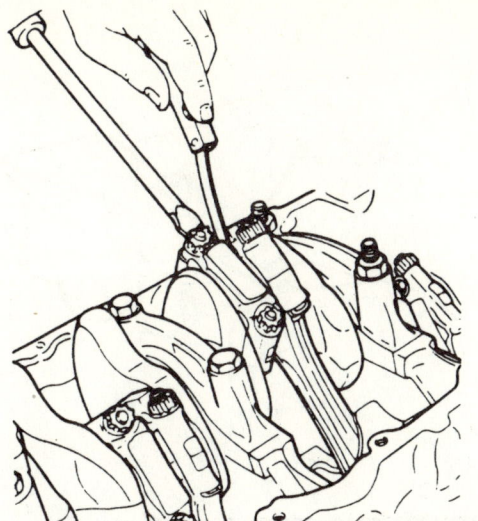

Checking connecting rod side clearance with a feeler gauge, use a small pry bar to carefully spread the connecting rods

CHECKING BEARING CLEARANCE AND REPLACING BEARINGS

NOTE: *Make sure connecting rods and their caps are kept together, and that the caps are installed in the proper direction.*

Replacement bearings are available in standard size, and in undersizes for reground crankshaft. Connecting rod-to-crankshaft bearing clearance is checked using Plastigage® at either the top or bottom of each crank journal. The Plastigage® has a range of 0.001–0.003″.

1. Remove the rod cap with the bearing shell. completely clean the bearing shell and the crank journal, and blow any oil from the oil hole in the crankshaft; Plastigage® is soluble in oil.
2. Place a piece of Plastigage® lengthwise along the bottom center of the lower bearing shell, then install the cap with shell and torque the bolt or nuts to specification. DO NOT turn the crankshaft with Plastigage® in the bearing.
3. Remove the bearing cap with the shell. the flattened Plastigage® will be found sticking to either the bearing shell or crank journal. Do not remove it yet.
4. Use the scale printed on the Plastigage® envelope to measure the flattened material to its widest point. The number within the scale which most closely corresponds to the width of the Plastigage® indicates bearing clearance in thousandths of an inch.
5. Check the specifications chart in this chapter for the desired clearance. It is advisable to install a new bearing if clearance exceeds 0.003″; however, if the bearing is in good condition and is not being checked because of bearing noise, bearing replacement is not necessary.
6. If you are installing new bearings, try a standard size, then each undersize in order until one is found that is within the specified limits when checked for clearance with Plastigage®. Each undersize shell has its size stamped on it.
7. When the proper size shell is found, clean off the Plastigage®, oil the bearing thoroughly, reinstall the cap with its shell and torque the rod bolt units to specification.

NOTE: *With the proper bearing selected and the nuts torqued, it should be possible to move the connecting rod back and forth freely on the crank journal as allowed by the specified connecting rod end clearance. If the rod cannot be moved, either the rod bearing is too far undersize or the rod is misaligned.*

PISTON AND CONNECTING ROD ASSEMBLY AND INSTALLATION

Install the connecting rod to the piston, making sure the piston installation notches

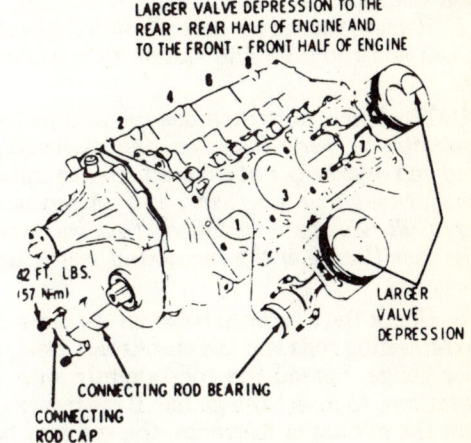

Piston locations in block, diesel shown

Using a wooden hammer handle, tap the piston down through the ring compressor and into the cylinder

and any marks on the rod are in proper relation to one another. Lubricate the wrist pin with clean engine oil, and install the pin into the rod and piston assembly, either by hand or by using a wrist pin press as required. Install snaprings if equipped, and rotate them in their grooves to make sure they are seated. To install the piston and connecting rod assembly:

1. Make sure connecting rod big-end bearings (including end cap) are of the correct size and properly installed.
2. Fit rubber hoses over the connecting rod bolts to protect the crankshaft journals, as in the Piston Removal procedure. Coat the rod bearings with clean oil.
3. Using the proper ring compressor, insert the piston assembly into the cylinder so that the notch in the top of the piston faces the front of the engine (this assumes that the dimple(s) or other markings on the connecting rods are in correct relation to the piston notch(s).
4. From beneath the engine, coat each crank journal with clean oil. Pull the connecting rod, with the bearing shell in place, into position against the crank journal.
5. Remove the rubber hoses. Install the bearing cap and cap nuts and torque to specification.

NOTE: *When more than one rod and piston assembly is being installed, the connecting rod cap attaching nuts should only be tightened enough to keep each rod in position until all have been installed. This will ease the installation of the remaining piston assemblies.*

6. Check the clearance between the sides of the connecting rods and the crankshaft using a feeler gauge. Spread the rods slightly with a screwdriver to insert the gauge. If clearance is below the minimum tolerance, the rod may be machined to provide adequate clearance. If clearance is excessive, substitute an unworn rod, and recheck. If clearance is still outside specifications, the crankshaft must be welded and reground, or replaced.
7. Replace the oil pump if removed and the oil pan.
8. Install the cylinder head(s) and intake manifold.

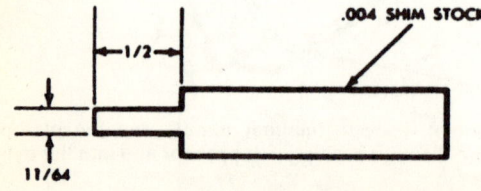

Dimensions for making an oil seal installation tool

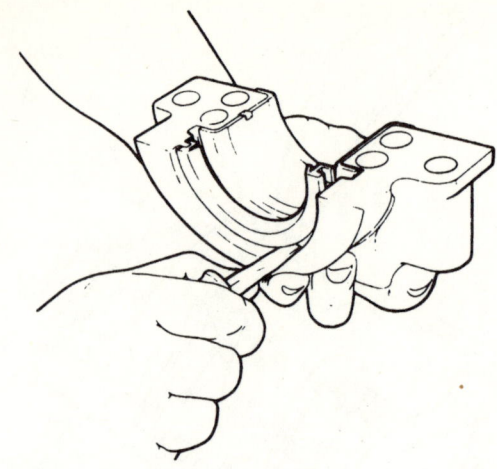

Removing the lower rear main seal

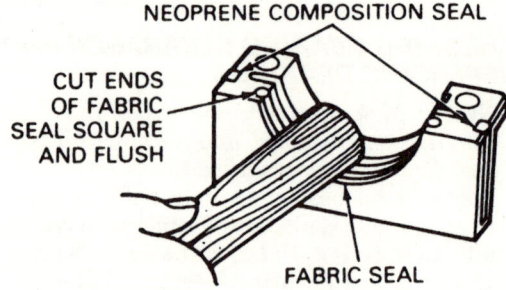

Rolling the new lower seal into place

Rear Main Oil Seal

REMOVAL AND INSTALLATION

V6 and V8 Gasoline Engines except V6-231 Engine and 8-307 Oldsmobile Produced Engine

TWO PIECE SEAL

1. Refer to the Oil Pan Removal and Installation procedures in this section and remove the oil pan.
2. Remove the oil pump and the rear main bearing cap.
3. Using a small pry bar, pry the oil seal from the rear main bearing cap.
4. Using a small hammer and a brass pin punch, drive the top half of the oil seal from the rear main bearing. Drive it out far enough, so it may be removed with a pair of pliers.
5. Using a non-abrasive cleaner, clean the rear main bearing cap and the crankshaft.
6. Fabricate an oil seal installation tool

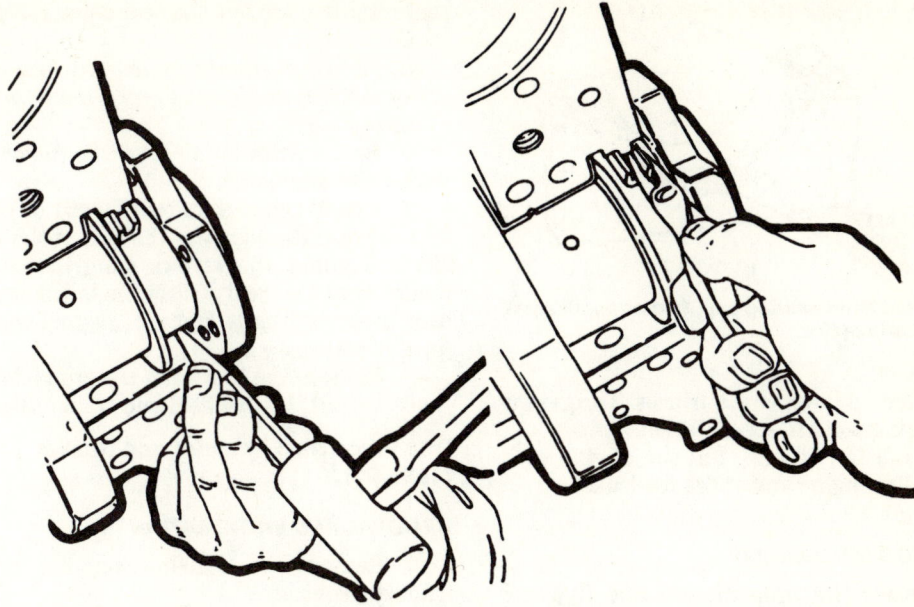

Removing the upper rear main seal

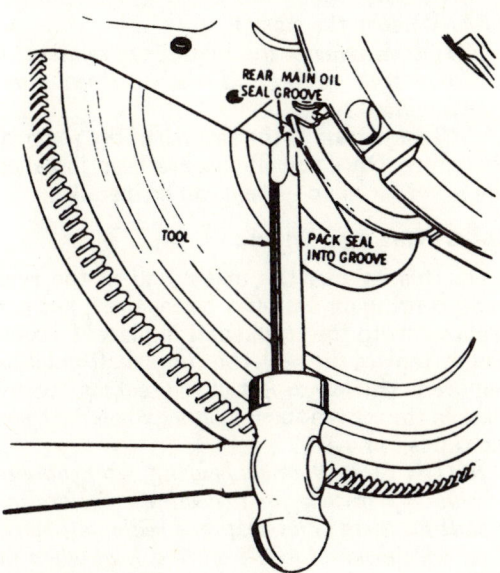

Packing the rear main oil seal

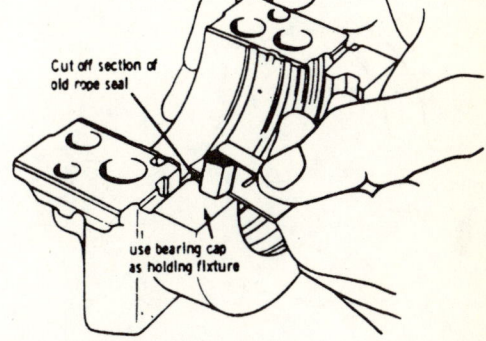

Cutting off the lower seal ends

from 0.004" shim stock, shape the end to ½" long by 11/64" wide.

7. Coat the new oil seal with engine oil; DO NOT coat the ends of the seal.

8. Position the fabricated tool between the crankshaft and the seal seat in the cylinder case.

9. Position the new half seal between the crankshaft and the top of the tool, so that the seal bead contacts the tip of the tool.

NOTE: *Make sure that the seal lip is positioned toward the front of the engine.*

10. Using the fabricated tool as a shoe horn, to protect the seal's bead from the sharp edge of the seal seat surface in the cylinder case, roll the seal around the crankshaft. when the seal's ends are flush with the engine block, remove the installation.

11. Using the same manner of installation, install the lower half onto the lower half of the rear main bearing cap.

12. Apply sealant to the cap-to-case mating surfaces and install the lower rear main bearing half to the engine; keep the sealant off of the seal's mating line.

13. Install the rear main bearing cap bolts and torque to 10-12 ft. lbs. Using a lead hammer, tap the crankshaft forward and rear-

ENGINE AND ENGINE OVERHAUL

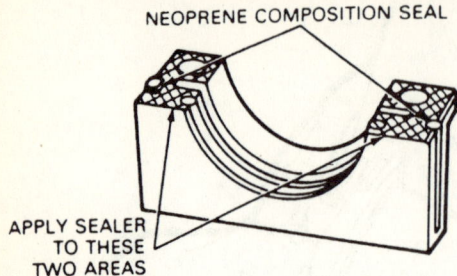

V6–231 lower main bearing cap. Sealer application, DO NOT over apply

ward, to line up the thrust surfaces. Torque the main bearing bolts to specification.

14. Install the oil pan. Fill the engine with oil. Start the engine and check for leaks.

ONE PIECE SEAL (1986-92)

1. Remove the transmission and flywheel from the vehicle.
2. Using the notches provided in the rear

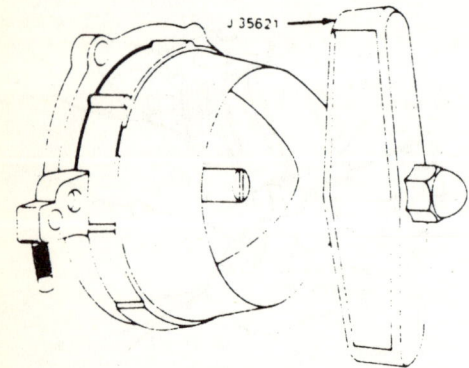

One-piece seal removal tool

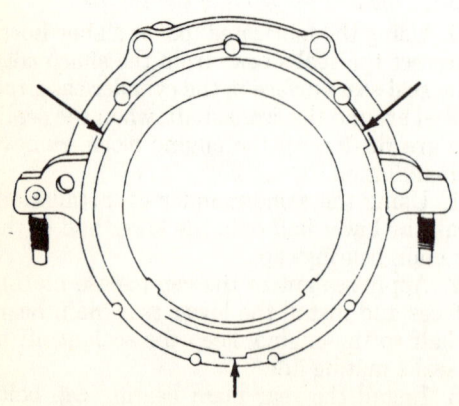

Seal removal notches

seal retainer, pry out the seal using the proper tool.

NOTE: *Care should be taken when removing the seal so as not to nick the crankshaft sealing surface.*

3. Before installation lubricate the new seal with clean engine oil.
4. Install the seal on tool J-3561 or equivalent. Thread the tool into the rear of the crankshaft. Tighten the screws snugly, this is to insure that the seal will be installed squarely over the crankshaft. Tighten the tool wing nut until it bottoms.
5. Remove the tool from the crankshaft.
6. Install the flywheel and transmission.

One Piece Seal Retainer and Gasket (1986-92)

REMOVAL AND INSTALLATION

1. Remove the transmission and flywheel from the vehicle.
2. Remove the oil pan bolts. Lower the oil pan.
3. Remove the retainer and seal assembly.
4. Remove the gasket.

NOTE: *Whenever the retainer is removed a new retainer gasket and rear main seal must be installed.*

5. Installation is the reverse of the removal procedure. Once the oil pan has been installed the new rear main oil seal can be installed.

V6-231 Gasoline Engine

On this engine, the upper half of the rear main bearing oil seal may be repaired, but not replaced, with the crankshaft in place. To completely replace the seal, the crankshaft must be removed. The lower part of the seal may be replace in the conventional manner when the bearing cap is removed.

NOTE: *Place the new bearing cap neoprene seals in kerosene for two minutes before installing them. The neoprene seals will swell up once exposed to the oil and heat when in the engine. It is normal for the seals to leak for a short time, until they become properly seated. The seals must NOT be cut to fit.*

1. Remove the oil pan. Remove the rear main bearing cap.
2. Using a blunt-edged tool, drive the upper seal into its groove until it is tightly packed. This is usually $1/4$-$3/4$".
3. Cut pieces of a new seal 1/16" longer than required to fill the grooves and install them, packing them into place.
4. Carefully trim any protruding edges of the seal.
5. Remove the old seal from the bearing cap, and install a new seal.

ENGINE AND ENGINE OVERHAUL

NOTE: *To help eliminate oil leakage at the joint where the cap meets the crankcase, apply RTV type sealer to the rear main bearing cap split line. When applying sealer, use only a thin coat as an over abundance will not allow the cap to seat properly.*

6. Install the main bearing cap. Install the oil pan.

8-307 Oldsmobile Produced Engine and Diesel Engine

1. Disconnect the negative battery cable. Drain the crankcase oil. Remove the oil pan. remove the rear main bearing cap.

CAUTION: *The EPA warns that prolonged contact with used engine oil may cause a number of skin disorders, including cancer! You should make every effort to minimize your exposure to used engine oil. Protective gloves should be worn when changing the oil. Wash your hands and any other exposed skin areas as soon as possible after exposure to used engine oil. Soap and water, or waterless hand cleaner should be used.*

NOTE: *A Chilton environmental tip. Used oil contains heavy metals, and has been determined to be hazardous to the environment. Recycling oil is the best way for disposal. Check local laws and recycle whenever possible.*

2. Using a special main seal tool or a tool that can be made from a dowel (see illustration), drive the upper seal into its groove on each side until it is tightly packed. This is usually 1/4-3/4".

3. Measure the amount the seal was driven up on one side; add 1/16", then cut this length from the old seal that was removed from the main bearing cap. Use a single-edge razor blade. Measure the amount the seal was driven up on the other side; add 11/16" and cut another length from the old seal. Use the main bearing cap as a holding fixture when cutting the seal as illustrated. Carefully trim the protruding seal.

4. Work these two pieces of seal up into the cylinder block on each side with two nail sets or small spry bars. Using the packing tool again, pack these pieces into the block, then trim them flush with a razor blade or hobby knife as shown. Do not scratch the bearing surfaces with the razor.

5. Install a new seal in the rear main bearing cap. Run a 1/16" bead of sealer onto the outer mating surface of the bearing cap. Assembly the cap to the block and torque to specification.

Crankshaft and Main Bearings

CRANKSHAFT REMOVAL

1. Drain the engine oil and remove the engine from the vehicle. Mount the engine on a work stand in a suitable working area. Invert the engine, so the oil pan is facing up.

CAUTION: *The EPA warns that prolonged contact with used engine oil may cause a number of skin disorders, including cancer! You should make every effort to minimize your exposure to used engine oil. Protective gloves should be worn when changing the oil. Wash your hands and any other exposed skin areas as soon as possible after exposure to*

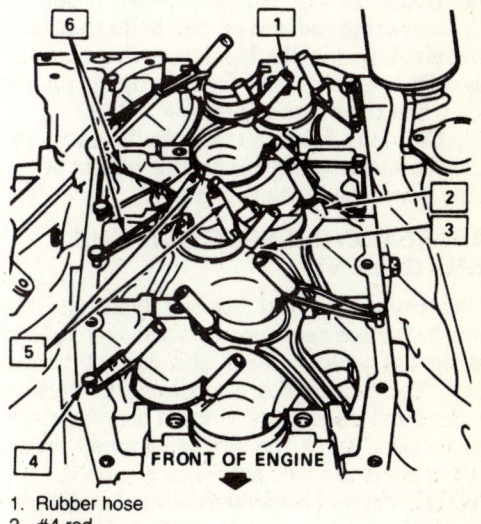

1. Rubber hose
2. #4 rod
3. #3 rod
4. Oil pan bolt
5. Note overlap of adjacent rods
6. Rubber bands

Crankshaft removal showing hose lengths on rod bolts

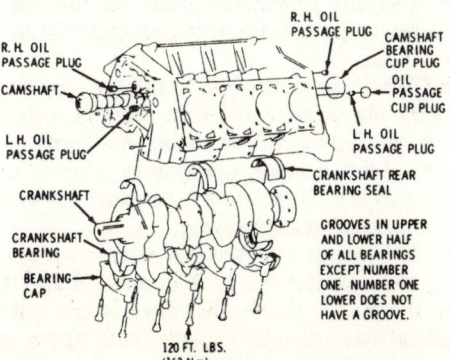

Diesel crankshaft, exploded view. Gasoline engines similar; bearing configuration may differ among engines

used engine oil. Soap and water, or waterless hand cleaner should be used.

NOTE: *A Chilton environmental tip. Used oil contains heavy metals, and has been determined to be hazardous to the environment. Recycling oil is the best way for disposal. Check local laws and recycle whenever possible.*

2. Remove the engine front (timing) cover.
3. Remove the timing chain and gears.
4. Remove the oil pan.
5. Remove the oil pump.
6. Stamp the cylinder number on the machined surfaces of the bolt bosses of the connecting rods and caps for identification when reinstalling. If the pistons are to be removed eventually from the connecting rod, mark the cylinder number on the pistons with silver paint or felt-tip pen for proper cylinder identification and cap-to-rod location.
7. Remove the connecting rod caps. Install lengths of rubber hose on each of the connecting rod bolts, to protect the crank journals when the crank is removed.
8. Mark the main bearing caps with a number punch or punch so that they can be reinstalled in their original positions. Remove all main bearing caps.
9. Remove the one piece rear crankshaft seal retainer (on 1986 and later V8).
10. Note the position of the keyway in the crankshaft so it can be installed in the same position.
11. Install rubber bands between a bolt on each connecting rod and oil pan bolts that have been reinstalled in the block (see illustration). This will keep the rods from banging on the block when the crank is removed.
12. Carefully lift the crankshaft out of the block. The rods will pivot to the center of the engine when the crank is removed.

MAIN BEARING INSPECTION AND REPLACEMENT

Like connecting rod big-end bearings, the crankshaft main bearings are shell-type inserts that do not utilize shims and cannot be adjusted. The bearings are available in various standard and undersizes; if main bearing clearance is found to be too sloppy, a new bearing (both upper and lower halves) is required.

NOTE: *Factory-undersized crankshafts are marked, sometimes with a "9" and/or a large spot of light green paint; the bearing caps also will have the paint on each side of the undersized journal.*

Generally, the lower half of the bearing shell (except No. 1 bearing) shows greater wear and fatigue. If the lower half only shows the effects of normal wear (no heavy scoring or discoloration), it can usually be assumed that the upper half is also in good shape; conversely, if the lower half is heavily worn or damaged, both halves should be replaced. Never replace one bearing half without replacing the other.

CHECKING CLEARANCE

Main bearing clearance can be checked both with the crankshaft in the vehicle and with the engine out of the vehicle. If the engine block is still in the vehicle, the crankshaft should be supported both front and rear (by the damper and flywheel ends) to remove clearance from the upper bearing. Total clearance can then be measured between the lower bearing and journal. If the block has been removed from the vehicle, and is inverted, the crank will rest on the upper bearings and the total clearance can be measured between the lower bearing and journal. Clearance is checked in the same manner as the connecting rod bearings, and Plastigage®.

NOTE: *Crankshaft bearing caps and bearing shells should NEVER be filed flush with the cap-to-block mating surface to adjust for wear in the old bearings. Always install new bearings.*

1. If the crankshaft has been removed, install it (block removed from vehicle). If the block is still in the vehicle, remove the oil pan and oil pump. Starting with the rear bearing cap, remove the cap and wipe all oil from the crank journal and bearing cap.
2. Place a strip of Plastigage® the full width of the bearing, (parallel to the crankshaft), on the journal.

WARNING: *Do not rotate the crankshaft while the gaging material is between the bearing and the journal.*

3. Install the bearing cap and evenly torque the cap bolts to specification.
4. Remove the bearing cap. The flattened Plastigage® will be sticking to either the bearing shell or the crank journal.
5. Use the graduated scale on the Plastigage® envelope to measure the material at its widest point.

NOTE: *If the flattened Plastigage® tapers towards the middle or ends. there is a difference in clearance indicating the bearing or journal has a taper, low spot or other irregularity. If this is indicated, measure the crank journal with a micrometer.*

6. If bearing clearance is within specifications, the bearing insert is in good shape. Replace the insert if the clearance is not within specifications. Always replace both upper and lower inserts as a unit.
7. Standard, 0.001" or 0.002" undersize bearings should produce the proper clearance. If these sizes still produce too sloppy a fit, the

ENGINE AND ENGINE OVERHAUL

crankshaft must be reground for use with the next undersize bearing. Recheck all clearances after installing new bearings.

8. Replace the rest of the bearings in the same manner. After all bearings have been checked, rotate the crankshaft to make sure there is no excessive drag. When checking the No. 1 main bearing, loosen the accessory drive belts (engine in vehicle) to prevent a tapered reading with Plastigage®.

MAIN BEARING REPLACEMENT

Engine Out of Vehicle

1. Remove and inspect the crankshaft.
2. Remove the main bearings from the bearing saddles in the cylinder block and main bearing caps.
3. Coat the bearing surfaces of the new, correct size main bearings with clean engine oil and install them in the bearing saddles in the block and in the main bearing caps.
4. Install the crankshaft. See Crankshaft Installation.

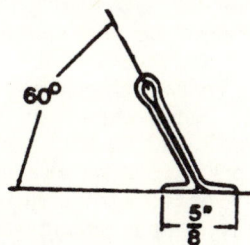

Home-made bearing roll-out pin

Engine in Vehicle

1. With the oil pan, oil pump and spark plugs removed, remove the cap from the main bearing needing replacement and remove the bearing from the cap.
2. Make a bearing roll-out pin, using a bent cotter pin as shown in the illustration. Install the end of the pin on the oil hole in the crankshaft journal.
3. Rotate the crankshaft clockwise as viewed from the front of the engine. This will roll the upper bearing out of the block.
4. Lube the new upper bearing with clean engine oil and insert the plain (unnotched) end between the crankshaft and the indented or notched side of the block. Roll the bearing into place, making sure that the oil holes are aligned. Remove the roll pin from the oil hole.
5. Lube the new lower bearing. Install the main bearing cap, making sure it is positioned in proper direction with the matchmarks in alignment.

6. Torque the main bearing cap bolts to specification.

NOTE: *See Crankshaft Installation for thrust bearing alignment.*

CRANKSHAFT END PLAY AND INSTALLATION

When main bearing clearance has been checked, bearings examined and/or replaced, the crankshaft can be installed. Thoroughly clean the upper and lower bearing surfaces, and lube them with clean engine oil. Install the crankshaft and main bearing caps.

Dip all main bearing cap bolts in clean oil, and torque all main bearing caps, excluding the thrust bearing cap, to specifications (see the Crankshaft and Connecting Rod chart in this chapter to determine which bearing is the thrust bearing). Tighten the thrust bearing bolts finger tight. To align the thrust bearing, pry the crankshaft to the extent of it axial travel several times, holding the last movement toward the front of the engine. Add thrust washers if required for proper alignment. Torque the thrust bearing cap to specifications.

To check crankshaft endplay, pry the crankshaft to the extreme rear of its axial travel,

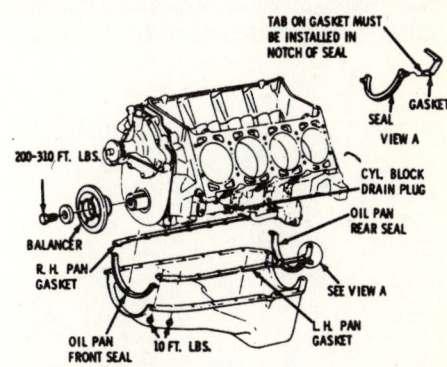

Oil pan installation. Gaskets and seals slightly differ among engines

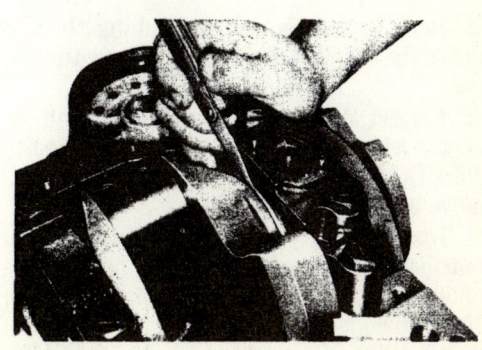

Measuring crankshaft-end-play with a feeler gauge

ENGINE AND ENGINE OVERHAUL

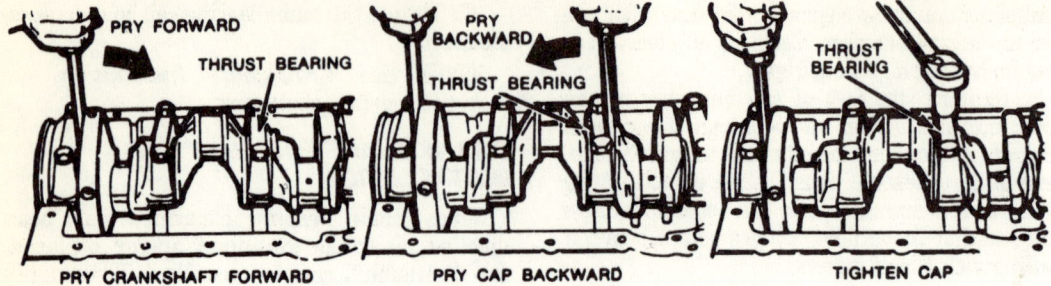

Aligning crankshaft thrust bearing

then to the extreme front of its travel. Using a feeler gauge, measure the endplay at the front of the rear main bearing. End play may also be measured at the thrust bearing. Install a new rear main bearing oil seal in the cylinder block and main bearing cap. Continue to reassemble the engine.

Flywheel and Ring Gear

REMOVAL AND INSTALLATION

The ring gear is an integral part of the flywheel and is not replaceable.

1. Disconnect the negative battery cable. Remove the transmission (on manual transmissions remove the clutch and pressure plate also).

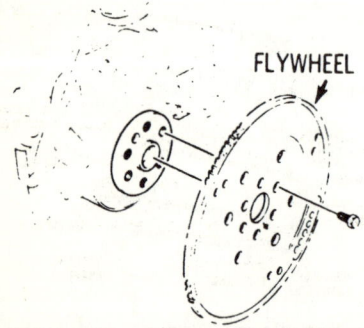

Flywheel installation. Follow torque specifications closely

2. Remove the six bolts attaching the flywheel to the crankshaft flange. Remove the flywheel.
3. Inspect the flywheel for cracks, and inspect the ring gear for burrs or worn teeth. Replace the flywheel if any damage is apparent. Remove burrs with a mill file.
4. Install the flywheel. The flywheel will only attach to the crankshaft in one position, as the bolt holes are unevenly spaced. Install the bolts and torque to specification.
5. Install the transmission and related components.

EXHAUST SYSTEM

Safety Precautions

For a number of reasons, exhaust system work can be the most dangerous type of work you can do on your car. Always observe the following precautions:

• Support the car extra securely. Not only will you often be working directly under it, but you'll frequently be using a lot of force, say, heavy hammer blows, to dislodge rusted parts. This can cause a car that's improperly supported to shift and possibly fall.

• Wear goggles. Exhaust system parts are always rusty. Metal chips can be dislodged, even when you're only turning rusted bolts. Attempting to pry pipes apart with a chisel makes the chips fly even more frequently.

• If you're using a cutting torch, keep it a great distance from either the fuel tank or lines. Stop what you're doing and feel the temperature of the fuel bearing pipes on the tank frequently. Even slight heat can expand and/or vaporize fuel, resulting in accumulated vapor, or even a liquid leak, near your torch.

• Watch where your hammer blows fall and make sure you hit squarely. You could easily tap a brake or fuel line when you hit an exhaust system part with a glancing blow. Inspect all lines and hoses in the area where you've been working.

CAUTION: *Be very careful when working on or near the catalytic converter. External temperatures can reach 1,500°F (816°C) and more, causing severe burns. Removal or installation should be performed only on a cold exhaust system.*

Special Tools

A number of special exhaust system tools can be rented from auto supply houses or local stores that rent special equipment. A common one is a tail pipe expander, designed to enable you to join pipes of identical diameter.

ENGINE AND ENGINE OVERHAUL   189

It may also be quite helpful to use solvents designed to loosen rusted bolts or flanges. Soaking rusted parts the night before you do the job can speed the work of freeing rusted parts considerably. Remember that these solvents are often flammable. Apply only to parts after they are cool!

Whenever working on the exhaust system please observe the following:

1. Check the complete exhaust system for open seams, holes loose connections, or other deterioration which could permit exhaust fumes to seep into the passenger compartment.
2. The exhaust system is supported by free-hanging rubber mountings which permit some movement of the exhaust system, but do not permit transfer of noise and vibration into the passenger compartment.
3. Before removing any component of the exhaust system, ALWAYS squirt a liquid rust dissolving agent onto the fasteners for ease of removal.
4. Annoying rattles and noise vibrations in the exhaust system are usually caused by misalignment of the parts. When aligning the system, leave all bolts and nuts loose until all parts are properly aligned, then tighten, working from front to rear.
5. When replacing a muffler and/or resonator, the tailpipe(s) should also be replaced.
6. When installing exhaust system parts, make sure there is enough clearance between the hot exhaust parts and pipes, and hoses that would be adversely affected by excessive heat. Also make sure there is adequate clearance from the floor pan to avoid possible overheating of the floor.
7. Exhaust pipe sealers should be used at all slip joint connections except at the catalytic convertor. Do not use any sealers at the convertor as the sealer will not withstand convertor temperatures.

Tailpipe

REMOVAL AND INSTALLATION

1. Raise and support the vehicle safely.
2. Remove the hanger clamps from the tail pipe. Remove the tailpipe-to-muffler clamp.
3. Disengage the tailpipe from the muffler and remove the tailpipe.
4. Inspect the tailpipe hangers; replace, if necessary.
5. Installation is the reverse of the removal procedure. Assemble the system, check the clearance and then tighten the components.

Crossover Pipe

REMOVAL AND INSTALLATION

1. Raise and support the vehicle safely.
2. Remove the exhaust pipe-to-exhaust manifolds nuts.
   NOTE: *The left side has an extension and packing; the right side has a heat riser valve assembly.*
3. Loosen the catalytic converter-to-transmission bracket.
4. Remove the crossover-to-catalytic converter clamp and the pipe from the converter.
   NOTE: *Check and lubricate the heat riser valve to make sure that it is operating properly.*
5. To install, add sealer to the connecting surfaces, assemble the system, check the clearances and tighten all the attachments.

Front Pipe

REMOVAL AND INSTALLATION

1. Raise and support the vehicle safely.
2. Remove the exhaust pipe-to-manifold nuts.
3. Support the catalytic converter and disconnect the pipe from the converter. Remove the pipe.
4. To install, use a new gasket, add sealer to the connecting surfaces, assemble the system, check the clearance and tighten the connectors.

Intermediate Pipe

The intermediate pipe is the section between the catalytic converter and the muffler.

REMOVAL AND INSTALLATION

1. Raise and support the vehicle safely.
2. Disconnect the intermediate pipe from the catalytic converter.
3. At the muffler, remove the clamp and the intermediate pipe.
4. To install, use a new clamp and nuts/bolts, assemble the system, check the clearances and tighten the connectors.

Muffler

The exhaust system pipes and resonators rearward of the mufflers, MUST BE replaced whenever a new muffler is installed.

REMOVAL AND INSTALLATION

1. Raise and support the vehicle safely.
2. On the single pipe system, cut the exhaust pipe near the front of the muffler. ON the dual pipe system, remove the U-bolt clamp at the front of the muffler and disengage the muffler from the exhaust pipe.

190 ENGINE AND ENGINE OVERHAUL

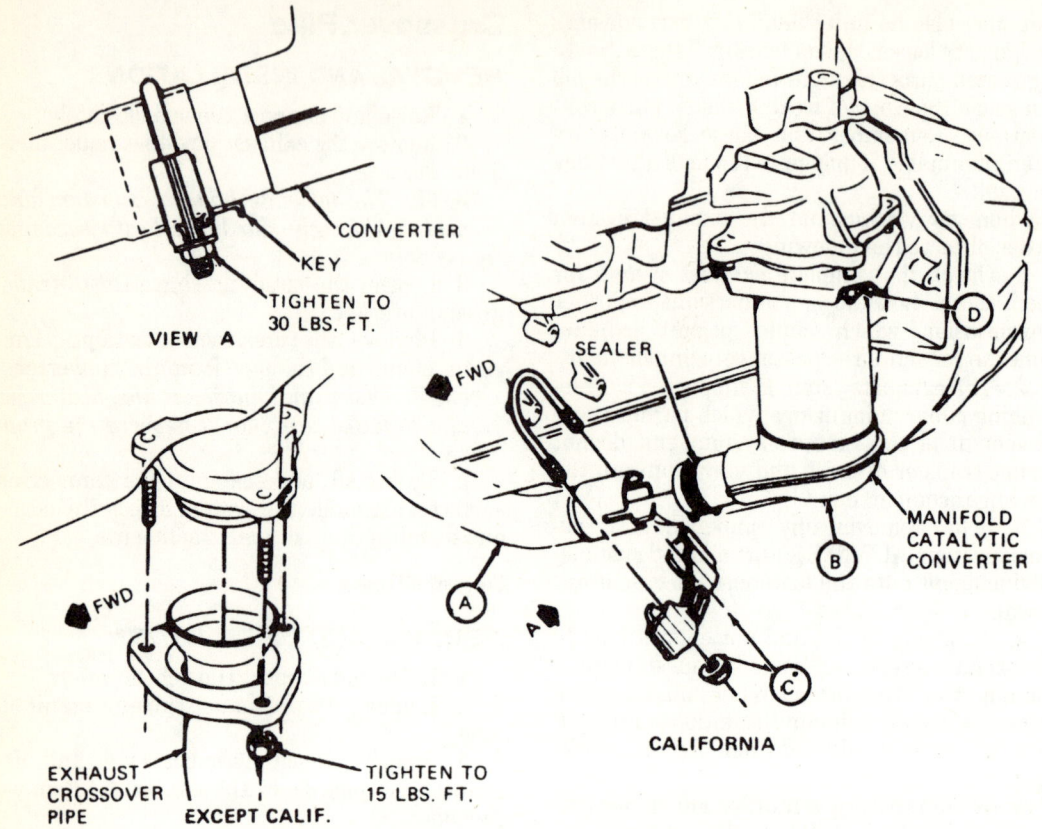

Replacing the manifold catalytic converter on the inline 6 cyl. engine

NOTE: *Before cutting the exhaust pipe, measure the service muffler exhaust pipe extension and make certain to allow $1^1/2"$ for the exhaust pipe-to-muffler extension engagement.*

3. At the rear of the muffler, remove the U-bolt clamp and disengage the muffler from the tailpipe.
4. Remove the tailpipe clamps and the tailpipe.
5. Inspect the muffler and the tailpipe hangers; replace if necessary.
6. To install, add sealer to the connecting surfaces, assemble the system, check the clearances and tighten all of the attachments.

Catalytic Converter

REMOVAL AND INSTALLATION

Except California Inline 6-Cylinder Gasoline Engine

1. Raise and support the vehicle safely.
2. Remove the clamp at the front of the converter, then cut the pipe at the front of the converter.

3. Remove the converter-to-intermediate pipe nuts/bolts.
4. Disconnect the converter-to-crossover pipe or front pipe.

NOTE: *Always install the catalytic converter into its original position in the exhaust system, the body has heat shields installed in those areas to protect the chassis from the extreme temperatures created by the converter.*

5. To install, add sealer to the connecting surfaces, use new clamps and nuts/bolts, assemble the system, check the clearances and tighten all of the attachments.

California Inline 6-Cylinder Gasoline Engine

This vehicle has a catalytic converter mounted to the exhaust manifold in addition to the one under the floor.

1. Raise and support the vehicle safely.
2. Remove the manifold converter-to-front pipe.
3. Remove the manifold converter-to-exhaust manifold nuts.
4. Remove the underfloor converter from the hanger and lower the front pipe with the manifold converter.

ENGINE AND ENGINE OVERHAUL

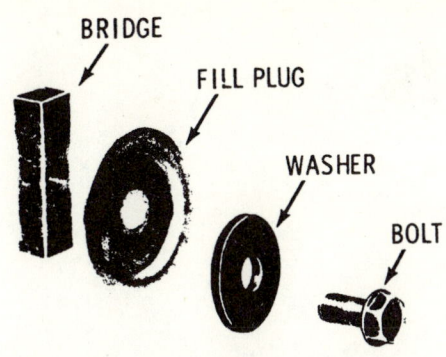

Exploded view of the replacement fill plug

5. Remove the manifold converter from the front pipe.
6. To install, add sealer to the connecting surfaces, use new claps and nuts/bolts, assemble the system, check the clearances and tighten all of the attachments.

Catalyst (Gasoline Engine)

On some vehicles the catalyst can be replaced while the converter is on the vehicle.

REMOVAL AND INSTALLATION

1. Raise and support the vehicle safely.
2. Connect an Aspirator tool No. J-25077 to the tailpipe.

NOTE: *If the vehicle has dual exhaust systems, attach the aspirator to one pipe and plug the other.*

3. Connect a 60 psi air hose to the aspirator, to hold the beads in place when the plug is removed.
4. To remove the converter fill plug, perform the following:
   a. Threaded Plug - Using a ¾" Allen wrench or tool No. J-25077-3, remove the plug from the bottom of the converter.
   b. Pressed Plug - Place a cold chisel between the plug and the converter shell. Deform the plug until it can be removed with a pair of pliers.

WARNING: *When using the chisel, be careful not to damage the converter shell. DO NOT pry on the plug, for it may damage the converter's sealing opening.*

5. Connect the Vibrator tool J-25077-6 to the converter so that the canister aligns with the converter opening.
6. Disconnect the air hose from the aspirator and allow the beads to drop into the collecting canister.
7. Connect a 60 psi air hose to the vibrator and allow the excess beads to be removed from the converter, then disconnect the air hose. Remove the canister from the vibrator and discard the used catalyst.
8. Fill the container with an approved replacement catalyst.
9. Install the fill tube extension to the vibrator fixture.
10. Connect the air hoses to the aspirator and the vibrator, then attach the canister.
11. After the catalyst stops flowing, disconnect the air hose from the vibrator. Remove the vibrator from the converter and check to see if the catalyst is flush with the fill plug hole (add catalyst if necessary).
12. To install the converter plug, perform the following:
    a. Threaded Plug - Apply an anti-seize compound to the threaded fill plug; install it and torque to 60 ft. lbs.
    b. Press Plug - Use a service plug. Install the bolt into the bridge, then the bridge into the converter's hole, by moving the bridge back-and-forth to dislodge the catalyst beads until the bridge is positioned. Remove the bolt from the bridge then position the washer and the fill plug (dished side out) over the bolt. Install the assembly to the bridge 4-5 turns and release the full plug (the aspirator will pull the plug into position). Torque the bolt to 28 ft. lbs.
13. Disconnect the air hose from the aspirator.
14. Start the vehicle and check for leaks.

Emission Controls

GASOLINE ENGINE EMISSION CONTROLS

There are three sources of automotive pollutants: Crankcase fumes, exhaust gases and gasoline evaporation. The pollutants formed from these substances all into three categories: unburnt hydrocarbons (HC), carbon monoxide (CO) and oxides of nitrogen (NOx). The equipment that is used to limit these pollutants is commonly called emission control equipment.

Positive Crankcase Ventilation

OPERATION

Until 1968, Chevrolets were available with two types of crankcase ventilation systems: The closed type found on all California vehicles and some non-California vehicles, or the open type. The open ventilation system (non-California vehicles) receives outside air through a vented oil filler cap. This filtered cap permits outside air to enter the valve cover and also allows crankcase vapors to escape from the valve cover into the atmosphere. The law requires that this filler cap be non-vented to prevent the emission of vapors into the air. To supply this closed system with fresh air, a hose runs from the carburetor air cleaner to an inlet hole in the valve cover. The carburetor end of the hose fits into a cup-shaped flame arrestor and filter in the air cleaner cover. In the event of a carburetor backfire, this arrestor prevents the spread of fire to the valve cover where it could create an explosion. Included in the system is a PCV (positive crankcase ventilation) valve that fits into an outlet hole in the top of the valve cover. A hose connects this valve to a vacuum outlet at the intake manifold. Contained within the valve housing is a valve (pointed at one end, flat at the other) positioned within a coiled spring. During idle or

Schematic of the PCV system

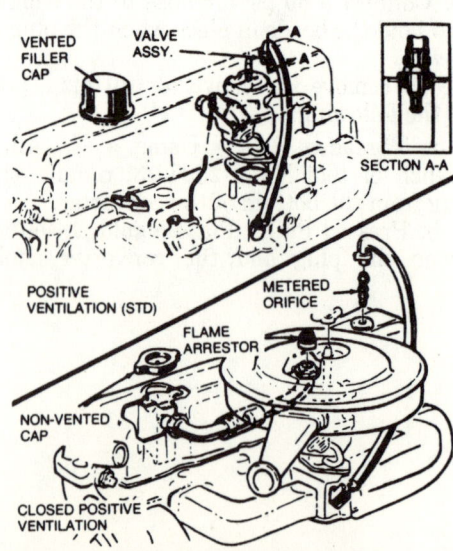

Closed and positive ventilation systems

EMISSION CONTROLS 193

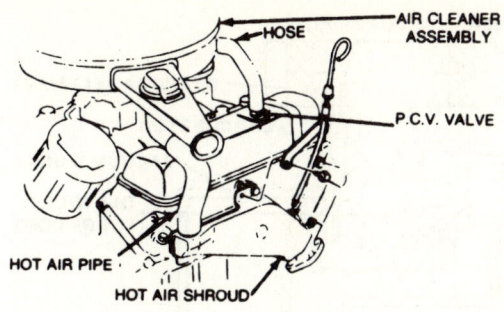

V6 PCV system

low speed operation, when manifold vacuum is highest, the valve spring tension is overcome by the high vacuum pull and, as a result, the valve is pulled up to very nearly seal off the manifold end of the valve housing. This restricts the flow of crankcase vapors to the intake manifold at a time when crankcase pressures are lowest and least disruptive to engine performance. At times of acceleration or constant speed, intake manifold vacuum is reduced to a point where it can no longer pull against the valve spring and so, spring force pulls the valve away from the housing outlet allowing crankcase vapors to escape through the hose to the intake manifold. Once inside the manifold, the gases enter the combustion chambers to be reburned. At times of engine backfire (at the carburetor) or when the engine is turned off, manifold vacuum ceases permitting the spring to pull the valve against the inlet (crankcase end) end of the valve housing. This seals off the inlet, thereby stopping the entrance of crankcase gases into the valve and preventing the possibility of a backfire spreading through the hose and valve to ignite these gases. The carburetor used with this system is set to provide a richer gas mixture to compensate for the additional air and gases going to the intake manifold. A valve that is clogged and stuck closed will not allow this extra air to reach the manifold. Consequently, the engine will run roughly and plugs will foul due to the creation, of an overly rich air/fuel mixture. It can be said that the PCV system performs three functions. It reduces air pollution by reburning the crankcase gases rather than releasing them to the atmosphere, and it increases engine life and gas economy. By recirculating crankcase gases, oil contamination that is harmful to engine parts is kept to a minimum. Recirculated gases returned to the intake manifold are combustible and, when combined with the air/fuel mixture from the carburetor, becomes fuel for operation, slightly increasing gas economy.

SERVICE

Inspect the PCV system hose and connections at each tune-up and replace any deteriorated hoses. Check the PCV valve at every tune-up and replace it every 24,000 miles on 1968-74 vehicles, and every 30,000 miles on later vehicles. See Chapter One for testing procedure.

REMOVAL AND INSTALLATION

The valve is inserted into a rubber grommet in the valve cover. At the narrow end, it is inserted into a hose and clamped. To remove it, gently pull it out of the valve cover, then open the clamp with a pair of pliers. Hold the clamp open while sliding it an inch or two down the hose (away from the valve), and then remove the valve. If the end of the hose is hard or cracked where it holds the valve, it may be feasible to cut the end off if there is plenty of extra hose. Otherwise, replace the hose. Replace the grommet in the valve cover if it is cracked or hard. Replace the clamp if it is broken or weak. In replacing the valve, make sure it is fully inserted in the hose, that the clamp is moved over the ridge on the valve so that the valve will not slip out of the hose, and that the valve is fully inserted into the grommet in the valve cover.

PCV Filter

REMOVAL AND INSTALLATION

1. Slide the rubber coupling that joins the tube coming from the valve cover to the filter off the filter nipple. Then, remove the top of the air cleaner. Slide the spring clamp off the filter, and remove the filter.
2. Inspect the rubber grommet in the valve cover and the rubber coupling for brittleness or cracking. Replace parts as necessary.
3. Insert the new PCV filter through the hole in the air cleaner with the open portion of the filter upward. Make sure that the square portion of filter behind the nipple fits into the (square) hole in the air cleaner.

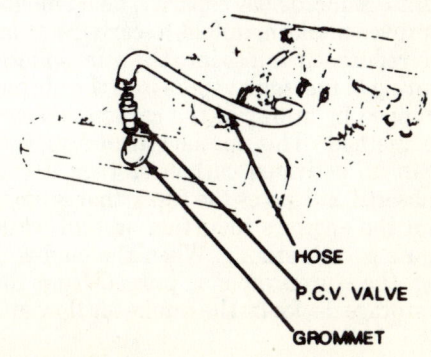

V8 PCV system

194 EMISSION CONTROLS

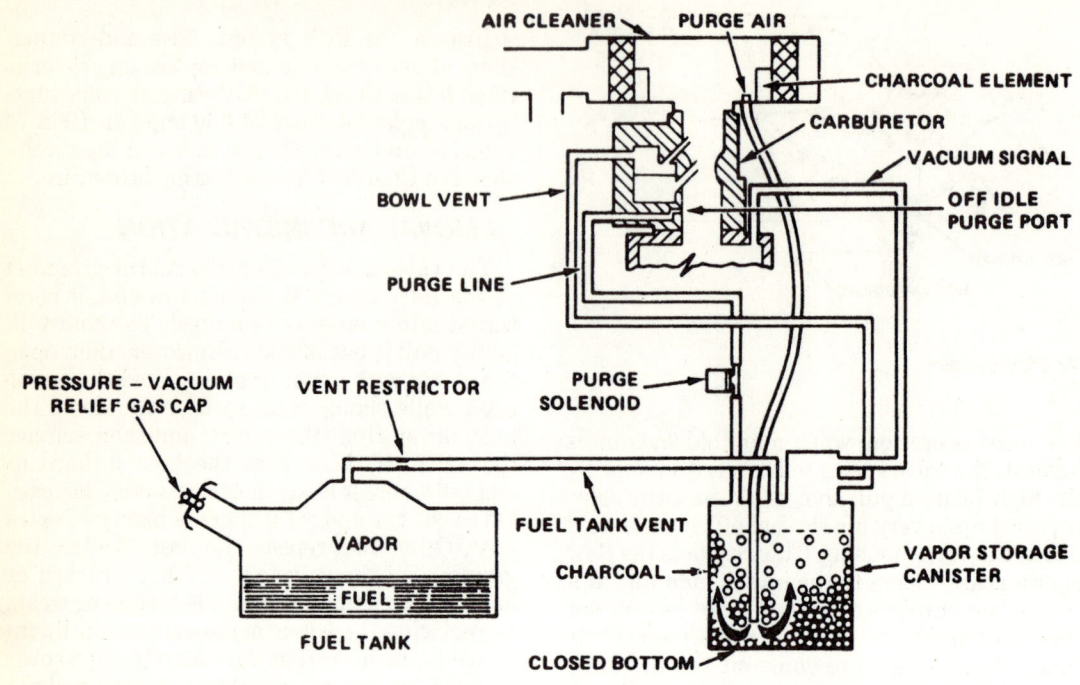

Closed canister EECS

4. Install a new spring clamp onto the nipple. Make sure the clamp goes under the ridge on the filter nipple all the way around. Then, reconnect the rubber coupling and install the air cleaner cover.

Evaporative Emission Control

OPERATION

This system which was introduced to California vehicles in 1970, and all other vehicles in 1971, reduces the amount of escaping gasoline vapors. Float bowl emissions are controlled by internal carburetor modifications. Redesigned bowl vents, reduced bowl capacity, heat shields, and improved intake manifold-to-carburetor insulation reduce vapor loss into the atmosphere. The venting of fuel tank vapors into the air has been stopped by means of the carbon canister storage method. This method transfers fuel vapors to an activated carbon storage device which absorbs and stores the vapor that is emitted from the engine's induction system while the engine is not running. When the engine is running, the stored vapor is purged from the carbon storage device by the intake air flow and then consumed in the normal combustion process. As the manifold vacuum reaches a certain point, it opens a purge control valve atop the charcoal storage canister. This allows air to be drawn into the canister, thus forcing the existing fuel vapors back into the engine to be burned normally.

The purge function on the 231 (1981) and 262 (1985-92) V6 engine is electronically controlled by a purge solenoid in the line which is itself controlled by the Electronic Control Module (ECM). When the system is in the Open Loop mode, the solenoid valve is energized, blocking all vacuum to the purge valve. When the system is in the Closed Loop mode, the solenoid is de-energized, thus allowing existing vacuum to operate the purge valve. This releases the trapped fuel vapor and it is forced into the induction system.

Most carbon canisters used are of the 'Open' design, meaning that air is drawn in through the bottom (filter) of the canister. Some 231 V6 canisters are of the 'Closed' design which means that the incoming air is drawn directly from the air cleaner.

EMISSION CONTROLS 195

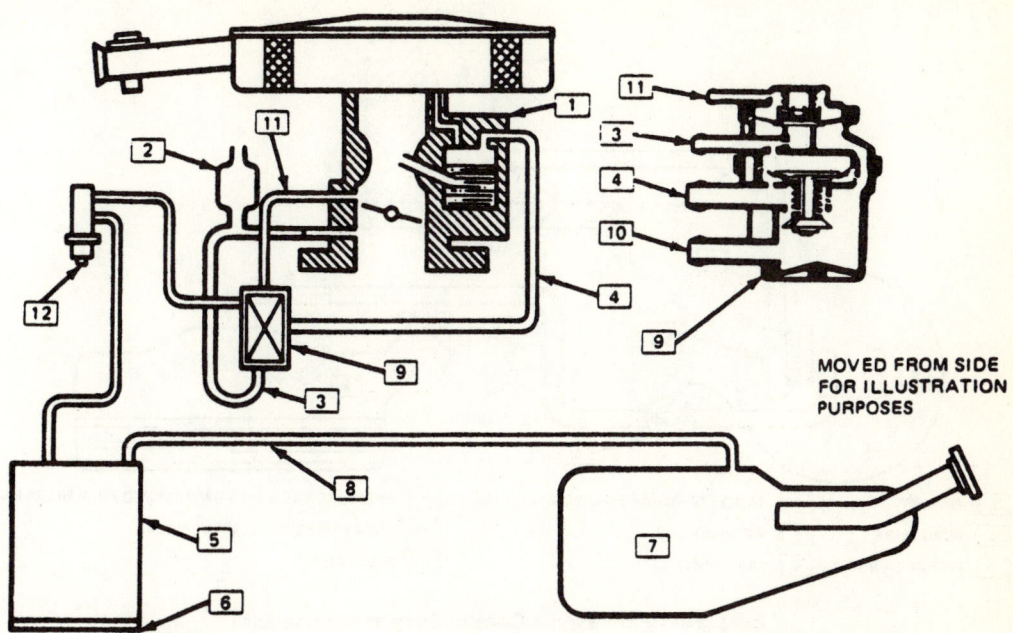

1. Carburetor
2. PCV valve
3. Manifold vacuum from PCV
4. Carburetor bowl tube
5. Vapor storage canister
6. Purge Air
7. Fuel tank
8. Fuel tank vent pipe
9. Canister control valve
10. Canister tube
11. Tube control vacuum
12. T.V.S. switch

EECS 5.0L engine except 1986–90 Oldsmobile produced engine

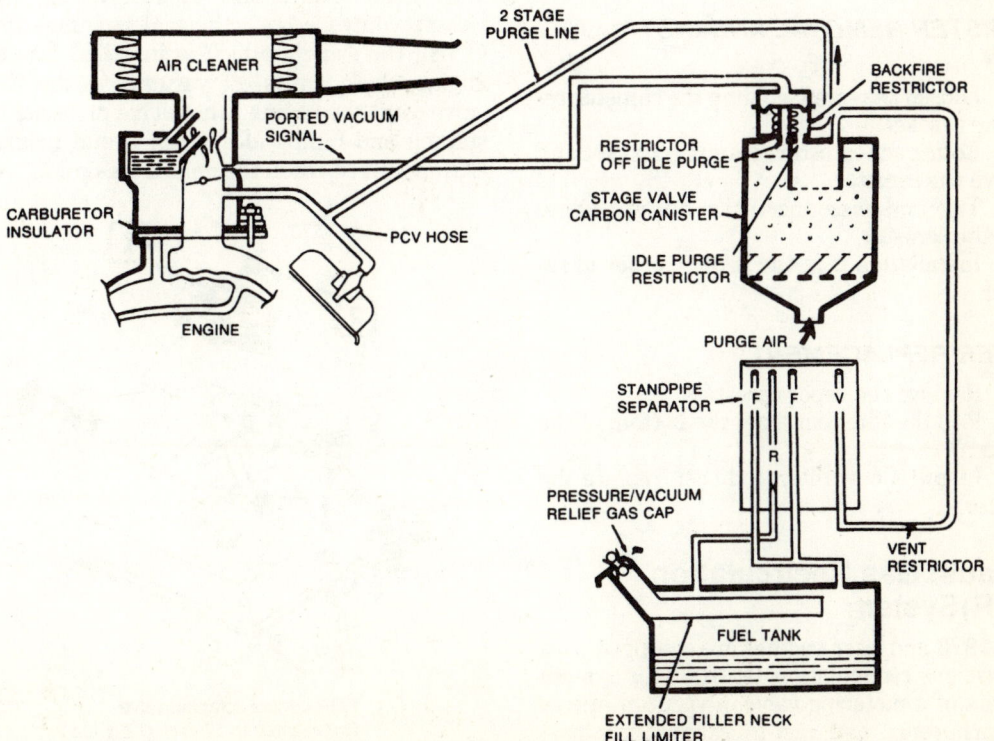

Schematic of the evaporative emissions system

196 EMISSION CONTROLS

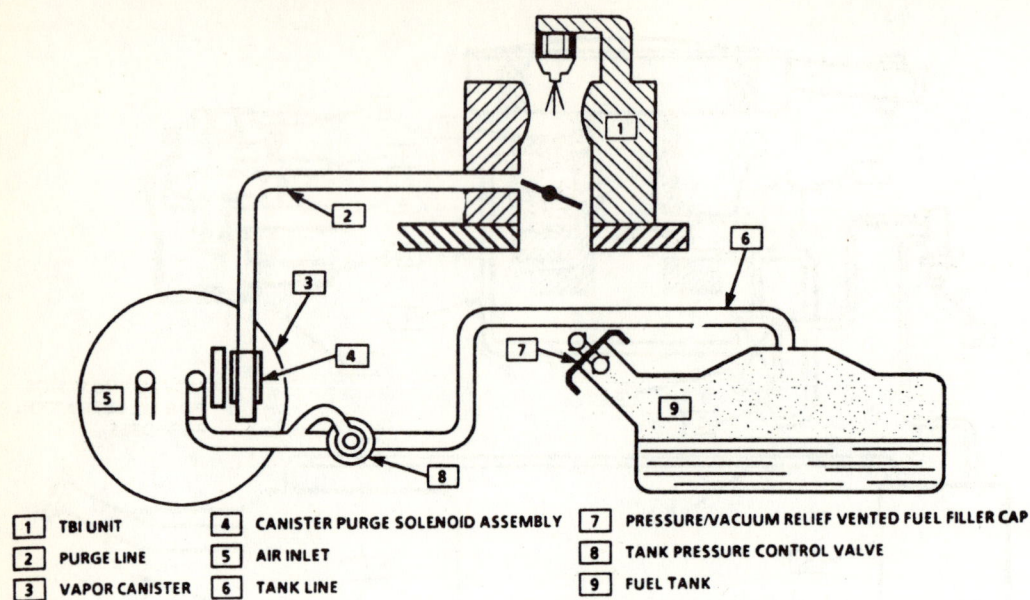

1	TBI UNIT	4	CANISTER PURGE SOLENOID ASSEMBLY	7	PRESSURE/VACUUM RELIEF VENTED FUEL FILLER CAP
2	PURGE LINE	5	AIR INLET	8	TANK PRESSURE CONTROL VALVE
3	VAPOR CANISTER	6	TANK LINE	9	FUEL TANK

Evaporative Emissions Control System schematic

SERVICE

The only service required is the periodic replacement of the canister filer (if so equipped). If the fuel tank cap on you vehicle ever requires replacement, make sure that it is of the same type as the original.

CANISTER REMOVAL AND INSTALLATION

1. Loosen the screw holding the canister retaining bracket.
2. Rotate the canister retaining bracket and remove the canister.
3. Tag and disconnect the hoses leading from the canister.
4. Installation is in the reverse order of removal.

FILTER REPLACEMENT

1. Remove the vapor canister.
2. Pull the filter out from the bottom of the canister.
3. Install a new filter and then replace the canister.

Exhaust Gas Recirculation (EGR) System

All 1973 and later engines are equipped with exhaust gas recirculation (EGR). This system consists of a metering valve, a vacuum line to the carburetor, and cast-in exhaust gas passages in the intake manifold. The EGR valve is controlled by carburetor vacuum, and accordingly opens and closes to admit exhaust gases into the fuel/air moisture. The exhaust gases lower the combustion temperature, and reduce the amount of oxides of nitrogen (NOx) produced. The valve is closed at idle between the two extreme throttle positions.

In most installations, vacuum to the EGR valve is controlled by a thermal vacuum switch (TVS); the switch, which is installed into the engine block, shuts off vacuum to the EGR valve until the engine is hot. This prevents the stalling and lumpy idle which would result if EGR occurred when the engine was cold.

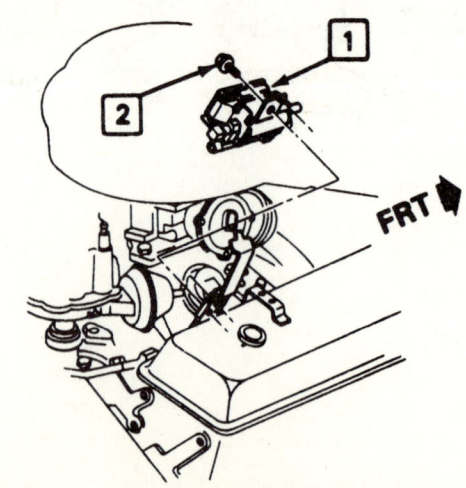

1. EGR control solenoid valve
2. Bolt (tighten to 17 N·m (12 ft. lbs.)

EGR vacuum controlled solenoid 305 V8 engine

EMISSION CONTROLS

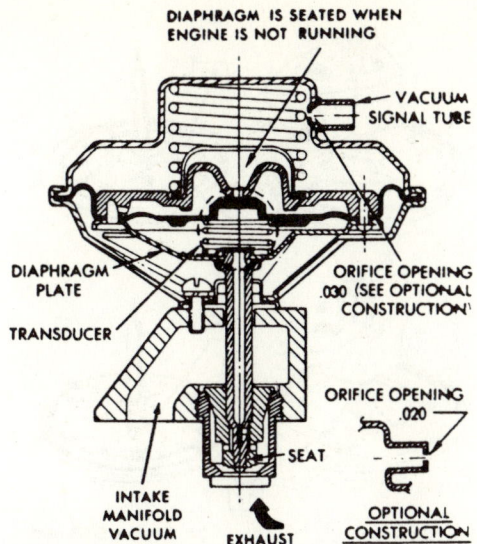

Negative backpressure EGR valve

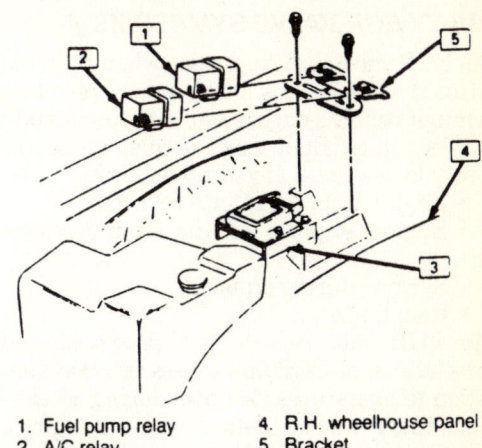

1. Fuel pump relay
2. A/C relay
3. E.S.C. module
4. R.H. wheelhouse panel
5. Bracket

Location of the EGR solenoid, 262 V6-1985 and later

EGR VACUUM CONTROL SOLENOID

To regulate EGR flow on 1981 and later vehicles, a solenoid is used in the vacuum line, and is controlled by the Electronic Control Module (ECM). The ECM uses information from the coolant temperature, throttle position, and manifold pressure sensors to regulate the vacuum solenoid. When the engine is cold, a signal from the ECM energizes the EGR solenoid, thus blocking vacuum to the EGR valve. The solenoid is also energized during cranking and wide-open throttle. When the engine warms up, the EGR solenoid is turned off by the ECM, and the EGR valve operates according to normal ported vacuum and exhaust backpressure signals.

As the vehicle accelerates, the carburetor throttle plate uncovers the vacuum port for the EGR valve. At 3-5 in.Hg, the EGR valve opens and then some of the exhaust gases are allowed to flow into the air/fuel mixture to lower the combustion temperature. At full throttle the valve closes again.

Some California engines are equipped with a dual diaphragm EGR valve. This valve further limits the exhaust gas opening (compared to the single diaphragm EGR valve) during high intake manifold vacuum periods, such as high speed cruising, and provides more exhaust gas recirculation during acceleration when manifold vacuum is low. In addition to the hose running to the thermal vacuum switch, a second hose is connected directly to the intake manifold.

For 1977, all California vehicles delivered in areas above 4000 ft. are equipped with back pressure EGR valves. This valve is also used on all 1978-81 vehicles. The EGR valve receives exhaust back pressure through its hollow shaft. This exerts a force on the bottom of the control valve diaphragm, opposed by a light spring. Under low exhaust pressure (low engine load and partial throttle), the EGR signal is reduced by an air bleed. Under conditions of high exhaust pressure (high engine load and large throttle opening), the air bleed is closed and the EGR valve responds to an unmodified vacuum signal. At wide open throttle, the EGR flow is reduced in proportion to the amount of vacuum signal available.

The 1979 and later vehicles have a ported signal vacuum EGR valve. The valve opening is controlled by the amount of vacuum obtained from a ported vacuum source on the carburetor and the amount of backpressure in the exhaust system.

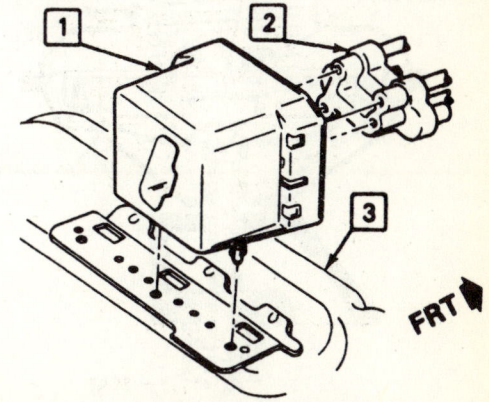

1. EGR/EFE solenoid valve
2. Harness
3. Rear rocker cover

231 V6 vacuum controlled solenoid locations

198 EMISSION CONTROLS

FAULTY EGR VALVE SYMPTOMS

An EGR valve that stays open when it should be closed causes weak combustion, resulting in a rough running engine and/or frequent stalling. Too much EGR flow at idle, cruise, or when cold can cause any of the following:
- Engine stopping after a cold start
- Engine stopping at idle after deceleration
- Surging during cruising
- Rough idle

An EGR valve which is tuck closed and allows little or no EGR flow causes extreme combustion temperatures (too hot) during acceleration. Spark knock (detonation or pinging), engine overheating and excess engine emissions can all be a result, as well as engine damage. See the accompanying EGR system diagnosis chart for possible cause and correction procedures.

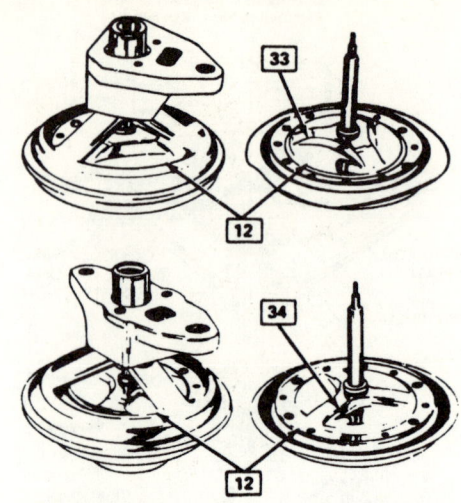

12. Diaphragm
33. Negative EGR web
34. Positive EGR web

Different EGR valve types

EGR VALVE REMOVAL AND INSTALLATION

1. Detach the vacuum lines from the EGR valve.
2. Unfasten the two bolts or bolt and clamp which attach the valve to the manifold. Withdraw the valve.
3. Installation is the reverse of removal. Always use a new gasket between the valve and the manifold. On dual diaphragm valves, attach the carburetor vacuum line to the tube at the top of the valve, and the manifold vacuum line to the tube at the center of the valve.

TVS SWITCH REMOVAL AND INSTALLATION

1. Drain the radiator.

CAUTION: *When draining the coolant, keep in mind that cats and dogs are attracted by the ethylene glycol antifreeze, and are quite likely to drink any that is left in an uncovered container or in puddles on the ground.*

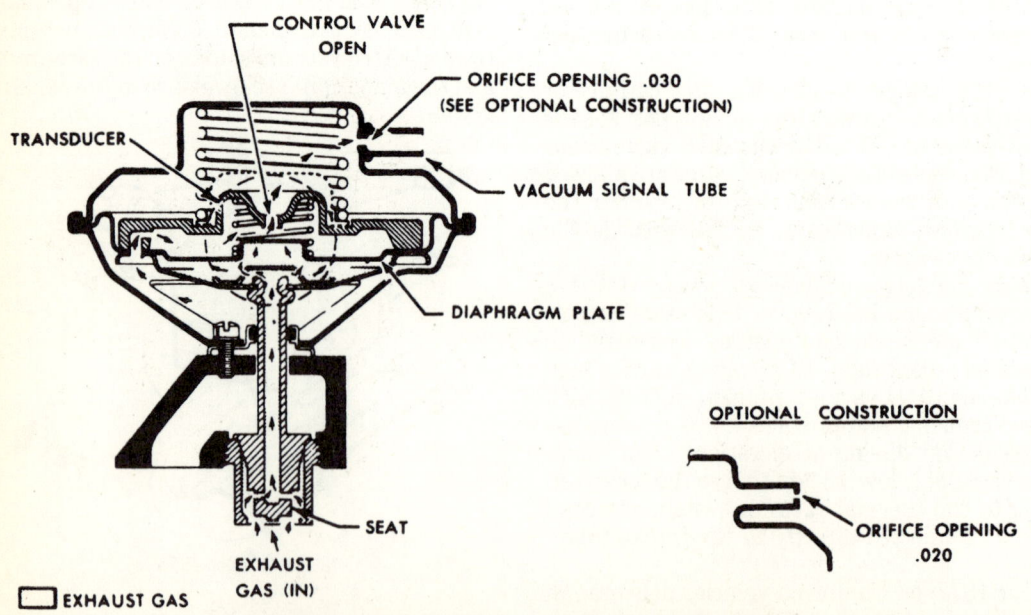

Cross section of a positive backpressure EGR valve

EMISSION CONTROLS 199

This will prove fatal in sufficient quantity. Always drain the coolant into a sealable container. Coolant should be reused unless it is contaminated or several years old.

2. Disconnect the vacuum lines from the switch noting their locations. Remove the switch.

3. Apply sealer to the threaded portion of the new switch, and install it, torquing to 15 ft. lbs.

4. Rotate the head of the switch to a position that will permit easy hookup of vacuum hoses. Then install the vacuum hoses to the proper connectors.

EGR VALVE CLEANING

Valves That Protrude from Mounting Face

WARNING: *Do not wash the valve assembly in solvents or degreasers. Permanent damage to the valve diaphragm may result.*

1. Remove the vacuum hose from the EGR valve assembly. Remove the two attaching bolts, remove the EGR valve from the intake manifold and discard the gasket.

2. Holding the valve assembly in hand, tap the valve lightly with a small plastic hammer to remove exhaust deposits from the valve seat. Shake out any loose particles. DO NOT put the valve in a vise.

3. Carefully remove any exhaust deposits from the mounting surface of the valve with a wire wheel or putty knife. Do not damage the mounting surface.

4. Depress the valve diaphragm and inspect the valve seating areas through the valve outlet for cleanliness. If the valve and/or seat are not completely clean, repeat Step 2.

5. Look for exhaust deposits in the valve outlet, and remove any deposits with an old screwdriver.

6. Clean the mounting surfaces of the intake manifold and valve assembly. Using a new gasket, install the valve assembly to the intake manifold. Torque the bolts to 25 ft. lbs. Connect the vacuum hose.

Shielded Valves or Valves That Do Not Protrude

1. Clean the base of the valve with a wire brush or wheel to remove exhaust deposits from the mounting surface.

2. Clean the valve seat and valve in an abrasive-type spark plug cleaning machine or sandblaster. Most machine shops provide this service. Make sure the valve portion is cleaned (blasted) for about 30 seconds, and that the valve is also cleaned with the diaphragm spring fully compressed (valve unseated). The cleaning should be repeated until all deposits are removed.

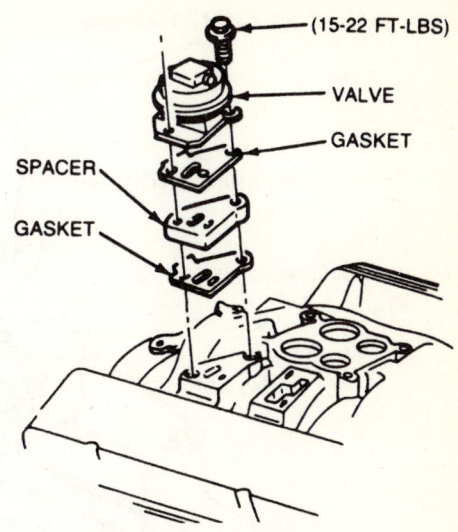

Typical EGR valve mounting location

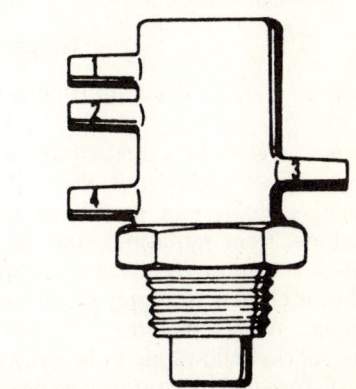

Thermostatic vacuum switch (TVS); Nipple 1 is to distributor, 2 is to TCS solenoid, 4 to intake manifold

3. The valve must be blown out with compressed air thoroughly to ensure all abrasive material is removed from the valve.

4. Clean the mounting surface of the intake manifold and valve assembly. Using a new gasket, install the valve assembly to the intake manifold. Torque the bolts to 25 ft. lbs. Connect the vacuum hose.

Thermostatic Air Cleaner

All engines utilize the THERMAC system (in 1978 it was called TAC, but was the same). This system is designed to warm the air entering the carburetor when underhood temperatures are low, and to maintain a controlled air temperature into the carburetor at all times. By allowing preheated air to enter the carburetor, the amount of time the choke is on is reduced, resulting in better fuel economy and

200 EMISSION CONTROLS

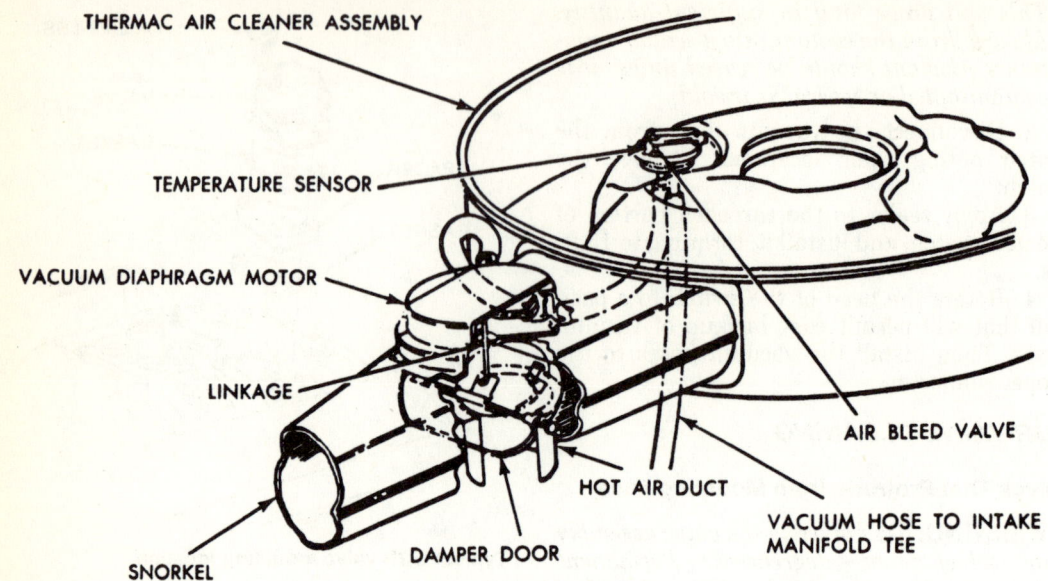

Typical THERMAC air cleaner

lower emissions. Engine warm-up time is also reduced.

The Thermac system is composed of the air cleaner body, a filter, sensor unit, vacuum diaphragm, damper door, and associated hoses and connections. Heat radiating from the exhaust manifold is trapped by a heat stove and is ducted to the air cleaner to supply heated air to the carburetor. A movable door in the air cleaner case snorkel allows air to be drawn in from the heat stove (cold operation) or from underhood air (warm operation). The door position is controlled by the vacuum motor, which receives intake manifold vacuum as modulated by the temperature sensor.

SYSTEM CHECKS

1. Check the vacuum hoses for leaks, kinks, breaks, or improper connections and correct any defects.

2. With the engine off, check the position of the damper door within the snorkel. A mirror can be used to make this job easier. The damper door should be open to admit outside air.

3. Apply at least 7 in.Hg of vacuum to the damper diaphragm unit. The door should close. If it doesn't, check the diaphragm linkage for binding and correct hookup.

4. With vacuum still applied and the door closed, clamp the tube to trap the vacuum. If the door doesn't remove closed, there is a leak in the diaphragm assembly.

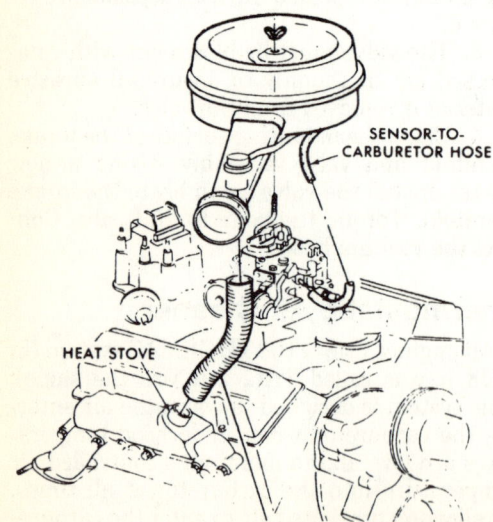

Thermostatic air cleaner (THERMAC)

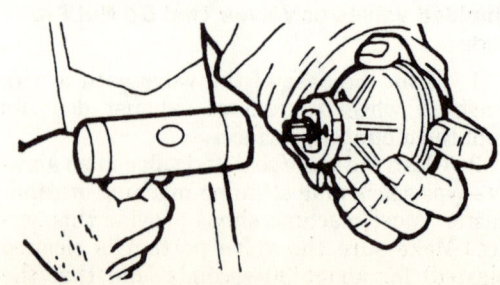

Cleaning the EGR (valve protruding). Tap lightly with hammer

EMISSION CONTROLS

EGR System Diagnosis

Condition	Possible Cause	Correction
Engine idles abnormally rough and/or stalls.	EGR valve vacuum hoses mis-routed.	Check EGR valve vacuum hose routing. Correct as required.
	Leaking EGR valve.	Check EGR valve for correct operation.
	EGR valve gasket failed or loose EGR attaching bolts.	Check EGR attaching bolts for tightness. Tighten as required. If not loose, remove EGR valve and inspect gasket. Replace as required.
	EGR control solenoid.	Check vacuum into control solenoid from carburetor EGR port with engine at normal operating temperature and at curb idle speed. Then check the vacuum out of the EGR control solenoid to EGR valve. If the two vacuum readings are not equal within ± 1/2 in. Hg. (1.7 kPa), then problem could be within EGR solenoid or ECM unit.
	Improper vacuum to EGR valve at idle.	Check vacuum from carburetor EGR port with engine at stabilized operating temperature and at curb idle speed. Vacuum should not exceed 1.0 in. Hg. If vacuum exceeds this, check carburetor idle.
Engine runs rough on light throttle acceleration and has poor part load performance.	EGR valve vacuum hose misrouted.	Check EGR valve vacuum hose routing. Correct as required.
	Check for loose valve.	Torque valve.
	Failed EGR control solenoid.	Same as listing in "Engine Idles Rough" condition.
	Sticky or binding EGR valve.	Clean EGR passage of all deposits.
		Remove EGR valve and inspect. Replace as required.
	Wrong or no EGR gasket(s) and/or Spacer.	Check and correct as required. Install new gasket(s), install spacer (if used), torque attaching parts.
Engine stalls on decelerations.	Control valve blocked or air flow restricted.	Check internal control valve function per service procedure.
	Restriction in EGR vacuum line or control solenoid signal tube.	Check EGR vacuum lines for kinks, bends, etc. Remove or replace hoses as required. Check EGR control solenoid function.
		Check EGR valve for excessive deposits causing sticky or binding operation. Replace valve.
	Sticking or binding EGR valve.	Remove EGR valve and replace valve.
Part throttle engine detonation.	Control solenoid blocked or air flow restricted.	Check control solenoid function per check chart.
	Insufficient exhaust gas recirculation flow during part throttle accelerations.	Check EGR valve hose routing. Check EGR valve operation. Repair or replace as required. Check EGR control solenoid as listed in "Engine Idles Rough" section. Check EGR passages and valve for excessive deposit. Clean as required.

(NOTICE: Non-Functioning EGR valve could contribute to part throttle detonation.)

		Check EGR per service procedure.

(NOTICE: Detonation can be caused by several other engine variables. Perform ignition and carburetor related diagnosis.)

Condition	Possible Cause	Correction
Engine starts but immediately stalls when cold.	EGR valve hoses misrouted.	Check EGR valve hose routings.
	EGR control solenoid system malfunctioning when engine is cold.	Perform check to determine if the EGR solenoid is operational. Replace as required.

(NOTICE: Stalls after start can also be caused by carburetor problems.)

202 EMISSION CONTROLS

REMOVAL AND INSTALLATION

Vacuum Motor

1. Remove the air cleaner.
2. Disconnect the vacuum hose from the motor.
3. Drill out the spot welds with a $1/8''$ hole, then enlarge as necessary to remove the retaining strap.
4. Remove the retaining strap.
5. Lift up the motor and cock it to one side to unhook the motor linkage at the control damper assembly.
6. To install the new vacuum motor, drill a $7/64''$ hole in the snorkel tube as the center of the vacuum motor retaining strap.
7. Insert the vacuum motor linkage into the control damper assembly.
8. Use the motor retaining strap and a sheet metal screw to secure the retaining strap and motor to the snorkel tube.

NOTE: *Make sure the screw does not interfere with the operation of the damper assembly. Shorten the screw if necessary.*

Temperature Sensor

1. Remove the air cleaner.
2. Disconnect the hoses at the air cleaner.
3. Pry up the tabs on the sensor retaining clip and remove the clip and sensor from the air cleaner.
4. Installation is the reverse of removal.

Air Injection Reactor System (A.I.R)

The AIR system injects compressed air into the exhaust system, near enough to the exhaust valves to continue the burning of the normally unburned segment of the exhaust gases. To do this it employs an air injection pump and a system of hoses, valves, tubes, etc., necessary to carry the compressed air from the pump to the exhaust manifolds. Carburetors and distributors for AIR engines have specific modifications to adapt them to the air injection system; those components should not be interchanged with those intended for use on engines that do not have the system.

A diverter valve is used to prevent backfiring. The valve senses sudden increases in manifold vacuum and ceases the injection of air during fuel-rich periods. During coasting, this valve di-

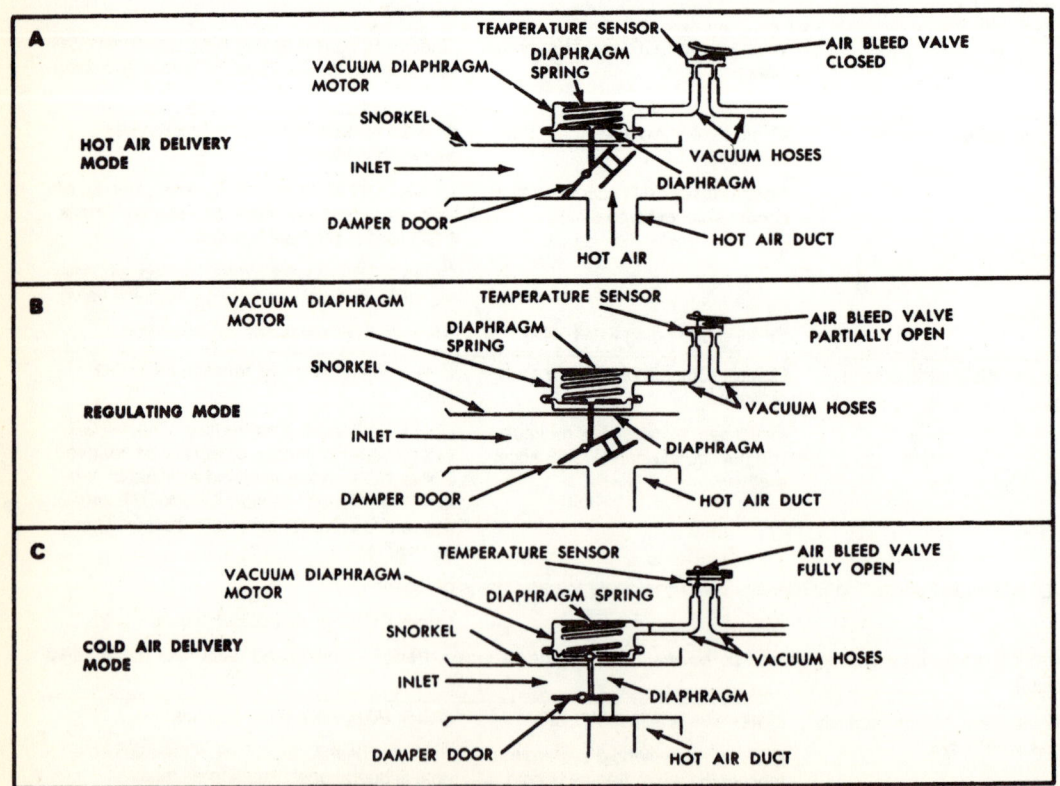

Schematic of the vacuum motor operation

EMISSION CONTROLS 203

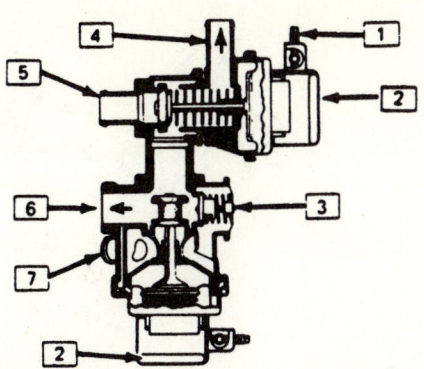

1. Vacuum signal
2. Solenoid
3. Relief valve
4. Converter air
5. Port air
6. Air from pump
7. Divert air

Electric divert/electric air switching valve (EDES)

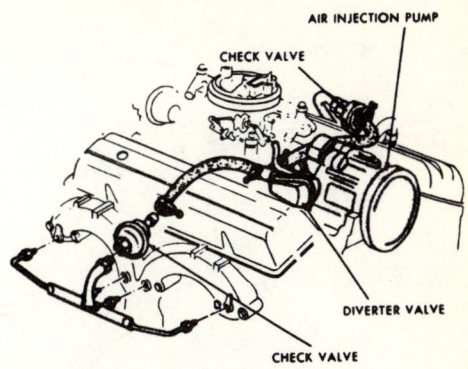

A.I.R. system components

verts the entire air flow through the pump muffler and during high engine speeds, expels it through a relief valve. Check valves in the system prevent exhaust gases from entering the pump.

NOTE: *The AIR system on the 231 V6 engine is slightly different, but its purpose remains the same.*

SERVICE

The AIR system's effectiveness depends on correct engine idle speed, ignition timing, and dwell. These settings should be strictly adhered to and checked frequently. All hoses and fittings should be inspected for condition and tightness of connections. Check the drive belt for wear and tension every 12 months or 12,000 miles.

COMPONENT REMOVAL AND INSTALLATION

Air Pump

WARNING: *Do not pry on the pump housing or clamp the pump in a vise; the housing is soft and may become distorted.*

1. Disconnect the air hoses at the pump.
2. Hold the pump pulley form turning and loosen the pulley bolts.
3. Loosen the pump mounting bolt and adjustment bracket bolt. Remove the drive belt.
4. Remove the mounting bolts, and then remove the pump.
5. Install the pump using a reverse of the removal procedure.

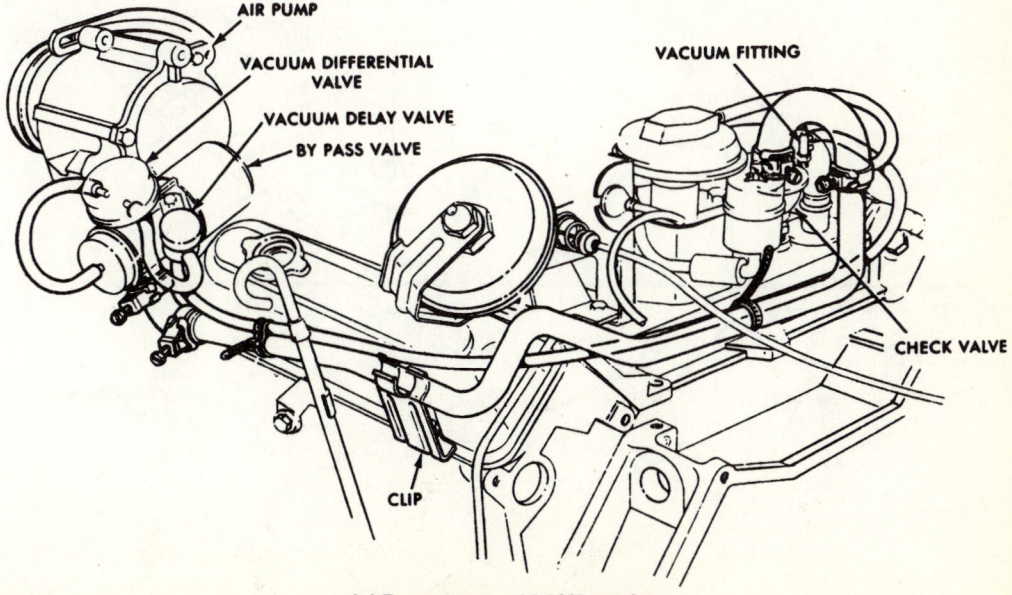

A.I.R. system — 231 V6 engine

EMISSION CONTROLS

A.I.R. system — inline six cylinder engine

Diverter (Anti-afterburn) Valve

1. Detach the vacuum sensing line from the valve.
2. Remove the other hose(s) from the valve.
3. Unfasten the diverter valve from the elbow or the pump body.
4. Installation is performed in the reverse order of removal. Always use a new gasket. Tighten the valve securing bolts to 85 inch lbs.

Air Management System

The Air Management System is used to provide additional oxygen to continue the combustion process after the exhaust gases leave the combustion chamber; much the same as the AIR system described earlier in this chapter. Air is injected into either the exhaust port(s), the exhaust manifold(s) or the catalytic con-

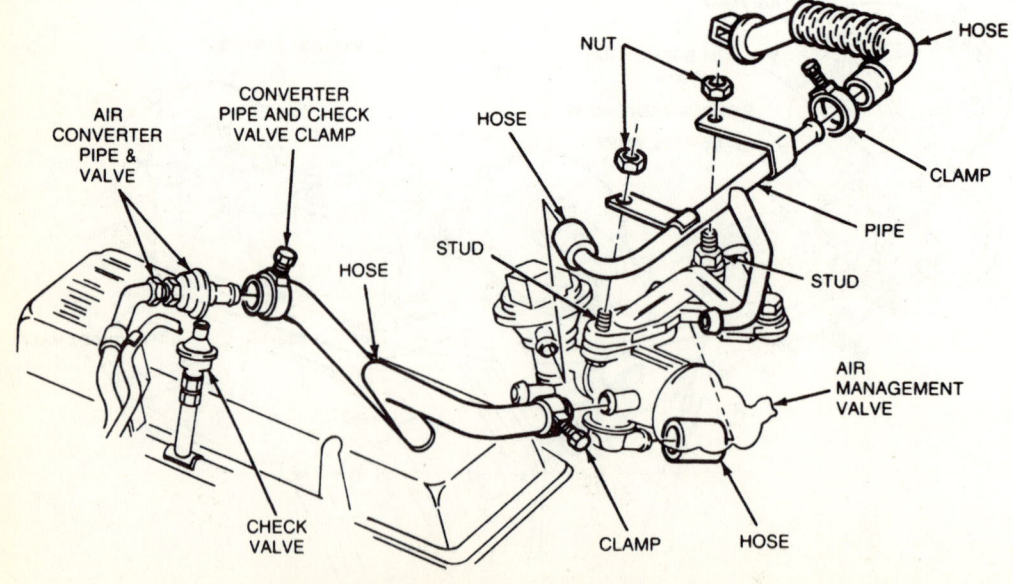

Check valve and hoses — 1981 A.I.R. system

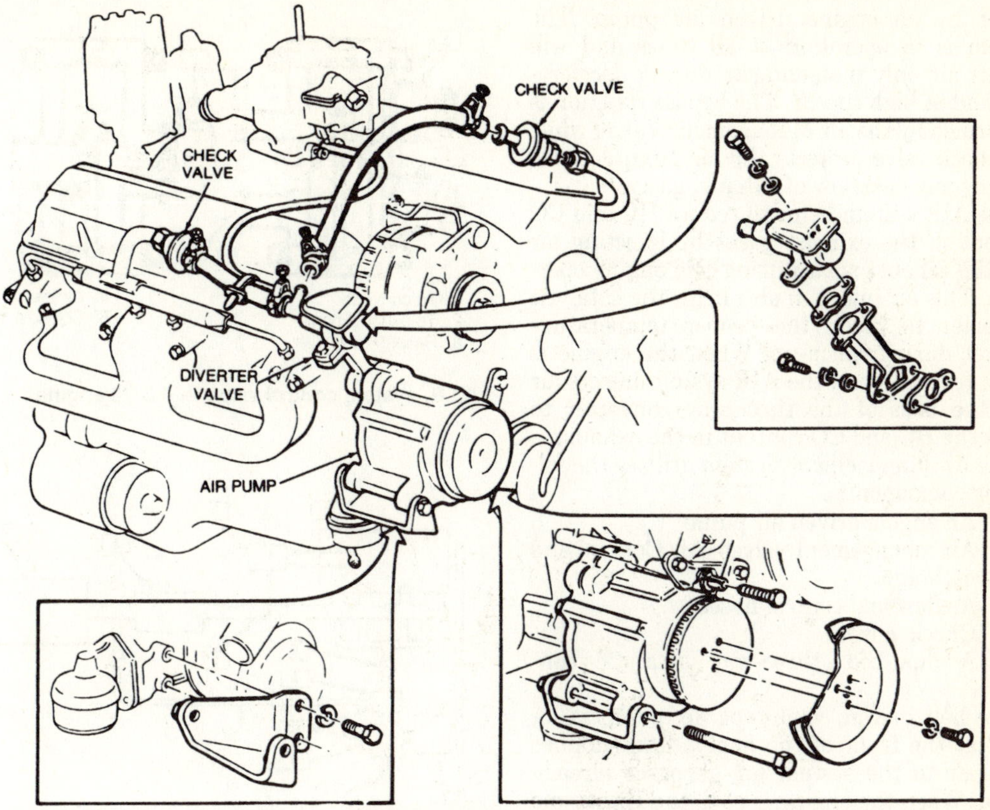

EMISSION CONTROLS 205

A.I.R. system — V8 engines

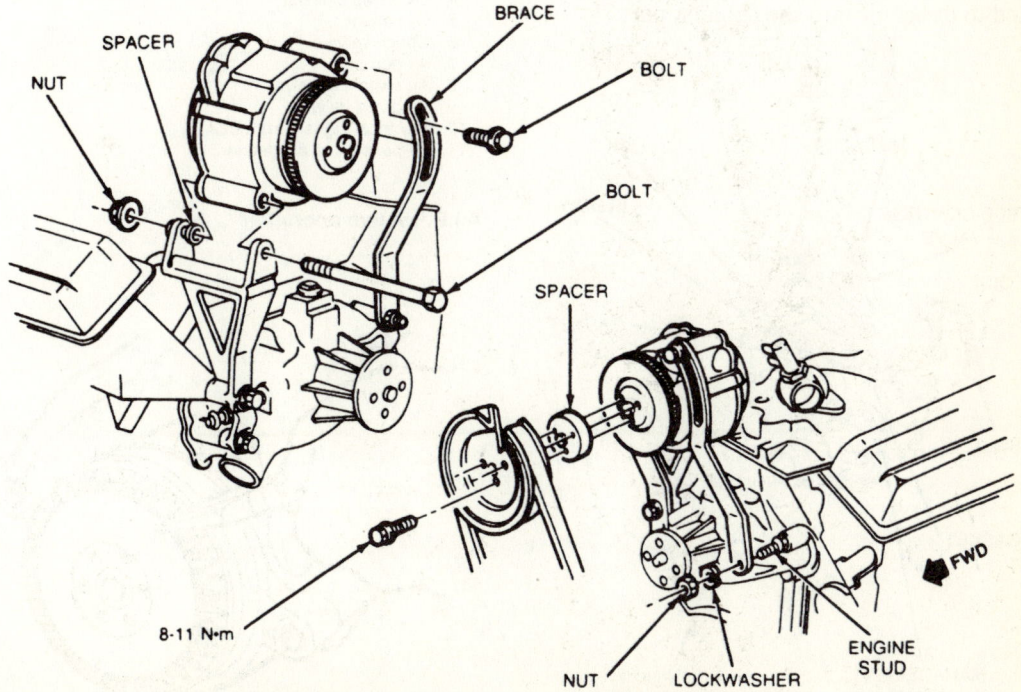

Removing the A.I.R. pump — 1981 and later A.I.R. systems

EMISSION CONTROLS

verter by an engine driven air pump. The system is in operation at all times and will bypass air only momentarily during deceleration and at high speeds. The bypass function is performed by the air Management Valve, while the check valve protects the air pump by preventing any backflow of exhaust gases.

The AIR system helps to reduce HC and CO content in the exhaust gases by injecting air into the exhaust ports during cold engine operation. This air injection also helps the catalytic converter to reach the proper temperature quicker during warm-up. When the engine is warm (closed loop), the AIR system injects air into the beds of the three-way converter to lower the HC and CO content in the exhaust.

The Air Management System utilizes the following components:
 1. An engine driven air pump
 2. Air management valves (Air Control and Air Switching)
 3. Airflow and control hoses
 4. Check valves
 5. A dual-bed, three-way catalytic converter.

The belt driven, vane-type air pump is located at the front of the engine and supplies clean air to the system for purposes already stated. When the engine is cold, the Electronic Control Module (ECM) energizes an air control solenoid. This allows air to flow to the air switching valve. The air switching valve is then energized to direct air into the exhaust port.

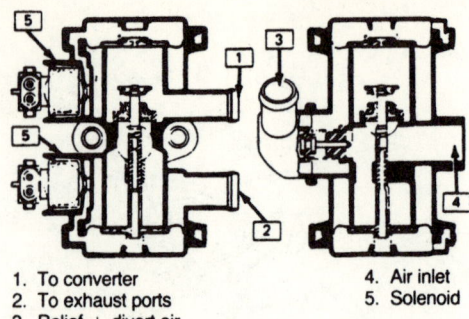

1. To converter
2. To exhaust ports
3. Relief + divert air
4. Air inlet
5. Solenoid

A.I.R. system control valve — 4.3 V6 engine

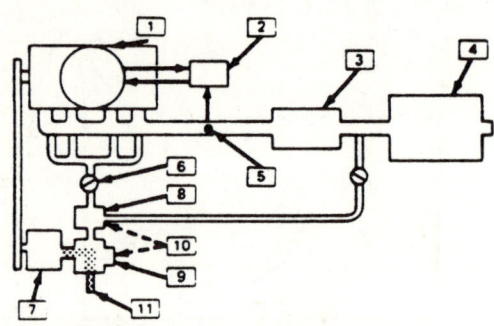

1. Closed loop fuel control
2. ECM
3. Reducing catalyst
4. Oxidizing catalyst
5. O_2 sensor
6. Check valve
7. Air pump
8. Air switching valve
9. Air divert valve
10. Electrical signals from ECM
11. By-pass air to air cleaner

A.I.R. system operation

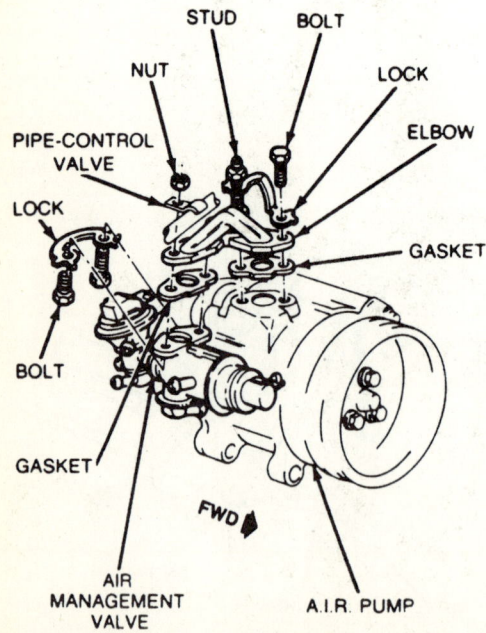

Removing the air management valve

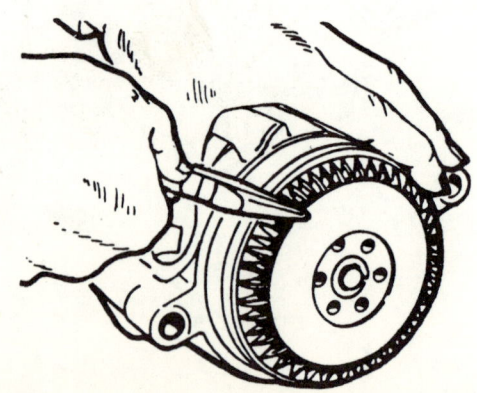

A.I.R. filter removal

EMISSION CONTROLS

When the engine is warm, the ECM de-energizes the air switching valve, thus directing the air between the beds of the catalytic converter. This then provides additional oxygen for the oxidizing catalyst in the second bed to decrease HC and CO levels, while at the same time keeping oxygen levels low in the first bed, enabling the reducing catalyst to effectively decrease the levels of NOx.

If the air control valve detects a rapid increase in manifold vacuum (deceleration), certain operating modes (wide open throttle, etc.) or if the ECM self-diagnostic system detects any problems in the system, air is diverted to the air cleaner or directly into the atmosphere.

The primary purpose of the ECM's divert mode is to prevent backfiring. Throttle closure at the beginning of deceleration will temporarily create air/fuel mixtures which are too rich to burn completely. These mixtures will become burnable when they reach the exhaust if they are combined with injection air. The next firing of the engine will ignite the mixture causing an exhaust backfire. Momentary diverting of the injection air from the exhaust prevents this.

The Air Management System check valves and hose should be checked periodically for any leaks, cracks or deterioration.

REMOVAL AND INSTALLATION

Air Pump

1. Remove the valves and/or adapter at the air pump.
2. Loosen the air pump adjustment bolt and remove the drive belt.
3. Unscrew the three mounting bolts and then remove the pump pulley.
4. Unscrew the pump mounting bolts and then remove the pump.
5. Installation is in the reverse order of removal. Be sure to adjust the drive belt tension after installing it.

Check Valve

1. Release the clamp and disconnect the air hoses from the valve.
2. Unscrew the check valve from the air injection pipe.
3. Installation is in the reverse order of removal.

Air Management Valve

1. Disconnect the negative battery cable.
2. Remove the air cleaner.
3. Tag and disconnect the vacuum hose from the valve.
4. Tag and disconnect the air outlet hoses from the valve.
5. Bend back the lock tabs and then remove the bolts holding the elbow to the valve.
6. Tag and disconnect any electrical connections at the valve and then remove the valve from the elbow.
7. Installation is in the reverse order of removal.

Pump Filter Removal

1. Remove the drive belt and pump pulley.
2. Using needlenosed pliers, pull the fan from the pump.

NOTE: *Use care to prevent any dirt or fragments from entering the air intake hole. DO NOT insert a screwdriver between the pump and the filter, and do not attempt to remove the metal hub. It is seldom possible to remove the filter without destroying it.*

3. To install a new filter, draw it on with the pulley and pulley bolts. DO NOT hammer or press the filter on the pump.
4. Draw the filter down evenly by torquing the bolts alternately. Make sure the outer edge of the filter slips into the housing. A slight amount of interference with the housing bore is normal.

NOTE: *The new filter may squeal initially until the sealing lip on the pump outer diameter has worn in.*

Anti-Dieseling Solenoid

Beginning in 1968 some vehicles may have an idle speed solenoid on the carburetor. All 1972-75 vehicles have idle solenoids. Due to the leaner carburetor settings required for emission control, the engine may have a tendency to diesel or run-on after the ignition is turned off. The carburetor solenoid, energized when the ignition is on, maintains the normal idle speed. When the ignition is turned off, the solenoid is de-energized and permits the throttle valves to fully close, thus preventing run-on. For adjustment of carburetors with idle solenoids see the section on carburetor adjustments later in this chapter.

Transmission Controlled Spark

1970-74

Introduced in 1970, this system controls exhaust emissions by eliminating vacuum advance in the lower forward gears.

The 1970 system consists of a transmission switch, solenoid vacuum switch, time delay relay, and a thermostatic water temperature switch. The solenoid vacuum switch is energized in the lower gears via the transmission switch and closes off distributor vacuum. The two-way transmission switch is activated by the shifter shaft on manual transmissions, and

208 EMISSION CONTROLS

Anti-dieseling solenoid

by oil pressure on automatic transmissions. The switch de-engerizes the solenoid in High gear, the plunger extends and uncovers the vacuum port, and the distributor receives full vacuum. The temperature switch overrides the system when engine temperature is below 63°F or above 232°F. This allows vacuum advance in all gears. A time delay relay opens 15 seconds after the ignition is switched on. Full vacuum advance during this delay eliminates the possibility of stalling.

The 1971 system is similar, except that the vacuum solenoid (now called a Combination Emissions Control or CEC solenoid) serves two functions. One function is to control distributor vacuum; the added function is to act as a deceleration throttle stop in High gear. This cuts down on emissions when the vehicle is coming to a stop in High gear. The CEC solenoid is controlled by a temperature switch, a transmission switch, and a 20 second time delay relay. This system also contains a reversing delay, which energizes the solenoid when the transmission switch, temperature switch or time delay completes the CEC circuit to ground. This system is directly opposite the 1970 system in operation. The 1970 vacuum solenoid was normally open to allow vacuum advance and when energized, closed to block vacuum. The 1971 system is normally closed blocking vacuum advance and when energized, opens to allow vacuum advance. The temperature switch completes the CEC circuit to ground when engine temperature is below 82°F. The time delay relay allows vacuum advance (and raised idle speed) for 200 seconds after the ignition key is turned to the ON position. Vehicles equipped with an automatic transmission and air conditioning also have a solid state timing device which engages the air conditioning compressor for three seconds after the ignition key is turned to the OFF position to prevent the engine from running on.

The 1972 6-cylinder system is similar to that used on 1971, except that an idle stop solenoid has been added to the system. In the energized position, the solenoid maintains engine speed at a predetermined fast idle. When the solenoid is de-energized by turning off the ignition, the solenoid allows the throttle plates to close beyond the normal idle position; thus cutting off their air supply and preventing engine run-on. The 6-cylinder is the only 1972 engine with a C.E.C. valve, which serves the same deceleration function as in 1971. The 1972 time delay relay delays full vacuum 20 seconds after the transmission is shifted into High gear. V8 engines use a vacuum advance solenoid similar to that used in 1970. this relay is normally closed to block vacuum and opens when energized to allow vacuum advance. The solenoid controls distributor vacuum advance and performs no throttle positioning function. The idle stop solenoid used operates in the same manner as the one on 6-cylinder engines. All air conditioned vehicles have an additional anti-diesel (run-on) solenoid which engages the compressor clutch for three seconds after the ignition is switched off. The 1973 TCS system differs from the 1972 system in three ways. The 23 second upshift delay has been replaced by a 20 second starting relay. This relay closes to complete the

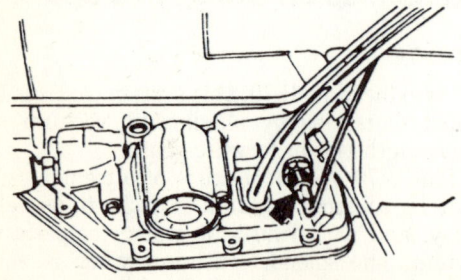

TCS switch location, THM 350. Location similar on other transmission (sides may vary)

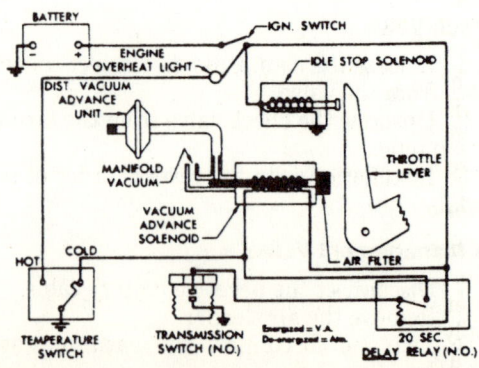

Small V8 TCS system, engine "OFF" mode shown. Large V8 and L6 similar

EMISSION CONTROLS

TCS circuit and open the TCS solenoid, allowing vacuum advance, for 20 seconds after the key is turned to the on position. The operating temperature of the temperature override switch has been raised to 93°F, and the switch which was used to engage the A/C compressor when the key was turned OFF has been eliminated. All vehicles are equipped with an electric throttle control solenoid to prevent run-on. The 1973 TCS system is used on all vehicles equipped with a 307 engine and all vehicles equipped with a V8 engine and a manual transmission.

The 1974 TCS system is used only on manual transmission equipped vehicles. System components remain unchanged from 1973. The vacuum advance solenoid is located on the coil bracket. The TCS system is not used in 1975 and later vehicles.

TESTING

If there is a TCS system malfunction, first connect a vacuum gauge in the hose between the solenoid valve and the distributor vacuum unit. Drive the vehicle or raise it on a frame lift and observe the vacuum gauge. If full vacuum is available in all gears, check for the following:
1. Blown fuse.
2. Disconnected wire at the solenoid-operated vacuum valve.
3. Disconnect wire at the transmission switch.
4. Temperature override switch energized due to low engine temperature.
5. Solenoid failure.

If no vacuum is available in any gear, check the following:
1. Solenoid valve vacuum lines switched.
2. Clogged solenoid vacuum valve.
3. Distributor or manifold vacuum lines leaking or disconnected.
4. Transmission switch or wire grounded.

Test for individual components are as follows:

Idle Stop Solenoid

This unit may be checked simply by observing it while an assistant switches the ignition on and off. It should extend further with the current switched on. The unit is not repairable.

Solenoid Vacuum Valve

Check that proper manifold vacuum is available. Connect the vacuum gauge in the line between the solenoid valve and the distributor. Apply 12 volts to the solenoid. If vacuum is still not available, the valve is defective, either mechanically or electrically. The unit is not repairable. If the valve is satisfactory, check the relay next.

Relay

1. With the engine at normal operating temperature and the ignition on, ground the solenoid vacuum valve terminal with the black lead. The solenoid should energize (no vacuum) if the relay is satisfactory.
2. With the solenoid energized as in Step 1, connect a jumper from the relay terminal with the green/white stripe lead to ground. The solenoid should de-energize (vacuum available) if the relay is satisfactory.
3. If the relay worked properly in Steps 1 and 2, check the temperature switch. The relay unit is not repairable.

Temperature Switch

The vacuum valve solenoid should be deenergized (vacuum available) with the engine cold. If it is not, ground the green/white stripe wire from the switch. If the solenoid now de-energizes, replace the switch. If the switch was satisfactory, check the transmission switch.

Transmission Switch

With the engine at normal operating temperature and the transmission in one of the no-vacuum gears, the vacuum valve solenoid should be energized (no vacuum). If not, remove the ground the switch electrical lead. If the solenoid energizes, replace the switch.

Early Fuel Evaporation System

The 1975 and later vehicles (except TBI) are equipped with this system to reduce engine-warm-up time, improve driveability, and reduce emissions. On start-up, a vacuum motor acts to close a heat valve in the exhaust manifold which causes exhaust gases to enter the intake manifold heat riser passages. Incoming fuel mixture is then heated and more complete fuel evaporation is provided during warm-up.

The system consists of a Thermal Vacuum Switch, and an Exhaust Heat Valve and actuator. The Thermal Vacuum Switch is located on the coolant outlet housing on V8s, and on the block on in-line 6-cylinder engines. When the engine is cold, the TVS conducts manifold vacuum to the actuator to close the valve. When engine coolant or, on 6-cylinder engines, oil warms up, vacuum is interrupted and the actuator should open the valve.

NOTE: *On the 231 V6 (1981) and 262 V6 (1985 and later) engines, the EFE system is controlled by the ECM.*

As of 1981, the 231 V6 Turbo utilizes a slightly different system. Although the function of this system remains the same, to reduce engine warm-up time, imporve driveability and to reduce emissions, the operation is entirely dif-

EMISSION CONTROLS

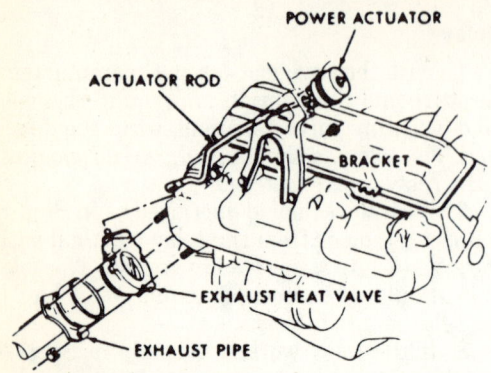

EFE system V8 engines, except 307 Oldsmobile produced engine

ferent. The new system is electric and uses a ceramic heater grid located underneath the primary bore of the carburetor as part of the carburetor insulator/gasket. When the engine coolant is below the specified calibration level, electrical current is supplied to the heater through an ECM controlled relay.

CHECKING THE EFE SYSTEM

1. With the engine overnight cold, have someone start the engine while you observe the Exhaust Heat Valve (on some V8s, the EFE valve actuator arm is covered by a two-piece metal cover, which must be removed for service). The valve should snap to the closed position.

2. Watch the valve as the engine warms up. By the time coolant starts circulating through the radiator (V-type engines) or oil is hot (in-line engines), the valve should snap open.

3. If the valve does not close, immediately disconnect the hose at the actuator, and check for vacuum by placing your finger over the end of the hose, or with a vacuum gauge. If there is no vacuum, immediately disconnect the hose leading to the TVS from the manifold at the TVS. If there is vacuum here, but not at the actuator, replace the TVS. If vacuum does not exist at the hose going to the TVS, check that the vacuum hose is free of cracks or breaks and tightly connected at the manifold, and that the manifold port is clear.

4. If the valve does not open when the engine coolant or oil warms up, disconnect the hose at the actuator, and check for vacuum by placing your finger over the end of the hose or using a vacuum gauge. If there is vacuum, replace the TVS. If there is no vacuum, replace the actuator.

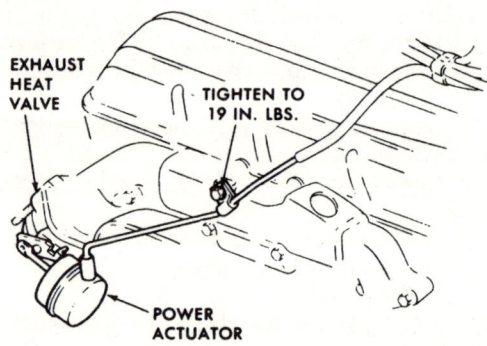

Exhaust heat valve and acuator V8, V6 similar

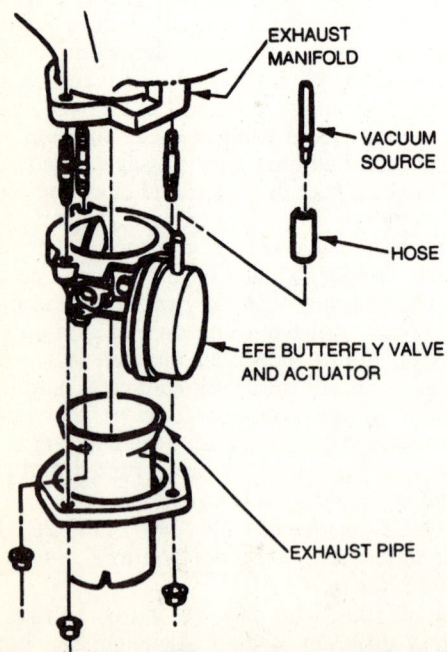

Vacuum servo type EFE assembly, except 307 Oldsmobile produced engine

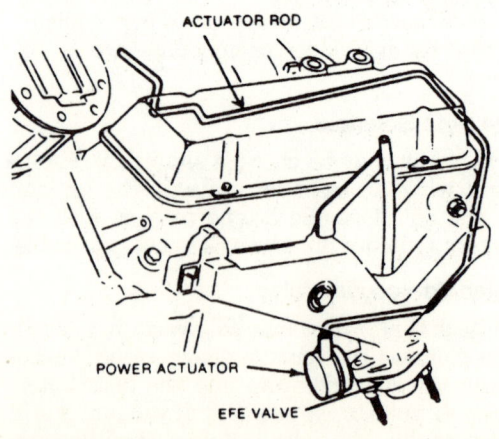

EFE system, 231 V6

EMISSION CONTROLS   211

TVS REMOVAL AND INSTALLATION

The Thermo Vacuum Switch (TVS) is located on the engine coolant outlet housing.

On V8 engines, drain coolant until the level is below the coolant outlet housing. No oil need be drained on 6-cylinder engines. Apply sealer to threads on V8 engines. Use no sealer on 6-cylinder engines. Note that the valve must be installed until just snug (120 inch lbs.) and then turned by hand just far enough to line up the fittings for hose connection.

HEATER GRID REMOVAL AND INSTALLATION

1. Disconnect the negative battery cable. Remove the air cleaner.
2. Tag and disconnect all electrical, vacuum and fuel connections from the carburetor.
3. Disconnect the EFE heater electrical connection.
4. Remove the carburetor as detailed later in this chapter.
5. Lift off the EFE heater.
6. Installation is in the reverse order of removal.
7. Start the engine and check for any leaks.

Controlled Combustion System

The CCS system relies upon leaner air/fuel mixtures and altered ignition timing to improve combustion efficiency. A special air cleaner with a thermostatically controlled opening is used on most CCS equipped vehicles to ensure that air entering the carburetor is kept at 100°F. This allows leaner carburetor settings and improves engine warm-up. A 15°F higher temperature thermostat is employed on CCS vehicles to further improve emission control.

SERVICE

Since the only extra component added with a CCS system is the thermostatically controlled air cleaner, there is no additional maintenance required; however, tune-up adjustments such as idle speed, ignition timing, and dwell become much more critical. Care must be taken to ensure that these settings are correct, both for trouble-free operation and a low emission level.

Computer Controlled Catalytic Converter System

The C-4 System, installed on certain 1979 and all 1980 vehicles sold in California, is an electronically controlled exhaust emission system. The purpose of the system is to maintain the ideal air/fuel ratio at which the catalytic converter is most effective.

Major components of the system include an Electronic Control Module (ECM), an oxygen sensor, an electronically controlled carburetor, and a three-way oxidation reduction catalytic converter. The system also includes a maintenance reminder flag connected to the odometer which becomes visible in the instrument cluster at regular intervals, signaling the need for oxygen sensor replacement.

The oxygen sensor, installed in the exhaust manifold, generates a voltage which varies with exhaust gas oxygen content. Lean mixtures (more oxygen) reduce voltage; rich mixtures (less oxygen) increase voltage. Voltage output is sent to the ECM.

An engine temperature sensor installed in the engine coolant outlet monitors engine coolant temperatures. Vacuum control switches and throttle position sensors also monitor engine conditions and supply signals to the ECM.

The Electronic Control Module receives input signals from all sensors. It processes these signals and generates a control signal sent to the carburetor. The control signal cycles between on (lean command) and off (rich command). The amount of on and off time is a function of the input voltage sent to the ECM by the oxygen sensor.

Rochester Dualjet (2-barrel) E2ME and E4ME (4-barrel) carburetors are used with the C-4 system. Basically, an electrically operated mixture control solenoid is installed in the carburetor float bowl. The solenoid controls the air/fuel mixture metered to the idle and main metering systems. Air metering to the idle system is controlled by an idle air bleed valve. It follows the movement of the mixture solenoid to control the amount of air bled into the idle system, enriching or leaning out the mixture as appropriate. Air/fuel mixture enrichment occurs when the fuel valve is open and the air bleed

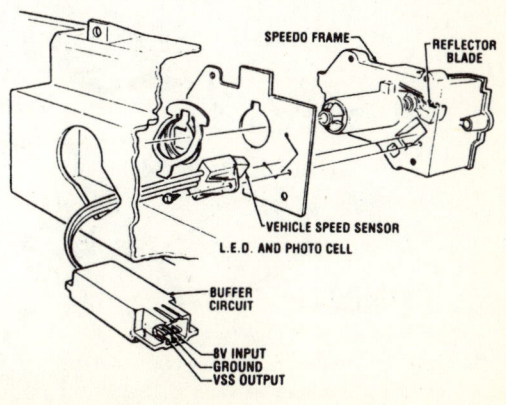

Vehicle speed sensor

212 EMISSION CONTROLS

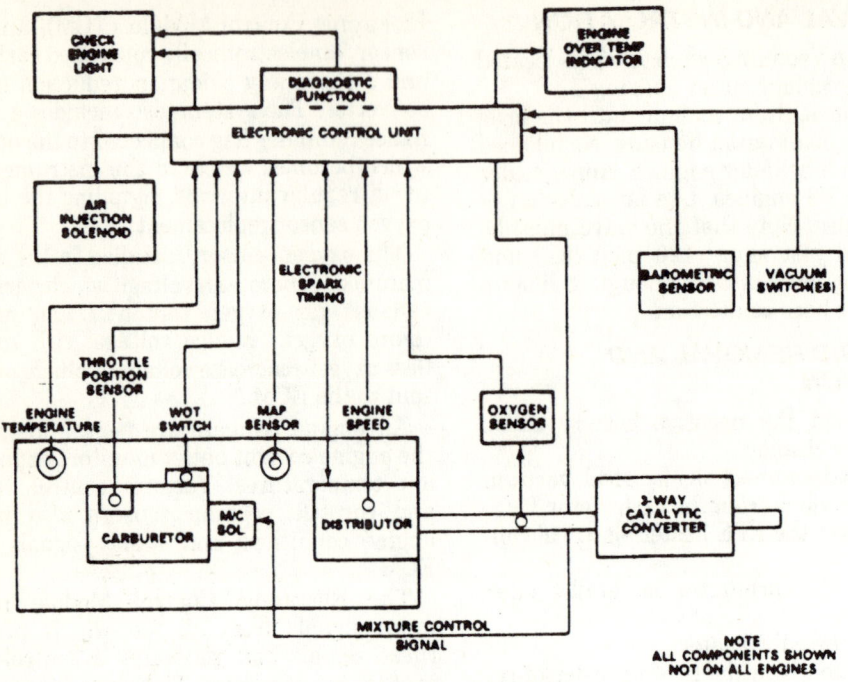

C-4 system schematic

valve is closed. All cycling of this system, which occurs ten times per second, is controlled by the ECM. A throttle position switch informs the ECM of the open or closed throttle operation. A number of different switches are used, varying with application. When the ECM receives a signal from the throttle switch, indicating a change of position, it immediately searches its memory for the least set of operating conditions that result in an ideal air/fuel ratio, and shifts to that set of conditions. The memory is continually updated during normal operation.

A Check-Engine light is included in the C-4 System installation. When a fault develops, the light comes on, and a trouble code is set into the ECM memory. However, if the fault is in-

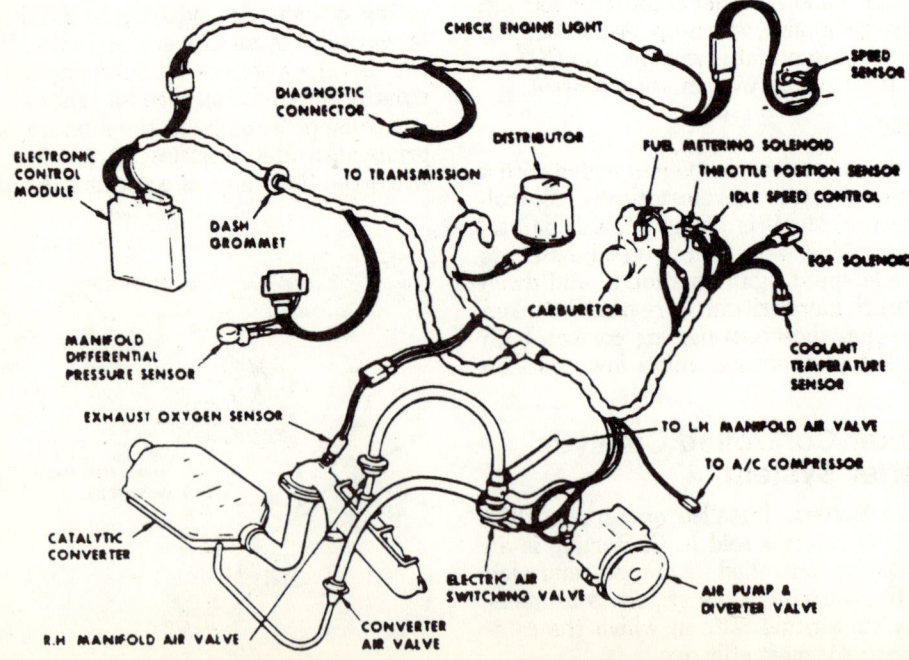

Computer Camand Control (CCC) system wiring

EMISSION CONTROLS 213

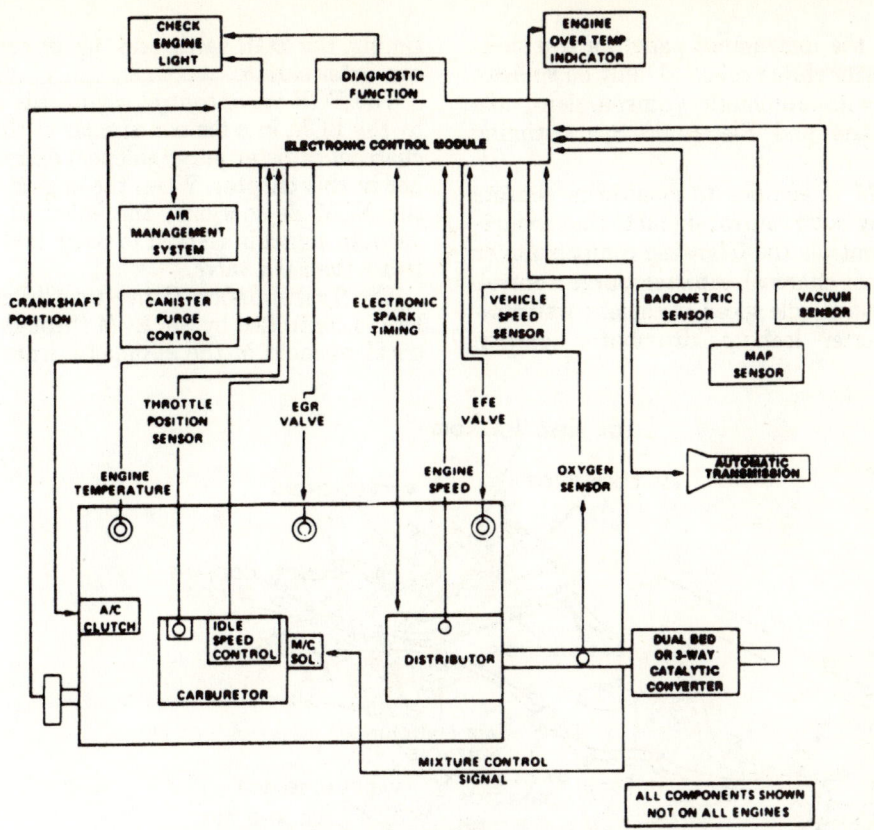

Computer Camand Control (CCC) system schematic

termittent, the light will go out, but the trouble code will remain in the ECM memory as long as the engine is running. The trouble codes are used as a diagnostic aid, and are pre-programmed.

Unless the required tools are available, troubleshooting the C-4 System should be confined to mechanical checks of electrical connectors, vacuum hoses and the like. All diagnosis and repair should be performed by a qualified mechanic.

Computer Command Control System

The Computer Command Control System, installed on all 1981 and later vehicles, is basically a modified version of the C-4 system. Its main advantage over its predecessor is that it can monitor and control a large number of interrelated emission control systems.

This new system can monitor up to 15 various engine/vehicle operating conditions and then use this information to control as many as 9 engine related systems. The System is thereby making constant adjustments to maintain good vehicle performance under all normal driving conditions while at the same time allowing the catalytic converter to effectively control the emissions of NOx, HC and CO.

In addition, the System has a built in diagnostic system that recognizes and identifies possible operational problems and alerts the driver through a Check Engine light in the instrument panel. The light will remain ON until the problem is corrected. The System also has built in back-up systems that in most cases of an operational problem will allow for the continued operation of the vehicle in a near normal manner until the repairs can be made.

The CCC system has some components in common with the C-4 system, although they are not interchangeable. These components include the Electronic Control Module (ECM), which, as previously stated, controls many more functions than does it predecessor, and oxygen sensor system, an electronically controlled variable mixture carburetor, a three-way catalytic converter, throttle position and coolant sensors, a Barometric Pressure Sensor (BARO), a Manifold Absolute Pressure Sensor (MAP) and a Check Engine light in the instrument panel.

Components unique to the CCC system include the Air Injection Reaction (AIR) management system, a charcoal canister purge solenoid, EGR valve controls, a vehicle speed

214 EMISSION CONTROLS

sensor (in the instrument panel), a transmission converter clutch solenoid (only on vehicles equipped with automatic transmission), idle speed control and Electronic Spark timing (EST).

The ECM, in addition to monitoring sensors and sending out a control signal to the carburetor, also controls the following components or sub-systems: charcoal canister purge control, the AIR system, idle speed, automatic transmission converter lock-up, distributor ignition timing, the EGR valve, and the air conditioner converter clutch.

The EGR valve control solenoid is activated by the ECM in a fashion similar to that of the charcoal canister purge solenoid described earlier in this chapter. When the engine is warm, the ECM de-energizes the solenoid and the vacuum signal is allowed to reach and then activate the EGR valve.

The Transmission Converter Clutch (TCC) lock is controlled by the ECM through an electrical solenoid in the automatic transmission.

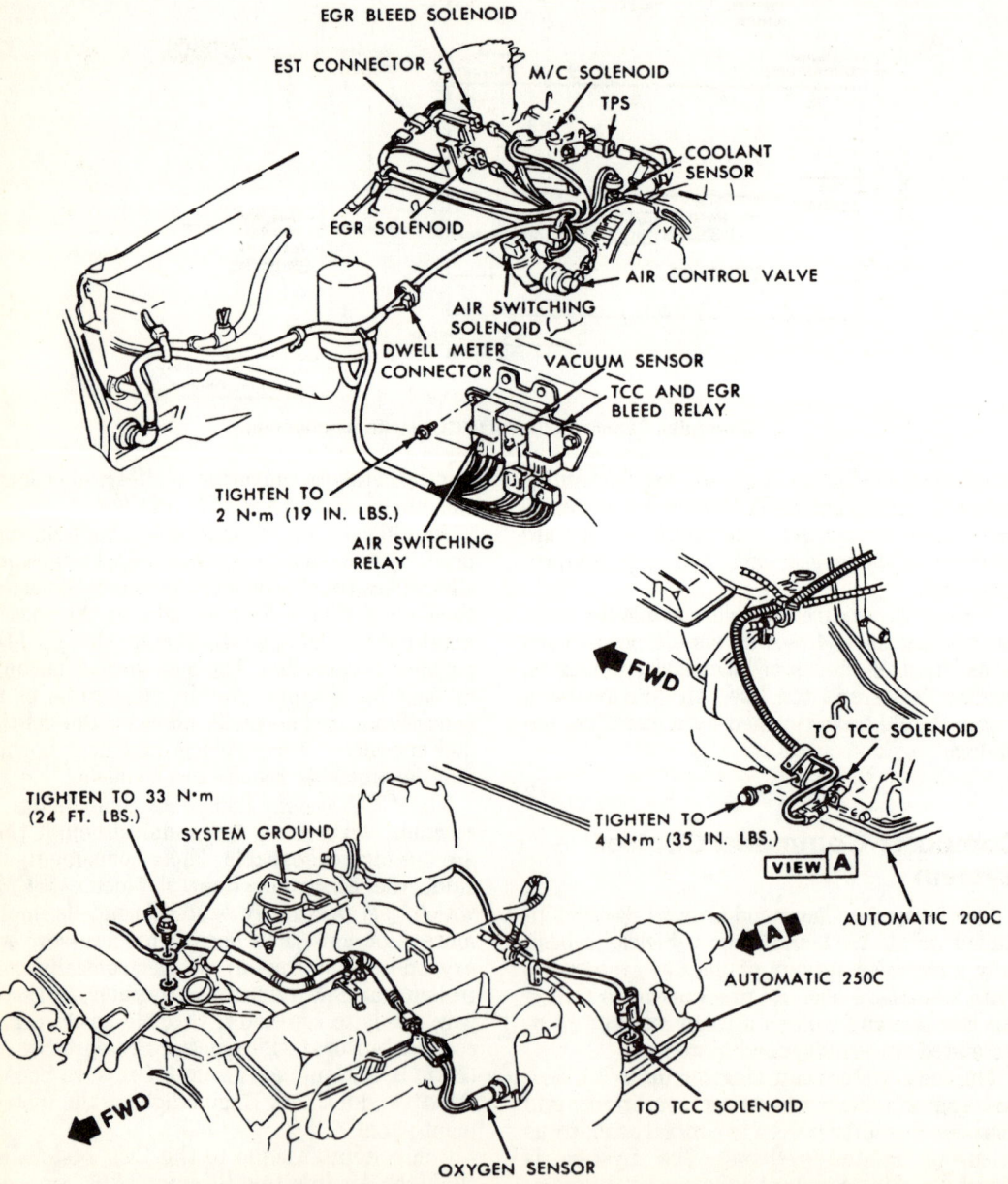

CCC component location, V8 engines except 307 Oldsmobile produced engine

EMISSION CONTROLS 215

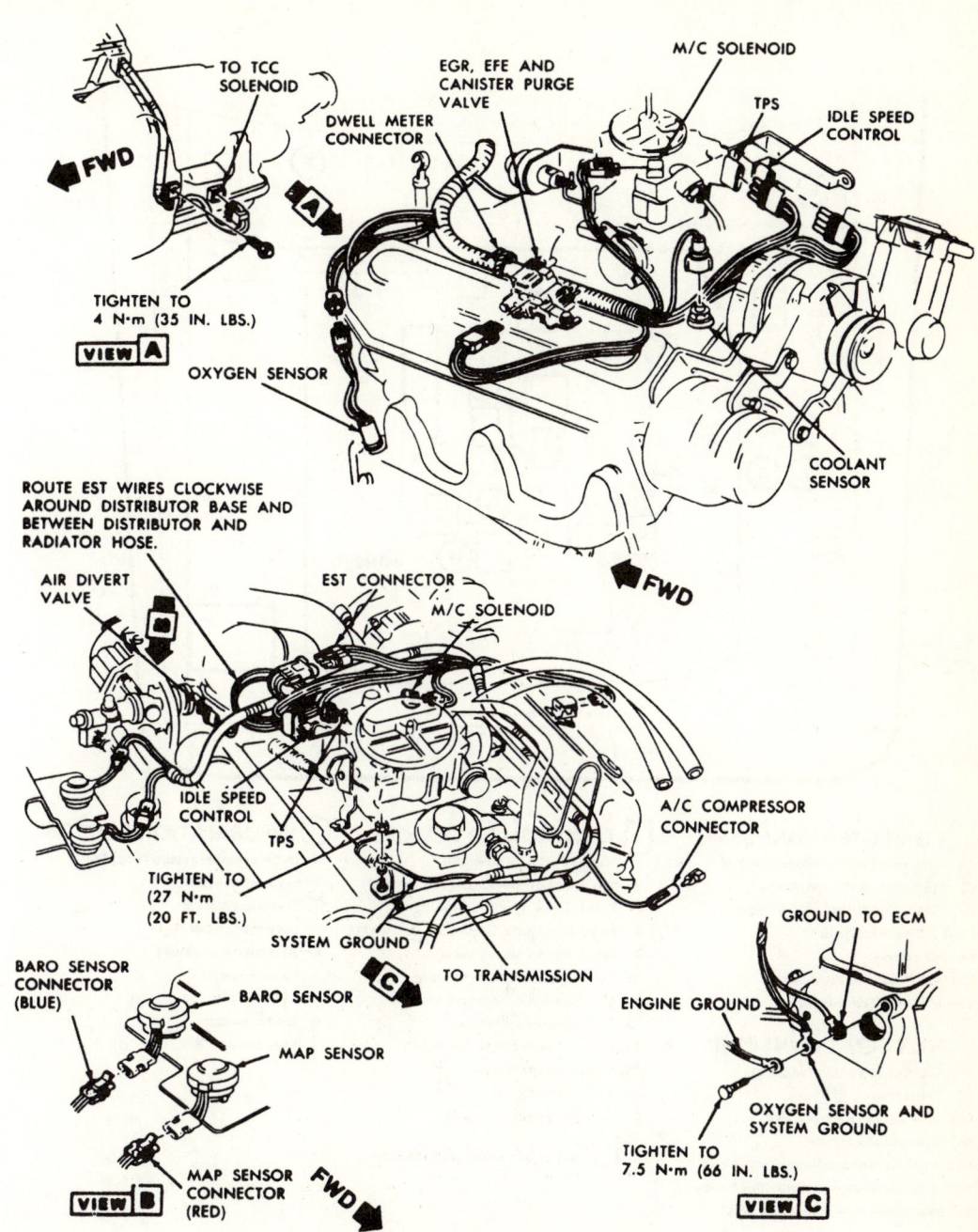

CCC component location, 231 V6 engine

When the vehicle speed sensor in the dash signals the ECM that the vehicle has attained the predetermined speed, the ECM energizes the solenoid which then allows the torque converter to mechanically couple the engine to the transmission. When the brake pedal is pushed, or during deceleration or passing, etc., the ECM returns the transmission to fluid drive.

The idle speed control adjusts the idle speed to all particular engine load conditions and will lower the idle under no-load or low-load conditions in order to converse fuel.

NOTE: *Not all engines use all systems. Control application may differ.*

The vehicle speed sensor (VSS) sends information to the ECM about vehicle speed to aid in controlling the transmission converter clutch.

The VSS is located in the speedometer frame. A reflective blade springs like a propeller, with

EMISSION CONTROLS

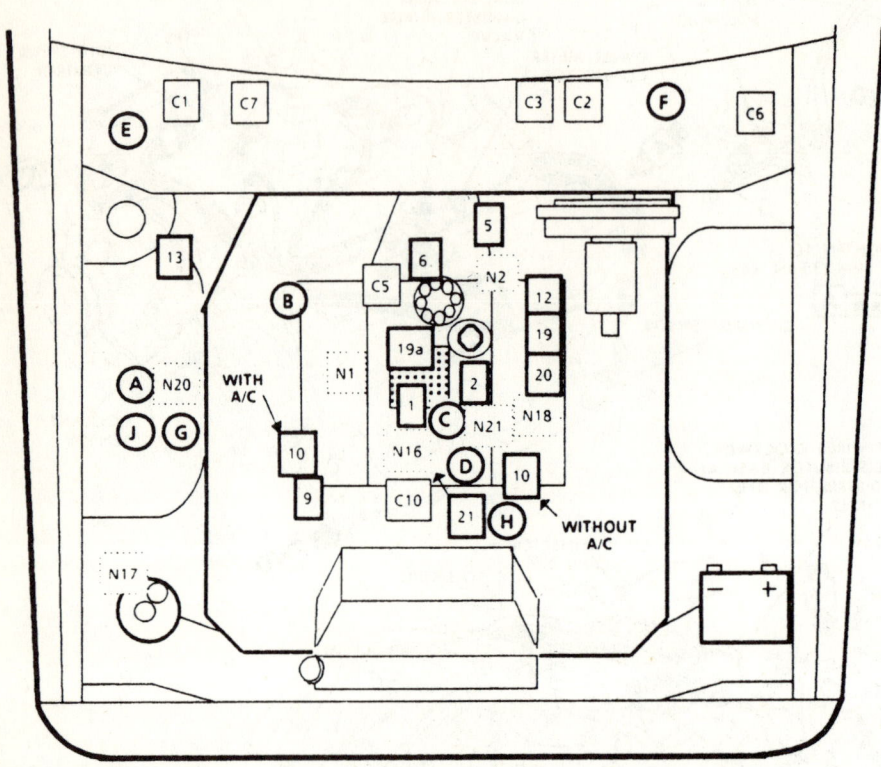

☐ COMPUTER HARNESS
- C1 Electronic Control Module (ECM)
- C2 ALDL diagnostic connector
- C3 "SERVICE ENGINE SOON" light
- C5 ECM harness ground
- C6 Fuse panel
- C7 "C.E./S.E.S." lamp driver
- C10 M/C dwell connector

☐ NOT ECM CONNECTED
- N1 Crankcase vent valve (PCV)
- N2 EFE valve
- N16 Fuel vapor canister valve
- N17 Fuel vapor canister
- N18 Anti-dieseling solenoid (VIN Y only)
- N20 Anti-dieseling vac. tank (VIN Y only)
- N21 Idle Speed Solenoid (VIN 9 only)

☐ CONTROLLED DEVICES
- 1   Mixture control solenoid
- 2   Idle load Compensator (VIN Y only)
- 5   Trans. Conv. Clutch connector
- 6   Electronic Spark Timing (EST) connector
- 9   Air injection divert valve
- 10 Air injection switching valve
- 12 Exh. Gas Recirc. vacuum solenoid
- 13 A/C compressor relay
- 19 Rear vacuum break solenoid
- 19a Rear vacuum break
- 20 ILC Solenoid
- 21 Canister purge solenoid
- ⓞ   Exhaust Gas Recirculation valve

◯ INFORMATION SENSORS
- A Differential Pressure (Vac) "G"
- B Exhaust oxygen
- C Throttle position
- D Coolant temperature
- E Barometric pressure
- F Vehicle speed
- G ESC module
- H Knock sensor
- J Map sensor "B"

Component location 1986–90 307 V8 Oldsmobile produced engines

EMISSION CONTROLS 217

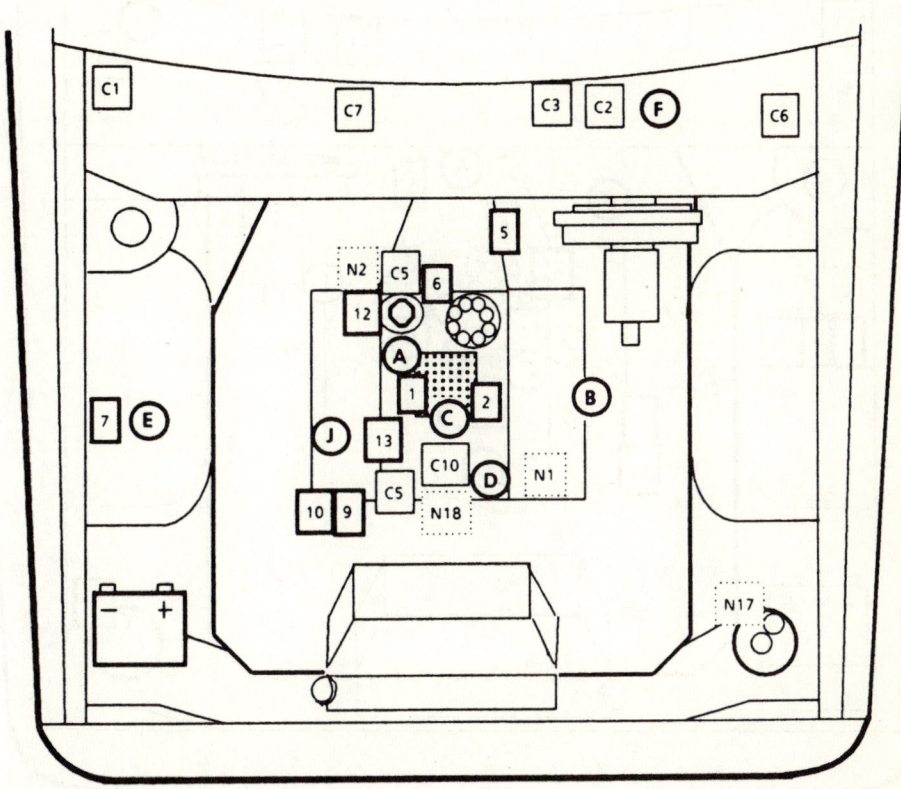

☐ COMPUTER HARNESS
- C1 Electronic Control Module
- C2 ALCL connector (Data Access)
- C3 "SERVICE ENGINE SOON" light
- C5 ECM harness ground
- C6 Fuse panel
- C7 "C.E./S.E.S." lamp driver
- C10 M/C dwell connector

⋮ NOT ECM CONNECTED
- N1 Crankcase vent (PCV)
- N2 EFE Valve
- N17 Fuel vapor canister
- N18 Purge Valve

☐ CONTROLLED DEVICES
- 1 Mixture control solenoid
- 2 Electric throttle kicker (idle solenoid)
- 5 Trans. Conv. Clutch connector
- 6 Electronic Spark Timing connector
- 7 Electronic Spark Control
- 9 Air divert solenoid
- 10 Air switching solenoid
- 12 Exh. Gas Recirc. vac. solenoid (PWM)
- 13 Fuel Vapor Canister Valve

 Exhaust Gas Recirculation valve

○ INFORMATION SENSORS
- A Pressure differential (vacuum)
- B Exhaust oxygen
- C Throttle position
- D Coolant temperature
- E Barometric pressure
- F Vehicle speed
- J Detonation (ESC)

Component location 1986–88 305 V8 engines

EMISSION CONTROLS

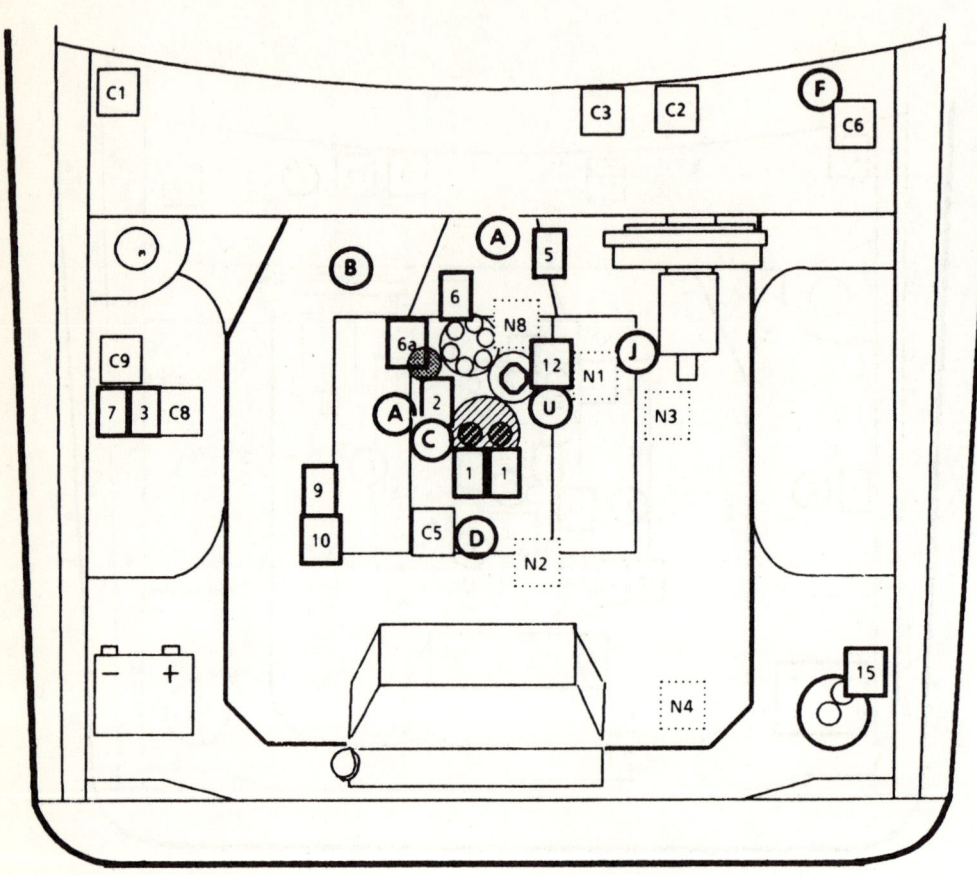

COMPUTER HARNESS
- C1 Electronic control module
- C2 ALCL diagnostic connector
- C3 "SERVICE ENGINE SOON" light
- C5 ECM harness ground
- C6 Fuse panel
- C8 Fuel pump test connector
- C9 Fuel pump fuse & ECM power

NOT ECM CONNECTED
- N1 Crankcase vent valve (PCV)
- N2 EFE/TVS Switch
- N3 EFE valve
- N4 P/S switch
- N8 Oil pressure switch

CONTROLLED DEVICES
1. Fuel injectors
2. Idle air control motor
3. Fuel pump relay
5. Trans. conv. clutch connector
6. EST distributor
6a. Remote ignition coil
7. Electronic spark control module
9. Air injection control solenoid
10. Air injection switching solenoid
12. Exh. gas recirc. vacuum solenoid
15. Fuel vapor canister solenoid

Exhaust gas recirculation valve

INFORMATION SENSORS
- A. Manifold absolute pressure
  (attached to air cleaner "G" series)
  (attached to cowl "B" series)
- B. Exhaust oxygen
- C. Throttle position
- D. Coolant temperature
- F. Vehicle speed
- J. ESC knock
- U. EGR temp. diagnostic switch on base of valve

Component location 1985–88 262 V6 engines

EMISSION CONTROLS 219

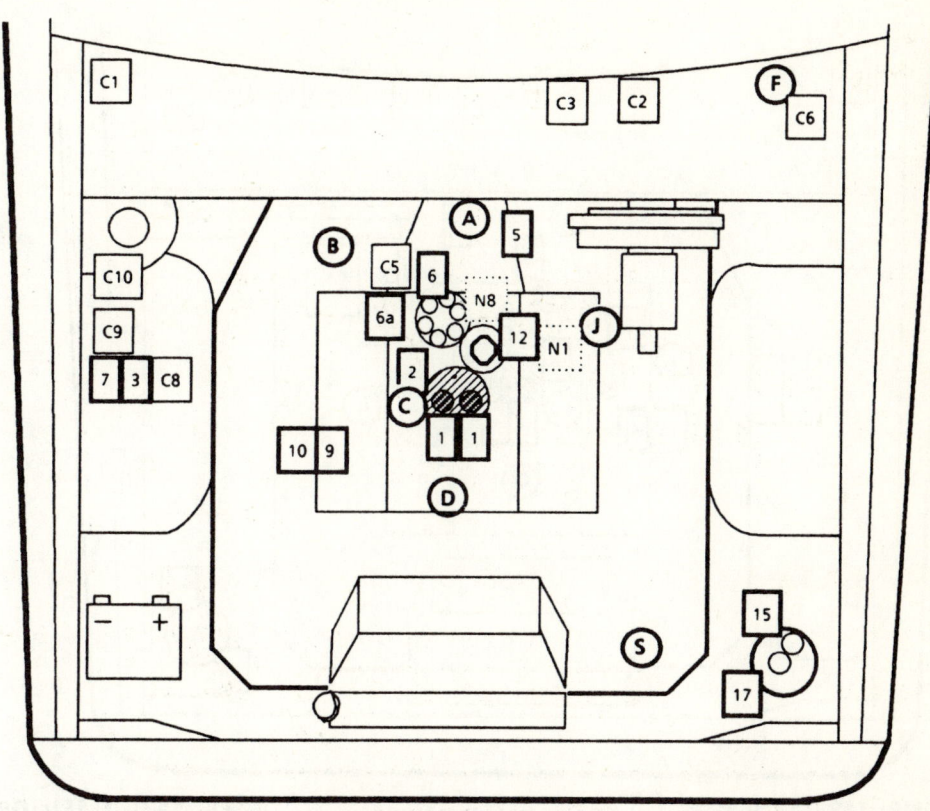

COMPUTER HARNESS
- C1 Electronic Control Module (ECM)
- C2 ALDL diagnostic connector
- C3 "SERVICE ENGINE SOON" light
- C5 ECM harness grounds
- C6 Fuse panel
- C8 Fuel pump test connector
- C9 Fuel pump/ECM fuse
- C10 Set timing connector

NOT ECM CONNECTED
- N1 Crankcase vent valve (PCV)
- N8 Oil pressure switch

CONTROLLED DEVICES
- 1   Fuel injector
- 2   Idle air control motor
- 3   Fuel pump relay
- 5   Torque converter clutch connector
- 6   EST distributor
- 6a Remote ignition coil
- 7   Electronic spark control module
- 9   AIR port solenoid
- 10 AIR converter solenoid
- 12 EGR solenoid
- 15 Fuel vapor canister solenoid
- 17 Fuel vapor canister

◯ Exhaust Gas Recirculation valve

INFORMATION SENSORS
- A   Manifold absolute pressure
- B   Exhaust oxygen
- C   Throttle position
- D   Coolant temperature
- F   Vehicle speed (buffer)
- J   ESC knock
- S   Power steering pressure switch

Component location 1989–90 262 V6 engines

EMISSION CONTROLS

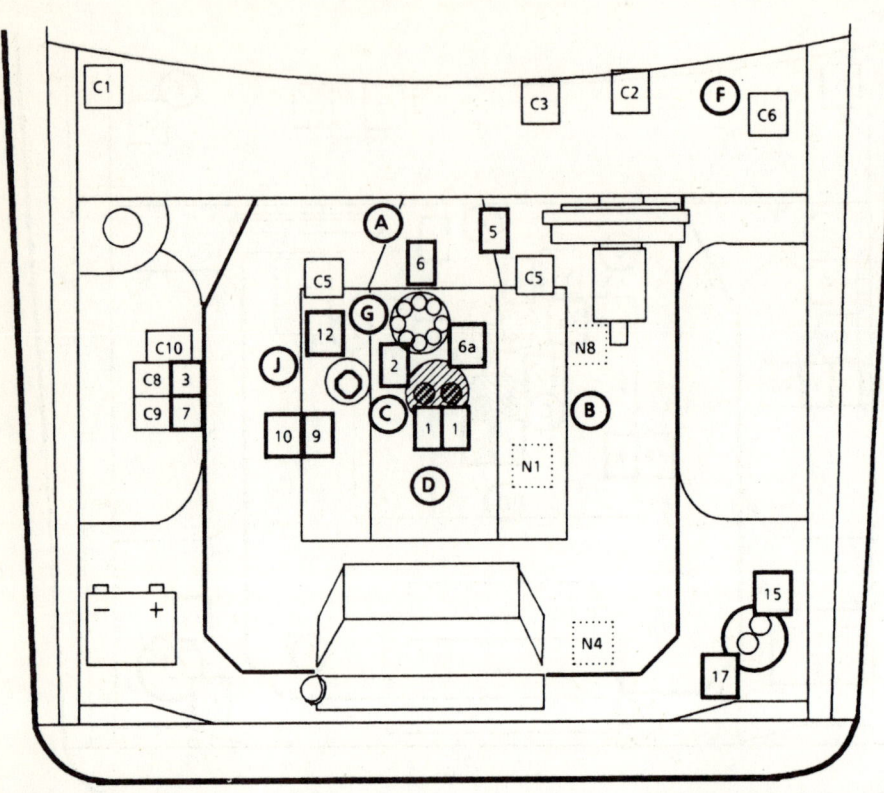

COMPUTER HARNESS
- C1 Electronic Control Module (ECM)
- C2 ALDL diagnostic connector
- C3 "SERVICE ENGINE SOON" light
- C5 ECM harness grounds
- C6 Fuse panel
- C8 Fuel pump test connector
- C9 Fuel pump/ECM fuse
- C10 Set timing connector

NOT ECM CONNECTED
- N1 Crankcase vent valve (PCV)
- N4 Power steering pressure switch
- N8 Oil pressure switch

CONTROLLED DEVICES
- 1   Fuel injector
- 2   Idle air control motor
- 3   Fuel pump relay
- 5   Torque converter clutch connector
- 6   EST distributor
- 6a Remote ignition coil
- 7   Electronic spark control module
- 9   AIR port solenoid
- 10 AIR converter solenoid
- 12 EGR solenoid
- 15 Fuel vapor canister solenoid
- 17 Fuel vapor canister

Exhaust Gas Recirculation valve

INFORMATION SENSORS
- A Manifold absolute pressure
- B Exhaust oxygen
- C Throttle position
- D Coolant temperature
- F Vehicle speed (buffer)
- G MAT (in air cleaner)
- J ESC knock

Component location 1989–90 305, 350 V8 engines

EMISSION CONTROLS 221

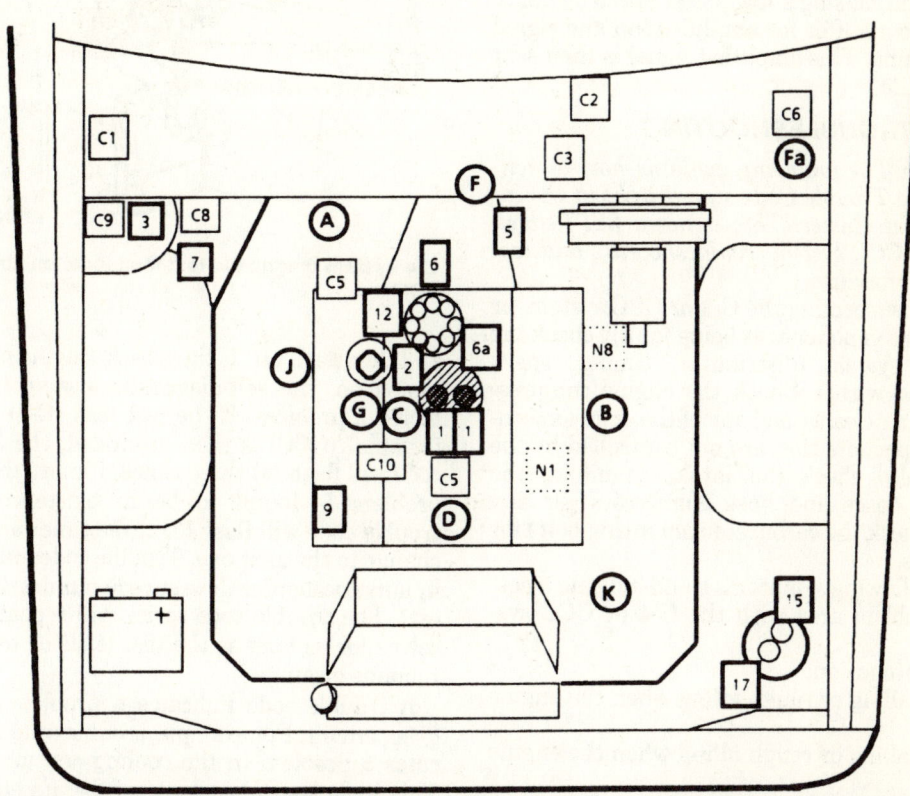

COMPUTER HARNESS
- C1 Electronic Control Module (ECM)
- C2 ALDL diagnostic connector
- C3 "SERVICE ENGINE SOON" light
- C5 ECM harness grounds
- C6 Fuse panel
- C8 Fuel pump "test" connector
- C9 Fuel pump/ECM fuse
- C10 Set timing connector

NOT ECM CONNECTED
- N1 Positive Crankcase Ventilation (PCV)
- N8 Oil pressure switch

CONTROLLED DEVICES
- 1 Fuel injector
- 2 Idle Air Control (IAC) motor
- 3 Fuel pump relay
- 5 Torque Converter Clutch (TCC) connector
- 6 EST distributor
- 6a Remote ignition coil
- 7 Electronic Spark Control (ESC) module
- 9 AIR Electric Diverter (EDV) valve solenoid
- 12 EGR solenoid
- 15 Fuel vapor canister solenoid
- 17 Fuel vapor canister

Exhaust Gas Recirculation valve

INFORMATION SENSORS
- A Manifold Absolute Pressure (MAP) sensor
- B Oxygen (O_2) sensor
- C Throttle Position Sensor (TPS)
- D Coolant Temperature Sensor (CTS)
- F Vehicle Speed Sensor (VSS)
- Fa Vehicle Speed Sensor (VSS) buffer
- G IAT (in air cleaner)
- J ESC knock
- K POWER STEERING PRESSURE SWITCH (WAGON ONLY)

Component location 1991–92 305, 350 V8 engines

EMISSION CONTROLS

its blades passing through a light beam from a light emitting diode (LED). As each blade enters the light beam, light is reflected back to a photocell causing a low power speed signal to be sent to a buffer for amplification and signal conditioning. This amplified signal is then sent to the ECM.

BASIC TROUBLESHOOTING

NOTE: *The following explains how to activate the Trouble Code signal light in the instrument cluster. This is not a full fledged C-4 or CCC system troubleshooting and isolation procedure.*

Before suspecting the C-4 or CCC system, or any of its components as being faulty, check the ignition system (distributor, timing, spark plugs and wires). Check the engine compression, the air cleaner and any of the emission control components that are not controlled by the ECM. Also check the intake manifold, the vacuum hoses and hose connectors for any leaks. Check the carburetor mounting bolts for tightness.

The following symptoms could indicate a possible problem area with the C-4 or CCC systems:
1. Detonation;
2. Stalling or rough idling when the engine is cold;
3. Stalling or rough idling when the engine is hot;
4. Missing;
5. Hesitation;
6. Surging;
7. Poor gasoline mileage;
8. Sluggish or spongy performance;
9. Hard starting when engine is cold;
10. Hard starting when the engine is hot;
11. Objectionable exhaust odors;
12. Engine cuts out;
13. Improper idle speed (CCC only).

As a bulb and system check, the Check Engine light will come on when the ignition switch is turned to the ON position but the engine is not started.

The Check Engine light will also produce the trouble code/codes by a series of flashes which translate as follows: When the diagnostic test lead (C-4) or terminal (CCC) under the instrument panel is grounded, with the ignition in the ON position and the engine not running, the Check Engine light will flash once, pause, and then flash twice in rapid succession. This is a Code 12, which indicates that the diagnostic system is working. After a long pause, the Code 12 will repeat itself until the engine is started or the ignition switch is turned OFF.

When the engine is started, the Check Engine light will remain on for a few seconds

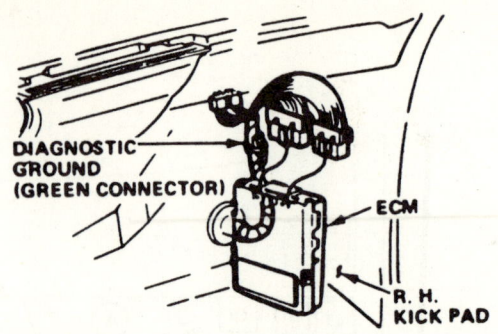

C-4 system diagnostic test lead location above the ECM

and then turn off. If the Check Engine light remains on, the self-diagnostic system has detected a problem. If the test lead (C-4) or test terminal (CCC) is then grounded, the trouble code will flash (3) three times. If more than one problem is found to be in existence, each trouble code will flash (3) three times and then change to the next one. Trouble codes will flash in numerical order (lowest code number to highest). The trouble code series will repeat themselves for as long as the test leads or terminal remains grounded.

A trouble code indicates a problem with a given circuit. For example, trouble code 14 indicates a problem in the cooling sensor circuit. This includes the coolant sensor, its electrical harness and the Electronic Control Module (ECM).

Since the self-diagnostic system cannot diagnose every possible fault in the system, the absence of a trouble code does not necessarily mean that the system is trouble free. To determine whether or not a problem with the system exists that does not activate a trouble code, a system performance check must be made. This job should be left to a qualified service technician.

In case of an intermittent fault in the

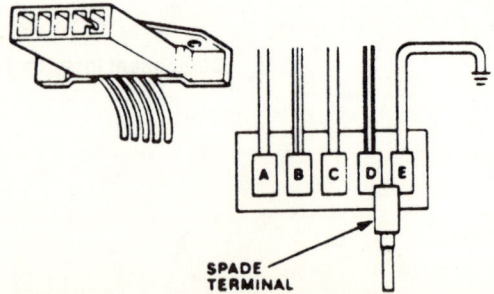

CCC system diagnostic test terminal located underneath the left side instrument panel

EMISSION CONTROLS 223

system, the Check Engine light will go out when the fault goes away, but the trouble code will remain in the memory of the ECM. Therefore, if a trouble code can be obtained even though the Check Engine light is not on, it must still be evaluated. It must be determined if the fault is intermittent or if the engine must be operating under certain conditions (acceleration, deceleration, etc.) before the Check Engine light will come on. In some cases, certain trouble codes will not be recorded in the ECM until the engine has been operated at part throttle for at least 5 to 8 minutes.

On the C-4 system, the ECM erases all trouble codes every time that the ignition is turned off. In the case of intermittent faults, a long term memory is desirable. This can be produced by connecting the orange connector/lead from terminal S of the ECM directly to the battery (or to a the CCC system, a trouble code will be stored until the terminal 'R' at the ECM has been disconnected from the battery for at least 10 seconds.

ACTIVATING THE TROUBLE CODE

On the C-4 system, activate the trouble code by grounding the trouble code test lead. Use the illustrations to help you locate the test lead under the instrument panel (usually a white and black wire from the lead to a suitable ground).

On the CCC system, locate the test terminal under the instrument panel (see illustration). Use a jumper wire and ground only the lead.

NOTE: *Ground the test lead/terminal according to the instructions given previously in the Basic Troubleshooting section.*

RESETTING THE TROUBLE CODE

To clear the codes from the memory of the ECM, either to determine if the malfunction will occur again or because the repair has been completed, the ECM power feed must be disconnected for at least 30 seconds. Depending on how the vehicle is equipped, the ECM power feed can be disconnected at the positive terminal, the inline fuseholder that originates at the positive connection at the battery, or the ECM fuse in the fuse block.

NOTE: *The negative battery terminal may also be disconnected to clear the codes, but other memory data will be lost, such as the clock and radio presets.*

CAUTION: *To prevent ECM damage, the ignition key must be in the **OFF** position when disconnecting or reconnecting the power source*

Mixture Control Solenoid (M/C)

The fuel flow through the carburetor idle main metering circuits is controlled by a mixture control (M/C) solenoid located in the car-

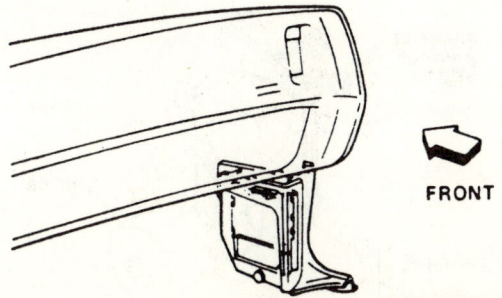

1. Jumper "B" to "A" to display diagnostic codes
2. Test terminal
3. Ground terminal

Under dash test terminal location

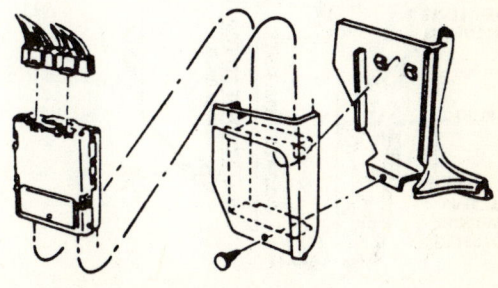

Electronic Control Module (ECM), all models

buretor. The M/C solenoid changes the air/fuel mixture to the engine by controlling the fuel flow through the carburetor. The ECM controls the solenoid by providing a ground. When the solenoid is energized, the fuel flow through the carburetor is reduced, providing a leaner mixture. When the ECM removes the ground, the solenoid is de-energized, increasing the fuel flow and providing a richer mixture. The M/C solenoid is energized and de-energized at a rate of 10 times per second.

Throttle Position Sensor (TPS)

The throttle position sensor is mounted in the carburetor body and is used to supply throttle position information to the ECM. The ECM memory stores an average of operation conditions with the ideal air/fuel ratios for each of those conditions. When the ECM receives a signal that indicates throttle position change, it immediately shifts to the last remembered set of operating conditions that resulted in an ideal air/fuel ratio control. The memory is continually being updated during normal operations.

Idle Speed Control (ISC)

231 V6

The idle speed control does just what its name implies - it controls the idle. the ISC is used to maintain low engine speeds while at the same time preventing stalling due to engine load changes. The system consists of a motor assembly mounted on the carburetor which moves the throttle lever so as to open or close the throttle blades.

The whole operation is controlled by the ECM. The ECM monitors engine load to determine the proper idle speed. To prevent stalling, it monitors the air conditioning compressor switch, the transmission, the park/neutral switch and the ISC throttle switch. The ECM processes all this information and then uses it to control the ISC motor which in turn will vary the idle speed as necessary.

Electronic Spark Timing (EST)

All 1980 vehicles equipped with the 231 V6 engine and all 1981 and later vehicles use EST. The EST distributor, as described in an earlier chapter, contains no vacuum or centrifugal advance mechanism and uses a seven terminal HEI module. It has four wires going to a four terminal connector in addition to the connectors normally found on the HEI distributors. A reference pulse, indicating engine rpm is sent to the ECM; terminal R on the 7-terminal HEI provides this pulse on all vehicles. The ECM determines the proper spark advance for the engine operating conditions and then sends an EST pulse back to the distributor.

NOTE: *The 1985 and later 6-262 distributor is equipped with a modified module which has eight terminals.*

Under most normal operating conditions, the ECM will control spark advance. However, under certain operating conditions such as cranking or when setting base timing, the distributor is capable of operating without ECM control. This condition is called BYPASS and is determined by the BYPASS lead which runs from the ECM to the distributor. When the BYPASS lead is at the proper voltage (5), the ECM will control the spark. If the lead is grounded or open circuited, the HEI module itself will control the spark. Disconnecting the 4-terminal EST connector will also cause the engine to operate in the BYPASS mode.

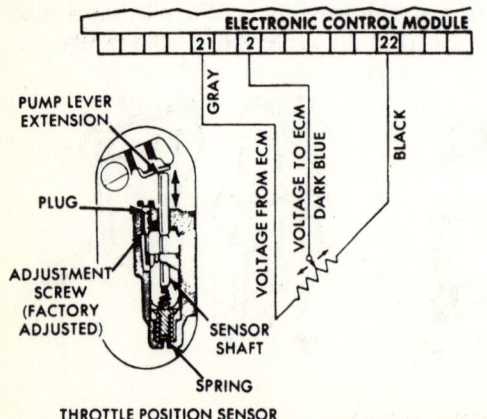

Throttle Position Sensor (TPS)

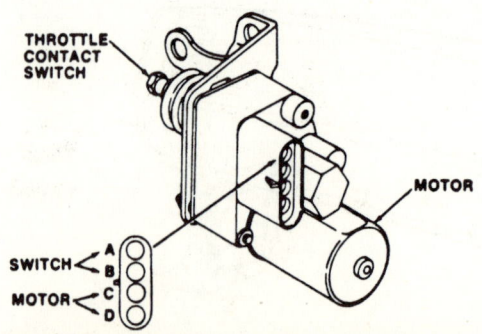

Idle Speed Control (ISC) motor is attached to the carburetor

EMISSION CONTROLS 225

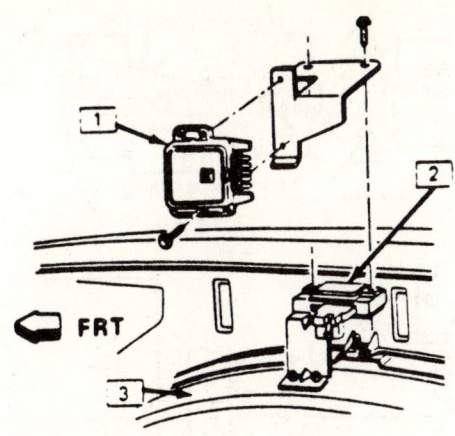

1. ESC module
2. Map sensor
3. Right wheelhouse

View of the electronic spark control module

Electronic Spark Control (ESC)

All 1984 and Later Engines

The Electronic Spark Control (ESC) system is a closed loop system that controls engine detonation by adjusting the spark timing. There are two basic components in this system, the controller and the sensor.

The controller processes the sensor signal and remodifies the EST signal to the distributor to adjust the spark timing. The process is continuous so that the presence of detonation is monitored and controlled. The controller is not capable of memory storage.

The sensor is a magnetorestrictive device, mounted in the engine block that detects the presence, or absence, and intensity of detonation according to the vibration characteristics of the engine. The output is an electrical signal which is sent to the controller.

Transmission Converter Clutch (TCC)

All 1981 vehicles equipped with an automatic transmission use TCC. The ECM controls the converter by means of a solenoid mounted in the transmission. When the vehicle speed reaches a certain level, the ECM energizes the solenoid and allows the torque converter to mechanically couple the transmission to the engine. When the operating conditions indicate that the transmission should operate as a normal fluid coupled transmission, the ECM will de-engerize the solenoid. Depressing the brake will also return the transmission to normal automatic operation.

Catalytic Converter

The catalytic converter is a muffler-life container built into the exhaust system to aid in the reduction of exhaust emissions. The catalyst element consists of individual pellets or a honeycomb monolithic substrate coated with a noble metal such as platinum, palladium, rho-

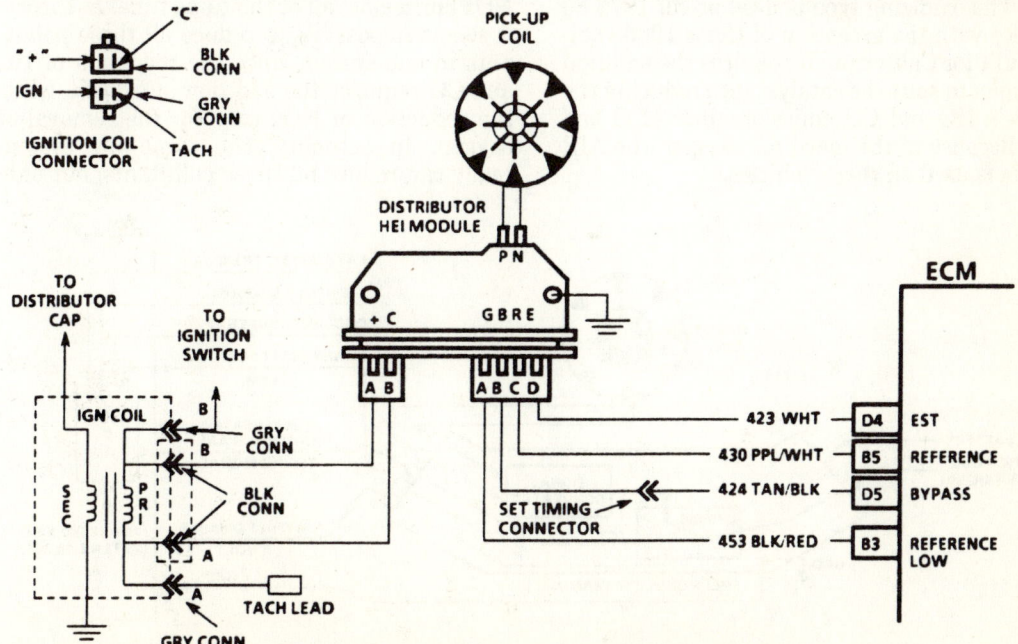

Remote coil/EST schematic, 305, 350 V8 engine

EMISSION CONTROLS

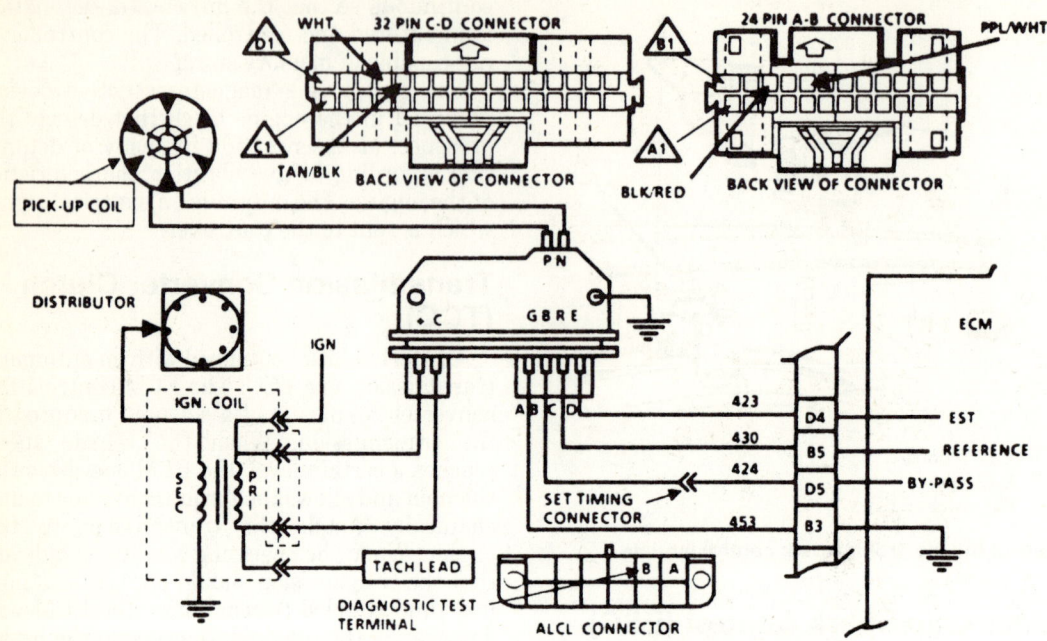

HEI system schematic with EST, 1985 and later 262 V6 engine

dium, or a combination. When the exhaust gases come into contact with the catalyst, a chemical reaction occurs which will reduce the pollutants into harmless substances like water and carbon dioxide.

There are essentially two types of catalytic converters: an oxidizing type and a three-way type. The oxidizing type is used on all 1975-80 vehicles with the exception of those 1980 vehicles built for California. It requires the addition of oxygen to spur the catalyst into reducing the engine's HC and CO emissions into H_2O and CO_2. Because of this need for oxygen, the AIR system is used on these vehicles.

The oxidizing catalytic converter, while effectively reducing HC and CO emissions, does little, if anything in the way of reducing NOx emissions. Thus, the three-way catalytic converter.

The three-way converter, unlike the oxidizing type, is capable of reducing HC, CO and NOx emissions; all at the same time. In theory, it seems impossible to reduce all three pollutants in one system since the reduction of HC and CO requires the addition of oxygen, while the reduction of NOx calls for the removal of oxygen. In actuality, the three-way system really can reduce all three pollutants, but only

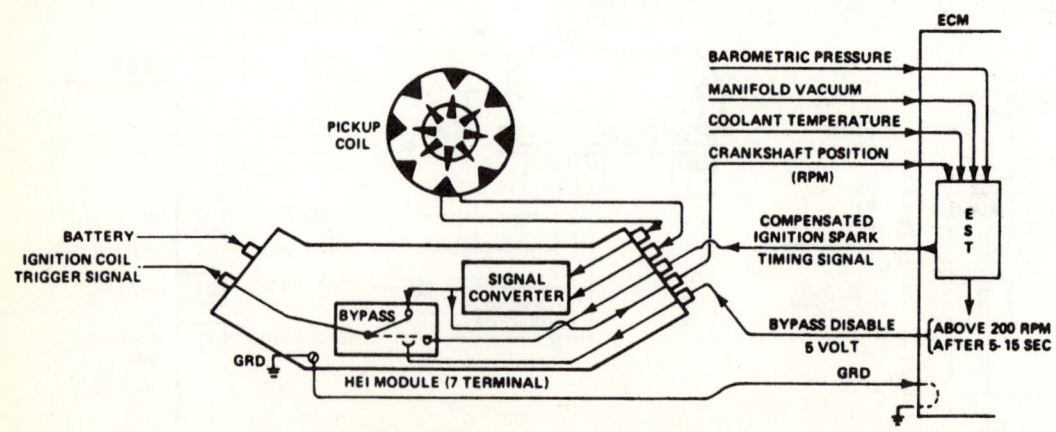

A schematic view of the EST circuitry

EMISSION CONTROLS 227

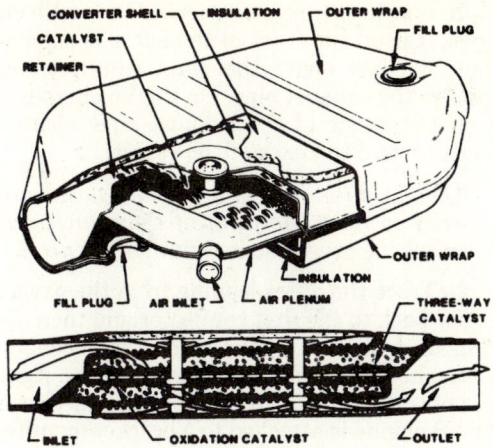

Dual bed type catalytic converter

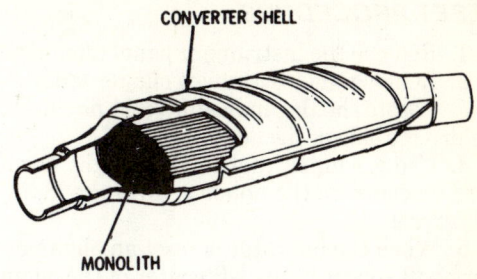

Single bed monolith catalytic converter

if the amount of oxygen in the exhaust system is precisely controlled. Due to this precise oxygen system is used only in vehicles equipped with an oxygen sensor system.

There are no service procedures required for the catalytic converter, although the converter body should be inspected occasionally for damage. Some vehicles equipped with the V6 engine require a catalyst charge at 30,000 mile intervals (consult your Owner's Manual).

PRECAUTIONS

1. Use only unleaded fuel.
2. Avoid prolonged idling; the engine should run no longer than 20 min. at curb idle and no longer than 10 min. at fast idle.
3. Do not disconnect any of the spark plug leads while the engine is running.
4. Make engine compression checks as quickly as possible.

CATALYST TESTING

At the present time there is no known way to reliably test catalytic converter operation in the field. The only reliable test is a 12 hour and 40 min. soak test (CVS) which must be done in a laboratory.

An infrared HC/CO tester is not sensitive enough to measure the higher tailpipe emissions from a failing converter. Thus, a bad converter may allow enough emissions to escape so that the vehicle is no longer in compliance with Federal or state standards, but will still not cause the needle on a tester to move off zero.

The chemical reactions which occur inside a catalytic converter generate a great deal of heat. Most converter problems can be traced to fuel or ignition system problems which cause unusually high emissions. As a result of the increased intensity of the chemical reactions, the converter literally burns itself up.

A completely failed converter might cause a tester to show a slight reading. As a result, it is occasionally possible to detect one of these.

As long as you avoid severe overheating and the use of leaded fuels it is reasonably safe to assume that the converter is working properly. If you are in doubt, take the vehicle to a diagnostic center that has a tester.

Maintenance Reminder System

An emissions indicator flag may appear in the odometer window of the speedometer on somevehicles. The flag could say "Sensor", "Emissions" or "Catalyst" depending on the part or assembly that is scheduled for regular emissions maintenance replacement. The word "Sensor" indicates a need for oxygen sensor replacement and the words "Emissions" or "Catalyst" indicate the need for catalytic converter catalyst replacement.

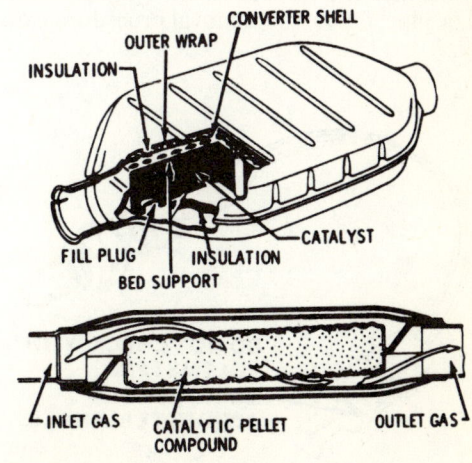

Bead type catalytic converter

228   EMISSION CONTROLS

RESET PROCEDURE

1. Remove the instrument panel trim plate.
2. Remove the instrument cluster lens.
3. Locate the flag indicator reset notches at the drivers side of the odometer.
4. Use a pointed tool to apply light downward pressure on the notches, until the indicator is reset.
5. When the indicator is reset an alignment mark will appear in the left center of the odometer window.

Oxygen Sensor

An oxygen sensor protrudes into the exhaust stream and monitors the oxygen content of the exhaust gases. The difference between the oxygen content of the exhaust gases and that of outside air generates a voltage signal to the ECM. The ECM monitors this voltage and, depending upon the value of the signal received, issues a command to adjust for a rich or a lean condition.

No attempt should ever be made to measure the voltage output of the sensor. The current drain of any conventional voltmeter would be such that it would permanently damage the sensor. Use these tools ONLY on the ECM side of the wiring harness connector AFTER disconnecting it from the sensor.

REMOVAL AND INSTALLATION

The oxygen sensor must be replaced every 30,000 miles (48,000 km.). The sensor may be difficult to remove when the engine temperature is below 120°F. Excessive removal force may damage the threads in the exhaust manifold or pipe; follow the removal procedure carefully.

1. Locate the oxygen sensor. On the V8 engines, it is on the front of the left side exhaust manifold, just above the point where it connects to the exhaust pipe. On the V6 engines, it is on the inside of the exhaust pipe where it bends toward the back of the vehicle.

NOTE: *On the V6 engine you may find it necessary to raise the front of the vehicle and remove the oxygen sensor from underneath.*

2. Trace the wires leading from the oxygen sensor back to the first connector and then disconnect them (the connector on the V6 engine is attached to a bracket mounted on the right, rear of the engine block, while the connector in the V8 engine is attached to a bracket mounted on the top of the left side exhaust manifold).
3. Spray a commercial heat riser solvent onto the sensor threads and allow it to soak in for at least five minutes.
4. Carefully unscrew and remove the sensor.
5. To install, first coat the new sensor's threads with G.M. anti-seize compound no. 5613695 or the equivalent. This is not a conventional anti-seize paste. The use of a regular compound may electrically insulate the sensor, rendering it inoperative. You must coat the threads with an electrically conductive anti-seize compound.
6. Installation torque is 30 ft. lbs. (42 Nm.). Do not overtighten.
7. Reconnect the electrical connector. Be careful not to damage the electrical pigtail. Check the sensor boot for proper fit and installation. Install the air cleaner, if removed.

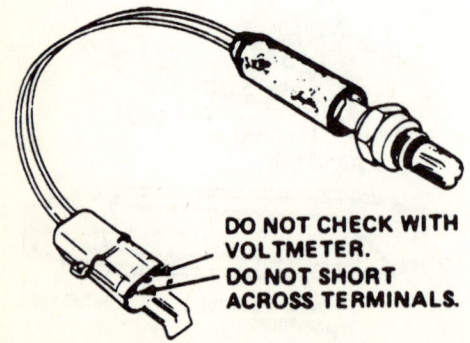

Oxygen sensor assembly

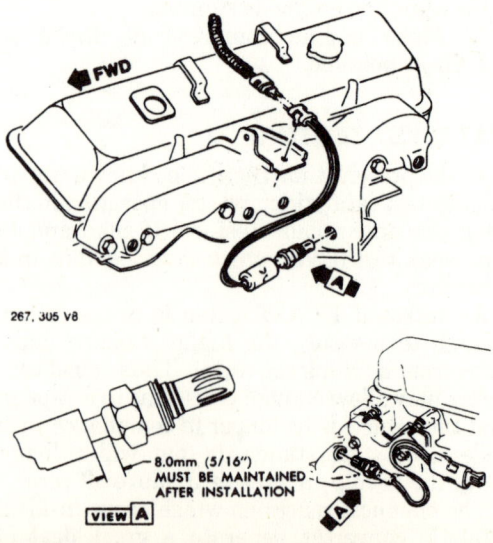

Oxygen sensor locations on all engines

EMISSION CONTROLS

DIESEL ENGINE EMISSIONS CONTROLS

Crankcase Ventilation

A Crankcase Depression Regulator Valve (CDRV) is used to regulate (meter) the flow of crankcase gases back into the engine to be burned. The CDRV is designed to limit vacuum in the crankcase as the gases are drawn from the valve covers through the CDRV and into the intake manifold (air crossover).

Fresh air enters the engine through the combination filter, check valve and oil fill cap. The fresh air mixes with blow-by gases and enters both valve covers. The gases pass through a filter installed on the valve covers and are drawn into connecting tubing.

Intake manifold vacuum acts against a spring loaded diaphragm to control the flow of crankcase gases. Higher intake vacuum levels pull the diaphragm closer to the top of the outlet tube. This reduces the amount of gases being drawn from the crankcase and decreases the vacuum level in the crankcase. As the intake vacuum decreases, the spring pushes the diaphragm away from the top of the outlet tube allowing more gases to flow to the intake manifold.

> NOTE: *Do not allow any solvent to come in contact with the diaphragm of the Crankcase Depression Regulator Valve because the diaphragm will fail.*

Exhaust Gas Recirculation (EGR)

To lower the formation of nitrogen oxides (NOx) in the exhaust, it is necessary to reduce combustion temperatures. This is done in the diesel, as in the gasoline engine, by introducing exhaust gases into the cylinders through the EGR valve.

FUNCTIONAL TESTS OR COMPONENTS

Vacuum Regulator Valve (VRV)

The Vacuum Regulator Valve is attached to the side of the injection pump and regulates vacuum in proportion to throttle angel. Vacuum from the vacuum pump is supplied to port A and vacuum at port B is reduced as the throttle is opened. At closed throttle, the vacuum is 15 inches; at half throttle, 6 inches; at wide open throttle there is 0 vacuum.

Exhaust Gas Recirculation (EGR) Valve

Apply vacuum to vacuum port. The valve should be fully open at 10.5 and closed below 6.

Response Vacuum Reducer (RVR)

Connect a vacuum gauge to the port marked To EGR valve or T.C.C solenoid. Connect a hand operated vacuum pump to the RVR port. Draw a 50.66 kPa (15 inch) vacuum on the pump and the reading on the vacuum gauge should be lower than the vacuum pump reading as follows:
- 0.75 Except High Altitude
- 2.5 High Altitude

Torque Converter Clutch Operated Solenoid

When the torque converter clutch is engaged, an electrical signal energizes the solenoid allowing ports 1 and 2 to be interconnected. When the solenoid is not energized, port 1 is closed and ports 2 and 3 are interconnected.

Solenoid Energized
- Ports 1 and 3 are connected.

Solenoid De-energized
- Ports 2 and 3 are connected.

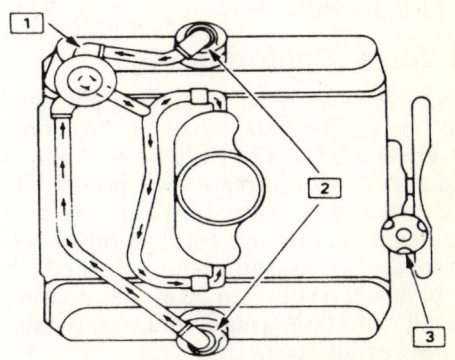

CRANKCASE VENTILATION SYSTEM SCHEMATIC
V-TYPE DIESEL ENGINE
WITH DEPRESSION REGULATOR VALVE
1. Crankcase depression regulator
2. Ventilation filter
3. Breather cap

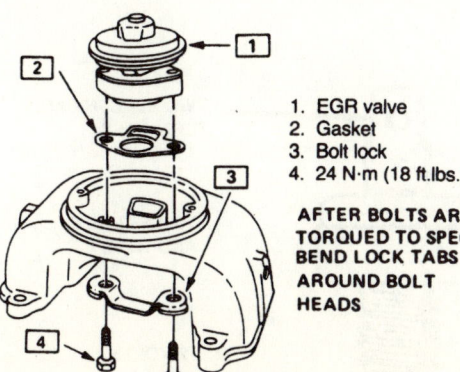

1. EGR valve
2. Gasket
3. Bolt lock
4. 24 N·m (18 ft.lbs.)

AFTER BOLTS ARE TORQUED TO SPECS BEND LOCK TABS AROUND BOLT HEADS

Diesel EGR valve location on top of intake manifold

230 EMISSION CONTROLS

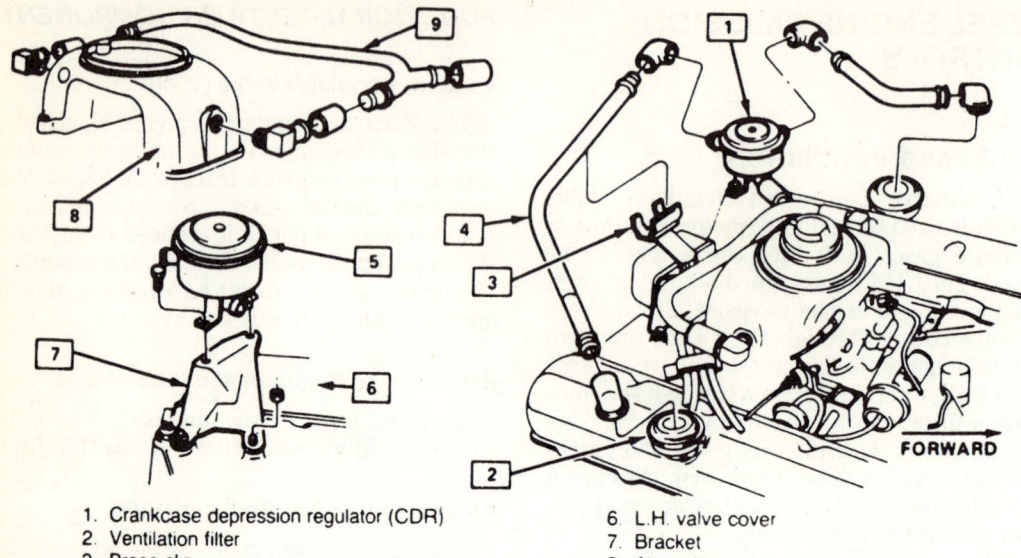

1. Crankcase depression regulator (CDR)
2. Ventilation filter
3. Brace clip
4. Ventilation pipes
5. Crankcase depression regulator (CDR)
6. L.H. valve cover
7. Bracket
8. Air crossover
9. Air crossover to regulator valve pipe

V8 diesel crankcase ventilation system

Engine Temperature Sensor (ETS)

OPERATION

The engine temperature sensor has two terminals. Twelve volts are applied to one terminal and the wire from the other terminal leads to the fast idle solenoid and Housing Pressure cold Advance solenoid that is part of the injection pump.

The switch contacts are closed below 125°F. At the calibration point, the contacts are open which turns off the solenoids.

Above Calibration
- Open circuit

Below Calibration
- Closed circuit

EPR Valve (California V6)

This valve is found between the right hand exhaust manifold and the exhaust pipe on California V6 diesel cars. The EPR valve is used in the exhaust flow to increase back pressure in the exhaust system, thus increasing exhaust flow through the EGR system. The valve operates from the same vacuum source as the EGR valve. It should be fully closed at idle, and will open as the throttle is opened until, at full throttle, it will be fully open.

TESTING THE EPR VALVE

To test the EPR valve, apply vacuum to the vacuum port on the valve. The valve should be fully closed at 12 in.Hg of vacuum, and open below 6 in.Hg. of vacuum.

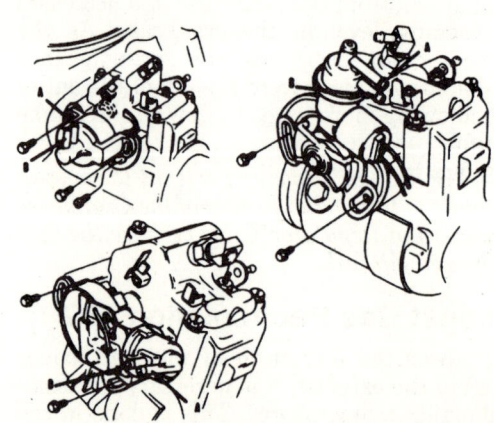

Vacuum regulator valve, mounted to diesel injection pumps

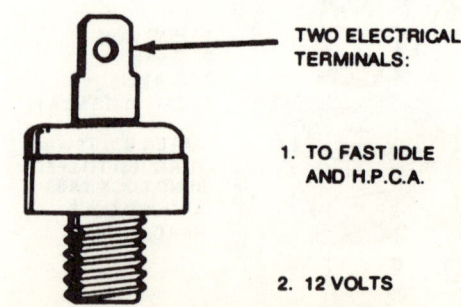

TWO ELECTRICAL TERMINALS:
1. TO FAST IDLE AND H.P.C.A.
2. 12 VOLTS

Engine Temperature Sensor (ETS), diesel engines

EMISSION CONTROLS 231

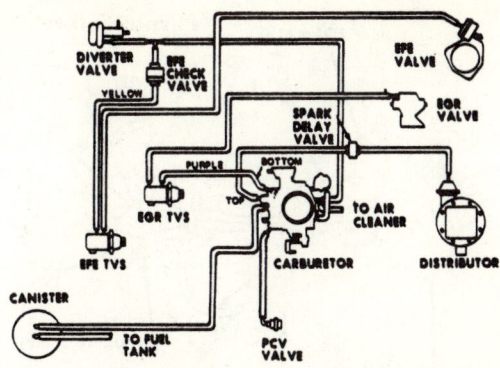

1977—305 V8 engine 2bbl. except California

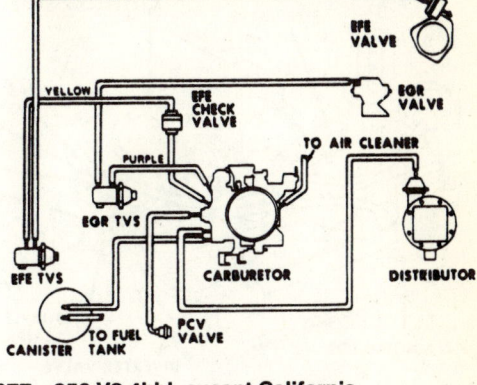

1977—350 V8 4bbl. except California

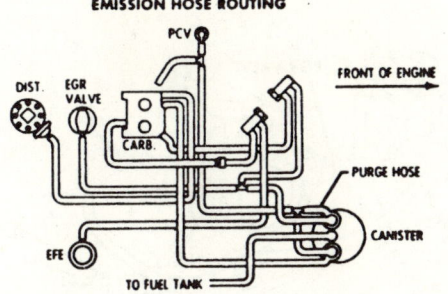

1979—350 V8 engine (910L4RU) except California

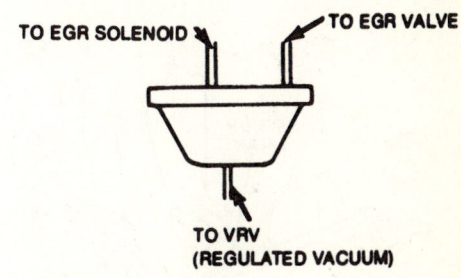

Diesel vacuum reducer, except California models

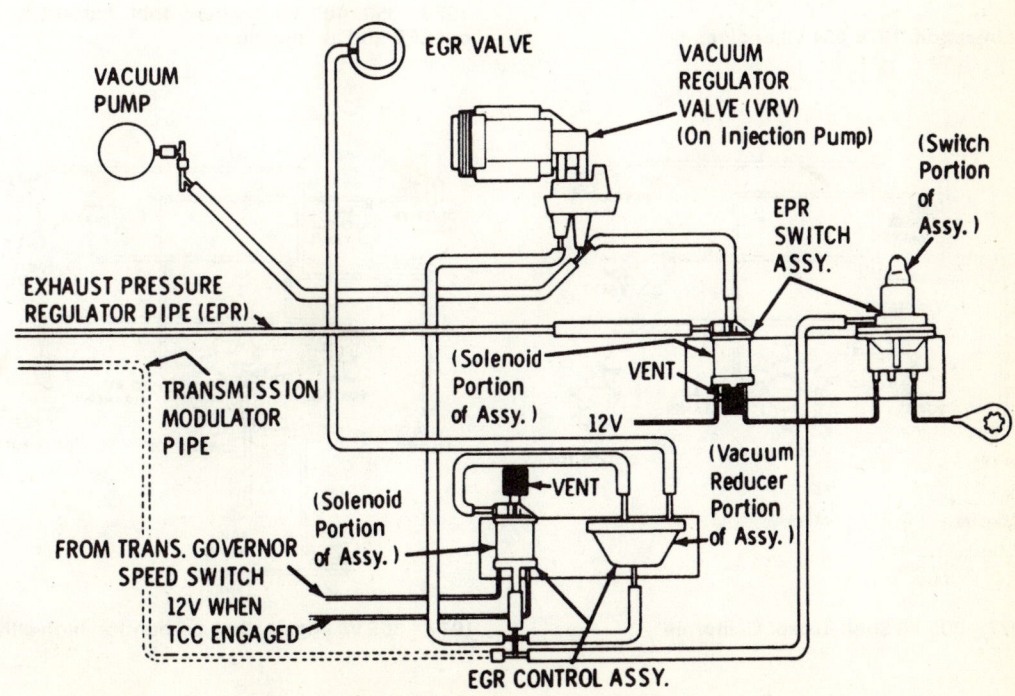

Diesel EGR system, except California models

232 EMISSION CONTROLS

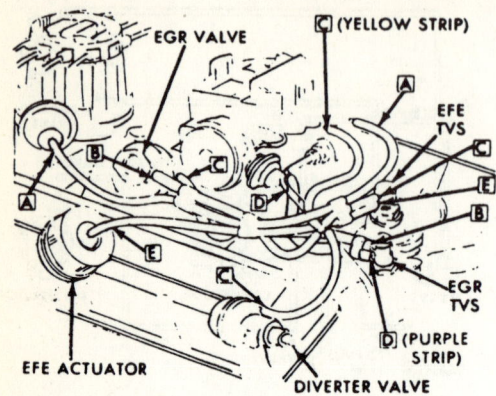

EFE and air injection, 1976—400 V8 engine

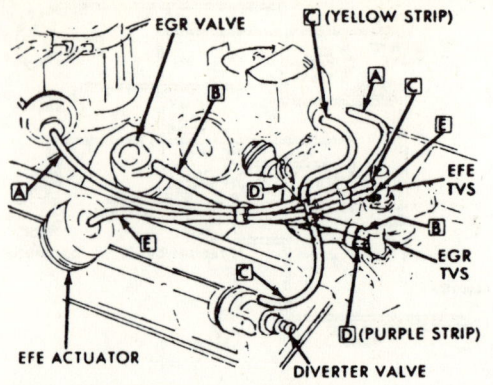

4bbl. Carburetor with EFE and air injection, 1976—350 V8 engine

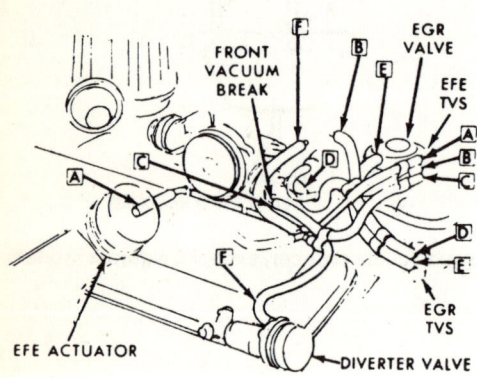

Air injection, 1976 454 V8 engines

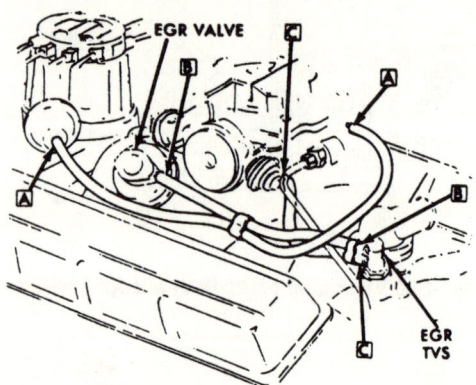

1976—350, 400 V8 engines 4bbl. carburetor without EFE and air injection

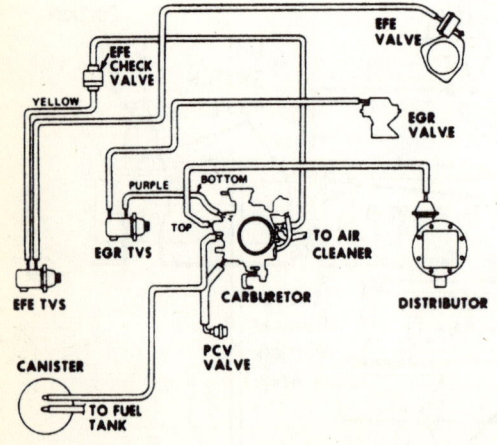

1977—305 V8 2bbl. except California

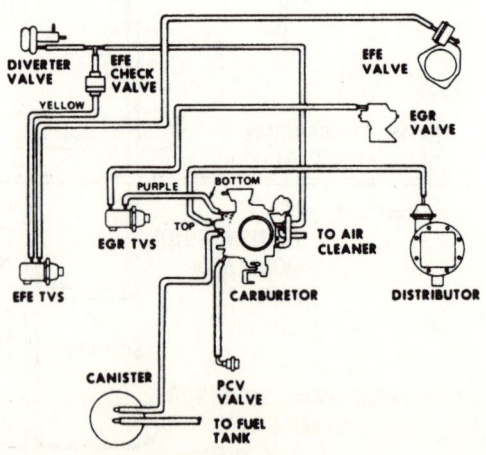

1977—305 V8 engine 2bbl. carburetor, high altitude

EMISSION CONTROLS 233

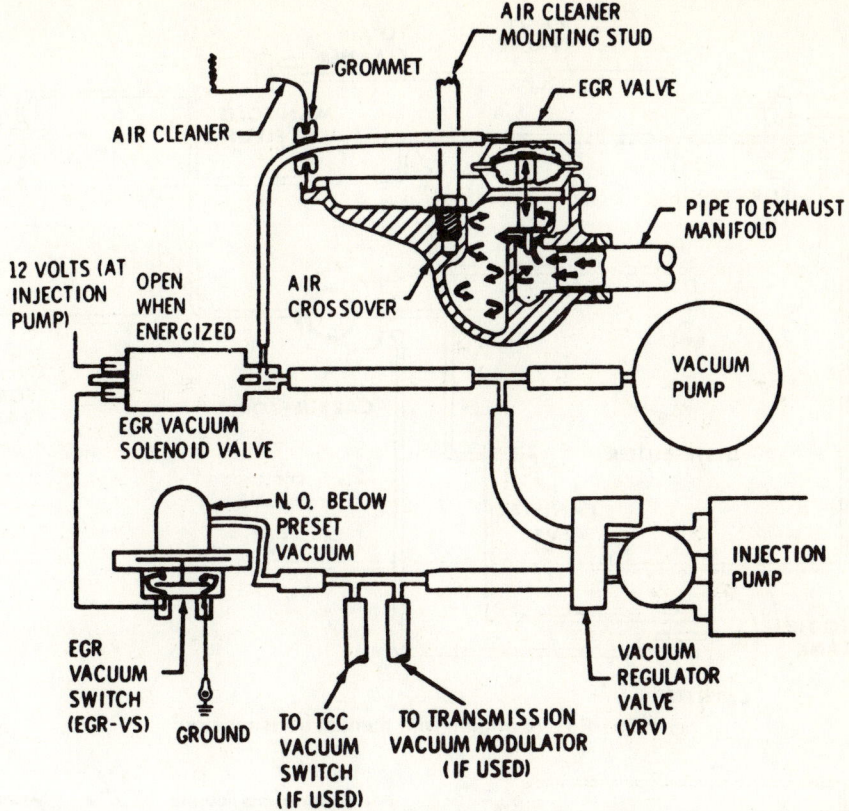

California EGR system, V8 diesel engines

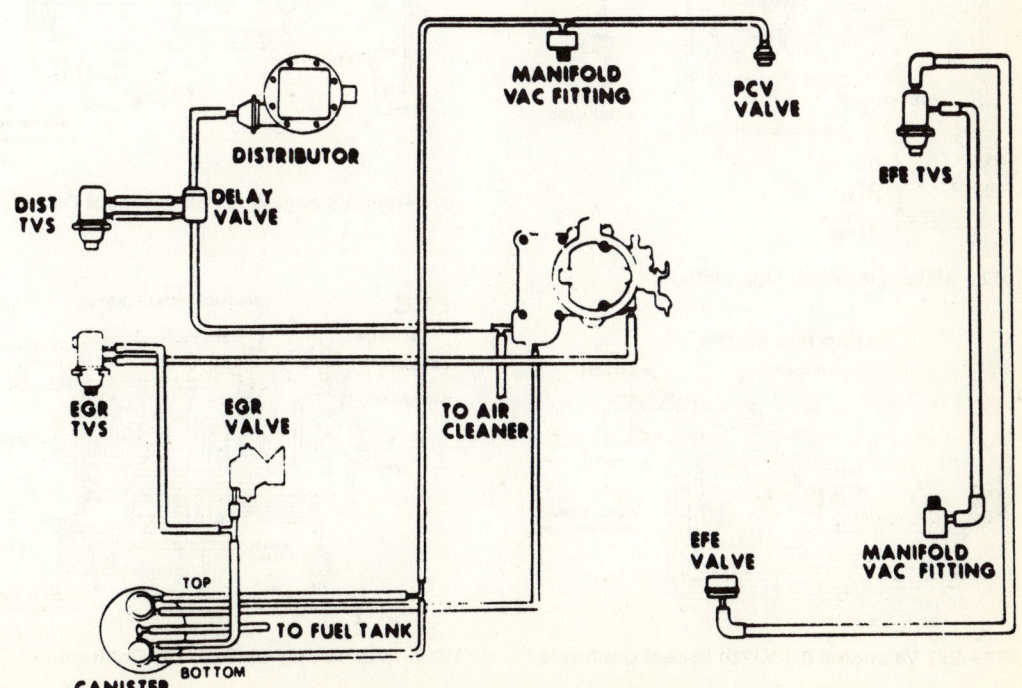

1978—250 L6 engine except California

234 EMISSION CONTROLS

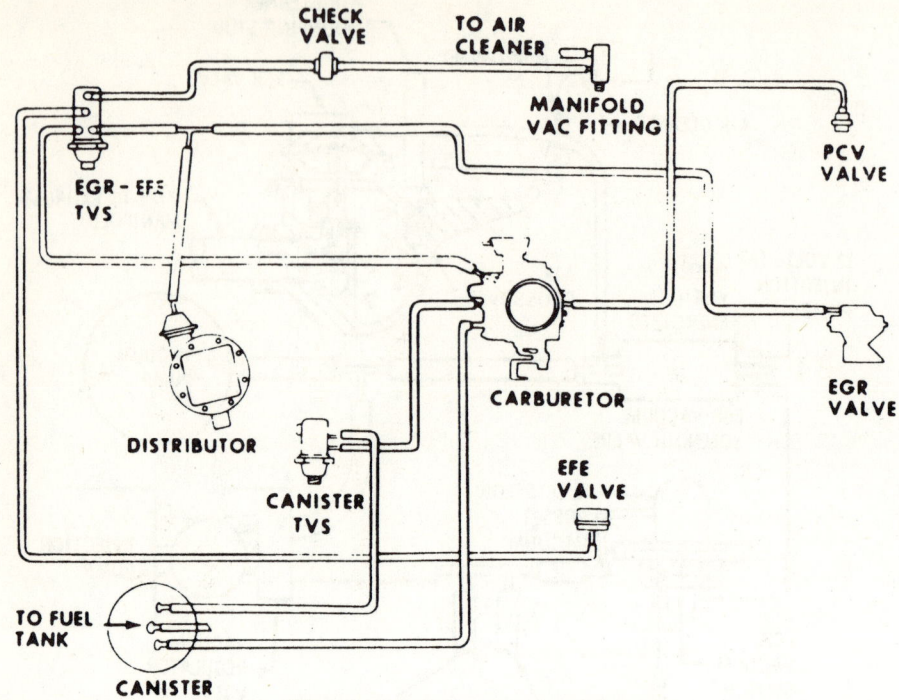
1978 – 231 V6 engine with manual transmission

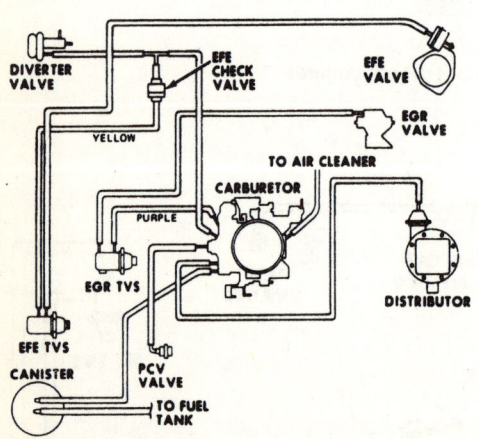
1977 – 350 engine 4bbl. high altitude

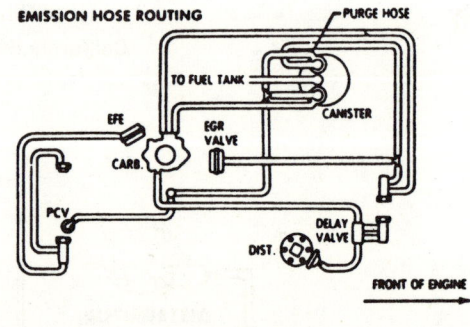
1979 – 305 V8 engine (91G2U) except California

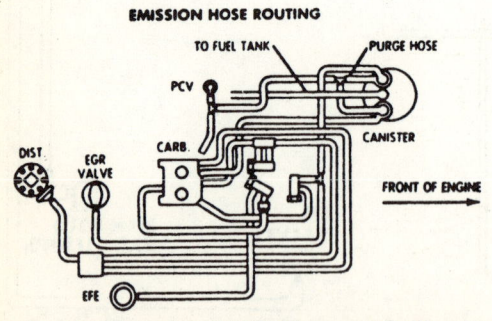
1979 – 267 V8 engine (910G2U) except California

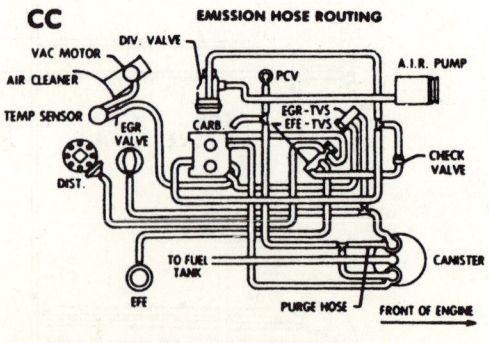
1979 – 305 V8 engine (910Y2V) California

EMISSION CONTROLS

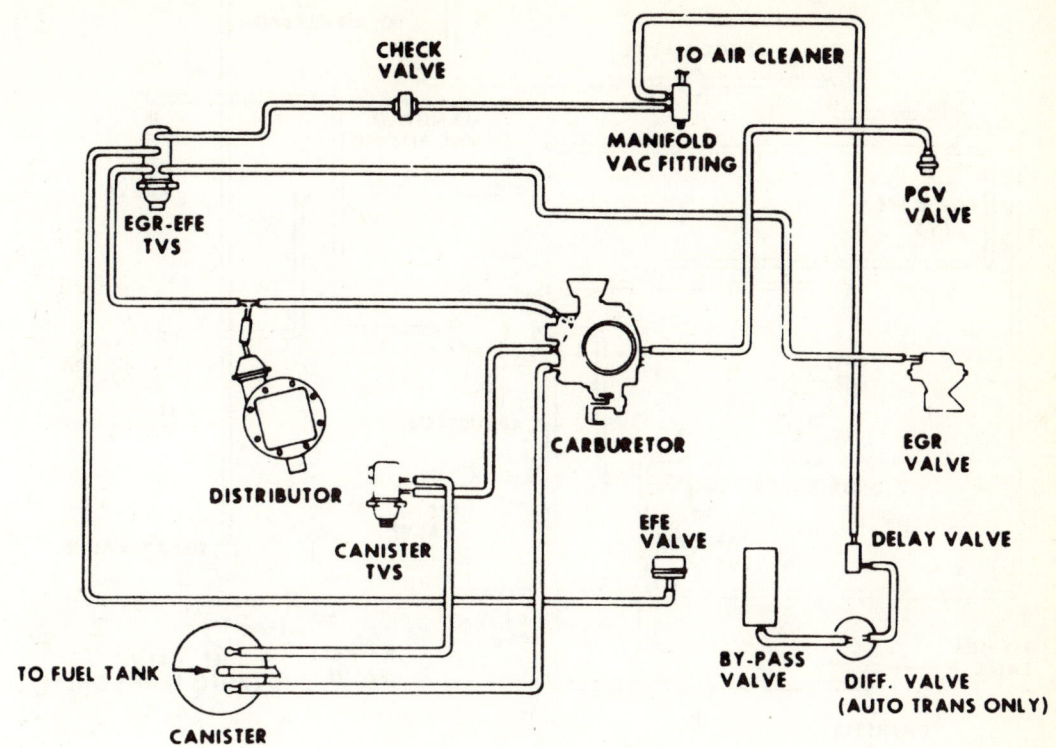

1978—231 V6 engine California

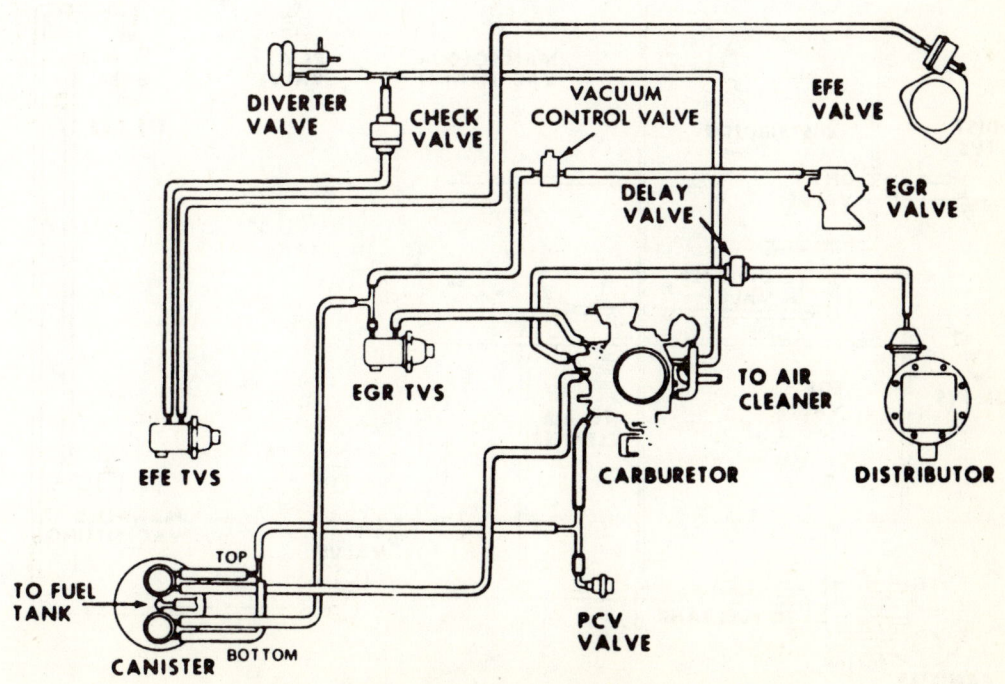

1978—305 V8 engine California

236 EMISSION CONTROLS

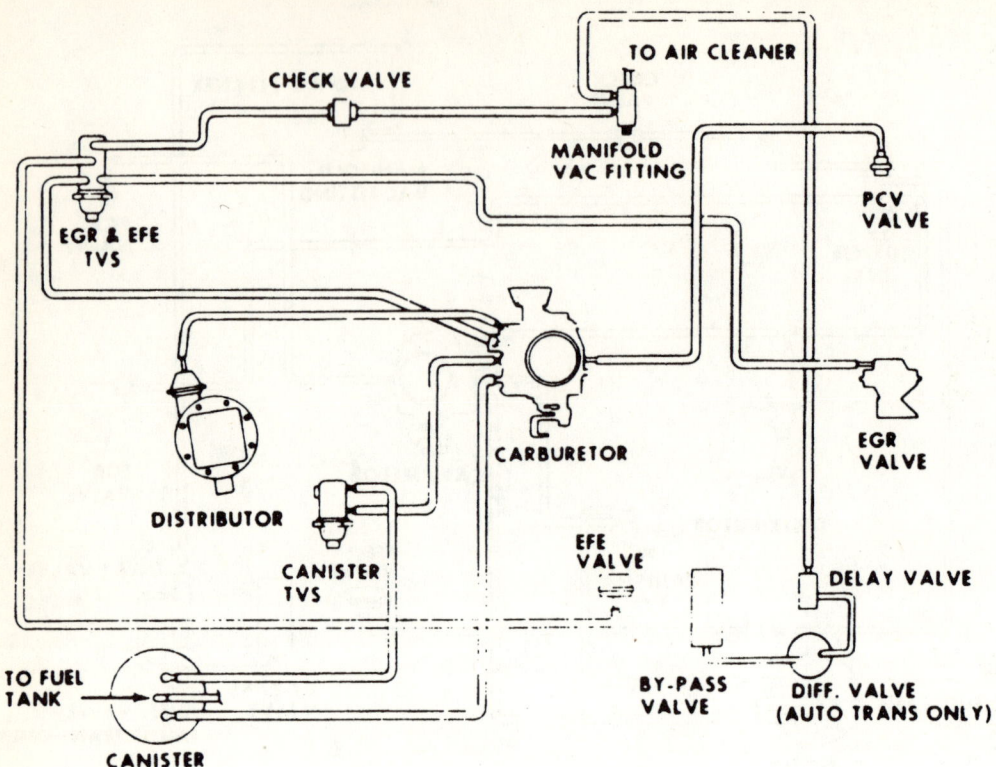

1978—231 V6 engine high altitude

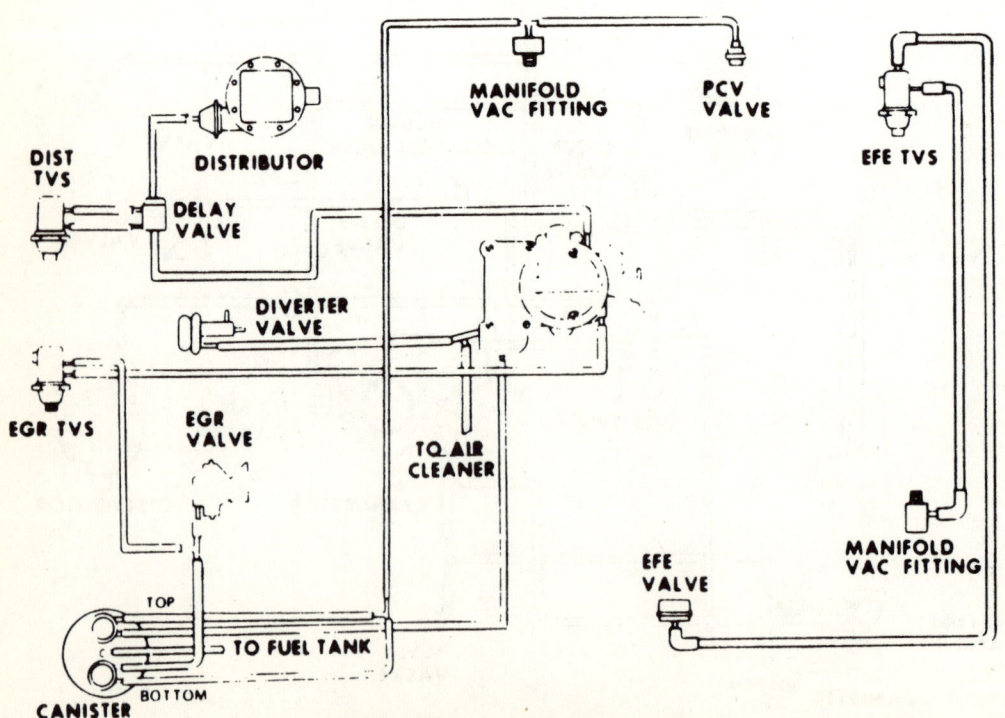

1978—250 L6 engine California

EMISSION CONTROLS

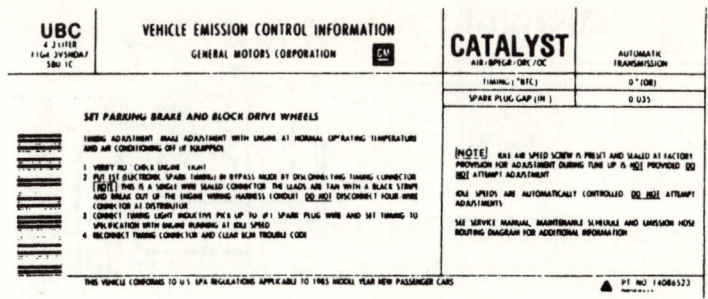 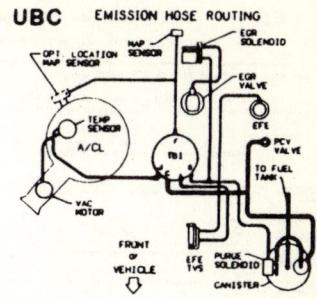

1985 – 262 V6 engine except California

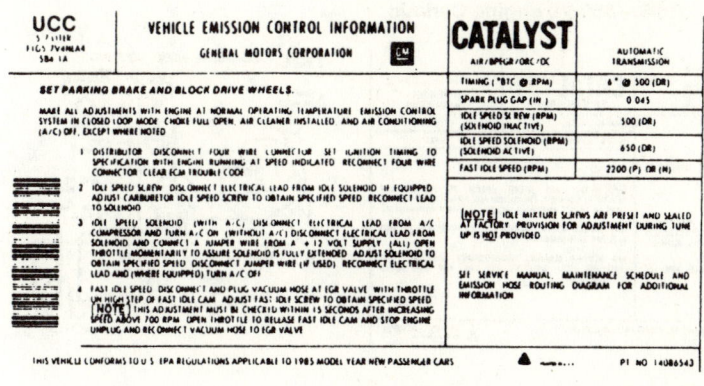 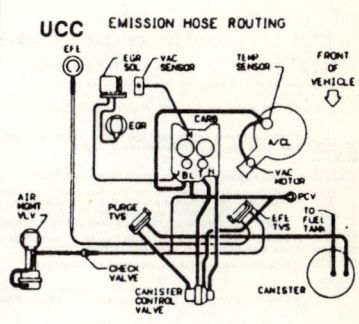

1985 – 350 V8 engine except California

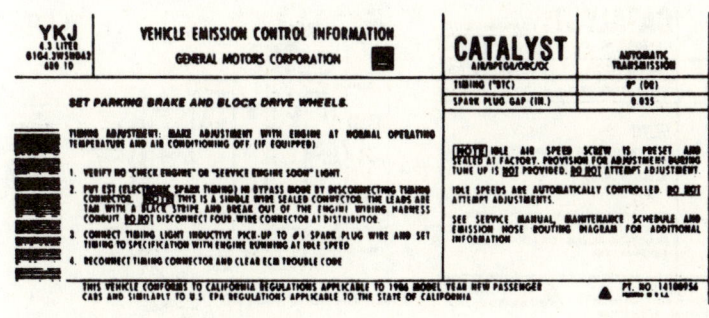 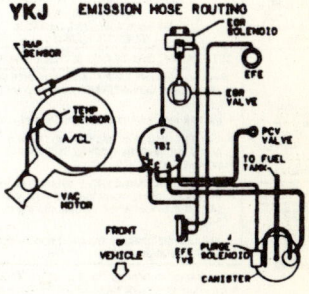

262 V6 engine with automatic trans, California

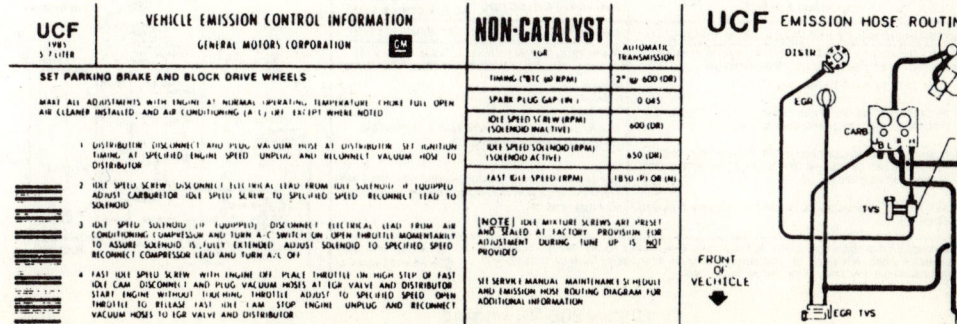

1985 – 350 V8 engine Export

238 EMISSION CONTROLS

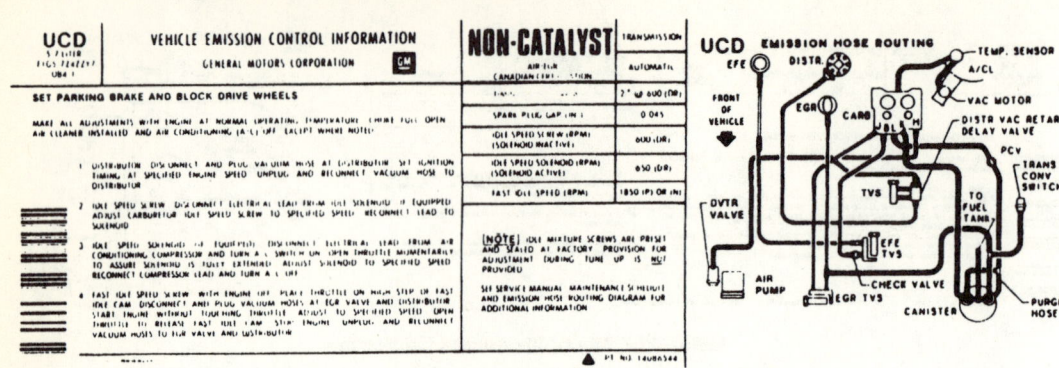

1985 – 350 V8 engine Canada

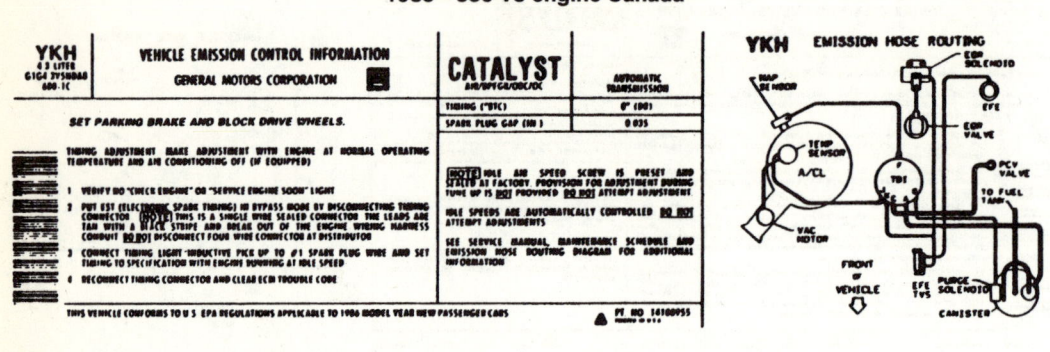

262 V6 engine with automatic trans, except California

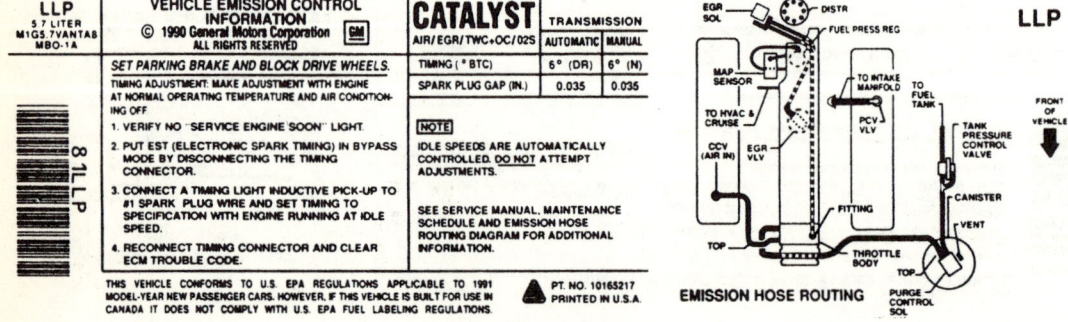

1991 – 350 V8 engine

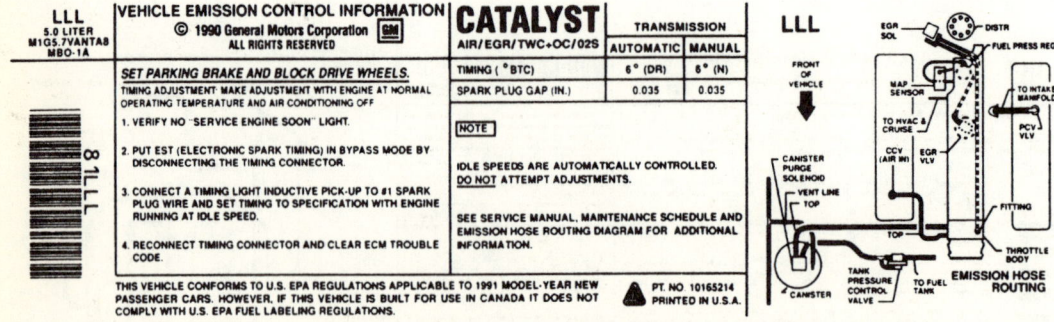

1991 – 305 V8 engine

EMISSION CONTROLS 239

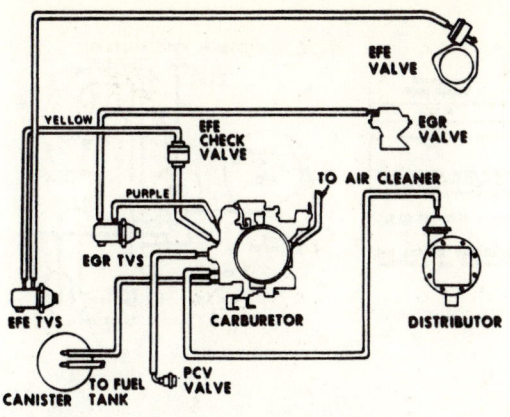
1978—350 V8 engine except California

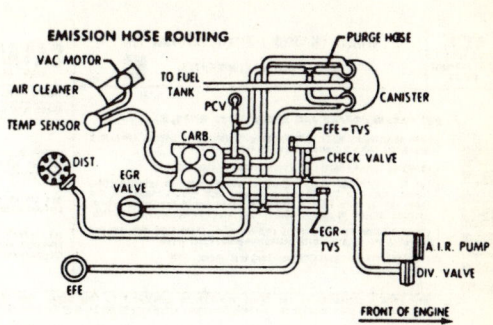
1979—350 V8 engine (910L4RU) California

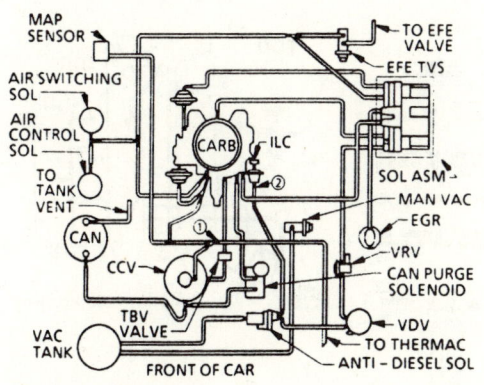
1991—350 V8 engine

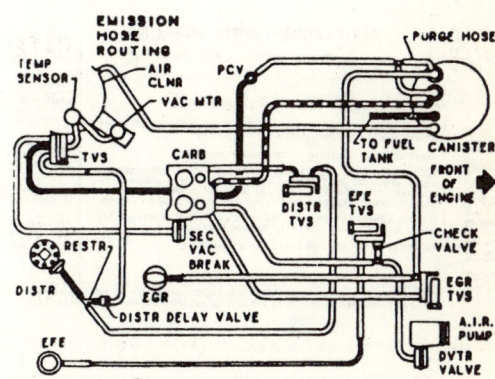
1980—305 V8 engine California

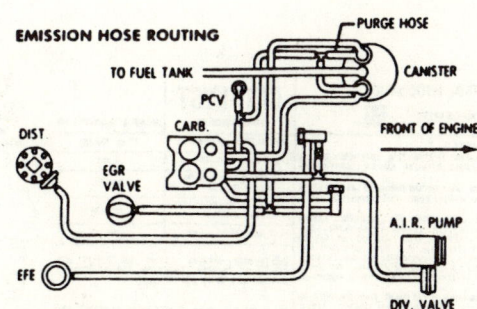
1979—350 V8 engine (910L4RU) high altitude

240 EMISSION CONTROLS

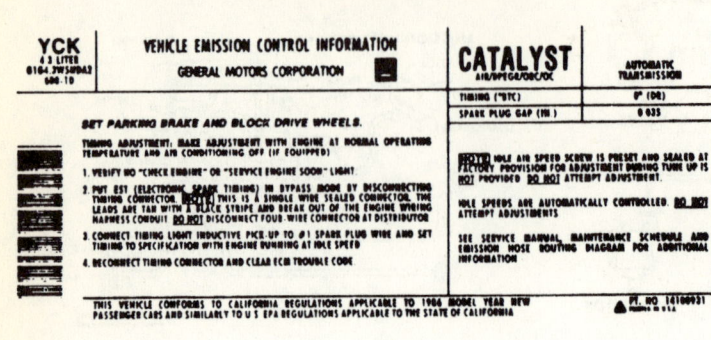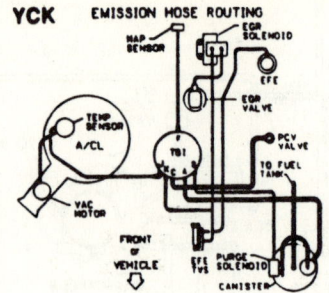

USAGE: 1-2BA00 & LB4 & NB2

262 V6 engine with automatic trans, except California

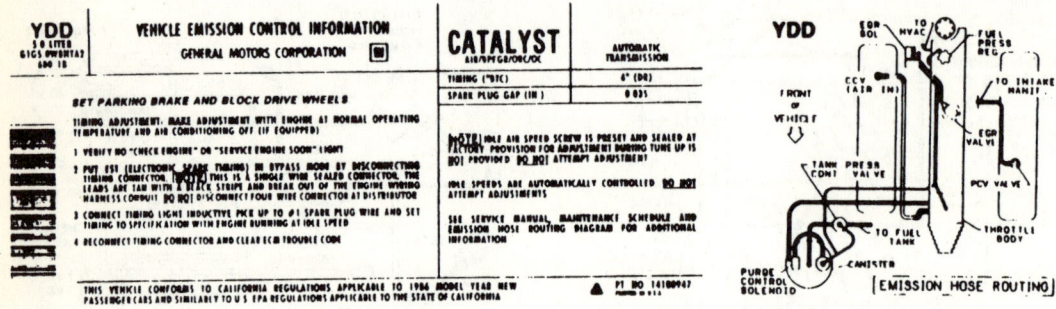

USAGE: 2FA00 & LB9 & NB2 (Pontiac only)

305 V8 engine except California

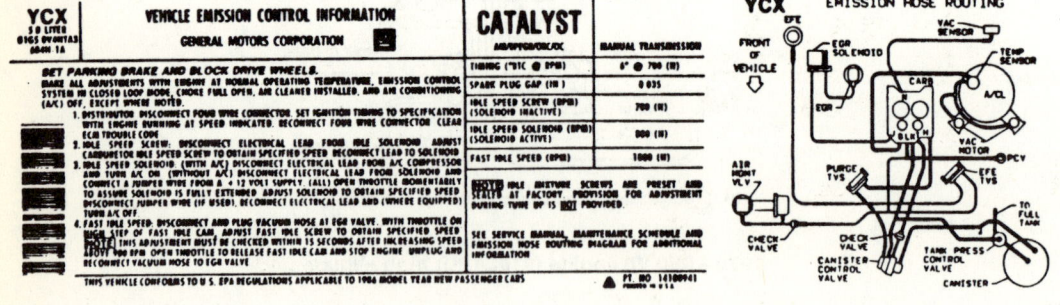

USAGE: 1-2FA00 & L69 & M39/MC4 & NA5

262 V6 engine with automatic trans, California

EMISSION CONTROLS 241

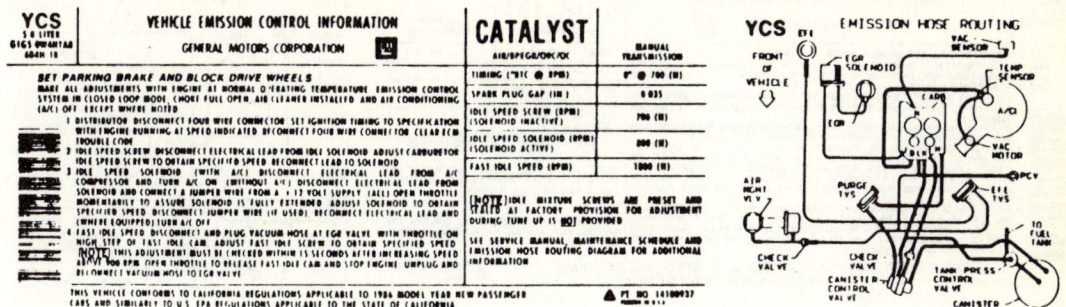

305 V8 engine with automatic trans California

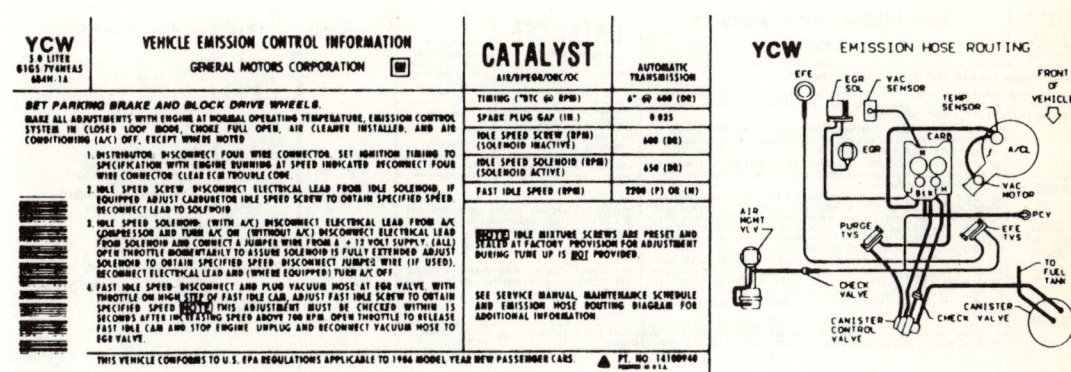

305 V8 engine California

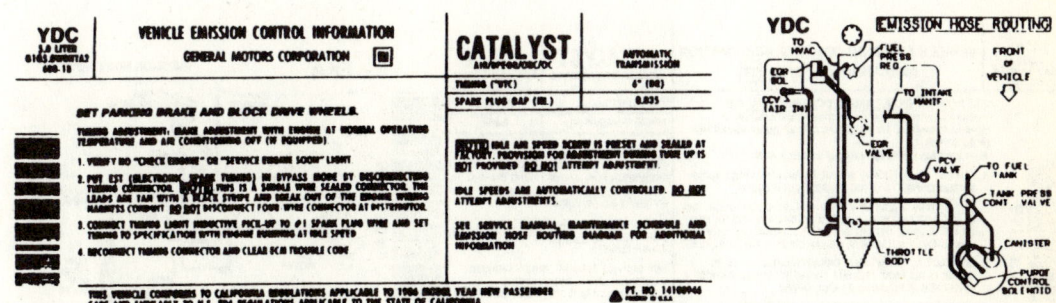

305 V8 engine with automatic trans except California

EMISSION CONTROLS

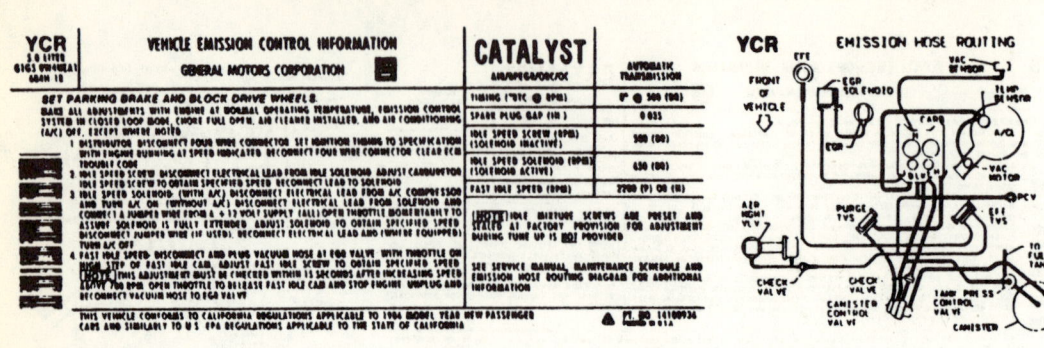

305 V8 engine California

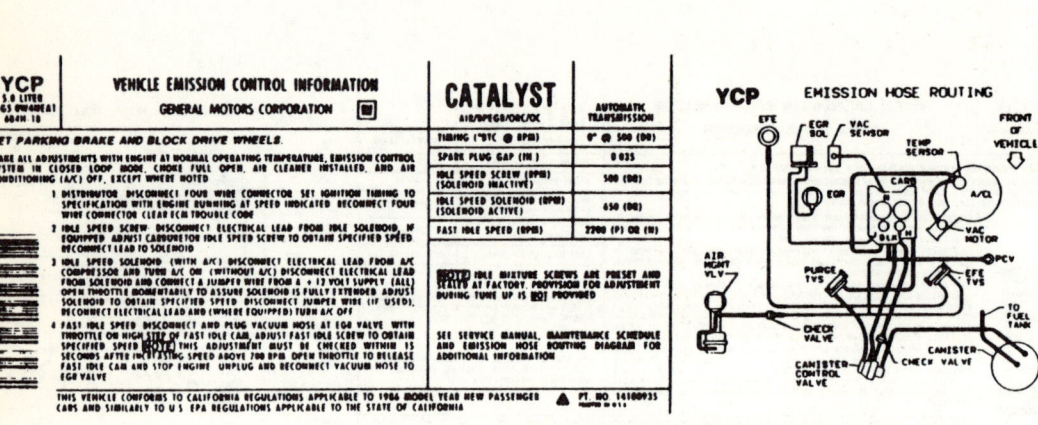

305 V8 engine California

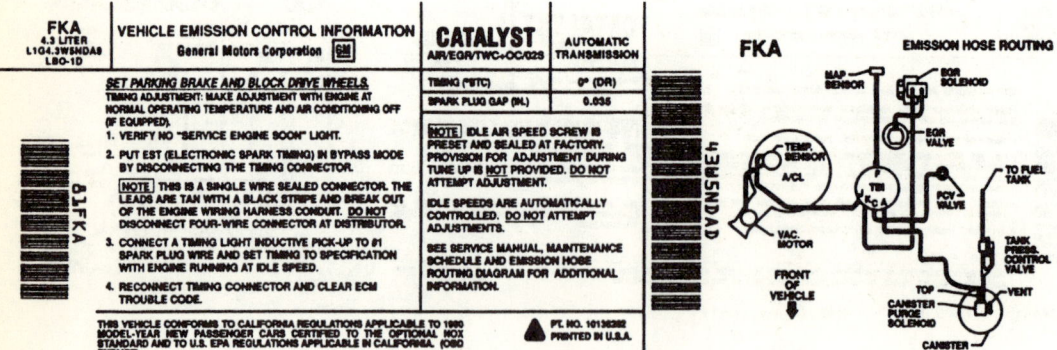

1990 – 262 V6 engine

EMISSION CONTROLS 243

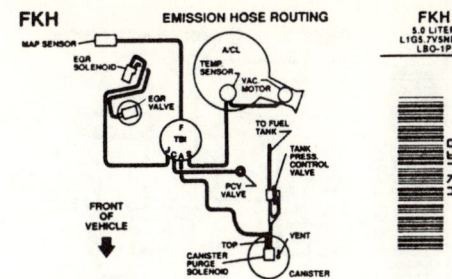

1990 — 305 V8 engine

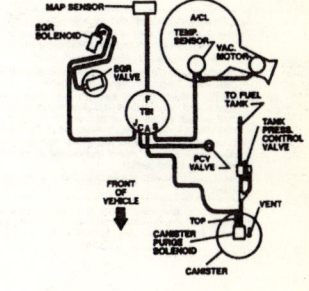

1990 — 305 V8 engine

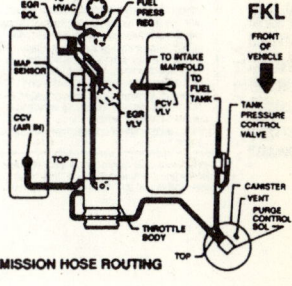

1990 — 305 V8 engine

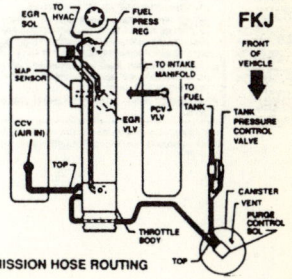

1990 — 305 V8 engine

EMISSION CONTROLS

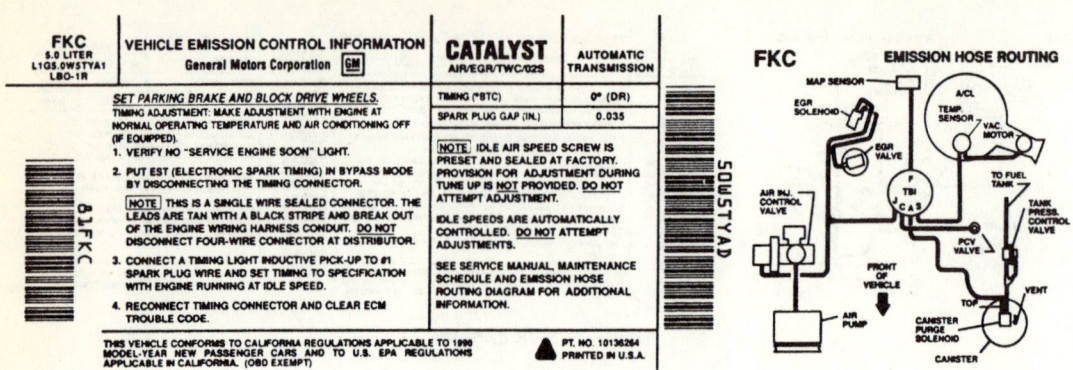

1990 – 305 V8 engine

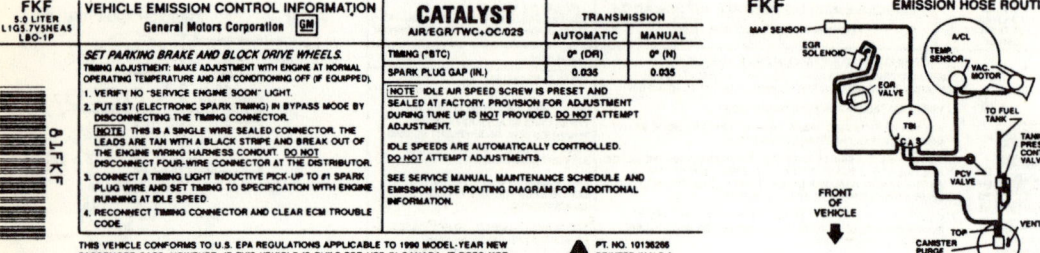

1990 – 305 V8 engine

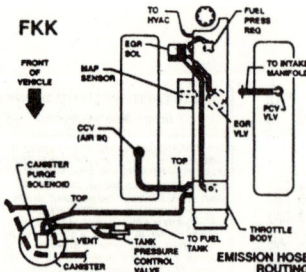

1990 – 305 V8 engine

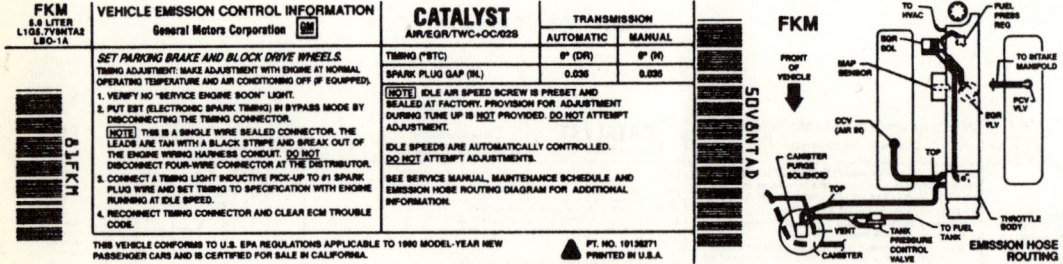

1990 – 305 V8 engine

EMISSION CONTROLS

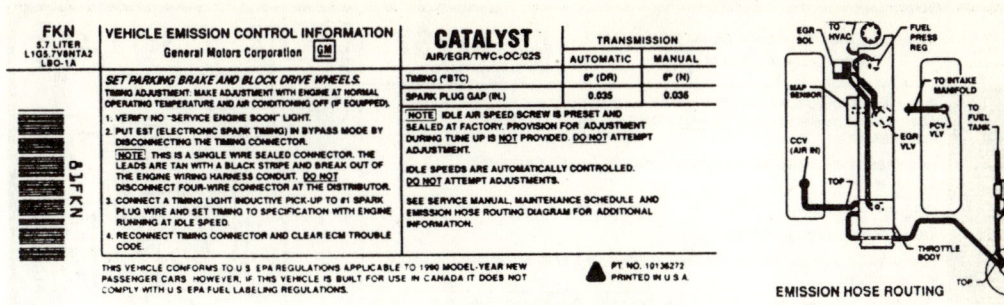

1990 – 350 V8 engine

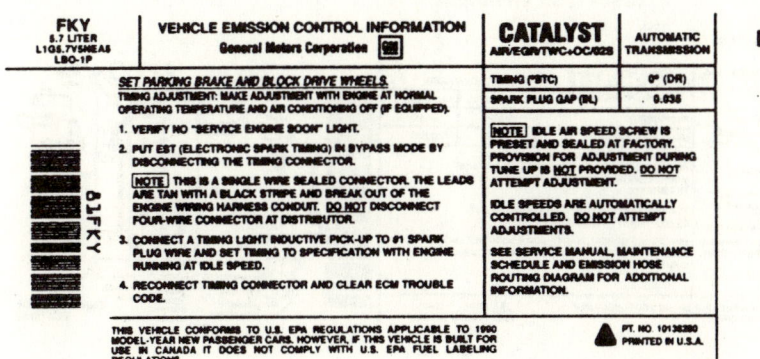 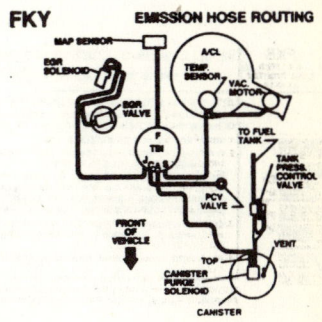

1990 – 350 V8 engine

1990 – 350 V8 engine

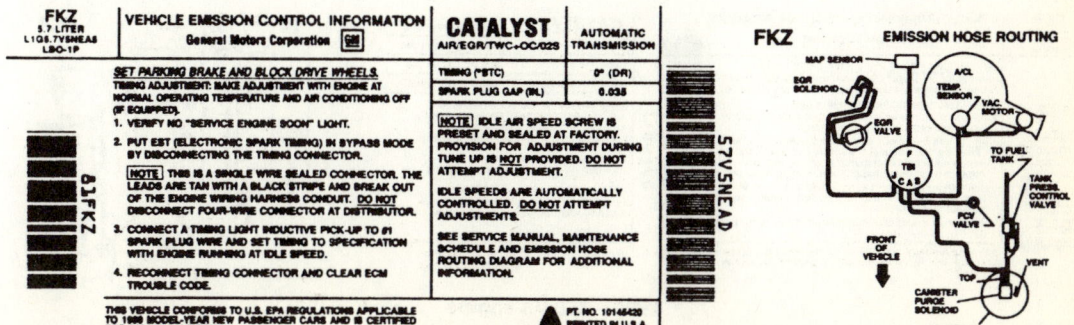

1990 – 350 V8 engine

EMISSION CONTROLS

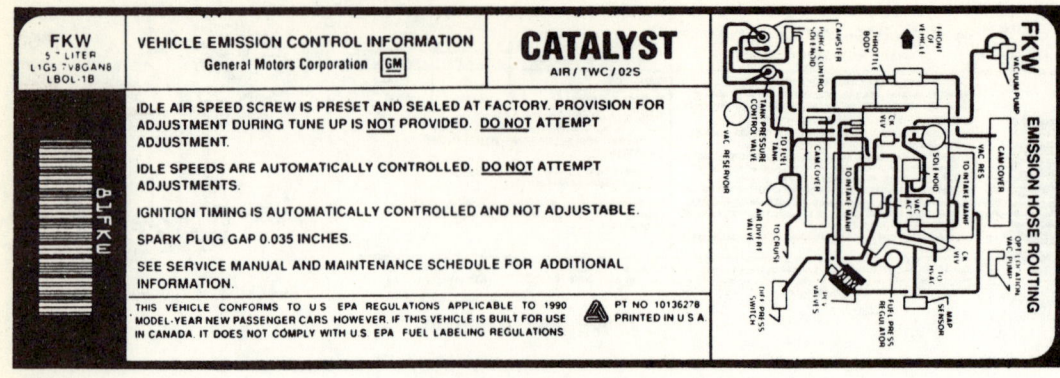

1990 – 350 V8 engine

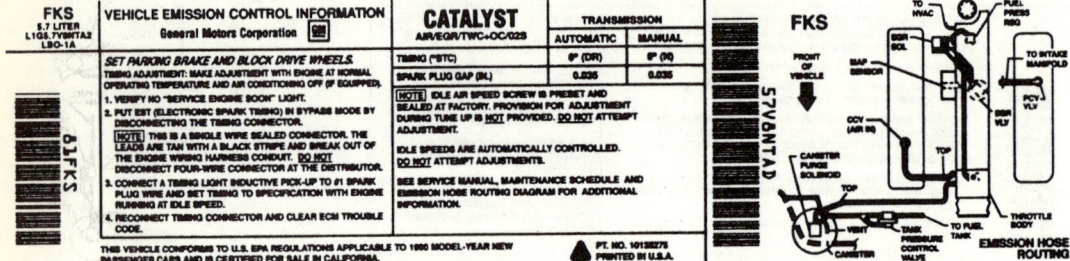

1990 – 350 V8 engine

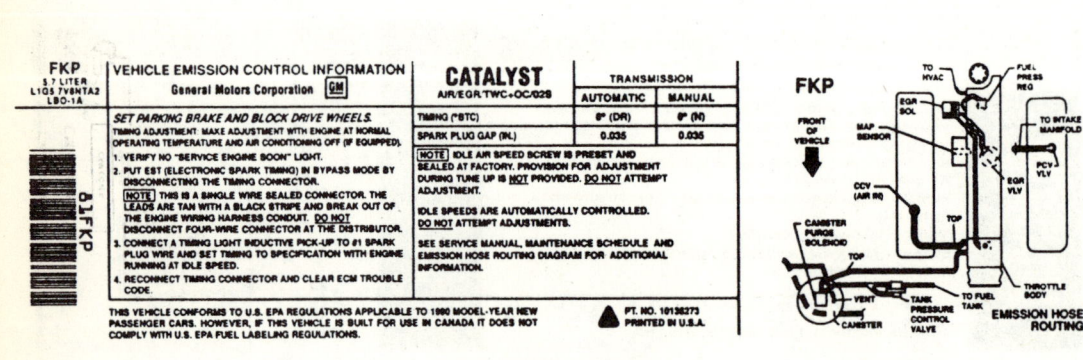

1990 – 350 V8 engine

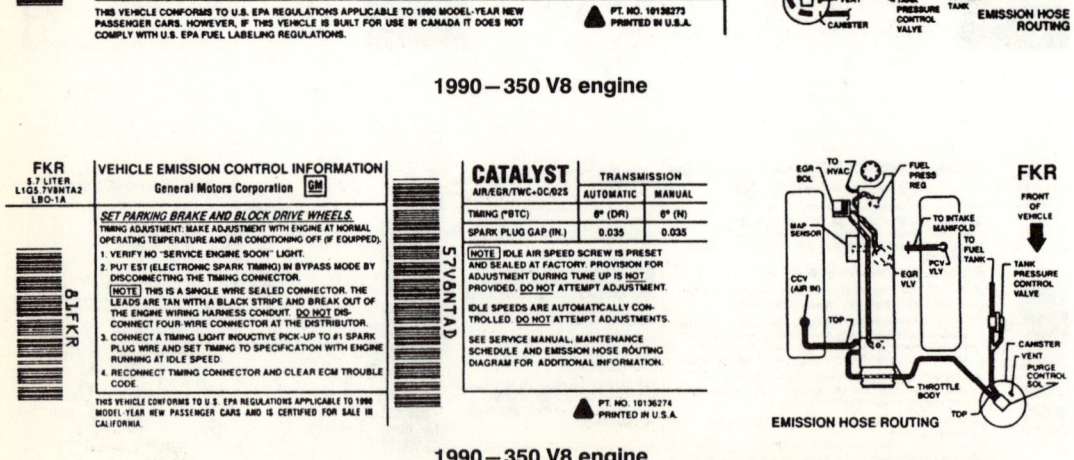

1990 – 350 V8 engine

EMISSION CONTROLS

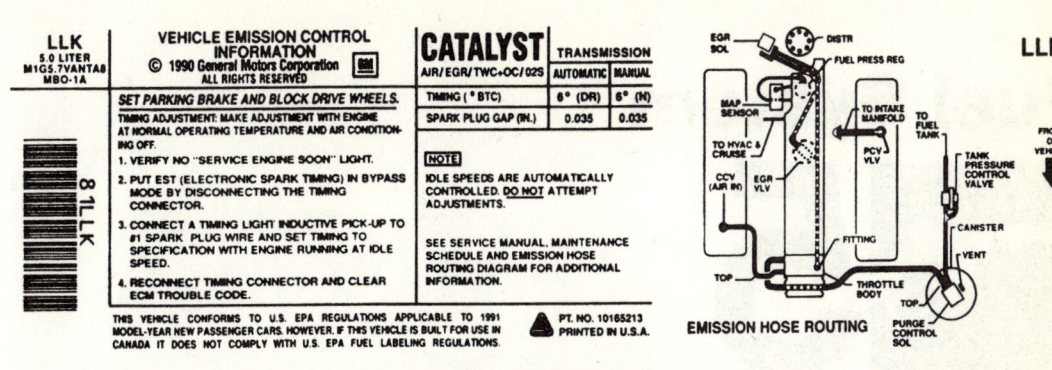

1991 – 305 V8 engine

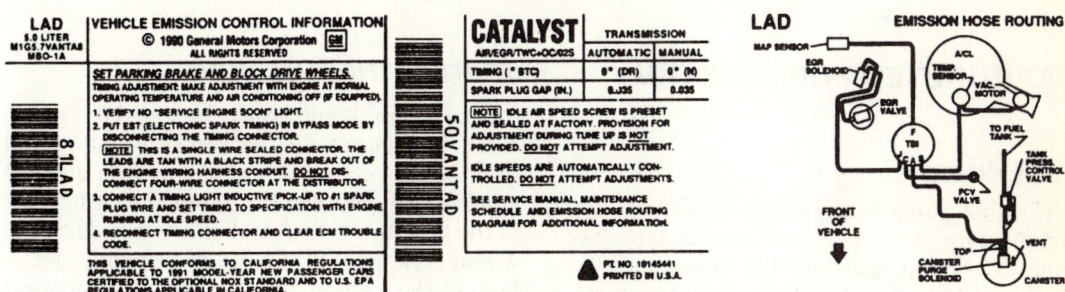

1991 – 305 V8 engine

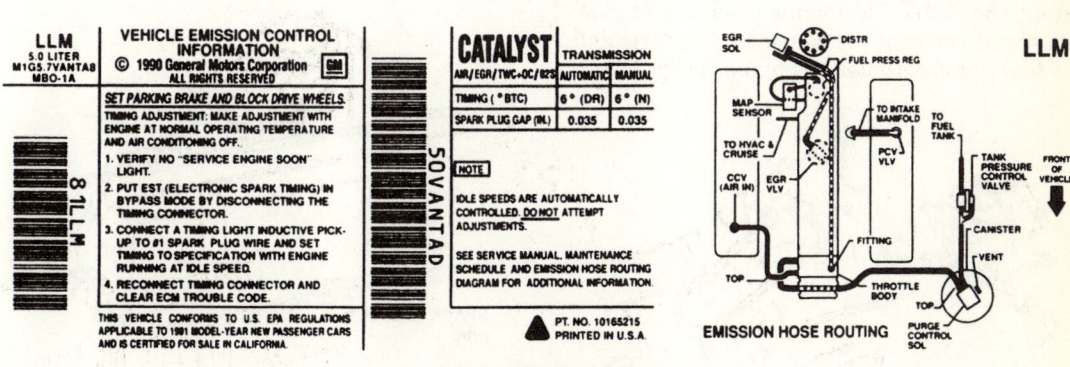

1991 – 305 V8 engine

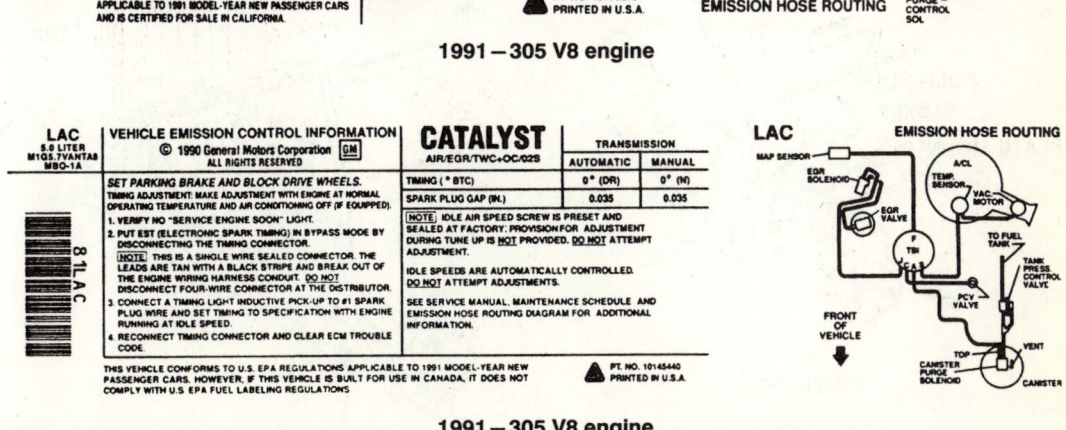

1991 – 305 V8 engine

Fuel System

CARBURETED FUEL SYSTEM

Fuel Pump

The fuel pump is the single action AC diaphragm type. It is a non-serviceable type fuel pump. The pump is actuated by an eccentric located on the engine camshaft. On inline 6-cylinder engines, the 231 V6 engine and the 307 Oldsmobile produced V8 engine the eccentric actuates the pump rocker arm. On V8 engines except the 8-307 Oldsmobile produced engine, a pushrod between the camshaft eccentric and the fuel pump actuates the pump rocker arm.

TESTING THE FUEL PUMP

Fuel pump should always be tested on the vehicle. The larger line between the pump and tank is the suction side of the system and the smaller line, between the pump and carburetor is the pressure side. A leak in the pressure side would be apparent because of dripping fuel. A leak in the suction side is often only apparent because of the reduced volume of fuel delivered to the pressure side. However, fuel may leak out on the suction side when the engine is off.

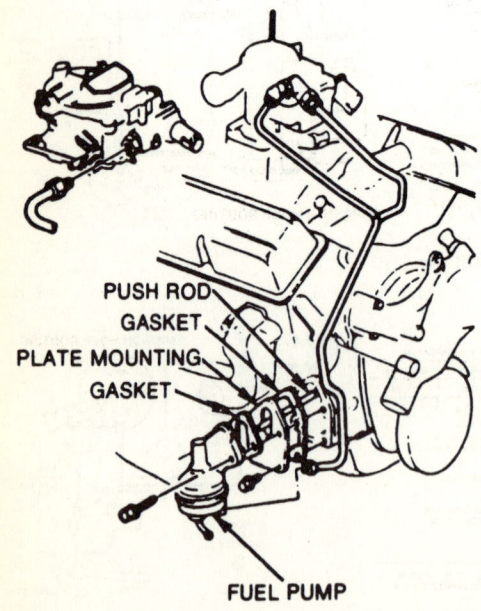

Fuel pump mounting V8 engine except 307 Oldsmobile produced engine

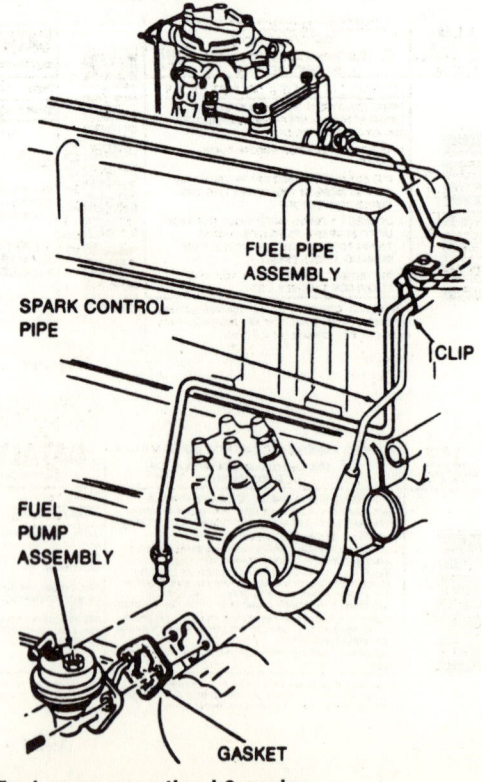

Fuel pump mounting L6 engine

FUEL SYSTEM

1. Tighten any loose line connections and look for any kinks or restrictions. Inspect rubber hoses for cracks or leaks and replace if necessary. Inspect the fuel filter for clogging and clean or replace it as necessary.

2. Disconnect the fuel line at the carburetor. Disconnect the distributor-to-coil primary wire or, on HEI systems, the distributor connector. Place a container at the end of the fuel line and crank the engine a few revolutions. If little or no gasoline flows from the line, either the fuel pump is inoperative or the line is plugged. Blow through the lines with compressed air and try the test again. Reconnect the line.

3. Attach a pressure gauge to the pressure side of the fuel line with a Tee fitting.

4. Run the engine and note the reading on the gauge. Stop the engine and compare the reading with the specifications listed in the Tune-Up Specifications chart. If the pump is operating properly, the pressure will be as specified and will be constant at idle speed. If pressure varies sporadically or is too high or low, the pump should be replaced.

5. Remove the pressure gauge.

REMOVAL AND INSTALLATION

NOTE: *When you connect the fuel pump outlet fitting, always use 2 wrenches to avoid damaging the pump.*

1. Disconnect the negative battery cable. Disconnect the fuel intake and outlet lines at the pump and plug the pump intake line.

2. On V6 and V8 engines, except 8-307 Oldsmobile produced engine, remove the upper bolt from the right front mounting boss. Insert a longer bolt ($3/8$-16 x 2") in this hole to hold the fuel pump pushrod.

3. Remove the two pump mounting bolts and lockwashers; remove the pump and its gasket.

4. If the rocker arm pushrod is to removed, remove the two adapter bolts and lockwashers and remove the adapter and its gasket.

5. Install the fuel pump with a new gasket reversing the removal procedure. Coat the mating surfaces with sealer.

6. Connect the fuel lines and check for leaks.

Carburetors

REMOVAL AND INSTALLATION

1. Disconnect the negative battery cable. Remove the air cleaner and its gasket.

2. Disconnect the fuel and vacuum lines from the carburetor. As required, disconnect the PCV line at the base of the carburetor.

3. Disconnect the choke coil rod, heated air line tube, or electrical connector.

4. If equipped, remove the cruise control. Disconnect the throttle linkage.

5. On automatic transmission equipped vehicles, disconnect the throttle valve linkage as required.

6. If CEC equipped, remove the CEC valve vacuum hose and electrical connector. Disconnect the EGR line, if so equipped.

7. Remove the idle stop solenoid, if equipped.

8. Remove the carburetor attaching nuts and/or bolts, gasket or insulator, and remove the carburetor.

To Install:

9. Position the carburetor on the intake manifold, using a new base gasket.

10. Install the carburetor retaining bolts. Connect the fuel line.

11. Connect the choke assembly. Connect the required linkage.

12. Connect the idle stop solenoid, the cruise control linkage and the CEC vacuum valve, as required.

13. Install the carburetor air cleaner assembly. Connect the negative battery cable.

14. Start the engine. Adjust the carburetor air/fuel mixture, if possible. Adjust the idle speed, as required. Roadtest the vehicle.

IDENTIFICATION

Carburetor identification numbers will generally be found in the following locations:

- MV, ME: Stamped on the vertical portion of the float bowl, adjacent to the fuel inlet nut.
- GV, GC: Stamped on the flat section of the float bowl next to the fuel inlet nut.
- E2SE, 2SE: Stamped on the vertical surface of the float bowl adjacent to the vacuum tube.
- M2MC: Stamped on the vertical surface of the left rear corner of the float bowl.
- E4ME, 4MV, M4MC, M4ME: Stamped on the vertical section of the float bowl, near the secondary throttle lever.

OVERHAUL

All Types

Efficient carburetion depends greatly on careful cleaning and inspection during overhaul, since dirt, gum, water, or varnish in or on the carburetor parts are often responsible for poor performance.

Overhaul your carburetor in a clean, dust free area. Carefully disassemble the carburetor, referring often to the exploded views and directions packaged with the rebuilding kit. Keep similar and look-alike parts segregated during

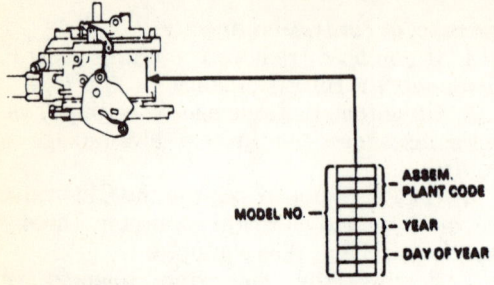

Carburetor I.D. location all models

disassembly and cleaning to avoid accidental interchange during assembly. Make a note of all jet sizes.

When the carburetor is disassembled, wash all parts (except diaphragms, electric choke units, pump plunger, and any other plastic, leather, fiber, or rubber parts) in clean carburetor solvent. do not leave parts in the solvent any longer than is necessary to sufficiently loosen the deposits. Excessive cleaning may remove the special finish from the float bowl and choke valve bodies, leaving these parts unfit for service. Rinse all parts in clean solvent and blow them dry with compressed air or allow them to air dry. Wipe clean all cork, plastic, leather, and fiber parts with clean, lint-free cloth.

Blow out all passages and jets with compressed air and be sure that there is no restrictions or blockages. Never use wire or similar tools to clean jets, fuel passages, or air bleeds. Clean all jets and valves separately to avoid accidental interchange.

Check all parts for wear or damage. If wear

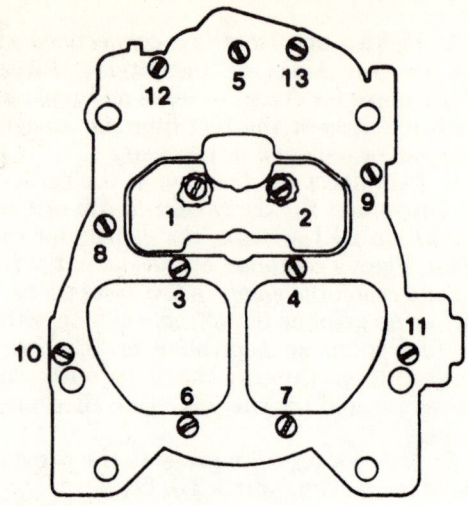

E4MC air horn tightening sequence

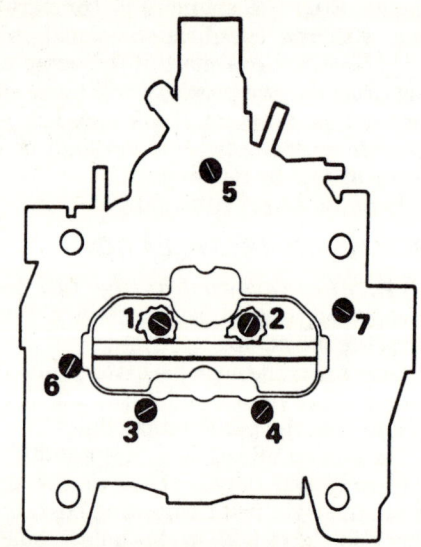

M2MC air horn tightening sequence

or damage is found, replaced the defective parts. Especially check the following:

1. Check the float needle and seat for wear. If wear is found, replace the complete assembly.

2. Check the float hinge pin for wear and the float(s) for dents or distortion. Replace the float if fuel has leaked into it.

3. Check the throttle and choke shaft bores for wear or an out-of-round condition. Damage or wear to the throttle arm, shaft, or shaft bore will often require replacement of the throttle body. These parts require a close tolerance of fit; wear may allow air leakage, which could affect starting and idling.

NOTE: *Throttle shafts and bushings are not included in overhaul kits. They can be purchased separately.*

4. Inspect the idle mixture adjusting needles for burrs or grooves. Any such condition

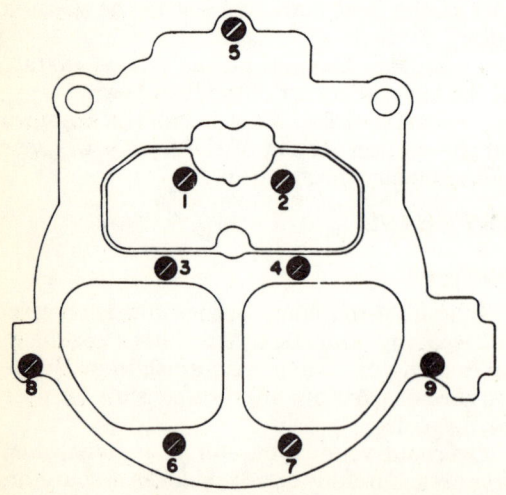

M4MC air horn tightening sequence

FUEL SYSTEM

requires replacement of the needle, since you will not be able to obtain a satisfactory idle.

5. Test the accelerator pump check valves. They should pass air one way but not the other. Test for proper seating by blowing and sucking on the valve. Replace the valve check ball and spring as necessary. If the valve is satisfactory wash the valve parts again to remove breath moisture.

6. Check the bowl cover for warped surfaces with a straightedge.

7. Closely inspect the accelerator pump plunger for wear and damage, replacing as necessary.

8. After the carburetor is assembled, check the choke valve for freedom of operation.

Carburetor overhaul kits are recommended for each overhaul. These kits contain all gaskets and new parts to replace those which deteriorate most rapidly. Failure to replace all parts supplied with the kit (especially gaskets) can result in poor performance later.

Some carburetor manufacturers supply overhaul kits of three basic types: minor repair; major repair; and gasket kits. Basically, they contain the following:

Minor Repair Kits:
- All gaskets
- Float needle valve
- All diagrams
- Spring for the pump diaphragm

Major Repair Kits:
- All jets and gaskets
- All diaphragms
- Float needle valve
- Pump ball valve
- Float
- Complete intermediate rod
- Intermediate pump lever
- Some cover holddown screws and washers

Gasket Kits:
- All gaskets

After cleaning and checking all components, reassemble the carburetor, using new parts and referring to the exploded view: When reassembling, make sure that all screws and jets are tight in their seats, but do not overtighten as the tips will be distorted. Tighten all screws gradually, in rotation. Do not tighten needle valves into their seats; uneven jetting will result. Always use new gaskets. Be sure to adjust the float level when reassembling.

PRELIMINARY CHECKS (ALL CARBURETORS)

The following should be observed before attempting any adjustments.

1. Thoroughly warm the engine. If the engine is cold, be sure that it reaches operating temperature.

2. Check the torque of all carburetor mounting nuts and assembly screws. Also check the intake manifold-to-cylinder head bolts. If air is leaking at any of these points, any attempts at adjustment will inevitably lead to frustration.

3. Check the manifold heat control valve (if used) to be sure that it is free.

4. Check and adjust the choke as necessary.

5. Adjust the idle speed and mixture. If the mixture screws are capped, don't adjust them unless all other causes of rough idle have been eliminated. If any adjustments are performed that might possibly change the idle speed or mixture, adjust the idle and mixture again when you are finished.

Before you make any carburetor adjustments make sure that the engine is in tune. Many problems which are thought to be carburetor related can be traced to an engine which is simply out-of-tune. Any trouble in these areas will have symptoms like those of carburetor problems.

Carter YF 1-BBL Carburetor

AUTOMATIC CHOKE ADJUSTMENT

1. Disconnect the choke rod from the choke lever.

2. Hold the choke valve closed and pull the rod up against the stop in the thermostat housing.

3. The top of the rod should be about one rod diameter above the top of the hole in the choke lever. If not, adjust the length of the rod by bending it at the bend.

4. Connect the choke rod at the lever.

HAND CHOKE ADJUSTMENT

1. Push in the hand choke knob until the knob is within 1/8" of the dash.

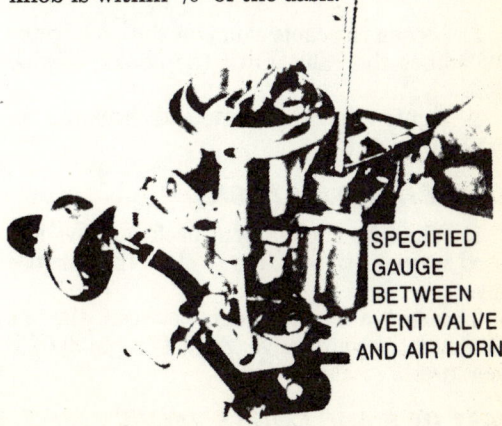

Adjusting the idle vent (Carter YF)

FUEL SYSTEM

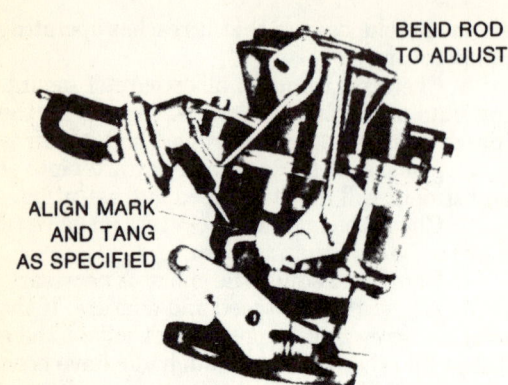

Choke rod adjustment (Carter YF)

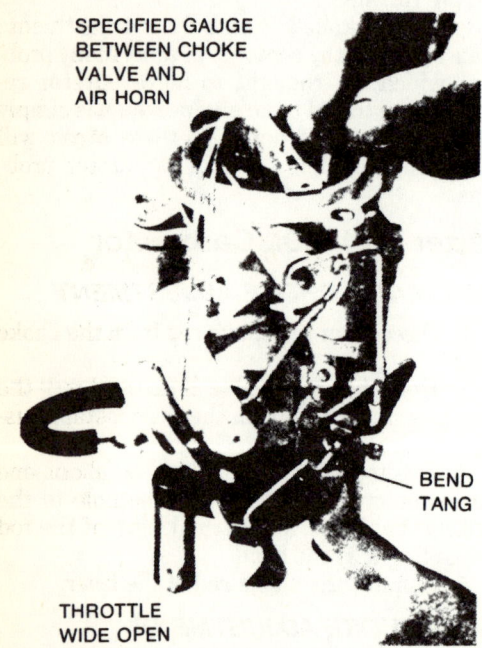

Choke unloader adjustment (Carter YF)

2. Loosen the cable clamp at the carburetor and adjust the cable until the choke is wide open.
3. Tighten the cable clamp and check the operation of the choke.

IDLE VENT ADJUSTMENT

1. With the choke open, back out the idle speed screw until it is free to close the throttle valve.
2. Insert a feeler gauge between the air horn and the vent valve. Adjust to get 0.065" clearance.

FAST IDLE AND CHOKE VALVE ADJUSTMENT

1. Hold the choke valve closed.
2. Close the throttle and mark the position of the throttle lever tang on the fast idle cam.
3. The mark on the fast idle cam should align with the upper edge of the tang on the throttle lever. If not, bend the choke rod as necessary.

CHOKE UNLOADER ADJUSTMENT

1. Open the throttle to the wide open position.
2. Using a rubber band, hold the choke valve closed.
3. Bend the unloader tang on the throttle lever to get the proper clearance (0.250") between the lower edge of the choke valve and the air horn wall.

VACUUM BREAK ADJUSTMENT

1. Hold the vacuum break arm against its stop and hold the choke closed with a rubber band.
2. Bend the vacuum break link to establish a clearance of 0.220" (automatic transmission) or 0.240" (manual transmission) between the lower edge of the choke valve and the air horn wall.

FLOAT LEVEL ADJUSTMENT

1. Turn the bowl cover upside down and measure the float level by measuring the distance between the float (free end) and cover. This distance should be $7/32"$; if not, bend the lip of the float, not the float arm.
2. Hold the cover in the proper position (not upside down) and measure the float drop from the cover to the float bottom at the end opposite the hinge. This distance should be $1 3/16"$.
3. Make any necessary adjustments with the stop tab on the float arm.

ACCELERATOR PUMP ADJUSTMENT

1. Seat the throttle valve by backing off the idle speed screw.
2. Hold the throttle valve closed.
3. Fully depress the diaphragm shaft and check the contact between the lower retainer and the lifter link.
4. This retainer (upper pump spring) should just contact the pump lifter link. Do not compress the spring.
5. Bend the pump connector link at its U-bend to make corrections.

METERING ROD ADJUSTMENT

1. Insert the metering rod through the metering jet and close the throttle valve. Press down on the upper pump spring until the pump bottoms.
2. The metering rod arm must rest on the

FUEL SYSTEM 253

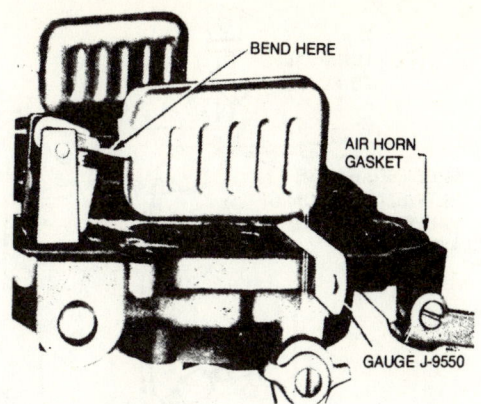

AFB float adjustment

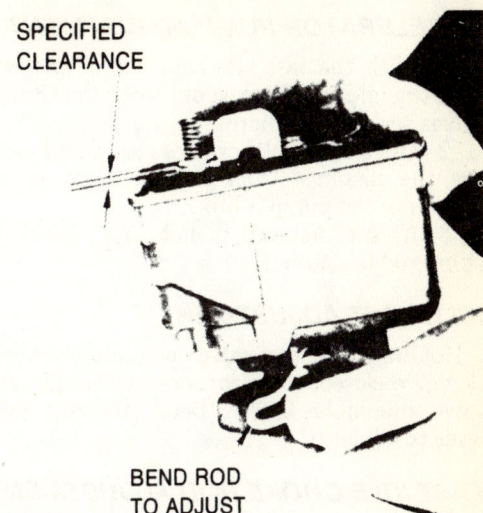

Accelerator pump adjustment (Carter YF)

pump lifter link; the rod eye should just slide over the arm pin.

3. To adjust, bend the metering rod arm.

Carter AFB 4-BBL Carburetor

AUTOMATIC CHOKE ADJUSTMENT

The automatic choke is correctly adjusted when the scribe mark on the coil housing is aligned with the center notch in the choke housing for automatic transmission vehicles and one notch lean for manual transmission vehicles.

FLOAT ADJUSTMENT

Remove the metering rods and the bowl cover. Align the float by sighting down its side to determine if it is parallel with the outer edge of the air horn. Bend the float to adjust. Float level is adjusted with the air horn inverted and the air horn gasket in place. Clearance between each float (at the outer end) and the air horn gasket should be $5/16''$. Bend to adjust.

FLOAT DROP ADJUSTMENT

Float drop is adjusted by holding the air horn in an upright position and bending the float arm until the vertical distance from the air horn gasket to the outer end of each float measures $3/4''$.

INTERMEDIATE CHOKE ROD ADJUSTMENT

Remove the choke coil housing assembly, gasket, and baffle plate. Position a 0.026" wire gauge between the bottom of the slot in the piston and the top of the float in the choke piston housing. Close the choke piston against the gauge and secure it with a rubber band. Bend the intermediate choke rod so that the distance between the top edge of the choke valve and the air horn divider measures 0.070".

ACCELERATOR PUMP ADJUSTMENT

The first step in adjusting the accelerator pump is to push aside the fast idle cam and firmly seat the throttle valves. Bend the pump rod at the lower angle to obtain a $1/2''$ clearance between the air horn and the top of the plunger shaft.

UNLOADED, CLOSING SHOE, AND SECONDARY THROTTLE ADJUSTMENT

To adjust the unloader, hold the throttle wide open and bend the unloaded tang to obtain a ú/$_{16}$" clearance between the upper edge of the choke valve and the inner wall of the air horn.

The clearance between the positive closing shoes on the primary and secondary throttle valves is checked with the valves closed. Bend the secondary closing shoe as required to obtain a clearance of 0.020".

The secondary throttle opening is governed by the pick-up lever on the primary throttle shaft. It has two points of contact with the loose lever on the primary shaft. If the contact points do not simultaneously engage, bend the pick-up lever to obtain proper engagement. The primary and secondary throttle valve opening must be synchronized.

Carter AVS 4-BBL Carburetor

AIR VALVE ADJUSTMENT

1. Turn the air valve bearing retainer until the air valve freely falls open.

2. Wind the air valve bearing counterclockwise until the air valve just starts to close.

3. Continue to wind the bearing an additional $2 1/8$ turns, and then tighten the retainer.

FUEL SYSTEM

ACCELERATOR PUMP ADJUSTMENT

1. With the fast idle cam out of the way, back the idle speed screw out until the throttle valves seat in their bores.
2. Hold the throttle valves closed and measure the distance from the air horn to the bottom of the pump S-link.
3. If this distance is not $^{11}/_{32}"$, bend the pump rod to obtain it.

IDLE VENT ADJUSTMENT

Holding the choke valve open and the throttle valve closed, the clearance at the idle vent valve should be 0.030". Bend the vent valve lever to adjust.

FAST IDLE CHOKE ROD ADJUSTMENT

Bend the fast idle rod at a lower angle until the fast idle cam index mark lines up with the fast idle adjustment screw. Perform this adjustment while holding the choke valve closed.

FLOAT LEVEL AND DROP ADJUSTMENT

These adjustments are made in the same manner as on the Carter AFB.

CHOKE UNLOADER ADJUSTMENT

1. Hold the throttle wide open and the choke valve toward the closed position with a rubber band.
2. Bend the unloader tang on the throttle shaft lever to obtain a clearance of 0.170". Between the upper edge of the choke valve and the dividing wall of the air horn.

CHOKE VACUUM BREAK ADJUSTMENT

1. Hold the vacuum break in against its stop and the choke valve toward the closed position with a rubber band.
2. Bend the vacuum break link at an offset

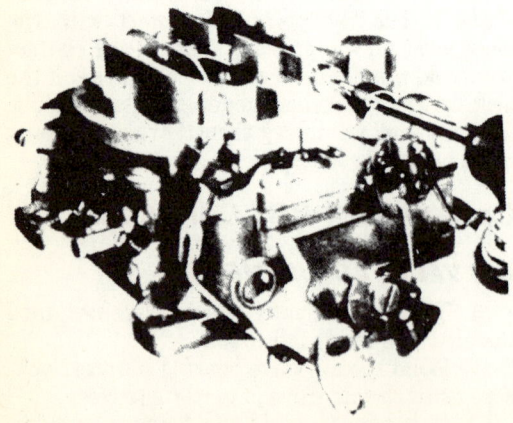

AVS air valve adjustment

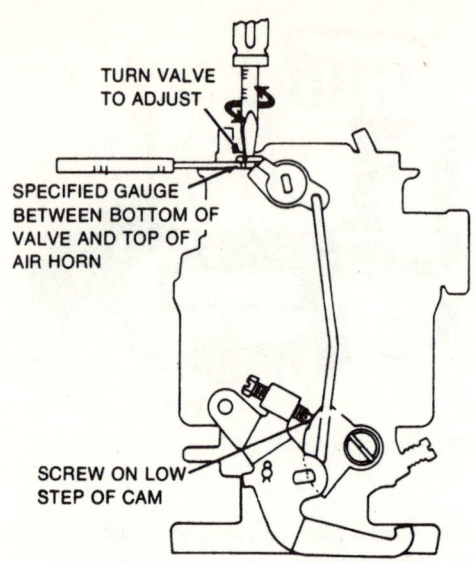

Idle vent adjustment (Rochester BV)

to obtain 0.120" clearance between the upper edge of the choke valve and the air horn.

CLOSING SHOE ADJUSTMENT

1. Hold the primary and secondary throttle valves closed.
2. Bend the secondary closing shoe to obtain 0.020" clearance between the positive closing shoes on the primary and secondary throttle levers.

SECONDARY THROTTLE OPENING

1. Check to see that the pickup lever contacts the loose lever on the primary shaft at both points simultaneously. Bend the pickup lever to obtain proper contact, if necessary.
2. If the primary and secondary throttle valve do not come to the wide open position simultaneously, bend the connecting link until they do.

SECONDARY LOCKOUT ADJUSTMENT

1. When the choke valve is closed, the lockout tang on the secondary throttle lever should engage the lockout dog. When the valve is open, the lockout dog should swing free of the tang.
2. Bend the lockout tang on the secondary throttle lever, if an adjustment is necessary.

Rochester BV 1-BBL Carburetor

AUTOMATIC CHOKE ADJUSTMENT

1. Disconnect the choke rod from the choke lever at the carburetor.
2. While holding the choke valve shut, pull the choke rod up against the stop in the thermostat housing.
3. Adjust the length of the choke rod so

FUEL SYSTEM 255

that the bottom edge of the choke rod is even with the top edge of the hole in the choke lever.

4. Check the linkage for freedom of operation.

IDLE VENT ADJUSTMENT

1. Position the carburetor lever on the low step of the fast idle cam.
2. The distance between the choke valve and the body casting should be 0.050″.
3. If an adjustment is necessary, turn the valve, using the proper tool.

FAST IDLE AND CHOKE VALVE ADJUSTMENT

1. Position the end of the idle adjusting screw on the next to highest step of the fast idle cam.
2. A 0.050″ feeler gauge should slide easily between the lower edge of the choke valve and the carburetor bore.
3. If necessary, bend the choke rod until the correct clearance is obtained.

UNLOADER ADJUSTMENT

1. Open the throttle to the wide open position.
2. A 0.230-0.270″ gauge should slide freely between the lower edge of the choke valve and the bore of the carburetor.
3. If necessary, bend the throttle tang to obtain the proper clearance.

VACUUM BREAK ADJUSTMENT

1. Hold the diaphragm lever against the diaphragm body.
2. The clearance between the lower edge of the choke valve and the air horn wall should be 0.136-0.154″ on Powerglide transmission equipped vehicles and 0.154-0.173″ on manual transmission equipped vehicles.
3. Bend the diaphragm link, if necessary.

FLOAT LEVEL ADJUSTMENT

1. Remove the air cleaner.
2. Disconnect the fuel line, fast idle rod, cam-to-choke kick lever, vacuum hose at the diaphragm, and the choke rod at the choke lever.
3. Remove the bowl cover screws and careful lift the cover off the carburetor.
4. Install a new gasket on the cover before making any adjustments.
5. Invert the cover assembly and measure the float level with a float gauge.

NOTE: *Rebuilding kits include a float level gauge.*

6. Check float centering while holding the cover sideways. Use the same gauge as in Step Five. The floats should not touch the gauge.
7. Hold the cover upright and measure the float drop. If the drop is more or less than $1^{3}/_{4}″$, bend the stop tang until the drop is correct.
8. Install the cover and reconnect the lines and linkage in the reverse order of removal.

Rochester MV 1-BBL Carburetor

The model MV carburetor is a single bore, downdraft carburetor with an aluminum throttle body, automatic choke, internally balanced venting, and a hot idle compensating system for vehicles equipped with automatic transmission. Newer models are also equipped with Combination Emission Control valves (CEC) and an Exhaust Gas Recirculation (EGR) system. An electrically operated idle stop solenoid replaced the idle stop screw of older models.

The MV carburetor is used on six cylinder vehicles from 1968 and service procedures apply to all MV carburetors.

FAST IDLE ADJUSTMENT

NOTE: *The fast idle adjustment must be made with the transmission in Neutral.*

1. Position the fast idle lever on the high step of the fast idle cam.
2. Be sure that the choke is properly ad-

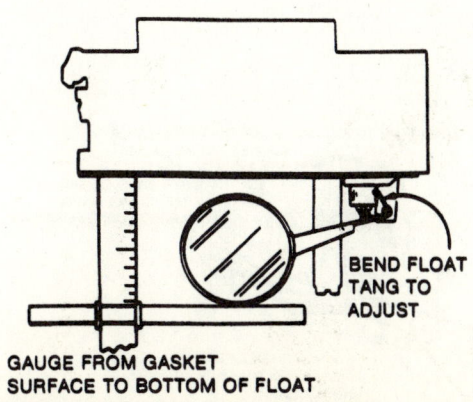

Adjusting the float drop (Rochester BV)

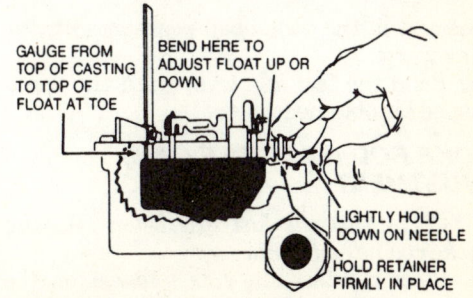

Adjusting the MV carburetor float level

MV Carburetor Specifications

Year	Carburetor Identification①	Float Level (in.)	Metering Rod (in.)	Pump Rod	Idle Vent (in.)	Vacuum Break (in.)	Auxiliary Vacuum Break (in.)	Fast Idle Off Car (in.)	Choke Rod (in.)	Choke Unloader (in.)	Fast Idle Speed (rpm)
1968	7028014	9/32	0.120	—	0.050	0.245	—	1½	0.150	0.350	2400②
	7028015	9/32	0.130	—	0.050	0.275	—	1½	0.150	0.350	2400②
	7028017	9/32	0.130	—	0.050	0.275	—	1½	0.150	0.350	2400②
1969	7029014	1/4	0.070	—	0.050	0.245	—	0.100	0.170	0.350	2400②
	7029015	1/4	0.090	—	0.050	0.275	—	0.100	0.200	0.350	2400②
	7029017	1/4	0.090	—	0.050	0.275	—	0.100	0.200	0.350	2400②
1970	7040014	1/4	0.070	—	—	0.200	—	0.110	0.170	0.350	2400②
	7040017	1/4	0.090	—	—	0.160	—	0.100	0.190	0.350	2400②
1971	7041014	1/4	0.080	—	—	0.200	—	0.100	0.160	0.350	—
	7041017	1/4	0.080	—	—	0.230	—	0.100	0.180	0.350	—
	7041023	1/16	—	—	—	0.200	—	0.110	0.120	0.350	—
1972	7042014	1/4	0.080	—	—	0.190	—	—	0.125	0.500	2400②
	7042017	1/4	0.078	—	—	0.225	—	—	0.150	0.500	2400②
	7042984	1/4	0.078	—	—	0.190	—	—	0.125	0.500	2400②
	7042987	1/4	0.076	—	—	0.225	—	—	0.150	0.500	2400②
1973	7043014	1/4	0.080	—	—	0.300	—	—	0.245	0.500	1800②
	7043017	1/4	0.080	—	—	0.350	—	—	0.275	0.500	1800②
1974	7044014	3/10	0.079	—	—	0.275	—	—	0.230	0.500	1800②③
	7044017	3/10	0.072	—	—	0.350	—	—	0.275	0.500	1800②③
	7044314	3/10	0.073	—	—	0.300	—	—	0.245	0.500	1800②③
1975	7045013	11/32	0.080	—	—	0.200	0.215	—	0.160	0.215	1800④
	7045012	11/32	0.080	—	—	0.350	0.312	—	0.275	0.275	1800④
	7045314	11/32	0.080	—	—	0.275	0.312	—	0.230	0.275	1800④
1976	17056012	11/32	0.084	—	—	0.140	0.265	—	0.100	0.260	2200⑤
	17066013	11/32	0.082	—	—	0.140	0.325	—	0.140	0.260	2100
	17056016	11/32	0.080	—	—	0.140	WFO	—	0.115	0.260	2200⑤
	17056018	11/32	0.084	—	—	0.140	0.265	—	0.100	0.260	2200⑤
	17056314	11/32	0.083	—	—	0.150	0.325	—	0.135	0.266	1700

① The carburetor identification tag is located at the rear of the carburetor on one of the air horn screws, or stamped on the float bowl, next to the fuel inlet nut.
② High step of cam.
③ Without vacuum advance.
④ 1700 rpm with automatic transmission in neutral.
⑤ 2100 rpm with integral intake manifold.

justed and in the wide open position with the engine warm.

3. Bend the fast idle lever until the specified speed is obtained.

CHOKE ROD (FAST IDLE CAM) ADJUSTMENT

NOTE: *Adjust the fast idle before making choke rod adjustments.*

1. Place the fast idle cam follower on the second step of the fast idle cam and hold it firmly against the rise to the high step.
2. Rotate the choke valve in the direction of

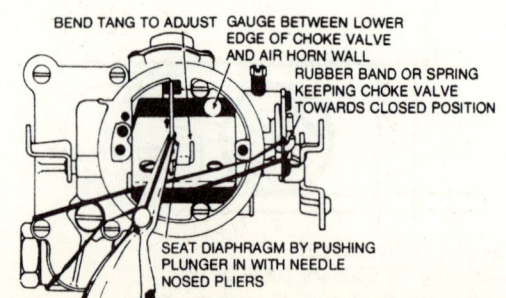

Vacuum break adjustment (MV)

FUEL SYSTEM

the closed choke by applying force to the choke coil lever.

3. Bend the choke rod, at the point shown in the illustration, to give the specified opening between the lower edge (upper edge-1976), of the choke valve and the inside air horn wall.

NOTE: *Measurement must be made at the center of the choke valve.*

CHOKE VACUUM BREAK ADJUSTMENT

The adjustment of the vacuum break diaphragm unit insures correct choke valve opening after engine starting.

1. Remove the air cleaner. On vehicles with the THERMAC air cleaner, plug the sensor's vacuum take off port.
2. Using an external vacuum source, apply vacuum to the vacuum break diaphragm until the plunger is fully seated.
3. When the plunger is seated, push the choke valve toward the closed position.

CHOKE UNLOADER ADJUSTMENT

1. While holding the choke valve closed, apply pressure to the choke operating lever.
2. Turn the throttle lever to the wide open position.
3. Measure the distance between the lower edge of the choke plate and the air horn wall.
4. If an adjustment is necessary, bend the unloader tang on the throttle lever.

FLOAT LEVEL ADJUSTMENT

1. While holding the float retainer in place, push down on the outer end of the float arm.
2. Measure the distance from the top of the float bowl casting (no gasket) and the toe of the float.
3. Bend the float as necessary to obtain the specified measurement.

METERING ROD ADJUSTMENT

1. Back out the idle adjusting screw or idle stop solenoid to close the throttle valve.
2. Apply pressure to the power piston hanger and hold the piston against its stop.
3. While holding the power piston down, turn the metering rod holder over the float surface of the bowl casting until the metering rod lightly touches the inside edge of the bowl.
4. Measure the space between the bowl and the bottom of the metering rod holder. This dimension should be 0.070-0.078".
5. Bend the metering rod holder if an adjustment is necessary.

CHOKE ROD ADJUSTMENT

1. Disconnect the choke rod from the upper choke lever and hold the choke valve closed.

2. Push the choke rod down to the bottom of its travel.
3. The top of the rod should be even with the bottom of the hole in the choke lever, band the rod if necessary.

Rochester 1ME 1-BBL Carburetors

FLOAT LEVEL ADJUSTMENT

1. Push down on the end of the float arm and against the top of the float needle to hold the retaining pin firmly in place.
2. While holding the position of the retaining pin, gauge from the top of the casting to the top of the index point at the toe of the float.
3. If float level needs to be changed do it by bending the float arm just on the float side of the float needle.

FAST IDLE ADJUSTMENT

1. If the carburetor has a stepped fast idle cam, put the cam follower on the high step of the cam. If the cam has a smooth contour, open the throttle slightly and rotate the cam to its highest position, then release throttle.
2. Support the lever with a pair of pliers, and bend the tang in or out to achieve specified rpm. Perform the adjustment with the engine hot and the choke open.

CHOKE COIL LEVER ADJUSTMENT

NOTE: *This adjustment requires a plug gauge or other metal rod of 0.120" diameter.*

1. Place the fast idle cam follower on the highest step of the cam or rotate the cam to its

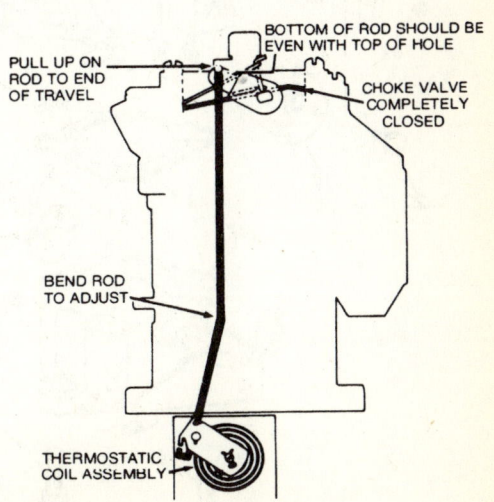

Adjusting the choke rod (MV)

FUEL SYSTEM

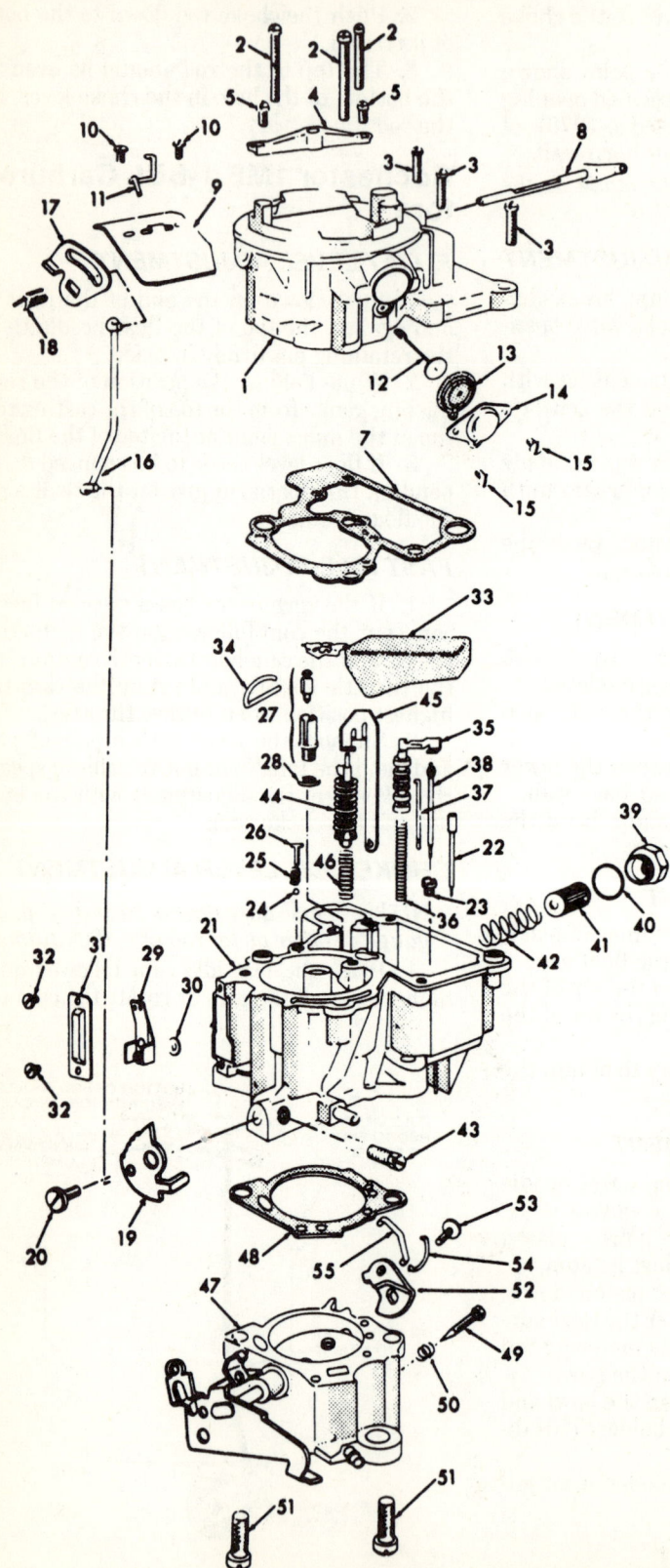

1. Air horn assembly
2. Screw—air horn—long
3. Screw—air horn—short
4. Bracket—air cleaner stud
5. Screw—bracket attaching
7. Gasket—air horn
8. Choke shaft and lever assembly
9. Choke valve
10. Screw—choke valve
11. Lever—vacuum break link
12. Vacuum break link assy.
13. Diaphragm—vacuum break
14. Cover—vacuum break
15. Screw—cover
16. Choke rod
17. Choke lever
18. Screw—choke lever
19. Cam—fast idle
20. Screw—cam attaching
21. Float bowl assembly
22. Idle tube assembly
23. Jet—main metering
24. Ball—pump discharge
25. Spring—pump discharge
26. Guide—pump discharge
27. Needle and seat assy.
28. Gasket—needle seat
29. Idle compensator assembly
30. Gasket—idle compensator
31. Cover—idle compensator
32. Screw—cover
33. Float assembly
34. Hinge pin—float
35. Power piston assembly
36. Spring—power piston
37. Rod—power piston
38. Metering rod and spring assembly
39. Filter nut—fuel inlet
40. Gasket—filter nut
41. Filter—fuel inlet
42. Spring—fuel filter
43. Screw—slow idle
44. Pump assembly
45. Lever—pump actuating
46. Spring—pump return
47. Throttle body assembly
48. Gasket—throttle body
49. Idle needle
50. Spring—idle needle
51. Screw—throttle body
52. Lever—pump and power rods—new
53. Screw—lever attaching
54. Link—power piston rod
55. Link—pump lever

Exploded view of the MV carburetor

1ME Carburetor Specifications

Year	Carburetor Identification[1] Number	Float Level (in.)	Metering Rod (in.)	Fast Idle Speed (rpm)	Fast Idle Cam (in.)	Vacuum Break (in.)	Choke Unloader (in.)	Choke Setting (notches)
1977	17057016	3/8	0.070	2000	0.095	0.125	0.325	1 Lean
	17057013	3/8	0.070	2000	0.100	0.120	0.270	3 Rich
	17057015	3/8	0.070	2000	0.100	0.125	0.325	1 Rich
	17057018	3/8	0.070	2000	0.085	0.120	0.325	1 Rich
	17057014	3/8	0.070	2000	0.100	0.125	0.325	2 Rich
	17057020	3/8	0.070	2000	0.085	0.120	0.120	2 Rich
	17057310	3/8	0.070	2000	0.085	0.125	0.200	Index
	17057312	3/8	0.070	1800	0.100	0.100	0.110	Index
	17057314	3/8	0.070	1800	0.100	0.110	0.225	Index
	17057318	3/8	0.070	1800	0.100	0.110	0.110	Index
	17057042, 17057044	5/32	0.080	2400	0.050	0.075	0.200	1 Rich
	17057045	5/32	0.080	2000	0.050	0.075	0.200	1 Rich
	17057332, 17057334	5/32	0.080	2400	0.050	0.075	0.200	2 Rich
	17057335	5/32	0.080	2300	0.050	0.080	0.200	2 Rich
	17057030	5/32	0.080	2400	0.050	0.080	0.200	2 Rich
	17057031	5/32	0.080	2300	0.050	0.080	0.200	2 Rich
	17057032, 17057034	5/32	0.080	2400	0.050	0.080	0.200	2 Rich
	17057035	5/32	0.080	2300	0.050	0.080	0.200	3 Rich
1978	17058013	3/8	0.080	2000	0.180	0.200	0.500	Index
	17058014	5/16	0.100	2100	0.180	0.200	0.500	Index
	17058020	5/16	0.100	2100	0.180	0.200	0.500	Index
	17058314	3/8	0.100	2000	0.190	0.245	0.400	Index
	17058031	5/32	0.080	2400	0.105	0.150	0.500	2 Rich
	17058032	5/32	0.080	2400	0.080	0.130	0.500	3 Rich
	17058033	5/32	0.080	2400	0.080	0.130	0.500	2 Rich
	17058034	5/32	0.080	2400	0.080	0.130	0.500	3 Rich
	17058035	5/32	0.080	2300	0.080	0.130	0.500	3 Rich
	17058036	5/32	0.080	2400	0.080	0.130	0.500	3 Rich
	17058037	5/32	0.080	2400	0.080	0.130	0.500	2 Rich
	17058038	5/32	0.080	2400	0.080	0.130	0.500	3 Rich
	17058042	5/32	0.080	2400	0.080	0.160	0.500	2 Rich
	17058044	5/32	0.080	2400	0.080	0.160	0.500	2 Rich
	17058045	5/32	0.080	2300	0.080	0.160	0.500	2 Rich
	17058332	5/32	0.080	2400	0.080	0.160	0.500	2 Rich
	17058334	5/32	0.080	2400	0.080	0.160	0.500	2 Rich
	17058335	5/32	0.080	2300	0.080	0.160	0.500	2 Rich
1979	17059014	3/8	0.095	2000	0.180	0.200	0.400	Index
	17059020	3/8	0.095	2000	0.180	0.200	0.400	Index
	17059013	3/8	0.095	1800	0.180	0.200	0.400	Index
	17059314	3/8	0.100	2000	0.190	0.245	0.400	Index

[1] Stamped on float bowl, next to fuel inlet nut
[2] 2200 rpm for the first two numbers
[3] 2200 rpm for the last two numbers

highest possible position. Close the choke all the way and hold.

2. Insert the plug gauge through the small hole in the out end of the choke lever. Bend the link at the lowest point of the curved portion until the gauge will enter the hole in the carburetor casting.

CHOKE ADJUSTMENT

1. Place the fast idle cam follower on the high step of the cam or rotate the cam until it is at the highest position.

2. Slightly loosen the three screws which retain the choke cover just enough to turn the cover - don't loosen them more than necessary, or the lever may slip out of the tan inside.

3. Turn the cover until the mark on the cover lines up with the appropriate mark on the choke housing. See specifications.

METERING ROD ADJUSTMENT

1. Hold the throttle wide open. Push downward on the metering rod until it can be slid out of the slot in the holder. Slide the rod out of

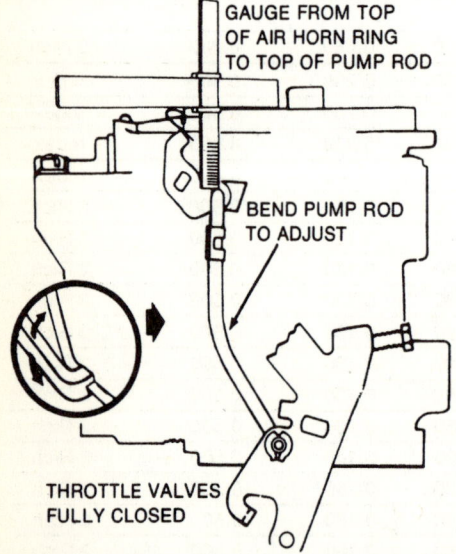

Adjusting the accelerator pump rod (2GV)

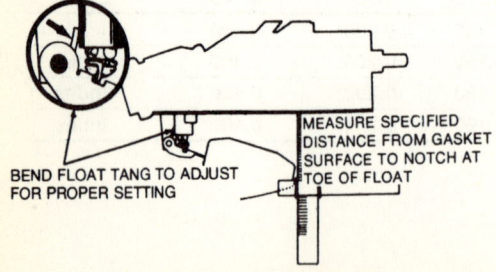

Adjusting the float drop-nitrophyl (2GV)

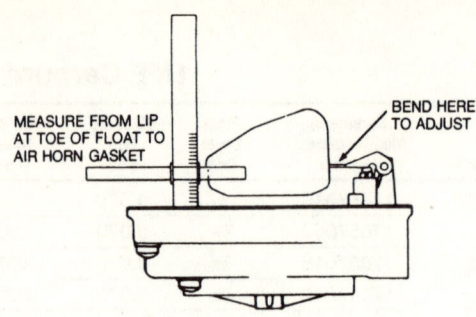

Adjusting the float level-nitrophyl (2GV)

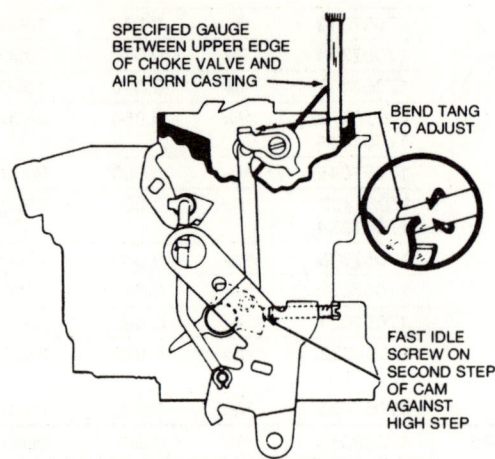

Adjusting the fast idle cam (2GV)

the holder and remove it from the main metering jet.

2. Back out the solenoid hex screw until the throttle can be closed all the way.

3. Remove the float bowl gasket.

4. Holding power piston down and throttle closed, swing the metering rod holder over the flat surface of the bowl casting next to the throttle bore. Measure the distance from the float surface to the outer end of the rod holder with the specified gauge (see specifications), or a metal rod of equivalent diameter.

5. Bend the horizontal portion of the rod holder where is joins the vertical portion until the gauge just passes between the holder and surface of the bowl casting with power piston bottomed.

Rochester 2GC, 2GV 2-BBL Carburetors

These procedures are for both the $1^1/_4$ and $1^1/_2$ models; where there are differences these are noted. The $1^1/_2$ model has larger throttle bores and an additional fuel feed circuit to make it suitable for use on the 350 V8s.

FUEL SYSTEM   261

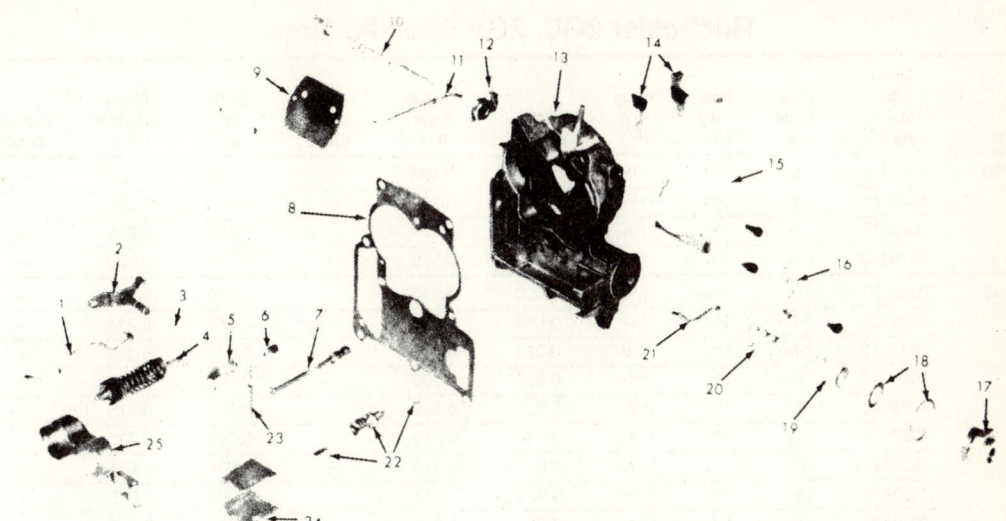

1. Pump rod
2. Pump outlet lever
3. Accelerator pump
4. Washer
5. Pump inner lever
6. Pump inner lever retainer
7. Power piston
8. Air horn-to-bowl gasket
9. Choke valve
10. Choke rod
11. Choke shaft
12. Choke kick lever
13. Air horn
14. Vent valve and shield
15. Vacuum diaphragm
16. Choke lever
17. Diaphragm link
18. Fuel inlet nut
19. Gaskets
20. Fuel filter
21. Filter spring
22. Float needle and seat
23. Float hinge pin
24. Splash shield
25. Float

Exploded view of the 2GV (1¼) air horn

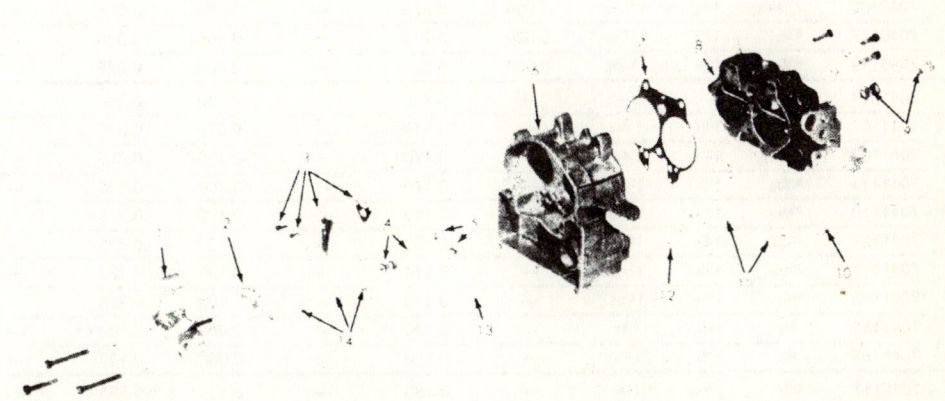

1. Cluster assembly
2. Cluster gasket
3. Hot idle compensator
4. Power valve assembly
5. Main jets
6. Bowl assembly
7. Throttle body-to-bowl gasket
8. Throttle body assembly
9. Idle speed screw
10. Choke rod
11. Idle mixture screws
12. Fast idle cam
13. Accelerator pump spring
14. Pump discharge check assembly

Exploded view of the 2GV (1½) float bowl

Rochester 2GC, 2GV Specifications

Year	Carburetor Identification ①	Float Level (in.)	Float Drop (in.)	Pump Rod (in.)	Idle Vent (in.)	Vacuum Break (in.)	Automatic Choke	Choke Rod (in.)	Choke Unloader (in.)	Fast Idle Speed
1968	7028110	3/4	1 3/4	1 1/8	1.000	0.100	—	0.060	0.200	—
	7028101	3/4	1 3/4	1 1/8	1.000	0.100	—	0.060	0.200	—
	7028112	3/4	1 3/4	1 1/8	1.000	0.100	—	0.060	0.200	—
	7028103	3/4	1 3/4	1 1/8	1.000	0.100	—	0.060	0.200	—
1969	7029101	27/32	1 3/4	1 1/8	0.020	0.100	—	0.060	0.215	—
	7029103	27/32	1 3/4	1 1/8	0.020	0.100	—	0.060	0.215	—
	7029110	27/32	1 3/4	1 1/8	0.020	0.100	—	0.060	0.215	—
	7029112	27/32	1 3/4	1 1/8	0.020	0.100	—	0.060	0.215	—
	7029102	3/4	1 3/4	1 13/32	0.020	0.215	—	0.085	0.275	—
	7029104	3/4	1 3/4	1 13/32	0.020	0.215	—	0.085	0.275	—
	7029127	3/4	1 3/4	1 13/32	0.020	0.215	—	0.085	0.275	—
	7029129	3/4	1 3/4	1 13/32	0.020	0.215	—	0.085	0.275	—
	7029117	3/4	1 3/4	1 13/32	0.020	0.215	—	0.085	0.275	—
	7029118	3/4	1 3/4	1 13/32	0.020	0.215	—	0.085	0.275	—
	7029119	5/8	1 3/4	1 13/32	0.020	0.215	—	0.085	0.275	—
	7029120	5/8	1 3/4	1 13/32	0.020	0.215	—	0.085	0.275	—
1970	7040110	27/32	1 3/4	1 1/8	0.020	0.100	—	0.060	0.215	—
	7040112	27/32	1 3/4	1 1/8	0.020	0.100	—	0.060	0.215	—
	7040101	27/32	1 3/4	1 1/8	0.020	0.125	—	0.060	0.160	—
	7040103	27/32	1 3/4	1 1/8	0.020	0.125	—	0.060	0.225	—
	7040114	23/32	1 3/8	1 17/32	0.020	0.200	—	0.085	0.325	—
	7040116	23/32	1 3/8	1 17/32	0.020	0.200	—	0.085	0.325	—
	7040113	23/32	1 3/8	1 17/32	0.020	0.215	—	0.085	0.275	—
	7040115	23/32	1 3/8	1 17/32	0.020	0.215	—	0.085	0.275	—
	7040118	23/32	1 3/8	1 17/32	0.020	0.215	—	0.085	0.325	—
	7040120	23/32	1 3/8	1 17/32	0.020	0.215	—	0.085	0.325	—
	7040117	23/32	1 3/8	1 17/32	0.020	0.215	—	0.085	0.325	—
	7040119	23/32	1 3/8	1 17/32	0.020	0.215	—	0.085	0.325	—
1971	7041024	1/16	—	—	—	0.140	—	0.080	0.350	—
	7041101	13/16	1 3/4	1 3/64	—	0.110	—	0.075	0.215	—
	7041102	25/32	1 3/4	1 5/32	—	0.170	—	0.100	0.325	—
	7041114	25/32	1 3/4	1 5/32	—	0.170	—	0.100	0.325	—
	7041113	25/32	1 3/4	1 5/32	—	0.180	—	0.100	0.325	—
	7041127	25/32	1 3/4	1 5/32	—	0.180	—	0.100	0.325	—
	7041117	25/32	1 3/4	1 5/32	—	0.170	—	0.100	0.325	—
	7041118	25/32	1 3/4	1 5/32	—	0.170	—	0.100	0.325	—
	7041181	5/8	1 3/4	1 3/8	—	0.120	—	0.080	0.180	—
	7041182	5/8	1 3/4	1 3/8	—	0.120	—	0.080	0.180	—
1972	7042111	23/32	1 9/32	1 1/2	—	0.180	—	0.100	0.325	—
	7042113	23/32	1 9/32	1 1/2	—	0.180	—	0.100	0.325	—
	7042831	23/32	1 9/32	1 1/2	—	0.180	—	0.100	0.325	—
	7042833	23/32	1 9/32	1 1/2	—	0.180	—	0.100	0.325	—
	7042112	23/32	1 9/32	1 1/2	—	0.170	—	0.100	0.325	—
	7042114	23/32	1 9/32	1 1/2	—	0.170	—	0.100	0.325	—
	7042118	23/32	1 9/32	1 1/2	—	0.190	—	0.100	0.325	—
	7042832	23/32	1 9/32	1 1/2	—	0.170	—	0.100	0.325	—
	7042834	23/32	1 9/32	1 1/2	—	0.170	—	0.100	0.325	—
	7042838	23/32	1 9/32	1 1/2	—	0.190	—	0.100	0.325	—
	7042100	23/32	1 31/32	1 5/16	—	0.080	—	0.040	0.215	—

Rochester 2GC, 2GV Specifications

Year	Carburetor Identification ①	Float Level (in.)	Float Drop (in.)	Pump Rod (in.)	Idle Vent (in.)	Vacuum Break (in.)	Automatic Choke	Choke Rod (in.)	Choke Unloader (in.)	Fast Idle Speed
	7042820	23/32	1 31/32	1 5/16	—	0.080	—	0.040	0.215	—
	7042101	23/32	1 31/32	1 5/16	—	0.110	—	0.075	0.215	—
	7042821	23/32	1 31/32	1 5/16	—	0.110	—	0.075	0.215	—
1973	7043100	21/32	1 9/32	1 5/16	—	0.080	—	0.150	0.215	—
	7043101	21/32	1 9/32	1 5/16	—	0.080	—	0.150	0.215	—
	7043120	21/32	1 9/32	1 5/32	—	0.080	—	0.150	0.215	—
	7043105	21/32	1 9/32	1 5/16	—	0.080	—	0.150	0.215	—
	7043114	19/32	1 9/32	1 7/16	—	0.130	—	0.245	0.325	—
	7043113	19/32	1 9/32	1 5/16	—	0.140	—	0.200	0.250	—
	7043112	19/32	1 9/32	1 5/16	—	0.130	—	0.245	0.325	—
	7043111	19/32	1 9/32	1 7/16	—	0.140	—	0.200	0.250	—
	7043182	19/32	1 9/32	1 7/16	—	0.130	—	0.245	0.325	—
1974	704411	19/32	1 9/32	1 9/32	—	0.140	—	0.200	0.250	1600 ②
	7044112	19/32	1 9/32	1 3/16	—	0.130	—	0.245	0.325	1600 ②
	7044113	19/32	1 9/32	1 9/32	—	0.140	—	0.200	0.250	1600 ②
	7044114	19/32	1 9/32	1 3/16	—	0.130	—	0.245	0.325	1600 ②
	7044115	19/32	1 9/32	1 9/32	—	0.140	—	0.200	0.250	1600 ②
	7044116	19/32	1 9/32	1 3/16	—	0.130	—	0.245	0.325	1600 ②
	7044118	19/32	1 9/32	1 3/16	—	0.130	—	0.245	0.325	1600 ②
	7044123	19/32	1 9/32	1 9/32	—	0.140	—	0.200	0.250	1600 ②
	7044124	19/32	1 9/32	1 3/16	—	0.130	—	0.245	0.325	1600 ②
1975	7045105	19/32	1 7/32	1 19/32	—	0.130	—	0.375	0.350	—
	7045106	19/32	1 7/32	1 19/32	—	0.130	—	0.380	0.350	—
	7045111	21/32	31/32	1 5/8	—	0.130	—	0.400	0.350	—
	7045112	21/32	31/32	1 5/8	—	0.130	—	0.400	0.350	—
	7045114	21/32	31/32	1 5/8	—	0.130	—	0.400	0.350	—
	7045115	21/32	31/32	1 5/8	—	0.130	—	0.400	0.350	—
	7045123	21/32	31/32	1 5/8	—	0.130	—	0.400	0.350	—
	7045124	21/32	31/32	1 5/8	—	0.130	—	0.400	0.350	—
	7045406	21/32	1 7/32	1 19/32	—	0.130	—	0.380	0.350	—
1976	17056108	9/16	1 9/32	1 21/32	—	0.140	Index	0.260	0.325	—
	17056110	9/16	1 9/32	1 21/32	—	0.140	Index	0.260	0.325	—
	17056111	9/16	1 9/32	1 21/32	—	0.140	Index	0.260	0.325	—
	17056112	9/16	1 9/32	1 21/32	—	0.140	Index	0.260	0.325	—
	17056113	9/16	1 9/32	1 21/32	—	0.140	Index	0.260	0.325	—
	17056114	21/32	31/32	1 11/16	—	0.130	1 Rich	0.260	0.325	—
	17056430, 17056432	9/16	1 9/32	1 21/32	—	0.140	Index	0.260	0.325	—
1977	17057108, 17057110, 17057111, 17057112, 17057113	9/16	1 9/32	1 21/32	—	0.140	Index	0.260	0.325	—
	17057114	21/32	31/32	1 11/16	—	0.130	1 Rich	0.260	0.325	—
	17057123	19/32	1 9/32	1 21/32	—	0.160	Index	0.260	0.325	—
	17057408, 17057410, 17057412, 17057414	21/32	1 9/32	1 21/32	—	0.160	1/2 Lean	0.260	0.325	—

① The carburetor identification tag is located at the rear of the carburetor on one of the air horn screws, or stamped on the float bowl, next to the fuel inlet nut.
② This setting is with the low idle at 500 rpm with the clutch fan disengaged.

264 FUEL SYSTEM

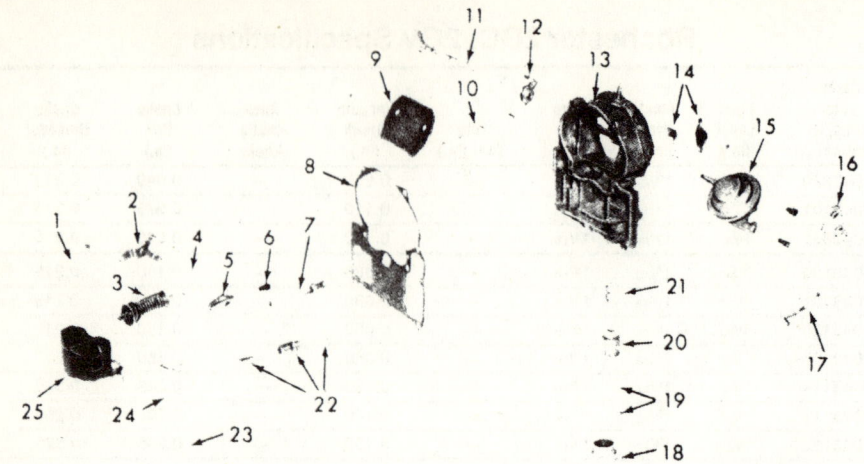

1. Pump rod
2. Pump outer lever
3. Washer
4. Accelerator pump
5. Pump inner lever
6. Pump inner lever retainer
7. Power piston
8. Air horn-to-bowl gasket
9. Choke valve
10. Choke shaft
11. Choke rod
12. Choke kick lever
13. Air horn
14. Vent valve and shield
15. Vacuum diaphragm
16. Choke lever
17. Fuel inlet nut
18. Gaskets
19. Fuel filter
20. Filter spring
21. Diaphragm link
22. Float needle and seat
23. Float hinge pin
24. Splash shields
25. Float

Exploded view of the 2GV (1½) air horn

FAST IDLE ADJUSTMENT

On 2GC and 2GV models the fast idle is set automatically when the curb idle and mixture is set.

FAST IDLE CAM ADJUSTMENT

1. Turn the idle screw onto the second step of the fast idle cam, abutting against the top step.
2. Hold the choke valve toward the closed position and check the clearance between the upper edge of the valve and the air horn wall.
3. If this measurement varies from specifications, bend the tang on the choke lever.

CHOKE UNLOADER ADJUSTMENT

1. Hold the throttle valves wide open and use a rubber band to hold the choke valve toward the closed position.
2. Measure the distance between the upper edge of the choke valve and the air horn wall.
3. If this measurement is not within specifications, bend the unloader tang on the throttle lever to correct it.

ACCELERATOR PUMP ROD ADJUSTMENT

1. Back out the idle stop screw and close the throttle valves in their bores.
2. Measure the distance from the top of the air horn to the top of the pump rod.
3. Bend the pump rod at angle to correct this dimension.

FLOAT LEVEL ADJUSTMENT

Invert the air horn and, with the gasket in place and the needle seated, measure the level as follows:
On nitrophyl floats, measure the air horn gasket to the lip on the toe of the float.
On brass floats, measure the air horn gasket to the lower edge of the float seam.
Bend the float tang to adjust the level.

FLOAT DROP ADJUSTMENT

Holding the air horn right side up, measure float drop as follows:
On nitrophyl floats, measure from the air horn gasket to the lip at the toe of the float.
On brass floats, measure from the air horn gasket to the bottom of the float.
Bend the float tang to adjust either type of float.

Rochester M2ME and E2ME 2-BBL Carburetors

FLOAT ADJUSTMENT

1. Remove the air horn from the throttle body.
2. Using your fingers, hold the retainer in place and then push the float down into light contact with the needle.
3. Measure the distance from the toe of the float (furthest from the hinge) to the top of the carburetor (gasket removed).

FUEL SYSTEM 265

4. To adjust, remove the float and gently bend the are to specifications. After adjustment, check the float alignment in the chamber. On engines equipped with the C-4 or the CCC system, where the float level varies more than $1/16''$ from specifications, adjust the float as follows:

FLOAT TOO HIGH:

Hold the retainer firmly in place and then push down the center of the float pontoon until the correct setting is obtained.

FLOAT TOO LOW:

1. Lift out the meter rods and remove the solenoid connector screw.
2. Turn the lean mixture solenoid screw clockwise until the screw is bottomed lightly in the float bowl. Count and record the number of turns before the screw is bottomed.
3. Turn the screw counterclockwise and remove it. Lift the solenoid and the connector from the float bowl.
4. Remove the float and bend the arm up to adjust it. Put the float back in and check its alignment.
5. Installation is in the reverse order of removal. Make sure that the solenoid lean mixture screw is backed out of the float bowl exactly the same number of turns as were recorded in Step 2.

PUMP ADJUSTMENT

NOTE: *All 1980 and later engines equipped with the C-4 or the CCC system have a non-adjustable pump lever. No adjustments are either necessary or possible.*

1. With the throttle closed and the fast idle screw off the steps of the fast idle cam, measure the distance from the air horn casting to the top of the pump stem.
2. To adjust the lever, support it firmly, using the proper tool and then bend it to obtain the proper specifications.
3. When the adjustment is correct, open and close the throttle a few times to check the linkage movement and alignment.

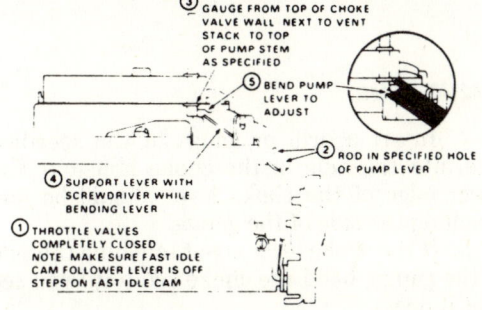

Pump adjustment—M2ME, E2ME (all except C-4 or CCC)

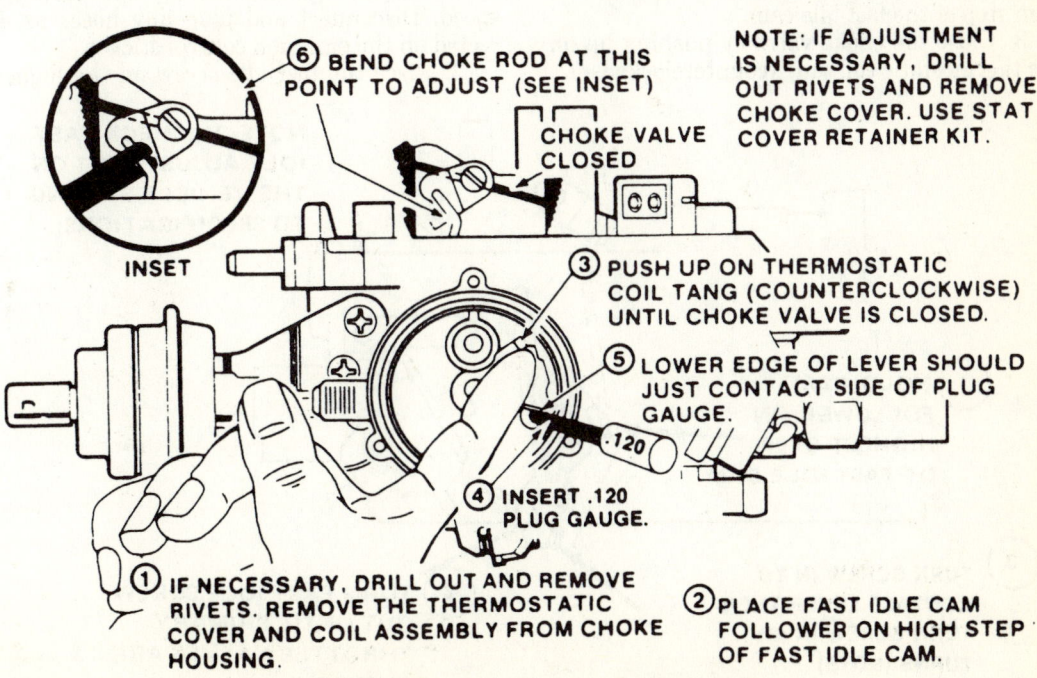

Adjusting the choke coil lever—M2ME, E2ME

266 FUEL SYSTEM

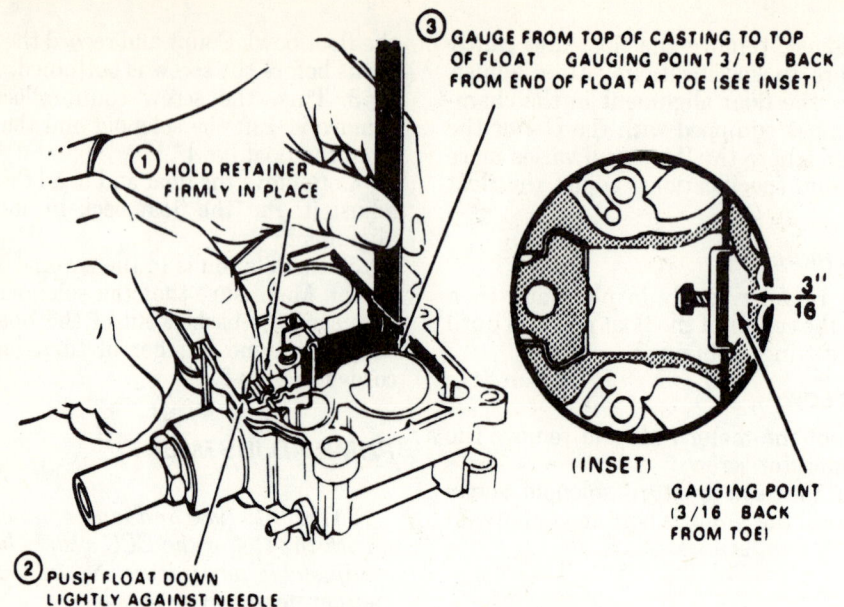

Adjusting the float — M2ME, E2ME

CHOKE COIL LEVER ADJUSTMENT

NOTE: *To complete this procedure you will need a Start Cover Retainer Kit; available at most auto parts suppliers.*

1. Drill out and remove the rivets. Retain the choke housing cover and then remove the thermostatic cover and coil assembly from the choke housing.
2. Place the fast idle cam follower on the high step of the fast idle cam.
3. Close the choke valve by pushing up on the thermostatic coil tang (counterclockwise).

4. Insert a drill or gauge of the specified size into the hole in the choke housing. The lower edge of the choke lever should be just touching the side of the gauge.
5. If the choke lever is not touching the side of the gauge, bend the choke rod until you see that it does.

FAST IDLE ADJUSTMENT

1. Set the ignition timing and curb idle speed. Disconnect and plug any hoses as directed on the emission control sticker.
2. Place the fast idle screw on the highest

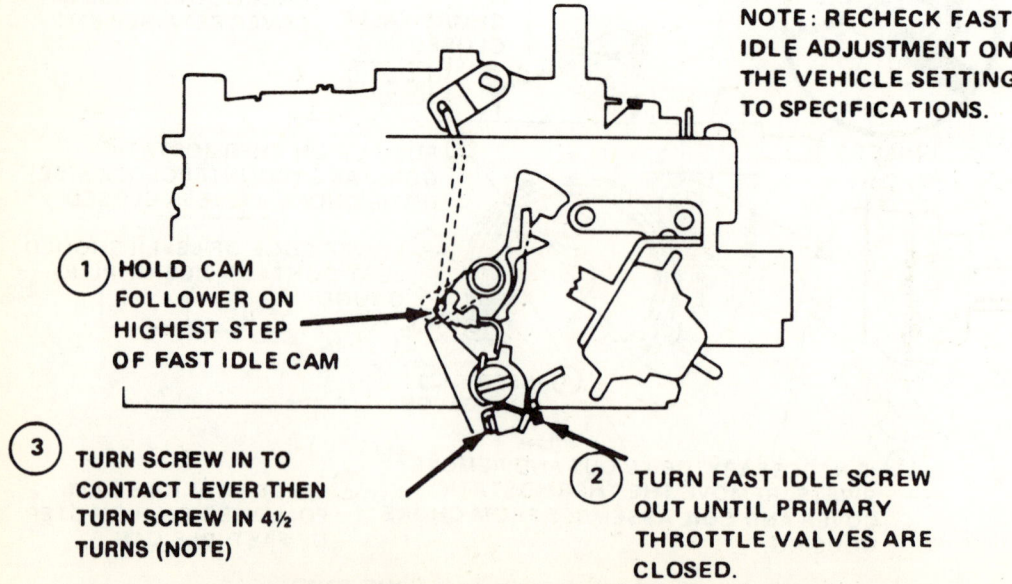

Adjusting the fast idle with the carburetor off the vehicle — M2ME, E2ME

FUEL SYSTEM

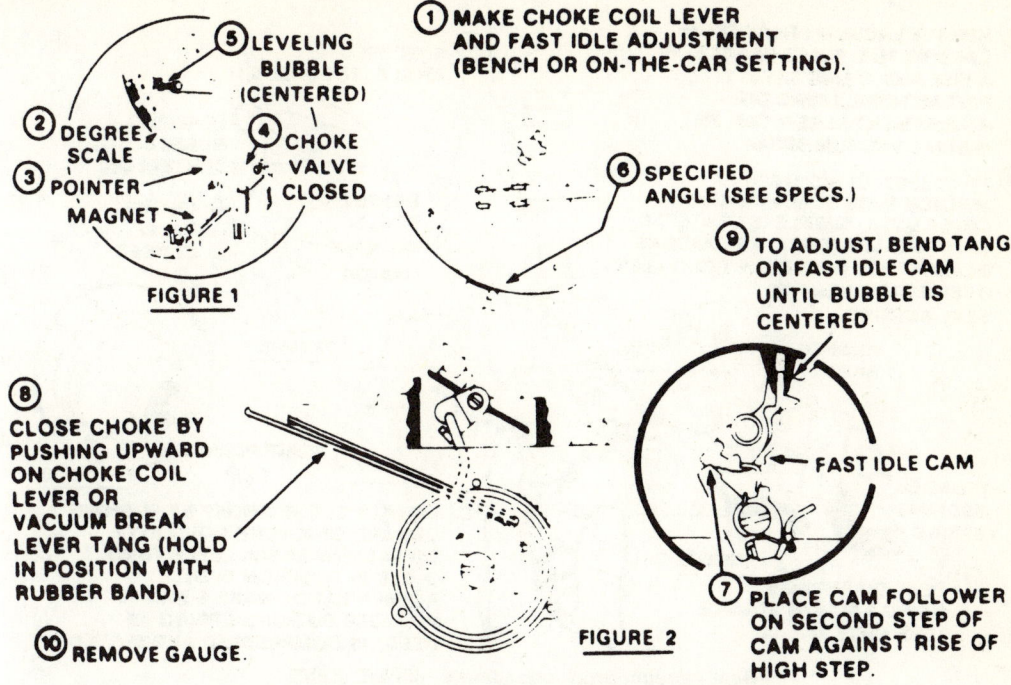

Fast idle cam (choke rod) adjustment—M2ME, E2ME

step of the fast idle cam.

3. Start the engine and adjust the engine speed to specifications with the fast idle screw.

FAST IDLE CAM (CHOKE ROD) ADJUSTMENT

NOTE: *A special angle gauge should be used. If it is not available, an inch measurement can be used.*

1. Adjust the choke coil lever and the fast idle speed.
2. Rotate the degree scale until it is zeroed.
3. Close the choke valve completely and place the magnet on top of it.
4. Center the bubble.
5. Rotate the scale so that the specified degree is opposite the pointer.
6. Place the fast idle screw on the second

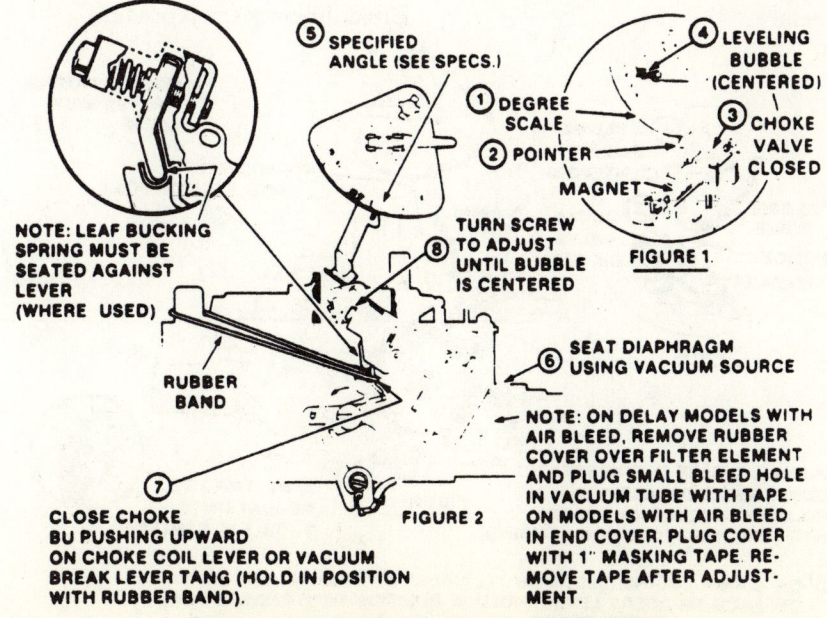

Front vacuum break adjustment—M2ME, E2ME

FUEL SYSTEM

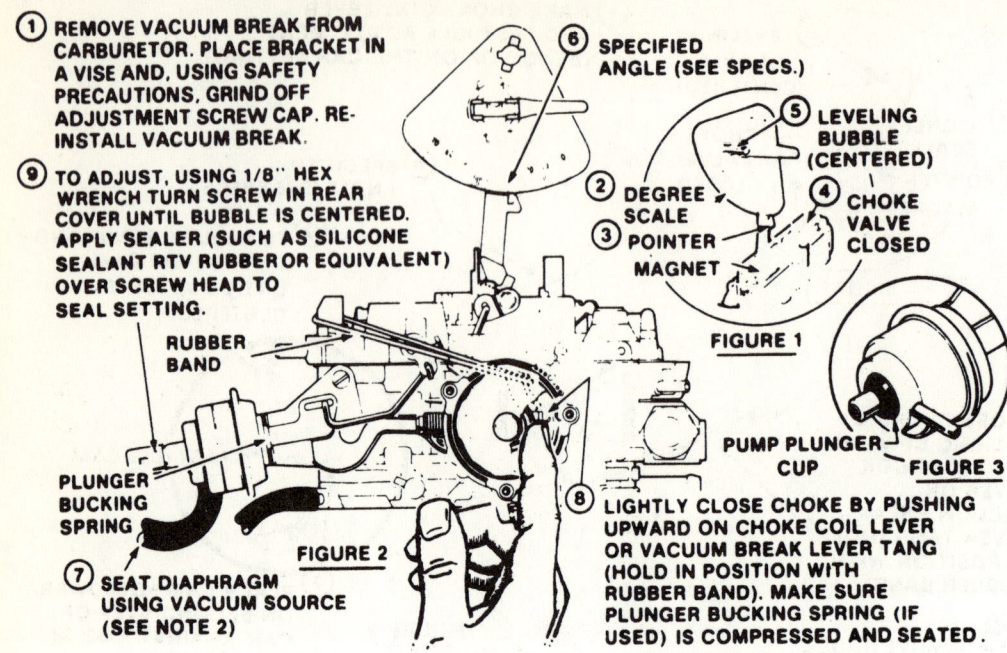

Rear vacuum break adjustment—M2ME, E2ME

step of the cam, against the rise of the high step.

7. Close the choke by pushing up on the choke coil lever or the vacuum break lever tang. You may hold it in position with a rubber band.

8. To adjust, bend the tang on the fast idle cam until the bubble is centered.

FRONT VACUUM BREAK ADJUSTMENT

1. Follow Steps 1-5 of the Fast Idle Cam Adjustment procedure.

2. Set the choke vacuum diaphragm using an outside vacuum source.

3. Close the choke valve by pushing up on the choke coil lever or the vacuum break lever. You may hold it in position with a rubber band.

4. To adjust, turn the screw in or out until the bubble in the gauge is centered.

REAR VACUUM BREAK ADJUSTMENT

1. Follow Steps 1-3 of the Front Vacuum Break Adjustment procedure.

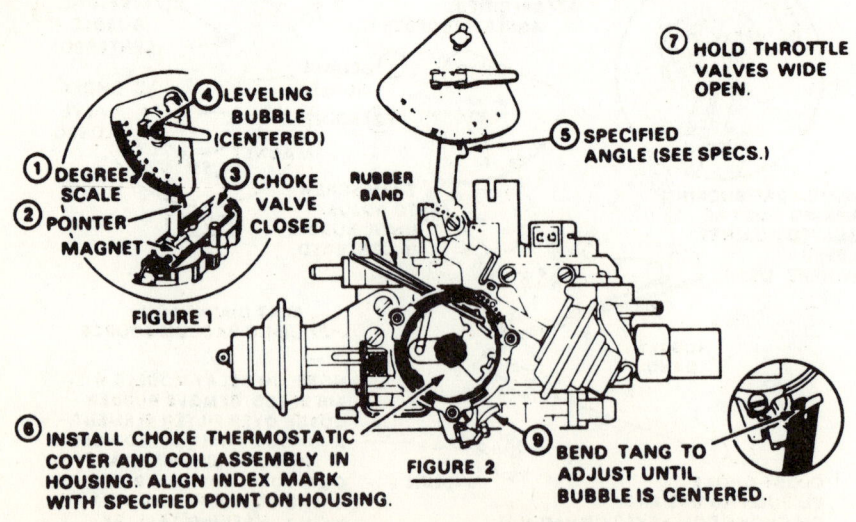

Unloader adjustment—M2ME, E2ME

Rochester M2ME, E2ME Specifications

Year	Carburetor Identification ①	Float Level (in.)	Choke Rod (deg./in.)	Choke Unloader (deg./in.)	Vacuum Break Lean or Front (deg./in.)	Vacuum Break Rich or Rear (deg./in.)	Pump Rod (in.)	Choke Coil Lever (in.)	Automatic Choke (notches)
1980	17080108	3/8	38/0.243	38/0.243	25/0.142	—	5/16 ②	0.120	Fixed
	17080110	3/8	38/0.243	38/0.243	25/0.142	—	5/16 ②	0.120	Fixed
	17080130	5/16	38/0.243	38/0.243	25/0.142	—	5/16 ②	0.120	Fixed
	17080131	5/16	38/0.243	38/0.243	25/0.142	—	5/16 ②	0.120	Fixed
	17080132	5/16	38/0.243	38/0.243	25/0.142	—	5/16 ②	0.120	Fixed
	17080133	5/16	38/0.243	38/0.243	25/0.142	—	5/16 ②	0.120	Fixed
	17080138	3/8	38/0.243	38/0.243	25/0.142	—	5/16 ②	0.120	Fixed
	17080140	3/8	38/0.243	38/0.243	25/0.142	—	5/16 ②	0.120	Fixed
	17080493	5/16	38/0.139	38/0.243	25/0.117	–/0.179	Fixed	0.120	Fixed
	17080495	5/16	38/0.139	38/0.243	25/0.117	–/0.179	Fixed	0.120	Fixed
	17080496	5/16	38/0.139	38/0.243	25/0.117	–/0.203	Fixed	0.120	Fixed
	17080498	5/16	38/0.139	38/0.243	25/0.117	–/0.203	Fixed	0.120	Fixed
1981	17080185	9/32	24.5/0.139	38/0.243	19/0.103	14/0.071	1/4 ②	0.120	Fixed
	17080187	9/32	24.5/0.139	38/0.243	19/0.103	14/0.071	1/4 ②	0.120	Fixed
	17080191	9/32	24.5/0.139	38/0.243	18/0.096	18/0.096	1/4 ②	0.120	Fixed
	17080491	5/16	24.5/0.139	38/0.243	21/0.117	35/0.220	Fixed	0.120	Fixed
	17080496	5/16	24.5/0.139	38/0.243	21/0.117	33/0.203	Fixed	0.120	Fixed
	17080498	5/16	24.5/0.139	38/0.243	21/0.117	33/0.203	Fixed	0.120	Fixed
	17081130	3/8	20/0.110	38/0.243	25/0.142	—	Fixed	0.120	Fixed
	17081131	3/8	20/0.110	38/0.243	25/0.142	—	Fixed	0.120	Fixed
	17081132	3/8	20/0.110	38/0.243	25/0.142	—	Fixed	0.120	Fixed
	17081133	3/8	20/0.110	38/0.243	25/0.142	—	Fixed	0.120	Fixed
	17081138	3/8	20/0.110	40/0.260	25/0.142	—	Fixed	0.120	Fixed
	17081140	3/8	20/0.110	40/0.260	25/0.142	—	Fixed	0.120	Fixed
	17081191	5/16	24.5/0.139	38/0.243	28/0.139	24/0.136	Fixed	0.120	Fixed
	17081192	5/16	24.5/0.139	38/0.243	28/0.139	24/0.136	Fixed	0.120	Fixed
	17081194	5/16	24.5/0.139	38/0.243	21/0.117	24/0.136	Fixed	0.120	Fixed
	17081196	5/16	24.5/0.139	38/0.243	28/0.139	24/0.136	Fixed	0.120	Fixed
	17081197	5/16	18/0.096	38/0.243	28/0.139	24/0.136	Fixed	0.120	Fixed
	17081198	5/16	24.5/0.139	38/0.243	28/0.139	24/0.136	Fixed	0.120	Fixed
	17081199	5/16	18/0.096	38/0.243	28/0.139	24/0.136	Fixed	0.120	Fixed
1982	17082130	3/8	20/–	38/0.243	27	—	Fixed	0.120	Fixed
	17082132	3/8	20/–	38/0.243	27	—	Fixed	0.120	Fixed
	17082138	3/8	20/–	38/0.243	27	—	Fixed	0.120	Fixed
	17082140	3/8	20/–	38/0.243	27	—	Fixed	0.120	Fixed
	17082497	5/16	24.5/0.139	32/–	28	24	Fixed	0.120	Fixed
1983	17082130	3/8	20/–	38/–	27	—	Fixed	0.120	Fixed
	17082132	3/8	20/–	38/–	27	—	Fixed	0.120	Fixed
	17083130	3/8	20/–	38/–	27	—	Fixed	0.120	Fixed
	17083132	3/8	20/–	38/–	27	—	Fixed	0.120	Fixed
	17083190	5/16	18/–	32/–	28	24	Fixed	0.120	Fixed
	17083192	5/16	18/–	32/–	28	24	Fixed	0.120	Fixed
	17083193	5/16	17/–	27/–	23	28	Fixed	0.120	Fixed
1984	17082130	3/8	20/–	38/–	27	—	Fixed	0.120	Fixed
	17082132	12/32	20/–	38/–	27	—	Fixed	0.120	Fixed
	17084191	10/32	18/–	32/–	28	24	Fixed	0.120	Fixed

① The carburetor identification number is stamped on the float bowl, next to the fuel inlet nut.   ② Inner hole.

FUEL SYSTEM

2. To adjust, use a 1/8" Allen wrench to turn the screw in the rear cover until the bubble is centered. After adjusting, apply silicone sealant RTV over the screw head to seal the setting.

UNLOADER ADJUSTMENT

1. Follow Steps 1-5 of the Fast Idle Cam Adjustment procedure.
2. If they have been previously removed, install the choke thermostatic cover and the coil assembly into the choke housing.
3. Close the choke valve by pushing up on the tang on the vacuum break lever (you may hold it with a rubber band).
4. Hold the primary throttle valves wide open.
5. To adjust, bend the tang on the fast idle lever until the bubble on the gauge is centered.

Rochester M4ME and E4ME 4-BBL Carburetors

NOTE: *Float, pump, choke coil lever, fast idle, fast idle cam (choke rod), front and rear vacuum break and unloader adjustments on these two carburetors are identical to those detailed in the M2ME and E2ME section. Please refer to them. There are, however, a number of procedures that apply only to the 4-bbl carburetors and to specific years. Please refer to the appropriate illustration.*

AIR VALVE ROD ADJUSTMENT

1. Using an outside vacuum source, seat the choke vacuum diaphragm. Put a piece of tape over the purge bleed hole if so equipped.
2. Close the air valve completely.
3. Insert the gauge between the rod and the end of the slot in the lever.
4. Bend the rod to adjust the clearance.

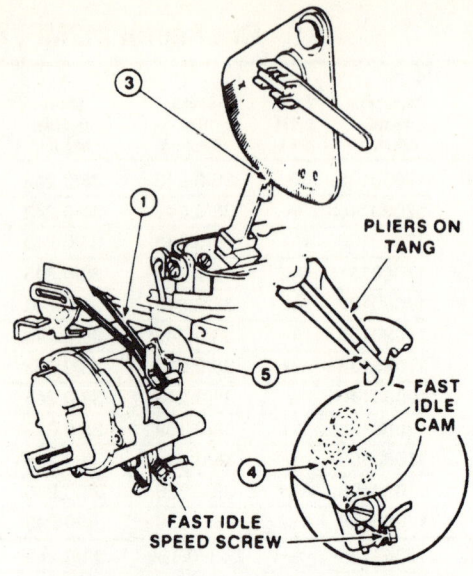

1. Attach rubber band to green tang of intermediate choke shaft
2. Open throttle to allow choke valve to close
3. Set up angle gage and set angle to specifications
4. Place cam follower on second step of cam, against rise of high step. If cam follower does not contact cam, turn in fast idle speed screw additional turn(s).
   Notice: Final fast idle speed adjustment must be performed according to under-hood emission control information label.
5. Adjust by bending tang of fast idle cam until bubble is centered

Choke rod fast idle cam adjustment E4ME 1983–88

SECONDARY LOCKOUT ADJUSTMENT

1. Pull the choke wide open by pushing out on the choke lever.
2. Open the throttle until the end of the secondary actuating lever is opposite the toe of the lockout lever.
3. Measure the clearance between the lock-

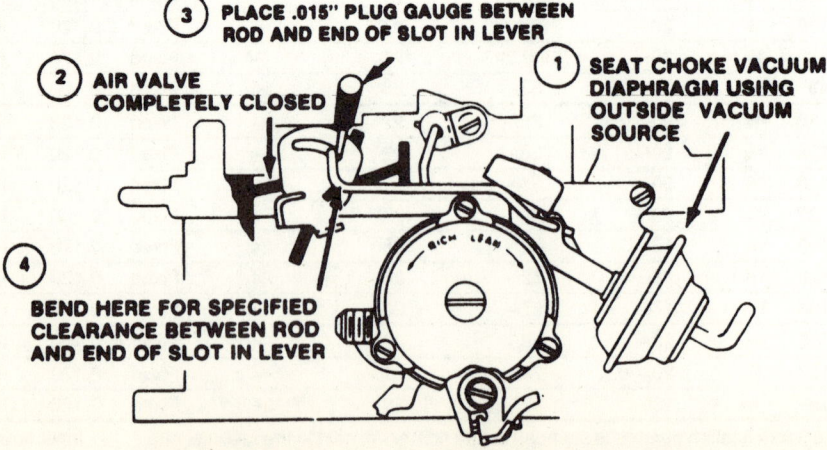

Adjusting the air valve rod—M4ME, E4ME (thru 1982)

FUEL SYSTEM

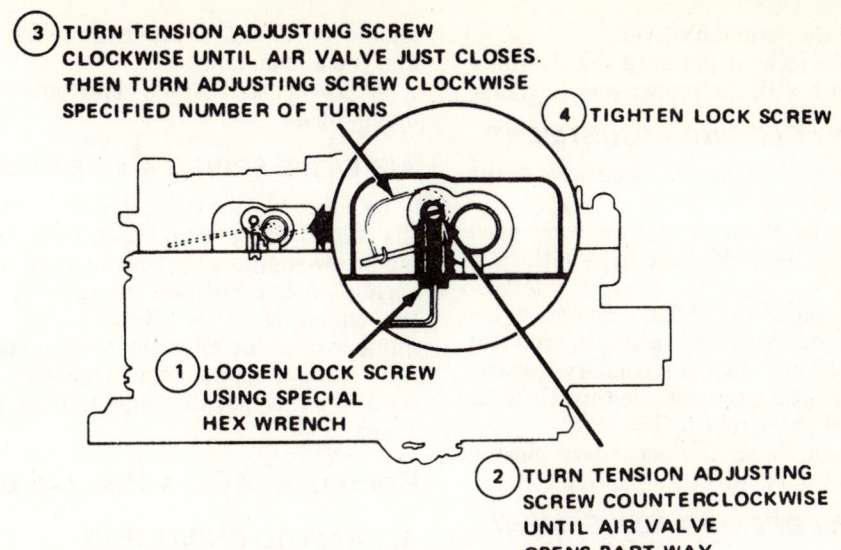

Adjusting the air valve spring—M4ME, E4ME

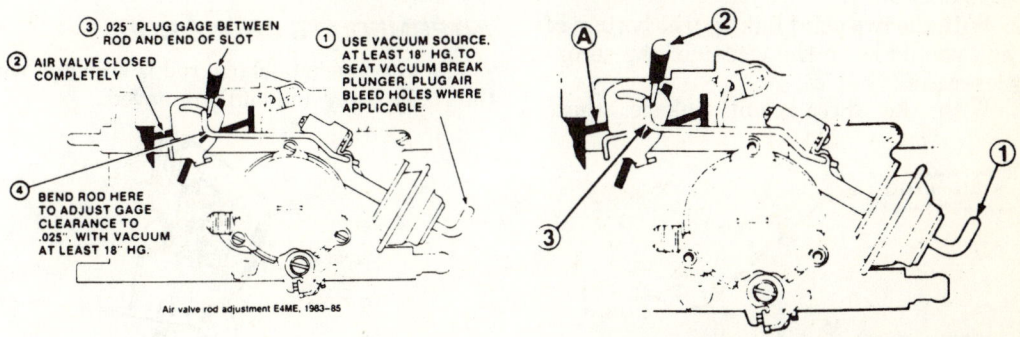

Air valve rod adjustment E4ME, 1983–85

1. Plug vacuum break bleed holes, if applicable. Air valves Ⓐ closed. Apply 15" Hg (51 k Pa) vacuum to seat vacuum break plunger.
2. Gage the clearance between air valve link and end of slot in lever.
3. Adjust, if necessary, by bending link.

Air valve rod adjustment E4ME 1986–88

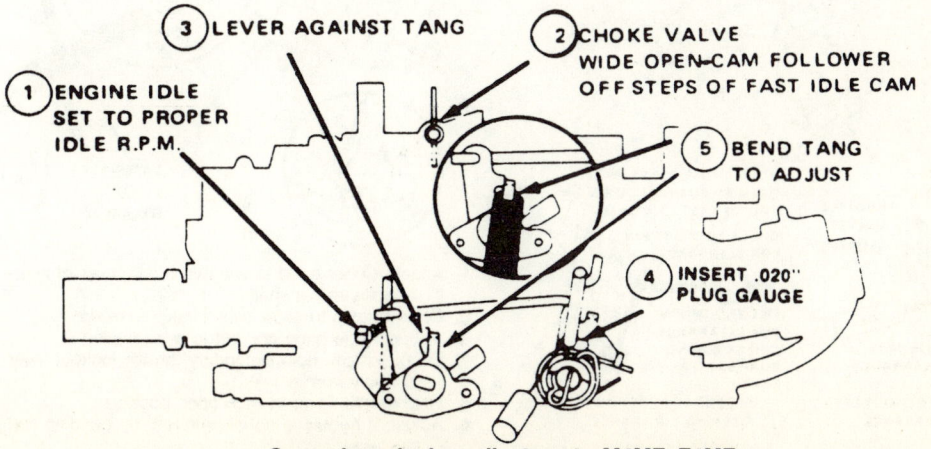

Secondary closing adjustment—M4ME, E4ME

FUEL SYSTEM

out lever and the secondary lever.

4. Bend the lockout pin until the clearance is in accordance with the proper specifications.

SECONDARY CLOSING ADJUSTMENT

1. Make sure that the idle speed is set to the proper specifications.
2. The choke valve should be wide open with the cam follower off of the steps of the fast idle cam.
3. There should be 0.020" clearance between the secondary throttle actuating rod and the front of the slot on the secondary throttle lever with the closing tang on the throttle lever resting against the actuating lever.
4. To adjust, bend the secondary closing tang on the primary throttle actuating rod.

SECONDARY OPENING ADJUSTMENT

1. Open the primary throttle valves until the actuating link contacts the upper tang on the secondary lever.
2. With the two point linkage, the bottom of the link should be in the center of the secondary lever slot.
3. With the three point linkage, there should be 0.070" clearance between the link and the middle tang.
4. To adjust, bend the upper tang on the secondary lever.

AIR VALVE SPRING ADJUSTMENT

To adjust the air valve spring windup, loosen the Allen lockscrew and then turn the adjusting screw counterclockwise so as to remove all spring tension. With the air valve closed, turn the adjusting screw clockwise the specified number of turns after the torsion spring contacts the pin on the shaft. Hold the adjusting screw in this position and tighten the lockscrew.

Rochester 4GC 4-BBL Carburetor

AUTOMATIC CHOKE ROD

Set the cover index mark on the one notch lean mark on the housing.

INTERMEDIATE CHOKE ROD

The intermediate choke rod is adjusted with the choke cover and baffle removed.

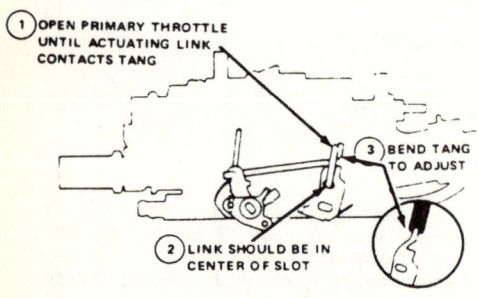

Secondary opening adjustment—M4ME, E4ME

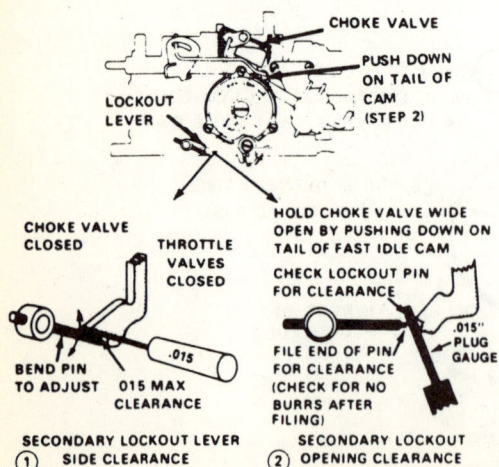

Secondary lockout adjustment—M4ME, E4ME

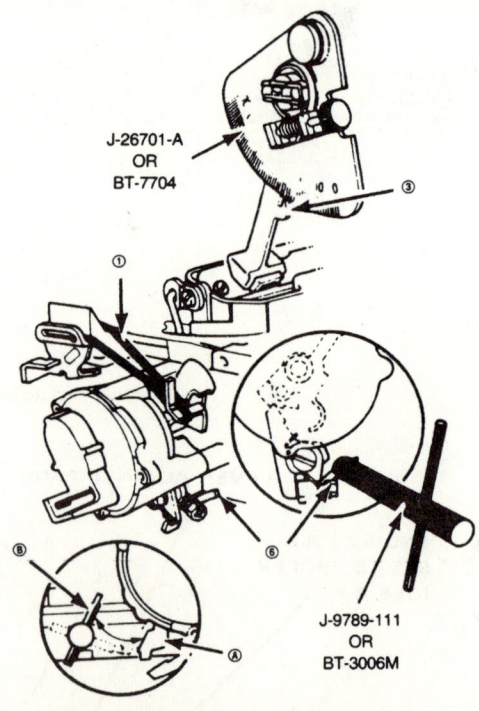

1. Attach rubber band to vacuum break lever of intermediate choke shaft
2. Open throttle to allow choke valve to close
3. Set up angle gage and set to specification
4. On quadrajet, hold secondary throttle lockout lever Ⓐ away from pin Ⓑ
5. Hold throttle lever in wide open position
6. Adjust, if bubble is not recentered, by bending fast idle lever

Unloader adjustment E4ME 1983–88

FUEL SYSTEM 273

1. Hold the choke valve closed, exert a light pressure on the choke piston to take up any lash, and then note if the choke piston is at the end of its sleeve.
2. If necessary, bend the intermediate choke rod for correct piston positioning.
3. Install the choke baffle and cover.

CHOKE ROD ADJUSTMENT

1. Turn the idle speed screw in until it just touches the second step of the fast idle cam.
2. Ensure that the choke trip lever is touching the choke counterweight lever.
3. While holding the idle speed screw on the second cam step and against the shoulder of the high step, there should be 0.043" clearance between the choke valve edge and the air horn dividing wall.
4. Bend the choke rod at a lower angle, if necessary.

CHOKE UNLOADER ADJUSTMENT

1. Hold the throttle valve wide open, while the choke trip lever touches the choke counterweight.
2. Clearance between the top of the choke valve and the dividing wall of the air horn should now be 0.235". Bend the fast idle cam tang, if necessary.

SECONDARY THROTTLE LOCKOUT ADJUSTMENT

1. Close the choke valve so that the secondary lockout tang is in the fast idle cam slot. Clearance between the fast idle cam and the tang should be 0.015".
2. Bend the tang horizontally as necessary to obtain the correct clearance.

FLOAT LEVEL AND DROP ADJUSTMENT

1. Remove the bowl cover.
2. Install a new gasket on the bowl cover surface.

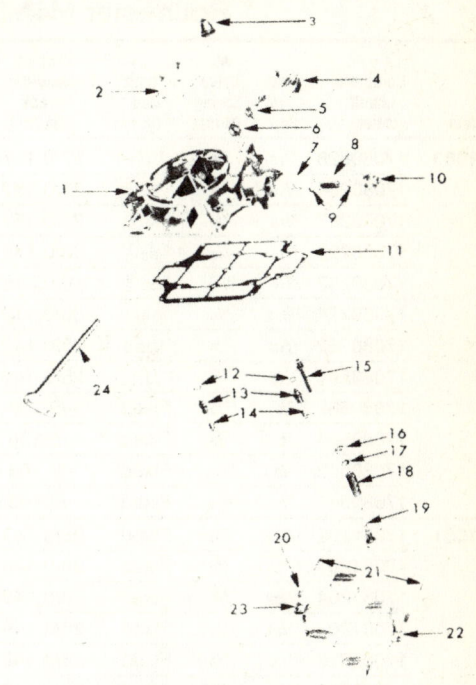

1. Bowl cover
2. Choke valve
3. Pump plunger rubber seal
4. Pump shaft and lever
5. Choke trip lever
6. Choke counterweight
7. Inlet filter spring
8. Inlet filter
9. Inlet gaskets
10. Inlet fitting
11. Bowl cover gasket
12. Float valve gasket
13. Float valve
14. Float valve needle
15. Power valve piston
16. Pump spring clip
17. Pump spring washer
18. Pump duration spring
19. Pump plunger
20. Float tension spring
21. Float hinge pin
22. Primary float
23. Secondary float
24. Choke valve shaft

Exploded view of 4GC air horn

3. Invert the cover and install the float lever gauges supplied with the carburetor rebuilding kit over the primary and secondary floats. The floats should just touch the gauges. The height from the bottom of the float to the bowl cover gasket is $1\frac{33}{64}$" for the primaries and $1\frac{37}{64}$" for the secondaries. Bend the float arms as necessary to obtain the correct level.
4. Center the floats in the level gauge, bending them to the left or right as necessary.

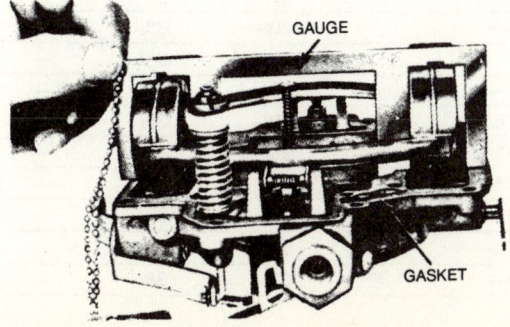

4GC float level adjustment

Rochester M4ME, E4ME Specifications

Year	Carburetor Identification ①	Float Level (in.)	Air Valve Spring (turn)	Pump Rod (in.)	Primary Vacuum Break (deg./in.)	Secondary Vacuum Break (deg./in.)	Secondary Opening (in.)	Choke Rod (deg./in.)	Choke Unloader (deg./in.)	Fast Idle Speed ② (rpm)
1980	17080828	7/16	7/8	1/4 ③	27/0.157	—	⑤	20/0.110	38/0.243	⑥
	17080204	7/16	7/8	1/4 ③	27/0.157	—	⑤	20/0.110	38/0.243	⑥
	17080207	7/16	7/8	1/4 ③	27/0.157	—	⑤	20/0.110	38/0.243	⑥
	17080228	7/16	7/8	9/32 ③	30/0.179	—	⑤	20/0.110	38/0.243	⑥
	17080243	3/16	9/16	9/32 ③	16/0.016	–/0.083	⑤	14.5/0.074	30/0.179	⑥
	17080274	15/32	5/8	5/16 ④	20/0.110	–/0.164	⑤	16/0.083	33/0.203	⑥
	17080282	7/16	7/8	11/32 ④	25/0.142	—	⑤	20/0.110	38/0.243	⑥
	17080284	7/16	7/8	11/32 ④	25/0.142	—	⑤	20/0.110	38/0.243	⑥
	17080502	1/2	7/8	Fixed	–/0.136	–/0.179	⑤	20/0.110	38/0.243	⑥
	17080504	1/2	7/8	Fixed	–/0.136	–/0.179	⑤	20/0.110	38/0.243	⑥
	17080542	3/8	9/16	Fixed	–/0.103	–/0.066	⑤	14.5/0.174	38/0.243	⑥
	17080543	3/8	9/16	Fixed	–/0.103	–/0.129	⑤	14.5/0.174	38/0.243	⑥
1981	17081202	11/32	7/8	Fixed	26/0.149	—	⑤	20/0.110	38/0.243	⑥
	17081203	11/32	7/8	Fixed	26/0.149	—	⑤	20/0.110	38/0.243	⑥
	17081204	11/32	7/8	Fixed	26/0.149	—	⑤	20/0.110	38/0.243	⑥
	17081207	11/32	7/8	Fixed	26/0.149	—	⑤	20/0.110	38/0.243	⑥
	17081216	11/32	7/8	Fixed	26/0.149	—	⑤	20/0.110	38/0.243	⑥
	17081217	11/32	7/8	Fixed	26/0.149	—	⑤	20/0.110	38/0.243	⑥
	17081218	11/32	7/8	Fixed	26/0.149	—	⑤	20/0.110	38/0.243	⑥
	17081242	5/16	7/8	Fixed	17/0.090	–/0.077	⑤	24.5/0.139	38/0.243	⑥
	17081243	1/4	7/8	Fixed	19/0.103	–/0.090	⑤	24.5/0.139	38/0.243	⑥
1982	17082202	11/32	7/8	Fixed	27	—	⑤	20/–	38/–	⑦
	17082204	11/32	7/8	Fixed	27	—	⑤	20/–	38/–	⑦
1983	17083202	11/32	7/8	Fixed	—	27/–	⑤	20/–	38/–	⑦
	17083203	11/32	7/8	Fixed	—	27/–	⑤	38/–	38/–	⑦
	17083204	11/32	7/8	Fixed	—	27/–	⑤	20/–	38/–	⑦
	17083207	11/32	7/8	Fixed	—	27/–	⑤	38/–	38/–	⑦
	17083216	11/32	7/8	Fixed	—	27/–	⑤	20/–	38/–	⑦
	17083218	11/32	7/8	Fixed	—	27/–	⑤	20/–	38/–	⑦
	17083236	11/32	7/8	Fixed	—	27/–	⑤	20/–	38/–	⑦
	17083506	7/16	7/8	Fixed	27/–	36/–	⑤	20/–	36/–	⑦
	17083508	7/16	7/8	Fixed	27/–	36/–	⑤	20/–	36/–	⑦
	17083524	7/16	7/8	Fixed	25/–	36/–	⑤	20/–	36/–	⑦
	17083526	7/16	7/8	Fixed	25/–	36/–	⑤	20/–	36/–	⑦
1984	17084201	11/32	7/8	Fixed	27/–	—	⑤	20/–	38/–	—
	17084205	11/32	7/8	Fixed	27/–	—	⑤	38/–	38/–	—
	17084208	11/32	7/8	Fixed	27/–	—	⑤	20/–	38/–	—
	17084209	11/32	7/8	Fixed	27/–	—	⑤	38/–	38/–	—
	17084210	11/32	7/8	Fixed	27/–	—	⑤	20/–	38/–	—
	17084507	7/16	1	Fixed	27/–	36/–	⑤	20/–	36/–	—
	17084509	7/16	1	Fixed	27/–	36/–	⑤	20/–	36/–	—
	17084525	7/16	1	Fixed	25/–	36/–	⑤	20/–	36/–	—
	17084527	7/16	1	Fixed	25/–	36/–	⑤	20/–	36/–	—
1985	17085202	11/32	7/8	Fixed	27/–	—	⑤	20/–	38/–	—
	17085203	11/32	7/8	Fixed	27/–	—	⑤	20/–	38/–	—

Rochester M4ME, E4ME Specifications

Year	Carburetor Identification ①	Float Level (in.)	Air Valve Spring (turn)	Pump Rod (in.)	Primary Vacuum Break (deg./in.)	Secondary Vacuum Break (deg./in.)	Secondary Opening (in.)	Choke Rod (deg./in.)	Choke Unloader (deg./in.)	Fast Idle Speed ② (rpm)
	17085204	11/32	7/8	Fixed	27/–	—	⑤	20/–	38/–	—
	17085207	11/32	7/8	Fixed	27/–	—	⑤	38/–	38/–	—
	17085218	11/32	7/8	Fixed	27/–	—	⑤	20/–	38/–	—
	17085502	7/16	7/8	Fixed	26/–	36/–	⑤	20/–	39/–	—
	17085503	7/16	7/8	Fixed	26/–	36/–	⑤	20/–	39/–	—
	17085506	7/16	1	Fixed	27/–	36/–	⑤	20/–	36/–	—
	17085508	7/16	1	Fixed	27/–	36/–	⑤	20/–	36/–	—
	17085524	7/16	1	Fixed	25/–	36/–	⑤	20/–	36/–	—
	17085526	7/16	1	Fixed	25/–	36/–	⑤	20/–	36/–	—
1986	17085502	7/16	7/8	Fixed	26/–	36/–	⑤	20/–	39/–	—
	17085503	7/16	7/8	Fixed	26/–	36/–	⑤	20/–	39/–	—
	17085506	7/16	1	Fixed	27/–	36/–	⑤	20/–	36/–	—
	17085508	7/16	1	Fixed	27/–	36/–	⑤	20/–	36/–	—
	17085524	7/16	1	Fixed	25/–	36/–	⑤	20/–	36/–	—
	17085526	7/16	1	Fixed	25/–	36/–	⑤	20/–	36/–	—
	17086003	11/32	7/8	Fixed	27/–	—	⑤	20/–	38/–	—
	17086004	11/32	7/8	Fixed	27/–	—	⑤	20/–	38/–	—
	17086005	11/32	7/8	Fixed	27/–	—	⑤	38/–	38/–	—
	17086006	11/32	7/8	Fixed	27/–	—	⑤	20/–	38/–	—
	17086040	11/32	7/8	Fixed	27/–	—	⑤	38/–	38/–	—
1987	17087129	11/32	7/8	Fixed	27/–	—	⑤	20/–	38/–	—
	17087130	11/32	7/8	Fixed	27/–	—	⑤	20/–	38/–	—
	17087132	11/32	7/8	Fixed	27/–	—	⑤	20/–	38/–	—
1988	17087306	11/32	7/8	Fixed	27/–	—	—	20/–	32/–	—
	17087129	11/32	7/8	Fixed	27/–	—	—	20/–	32/–	—
	17087132	11/32	7/8	Fixed	27/–	—	—	20/–	32/–	—

① The carburetor identification number is stamped on the float bowl, near the secondary throttle lever.
② With manual transmission; w/o vacuum advance and the throttle positioned on the high step of the cam
③ Inner hole
④ Outer hole
⑤ No measurement necessary on two point linkage; see text
⑥ 4 turns after contacting lever for preliminary setting
⑦ 4½ turns after contacting lever for preliminary setting

5. While holding the bowl cover in an upright position, measure the distance from the bowl cover gasket to the bottom of the float. This distance, float drop, should be $2^{1}/_{4}''$. Bend the float tang on the end of the hinge arm to correct the drop.

6. Install the bowl cover as outlined under overhaul.

Rochester 4MC, 4MV, M4MC 4-BBL Carburetors

The Rochester Quadrajet carburetor is a two stage, 4-barrel downdraft carburetor. The designation MC or MV refers to the type of choke system the carburetor is designed for. The MV model is equipped with a manifold thermostatic choke coil. The MC model has a choke housing and coil mounted on the side of the float bowl.

The primary side of the carburetor is equipped with $1^{3}/_{8}''$ diameter bores and a triple venturi with plain tube nozzles. During off idle and part throttle operation, the fuel is metered through tapered metering rods operating in specially designed jets positioned by a manifold vacuum responsive piston.

The secondary side of the carburetor con-

FUEL SYSTEM

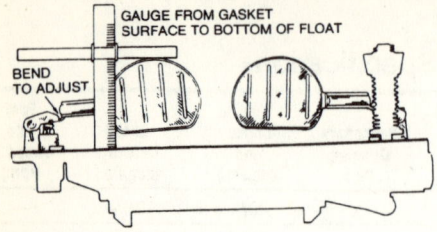

4GC float level adjustment

tains two 2¼" bores. An air valve is used on the secondary side for metering control and supplements the primary bores.

The secondary air valve operates tapered metering rods which regulate the fuel in constant proportion to the air being supplied.

ACCELERATOR PUMP

1. close the primary throttle valves by backing out the slow idle screw and making sure that the fast idle cam follower is off the steps of the fast idle cam.
2. Bend the secondary throttle closing tang away from the primary throttle lever.
3. With the pump in the appropriate hole in the pump lever, measure from the top of the choke valve wall to the top of the pump stem.
4. To adjust, bend the pump lever while supporting it, using the proper tool.
5. After adjusting, reading the secondary throttle tang and the slow idle screw.

IDLE VENT ADJUSTMENT

NOTE: *This adjustment is not required on 1977-79 carburetors.*

After adjusting the accelerator pump rod as specified above, open the primary throttle valve enough to just close the idle vent. Measure from the top of the choke valve wall to the top of the pump plunger stem. If adjustment is necessary, bend the wire tang on the pump lever.

FLOAT LEVEL

With the air horn assembly upside down, measure the distance from the air horn gasket surface (gasket removed) to the top of the float at the toe. Measure at a point $3/16$" back from the top of the float on 1977-79 carburetors.

NOTE: *Make sure the retaining pin is firmly held in place and that the tang of the float is firmly against the needle and seat assembly.*

FAST IDLE

1. Position the fast idle lever on the high step of the fast idle cam. Disconnect and plug the vacuum hose at the EGR valve.
2. Be sure that the choke is wide open and the engine warm.
3. Turn the fast idle screw to gain the proper fast idle rpm.

CHOKE ROD ADJUSTMENT

Position the cam follower on the second step of the fast idle cam, touching the high step. Close the choke valve directly on models up to 1976. On 1977 models, remove the choke thermostatic cover, and then hold the choke closed by pushing upward on the choke coil lever. Gauge the clearance between the lower edge of the choke valve and the carburetor body on models to 1975, and between the upper edge of the choke valve and the carburetor body on 1976 and 1977 models. Bend the choke rod to obtain the specified clearance. On 1977 models, install the choke thermostatic cover and adjust it to specification when the adjustment is complete. This adjustment requires sophisticated special tools on 1978 and later models, and so is not included here.

AIR VALVE DASHPOT ADJUSTMENT

Set the vacuum break diaphragm. On 1977-79 models, this requires plugging the bleed

QUADRAJET CARBURETOR SPECIFICATIONS
Chevrolet

Year	Carburetor Identification	Float Level (in.)	Air Valve Spring (turn)	Pump Rod (in.)	Primary Vacuum (deg./in.)	Secondary Vacuum (deg./in.)	Secondary Opening (in.)	Choke Rod (deg./in.)	Choke Unloader (deg./in.)	Fast Idle Speed (rpm)
1988	17087306	11/32	7/8	Fixed	27°	—	①	20°	32°	②
	17087129	11/32	7/8	Fixed	27°	—	①	20°	32°	②
	17087132	11/32	7/8	Fixed	27°	—	①	20°	32°	②
1989	17088115	11/32	1/2	Fixed	25°	43°	①	14°	35°	②
1990–92	17088115	11/32	1/2	Fixed	25°	43°	①	14°	35°	②
	17086008	11/32	1/2	Fixed	25°	43°	①	14°	35°	②

① No measurement necessary on two point linkage
② See underhood decal

FUEL SYSTEM 277

1. Air horn-to-bowl gasket
2. Air horn assembly
3. Air horn-to-bowl retaining screws
4. Idle vent valve lever
5. Idle vent valve
   a. Bimetal
   b. Spring
6. Choke shaft and lever
7. Choke valve
8. Idle vent shield
9. Countersunk air horn retaining screws
10. Secondary metering rods
11. Metering rod hanger

Exploded view of the air horn (4MV)

purge hole on the back of the diaphragm with tape, and the use of an external vacuum source. Hold the air valve tightly closed on all models. Gauge the clearance between the dashpot rod and the end of the slot in the air valve lever. Bend the rod to adjust. Remove the tape from the bleed purge hole.

CHOKE COIL ROD

1968-76

1. Close the choke valve by rotating the choke coil lever counterclockwise.
2. Disconnect the thermostatic coil rod

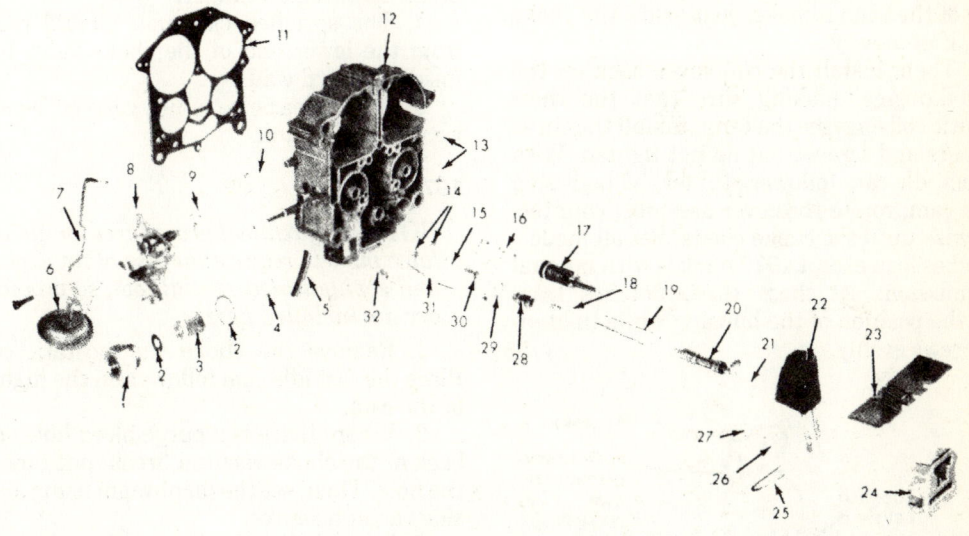

1. Fuel inlet nut
2. Gasket
3. Fuel filter
4. Fuel filter spring
5. Vacuum break hose
6. Vacuum diaphragm
7. Air valve dashpot
8. Choke control bracket
9. Fast idle cam
10. Secondary throttle lockout
11. Throttle body-to-bowl gasket
12. Float bowl assembly
13. Idle speed screw
14. Primary jets
15. Pump discharge ball
16. Pump return spring
17. Accelerator pump
18. Power piston spring
19. Primary metering rods
20. Power piston
21. Metering rod retainer
22. Float
23. Secondary air baffle
24. Float bowl insert
25. Float hinge pin
26. Float needle pull clip
27. Float needle
28. Float needle seat
29. Needle seat gasket
30. Discharge ball retainer
31. Choke rod
32. Choke lever

Exploded view of the float bowl (4MV)

from the upper lever.

3. Push down on the rod until it contacts the bracket of the coil.

4. The rod must fit in the notch of the upper lever.

5. If it does not, it must be bent on the curved portion just below the upper lever.

CHOKE COIL LEVER AND CHOKE THERMOSTATIC COIL

1977-79

1. Remove the three mounting screws and retainers, and pull the thermostatic coil cover assembly off the choke housing and set it aside.

2. Place the fast idle cam follower on the high step of the cam and then push up on the thermostatic coil tang in the choke housing until the choke is closed.

3. Insert a 0.120" plug gauge or rod of that diameter into the hole in the housing located just below the lever. With the choke closed, the lever should just touch the gauge.

4. Adjust the choke rod by changing the angle of the bend it makes just below the choke itself, if necessary.

5. Then, install the coil cover back on the choke housing, making sure that the thermostatic coil engages the tang. Install the three retainers and screws, but do not tighten. With the fast idle cam follower still on the high step of the cam, rotate the cover assembly counterclockwise until the choke closes. Set all models 2 notches lean except 1977 models with manual transmission; set these three notches lean. Hold the position of the housing while tightening screws evenly.

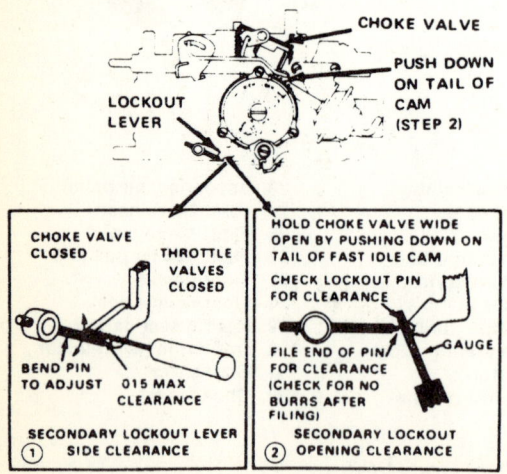

Adjusting the secondary lockout (1977–79)

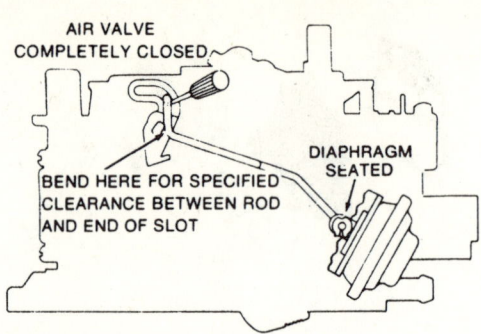

Adjusting the air valve dashpot (4MV)

VACUUM BREAK

1968-76

1. Fully seat the vacuum break diaphragm using an outside vacuum source.

2. Open the throttle valve enough to allow the fast idle cam follower to clear the fast idle cam.

3. The end of the vacuum break rod should be at the outer end of the slot in the vacuum break diaphragm plunger.

4. The specified clearance should register from the lower end of the choke valve to the inside air horn wall.

5. If the clearance is not correct, bend the vacuum break link.

1977

NOTE: *Adjustment procedures for 1978 and later models require the use of an expensive and sophisticated special tool, so procedures are not included here.*

1. Remove the choke thermostatic cover. Place the fast idle cam follower on the high step of the cam.

2. Where there is a purge bleed hole on the back of the choke vacuum break, put tape over the hole. Then, set the diaphragm using an outside vacuum source.

3. Push the inside choke coil lever counterclockwise until the tang on the vacuum break lever touches the tang on the vacuum break plunger stem.

4. Place a gauge of the following diameter between the upper edge of the choke butterfly and the inside wall of the air horn:

- California Engines: 0.165"
- All Other Engines: 0.160"

5. If the dimension is incorrect, adjust the screw on the vacuum break plunger stem until all play is taken up and choke butterfly just touches the gauge when it's held vertically.

6. Reconnect the vacuum line to vacuum break port of carburetor, remove tape from

FUEL SYSTEM 279

purge hole, if applied and reinstall and adjust choke thermostat.

CHOKE UNLOADER ADJUSTMENT

NOTE: *Performing this adjustment on 1978 and later models requires sophisticated special tools, so the procedure is not included here.*

On 1977 models, make sure the choke thermostatic spring is properly adjusted (see above). Close the choke valve and secure it with a rubber band hooked to the vacuum break lever. Open the primary throttles all the way. Then measure the distance between the air horn and edge of the choke butterfly. On models up to and including 1976, use the bottom side of the butterfly for this measurement; on 1977 models, use the top side. Bend the fast idle lever tang to achieve the proper opening of the choke.

SECONDARY LOCKOUT

1968-76

Completely open the choke valve and rotate the vacuum break lever clockwise. Bend the lever if the measurement between the lever and the secondary throttle exceeds specifications. Close the choke and gauge the distance between the lever and the secondary throttle shaft pin. Bend the lever to adjust.

1977-79

Please refer to the illustration that follows for this procedure.

AIR VALVE SPRING ADJUSTMENT

NOTE: *Loosening and tightening the locking screw to adjust the air valve spring requires a hex wrench on 1977 and later models carburetors.*

Remove all spring tension by loosening the locking screw and backing out the adjusting screw. Close the air valve and turn the adjusting screw in until the torsion spring touches the pin on the shaft, and then turn it the additional number of turns specified. Secure the locking screw.

SECONDARY CLOSING ADJUSTMENT

This adjustment assures proper closing of the secondary throttle plates.

1. Set the idle as per instructions in the appropriate section. Make sure that the fast idle cam follower is not resting on the fast idle cam.
2. There should be 0.020" clearance between the secondary throttle actuating rod and the front of the slot on the secondary throttle lever with the closing tang on the throttle lever resting against the actuating lever.
3. Bend the tang on the primary throttle actuating rod to adjust.

SECONDARY OPENING ADJUSTMENT

1. Open the primary throttle valves until the actuating link contacts the upper tang on the secondary lever.
2. With 2-point linkage, the bottom of the link should be in the center of the secondary lever slot.
3. With three point linkage, there should be 0.070" clearance between the link and the middle tang.
4. Bend the upper tang on the secondary lever to adjust as necessary.

Rochester E4MC 4-BBL Carburetor

FLOAT UNLOADER ADJUSTMENT

1. Remove the air horn and gasket.
2. Remove the solenoid plunger, metering rods and float bowl insert.

NOTE: *If necessary to remove the lean mixture solenoid adjusting screw, count and record the number of turns it takes to lightly bottom the screw, using tool J-28696 or BT-7928.*

3. Attach tool J-34817-1 or tool BT-8227-A to the float bowl.
4. Place tool J-34817-3 or tool BT-8227A-1 in the base with the contact pin resting on the outer edge of the float lever.
5. Measure the distance from the top of the casting to the top of the float, at a point $3/16$" from the large end of the float. Use tool J-9789-90 or tool BT-8037.
6. If more than $2/32$" from specification, use tool J-9789-90 or tool BT-8037 to bend the lever up or down. Remove the bending tool and check for proper specification. Repeat the process, as required.
7. Check for proper float alignment. Reassemble the carburetor.

AIR VALVE RETURN SPRING ADJUSTMENT

1. Loosen the set screw.
2. Turn the spring fulcrum pin counterclockwise until the air valves open.
3. Turn the pin clockwise until the air valves close, then any additional turns as specified.
4. Tighten the set screw. Apply lithium grease to the spring contact area.

CHOKE STAT LEVER ADJUSTMENT

1. If riveted, drill out and remove the rivets. Remove the choke cover and stat assembly.

Rochester 4MV, 4MC, M4MC Specifications

Year	Carburetor Identification ①	Float Level (in.)	Air Valve Spring	Pump Rod (in.)	Idle Vent (in.)	Vacuum Break (in.)	Secondary Opening (in.)	Choke Rod (rpm)	Choke Unloader Speed	Fast Idle
1968	7028212	9/32	3/8 turn	9/32	3/8	0.160	0.010	0.100	0.260	—
	7028213	9/32	3/8 turn	9/32	3/8	0.245	0.010	0.100	0.300	—
	7028229	9/32	7/8 turn	9/32	3/8	0.245	0.010	0.100	0.300	—
	7028208	9/32	3/8 turn	9/32	3/8	0.160	0.010	0.100	0.260	—
	7028207	9/32	3/8 turn	9/32	3/8	0.245	0.010	0.100	0.300	—
	7028219	9/32	7/8 turn	9/32	3/8	0.245	0.010	0.100	0.300	—
	7028218	3/16	7/8 turn	9/32	3/8	0.160	0.010	0.100	0.300	—
	7028217	3/16	7/8 turn	9/32	3/8	0.245	0.010	0.100	0.300	—
	7028210	3/16	7/8 turn	9/32	3/8	0.160	0.010	0.100	0.300	—
	7028211	3/16	7/8 turn	9/32	3/8	0.245	0.010	0.100	0.300	—
	7028216	3/16	7/8 turn	9/32	3/8	0.160	0.010	0.100	0.300	—
	7028209	3/16	7/8 turn	9/32	3/8	0.245	0.010	0.100	0.300	—
1969	7029203	7/32	7/16 turn	5/16	3/8	0.245	0.015	0.100	0.450	—
	7029202	7/32	7/16 turn	5/16	3/8	0.180	0.015	0.100	0.450	—
	7029207	3/16	13/16 turn	5/16	3/8	0.245	0.015	0.100	0.450	—
	7029215	1/4	13/16 turn	5/16	3/8	0.245	0.015	0.100	0.450	—
	7029204	1/4	13/16 turn	5/16	3/8	0.180	0.015	0.100	0.450	—
1970	7040202	1/4	7/16 turn	5/16	—	0.245	—	0.100	0.450	—
	7040203	1/4	7/16 turn	5/16	—	0.275	—	0.100	0.450	—
	7040207	1/4	13/16 turn	5/16	—	0.275	—	0.100	0.450	—
	7040200	1/4	13/16 turn	5/16	—	0.245	—	0.100	0.450	—
	7040201	1/4	13/16 turn	5/16	—	0.275	—	0.100	0.450	—
	7040204	1/4	13/16 turn	5/16	—	0.245	—	0.100	0.450	—
	7040205	1/4	13/16 turn	5/16	—	0.275	—	0.100	0.450	—
1971	7041200	1/4	7/16 turn	—	—	0.260	—	0.100	—	—
	7041202	1/4	7/16 turn	—	—	0.260	—	0.100	—	—
	7041204	1/4	7/16 turn	—	—	0.260	—	0.100	—	—
	7041212	1/4	7/16 turn	—	—	0.260	—	0.100	—	—
	7041201	1/4	7/16 turn	—	—	0.275	—	0.100	—	—
	7041203	1/4	7/16 turn	—	—	0.275	—	0.100	—	—
	7041205	1/4	7/16 turn	—	—	0.275	—	0.100	—	—
	7041213	1/4	7/16 turn	—	—	0.275	—	0.100	—	—
1972	7042202	1/4	1/2 turn	3/8	—	0.215	—	0.100	0.450	—
	7042203	1/4	1/2 turn	3/8	—	0.215	—	0.100	0.450	—
	7042902	1/4	1/2 turn	3/8	—	0.215	—	0.100	0.450	—
	7042903	1/4	1/2 turn	3/8	—	0.215	—	0.100	0.450	—
1973	7043212	7/32	1 turn	13/32	—	0.250	—	0.430	0.450	—
	7043213	7/32	1 turn	13/32	—	0.250	—	0.430	0.450	—
1974	7044202	1/4	7/8 turn	13/32 ②	—	0.230	—	0.430	0.450	1600③-1300④
	7044203	1/4	7/8 turn	13/32 ②	—	0.230	—	0.430	0.450	1600③-1300④
	7044208	1/4	1 turn	13/32 ②	—	0.230	—	0.430	0.450	1600③-1300④
	7044209	1/4	1 turn	13/32 ②	—	0.230	—	0.430	0.450	1600③-1300④

Rochester 4MV, 4MC, M4MC Specifications

Year	Carburetor Identification ①	Float Level (in.)	Air Valve Spring	Pump Rod (in.)	Idle Vent (in.)	Vacuum Break (in.)	Secondary Opening (in.)	Choke Rod (rpm)	Choke Unloader Speed	Fast Idle
	7044502	1/4	7/8 turn	13/32 ②	—	0.230	—	0.430	0.450	1600③–1300④
	7044503	1/4	7/8 turn	13/32 ②	—	0.230	—	0.430	0.450	1600③–1300④
1975	7045202	15/32	7/8 turn	0.275 ⑤	—	0.180 ⑥	—	0.300	0.325	—
	7045203	15/32	7/8 turn	0.275 ⑤	—	0.180 ⑥	—	0.300	0.325	—
	7045208	15/32	7/8 turn	0.275 ⑤	—	0.180 ⑥	—	0.300	0.325	—
	7045209	15/32	7/8 turn	0.275 ⑤	—	0.180 ⑥	—	0.300	0.325	—
1976	17056202	13/32	7/8 turn	9/32	—	0.185	—	0.325	0.325	—
	17056203	13/32	7/8 turn	9/32	—	0.170	—	0.325	0.325	—
	17056528	13/32	7/8 turn	9/32	—	0.185	—	0.325	0.325	—
1977	17057203	15/32	7/8	9/32 ⑤	—	0.160	—	0.325	0.280	1300
	17057202	15/32	—	9/32 ⑤	—	0.160	—	0.325	0.280	1600 ⑤
	17057502	15/32	7/8	9/32 ⑤	—	0.165	—	0.325	0.280	1600 ⑤
1978	17058203	15/32	7/8	9/32	—	0.179	—	0.314	0.277	⑦
	17058202	15/32	7/8	9/32	—	0.179	—	0.314	0.277	⑦
	17058502	15/32	7/8	9/32	—	0.187	—	0.314	0.277	⑦
1979	17059203	15/32	7/8	1/4	—	0.157	—	0.243	0.243	⑦
	17059207	15/32	7/8	1/4	—	0.157	—	0.243	0.243	⑦
	17059216	15/32	7/8	1/4	—	0.157	—	0.243	0.243	⑦
	17059217	15/32	7/8	1/4	—	0.157	—	0.243	0.243	⑦
	17059218	15/32	7/8	1/4	—	0.164	—	0.243	0.243	⑦
	17059222	15/32	7/8	1/4	—	0.164	—	0.243	0.243	⑦
	17059502	15/32	7/8	1/4	—	0.164	—	0.243	0.243	⑦
	17059504	15/32	7/8	1/4	—	0.164	—	0.243	0.243	⑦
	17059582	15/32	7/8	11/32	—	0.203	—	0.243	0.314	⑦
	17059584	15/32	7/8	11/32	—	0.203	—	0.243	0.314	⑦
	17059210	15/32	1	9/32	—	0.157	—	0.243	0.243	⑦
	17059211	15/32	1	9/32	—	0.157	—	0.243	0.243	⑦
	17059228	15/32	1	9/32	—	0.157	—	0.243	0.243	⑦

① The carburetor identification tag is located at the rear of the carburetor on one of the air horn screws
② Without vacuum advance
③ With automatic transmission; vacuum advance connected and EGR disconnected after the throttle positioned on the high step of cam
④ With manual transmission; without vacuum advance and the throttle positioned on the high step of cam
⑤ Inner pump rod location
⑥ Front vacuum break given; rear—0.170 in.
⑦ See engine compartment sticker

2. Place the fast idle cam on the high step against the cam follower lever.

3. Push up on the choke stat lever to close the choke valve.

4. Check the stat lever for correct orientation by inserting a 0.120" plug gauge in the hole.

5. The gauge should fit in the hole and touch the edge of the lever. Adjust, as required by bending the choke link.

CHOKE VALVE ANGLE GAUGE

1. Attach the angle gauge magnet to the choke valve.

2. Rotate the degree scale, of the tool, until zero is opposite the pointer.

3. Center the leveling bubble. Rotate the scale to the specified angle.

4. Open the choke valve. Adjust the linkage if the bubble is not recentered.

FUEL SYSTEM

FAST IDLE CAM ADJUSTMENT

1. Attach a rubber band to the vacuum brake lever of the intermediate choke shaft.
2. Open the throttle and allow the choke valve to close.
3. Set up the angle gauge tool to specification.
4. Place the fast idle cam **A** on the second step against cam follower lever **B**, with the lever contacting the rise of the high step. If the lever does not contact the cam, turn the fast idle adjusting screw **C** in an additional turn until contact is made.
5. Adjust, if the bubble is not recentered by bending the fast idle cam kick lever.

VACUUM BRAKE ADJUSTMENT DATA

FRONT VACUUM BRAKE ADJUSTMENT

1. Attach a rubber band to the vacuum brake lever of the intermediate choke shaft.
2. Open the throttle and allow the choke valve to close.
3. Set up the angle gauge tool to specification.
4. Plug the vacuum brake bleed holes, as required.
5. Apply 15 in.Hg of vacuum to the vacuum break plunger. As required, seat bucking spring **A**.
6. If necessary, bend the air valve link **B** to permit full plunger travel. Reapply the vacuum source.
7. Adjust as necessary if the bubble on the gauge is not recentered, by turning the screw.

REAR VACUUM BREAK ADJUSTMENT

1. Attach a rubber band to the vacuum brake lever of the intermediate choke shaft.
2. Open the throttle and allow the choke valve to close.
3. Set up the angle gauge tool to specification.
4. Plug the vacuum brake bleed holes, as required.
5. Apply 15 in.Hg of vacuum to the vacuum break plunger. As required, compress the plunger bucking spring.
6. If necessary, bend the air valve link **A** to permit full plunger travel. Reapply the vacuum source.
7. Adjust as necessary if the bubble on the gauge is not recentered, by supporting the "S" and bending the vacuum break link or by turning the screw with a hex head wrench.

FRONT AIR VALVE LINK ADJUSTMENT

1. As required, plug the vacuum break bleed holes. Close the air valves, **A**.
2. Apply 15 in.Hg of vacuum to seat the vacuum break plunger.
3. Gauge the clearance between the air valve link and the end of the slot in the lever.
4. Adjust, as required by bending the link.

1. REMOVE AIR HORN & GASKET.
2. REMOVE SOLENOID PLUNGER, METERING RODS, FLOAT BOWL INSERT. IF NECESSARY TO REMOVE SOLENOID (LEAN MIXTURE) ADJUSTING SCREW. COUNT AND MAKE RECORD OF NUMBER OF TURNS IT TAKES TO LIGHTLY BOTTOM SCREW, USING J-28696-10 OR BT-7928. (RETURN TO EXACT POSITION WHEN REASSEMBLING.)
3. ATTACH J-34817-1 OR BT-8227A-1 TO FLOAT BOWL.
4. PLACE J-34817-3 OR BT-8227A IN BASE WITH CONTACT PIN RESTING ON OUTER EDGE OF FLOAT LEVER.
5. MEASURE DISTANCE FROM TOP OF CASTING TO TOP OF FLOAT, AT POINT 3/16" FROM LARGE END OF FLOAT. USE J-9789-90 OR BT-8037.
6. IF MORE THAN ±2/32" FROM SPECIFICATION. USE J-34817-15 OR BT-8233 TO BEND LEVER UP OR DOWN. REMOVE BENDING TOOL AND MEASURE, REPEATING UNTIL WITHIN SPECIFICATION.
7. CHECK FLOAT ALIGNMENT.
8. REASSEMBLE CARBURETOR.

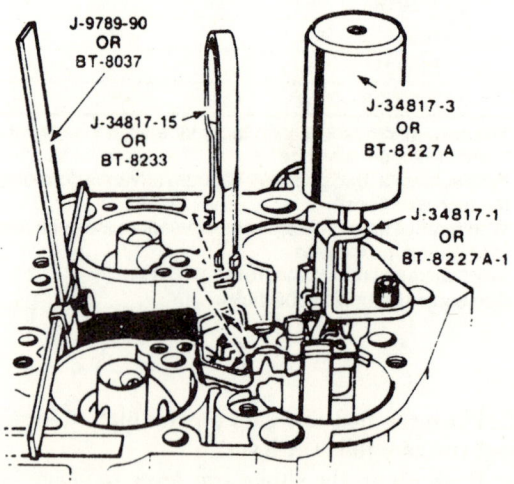

Float adjustment E4MC

FUEL SYSTEM 283

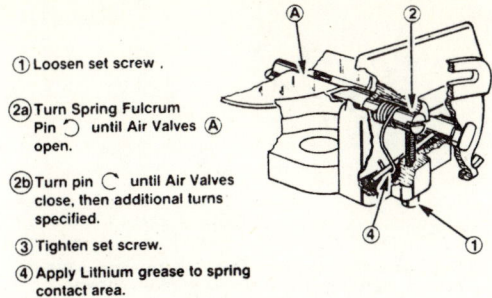

① Loosen set screw.
②a Turn Spring Fulcrum Pin ↺ until Air Valves Ⓐ open.
②b Turn pin ↻ until Air Valves close, then additional turns specified.
③ Tighten set screw.
④ Apply Lithium grease to spring contact area.

Air valve return spring adjustment E4MC

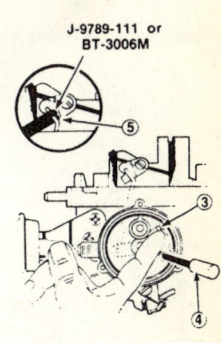

① If riveted, drill out and remove rivets. Remove Choke Cover and Stat Assembly.
② Place Fast Idle Cam on high step against Cam Follower Lever.
③ Push up on Choke Stat Lever to close Choke Valve.
④ Check Stat Lever for correct orientation by inserting .120" plug gage in hole.
Gage should fit in hole and touch edge of lever.
⑤ Adjust, if necessary, by bending Choke Link.

Choke stat lever adjustment E4MC

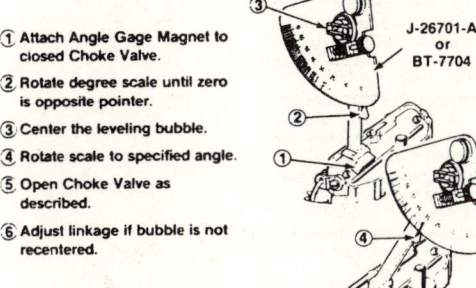

① Attach Angle Gage Magnet to closed Choke Valve.
② Rotate degree scale until zero is opposite pointer.
③ Center the leveling bubble.
④ Rotate scale to specified angle.
⑤ Open Choke Valve as described.
⑥ Adjust linkage if bubble is not recentered.

Choke valve angle gauge E4MC

UNLOADER ADJUSTMENT

1. Attach a rubber band to the vacuum brake lever of the intermediate choke shaft.
2. Open the throttle and allow the choke valve to close.
3. Set up the angle gauge tool to specification.
4. Hold the secondary throttle lockout lever **A** away from pin **B**.
5. Hold the throttle lever in the wide open position. Adjust the bubble if it is not recentered by bending the fast idle lever.

SECONDARY THROTTLE LOCKOUT ADJUSTMENT

1. Place the fast idle cam **A** on the high step against the cam follower lever.
2. Hold the throttle lever closed. Gauge the

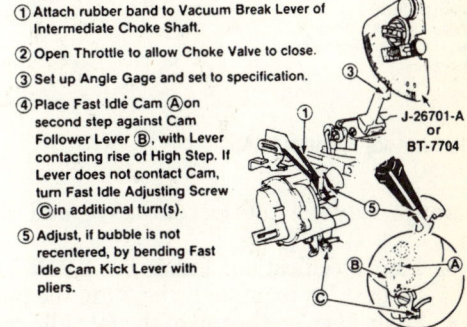

① Attach rubber band to Vacuum Break Lever of Intermediate Choke Shaft.
② Open Throttle to allow Choke Valve to close.
③ Set up Angle Gage and set to specification.
④ Place Fast Idle Cam Ⓐ on second step against Cam Follower Lever Ⓑ, with Lever contacting rise of High Step. If Lever does not contact Cam, turn Fast Idle Adjusting Screw Ⓒ in additional turn(s).
⑤ Adjust, if bubble is not recentered, by bending Fast Idle Cam Kick Lever with pliers.

Fast idle cam adjustment E4MC

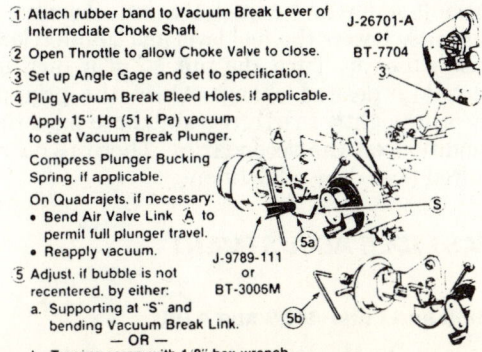

① Attach rubber band to Vacuum Break Lever of Intermediate Choke Shaft.
② Open Throttle to allow Choke Valve to close.
③ Set up Angle Gage and set to specification.
④ Plug Vacuum Break Bleed Holes, if applicable.
Apply 15" Hg (51 k Pa) vacuum to seat Vacuum Break Plunger.
Compress Plunger Bucking Spring, if applicable.
On Quadrajets, if necessary:
• Bend Air Valve Link Ⓐ to permit full plunger travel.
• Reapply vacuum.
⑤ Adjust, if bubble is not recentered, by either:
   a. Supporting at "S" and bending Vacuum Break Link.
   — OR —
   b. Turning screw with 1/8" hex wrench.

Rear vacuum break adjustment E4MC

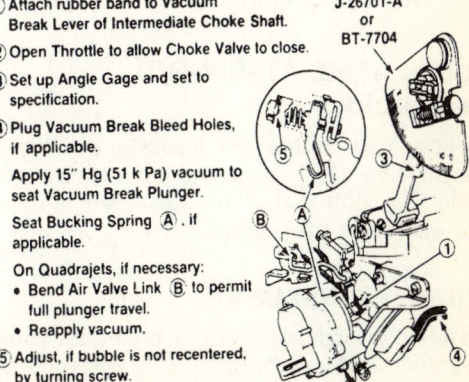

① Attach rubber band to Vacuum Break Lever of Intermediate Choke Shaft.
② Open Throttle to allow Choke Valve to close.
③ Set up Angle Gage and set to specification.
④ Plug Vacuum Break Bleed Holes, if applicable.
Apply 15" Hg (51 k Pa) vacuum to seat Vacuum Break Plunger.
Seat Bucking Spring Ⓐ, if applicable.
On Quadrajets, if necessary:
• Bend Air Valve Link Ⓑ to permit full plunger travel.
• Reapply vacuum.
⑤ Adjust, if bubble is not recentered, by turning screw.

Front vacuum break adjustment E4MC

FUEL SYSTEM

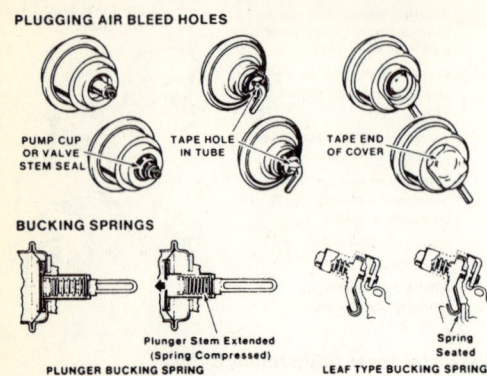

Vacuum break assemblies E4MC

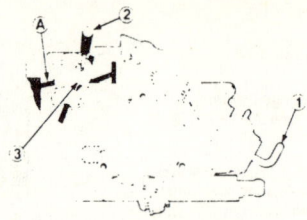

1. Plug Vacuum Break bleed holes, if applicable. Air Valves A closed. Apply 15" Hg (51 k Pa) vacuum to seat Vacuum Break Plunger.
2. Gage the clearance between Air Valve Link and end of slot in lever.
3. Adjust, if necessary, by bending link.

Air valve link adjustment E4MC

1. Attach rubber band to Vacuum Break Lever of Intermediate Choke Shaft.
2. Open Throttle to allow Choke Valve to close.
3. Set up Angle Gage and set to specification.
4. On Quadrajet, hold Secondary Throttle Lockout Lever Ⓐ away from pin Ⓑ.
5. Hold Throttle Lever in wide open position.
6. Adjust, if bubble is not recentered, by bending Fast Idle Lever.

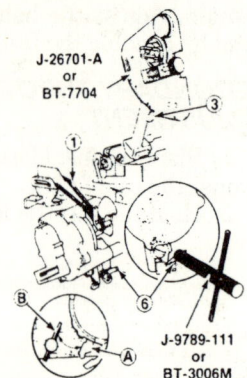

Unloader adjustment E4MC

1. Place Fast Idle Cam Ⓐ on high step against Cam Follower Lever.
2. Hold Throttle Lever closed.
3. Gage the clearance between Lockout Lever and pin. It must be .015" ±.005".
4. Adjust, if necessary, by bending pin.
5. Push down on tail of Fast Idle Cam Ⓐ to move Lockout Lever away from pin.
6. Rotate Throttle Lever to bring Lockout Pin to position of minimum clearance with Lockout Lever.
7. Gage the clearance between Lockout Lever and pin. Minimum must be .015".
8. Adjust, if necessary, by filing end of pin.

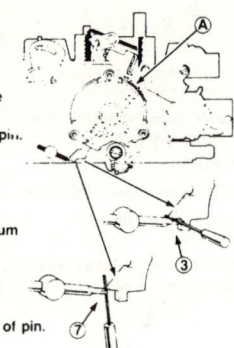

Secondary lockout adjustment E4MC

clearance between the lockout lever and the pin.

3. The specification must be 0.015" ± 0.005". Adjust, as required by bending the pin.
4. Push down on the tail of the fast idle cam **A** to move the lockout lever away from the pin.
5. Gauge the clearance between the lockout lever and the pin. Minimum clearance must be 0.015".
6. Adjust, as required by bending the pin.

Holley 4150, 4160 4-BBL Carburetors

These carburetors are basically similar in design. The 4160 is an end-inlet carburetor, while the 4150 carburetor has been both an end and center inlet design.

CHOKE ADJUSTMENT

The early model 4150 uses a bimetallic choke mounted on the carburetor. It is correctly set when the cover scribe mark aligns with the specified notch mark. The later model 4150 and 4160 employ a remotely located choke. To adjust, disconnect the choke rod at the choke lever and secure the choke lever shut. Bend the rod so that when the rod is depressed to the contact stop, the top is even with the bottom of the hole in the choke lever.

FLOAT LEVEL ADJUSTMENT

Position the vehicle on a flat, level surface and start the engine. Remove the sight plugs and check to see that the fuel level reaches the bottom threads of the sight plug port. A plus or minus tolerance of $1/32$". is acceptable. To change the level, loosen the fuel inlet needle locking screw and adjust the nut. Turning it clockwise lowers the fuel level and counterclockwise raises it. Turn the nut $1/6$ of a turn for each $1/16$" desired change. Open the primary throttle slightly to assure a stabilized adjusting condition on the secondaries. There is no required float drop adjustment.

FAST IDLE ADJUSTMENT

1968 and Later 4150 and 4160

Open the throttle and place the choke plate fast idle lever against the top step of the fast idle cam. Bend the fast idle lever to obtain the specified throttle plate opening.

FUEL SYSTEM 285

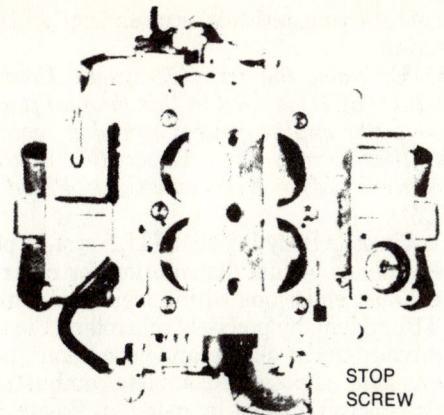

Adjusting the secondary throttle valve (HOLLEY)

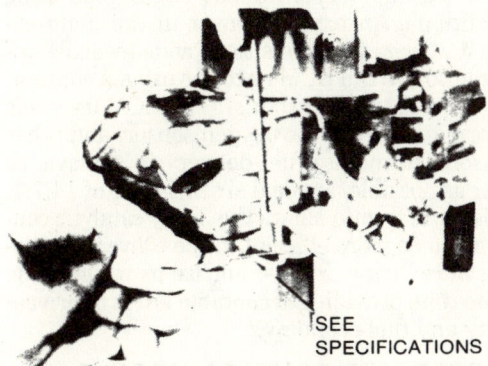

Adjusting the accelerator pump (HOLLEY)

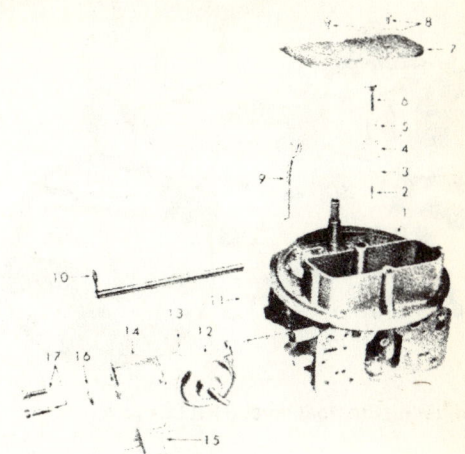

1. Main body assembly
2. Pump discharge needle
3. Pump discharge nozzle gasket
4. Pump discharge nozzle
5. Pump discharge nozzle screw gasket
6. Pump discharge nozzle screw
7. Choke valve
8. Choke valve screw
9. Choke rod
10. Choke shaft and lever
11. Choke rod seal
12. Vacuum break
13. Vacuum break link
14. Choke lever
15. Fast idle cam
16. Choke lever retainer
17. Vacuum break screw

Exploded view of the main body assembly (HOLLEY)

CHOKE UNLOADER ADJUSTMENT

Adjustment should be make with the engine not running. Fully open and secure the throttle plate. Force the choke valve toward a closed position, so that contact is made with the unloader tang. Bend the choke rod to gain the specified clearance between the main body and the lower edge of the choke valve.

ACCELERATOR PUMP ADJUSTMENT

With the engine off, block the throttle open and push the pump lever down, Clearance between the pump lever arm and the spring ad-

Holley 4150, 4160 Specifications

Year	Model or Type	Float Level (in.) Prim	Float Level (in.) Sec	Float Drop (in.) Prim	Float Drop (in.) Sec	Pump Travel Setting (in.)	Choke Setting Unloader (in.)	Choke Setting Housing	Secondary Locknut Adj.
1968	327-325 hp—4 bbl (4150)	A①		A①	0.066	0.015	0.265	—	—
1968–69	V8-396 (4150)	B①		B①	0.066	0.015	0.350	—	—
1970	454 (4150)	0.350		—	—	0.015	0.350	—	—
1971	454 (4160)	②		①	—	0.015	0.350	—	—

A—Primary 0.170, Secondary 0.300
B—Primary 0.350, Secondary 0.500
① Float adjustment: Fuel level should be plus or minus 1/32 in. with threads at bottom of sight holes. To adjust turn adjusting nut on top of bowl clockwise, to lower, counterclockwise to raise.
② Float centered in bowl

286 FUEL SYSTEM

Adjusting the float level (HOLLEY)

justing nut should be 0.015″ minimum. Turn the screw or nut to adjust this clearance.

SECONDARY THROTTLE VALVE ADJUSTMENT

Close the throttle plates, and then turn the adjustment screw until it contacts the throttle lever. Advance the screw 1/2 turn more.

AIR VENT VALVE ADJUSTMENT

Close the throttle valves and open the choke valve so that the throttle arm is free of the idle screw. Bend the air vent valve rod to obtain the specified clearance between the choke valve and seat. Advance the idle speed screw until it touches the throttle lever, and then advance it 1 1/2 turns.

VACUUM BREAK ADJUSTMENT

Secure the choke valve closed and the vacuum break against the stop. Bend the vacuum break link to gain the specified clearance between the main body and the lower edge of the choke valve.

GASOLINE FUEL INJECTION SYSTEM

General Information

The electronic throttle body fuel injection system is a fuel metering system with the amount of fuel delivered by the throttle body injector(s) (TBI) determined by an electronic signal supplied by the Electronic Control Module (ECM) or Powertrain Control Module (PCM). The ECM monitors various engine and vehicle conditions to calculate the fuel delivery time (pulse width) of the injector(s). The fuel pulse may be modified by the ECM to account for special operating conditions, such as cranking, cold starting, altitude, acceleration, and deceleration.

NOTE: *When the term Electronic Control Module (ECM) is used in this manual it will refer to the engine control computer regardless that it may be a Powertrain Control Module (PCM) or Electronic Control Module (ECM).*

The Throttle Body Injection (TBI) system provides a means of fuel distribution for controlling exhaust emissions within legislated limits. The TBI system, by precisely controlling the air/fuel mixture under all operating conditions, provides as near as possible complete combustion.

This is accomplished by using an Electronic Control Module (ECM) (a small on-board microcomputer) that receives electrical inputs from various sensors about engine operating conditions. An oxygen sensor in the main exhaust stream functions to provide feedback information to the ECM as to the oxygen content, lean or rich, in the exhaust. The ECM uses this information from the oxygen sensor, and other sensors, to modify fuel delivery to achieve, as near as possible, an ideal air/fuel ratio of 14.7:1. This air/fuel ratio allows the 3-way catalytic converter to be more efficient in the conversion process of reducing exhaust emissions while at the same time providing acceptable levels of driveability and fuel economy.

ELECTRONIC CONTROL MODULE

The ECM program electronically signals the fuel injector in the TBI assembly to provide the correct quantity of fuel for a wide range of operating conditions. Several sensors are used to determine existing operating conditions and the ECM then signals the injector to provide the precise amount of fuel required.

The ECM used on EFI vehicles has a learning capability. If the battery is disconnected to clear diagnostic codes, or for repair, the learning process has to begin all over again. A change may be noted in vehicle performance. To teach the vehicle, make sure the vehicle is at operating temperature and drive at part throttle, under moderate acceleration and idle conditions, until performance returns.

With the EFI system, the TBI assembly is centrally located on the intake manifold where air and fuel are distributed through a single bore in the throttle body, similar to a carbureted engine. Air for combustion is controlled by a single throttle valve which is connected to the accelerator pedal linkage by a throttle shaft and lever assembly. A special plate is located directly beneath the throttle valve to aid in mixture distribution.

Fuel for combustion is supplied by 1 or 2 fuel injector(s), mounted on the TBI assembly, whose metering tip is located directly above the throttle valve. The injector is pulsed or timed open or closed by an electronic output signal received from the ECM. The ECM receives inputs concerning engine operating conditions from the various sensors (coolant temperature sensor, oxygen sensor, etc.). The ECM, using this information, performs high speed calculations of engine fuel requirements and pulses or times the injector, open or closed, thereby controlling fuel and air mixtures to achieve, as near as possible, ideal air/fuel mixture ratios.

When the ignition key is turned **ON**, the ECM will initialize (start program running) and energize the fuel pump relay. The fuel pump pressurizes the system to approximately 10 psi. If the ECM does not receive a distributor reference pulse (telling the ECM the engine is turning) within 2 seconds, the ECM will then de-energize the fuel pump relay, turning off the fuel pump. If a distributor reference pulse is later received, the ECM will turn the fuel pump back on.

The ECM controls the exhaust emissions by modifying fuel delivery to achieve, as near as possible, and air/fuel ratio of 14.7:1. The injector on-time is determined by various inputs to the ECM. By increasing the injector pulse, more fuel is delivered, enriching the air/fuel ratio. Decreasing the injector pulse, leans the air/fuel ratio. Pulses are sent to the injector in 2 different modes: synchronized and nonsynchronized.

Synchronized Mode

In synchronized mode operation, the injector is pulsed once for each distributor reference pulse. In dual injector throttle body systems, the injectors are pulse alternately.

Nonsynchronized Mode

In nonsynchronized mode operation, the injector is pulsed once every 12.5 milliseconds or 6.25 milliseconds depending on calibration. This pulse time is totally independent of distributor reference pulses. Nonsynchronized mode results only under the following conditions:
  1. The fuel pulse width is too small to be delivered accurately by the injector (approximately 1.5 milliseconds)
  2. During the delivery of prime pulses (prime pulses charge the intake manifold with fuel during or just prior to engine starting)
  3. During acceleration enrichment
  4. During deceleration leanout

The basic TBI unit is made up of 2 major casting assemblies: (1) a throttle body with a valve to control airflow and (2) a fuel body assembly with an integral pressure regulator and fuel injector to supply the required fuel. An electronically operated device to control the idle speed and a device to provide information regarding throttle valve position are included as part of the TBI unit.

The fuel injector(s) is a solenoid-operated device controlled by the ECM. The incoming fuel is directed to the lower end of the injector assembly which has a fine screen filter surrounding the injector inlet. The ECM actuates the solenoid, which lifts a normally closed ball valve off a seat. The fuel under pressure is injected in a conical spray pattern at the walls of the throttle body bore above the throttle valve. The excess fuel passes through a pressure regulator before being returned to the vehicle's fuel tank.

The pressure regulator is a diaphragm-operated relief valve with injector pressure on one side and air cleaner pressure on the other. The function of the regulator is to maintain a constant pressure drop across the injector throughout the operating load and speed range of the engine.

The throttle body portion of the TBI may contain ports located at, above, or below the throttle valve. These ports generate the vacuum signals for the EGR valve, MAP sensor, and the canister purge system.

The Throttle Position Sensor (TPS) is a variable resistor used to convert the degree of throttle plate opening to an electrical signal to the ECM. The ECM uses this signal as a reference point of throttle valve position. In addition, an Idle Air Control (IAC) assembly, mounted in the throttle body s used to control idle speeds. A cone-shaped valve in the IAC assembly is located in an air passage in the throttle body that leads from the point beneath the air cleaner to below the throttle valve. The ECM monitors idle speeds and, depending on engine load, moves the IAC cone in the air passage to increase or decrease air bypassing the throttle valve to the intake manifold for control of idle speeds.

Cranking Mode

During engine crank, for each distributor reference pulse the ECM will deliver an injector pulse (synchronized). The crank air/fuel ratio will be used if the throttle position is less than 80% open. Crank air fuel is determined by the ECM and ranges from 1.5:1 at –33°F (–36°C) to 14.7:1 at 201°F (94°C).

The lower the coolant temperature, the longer the pulse width (injector on-time) or richer the air/fuel ratio. The higher the coolant temperature, the less pulse width (injector on-time) or the leaner the air/fuel ratio.

FUEL SYSTEM

Clear Flood Mode

If for some reason the engine should become flooded, provisions have been made to clear this condition. To clear the flood, the driver must depress the accelerator pedal enough to open to wide-open throttle position. The ECM then issues injector pulses at a rate that would be equal to an air/fuel ratio of 20:1. The ECM maintains this injector rate as long as the throttle remains wide open and the engine rpm is below 600. If the throttle position becomes less than 80%, the ECM then would immediately start issuing crank pulses to the injector calculated by the ECM based on the coolant temperature.

Run Mode

There are 2 different run modes. When the engine rpm is above 400, the system goes into open loop operation. In open loop operation, the ECM will ignore the signal from the oxygen (O_2) sensor and calculate the injector on-time based upon inputs from the coolant and manifold absolute pressure sensors.

During open loop operation, the ECM analyzes the following items to determine when the system is ready to go to the closed loop mode:

1. The oxygen sensor varying voltage output. (This is dependent on temperature).
2. The coolant sensor must be above specified temperature.
3. A specific amount of time must elapse after starting the engine. These values are stored in the PROM.

When these conditions have been met, the system goes into closed loop operation In closed loop operation, the ECM will modify the pulse width (injector on-time) based upon the signal from the oxygen sensor. The ECM will decrease the on-time if the air/fuel ratio is too rich, and will increase the on-time if the air/fuel ratio is too lean.

The pulse width, thus the amount of enrichment, is determined by manifold pressure change, throttle angle change, and coolant temperature. The higher the manifold pressure and the wider the throttle opening, the wider the pulse width. The acceleration enrichment pulses are delivered nonsynchronized.

Any reduction in throttle angle will cancel the enrichment pulses. This way, quick movements of the accelerator will not over-enrich the mixture.

Acceleration Enrichment Mode

When the engine is required to accelerate, the opening of the throttle valve(s) causes a rapid increase in Manifold Absolute Pressure (MAP). This rapid increase in the manifold pressure causes fuel to condense on the manifold walls. The ECM senses this increase in throttle angle and MAP, and supplies additional fuel for a short period of time. This prevents the engine from stumbling due to too lean a mixture.

Deceleration Leanout Mode

Upon deceleration, a leaner fuel mixture is required to reduce emission of hydrocarbons (HC) and carbon monoxide (CO). To adjust the injection on-time, the ECM uses the decrease in manifold pressure and the decrease in throttle position to calculate a decrease in pulse width. To maintain an idle fuel ratio of 14.7:1, fuel output is momentarily reduced. This is done because of the fuel remaining in the intake manifold during deceleration.

Deceleration Fuel Cut-Off Mode

The purpose of deceleration fuel cut-off is to remove fuel from the engine during extreme deceleration conditions. Deceleration fuel cut-off is based on values of manifold pressure, throttle position, and engine rpm stored in the calibration PROM. Deceleration fuel cut-off overrides the deceleration enleanment mode.

NOTE: *This book contains testing and service procedures for your vehicles fuel injection system. More comprehensive testing and diagnosis procedures may be found in CHILTON'S GUIDE TO FUEL INJECTION AND FEEDBACK CARBURETORS, book part number 7488, 7768, or 8173 dependent upon wich model year of your vehicle, available at your local retailer.*

The 4.3L engine equipped with fuel injection has a bleed in the pressure regulator to relieve pressure any time the engine is turned off, however a small amount of fuel may be released when the fuel line is disconnected. As a precaution, cover the fuel line with a cloth and dispose of properly.

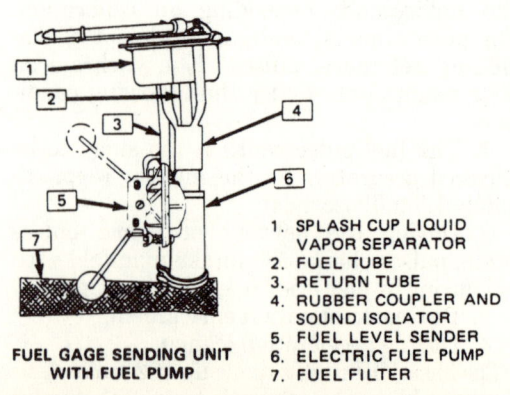

FUEL GAGE SENDING UNIT WITH FUEL PUMP

1. SPLASH CUP LIQUID VAPOR SEPARATOR
2. FUEL TUBE
3. RETURN TUBE
4. RUBBER COUPLER AND SOUND ISOLATOR
5. FUEL LEVEL SENDER
6. ELECTRIC FUEL PUMP
7. FUEL FILTER

Electric fuel pump assembly

FUEL SYSTEM   289

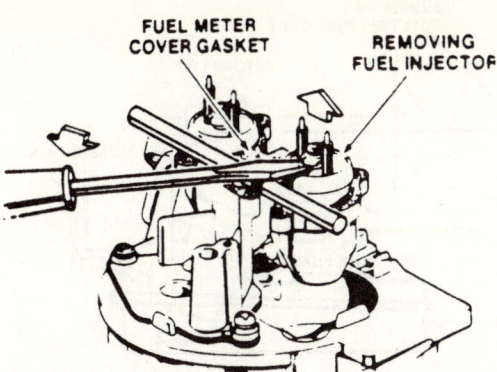

Removing the fuel injector from the throttle body

Electric Fuel Pump

REMOVAL AND INSTALLATION

4.3L V6 Engine

1. With the engine turned OFF, relieve the fuel pressure at the pressure regulator. See the Warning above.
2. Disconnect the negative battery cable.
3. Raise and support the rear of the vehicle safely.
4. Drain the fuel tank, then remove it.
5. Remove the fuel lever sending device and pump assembly locking ring, which is located on top of the fuel tank. Lift the assembly from the tank and remove the pump from the fuel lever sending device.
6. Pull the pump up into the attaching hose while pulling it outward away from the bottom support. Be careful not to damage the rubber insulator and strainer during removal. After the pump assembly is clear of the bottom support, pull it out of the rubber connector.

To Install:

7. Install the pump to the fuel sending gauge unit. Position the assembly in the fuel tank.
8. Install the locking ring. Install the fuel tank. Connect the fuel lines.
9. Lower the vehicle. Fill the tank with gasoline. Connect the battery cable.
10. Start the engine and check for leaks.

Testing

1. Secure two sections of $^3/_8"$ x 10" steel tubing, with a double flare on one end of each section.
2. Install a flare nut on each section of tubing, then connect each of the sections into the flare nut-to-flare nut adapter, included in Gauge Adapter tool No. J-29658-82.
3. Attach the pipe and the adapter assembly to the Gage tool No. J-29658.
4. Raise and support the vehicle on jackstands.
5. Remove the air cleaner and plug the THERMAC vacuum port on the TBI.
6. Disconnect the fuel feed hose between the fuel tank and the filter, then secure the other ends of the $^3/_8"$ tubing into the fuel hoses with hose clamps.

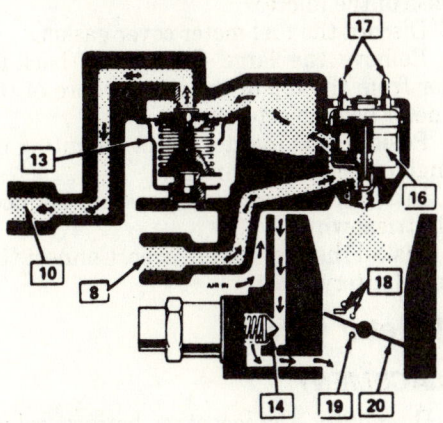

8. Fuel supply
10. Fuel return
13. Pressure regulator (part of fuel meter cover)
14. Idle air control (IAC) valve (shown open)
16. Fuel injector
17. Fuel injector terminals
18. Ported vacuum sources*
19. Manifold vacuum source*
20. Throttle valve

*May Be Different on some Models.

Cross sectional view of the TBI operation

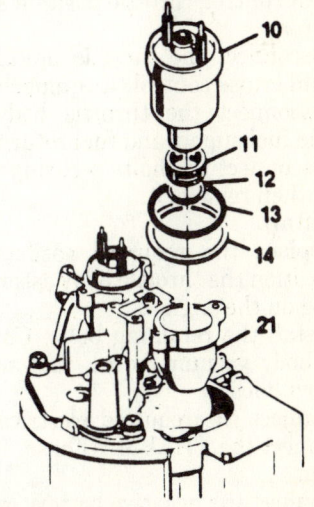

10. Injector-fuel
11. Filter-fuel injector inlet
12. "O" ring-fuel injector-lower
13. "O" ring-fuel injector-upper
14. Washer-fuel injector
21. Fuel meter body assembly

Exploded view of the fuel injector

FUEL SYSTEM

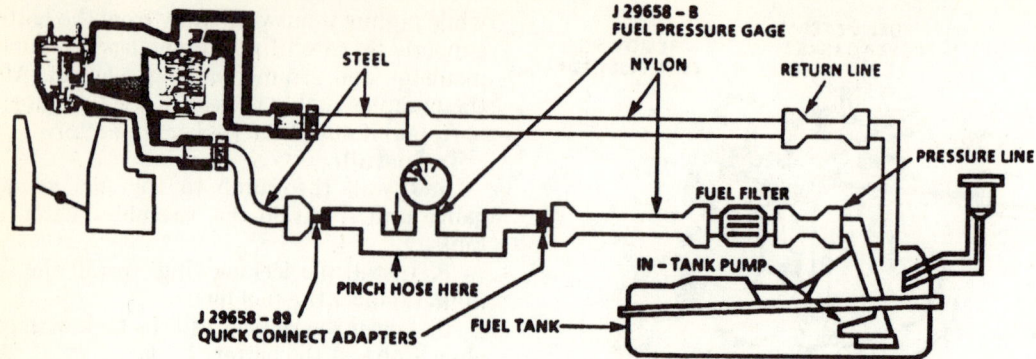

Testing the fuel pump on TBI, 1991 shown

7. Start the engine, check for leaks and observe the fuel pressure, it should be 9-13 psi.

8. Depressurize the fuel system, remove the testing tool, remove the plug from the THERMAC vacuum port, reconnect the fuel line, start the engine and check for fuel leaks.

Throttle Body

REMOVAL AND INSTALLATION

4.3L V6 Engine

1. Release the fuel pressure at the pressure regulator (see Warning above). Disconnect the negative battery cable.
2. Disconnect the THERMAC hose from the engine fitting and remove the air cleaner.
3. Disconnect the electrical connectors at the idle air control, throttle position sensor and the injector.
4. Disconnect the throttle linkage, return spring and cruise control (if equipped).
5. Disconnect the throttle body vacuum hoses, the fuel supply and fuel return lines.
6. Disconnect the bolts securing the throttle body, then remove it.

To Install:

7. Replace the required gaskets and O-rings. Position the throttle body assembly to its mounting on the engine.
8. Install the retaining bolts. Connect the throttle body vacuum hoses, fuel supply and fuel return lines.
9. Connect the required electrical connectors. Connect the THERMAC hose. Install the air cleaner.
10. Connect the negative battery cable. start the engine and check for leaks.

Injector

REPLACEMENT

WARNING: *When removing the injectors, be careful not to damage the electrical connector pins (on top of the injector), the injector fuel filter and the nozzle. The fuel injector is serviced as a complete assembly. The injector is an electrical component and should not be immersed in any kind of cleaner.*

1. Remove the air cleaner assembly. Relieve the fuel pressure (see Warning above). Disconnect the negative battery cable.
2. At the injector connector, squeeze the two tabs together and pull straight up.
3. Remove the fuel meter cover and leave the cover gasket in place.
4. Using a small pry bar or tool No. J-26868, carefully lift the injector until it is free from the fuel meter body.
5. Remove the small O-ring from the nozzle end of the injector. Carefully rotate the injector's fuel filter back and forth to remove it from the base of the injector.
6. Discard the fuel meter cover gasket.
7. Remove the large O-ring and back-up washer from the top of the counterbore of the fuel meter body injector cavity.
8. Position the injector to its mounting on the engine, using new O-rings.
9. Install the injector into position. Connect the electrical wire.
10. Install the fuel meter cover. Connect the negative battery cable.

Fuel Meter Cover

REPLACEMENT

1. Disconnect the negative battery cable. Remove the air cleaner assembly.
2. Disconnect the electrical connector from the fuel injector.
3. Remove the fuel meter-to-fuel meter body screws and lockwashers.

NOTE: *When removing the fuel meter cover screws note the location of the two short screws.*

4. Remove the fuel meter cover and discard the gasket.

FUEL SYSTEM 291

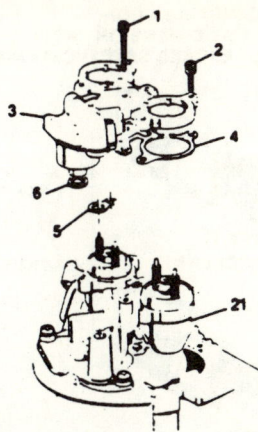

1. Screw assembly–fuel meter cover attaching–long
2. Screw assembly–fuel meter cover attaching–short
3. Fuel meter cover assembly
4. Gasket–fuel meter cover
5. Gasket–fuel meter outlet
6. Seal–pressure regulator
21. Fuel meter body assembly

Exploded view of the fuel meter cover

5. To install, use a new gasket and reverse the removal procedures.

Idle Air Control Valve

REPLACEMENT

1. Disconnect the negative battery cable. Remove the air cleaner assembly.
2. Disconnect the electrical connector from the idle air control valve.
3. Using a $1/4''$ wrench or tool J-33031, remove the idle air control valve.

WARNING: *Before install a new idle air control valve, measure the distance that the valve extends (from the motor housing to the end of the cone); the distance should be no greater than $1/8''$. If it extends too far, damage will occur to the valve when it is installed.*

4. To complete the installation, use a new gasket and reverse the removal procedures.

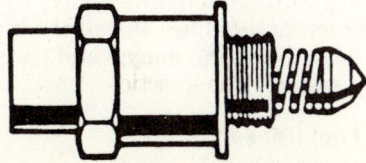

DUAL TAPER VALVE

View of the idle air control valve used with automatic transmissions

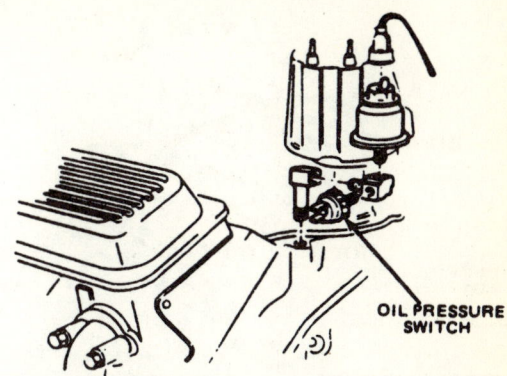

View of the oil pressure switch

Start the engine and allow it to reach operating temperature.

NOTE: *The ECM will reset the idle speed when the vehicle is driven at 30 mph.*

Fuel Pump Relay

REPLACEMENT

The fuel pump relay is located in the engine compartment. Other than checking for loose electrical connections, the only service necessary is to replace the relay. Before replacing the relay, disconnect the negative battery cable.

Oil Pressure Switch

REPLACEMENT

1. Disconnect the negative battery. Remove the electrical connector from the switch.
2. Remove the oil pressure switch.
3. To install, reverse the removal procedures.

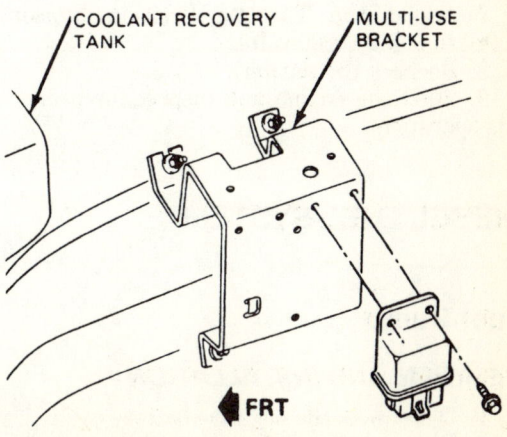

Fuel pump relay location

FUEL SYSTEM

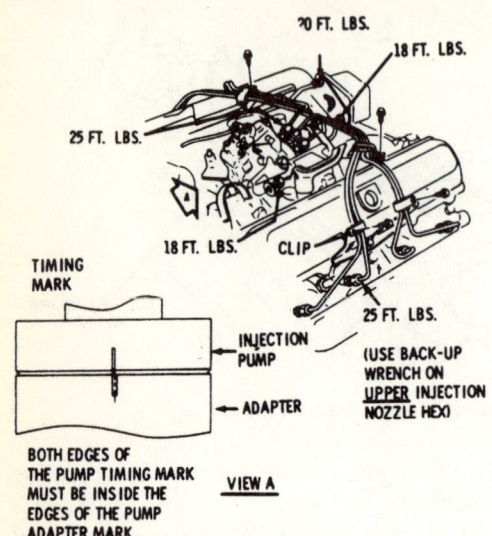

Diesel engine timing marks and injector lines

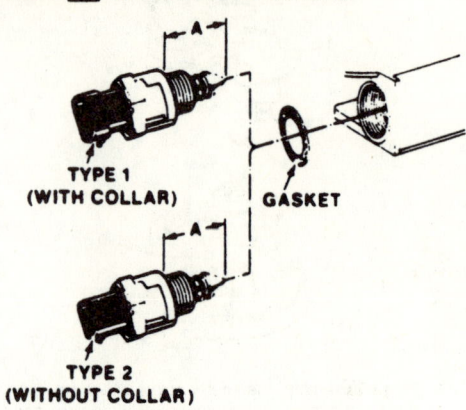

Installing the idle air control valve

Minimum Idle Speed

ADJUSTMENT

Only if parts of the throttle body have been replaced should this procedure be performed; the engine should be at operating temperature.

1. Ground the diagnostic lead of the IAC motor.
2. Turn the ignition ON, BUT DO NOT start the engine and wait for 30 seconds.
3. With the ignition ON, disconnect the electrical connector from the IAC motor.
4. Remove the ground from the diagnostic lead and start the engine.
5. Adjust the idle set screw to 500-600 rpm with the transmission in Drive.
6. Turn the ignition OFF and reconnect the electrical to the IAC motor.
7. Adjust the Throttle Position Sensor (TPS) to 0.450-0.600 volts.
8. Recheck the setting.
9. Start the engine and inspect for proper idle operation.

DIESEL FUEL SYSTEM

Fuel Pump

REMOVAL AND INSTALLATION

1. Disconnect the negative battery cable.
2. Disconnect and plug the fuel lines at the fuel pump assembly.
3. As required to gain working clearance,

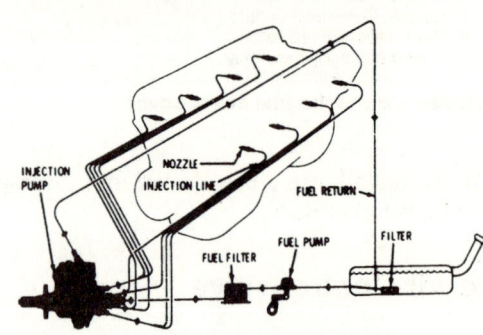

Diesel engine fuel system schematic

move the air conditioning compressor out of your way.

4. Remove the fuel pump retaining bolts. Remove the fuel pump from the engine block. Discard the fuel pump mounting gasket.
5. Installation is the reverse of the removal procedure.
6. Connect the negative battery cable. Start the engine and check for leaks.

Fuel Filter

REMOVAL AND INSTALLATION

The fuel filter is located at the back of the engine above the intake manifold. Disconnect the negative battery cable. Disconnect the fuel lines and remove the filter. Install the lines to the new filter. Start the engine and check for leaks.

Water in Fuel (Diesel)

Water is the worst enemy of the diesel fuel injection system. The injection pump, which is designed and constructed to extremely close tolerances, and the injectors can be easily damaged if enough water is forced through them in the fuel. Engine performance will also be drastically affected, and engine damage can occur.

Diesel fuel is much more susceptible than gasoline to water contamination. Diesel engine cars are equipped with an indicator lamp system that turns on an instrument panel lamp if water (1 to 2½ gallons) is detected in the fuel tank. The lamp will come on for 2 to 5 seconds each time the ignition is turned ON, assuring the driver the lamp is working. If there is water in the fuel, the light will come back on after a 15 to 20 second off delay, and then remain ON.

Purging the Fuel Tank

Vehicles which have a Water in Fuel light may have the water removed from the tank with a siphon pump. The pump hose should be hooked up to the ¼" fuel return hose (the smaller of the two hoses) above the rear axle or under the hood near the fuel pump. Siphoning should continue until all water is removed from the tank. Use a clear plastic hose or observe the filter bowl on the siphon pump (if equipped) to determine when clear fuel begins to flow. Be sure to remove the cap on the fuel tank while purging. Replace the cap when finished. Discard the fuel filter and replace with a new filter.

Fuel Injection Pump and Lines

REMOVAL AND INSTALLATION

1. Disconnect the negative battery cable. Remove the air cleaner.
2. Remove the filters and pipes from the valve covers and air crossover.
3. Remove the air crossover, then cap the intake manifold with screen covers (tool J-26996-1).
4. Disconnect the throttle rod and return spring.
5. Remove the bellcrank.
6. Remove the throttle and transmission cables from the intake manifold brackets.
7. Disconnect the fuel lines from the filter and remove the filter.
8. Disconnect the fuel inlet line from the pump.
9. Remove the rear air conditioning compressor brace and the fuel line.
10. Disconnect the fuel return line from the injection pump.
11. Remove the clamps and pull the fuel return line from each injection nozzle.
12. Using two wrenches, disconnect the high pressure line from the nozzles.
13. Remove the three injection pump retaining nuts with tool J-26987 or its equivalent.
14. Remove the injection pump. Be sure to cap all lines and nozzles.

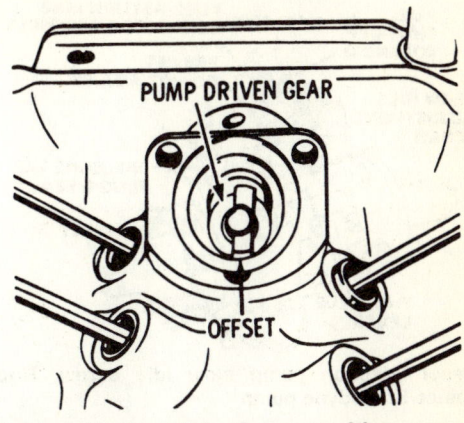

Offset on the diesel injection pump driven gear

To install:
15. Remove the protective caps.
16. Line up the offset tang on the pump driveshaft with the pump driven gear and install the pump.
17. Install, but do not tighten the pump retaining nuts.
18. Connect the high pressure lines to the nozzles.
19. Using two wrenches, torque the high pressure line nuts to 25 ft. lbs.
20. Connect the fuel return lines to the nozzles and pump.
21. Align the timing mark on the injection pump with the line on the timing mark adaptor and torque the mounting nuts to 35 ft. lbs.

NOTE: *A ¾" open end wrench on the boss at the front of the injection pump will aid in rotating the pump to align the marks.*

22. Adjust the throttle rod:
    a. Remove the clip from the cruise control rod and the rod from the bellcrank.
    b. Loosen the locknut on the throttle rod a few turns, then shorten the rod several turns.
    c. Rotate the bellcrank to the full throttle stop, then lengthen the throttle rod until the unjection pump lever contacts the injection pump full throttle stop, then release the bellcrank.
    d. Tighten the throttle rod locknut.
23. Install the fuel inlet line between the transfer pump and the filter.
24. Install the rear air conditioning compressor brace.
25. Install the bellcrank and clip.
26. Connect the throttle rod and return spring.
27. To adjust the transmission cable:
    a. Push the snap-lock to the disengaged position.
    b. Rotate the injection pump lever to the full throttle stop and hold it there.

294 FUEL SYSTEM

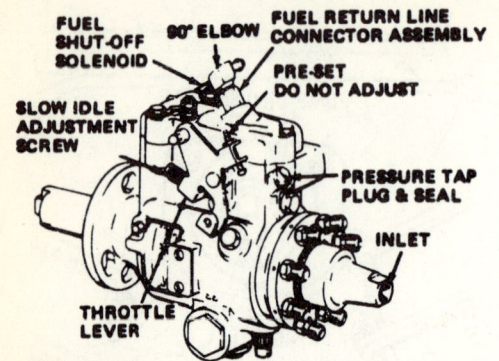

Diesel injection pump slow idle screw, Roosa Master/Stanodyne pump

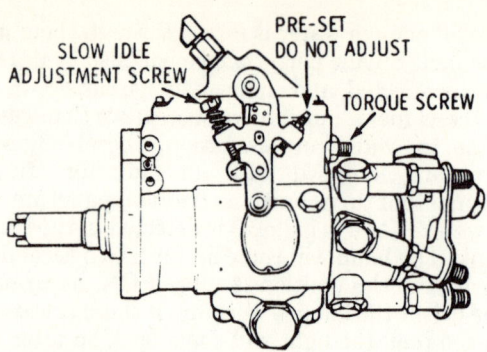

CAV injection pump slow idle screw

Fast idle solenoid adjustment, solenoid mounted on side of injection pump

c. Push the snap-lock until it is flush.
d. Release the injection pump lever.
28. Start the engine and check for fuel leaks.
29. Remove the screened covers and install the air crossover.
30. Install the tubes in the air flow control valve in the air crossover and install the ventilation filter in the valve covers.
31. Install the air cleaner.
32. Start the engine and allow it to run for two minutes. Stop the engine, let it stand for two minutes, then restart. This permits the air to bleed off within the pump.

SLOW IDLE SPEED ADJUSTMENT

1. Run the engine to normal operating temperature.
2. Insert the probe of a magnetic pickup tachometer into the timing indicator hole.
3. Set the parking brake and block the drive wheels.
4. Place the transmission in Drive and turn the air conditioning off.
5. Turn the slow idle screw on the injection pump to obtain the idle specification on the emission control label.

FAST IDLE SOLENOID ADJUSTMENT

1. With the ignition OFF, disconnect the single green wire from the fast idle relay located on the front of the firewall.
2. Set the parking brake and block the drive wheels.
3. Start the engine and adjust the solenoid (energized) to the specifications on the underhood emission control label.
4. Turn the ignition switch OFF and reconnect the green wire.

CRUISE CONTROL SERVO RELAY ROD ADJUSTMENT

1. Turn the engine Off.
2. Adjust the rod to minimum slack then put the clip in the first free hole closest to the bellcrank, but within the servo bail.

Injection Timing

CHECKING

NOTE: *A special diesel timing meter is needed to check injection timing. There are a few variations of this meter, but the type desirable here uses a signal through a glow plug probe to determine combustion timing. The meter picks up the engine speed in RPM and the crankshaft position from the crankshaft balancer. this tool is available at automotive supply houses and from tool jobbers; it is the counterpart to a gasoline engine timing light, coupled with a tachometer. An intake manifold cover is also needed. The marks on the pump and adapter flange will normally be aligned within 0.030".*

1. Place the transmission shift lever in PARK, apply the parking brake and block the rear wheels.
2. Start the engine and let it run at idle until fully warm. Shut off the engine.

NOTE: *If the engine is not allowed to completely warm up, the probe may soot up, causing incorrect timing readings.*

3. Remove the air cleaner assembly and carefully install cover J-26996-1. This cover over the intake is important. Disconnect the EGR valve hose.

4. Clean away all dirt from the engine probe holder (RPM counter) and the crankshaft balancer rim.

5. Clean the lens on both ends of the glow plug probe and clean the lens in the photoelectric pick-up. Use a tooth pick to scrape the carbon from the combustion chamber side of the glow plug probe, then look through the probe to make sure it's clean. Cleanliness is crucial for accurate readings.

6. Install the probe into the crankshaft RPM counter (probe holder) on the engine front cover.

7. Remove the glow plug from No. 3 cylinder. Install the glow plug probe in the glow plug opening and torque to 8 ft. lbs.

8. Set the timing meter offset selector to **B**.

9. Connect the battery leads, red to positive, black to negative.

10. Disconnect the two-lead connector from the alternator.

11. Start the engine. Adjust the engine RPM to the speed specified on the emissions control decal.

12. Observe the timing reading, then observe it again in 2 minutes. When the readings stabilize over the 2 minutes intervals, compare that final stabilized reading to the one specified on the emissions control decal. The timing reading will be at ATDC (After Top Dead Center) reading when set to specifications.

13. Disconnect the timing meter and install the removed glow plug, torquing it to 12 ft. lbs.

14. Connect the generator two-lead connection.

15. Install the air cleaner assembly and connect the EGR valve hose.

ADJUSTMENT

1. Shut off the engine.

2. Note the relative position of the marks on the pump flange and the pump adapter.

3. Loosen the nuts or bolts holding the pump to a point where the pump can just be rotated. Use a ¾" open-end wrench on the boss at the front of the injection pump. You may need a wrench with a slight offset to clear the fuel return.

4. Rotate the pump to the left to advance the timing and to the right to retard the timing. The width of the mark on the adaptor is equal to about 1° of timing. Move the pump the amount that is needed and tighten the pump retaining nuts to 18 ft. lbs.

5. Start the engine and recheck the timing as described earlier. Reset the timing if necessary.

6. Adjust the injection pump rod. Reset the fast and curb idle speeds.

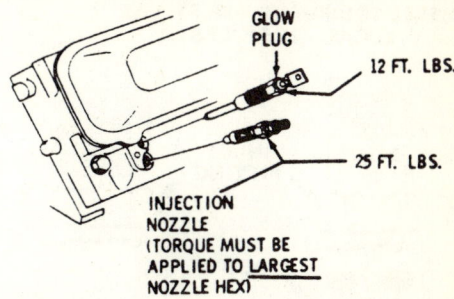

Injection nozzle and glow plug installation

NOTE: *Wild needle fluctuations on the timing meter indicate a cylinder not firing properly. Correction of this condition must be made prior to adjusting the timing.*

7. If after resetting the timing, the timing marks are far apart and the engine still runs poorly, the dynamic timing could still be off. It is possible that a malfunctioning cylinder will cause incorrect timing. If this occurs, it is essential that timing be checked in cylinders 2 or 3. If different timing exists between cylinders, try both positions to determine which timing works best.

Injection Nozzle

REMOVAL AND INSTALLATION

The injection nozzles on these engines are simply unbolted from the cylinder head, after the fuel lines are removed, in similar fashion to

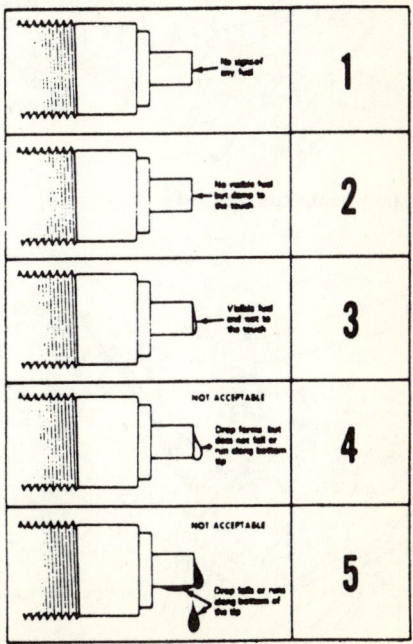

Checking injection nozzle seat tightness

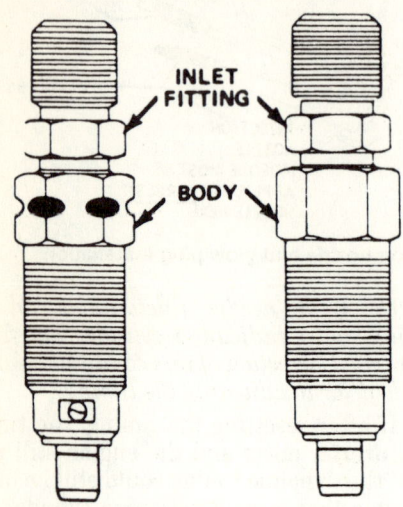

Injection nozzles, two types

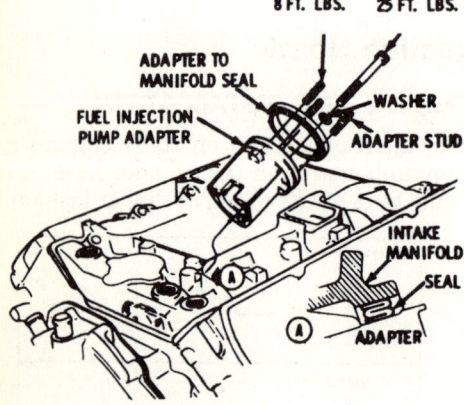

Injection pump adaptor bolts

a spark plug. Be careful not to damage the nozzle end and make sure you remove the copper nozzle gasket from the cylinder head if it does not come off with the nozzle. Clean the carbon off the tip of the nozzle with a soft brass wire brush and install the nozzles, with gaskets.

NOTE: *1981 and later models use two types of injectors, CAV Lucas and Diesel Equipment. When installing the inlet fittings, torque the Diesel Equipment injector fitting to 45 ft. lbs. and the CAV Lucas to 25 ft. lbs.*

Injection Pump Adapter, Adapter Seal, and New Adapter Timing Mark

REMOVAL AND INSTALLATION

NOTE: *Skip Steps 4 and 9 if a new adapter is not being installed.*

1. Disconnect the negative battery cable. Remove injection pump and lines as described earlier.
2. Remove the injection pump adapter.
3. Remove the seal from the adapter.
4. File the timing mark from the adapter. Do not file the mark off the pump.
5. Position the engine at TDC of No. 1 cylinder. Align the mark on the balancer with the zero mark on the indicator. The index is offset to the right when No. 1 is at TDC.
6. Apply a chassis lube to the seal areas. Install, but do not tighten the injection pump.
7. Install the new seal on the adapter using tool J-28425, or its equivalent.
8. Torque the adapter bolts to 25 ft. lbs.
9. Install timing tool J-26896 into the injection pump adapter. Torque the tool, toward No. 1 cylinder to 50 ft. lbs. Mark the injection pump adapter. Remove the tool.
10. Install the injection pump.

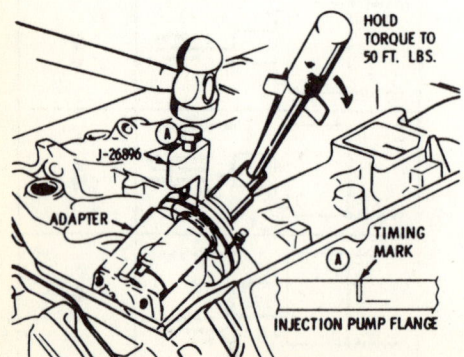

Marking the injection pump adaptor

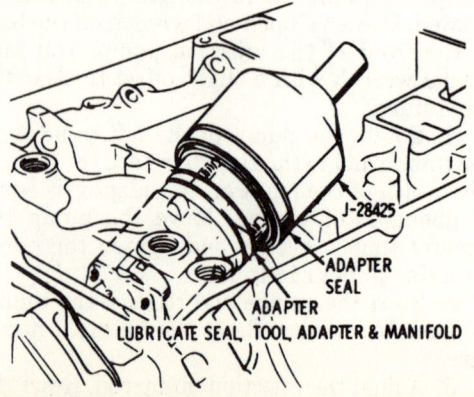

Installing the adaptor seal

Glow Plugs

There are two types of glow plugs used on General Motors Corp. diesels; the fast glow type and the slow glow type. The fast flow type use pulsing current applied to 6 volt glow plug while the slow glow type use continuous current applied to 12 volt glow plugs.

An easy way to tell the plugs apart is that the fast glow (6 volt) plugs have an $5/16''$ wide electrical connector plug while the slow glow (12 volt) connector plug is $1/4''$ wide. Do not attempt to interchange any parts of these two glow plug systems.

FUEL TANK

DRAINING

1. Remove the fuel tank cap.
2. Connect a siphon pump to the $1/4''$ fuel return hose (the smaller of the two hoses) above the rear axle, or under the hood near the fuel pump on the passenger's side of the engine, near the front.
3. Operate the siphon pump until all fuel is removed from the fuel tank. Be sure to reinstall the fuel return hose and the fuel cap.

REMOVAL AND INSTALLATION

1. Disconnect the negative battery cable. Siphon the fuel from the tank.
2. Raise and support the vehicle safely. On 1991-92 models, remove the fuel tank shield attaching bolts and lower the shield.
3. Disconnect the fuel line and the gauge sending unit wire from the tank.
4. Disconnect the black ground wire, as required. Disconnect the vent and fuel filler hose from the tank.
5. Remove the retaining bolts from the tank straps, lower the support straps, and then carefully lower the tank from the vehicle.
6. Reverse the removal procedure to install the fuel tank.

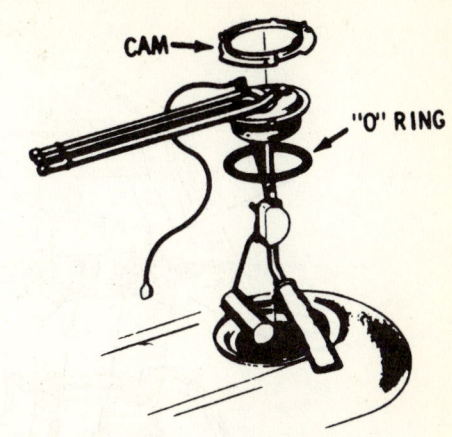

Installing fuel gauge unit into fuel tank

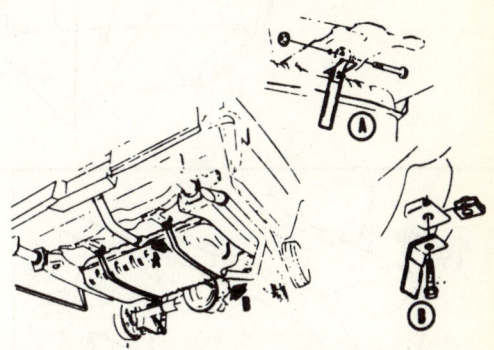

Fuel tank, except station wagon

298 FUEL SYSTEM

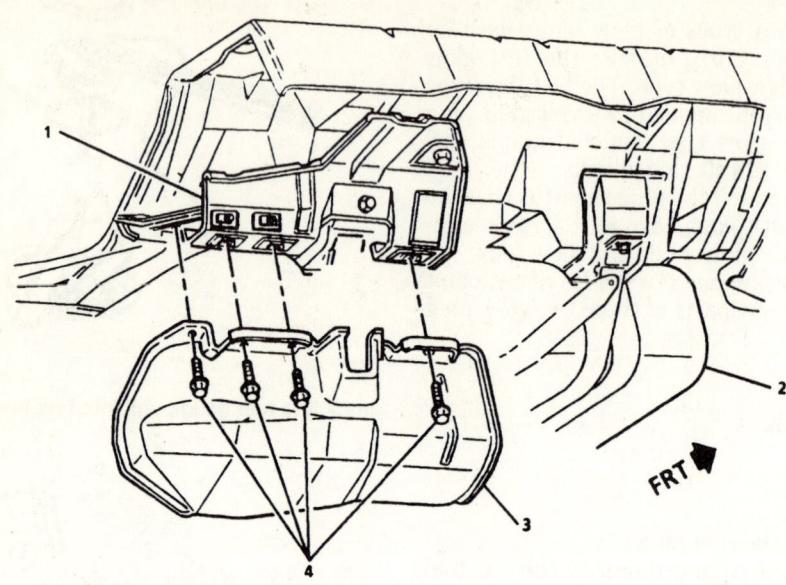
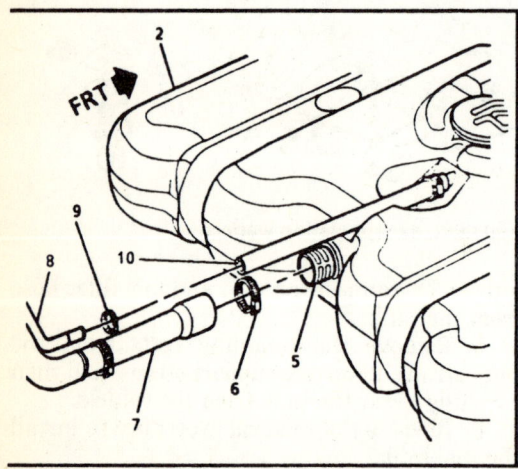
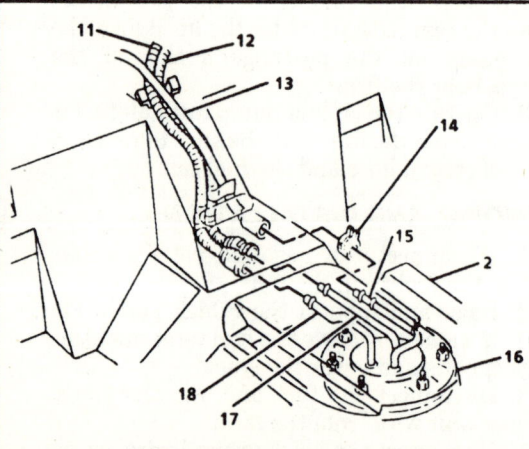

#	Component	#	Component
1	UPPER FUEL TANK SHIELD	10	VENT HOSE
2	FUEL TANK	11	FUEL FEED LINE
3	LOWER FUEL TANK SHIELD	12	FUEL RETURN LINE
4	LOWER FUEL TANK SHIELD ATTACHING SCREWS (4)	13	VAPOR LINE
5	FUEL TANK FILLER TUBE NIPPLE	14	VAPOR LINE CLAMP
6	FUEL FILLER TUBE CLAMP	15	VAPOR TUBE
7	FUEL FILLER CONNECTING TUBE	16	FUEL SENDER ASSEMBLY
8	VENT TUBE	17	FUEL RETURN TUBE
9	VENT TUBE CLAMP	18	FUEL FEED TUBE

Fuel tank mounting components, 1991–92 Sedan

FUEL SYSTEM 299

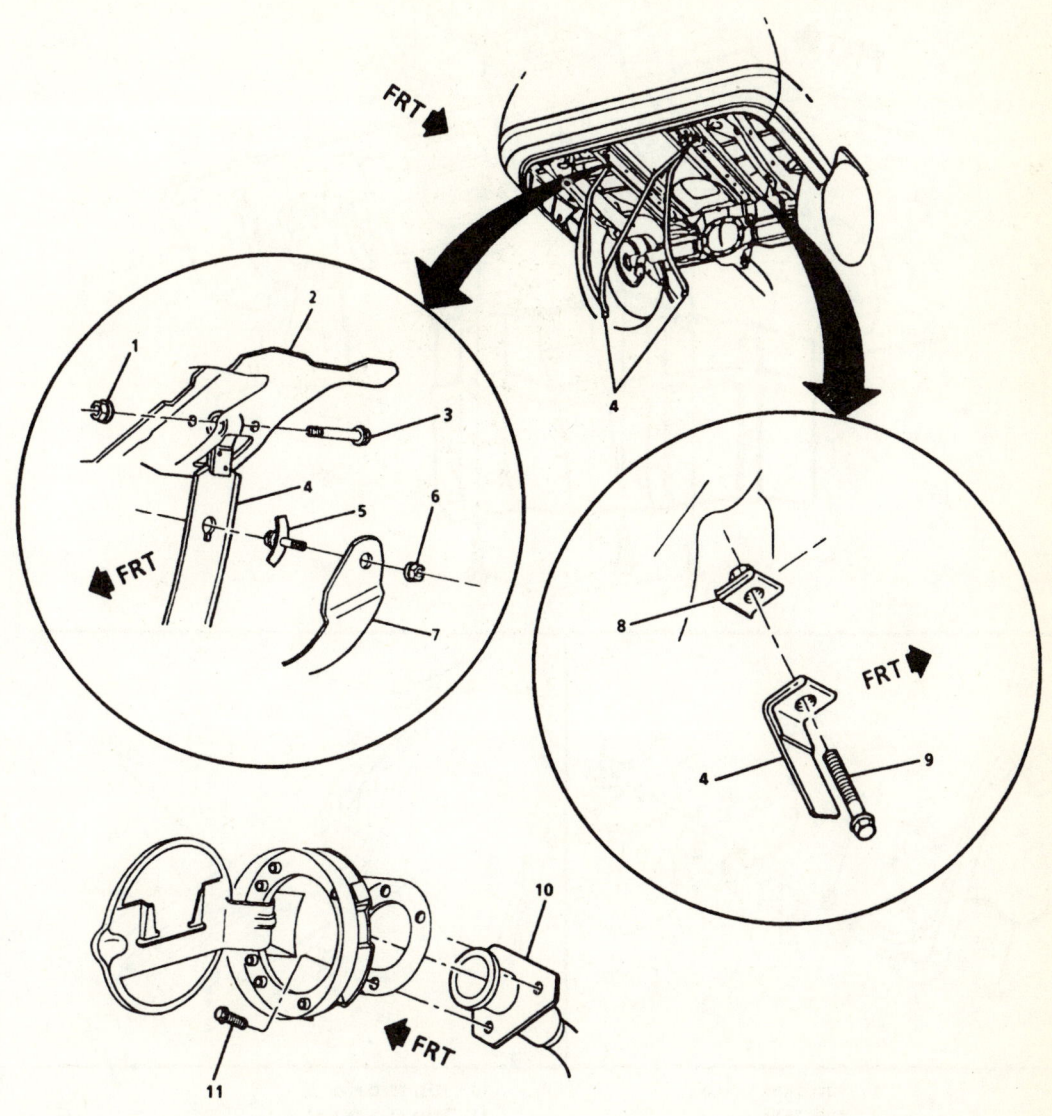

1. REAR FUEL TANK RETAINING STRAP NUTS (2)
2. UNDER BODY
3. REAR FUEL TANK RETAINING STRAP BOLTS (2)
4. FUEL TANK RETAINING STRAPS (2)
5. FUEL TANK CROSS STRAP ATTACHING SCREW
6. FUEL TANK CROSS STRAP ATTACHING NUT
7. FUEL TANK CROSS STRAP
8. FRONT FUEL TANK RETAINING STRAP BODY NUTS (2)
9. FRONT FUEL TANK RETAINING STRAP BOLTS (2)
10. FUEL FILLER TUBE
11. FUEL FILLER TUBE ATTACHING SCREWS (3)

Fuel tank mounting, 1991–92 Wagon

300 FUEL SYSTEM

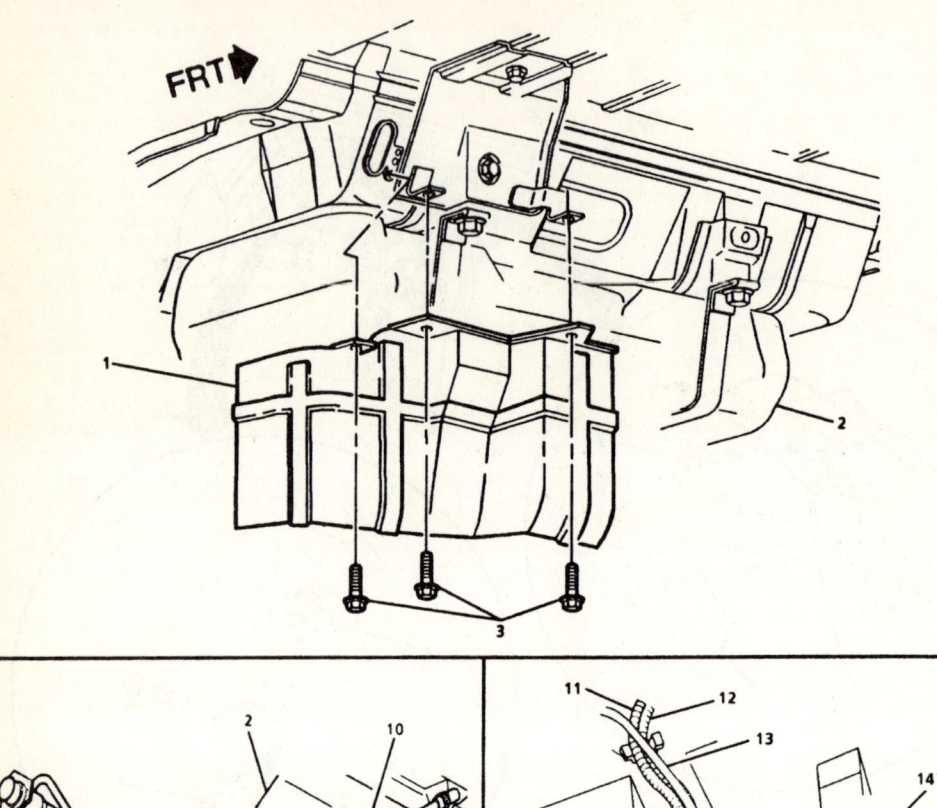
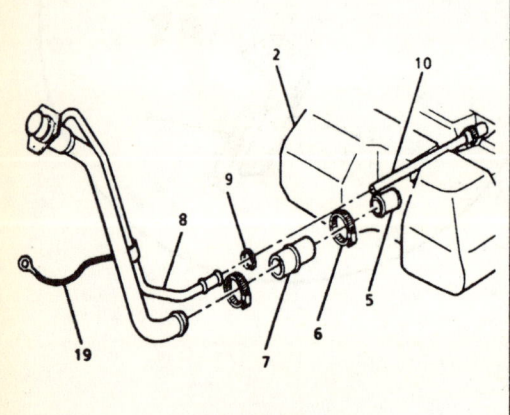
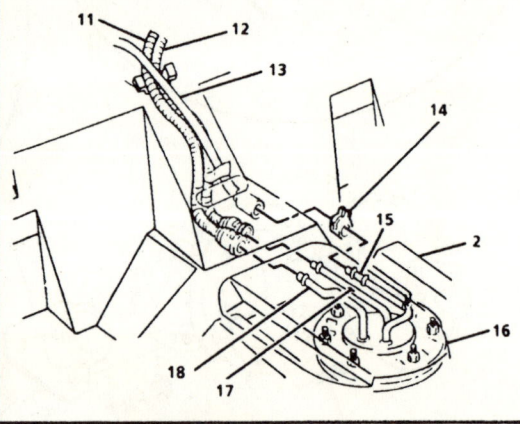

1	FUEL TANK SHIELD	10	FUEL FEED PIPE
2	FUEL TANK	11	FUEL RETURN PIPE
3	FUEL TANK SHIELD ATTACHING SCREWS (3)	12	VAPOR PIPE
4	FUEL TANK FILLER TUBE NIPPLE	13	VAPOR PIPE CLAMP
5	FUEL FILLER TUBE CLAMP	14	VAPOR TUBE
6	FUEL FILLER CONNECTING TUBE	15	FUEL SENDER ASSEMBLY
7	VENT TUBE	16	FUEL RETURN TUBE
8	VENT TUBE CLAMP	17	FUEL FEED TUBE
9	VENT HOSE	18	GROUNDING STRAP

Fuel tank mounting components, 1991–92 Wagon

CHILTON'S
FUEL ECONOMY & TUNE-UP TIPS

55 WAYS TO IMPROVE FUEL ECONOMY

Tune-up • Spark Plug Diagnosis • Emission Controls
Fuel System • Cooling System • Tires and Wheels
General Maintenance

CHILTON'S FUEL ECONOMY & TUNE-UP TIPS

Fuel economy is important to everyone, no matter what kind of vehicle you drive. The maintenance-minded motorist can save both money and fuel using these tips and the periodic maintenance and tune-up procedures in this Repair and Tune-Up Guide.

There are more than 130,000,000 cars and trucks registered for private use in the United States. Each travels an average of 10-12,000 miles per year, and, and in total they consume close to 70 billion gallons of fuel each year. This represents nearly ⅔ of the oil imported by the United States each year. The Federal government's goal is to reduce consumption 10% by 1985. A variety of methods are either already in use or under serious consideration, and they all affect you driving and the cars you will drive. In addition to "down-sizing", the auto industry is using or investigating the use of electronic fuel delivery, electronic engine controls and alternative engines for use in smaller and lighter vehicles, among other alternatives to meet the federally mandated Corporate Average Fuel Economy (CAFE) of 27.5 mpg by 1985. The government, for its part, is considering rationing, mandatory driving curtailments and tax increases on motor vehicle fuel in an effort to reduce consumption. The government's goal of a 10% reduction could be realized — and further government regulation avoided — if every private vehicle could use just 1 less gallon of fuel per week.

How Much Can You Save?

Tests have proven that almost anyone can make at least a 10% reduction in fuel consumption through regular maintenance and tune-ups. When a major manufacturer of spark plugs sur-

TUNE-UP

1. Check the cylinder compression to be sure the engine will really benefit from a tune-up and that it is capable of producing good fuel economy. A tune-up will be wasted on an engine in poor mechanical condition.

2. Replace spark plugs regularly. New spark plugs alone can increase fuel economy 3%.

3. Be sure the spark plugs are the correct type (heat range) for your vehicle. See the Tune-Up Specifications.

Heat range refers to the spark plug's ability to conduct heat away from the firing end. It must conduct the heat away in an even pattern to avoid becoming a source of pre-ignition, yet it must also operate hot enough to burn off conductive deposits that could cause misfiring.

The heat range is usually indicated by a number on the spark plug, part of the manufacturer's designation for each individual spark plug. The numbers in bold-face indicate the heat range in each manufacturer's identification system.

Manufacturer	Typical Designation
AC	R **45** TS
Bosch (old)	WA **145** T30
Bosch (new)	HR **8** Y
Champion	RBL **15** Y
Fram/Autolite	41**5**
Mopar	P-**62** PR
Motorcraft	BRF-**42**
NGK	BP **5** ES-15
Nippondenso	W **16** EP
Prestolite	14GR **5** 2A

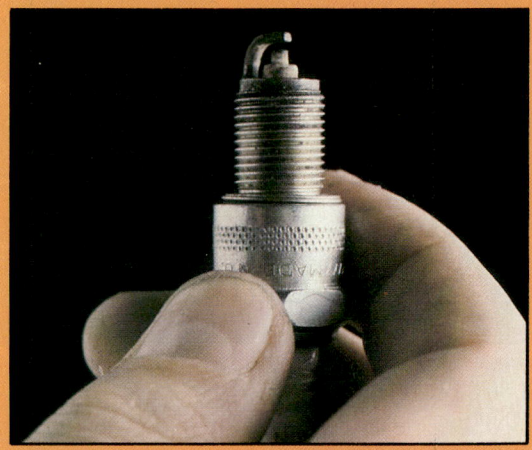

Periodically, check the spark plugs to be sure they are firing efficiently. They are excellent indicators of the internal condition of your engine.

On AC, Bosch (new), Champion, Fram/Autolite, Mopar, Motorcraft and Prestolite, a higher number indicates a hotter plug. On Bosch (old), NGK and Nippondenso, a higher number indicates a colder plug.

4. Make sure the spark plugs are properly gapped. See the Tune-Up Specifications in this book.

5. Be sure the spark plugs are firing efficiently. The illustrations on the next 2 pages show you how to "read" the firing end of the spark plug.

6. Check the ignition timing and set it to specifications. Tests show that almost all cars have incorrect ignition timing by more than 2°.

veyed over 6,000 cars nationwide, they found that a tune-up, on cars that needed one, increased fuel economy over 11%. Replacing worn plugs alone, accounted for a 3% increase. The same test also revealed that 8 out of every 10 vehicles will have some maintenance deficiency that will directly affect fuel economy, emissions or performance. Most of this mileage-robbing neglect could be prevented with regular maintenance.

Modern engines require that all of the functioning systems operate properly for maximum efficiency. A malfunction anywhere wastes fuel. You can keep your vehicle running as efficiently and economically as possible, by being aware of your vehicle's operating and performance characteristics. If your vehicle suddenly develops performance or fuel economy problems it could be due to one or more of the following:

PROBLEM	POSSIBLE CAUSE
Engine Idles Rough	Ignition timing, idle mixture, vacuum leak or something amiss in the emission control system.
Hesitates on Acceleration	Dirty carburetor or fuel filter, improper accelerator pump setting, ignition timing or fouled spark plugs.
Starts Hard or Fails to Start	Worn spark plugs, improperly set automatic choke, ice (or water) in fuel system.
Stalls Frequently	Automatic choke improperly adjusted and possible dirty air filter or fuel filter.
Performs Sluggishly	Worn spark plugs, dirty fuel or air filter, ignition timing or automatic choke out of adjustment.

Check spark plug wires on conventional point type ignition for cracks by bending them in a loop around your finger.

Be sure that spark plug wires leading to adjacent cylinders do not run too close together. (Photo courtesy Champion Spark Plug Co.)

7. If your vehicle does not have electronic ignition, check the points, rotor and cap as specified.

8. Check the spark plug wires (used with conventional point-type ignitions) for cracks and burned or broken insulation by bending them in a loop around your finger. Cracked wires decrease fuel efficiency by failing to deliver full voltage to the spark plugs. One misfiring spark plug can cost you as much as 2 mpg.

9. Check the routing of the plug wires. Misfiring can be the result of spark plug leads to adjacent cylinders running parallel to each other and too close together. One wire tends to pick up voltage from the other causing it to fire "out of time".

10. Check all electrical and ignition circuits for voltage drop and resistance.

11. Check the distributor mechanical and/or vacuum advance mechanisms for proper functioning. The vacuum advance can be checked by twisting the distributor plate in the opposite direction of rotation. It should spring back when released.

12. Check and adjust the valve clearance on engines with mechanical lifters. The clearance should be slightly loose rather than too tight.

SPARK PLUG DIAGNOSIS

Normal

APPEARANCE: This plug is typical of one operating normally. The insulator nose varies from a light tan to grayish color with slight electrode wear. The presence of slight deposits is normal on used plugs and will have no adverse effect on engine performance. The spark plug heat range is correct for the engine and the engine is running normally.
CAUSE: Properly running engine.
RECOMMENDATION: Before reinstalling this plug, the electrodes should be cleaned and filed square. Set the gap to specifications. If the plug has been in service for more than 10-12,000 miles, the entire set should probably be replaced with a fresh set of the same heat range.

Oil Deposits

APPEARANCE: The firing end of the plug is covered with a wet, oily coating.
CAUSE: The problem is poor oil control. On high mileage engines, oil is leaking past the rings or valve guides into the combustion chamber. A common cause is also a plugged PCV valve, and a ruptured fuel pump diaphragm can also cause this condition. Oil fouled plugs such as these are often found in new or recently overhauled engines, before normal oil control is achieved, and can be cleaned and reinstalled.
RECOMMENDATION: A hotter spark plug may temporarily relieve the problem, but the engine is probably in need of work.

Incorrect Heat Range

APPEARANCE: The effects of high temperature on a spark plug are indicated by clean white, often blistered insulator. This can also be accompanied by excessive wear of the electrode, and the absence of deposits.
CAUSE: Check for the correct spark plug heat range. A plug which is too hot for the engine can result in overheating. A car operated mostly at high speeds can require a colder plug. Also check ignition timing, cooling system level, fuel mixture and leaking intake manifold.
RECOMMENDATION: If all ignition and engine adjustments are known to be correct, and no other malfunction exists, install spark plugs one heat range colder.

Photos Courtesy Fram Corporation

Carbon Deposits

APPEARANCE: Carbon fouling is easily identified by the presence of dry, soft, black, sooty deposits.
CAUSE: Changing the heat range can often lead to carbon fouling, as can prolonged slow, stop-and-start driving. If the heat range is correct, carbon fouling can be attributed to a rich fuel mixture, sticking choke, clogged air cleaner, worn breaker points, retarded timing or low compression. If only one or two plugs are carbon fouled, check for corroded or cracked wires on the affected plugs. Also look for cracks in the distributor cap between the towers of affected cylinders.
RECOMMENDATION: After the problem is corrected, these plugs can be cleaned and reinstalled if not worn severely.

MMT Fouled

APPEARANCE: Spark plugs fouled by MMT (Methycyclopentadienyl Maganese Tricarbonyl) have reddish, rusty appearance on the insulator and side electrode.
CAUSE: MMT is an anti-knock additive in gasoline used to replace lead. During the combustion process, the MMT leaves a reddish deposit on the insulator and side electrode.
RECOMMENDATION: No engine malfunction is indicated and the deposits will not affect plug performance any more than lead deposits (see Ash Deposits). MMT fouled plugs can be cleaned, regapped and reinstalled.

High Speed Glazing

APPEARANCE: Glazing appears as shiny coating on the plug, either yellow or tan in color.
CAUSE: During hard, fast acceleration, plug temperatures rise suddenly. Deposits from normal combustion have no chance to fluff-off; instead, they melt on the insulator forming an electrically conductive coating which causes misfiring.
RECOMMENDATION: Glazed plugs are not easily cleaned. They should be replaced with a fresh set of plugs of the correct heat range. If the condition recurs, using plugs with a heat range one step colder may cure the problem.

Ash (Lead) Deposits

APPEARANCE: Ash deposits are characterized by light brown or white colored deposits crusted on the side or center electrodes. In some cases it may give the plug a rusty appearance.
CAUSE: Ash deposits are normally derived from oil or fuel additives burned during normal combustion. Normally they are harmless, though excessive amounts can cause misfiring. If deposits are excessive in short mileage, the valve guides may be worn.
RECOMMENDATION: Ash-fouled plugs can be cleaned, gapped and reinstalled.

Detonation

APPEARANCE: Detonation is usually characterized by a broken plug insulator.
CAUSE: A portion of the fuel charge will begin to burn spontaneously, from the increased heat following ignition. The explosion that results applies extreme pressure to engine components, frequently damaging spark plugs and pistons.
  Detonation can result by over-advanced ignition timing, inferior gasoline (low octane) lean air/fuel mixture, poor carburetion, engine lugging or an increase in compression ratio due to combustion chamber deposits or engine modification.
RECOMMENDATION: Replace the plugs after correcting the problem.

Photos Courtesy Champion Spark Plug Co.

EMISSION CONTROLS

13. Be aware of the general condition of the emission control system. It contributes to reduced pollution and should be serviced regularly to maintain efficient engine operation.
14. Check all vacuum lines for dried, cracked or brittle conditions. Something as simple as a leaking vacuum hose can cause poor performance and loss of economy.
15. Avoid tampering with the emission control system. Attempting to improve fuel econ-

FUEL SYSTEM

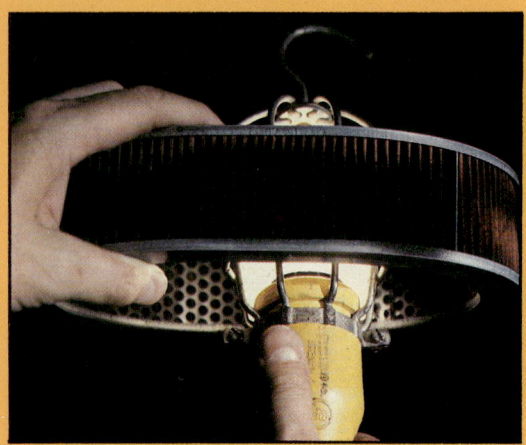

Check the air filter with a light behind it. If you can see light through the filter it can be reused.

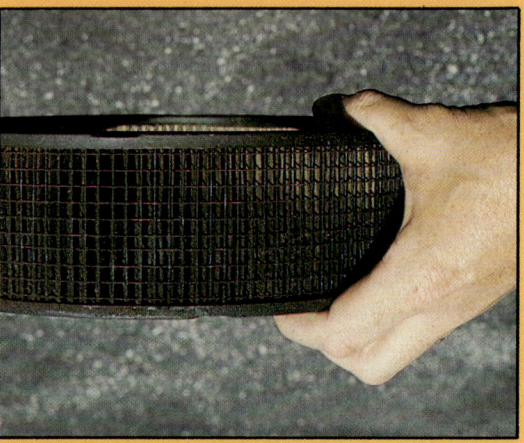

Extremely clogged filters should be discarded and replaced with a new one.

18. Replace the air filter regularly. A dirty air filter richens the air/fuel mixture and can increase fuel consumption as much as 10%. Tests show that 1/3 of all vehicles have air filters in need of replacement.
19. Replace the fuel filter at least as often as recommended.
20. Set the idle speed and carburetor mixture to specifications.
21. Check the automatic choke. A sticking or malfunctioning choke wastes gas.
22. During the summer months, adjust the automatic choke for a leaner mixture which will produce faster engine warm-ups.

COOLING SYSTEM

29. Be sure all accessory drive belts are in good condition. Check for cracks or wear.
30. Adjust all accessory drive belts to proper tension.
31. Check all hoses for swollen areas, worn spots, or loose clamps.
32. Check coolant level in the radiator or expansion tank.
33. Be sure the thermostat is operating properly. A stuck thermostat delays engine warm-up and a cold engine uses nearly twice as much fuel as a warm engine.
34. Drain and replace the engine coolant at least as often as recommended. Rust and scale

TIRES & WHEELS

38. Check the tire pressure often with a pencil type gauge. Tests by a major tire manufacturer show that 90% of all vehicles have at least 1 tire improperly inflated. Better mileage can be achieved by over-inflating tires, but never exceed the maximum inflation pressure on the side of the tire.
39. If possible, install radial tires. Radial tires deliver as much as ½ mpg more than bias belted tires.
40. Avoid installing super-wide tires. They only create extra rolling resistance and decrease fuel mileage. Stick to the manufacturer's recommendations.
41. Have the wheels properly balanced.

omy by tampering with emission controls is more likely to worsen fuel economy than improve it. Emission control changes on modern engines are not readily reversible.

16. Clean (or replace) the EGR valve and lines as recommended.

17. Be sure that all vacuum lines and hoses are reconnected properly after working under the hood. An unconnected or misrouted vacuum line can wreak havoc with engine performance.

---

23. Check for fuel leaks at the carburetor, fuel pump, fuel lines and fuel tank. Be sure all lines and connections are tight.

24. Periodically check the tightness of the carburetor and intake manifold attaching nuts and bolts. These are a common place for vacuum leaks to occur.

25. Clean the carburetor periodically and lubricate the linkage.

26. The condition of the tailpipe can be an excellent indicator of proper engine combustion. After a long drive at highway speeds, the inside of the tailpipe should be a light grey in color. Black or soot on the insides indicates an overly rich mixture.

27. Check the fuel pump pressure. The fuel pump may be supplying more fuel than the engine needs.

28. Use the proper grade of gasoline for your engine. Don't try to compensate for knocking or "pinging" by advancing the ignition timing. This practice will only increase plug temperature and the chances of detonation or pre-ignition with relatively little performance gain.

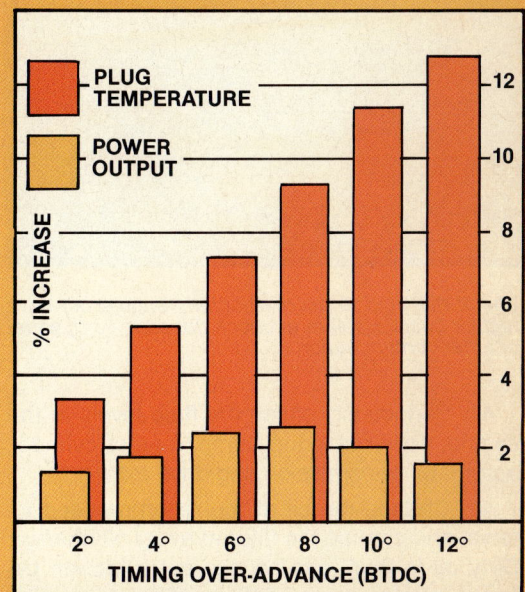

Increasing ignition timing past the specified setting results in a drastic increase in spark plug temperature with increased chance of detonation or preignition. Performance increase is considerably less. (Photo courtesy Champion Spark Plug Co.)

---

that form in the engine should be flushed out to allow the engine to operate at peak efficiency.

35. Clean the radiator of debris that can decrease cooling efficiency.

36. Install a flex-type or electric cooling fan, if you don't have a clutch type fan. Flex fans use curved plastic blades to push more air at low speeds when more cooling is needed; at high speeds the blades flatten out for less resistance. Electric fans only run when the engine temperature reaches a predetermined level.

37. Check the radiator cap for a worn or cracked gasket. If the cap does not seal properly, the cooling system will not function properly.

---

42. Be sure the front end is correctly aligned. A misaligned front end actually has wheels going in differed directions. The increased drag can reduce fuel economy by .3 mpg.

43. Correctly adjust the wheel bearings. Wheel bearings that are adjusted too tight increase rolling resistance.

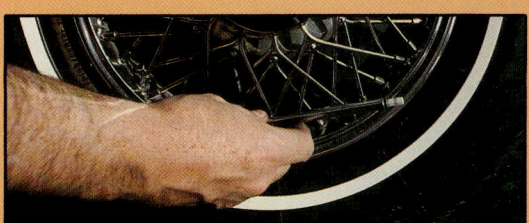

Check tire pressures regularly with a reliable pocket type gauge. Be sure to check the pressure on a cold tire.

GENERAL MAINTENANCE

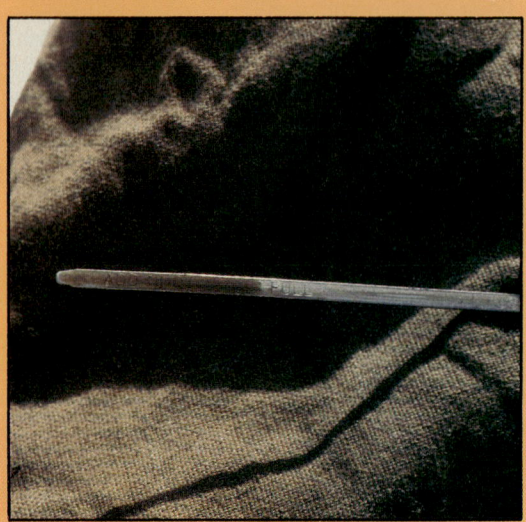

Check the fluid levels (particularly engine oil) on a regular basis. Be sure to check the oil for grit, water or other contamination.

A vacuum gauge is another excellent indicator of internal engine condition and can also be installed in the dash as a mileage indicator.

44. Periodically check the fluid levels in the engine, power steering pump, master cylinder, automatic transmission and drive axle.

45. Change the oil at the recommended interval and change the filter at every oil change. Dirty oil is thick and causes extra friction between moving parts, cutting efficiency and increasing wear. A worn engine requires more frequent tune-ups and gets progressively worse fuel economy. In general, use the lightest viscosity oil for the driving conditions you will encounter.

46. Use the recommended viscosity fluids in the transmission and axle.

47. Be sure the battery is fully charged for fast starts. A slow starting engine wastes fuel.

48. Be sure battery terminals are clean and tight.

49. Check the battery electrolyte level and add distilled water if necessary.

50. Check the exhaust system for crushed pipes, blockages and leaks.

51. Adjust the brakes. Dragging brakes or brakes that are not releasing create increased drag on the engine.

52. Install a vacuum gauge or miles-per-gallon gauge. These gauges visually indicate engine vacuum in the intake manifold. High vacuum = good mileage and low vacuum = poorer mileage. The gauge can also be an excellent indicator of internal engine conditions.

53. Be sure the clutch is properly adjusted. A slipping clutch wastes fuel.

54. Check and periodically lubricate the heat control valve in the exhaust manifold. A sticking or inoperative valve prevents engine warm-up and wastes gas.

55. Keep accurate records to check fuel economy over a period of time. A sudden drop in fuel economy may signal a need for tune-up or other maintenance.

© 1980 Chilton Book Company, Radnor, PA 19089

FUEL SYSTEM 301

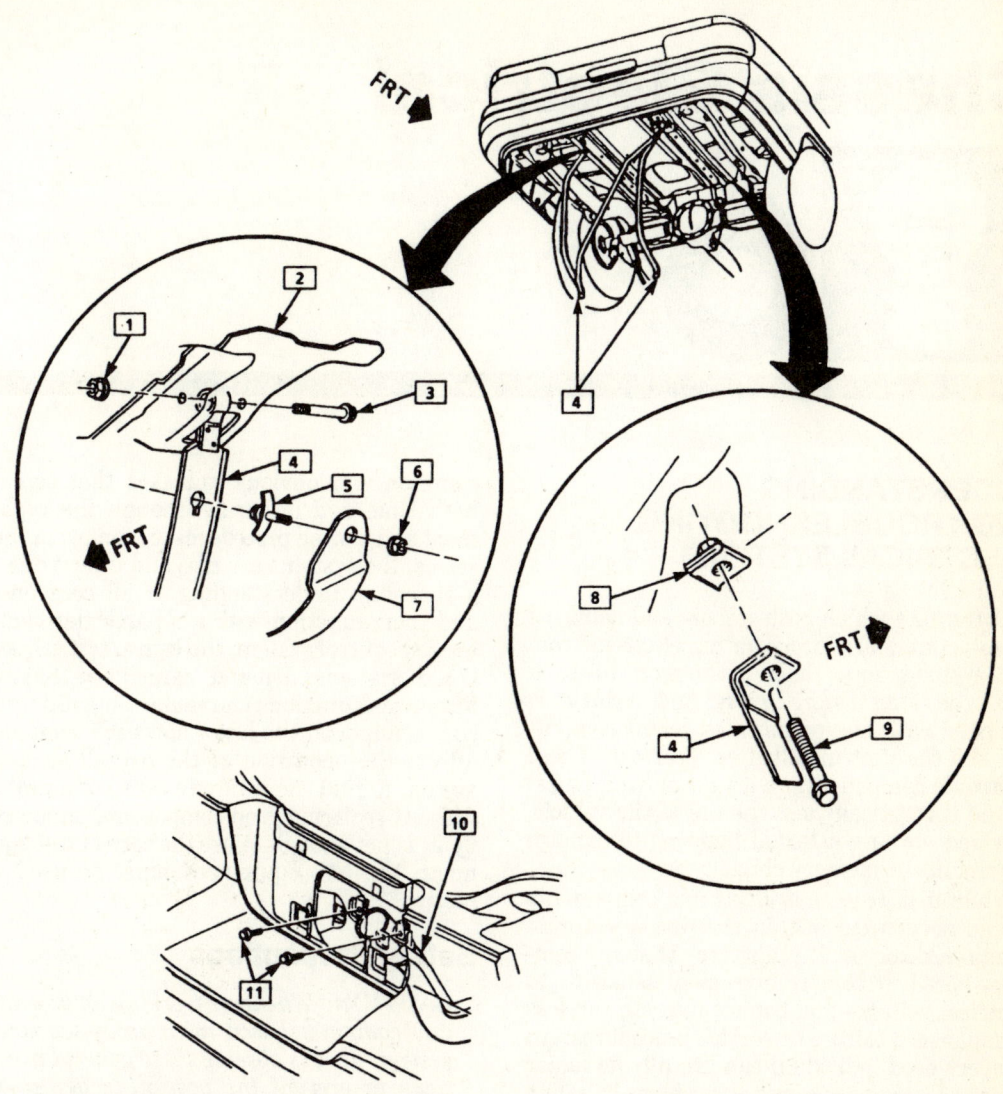

1	REAR FUEL TANK RETAINING STRAP NUTS (2) - 25 N·m (18 lb. ft.)
2	UNDER BODY
3	REAR FUEL TANK RETAINING STRAP BOLTS (2)
4	FUEL TANK RETAINING STRAPS (2)
5	FUEL TANK CROSS STRAP ATTACHING SCREW
6	FUEL TANK CROSS STRAP ATTACHING NUT - 15 N·m (11 lb. ft.)
7	FUEL TANK CROSS STRAP
8	FRONT FUEL TANK RETAINING STRAP BODY NUTS (2)
9	FRONT FUEL TANK RETAINING STRAP BOLTS (2) - 32 N·m (24 lb. ft.)
10	FUEL FILLER TUBE
11	FUEL FILLER TUBE ATTACHING SCREWS (2) - 2 N·m (18 lb. in.)

Fuel tank mounting components, 1991–92 Sedan

Chassis Electrical

UNDERSTANDING AND TROUBLESHOOTING ELECTRICAL SYSTEMS

At the rate which both import and domestic manufacturers are incorporating electronic control systems into their production lines, it won't be long before every new vehicle is equipped with one or more on-board computers, like the unit installed on your car. These electronic components (with no moving parts) should theoretically last the life of the vehicle, provided nothing external happens to damage the circuits or memory chips.

While it is true that electronic components should never wear out, in the real world malfunctions do occur. It is also true that any computer-based system is extremely sensitive to electrical voltages and cannot tolerate careless or haphazard testing or service procedures. An inexperienced individual can literally do major damage looking for a minor problem by using the wrong kind of test equipment or connecting test leads or connectors with the ignition switch ON. When selecting test equipment, make sure the manufacturers instructions state that the tester is compatible with whatever type of electronic control system is being serviced. Read all instructions carefully and double check all test points before installing probes or making any test connections.

The following section outlines basic diagnosis techniques for dealing with computerized automotive control systems. Along with a general explanation of the various types of test equipment available to aid in servicing modern electronic automotive systems, basic repair techniques for wiring harnesses and connectors is given. Read the basic information before attempting any repairs or testing on any computerized system, to provide the background of information necessary to avoid the most common and obvious mistakes that can cost both time and money. Although the replacement and testing procedures are simple in themselves, the systems are not, and unless one has a thorough understanding of all components and their function within a particular computerized control system, the logical test sequence these systems demand cannot be followed. Minor malfunctions can make a big difference, so it is important to know how each component affects the operation of the overall electronic system to find the ultimate cause of a problem without replacing good components unnecessarily. It is not enough to use the correct test equipment; the test equipment must be used correctly.

Safety Precautions

CAUTION: *Whenever working on or around any computer based microprocessor control system, always observe these general precautions to prevent the possibility of personal injury or damage to electronic components.*

• Never install or remove battery cables with the key ON or the engine running. Jumper cables should be connected with the key OFF to avoid power surges that can damage electronic control units. Engines equipped with computer controlled systems should avoid both giving and getting jump starts due to the possibility of serious damage to components from arcing in the engine compartment when connections are made with the ignition ON.

• Always remove the battery cables before charging the battery. Never use a high output charger on an installed battery or attempt to use any type of "hot shot" (24 volt) starting aid.

• Exercise care when inserting test probes into connectors to insure good connections without damaging the connector or spreading the

pins. Always probe connectors from the rear (wire) side, NOT the pin side, to avoid accidental shorting of terminals during test procedures.
• Never remove or attach wiring harness connectors with the ignition switch ON, especially to an electronic control unit.
• Do not drop any components during service procedures and never apply 12 volts directly to any component (like a solenoid or relay) unless instructed specifically to do so. Some component electrical windings are designed to safely handle only 4 or 5 volts and can be destroyed in seconds if 12 volts are applied directly to the connector.
• Remove the electronic control unit if the vehicle is to be placed in an environment where temperatures exceed approximately 176°F (80°C), such as a paint spray booth or when arc or gas welding near the control unit location in the car.

ORGANIZED TROUBLESHOOTING

When diagnosing a specific problem, organized troubleshooting is a must. The complexity of a modern automobile demands that you approach any problem in a logical, organized manner. There are certain troubleshooting techniques that are standard:

1. Establish when the problem occurs. Does the problem appear only under certain conditions? Were there any noises, odors, or other unusual symptoms?
2. Isolate the problem area. To do this, make some simple tests and observations; then eliminate the systems that are working properly. Check for obvious problems such as broken wires, dirty connections or split or disconnected vacuum hoses. Always check the obvious before assuming something complicated is the cause.
3. Test for problems systematically to determine the cause once the problem area is isolated. Are all the components functioning properly? Is there power going to electrical switches and motors? Is there vacuum at vacuum switches and/or actuators? Is there a mechanical problem such as bent linkage or loose mounting screws? Doing careful, systematic checks will often turn up most causes on the first inspection without wasting time checking components that have little or no relationship to the problem.
4. Test all repairs after the work is done to make sure that the problem is fixed. Some causes can be traced to more than one component, so a careful verification of repair work is important to pick up additional malfunctions that may cause a problem to reappear or a different problem to arise. A blown fuse, for example, is a simple problem that may require more than another fuse to repair. If you don't look for a problem that caused a fuse to blow, for example, a shorted wire may go undetected.

Experience has shown that most problems tend to be the result of a fairly simple and obvious cause, such as loose or corroded connectors or air leaks in the intake system; making careful inspection of components during testing essential to quick and accurate troubleshooting. Special, hand held computerized testers designed specifically for diagnosing the system are available from a variety of aftermarket sources, as well as from the vehicle manufacturer, but care should be taken that any test equipment being used is designed to diagnose that particular computer controlled system accurately without damaging the control unit (ECU) or components being tested.

NOTE: *Pinpointing the exact cause of trouble in an electrical system can sometimes only be accomplished by the use of special test equipment. The following describes commonly used test equipment and explains how to put it to best use in diagnosis. In addition to the information covered below, the manufacturer's instructions booklet provided with the tester should be read and clearly understood before attempting any test procedures.*

TEST EQUIPMENT

Jumper Wires

Jumper wires are simple, yet extremely valuable, pieces of test equipment. Jumper wires are merely wires that are used to bypass sections of a circuit. The simplest type of jumper wire is merely a length of multi-strand wire with an alligator clip at each end. Jumper wires are usually fabricated from lengths of standard automotive wire and whatever type of connector (alligator clip, spade connector or pin connector) that is required for the particular vehicle being tested. The well equipped tool box will have several different styles of jumper wires in several different lengths. Some jumper wires are made with three or more terminals coming from a common splice for special purpose testing. In cramped, hard-to-reach areas it is advisable to have insulated boots over the jumper wire terminals in order to prevent accidental grounding, sparks, and possible fire, especially when testing fuel system components.

Jumper wires are used primarily to locate open electrical circuits, on either the ground (−) side of the circuit or on the hot (+) side. If an electrical component fails to operate, connect the jumper wire between the component and a good ground. If the component operates only with the jumper installed, the ground circuit is open. If the ground circuit is good, but the component does not operate, the circuit between the power feed and component is open. You can sometimes connect the jumper wire directly from the battery to the hot terminal of the component, but first make sure the component uses 12 volts in operation. Some electrical components, such as fuel injectors, are designed to operate on about 4 volts and running 12 volts directly to the injector terminals can burn out the wiring. By inserting an inline fuseholder between a set of test leads, a fused jumper wire can be used for bypassing open circuits. Use a 5 amp fuse to provide protection against voltage spikes. When in doubt, use a voltmeter to check the voltage input to the component and measure how much voltage is being applied normally. By moving the jumper wire successively back from the lamp toward the power source, you can isolate the area of the circuit where the open is located. When the component stops functioning, or the power is cut off, the open is in the segment of wire between the jumper and the point previously tested.

CAUTION: *Never use jumpers made from wire that is of lighter gauge than used in the circuit under test. If the jumper wire is of too small gauge, it may overheat and possibly melt. Never use jumpers to bypass high resistance loads (such as motors) in a circuit. Bypassing resistances, in effect, creates a short circuit which may, in turn, cause damage and fire. Never use a jumper for anything other than temporary bypassing of components in a circuit.*

12 Volt Test Light

The 12 volt test light is used to check circuits and components while electrical current is flowing through them. It is used for voltage and ground tests. Twelve volt test lights come in different styles but all have three main parts; a ground clip, a probe, and a light. The most commonly used 12 volt test lights have pick-type probes. To use a 12 volt test light, connect the ground clip to a good ground and probe wherever necessary with the pick. The pick should be sharp so that it can penetrate wire insulation to make contact with the wire, without making a large hole in the insulation. The wrap-around light is handy in hard to reach areas or where it is difficult to support a wire to push a probe pick into it. To use the wrap around light, hook the wire to probed with the hook and pull the trigger. A small pick will be forced through the wire insulation into the wire core.

CAUTION: *Do not use a test light to probe electronic ignition spark plug or coil wires. Never use a pick-type test light to probe wiring on computer controlled systems unless specifically instructed to do so. Any wire insulation that is pierced by the test light probe should be taped and sealed with silicone after testing.*

Like the jumper wire, the 12 volt test light is used to isolate opens in circuits. But, whereas the jumper wire is used to bypass the open to operate the load, the 12 volt test light is used to locate the presence of voltage in a circuit. If the test light glows, you know that there is power up to that point; if the 12 volt test light does not glow when its probe is inserted into the wire or connector, you know that there is an open circuit (no power). Move the test light in successive steps back toward the power source until the light in the handle does glow. When it does glow, the open is between the probe and point previously probed.

NOTE: *The test light does not detect that 12 volts (or any particular amount of voltage) is present; it only detects that some voltage is present. It is advisable before using the test light to touch its terminals across the battery posts to make sure the light is operating properly.*

Self-Powered Test Light

The self-powered test light usually contains a 1.5 volt penlight battery. One type of self-powered test light is similar in design to the 12 volt test light. This type has both the battery and the light in the handle and pick-type probe tip. The second type has the light toward the open tip, so that the light illuminates the contact point. The self-powered test light is dual purpose piece of test equipment. It can be used to test for either open or short circuits when power is isolated from the circuit (continuity test). A powered test light should not be used on any computer controlled system or component unless specifically instructed to do so. Many engine sensors can be destroyed by even this small amount of voltage applied directly to the terminals.

Open Circuit Testing

To use the self-powered test light to check for open circuits, first isolate the circuit from the vehicle's 12 volt power source by disconnecting the battery or wiring harness connector. Connect the test light ground clip to a good ground and probe sections of the circuit sequentially with the test light. (start from either end of the circuit). If the light is out, the open is between the probe and the circuit ground. If the light is on, the open is between the probe and end of the circuit toward the power source.

Short Circuit Testing

By isolating the circuit both from power and from ground, and using a self-powered test light, you can check for shorts to ground in the circuit. Isolate the circuit from power and ground. Connect the test light ground clip to a good ground and probe any easy-to-reach test point in the circuit. If the light comes on, there is a short somewhere in the circuit. To isolate the short, probe a test point at either end of the isolated circuit (the light should be on). Leave the test light probe connected and open connectors, switches, remove parts, etc., sequentially, until the light goes out. When the light goes out, the short is between the last circuit component opened and the previous circuit opened.

NOTE: *The 1.5 volt battery in the test light does not provide much current. A weak battery may not provide enough power to illuminate the test light even when a complete circuit is made (especially if there are high resistances in the circuit). Always make sure that the test battery is strong. To check the battery, briefly touch the ground clip to the probe; if the light glows brightly the battery is strong enough for testing. Never use a self-powered test light to perform checks for opens or shorts when power is applied to the electrical system under test. The 12 volt vehicle power will quickly burn out the 1.5 volt light bulb in the test light.*

Voltmeter

A voltmeter is used to measure voltage at any point in a circuit, or to measure the voltage drop across any part of a circuit. It can also be used to check continuity in a wire or circuit by indicating current flow from one end to the other. Voltmeters usually have various scales on the meter dial and a selector switch to allow the selection of different voltages. The voltmeter has a positive and a negative lead. To avoid damage to the meter, always connect the negative lead to the negative (−) side of circuit (to ground or nearest the ground side of the circuit) and connect the positive lead to the positive (+) side of the circuit (to the power source or the nearest power source). Note that the negative voltmeter lead will always be black and that the positive voltmeter will always be some color other than black (usually red). Depending on how the voltmeter is connected into the circuit, it has several uses.

A voltmeter can be connected either in parallel or in series with a circuit and it has a very high resistance to current flow. When connected in parallel, only a small amount of current will flow through the voltmeter current path; the rest will flow through the normal circuit current path and the circuit will work normally. When the voltmeter is connected in series with a circuit, only a small amount of current can flow through the circuit. The circuit will not work properly, but the voltmeter reading will show if the circuit is complete or not.

Available Voltage Measurement

Set the voltmeter selector switch to the 20V position and connect the meter negative lead to the negative post of the battery. Connect the positive meter lead to the positive post of the battery and turn the ignition switch ON to provide a load. Read the voltage on the meter or digital display. A well charged battery should register over 12 volts. If the meter reads below 11.5 volts, the battery power may be insufficient to operate the electrical system properly. This test determines voltage available from the battery and should be the first step in any electrical trouble diagnosis procedure. Many electrical problems, especially on computer controlled systems, can be caused by a low state of charge in the battery. Excessive corrosion at the battery cable terminals can cause a poor contact that will prevent proper charging and full battery current flow.

Normal battery voltage is 12 volts when fully charged. When the battery is supplying current to one or more circuits it is said to be "under load". When everything is off the electrical system is under a "no-load" condition. A fully charged battery may show about 12.5 volts at no load; will drop to 12 volts under medium load; and will drop even lower under heavy load. If the battery is partially discharged the voltage decrease under heavy load may be excessive, even though the battery shows 12 volts or more at no load. When allowed to discharge further, the battery's available voltage under load will decrease more severely. For this reason, it is important that the battery be fully charged during all testing procedures to avoid errors in diagnosis and incorrect test results.

Voltage Drop

When current flows through a resistance, the voltage beyond the resistance is reduced (the larger the current, the greater the reduction in voltage). When no current is flowing, there is no voltage drop because there is no current flow. All points in the circuit which are connected to the power source are at the same voltage as the power source. The total voltage drop always equals the total source voltage. In a long circuit with many connectors, a series of small, unwanted voltage drops due to corrosion at the connectors can add up to a total loss of voltage which impairs the operation of the normal loads in the circuit.

INDIRECT COMPUTATION OF VOLTAGE DROPS

1. Set the voltmeter selector switch to the 20 volt position.
2. Connect the meter negative lead to a good ground.
3. Probe all resistances in the circuit with the positive meter lead.
4. Operate the circuit in all modes and observe the voltage readings.

DIRECT MEASUREMENT OF VOLTAGE DROPS

1. Set the voltmeter switch to the 20 volt position.
2. Connect the voltmeter negative lead to the ground side of the resistance load to be measured.
3. Connect the positive lead to the positive side of the resistance or load to be measured.
4. Read the voltage drop directly on the 20 volt scale.

Too high a voltage indicates too high a resistance. If, for example, a blower motor runs too slowly, you can determine if there is too high a resistance in the resistor pack. By taking voltage drop readings in all parts of the circuit, you can isolate the problem. Too low a voltage drop indicates too low a resistance. If, for example, a blower motor runs too fast in the MED and/or LOW position, the problem can be isolated in the resistor pack by taking voltage drop readings in all parts of the circuit to locate a possibly shorted resistor. The maximum allowable voltage drop under load is critical, especially if there is more than one high resistance problem in a circuit because all voltage drops are cumulative. A small drop is normal due to the resistance of the conductors.

HIGH RESISTANCE TESTING

1. Set the voltmeter selector switch to the 4 volt position.
2. Connect the voltmeter positive lead to the positive post of the battery.
3. Turn on the headlights and heater blower to provide a load.
4. Probe various points in the circuit with the negative voltmeter lead.
5. Read the voltage drop on the 4 volt scale. Some average maximum allowable voltage drops are:
   FUSE PANEL – 7 volts
   IGNITION SWITCH – 5 volts
   HEADLIGHT SWITCH – 7 volts
   IGNITION COIL (+) – 5 volts
   ANY OTHER LOAD – 1.3 volts

NOTE: *Voltage drops are all measured while a load is operating; without current flow, there will be no voltage drop.*

Ohmmeter

The ohmmeter is designed to read resistance (ohms) in a circuit or component. Although there are several different styles of ohmmeters, all will usually have a selector switch which permits the measurement of different ranges of resistance (usually the selector switch allows the multiplication of the meter reading by 10, 100, 1000, and 10,000). A calibration knob allows the meter to be set at zero for accurate measurement. Since all ohmmeters are powered by an internal battery (usually 9 volts), the ohmmeter can be used as a self-powered test light. When the ohmmeter is connected, current from the ohmmeter flows through the circuit or component being tested. Since the ohmmeter's internal resistance and voltage are known values, the amount of current flow through the meter depends on the resistance of the circuit or component being tested.

The ohmmeter can be used to perform continuity test for opens or shorts (either by observation of the meter needle or as a self-powered test light), and to read actual resistance in a circuit. It should be noted that the ohmmeter is used to check the resistance of a component or wire while there is no voltage applied to the circuit. Current flow from an outside voltage source (such as the vehicle battery) can damage the ohmmeter, so the circuit or component should be isolated from the vehicle electrical system before any testing is done. Since the ohmmeter uses its own voltage source, either lead can be connected to any test point.

NOTE: *When checking diodes or other solid state components, the ohmmeter leads can only be connected one way in order to measure current flow in a single direction. Make sure the positive (+) and negative (–) terminal connections are as described in the test procedures to verify the one-way diode operation.*

In using the meter for making continuity checks, do not be concerned with the actual resistance readings. Zero resistance, or any resistance readings, indicate continuity in the circuit. Infinite resistance indicates an open in the circuit. A high resistance reading where there should be none indicates a problem in the circuit. Checks for short circuits are made in the same manner as checks for open circuits except that the circuit must be isolated from both power and normal ground. Infinite resistance indicates no continuity to ground, while zero resistance indicates a dead short to ground.

RESISTANCE MEASUREMENT

The batteries in an ohmmeter will weaken with age and temperature, so the ohmmeter must be calibrated or "zeroed" before taking measurements. To zero the meter, place the selector switch in its lowest range and touch the two ohmmeter leads together. Turn the calibration knob until the meter needle is exactly on zero.

NOTE: *All analog (needle) type ohmmeters must be zeroed before use, but some digital ohmmeter models are automatically calibrated when the switch is turned on. Self-calibrating digital ohmmeters do not have an adjusting knob, but its a good idea to check for a zero readout before use by touching the leads together. All computer controlled systems require the use of a digital ohmmeter with at least 10 megohms impedance for testing. Before any test procedures are attempted, make sure the ohmmeter used is compatible with the electrical system or damage to the on-board computer could result.*

To measure resistance, first isolate the circuit from the vehicle power source by disconnecting the battery cables or the harness connector. Make sure the key is OFF when disconnecting any components or the battery. Where necessary, also isolate at least one side of the circuit to be checked to avoid reading parallel resistances. Parallel circuit resistances will always give a lower reading than the actual resistance of either of the branches. When measuring the resistance of parallel circuits, the total resistance will always be lower than the smallest resistance in the circuit. Connect the meter leads to both sides of the circuit (wire or component) and read the actual measured ohms on the meter scale. Make sure the selector switch is set to the proper ohm scale for the circuit being tested to avoid misreading the ohmmeter test value.

CAUTION: *Never use an ohmmeter with power applied to the circuit. Like the self-powered test light, the ohmmeter is designed to operate on its own power supply. The normal 12 volt automotive electrical system current could damage the meter.*

Ammeters

An ammeter measures the amount of current flowing through a circuit in units called amperes or amps. Amperes are units of electron flow which indicate how fast the electrons are flowing through the circuit. Since Ohms Law dictates that current flow in a circuit is equal to the circuit voltage divided by the total circuit resistance, increasing voltage also increases the current level (amps). Likewise, any decrease in resistance will increase the amount of amps in a circuit. At normal operating voltage, most circuits have a characteristic amount of amperes, called "current draw" which can be measured using an ammeter. By referring to a specified current draw rating, measuring the amperes, and comparing the two values, one can determine what is happening within the circuit to aid in diagnosis. An open circuit, for example, will not allow any current to flow so the ammeter reading will be zero. More current flows through a heavily loaded circuit or when the charging system is operating.

An ammeter is always connected in series with the circuit being tested. All of the current that normally flows through the circuit must also flow through the ammeter; if there is any other path for the current to follow, the ammeter reading will not be accurate. The ammeter itself has very little resistance to current flow and therefore will not affect the circuit, but it will measure current draw only when the circuit is closed and electricity is flowing. Excessive current draw can blow fuses and drain the battery, while a reduced current draw can cause motors to run slowly, lights to dim and other components to not operate properly. The ammeter can help diagnose these conditions by locating the cause of the high or low reading.

Multimeters

Different combinations of test meters can be built into a single unit designed for specific tests. Some of the more common combination test devices are known as Volt/Amp testers, Tach/Dwell meters, or Digital Multi-meters. The Volt/Amp tester is used for charging system, starting system or battery tests and consists of a voltmeter, an ammeter and a variable resistance carbon pile. The voltmeter will usually have at least two ranges for use with 6, 12 and 24 volt systems. The ammeter also has more than one range for testing various levels of battery loads and starter current draw and the carbon pile can be adjusted to offer different amounts of resistance. The Volt/Amp tester has heavy leads to carry large amounts of current and many later models have an inductive ammeter pickup that clamps around the wire to simplify test connections. On some models, the ammeter also has a zero-center scale to allow testing of charging and starting systems without switching leads or polarity. A digital multi-meter is a voltmeter, ammeter and ohmmeter combined in an instrument which gives a digital readout. These are often used when testing solid state circuits because of their high input impedance (usually 10 megohms or more).

The tach/dwell meter combines a tachometer and a dwell (cam angle) meter and is a specialized kind of voltmeter. The tachometer scale is marked to show engine speed in rpm and the dwell scale is marked to show degrees of distributor shaft rotation. In most electronic ignition systems, dwell is determined by the control unit, but the dwell meter can also be used to check the duty cycle (operation) of some electronic engine control systems. Some tach/dwell meters are powered by an internal battery, while others take their power from the car battery in use. The battery powered testers usually require calibration much like an ohmmeter before testing.

Special Test Equipment

A variety of diagnostic tools are available to help troubleshoot and repair computerized engine control systems. The most sophisticated of these devices are the console type engine analyzers that usually occupy a garage service bay, but there are several types of aftermarket electronic testers available that will allow quick circuit tests of the engine control system by plugging directly into a special connector located in the engine compartment or under the dashboard. Several tool and equipment manufacturers offer simple, hand held testers that measure various circuit voltage levels on command to check all system components for proper operation. Although these testers usually cost about $300-$500, consider that the average computer control unit (or ECM) can cost just as much and the money saved by not replacing perfectly good sensors or components in an attempt to correct a problem could justify the purchase price of a special diagnostic tester the first time it's used.

These computerized testers can allow quick and easy test measurements while the engine is operating or while the car is being driven. In addition, the on-board computer memory can be read to access any stored trouble codes; in effect allowing the computer to tell you where it hurts and aid trouble diagnosis by pinpointing exactly which circuit or component is malfunctioning. In the same manner, repairs can be tested to make sure the problem has been corrected. The biggest advantage these special testers have is their relatively easy hookups that minimize or eliminate the chances of making the wrong connections and getting false voltage readings or damaging the computer accidentally.

NOTE: *It should be remembered that these testers check voltage levels in circuits; they don't detect mechanical problems or failed components if the circuit voltage falls within the preprogrammed limits stored in the tester PROM unit. Also, most of the hand held testers are designed to work only on one or two systems made by a specific manufacturer.*

A variety of aftermarket testers are available to help diagnose different computerized control systems. Owatonna Tool Company (OTC), for example, markets a device called the OTC Monitor which plugs directly into the assembly line diagnostic link (ALDL). The OTC tester makes diagnosis a simple matter of pressing the correct buttons and, by changing the internal PROM or inserting a different diagnosis cartridge, it will work on any model from full size to subcompact, over a wide range of years. An adapter is supplied with the tester to allow connection to all types of ALDL links, regardless of the number of pin terminals used. By inserting an updated PROM into the OTC tester, it can be easily updated to diagnose any new modifications of computerized control systems.

Wiring Harnesses

The average automobile contains about $1/2$ mile of wiring, with hundreds of individual connections. To protect the many wires from damage and to keep them from becoming a confusing tangle, they are organized into bundles, enclosed in plastic or taped together and called wire harnesses. Different wiring harnesses serve different parts of the vehicle. Individual wires are color coded to help trace them through a harness where sections are hidden from view.

A loose or corroded connection or a replacement wire that is too small for the circuit will add extra resistance and an additional voltage drop to the circuit. A ten percent voltage drop can result in slow or erratic motor operation, for example, even though the circuit is complete. Automotive wiring or circuit conductors can be in any one of three forms:

1. Single strand wire
2. Multi-strand wire
3. Printed circuitry

Single strand wire has a solid metal core and is usually used inside such components as alternators, motors, relays and other devices. Multi-strand wire has a core made of many small strands of wire twisted together into a single conductor. Most of the wiring in an automotive electrical system is made up of multi-strand wire, either as a single conductor or grouped together in a harness. All wiring is color coded on the insulator, either as a solid color or as a colored wire with an identification stripe. A printed circuit is a thin film of copper or other conductor that is printed on an insulator backing. Occasionally, a printed circuit is sandwiched between two sheets of plastic for more protection and flexibility. A complete printed circuit, consisting of conductors, insulating material and connectors for lamps or other components is called a printed circuit board. Printed circuitry is used in place of individual wires or harnesses in places where space is limited, such as behind instrument panels.

Wire Gauge

Since computer controlled automotive electrical systems are very sensitive to changes in resistance, the selection of properly sized wires is critical when systems are repaired. The wire gauge number is an expression of the cross section area of the conductor. The most common system for expressing wire size is the American Wire Gauge (AWG) system.

Wire cross section area is measured in circular mils. A mil is $1/_{1000}$" (0.001"); a circular mil is the area of a circle one mil in diameter. For example, a conductor 1/4" in diameter is 0.250 in. or 250 mils. The circular mil cross section area of the wire is 250 squared (250²) or 62,500 circular mils. Imported car models usually use metric wire gauge designations, which is simply the cross section area of the conductor in square millimeters (mm²).

Gauge numbers are assigned to conductors of various cross section areas. As gauge number increases, area decreases and the conductor becomes smaller. A 5 gauge conductor is smaller than a 1 gauge conductor and a 10 gauge is smaller than a 5 gauge. As the cross section area of a conductor decreases, resistance increases and so does the gauge number. A conductor with a higher gauge number will carry less current than a conductor with a lower gauge number.

NOTE: *Gauge wire size refers to the size of the conductor, not the size of the complete wire. It is possible to have two wires of the same gauge with different diameters because one may have thicker insulation than the other.*

12 volt automotive electrical systems generally use 10, 12, 14, 16 and 18 gauge wire. Main power distribution circuits and larger accessories usually use 10 and 12 gauge wire. Battery cables are usually 4 or 6 gauge, although 1 and 2 gauge wires are occasionally used. Wire length must also be considered when making repairs to a circuit. As conductor length increases, so does resistance. An 18 gauge wire, for example, can carry a 10 amp load for 10 feet without excessive voltage drop; however if a 15 foot wire is required for the same 10 amp load, it must be a 16 gauge wire.

An electrical schematic shows the electrical current paths when a circuit is operating properly. It is essential to understand how a circuit works before trying to figure out why it doesn't. Schematics break the entire electrical system down into individual circuits and show only one particular circuit. In a schematic, no attempt is made to represent wiring and components as they physically appear on the vehicle; switches and other components are shown as simply as possible. Face views of harness connectors show the cavity or terminal locations in all multi-pin connectors to help locate test points.

If you need to backprobe a connector while it is on the component, the order of the terminals must be mentally reversed. The wire color code can help in this situation, as well as a keyway, lock tab or other reference mark.

NOTE: *Wiring diagrams are not included in this book. As trucks have become more complex and available with longer option lists, wiring diagrams have grown in size and complexity. It has become almost impossible to provide a readable reproduction of a wiring diagram in a book this size. Information on ordering wiring diagrams from the vehicle manufacturer can be found in the owner's manual.*

WIRING REPAIR

Soldering is a quick, efficient method of joining metals permanently. Everyone who has the occasion to make wiring repairs should know how to solder. Electrical connections that are soldered are far less likely to come apart and will conduct electricity much better than connections that are only "pig-tailed" together. The most popular (and preferred) method of soldering is with an electrical soldering gun. Soldering irons are available in many sizes and wattage ratings. Irons with higher wattage ratings deliver higher temperatures and recover lost heat faster. A small soldering iron rated for no more than 50 watts is recommended, especially on electrical systems where excess heat can damage the components being soldered.

There are three ingredients necessary for successful soldering; proper flux, good solder and sufficient heat. A soldering flux is necessary to clean the metal of tarnish, prepare it for soldering and to enable the solder to spread into tiny crevices. When soldering, always use a resin flux or resin core solder which is non-corrosive and will not attract moisture once the job is finished. Other types of flux (acid core) will leave a residue that will attract moisture and cause the wires to corrode. Tin is a unique metal with a low melting point. In a molten state, it dissolves and alloys easily with many metals. Solder is made by mixing tin with lead. The most common proportions are 40/60, 50/50 and 60/40, with the percentage of tin listed first. Low priced solders usually contain less tin, making them very difficult for a beginner to use because more heat is required to melt the solder. A common solder is 40/60 which is well suited for all-around general use, but 60/40 melts easier, has more tin for a better joint and is preferred for electrical work.

Soldering Techniques

Successful soldering requires that the metals to be joined be heated to a temperature that will melt the solder—usually 360–460°F (182–238°C). Contrary to popular belief, the purpose of the soldering iron is not to melt the solder itself, but to heat the parts being soldered to a temperature high enough to melt the solder when it is touched to the work. Melting flux-cored solder on the soldering iron will usually destroy the effectiveness of the flux.

NOTE: *Soldering tips are made of copper for good heat conductivity, but must be "tinned" regularly for quick transference of heat to the project and to prevent the solder from sticking to the iron. To "tin" the iron, simply heat it and touch the flux-cored solder to the tip; the solder will flow over the hot tip. Wipe the excess off with a clean rag, but be careful as the iron will be hot.*

After some use, the tip may become pitted. If so, simply dress the tip smooth with a smooth file and "tin" the tip again. An old saying holds that "metals well cleaned are half soldered." Flux-cored solder will remove oxides but rust, bits of insulation and oil or grease must be removed with a wire brush or emery cloth. For maximum strength in soldered parts, the joint must start off clean and tight. Weak joints will result in gaps too wide for the solder to bridge.

If a separate soldering flux is used, it should be brushed or swabbed on only those areas that are to be soldered. Most solders contain a core of flux and separate fluxing is unnecessary. Hold the work to be soldered firmly. It is best to solder on a wooden board, because a metal vise will only rob the piece to be soldered of heat and make it difficult to melt the solder. Hold the soldering tip with the broadest face against the work to be soldered. Apply solder under the tip close to the work, using enough solder to give a heavy film between the iron and the piece being soldered, while moving slowly and making sure the solder melts properly. Keep the work level or the solder will run to the lowest part and favor the thicker parts, because these require more heat to melt the solder. If the soldering tip overheats (the solder coating on the face of the tip burns up), it should be retinned. Once the soldering is completed, let the soldered joint stand until cool. Tape and seal all soldered wire splices after the repair has cooled.

Wire Harness and Connectors

The on-board computer (ECM) wire harness electrically connects the control unit to the various solenoids, switches and sensors used by the control system. Most connectors in the engine compartment or otherwise exposed to the elements are protected against moisture and dirt which could create oxidation and deposits on the terminals. This protection is important because of the very low voltage and current levels used by the computer and sensors. All connectors have a lock which secures the male and female terminals together, with a secondary lock holding the seal and terminal into the connector. Both terminal locks must be released when disconnecting ECM connectors.

These special connectors are weather-proof and all repairs require the use of a special terminal and the tool required to service it. This tool is used to remove the pin and sleeve terminals. If removal is attempted with an ordinary pick, there is a good chance that the terminal will be bent or deformed. Unlike standard blade type terminals, these terminals cannot be straightened once they are bent. Make certain that the connectors are properly seated and all of the sealing rings in place when connecting leads. On some models, a hinge-type flap provides a backup or secondary locking feature for the terminals. Most secondary locks are used to improve the connector reliability by retaining the terminals if the small terminal lock tangs are not positioned properly.

Molded-on connectors require complete replacement of the connection. This means splicing a new connector assembly into the harness. All splices in on-board computer systems should be soldered to insure proper contact. Use care when probing the connections or replacing terminals in them as it is possible to short between opposite terminals. If this happens to the wrong terminal pair, it is possible to damage certain components. Always use jumper wires between connectors for circuit checking and never probe through weather-proof seals.

Open circuits are often difficult to locate by sight because corrosion or terminal misalignment are hidden by the connectors. Merely wiggling a connector on a sensor or in the wiring harness may correct the open circuit condition. This should always be considered when an open circuit or a failed sensor is indicated. Intermittent problems may also be caused by oxidized or loose connections. When using a circuit tester for diagnosis, always probe connections from the wire side. Be careful not to damage sealed connectors with test probes.

All wiring harnesses should be replaced with

CHASSIS ELECTRICAL 311

identical parts, using the same gauge wire and connectors. When signal wires are spliced into a harness, use wire with high temperature insulation only. With the low voltage and current levels found in the system, it is important that the best possible connection at all wire splices be made by soldering the splices together. It is seldom necessary to replace a complete harness. If replacement is necessary, pay close attention to insure proper harness routing. Secure the harness with suitable plastic wire clamps to prevent vibrations from causing the harness to wear in spots or contact any hot components.

> NOTE: *Weatherproof connectors cannot be replaced with standard connectors. Instructions are provided with replacement connector and terminal packages. Some wire harnesses have mounting indicators (usually pieces of colored tape) to mark where the harness is to be secured.*

In making wiring repairs, it's important that you always replace damaged wires with wires that are the same gauge as the wire being replaced. The heavier the wire, the smaller the gauge number. Wires are color-coded to aid in identification and whenever possible the same color coded wire should be used for replacement. A wire stripping and crimping tool is necessary to install solderless terminal connectors. Test all crimps by pulling on the wires; it should not be possible to pull the wires out of a good crimp.

Wires which are open, exposed or otherwise damaged are repaired by simple splicing. Where possible, if the wiring harness is accessible and the damaged place in the wire can be located, it is best to open the harness and check for all possible damage. In an inaccessible harness, the wire must be bypassed with a new insert, usually taped to the outside of the old harness.

When replacing fusible links, be sure to use fusible link wire, NOT ordinary automotive wire. Make sure the fusible segment is of the same gauge and construction as the one being replaced and double the stripped end when crimping the terminal connector for a good contact. The melted (open) fusible link segment of the wiring harness should be cut off as close to the harness as possible, then a new segment spliced in as described. In the case of a damaged fusible link that feeds two harness wires, the harness connections should be replaced with two fusible link wires so that each circuit will have its own separate protection.

> NOTE: *Most of the problems caused in the wiring harness are due to bad ground connections. Always check all vehicle ground connections for corrosion or looseness before performing any power feed checks to eliminate the chance of a bad ground affecting the circuit.*

Repairing Hard Shell Connectors

Unlike molded connectors, the terminal contacts in hard shell connectors can be replaced. Weatherproof hard-shell connectors with the leads molded into the shell have non-replaceable terminal ends. Replacement usually involves the use of a special terminal removal tool that depress the locking tangs (barbs) on the connector terminal and allow the connector to be removed from the rear of the shell. The connector shell should be replaced if it shows any evidence of burning, melting, cracks, or breaks. Replace individual terminals that are burnt, corroded, distorted or loose.

> NOTE: *The insulation crimp must be tight to prevent the insulation from sliding back on the wire when the wire is pulled. The insulation must be visibly compressed under the crimp tabs, and the ends of the crimp should be turned in for a firm grip on the insulation.*

The wire crimp must be made with all wire strands inside the crimp. The terminal must be fully compressed on the wire strands with the ends of the crimp tabs turned in to make a firm grip on the wire. Check all connections with an ohmmeter to insure a good contact. There should be no measurable resistance between the wire and the terminal when connected.

Mechanical Test Equipment

Vacuum Gauge

Most gauges are graduated in inches of mercury (in.Hg), although a device called a manometer reads vacuum in inches of water (in. H_2O). The normal vacuum reading usually varies between 18 and 22 in.Hg at sea level. To test engine vacuum, the vacuum gauge must be connected to a source of manifold vacuum. Many engines have a plug in the intake manifold which can be removed and replaced with an adapter fitting. Connect the vacuum gauge to the fitting with a suitable rubber hose or, if no manifold plug is available, connect the vacuum gauge to any device using manifold vacuum, such as EGR valves, etc. The vacuum gauge can be used to determine if enough vacuum is reaching a component to allow its actuation.

Hand Vacuum Pump

Small, hand-held vacuum pumps come in a variety of designs. Most have a built-in vacuum gauge and allow the component to be tested without removing it from the vehicle. Operate the pump lever or plunger to apply the correct

312 CHASSIS ELECTRICAL

amount of vacuum required for the test specified in the diagnosis routines. The level of vacuum in inches of Mercury (in.Hg) is indicated on the pump gauge. For some testing, an additional vacuum gauge may be necessary.

Intake manifold vacuum is used to operate various systems and devices on late model vehicles. To correctly diagnose and solve problems in vacuum control systems, a vacuum source is necessary for testing. In some cases, vacuum can be taken from the intake manifold when the engine is running, but vacuum is normally provided by a hand vacuum pump. These hand vacuum pumps have a built-in vacuum gauge that allow testing while the device is still attached to the component. For some tests, an additional vacuum gauge may be necessary.

HEATING AND AIR CONDITIONING

Refer to Chapter 1 for discharging, charging and safety precautions pertaining to the service and repair of the A/C system.

Blower Motor

REMOVAL AND INSTALLATION

1968–76

1. Disconnect the negative battery cable.
2. Disconnect the hoses and wiring from the fender skirt.
3. Remove all fender skirt attaching bolts except those attaching the skirt to the radiator support.
4. Pull out, then down, on the skirt. Place a block between the skirt and the fender.
5. Remove the blower-to-case attaching screw. Remove the blower assembly.
6. Remove the blower wheel retaining nut and separate the motor and the wheel.
7. Install the blower wheel on the motor and secure the retaining nut.
8. Position the blower motor assembly to its mounting. The open end of the blower should be away from the motor. Install the retaining screws.
9. Install the fender skirt. Connect the required hoses and electrical wiring. Connect the negative battery cable.

1977–90

1. Disconnect the negative battery cable.
2. Disconnect the electrical connections from the blower motor assembly. Remove the blower motor vent tube.
3. Remove the blower motor retaining screws. Remove the blower motor from its mounting.
4. Remove the blower wheel retaining nut and separate the motor and the wheel.
5. Installation is the reverse of the removal procedure. The open end of the blower wheel should be away from the motor.

1991–92

1. Disconnect the battery ground cable.
2. From inside, remove the four retaining screws from the upper rear edge of the sound insulator.
3. Pull the sound insulator straight back until the two locator studs at the forward edge are disengaged.
4. Disconnect the blower motor electrical connector.
5. Pull the right side hinge pillar trim finish (kick) panel away from the body hinge pillar.
6. Remove the ECM retainer attaching screw, swing the ECM module and retainer far enough to provide clearance for the removal and installation of the blower motor assembly.
7. Remove two of the three blower motor cover attaching screws (leave the one screw nearest the right side rear), while supporting the blower motor remove the third attaching screw.
8. Carefully lower the blower motor and fan assembly until the rubber mounting grommets on the motor are clear of the locating bosses, then remove the assembly from vehicle.
9. Remove the blower motor cooling tube seal from the bottom of module or the blower motor mounting flange, if equipped.
10. Using a heat knife, make three cuts through the plastic fan sleeve, approximately $1/2$ in. from the base of the fan hub and should run lengthwise to the shaft to the end of the plastic sleeve, on the motor shaft. The three cuts should be evenly spaced around the sleeve.

NOTE: *Do not hammer on or place any heavy side force on the fan, damage to the motor could result.*

11. Carefully split the plastic sleeve from the blower motor shaft and pull the fan straight off the shaft. Discard the fan.

To Install:

12. Place the new fan on the blower motor shaft. Apply hand pressure to the dome shaped central portion of the fan to press the fan sleeve until it becomes lightly seated on the shaft. Do not use excessive force.
13. Connect a $1/2$ in. drive to $3/8$ in. drive socket adapter to a $1/2$ in. socket with $3/8$ in. drive. Slip the end of the fan sleeve into the $1/2$ in. socket until it bottoms out in the socket.

NOTE: *The socket must be long enough to extend past the rim of the fan so as to be able* to support the blower motor and fan during the next step. If adapter does not extend past that point use a longer socket.

14. Invert the blower assembly, using a small arbor press and short 1/4 in. diameter screw or rod, press the motor shaft into the fan sleeve until there is 19/64 in. clearance between the motor flange and the fan rim.

15. Reverse the removal procedure to install the blower assembly into vehicle.

Heater Core

REMOVAL AND INSTALLATION

1968-71

NOTE: *It is not necessary to discharge the air conditioning system to replace the heater core.*

1. Disconnect the negative battery cable.
2. Drain the cooling system.

CAUTION: *When draining the coolant, keep in mind that cats and dogs are attracted by the ethylene glycol antifreeze, and are quite likely to drink any that is left in an uncovered container or in puddles on the ground. This will prove fatal in sufficient quantity. Always drain the coolant into a sealable container. Coolant should be reused unless it is contaminated or several years old.*

3. Disconnect the heater hoses from the heater core at the firewall.

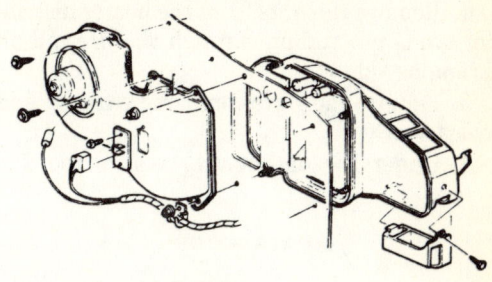

Heater blower mounting

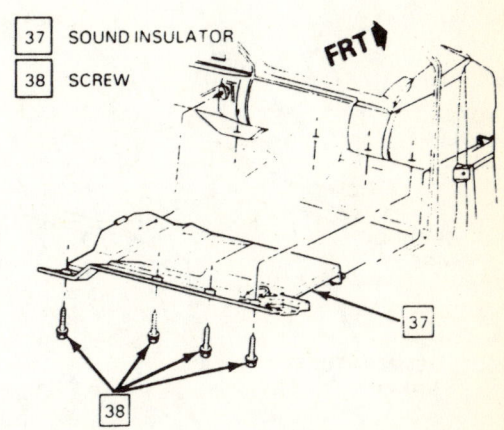

Sound insulator removal, 1991–92

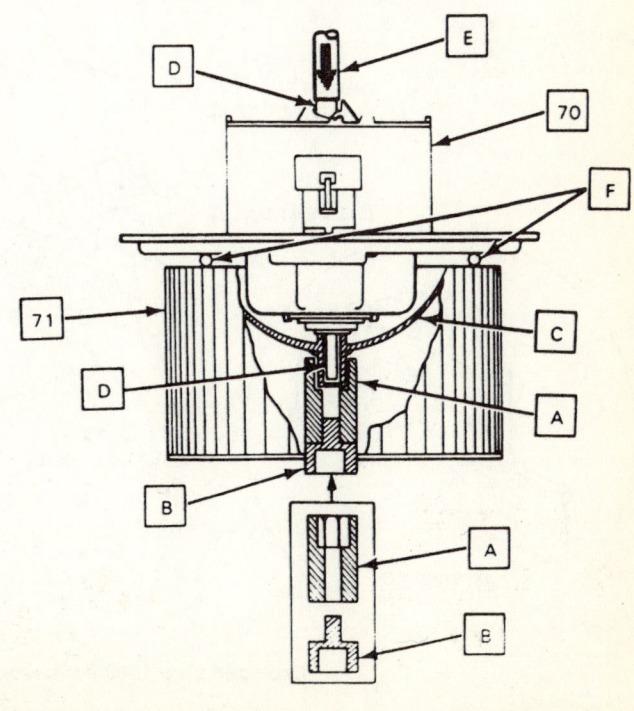

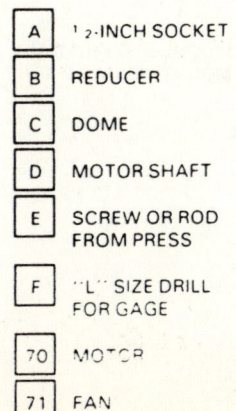

Blower motor fan installation, 1991–92

314 CHASSIS ELECTRICAL

4. Remove the nuts from the heater distributor studs protruding through the firewall on the engine side.

5. Remove the glove compartment door and inner tray.

6. Under the dashboard, remove the five center duct hoses, duct cables, center duct-to-selector duct screw and the center duct.

7. From inside the vehicle, drill out the lower case stud using a $1/4''$ drill.

8. Remove the center floor air duct.

9. Remove the firewall screws and pull the selector from the firewall.

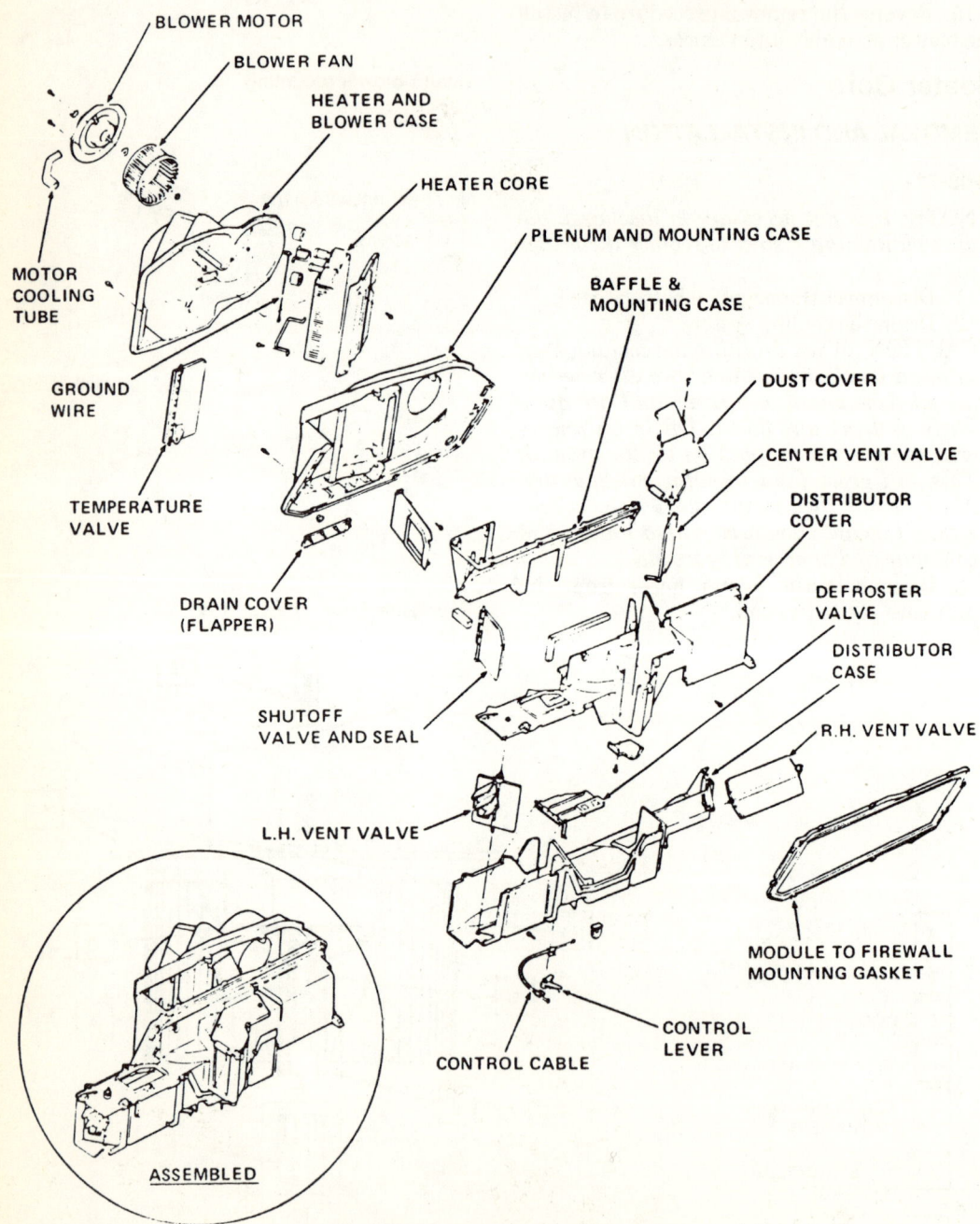

Exploded view 1982 heater module

CHASSIS ELECTRICAL 315

10. Disconnect and tag all wires, vacuum lines and cables attached to the assembly. Remove the heater assembly from the vehicle.

NOTE: *It is very important to tag all cables and vacuum lines during removal. Failure to do so, may cause improper operation of the heater and/or air conditioning systems.*

11. Scribe the temperature door camming plate-to-selector duct relationship and remove the plate.
12. Remove the heater core and core housing from the selector duct.

To Install:

13. Before installation, clean out all debris and coolant from the heater casing.
14. Match the new heater core to the old one to insure a proper fit.
15. Attach the heater core and housing to the selector duct.
16. Align the marks on the temperature door plate and selector duct and install the plate.
17. Install the heater assembly in the vehicle and connect all of the hoses, cables and vacuum lines according the way they are tagged.
18. Install the inside firewall mounting

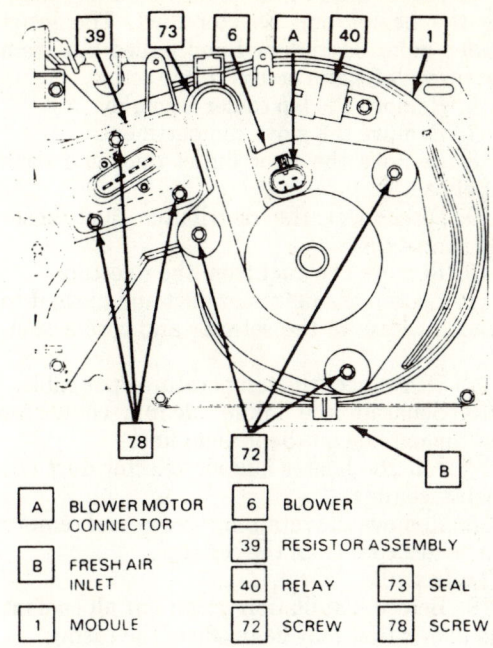

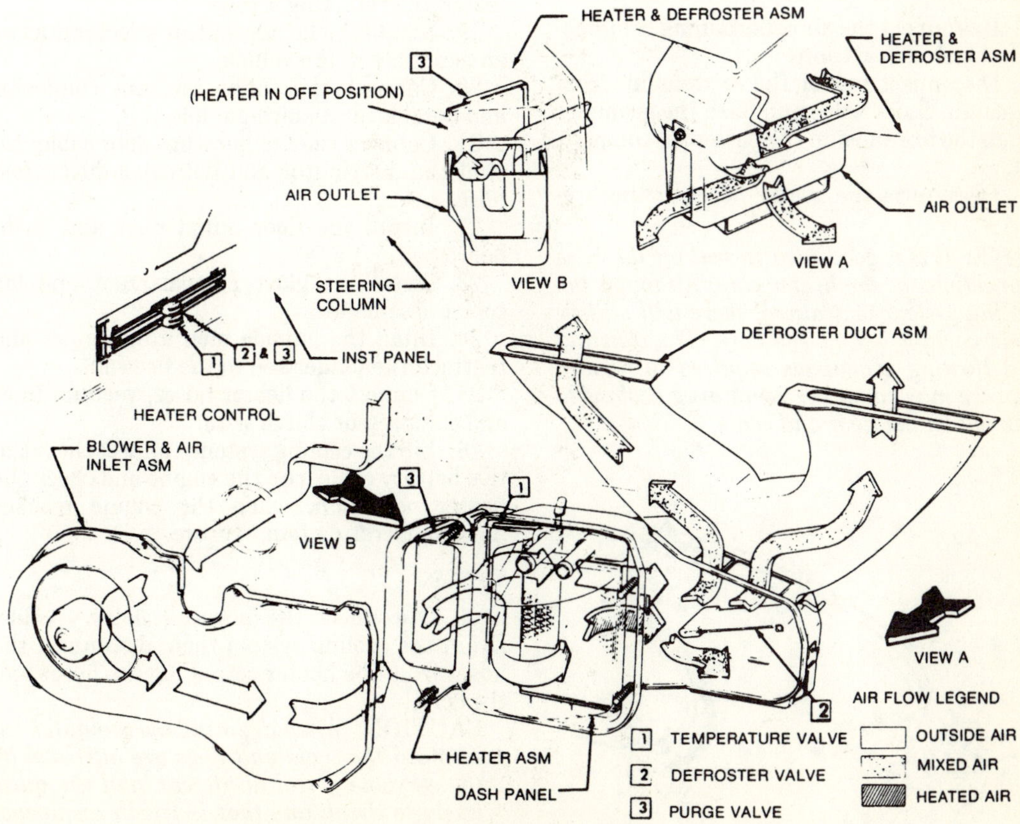

Blower housing area, 1991–92

Heater airflow diagram

screws and center floor air duct.

19. Install the remaining hoses and air ducts.

20. Install the glove compartment tray and door.

21. On the engine side of the firewall, install the two outer heater assembly mounting screws.

22. Check the condition of the heater hoses and replace them if necessary. Connect the hoses to the heater core at the firewall.

23. Fill the cooling system, run the engine and check the heater operation.

1972-76

1. Disconnect the negative battery cable. Drain the cooling system. It is not necessary to evacuate the air conditioning refrigerant.

CAUTION: When draining the coolant, keep in mind that cats and dogs are attracted by the ethylene glycol antifreeze, and are quite likely to drink any that is left in an uncovered container or in puddles on the ground. This will prove fatal in sufficient quantity. Always drain the coolant into a sealable container. Coolant should be reused unless it is contaminated or several years old.

2. Disconnect the air conditioning compressor clutch electrical connector.

3. Disconnect and tag the vacuum line from the vacuum check valve and push the grommet through the firewall into the passenger compartment.

4. Disconnect the heater hoses at the firewall.

NOTE: It is a good idea to seal off the hose connections at the heater core. Although the cooling system is drained, there will still be some coolant left in the core. The twisting and turning sometimes required to remove the core may cause the remaining coolant to leak out on the floor and seats.

5. Remove the three screws and nuts retaining the heater and selector duct. The inner fender must be loosened and pulled out from the firewall to gain access to one screw.

6. Remove the lap cooler assembly.

7. Remove the glove compartment.

8. Remove the floor outlet duct and dash panel pad.

9. Disconnect the distributor duct hoses and connector.

10. Remove the duct from the selector.

11. Loosen the defroster duct and move it to provide access to the selector and core assembly.

12. Disconnect the temperature door cable.

13. Separate the inline vacuum connector and the outside air diaphragm line.

14. Lift the heater and air selector duct out as an assembly.

15. Remove the retaining screws and remove the heater core from the selector.

To Install:

16. Before installation, clean out all coolant and debris that may be inside of the casing.

17. Match up the new and old heater cores to insure a proper fit.

18. Install the heater core in the selector and install the retaining screws.

19. Install the heater and air selector duct as an assembly in the vehicle.

20. Connect the inline vacuum connector and outside air diaphragm line.

21. Connect the temperature door cable. Install the distributor and defroster ducts, and hoses.

22. Install the floor outlet duct and dashboard pad.

23. Install,the glove compartment and lap cooler assemblies.

24. Intall the outside mounting screws and reattach the fenderwell to the firewall.

25. Connect the heater hoses, vacuum lines and compressor clutch wire.

26. Fill the cooling system, connect the negative battery cable, run the engine and check the heater operation, when the engine reaches normal operating temperature.

1977-90

1. Disconnect the negative battery cable. Drain the cooling system then, disconnect the hoses from the heater core. Plug the hoses and the core.

CAUTION: When draining the coolant, keep in mind that cats and dogs are attracted by the ethylene glycol antifreeze, and are quite likely to drink any that is left in an uncovered container or in puddles on the ground. This will prove fatal in sufficient quantity. Always drain the coolant into a sealable con-

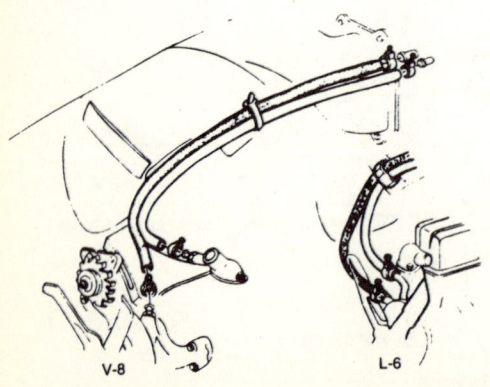

Heater hose routing

CHASSIS ELECTRICAL 317

tainer. *Coolant should be reused unless it is contaminated or several years old.*

2. Remove the diagnostic connector plug to heater case screws.

3. Disconnect and tag the connectors from the blower motor, resistor, blower relay and the thermostatic switch.

4. Remove the air conditioning wiring harness retainer from the blower case shroud.

5. Remove the module screen (leaf screen) assembly screws and remove the screen.

6. Remove the upper case half to lower case half cowl attaching screws. Two of these screws are inside the air intake at the case separation point.

7. Remove the thermostatic switch by removing the electrical connector, the switch attaching screws and the small tube running from the switch to the evaporator inlet pipe. To remove the small pipe, remove the insulation at the evaporator inlet pipe and loosen the two clamps. When you reinstall this line make sure the line is replaced in its original position.

8. Remove the evaporator inlet pipe support bracket.

9. Remove the heater-evaporator core case cover, making sure you do not damage the sealer.

10. Remove the heater core to case attaching screw and pull up firmly to release the core from the spring loaded clip. Remove the heater core from the vehicle.

To Install:

11. Line up the core base with the clip at the bottom of the case before you insert the core. The top retaining bracket will line up with the hole at the top of the core when properly seated.

12. Install the new core in the spring loaded clip and install the core to case mounting screws.

13. Install the case cover making sure that cover is completely seated in the sealer.

14. Install the evaporator inlet pipe support bracket.

15. Install the small pipe from the thermostatic switch to the evaporator pipe in the exact position as before removal. Wrap the sealer around the pipe.

16. Install the thermostatic switch and mounting screws. Connect the switch wiring.

17. Install the upper case half to the lower half and tighten the mounting screws.

18. Install the leaf screen. Install the wiring harness in the retainer on the casing.

19. Connect the wiring and diagnostic connector at the casing.

20. Connect the heater hoses to the core and fill the cooling system.

21. Connect the negative battery cable, run the engine and check the heater operation.

1991-92

1. Disconnect the negative battery cable. Drain the cooling system.

CAUTION: *When draining the coolant, keep in mind that cats and dogs are attracted by the ethylene glycol antifreeze, and are quite likely to drink any that is left in an uncovered container or in puddles on the ground. This will prove fatal in sufficient quantity. Always drain the coolant into a sealable container. Coolant should be reused unless it is contaminated or several years old.*

2. Remove the screw holding the hose assembly to the cowl panel.

3. Release the quick connect fittings, on the inlet and outlet pipes of the heater core, by squeezing both release tabs at the base of the heater core tube and pulling on the pipe to disengage the fitting.

4. From inside vehicle, remove the four retaining screws on the upper right side sound insulator. Pull the insulator straight back until the two locator studs are disengaged.

5. Remove the nut from the shroud panel stud on the Instrument panel lower reinforcement.

6. Remove the screw from the instrument panel carrier and remove the instrument panel lower reinforcement.

7. Disconnect both vacuum harness connectors, then remove the connector halves from the lower evaporator case and position the harnesses out of the way.

8. Remove the right side hinge pillar finish (kick) panel by pulling it away from the body pillar.

9. Roll the carpeting back enough to provide access to the lower area of the A/C module.

10. Remove the seven screws attaching the lower case to the module.

NOTE: *The forward center attaching screw may be difficult find. It is located almost directly below the lower right side heater core tube, the screw head is straight down from the A/C tank, using a $1/4$ in. drive 7mm socket with a swivel adaptor and short extension on a ratchet handle will aide in the removal.*

11. Remove the lower evaporator case.

12. Remove the heater core mounting straps and screws.

13. Carefully pull the heater core rearward, working the heater core tubes out of the seal.

To Install:

14. Transfer the quick connect tabs to the new heater core.

15. Insert heater core into position, carefully pushing the tubes through the seal. Install the mounting straps and screws.

318 CHASSIS ELECTRICAL

16. Install the lower evaporator case and tighten attaching screws evenly.
17. Fit the carpeting back into place.
18. Connect both vacuum connectors and install assembled connectors to the lower case.
19. With the instrument panel lower reinforcement held into position, install the nut to the shroud panel stud (tighten to 89 inch lbs.) and install the screw to the instrument panel carrier (tighten to 89 inch lbs.).
20. Snap the right side hinge pillar finish trim (kick) panel into place.
21. Slide the sound insulator forward into position engaging both locator tabs, and install the four attaching screws.
22. Hold the control valve and hose assembly into position, align the quick connect fitting tabs with the grooves in the fitting sleeve.
23. Push the sleeve into place on the heater core tube, then pull back on the sleeve to check for proper connection. Repeat for other connector.
24. Install the heater outlet pipe retaining screw. Fill with coolant, start engine and check for leaks and proper heater operation.

Control Head

REMOVAL AND INSTALLATION

1968-90

1. Disconnect the negative battery cable.
2. Remove the radio knobs and the clock set knob, as required.
3. Remove the instrument bezel retaining screws.
4. Pull the bezel out to disconnect the rear defogger switch and the remote mirror control, as required.
5. Remove the instrument panel retaining screws.
6. Remove the control head to dash screws and pull the head out.
7. Disconnect the electrical connectors and/or control cables. Remove the control head.
8. Installation is the reverse of the removal procedure.

1991

1. Remove the steering column opening filler by removing the two screws near the lower edge of the filler. Pull down to unsnap the four integral clips holding the filler to instrument panel.
2. Loosen the steering column attaching nuts and lower the steering column.
3. Remove the eight trim plate attaching screws. Pull the trim plate straight away from the instrument panel carrier, snapping the seven integral clips out of the slots in the instrument panel.
4. Remove the temperature control cable push-on retainer and cable loop from the pin at the bottom of the control assembly.

NOTE: *Do not attempt to remove the temperature control cable by force without releasing the lock tab in the slot of the control assembly. Damage to the lock tab will result.*

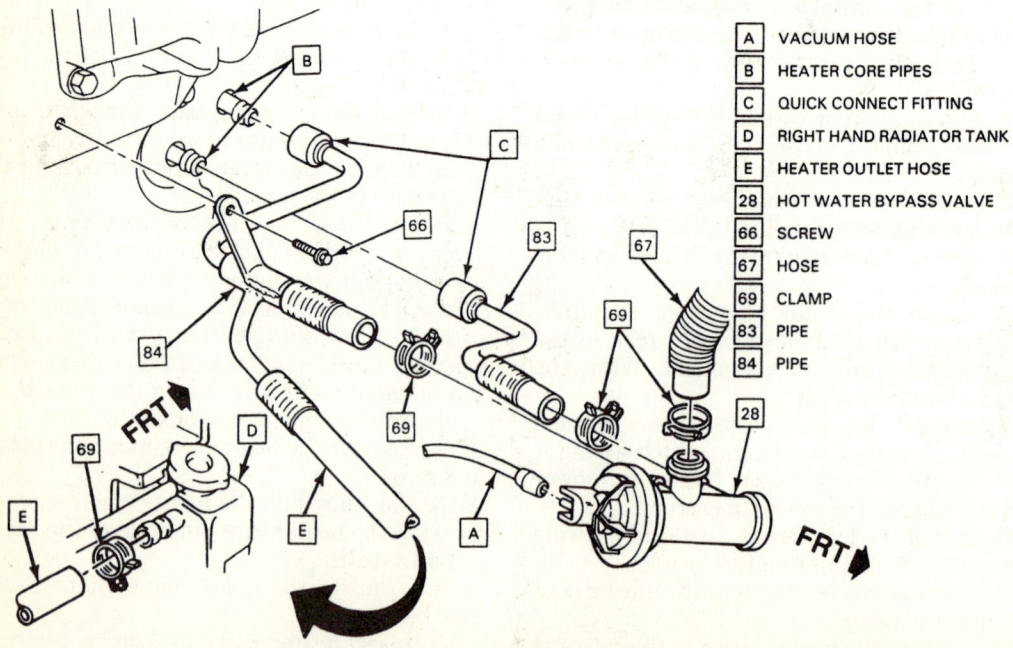

Hot water bypass valve and pipe assembly, 1992

CHASSIS ELECTRICAL 319

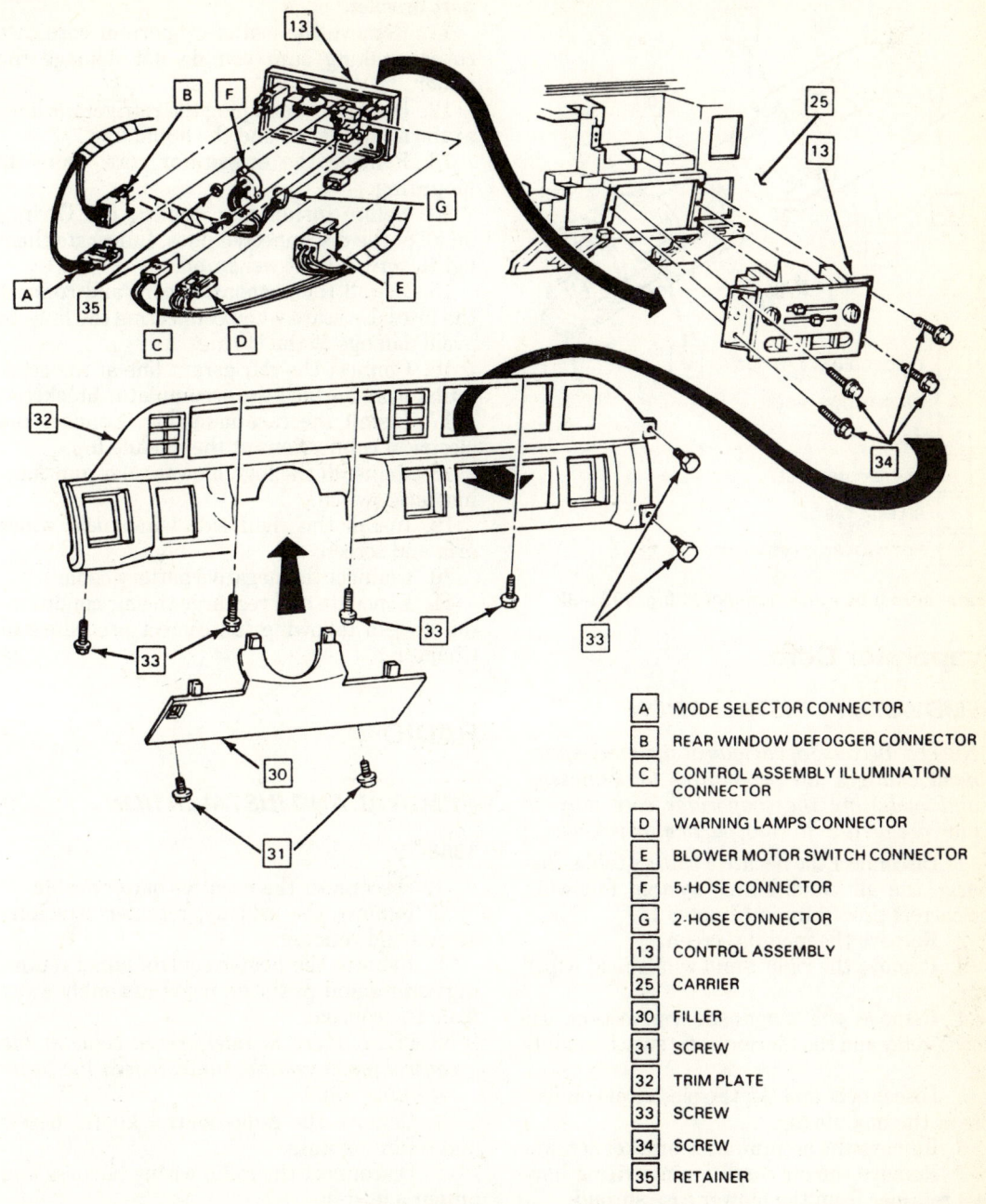

A	MODE SELECTOR CONNECTOR
B	REAR WINDOW DEFOGGER CONNECTOR
C	CONTROL ASSEMBLY ILLUMINATION CONNECTOR
D	WARNING LAMPS CONNECTOR
E	BLOWER MOTOR SWITCH CONNECTOR
F	5-HOSE CONNECTOR
G	2-HOSE CONNECTOR
13	CONTROL ASSEMBLY
25	CARRIER
30	FILLER
31	SCREW
32	TRIM PLATE
33	SCREW
34	SCREW
35	RETAINER

Control assembly, 1991–92

5. Squeeze the lock tab toward the left side of vehicle to release temperature control cable retainer and clip the retainer down and out of the slot in the control assembly.

6. Remove the control assembly attaching screws and pull assembly out just far enough to reach the electrical connectors, vacuum harness connectors and temperature control cable end at the back of the control assembly.

7. Disconnect the electrical connectors.

8. Remove the two push-on retainers holding the circular 5-hose connector, and pull the connector off the vacuum switch.

9. Pull the 2-hose connector off the hot water valve vacuum switch, and remove the control assembly.

10. Installation is the reverse of the removal procedure.

320 CHASSIS ELECTRICAL

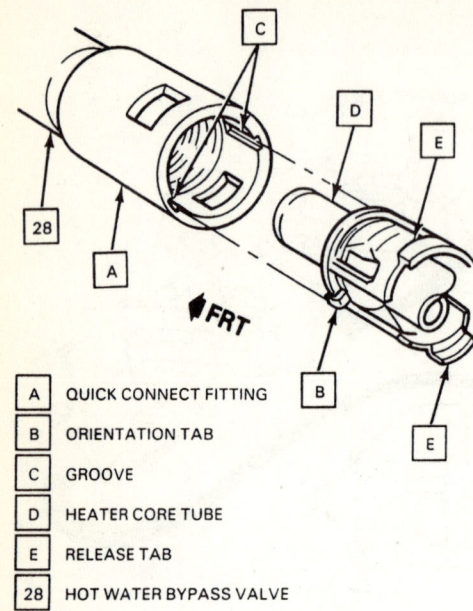

A	QUICK CONNECT FITTING
B	ORIENTATION TAB
C	GROOVE
D	HEATER CORE TUBE
E	RELEASE TAB
28	HOT WATER BYPASS VALVE

Heater core tube quick connect fitting, 1991–92

Evaporator Core

REMOVAL AND INSTALLATION

NOTE: *Because of various design and equipment changes, all procedures for removing and installing the evaporator core may or may not pertain to your particular vehicle.*

1. Disconnect the negative battery cable. Discharge the air conditioning system following the correct procedure in Chapter 1.
2. Remove the module screen.
3. Remove the right hand windshield wiper arm.
4. Remove the diagnostic connection, Hi blower relay and the thermostatic switch mounting.
5. Disconnect and tag the electrical connectors at the module top.
6. Remove the accumulator bracket screws.
7. Remove the air conditioning wiring harness retainer from the blower case shroud.
8. Remove the upper case half to lower case half cowl attaching screws. Two of these screws are inside the air intake at the case separation point.
9. Remove the thermostatic switch by removing the electrical connector, the switch attaching screws and the small tube running from the switch to the evaporator inlet pipe. To remove the small pipe, remove the insulation at the evaporator inlet pipe and loosen the two clamps. When you reinstall this line make sure the line is replaced in its original position.
10. Remove the evaporator inlet pipe support bracket.
11. Remove the heater-evaporator core case cover, making sure you do not damage the sealer.
12. Disconnect and plug the refrigerant lines at the accumulator and the liquid line.
13. Remove the evaporator core, from its mounting.
14. Before installation, replace the O-rings on all of the disconnected lines. Lubricate them lightly with a light weight oil.
15. Install the evaporator core and connect the lines. Use care when connecting the lines to avoid damage to the O-rings.
16. Connect the refrigerant line at the accumulator and attach the accumulator bracket.
17. Install the case assembly. Connect the electrical connections at the module top.
18. Reinstall the high blower relay and thermostatic switch.
19. Install the right side windshield wiper arm and screen.
20. Connect the negative battery cable.
21. Evacuate and recharge the air conditioning system following the correct procedure in Chapter 1.

RADIO

REMOVAL AND INSTALLATION

1968-72

1. Disconnect the negative battery cable.
2. Remove the ashtray, retainer attaching screws and retainer.
3. Remove the heater control panel retaining screws and push the panel assembly away from the console.

NOTE: *If there is interference between the control panel and the radio, loosen the radio retaining nuts.*

4. Remove the radio control knobs, bezels and retaining nuts.
5. Disconnect the radio wiring harness and antenna lead-in.
6. Remove the radio rear brace attaching screw and remove the radio.

To Install:

7. To install, insert the radio into the rear brace part of the way and connect the antenna and electrical connectors.
8. Slide the radio the rest of the way in and install the rear brace mounting screw.
9. Install the front retaining nuts, washers, bezels and control knobs.
10. Install the heater control panel and bezel.
11. Install the ash tray and retainer.

CHASSIS ELECTRICAL   321

12. Connect the negative battery cable and check the radio operation.

1973-76

1. Disconnect the negative battery cable.
2. On cars with air conditioning, remove the lap cooler duct.
3. Turn the radio control knobs until the slots in the bottom of the knobs are visible. Depress the metal retainers and remove the knobs and bezels.
4. Remove the radio control shaft nuts and washers.
5. Remove the right side bracket-to-instrument panel bolt and stud nut on the left side of the radio.
6. Pull the radio forward and disconnect the wiring from the radio. Remove the radio from the car.

To Install:

7. Position the radio close enough to connect the wiring and antenna.
8. Slide the radio inward and install the right and left side mounting screws.
9. Install and tighten the radio shaft nuts and washers. Install the radio knobs and bezels.
10. Install the lap cooler duct and connect the negative battery cable.
11. Check the radio operation.

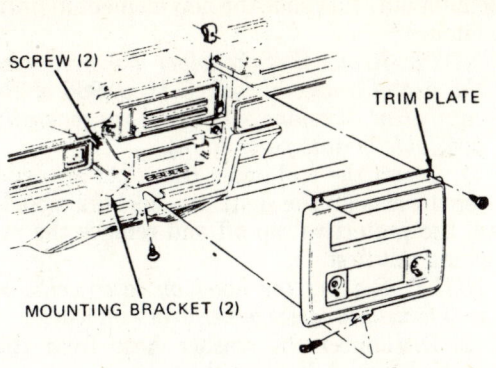

Radio installation, 1982

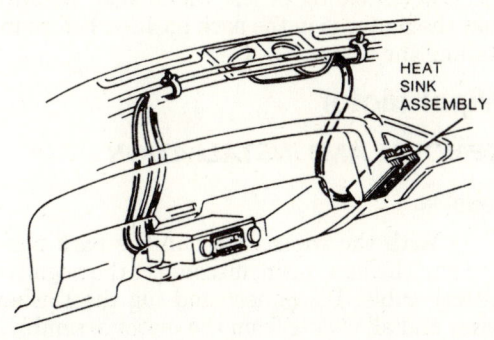

Radio heat sink location

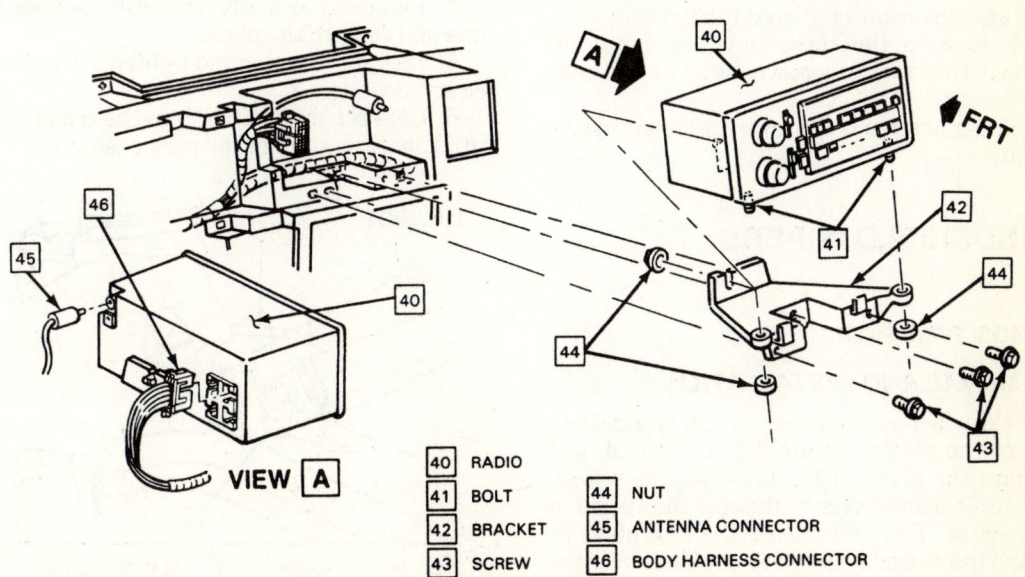

Radio installation, 1991–92

322 CHASSIS ELECTRICAL

1977-90

1. Disconnect the negative battery cable. Remove the radio knobs from the shafts.
2. Remove the three trim plate mounting screws and trim plate.
3. From under the dashboard, remove the two screws and bottom mounting nut holding the radio mounting bracket to the instrument panel.
4. Slide the radio outwards slightly and disconnect the electrical and antenna connectors from the back of the radio.
5. Remove the radio with the mounting bracket still attached.

To Install:

6. Insert the radio and mounting bracket in far enough to connect the antenna and wiring to the rear of the radio.
7. Slide the radio in all of the way and install the mounting nut and bolt.
8. Install the front trim plate and mounting screws.
9. Install the radio knobs on the shafts, connect the negative battery cable and check the radio operation.

1991-92

1. Remove the left side trim plate
2. Remove the three screws attaching the bracket to the instrument panel carrier.
3. Remove the bracket with attached radio from the instrument panel.
4. Disconnect the body harness connector and antenna connector from the radio.
5. Remove the three nuts attaching the radio to the bracket, remove the radio and bolts if necessary.
6. Installation is the reverse of removal procedure.

WINDSHIELD WIPERS

Blade and Arm

REMOVAL AND INSTALLATION

If the wiper assembly has a press type release tab at the center, simply depress the tab and remove the blade. If the blade has no release tab, use a screwdriver to depress the spring at the center. This will release the assembly. To install the assembly, position the blade over the pin at the tip of the arm and press until the spring retainer engages the groove in the pin.

To remove the element, either depress the release button or squeeze the spring type retainer clip at the out end together, and slide the blade element out. Just slide the new element in until it latches.

NOTE: *Removal of the wiper arms requires the use of a special tool, G.M. J8966 or its equivalent. Versions of this tool are generally available in auto parts stores.*

1. Insert the tool under the wiper arm and lever the arm off the shaft. On 1991-92 models pull the protective cap off and remove the attaching nut first.

NOTE: *Raising the hood on later vehicles will facilitate easier wiper arm removal.*

2. Disconnect the washer hose from the arm, if equipped. Remove the arm.
3. Installation is in the reverse order of removal. The proper park position for the arms is with the blades approximately 2″ (50mm) above the lower molding of the windshield. Be sure that the motor is in the park position before installing the arms.

Wiper Motor

REMOVAL AND INSTALLATION

1968-90

1. With the wiper motor in the park position and the hood open; disconnect the negative battery cable. Disconnect and tag the washer hoses and all wiring from the motor assembly.
2. Remove the wiper motor access cover.
3. Loosen the nuts which retain the drive link to the crank arm ball stud.
4. Remove the motor mounting screws or nuts and remove the motor.
5. Install the motor and tighten the mounting screws.
6. Connect the drive link to the crank arm with the wipers still in the park position.

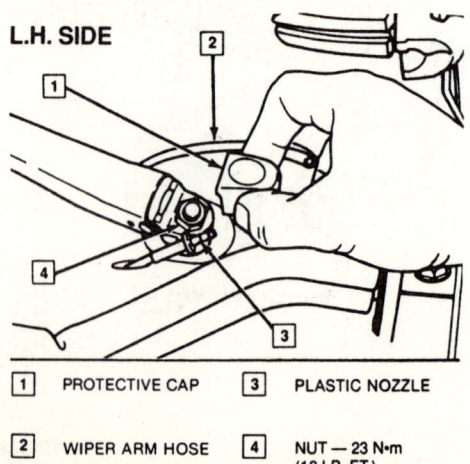

| 1 | PROTECTIVE CAP | 3 | PLASTIC NOZZLE |
| 2 | WIPER ARM HOSE | 4 | NUT — 23 N•m (18 LB. FT.) |

Protective cap, fluid hose and nut removal, 1991-92

7. Install the access cover and connect the hoses and wiring to the motor.
8. Connect the negative battery cable.
9. Before testing the motor, make sure that all tools and rags have been removed from the area.
10. Test the motor to insure that it returns to the park position and that the washers work. Make sure that there is no interference on either side of the windshield and that the wiper arms do not bind.

1991–92

1. Disconnect the battery ground cable.
2. Remove the right side wiper arm and hose.

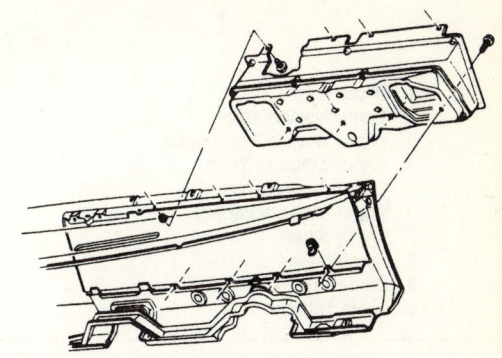

Instrument cluster mounting, 1968

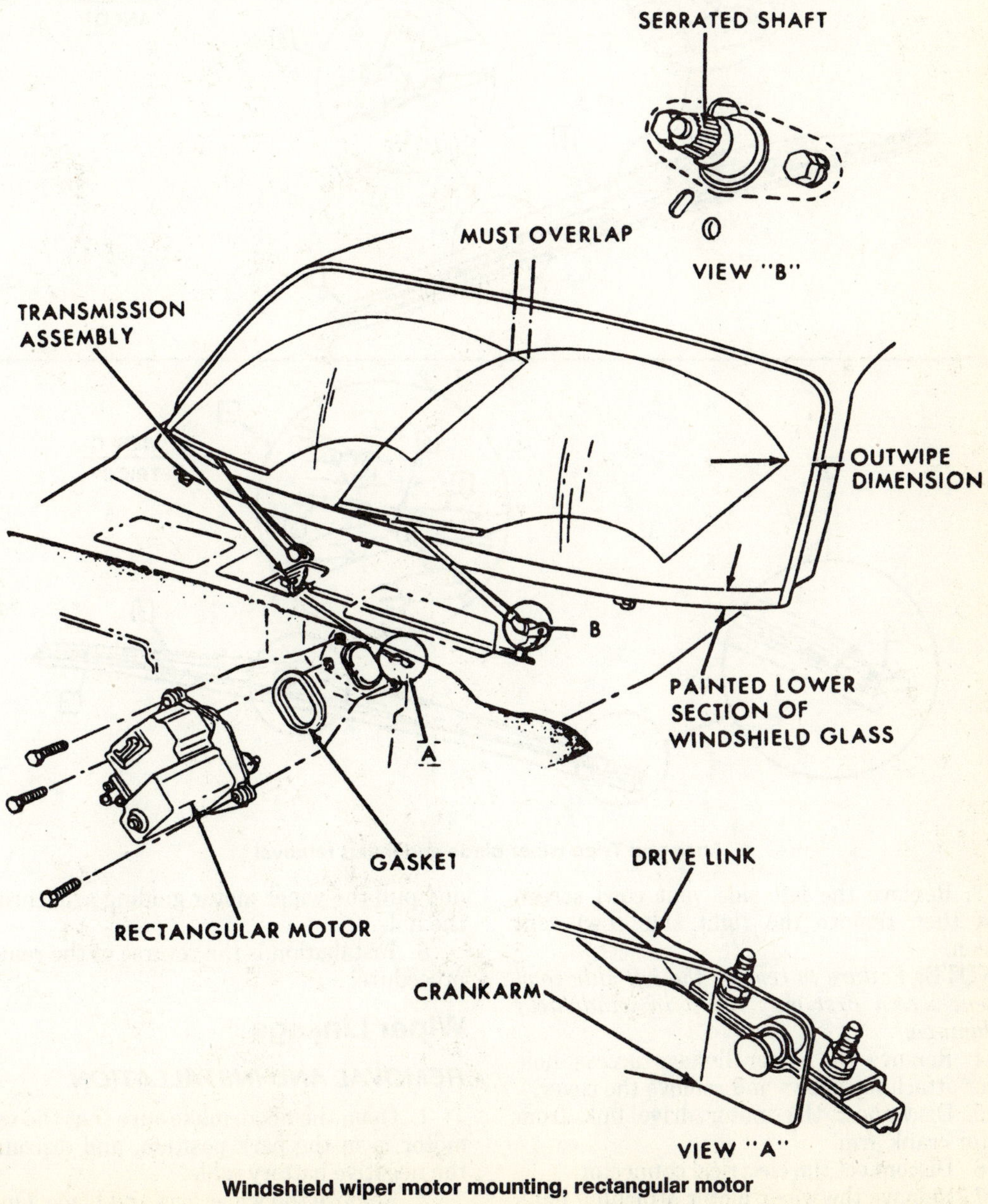

Windshield wiper motor mounting, rectangular motor

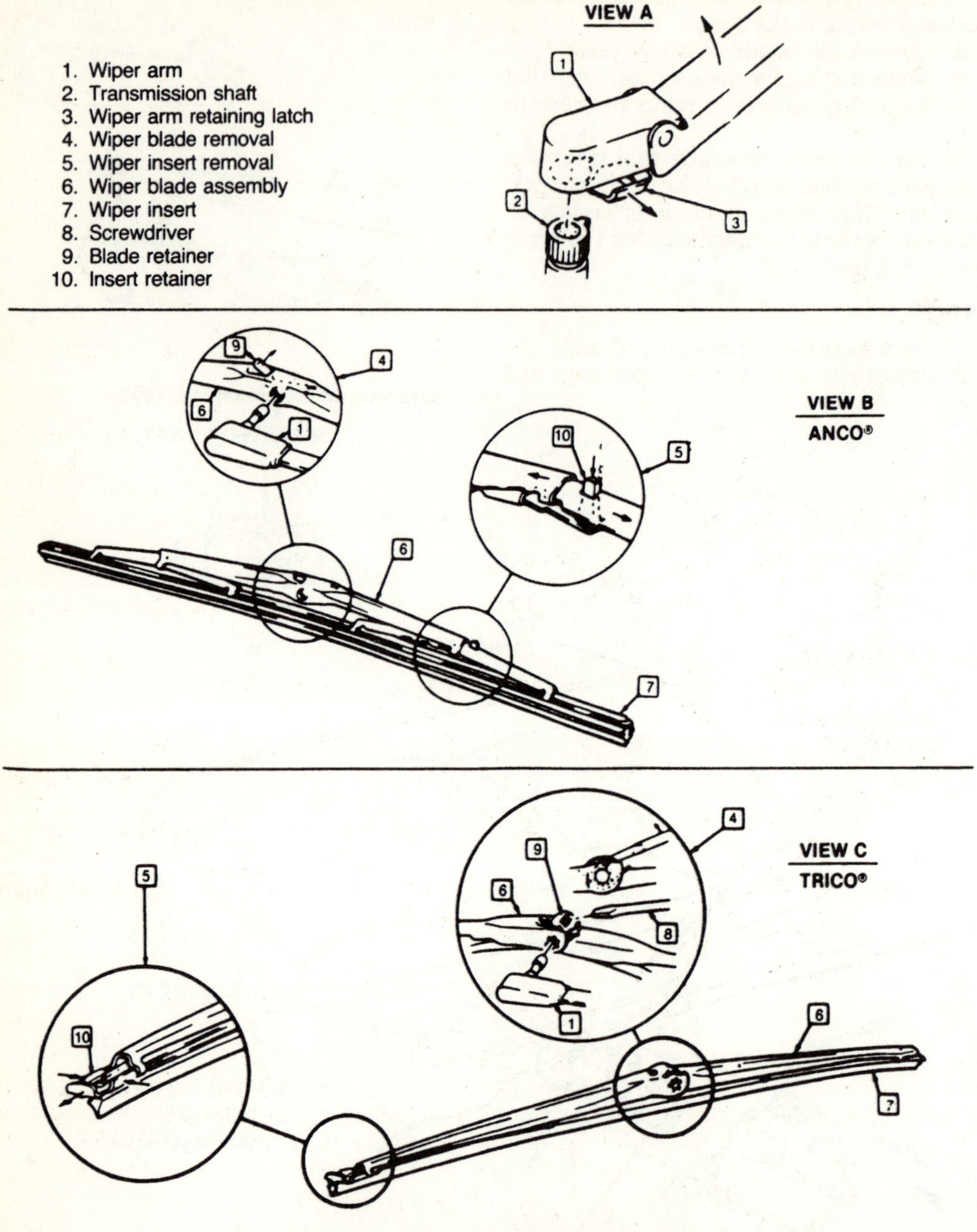

324 CHASSIS ELECTRICAL

1. Wiper arm
2. Transmission shaft
3. Wiper arm retaining latch
4. Wiper blade removal
5. Wiper insert removal
6. Wiper blade assembly
7. Wiper insert
8. Screwdriver
9. Blade retainer
10. Insert retainer

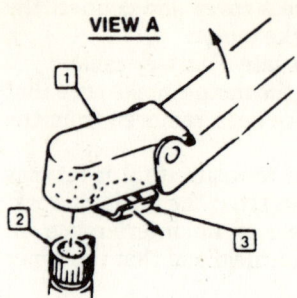

Anco and Trico wiper blade and insert removal

3. Remove the left side vent cowl screen first then remove the right side cowl vent screen.

NOTE: *Failure to remove the left side cowl vent screen first may result in windshield damage.*

4. Remove the wiper linkage access hole cover attaching screws and remove the cover.
5. Disconnect the motor drive link from motor crank arm.
6. Disconnect the electrical connector.
7. Remove the wiper motor attaching bolts and pull the wiper motor guiding arm through the hole.
8. Installation is the reverse of the removal procedure.

Wiper Linkage

REMOVAL AND INSTALLATION

1. Open the hood, make sure that the wiper motor is in the park position, and disconnect the negative battery cable.
2. Remove the wiper arm and blade. On the

left hand arm assembly, remove the retaining clip from the pin on the drive arm.

3. Remove the air intake grill or screen. On 1991–92 models you must remove the left side first.

4. Loosen the nuts which retain the drive rod ball stud to the crank arm and detach the drive rod from the crank arm.

5. Remove the transmission retaining screws or nuts, then lower the drive rod assemblies into the cowl.

6. Remove the transmission and linkage from the cowl through the cowl opening.

To Install:

7. Place the new linkage in the cowling and pull the drive rod assembly into the transmission. Install the retaining nut.

8. Make sure that the wipers are still in the

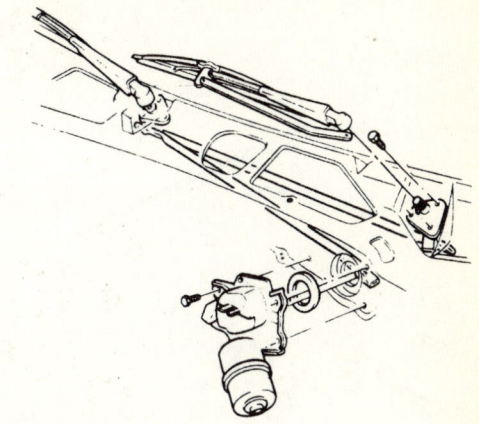

Wiper motor mounting

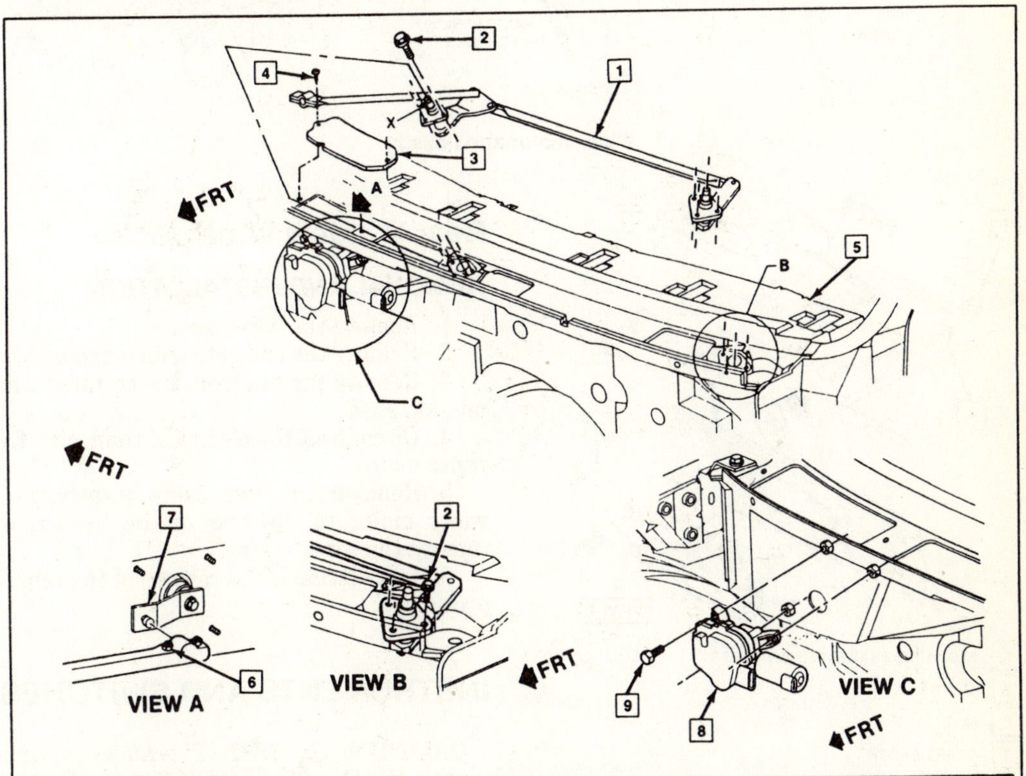

Wiper motor and linkage placement, 1991–92

1 LINKAGE ASSEMBLY	4 SCREW (2) — 1.5 N·m (13 LB. IN.)
2 SCREW (6) — 8 N·m (71 LB. IN.)	5 DASH PANEL
3 WIPER LINKAGE ACCESS HOLE COVER	6 NUT (2) — 3 N·m (27 LB. IN.)
7 MOTOR CRANK ARM	
8 WIPER MOTOR	
9 BOLT (3) — 9 N·m (80 LB. IN.)	

326 CHASSIS ELECTRICAL

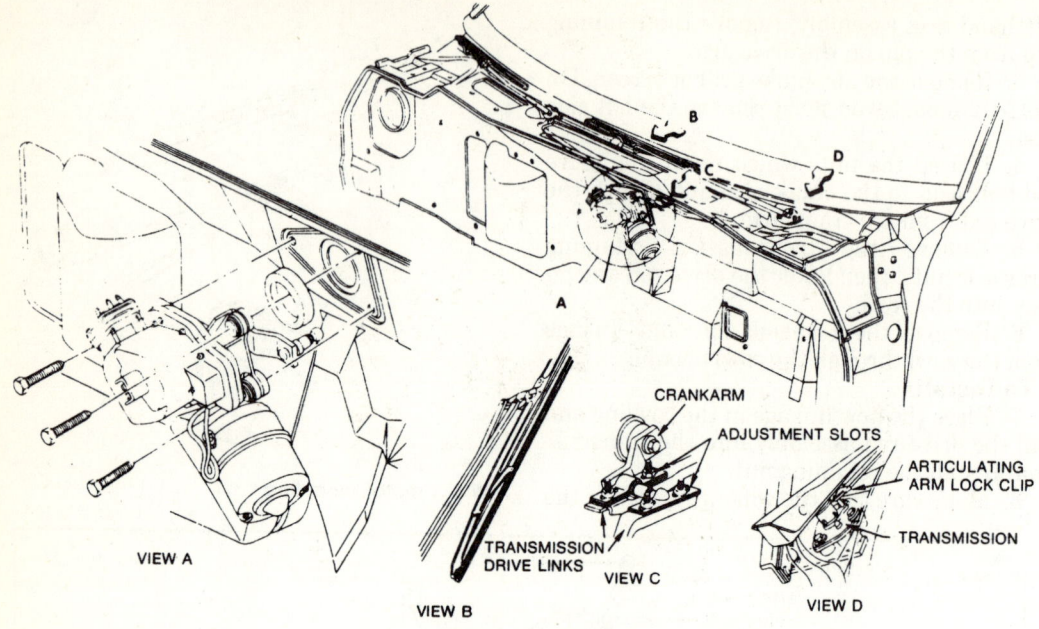

Wiper motor and linkage

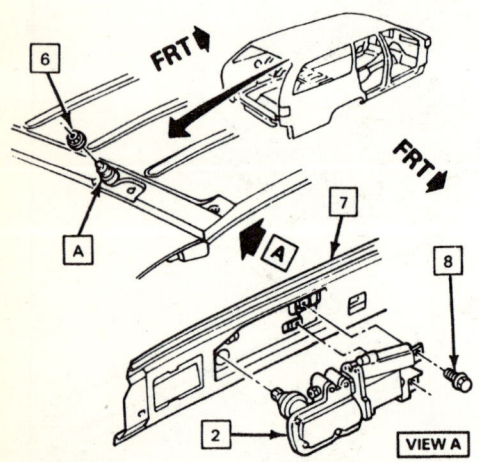

A	SERRATED TRANSMISSION SHAFT
2	MOTOR
6	NUT
7	REAR HEADER
8	BOLT

Rear window wiper motor installation

park position and connect the drive arm to the crank arm. Tighten the retaining nut.

9. Install the wiper arm assemblies and retaining clip(s).

10. Check that no tools are left in the air intake and install the grille or screen.

11. Connect the negative battery cable and check the operation of the wiper to insure full, smooth travel and a return to the park position.

Rear Window Wiper Motor

REMOVAL AND INSTALLATION

1. Remove the wiper arm.
2. Remove the end gate trim panel molding.
3. Remove the nut from the serrated transmission shaft.
4. Disconnect the electrical connector from wiper motor.
5. Remove the two bolts attaching the wiper motor to the rear of the header, and remove the wiper motor.
6. Installation is the reverse of the removal procedure.

INSTRUMENTS AND SWITCHES

CAUTION: *On 1991-92 vehicles equipped with the Supplemental Inflatable Restraint (SIR) System, refer to Steering in chapter 8.*

Instrument Cluster

REMOVAL AND INSTALLATION

1968

NOTE: *Before removing any instrument cluster, check to see if any light bulbs are burned out. If so, note their positions and replace them as soon as the cluster is removed. Replacement is much easier with the cluster removed than with it installed.*

CHASSIS ELECTRICAL 327

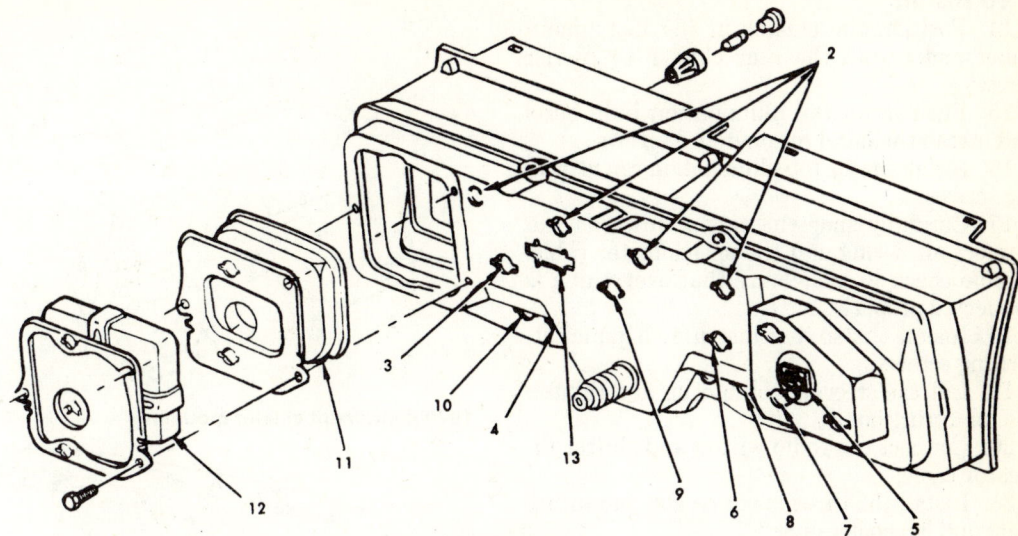

1. Clock knob assembly
2. Instrument cluster lamps
3. R.H. directional indicator
4. "Gen" indicator
5. Fuel gauge
6. L.H. directional indicator
7. "Hot" indicator
8. "Brake" indicator
9. "Bright" indicator
10. "Oil Press" indicator
11. Clock cover
12. Clock assembly
13. Instrument panel connector

1968 instrument cluster connectors

1. Disconnect the negative battery cable.
2. Carefully unplug forward wiring harness from the fuse panel and remove panel from firewall.
3. Remove the screws retaining the gauge cluster to instrument panel.
4. From the mast jacket, remove the screws retaining the column-mounted automatic transmission pointer cable.
5. From behind the cluster, disconnect the speedometer cable, harness connector, clock, speed warning device, defogger, convertible top or tailgate switches and vacuum hoses, if so equipped. If equipped with gauge pack, disconnect the oil pressure line.
6. Using care to prevent scratching the mast jacket, tip the top of the cluster forward and remove it from the vehicle.

To Install:

7. Carefully position the cluster close to the wiring and cables and connect it.
8. Double check to make sure that every wire disconnected is in the proper position.
9. Connect the transmission pointer cable and install the cluster to the instrument panel.
10. Connect the fuse panel to the firewall and plug in the forward wiring harness.
11. Connect the negative battery cable and check to make sure that all bulbs, gauges and switches affected by the removal are working properly.

1969

1. Disconnect the negative battery cable.
2. Remove the glove box and air conditioning center dash outlet.
3. Remove the four screws above the instruments and gently pull instrument panel pad loose from dash clips.
4. If equipped with air conditioning, remove the lap cooler from under the steering column.
5. Remove the three cluster mounting bolts from the underside of the dash.
6. Disconnect the shift indicator cable wire on the steering column.
7. Disconnect the radio wiring.
8. Loosen, but do not remove, the steering column mounting bolts and lower the steering column onto the front seat.
9. Remove the instrument panel top attaching screws and gently lift and tilt the panel forward.
10. Disconnect and tag the speedometer cable (press snap retainer) and all electrical connections.
11. Remove the six top illumination can attaching screws.
12. Remove the indicator bulb bezel by pushing in on the right-side of each cover to expose the screws, then remove the bezel retainer screws (4).
13. Remove the rear cluster to carrier attaching screws and remove the cluster.

CHASSIS ELECTRICAL

To Install:

14. Position the cluster in the instrument panel and install the rear cluster to carrier screws.
15. Push in on the illumination bulb bezel and install the bezel retainer and screws.
16. Install the six top illumination can attaching screws.
17. Carefully slide the cluster inwards and connect all wiring and the speedometer cable. Double check to make sure that everything is connected properly.
18. Install the six top instrument panel attaching screws.
19. Lift the steering column up and tighten the mounting bolts.
20. Connect the radio wiring and shifter indicator cable.
21. Install the three lower cluster mounting bolts and lap cooler duct.
22. Install the instrument panel pad in the dash clips and then install the top four pad mounting bolts.
23. Install the air conditioning center outlet and glove box.
24. Connect the negative battery cable and check to make sure that all instruments and lights are functioning properly.

1970

1. Disconnect the battery ground cable.
2. Remove the air conditioning lap cooler under the steering column.
3. Remove the steering column mounting bolts. Lower and support the steering column.
4. Remove the dash pad and if applicable, disconnect the center air conditioning outlet hose.
5. Disconnect the shift indicator cable on the steering column and remove the indicator bulb bezel.
6. Remove shift indicator lamp housing.
7. Remove the radio, as required.
8. Remove the instrument panel trim plate. The plate is held in place with snap-in studs on the left-side, right-side and to the right of the steering column. You can use a hooked tool to pull the trim plate free.
9. Remove the air conditioner/heater control assembly.
10. Remove the lower instrument cluster-to-bracket and parking brake pedal bracket screws.
11. Remove the ash tray and its retaining bracket.
12. Remove the three screws at the top of the instrument cluster and tilt the cluster forward.
13. Disconnect and tag the speedometer cable and instrument wiring harness. The illu-

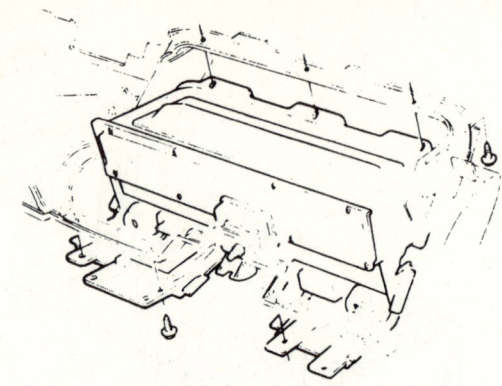

1970 instrument cluster mounting

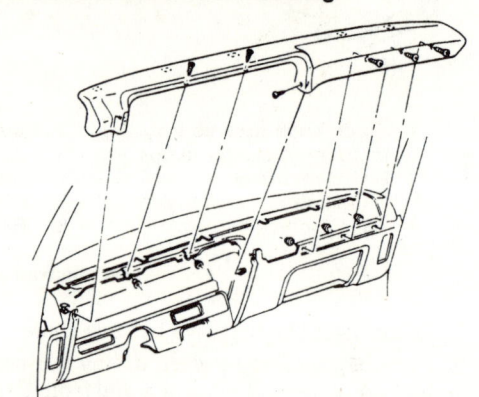

1970 dash pad

mination cover must be removed to get at the wiring.

14. Remove the instrument cluster.

To Install:

15. Position the cluster close to the wiring and connect the speedometer cable and wiring harness to the cluster.
16. Install the illumination cover after connecting the wiring.
17. Install the top three instrument cluster mounting screws.
18. Install the ash tray and retaining bracket.
19. Install the lower instrument cluster mounting screws and then the parking brake pedal bracket.
20. Install the heater/air conditioner control assembly.
21. Install the instrument panel trim plate into the snap in studs.
22. Install the radio, shift indicator housing and indicator cable.
23. Install the center air conditioning outlet and dash pad.
24. Install the steering column into the mounting bracket and tighten the bolts.
25. Install the center lap cooler on vehicles with air conditioning.

CHASSIS ELECTRICAL

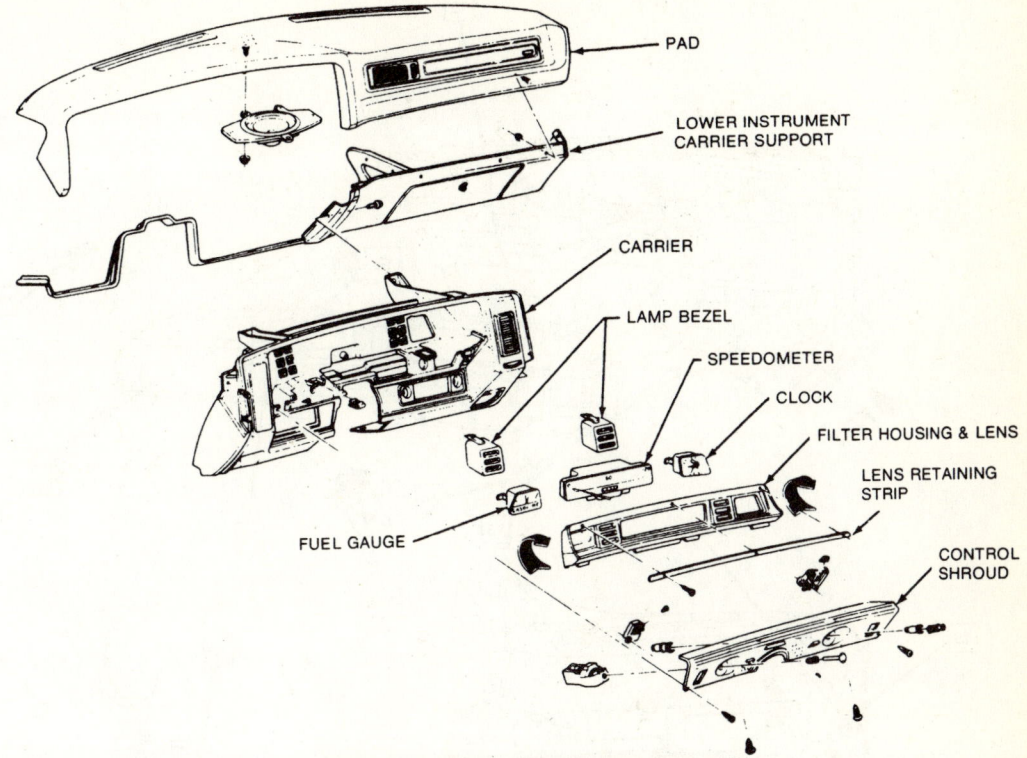

1972 instrument panel components

26. Connect the negative battery cable and check that all instrument and lights are working correctly.

1971-76

1. Disconnect the negative battery cable.
2. Remove the cigar lighter knob and then the screw under it.
3. Pull out the headlight switch and remove screw in middle of shaft.
4. Remove the two screws in the lower corners and remove the shroud.
5. Remove the clock stem set screw and knob.
6. Remove the instrument cluster lens and retaining strip. There are three screws holding the lens retaining strip.
7. Tilt the filter housing back and remove it.
8. The speedometer, fuel gauge, or clock may now be removed.

To Install:

9. Replace the necessary gauge(s) and reinstall the filter housing.
10. Install the lens and retaining strip.
11. Install the clock stem knob and set screw.

WARNING: *The set screw for the clock stem must not be over tightened. The stem is very small and can break very easily requiring the clock to be replaced.*

12. Install the shroud and the two lower corner mounting screws.
13. Install the headlight switch and the screw in the middle of the shaft.
14. Install the cigar lighter knob and screw.
15. Connect the negative battery cable and check the all lights and instruments affected are functioning properly.

1977-90

1. Disconnect the negative battery cable. Remove the steering column lower cover screw and remove the cover.
2. Disconnect the shift indicator cable from the steering column.
3. As required, remove the two screws mounting the steering column. Lower the steering column onto the front seat. Make sure that the column is adequately supported on the seat.

NOTE: *On vehicles equipped with tilt steering wheel it may be possible to position the wheel in the full down detent in order to remove the cluster assembly from the vehicle.*

4. Remove the six screws and three snap-in fasteners from around the edge of the instrument cluster lens.
5. Remove two screws from the sheet metal

330 CHASSIS ELECTRICAL

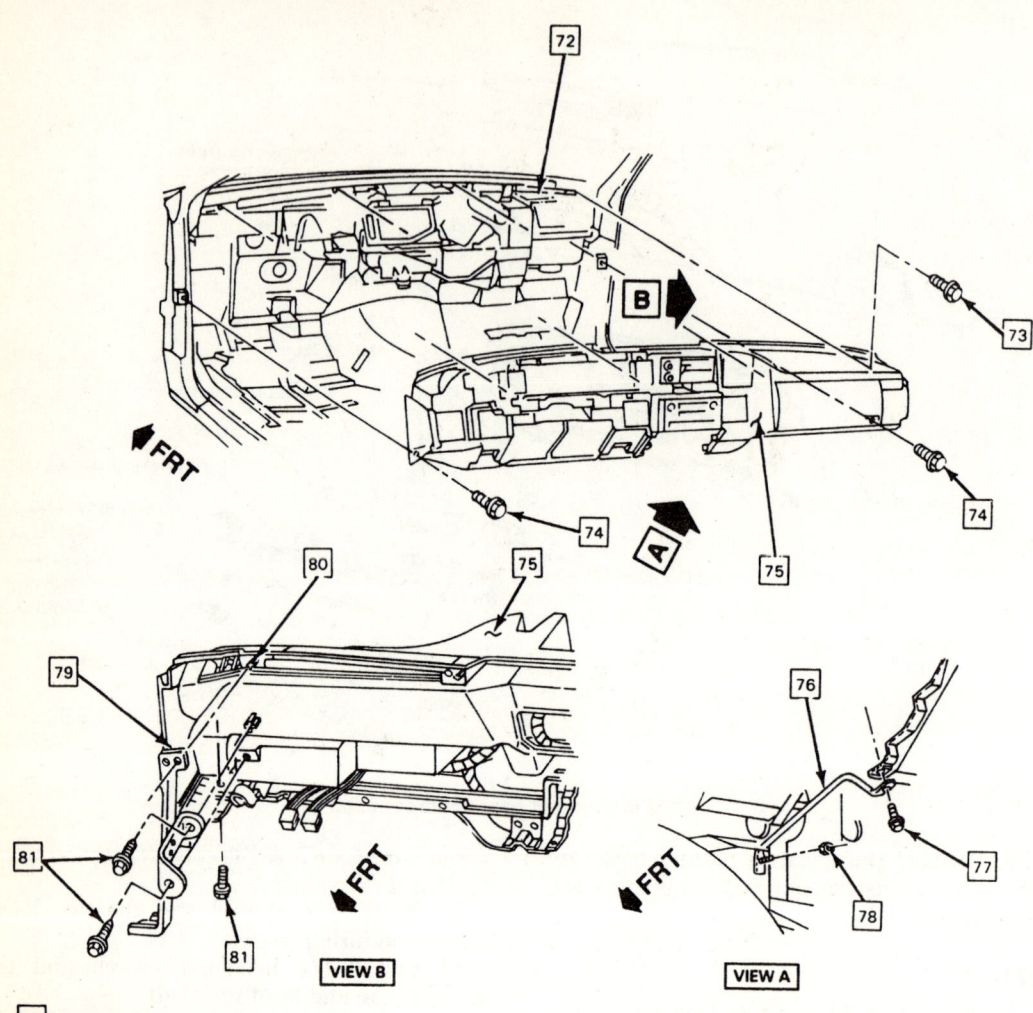

72	WINDSHIELD LOWER FRAME FLANGE
73	BOLT
74	BOLT
75	CARRIER
76	LOWER REINFORCEMENT
77	BOLT
78	NUT
79	OUTER REINFORCEMENT
80	LOCATOR PIN
81	SCREW

Instrument panel carrier mounting, 1991–92

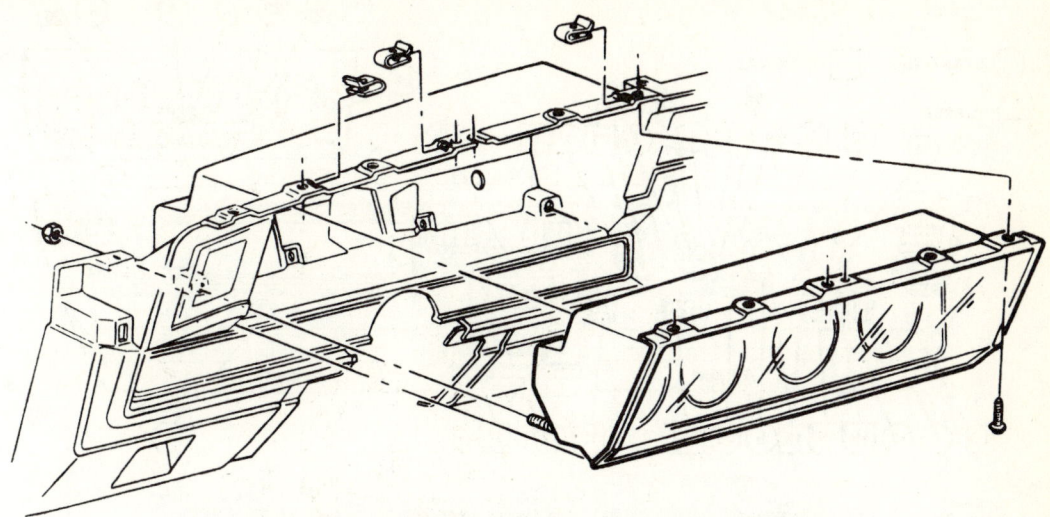

Mounting detail of the 1977 and later instrument cluster

trim plate, then remove the two nuts from the studs at the lower corner of the cluster.

6. Reach up behind the cluster and disconnect the speedometer cable, then remove the cluster by pulling outwards on it.

To Install:

7. Slide the cluster inwards enough to connect the wiring and speedometer cable to the rear of the cluster.

8. Install the two mounting nuts on the lower corner of the cluster, then install the trim plate and screws.

9. Install the lens over the cluster, then install the six screws and three snap-in fasteners around the edge of the cluster.

10. As required, raise the steering column and install the mounting bolts.

11. Connect the shifter indicator cable at the steering column.

12. Install the steering column lower cover and connect the negative battery cable.

13. Check that all of the lights and gauges work correctly.

1991–92

1. Disconnect the battery ground cable.

2. Remove the two screws and pull the steering column filler down.

3. Open the instrument panel compartment door and gently unsnap the right side molding from the instrument panel carrier.

4. Loosen the capsule nuts attaching the steering column support bracket to the instrument panel carrier. Only turn the nuts to the end of their threads do not remove the nuts, then lower the steering column down.

5. Remove the six screws attaching the left

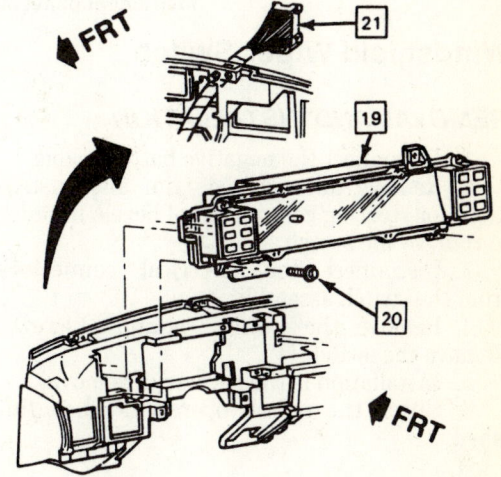

19	CLUSTER
20	SCREW
21	CONNECTOR

1991–92 instrument cluster

side trim plate to the instrument panel, carefully unsnap and remove the trim plate.

6. Remove the five screws attaching the cluster to the instrument panel.

7. Unclip the shift indicator cable from the steering column.

8. Gently pull the cluster from the connector and remove from vehicle.

9. Installation is the reverse of the removal procedure.

332 CHASSIS ELECTRICAL

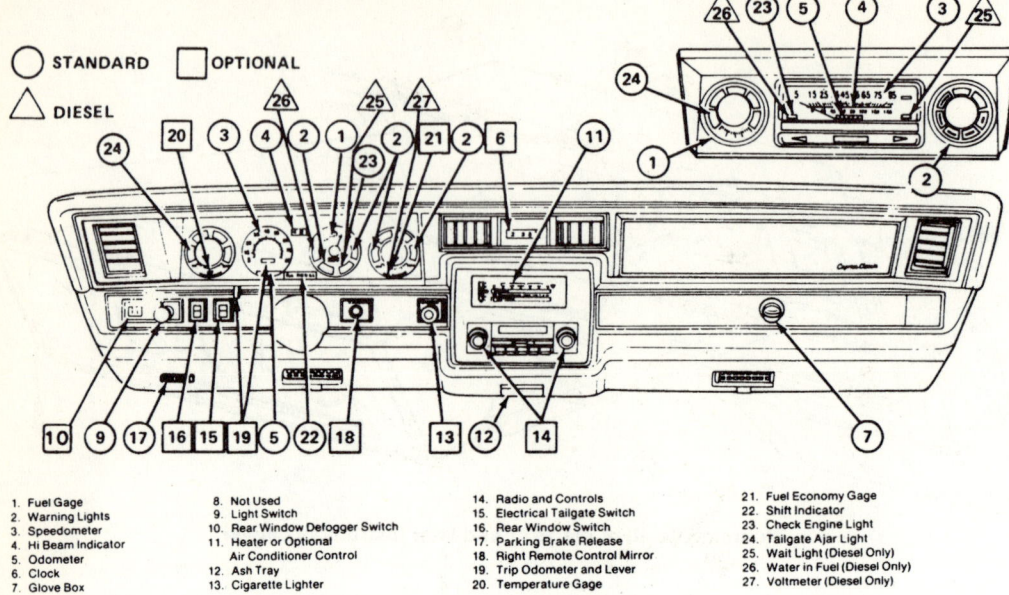

Instrument panel cluster assembly, 1977–90

1. Fuel Gage
2. Warning Lights
3. Speedometer
4. Hi Beam Indicator
5. Odometer
6. Clock
7. Glove Box
8. Not Used
9. Light Switch
10. Rear Window Defogger Switch
11. Heater or Optional Air Conditioner Control
12. Ash Tray
13. Cigarette Lighter
14. Radio and Controls
15. Electrical Tailgate Switch
16. Rear Window Switch
17. Parking Brake Release
18. Right Remote Control Mirror
19. Trip Odometer and Lever
20. Temperature Gage
21. Fuel Economy Gage
22. Shift Indicator
23. Check Engine Light
24. Tailgate Ajar Light
25. Wait Light (Diesel Only)
26. Water in Fuel (Diesel Only)
27. Voltmeter (Diesel Only)

Windshield Wiper Switch

REMOVAL AND INSTALLATION

1. Disconnect the negative battery cable.
2. Remove all required trim and instrument, or steering column panel bezels in order to remove the switch.
3. Disconnect the electrical connectors from the switch assembly.
4. Remove the switch retaining screws. Remove the switch.
5. Installation is the reverse of removal.
6. Check the switch operation when finished.

Headlight Switch

REMOVAL AND INSTALLATION

1968–90

1. Disconnect the negative battery cable. As required, remove the steering column trim cover.
2. Pull the headlight control knob to the ON position.
3. Reaching up under the instrument panel, depress the switch shaft release button, then pull the shaft and the control knob out.
4. Remove the ferrule nut and the switch from the instrument panel.
5. Disconnect the multi-contact electrical connector from the switch.
6. Installation is the reverse of the removal procedure.

1991–92

1. Disconnect the battery ground cable.
2. Remove the left side trim plate from the instrument panel.
3. Remove the three screws attaching the switch to instrument panel and remove the switch.
4. Disconnect the panel lamp dimmer switch connector, headlamp switch connector and the twilight sentinel switch connector, if so equipped, from the switch.
5. Disconnect the headlamp switch indicator lamp, if so equipped.
6. Installation is the reverse of the removal procedure.

Clock

REMOVAL AND INSTALLATION

1. Disconnect the negative battery cable.
2. Remove the clock panel cluster bezel.
3. Remove the clock retaining screws and remove the clock.
4. Disconnect the electrical connectors from the clock assembly.
5. Installation is the reverse of the removal procedure.

Speedometer Cable

REPLACEMENT

NOTE: *Although not necessary, removing the instrument cluster will give better access to the speedometer cable.*

1. Reach behind the instrument cluster and push the speedometer cable casing toward the

CHASSIS ELECTRICAL 333

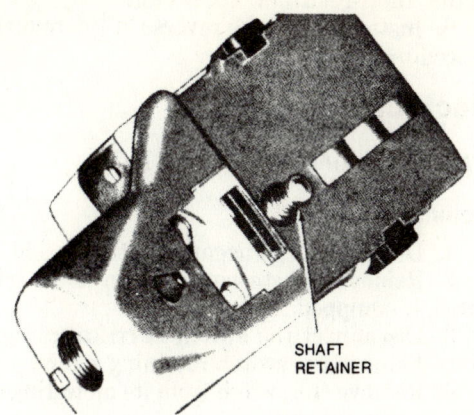

Headlight switch, early model

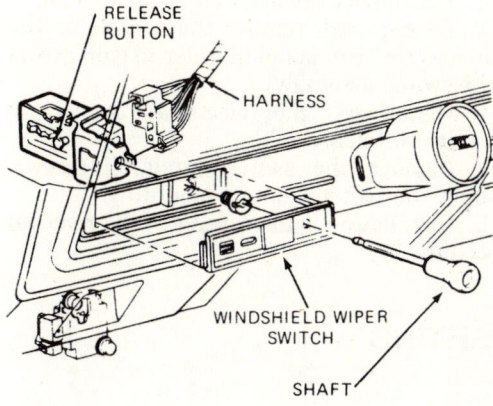

1977–90 headlight switch removal

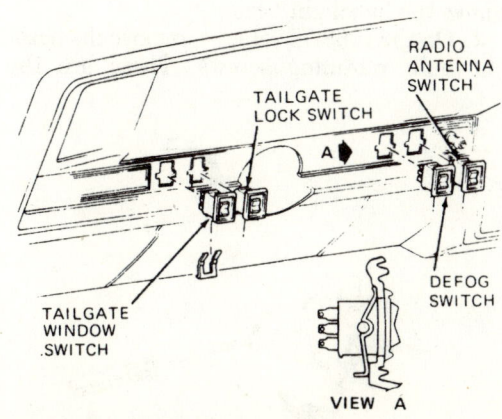

1977–90 auxiliry switch removal

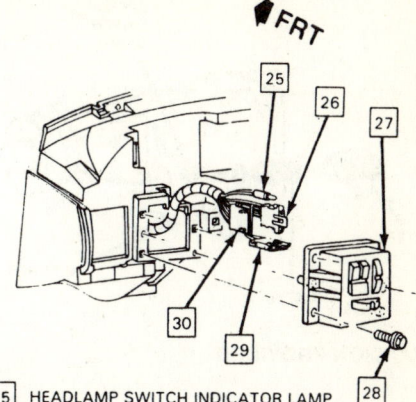

25	HEADLAMP SWITCH INDICATOR LAMP (WITH TWILIGHT SENTINEL)
26	PANEL LAMP DIMMER SWITCH CONNECTOR
27	SWITCH
28	SCREW
29	TWILIGHT SENTINEL SWITCH CONNECTOR (RPO T82)
30	HEADLAMP SWITCH CONNECTOR

Headlight and instrument panel dimmer switch, 1991–92

speedometer while depressing the retaining spring on the back of the instrument cluster case. Once the retaining spring has released, hold it in while pulling outward on the casing to disconnect the casing from the speedometer.

2. Remove the cable casing sealing plug from the dash panel. Then, pull the casing down from behind the dash and remove the cable.

NOTE: *If the vehicle is equipped with cruise control, the speedometer cable can be disconnected at the transducer which is mounted on the right fender inner panel.*

3. If the vehicle is not equipped with cruise control and the cable is broken and cannot be entirely removed from the top, support the vehicle securely, and then unscrew the cable casing connector at the transmission. Pull the

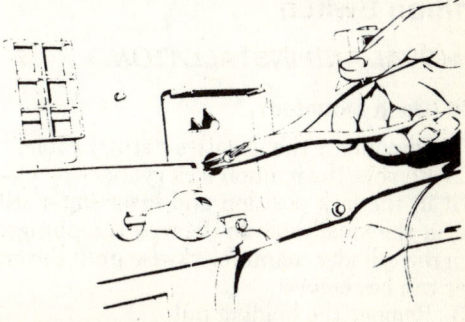

Speedometer cable removal

334 CHASSIS ELECTRICAL

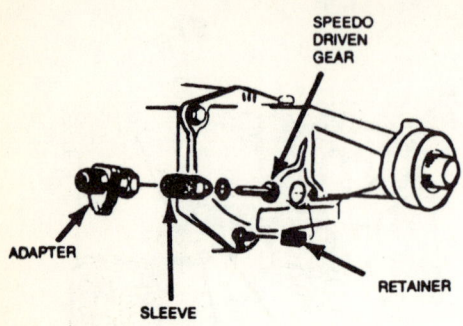

NO CAPTION PROVIDED

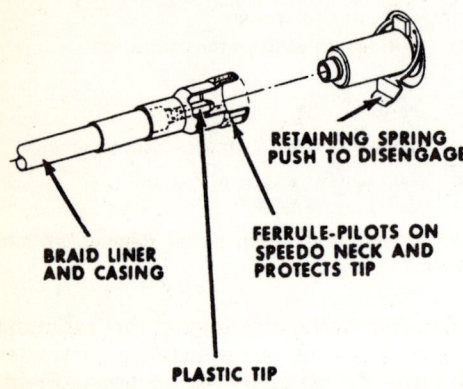

Speedometer cable attachment at instrument cluster

bottom part of the cable out, and then screw the connector back onto the transmission.

4. Lubricate the new cable with a speedometer cable grease. Insert it into the casing until it bottoms. Push inward while rotating it until the square portion at the bottom engages with the coupling in the transmission or cruise control transducer, permitting the cable to move in another inch or so. Then, reconnect the cable casing to the speedometer and install the sealing plug into the dash panel.

Ignition Switch

REMOVAL AND INSTALLATON

1968 (Dash Mounted)

1. Disconnect the negative battery cable.
2. Remove the ignition lock cylinder by placing it in the lock position and inserting a stiff wire in the small hole to depress the plunger. Turn the cylinder counterclockwise until the cylinder can be removed.
3. Remove the holding nut.
4. Pull the switch from under the dash and remove the connectors.
5. Using an awl, un-snap the locking tangs of the "theft-resistant" connector.
6. Installation is the reverse of the removal procedure.

Back-up Light switch

REMOVAL AND INSTALLATION

Column Mounted

1. Disconnect the negative battery cable.
2. Remove the steering column lower trim panel, if equipped.
3. Disconnect the switch electrical connections. Remove the switch retaining screws.
4. Remove the switch from its mounting on the steering column.
5. Installation is the reverse of the removal procedure.

Console Mounted

1. Disconnect the negative battery cable.
2. As required, remove the console or the shift selector trim panel in order to gain access to the switch assembly.
3. Disconnect the electrical connectors from the switch assembly.
4. Remove the switch retaining screws. Remove the switch from its mounting.
5. Installation is the reverse of the removal procedure.

LIGHTING

Headlights

REMOVAL AND INSTALLATION

1. Unscrew the four retaining screws and remove the headlight bezel.
2. On the 1968-75 vehicles, remove the headlight bulb retaining screws. These are the

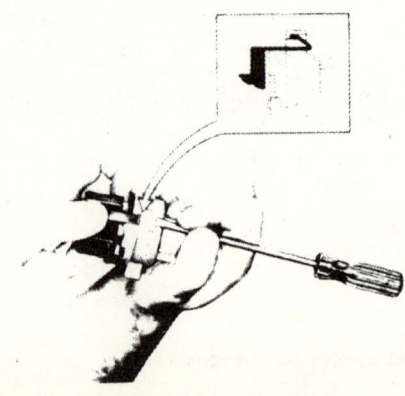

Unlocking the ignition switch connector, 1968

CHASSIS ELECTRICAL 335

Exploded view of headlight assembly, 1968–76

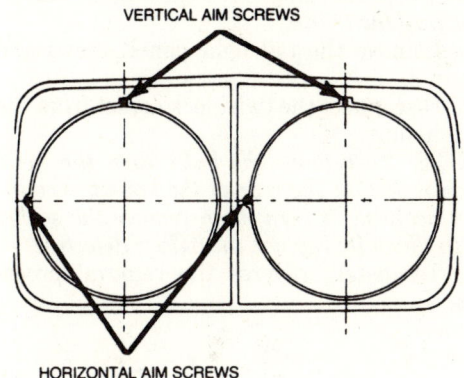

Dual headlight adjustment screw location

screws which hold the retaining ring for the bulb in the front of the vehicle. Do not touch the two headlight aiming screws, at the top and the side of the retaining ring (these screws will have different heads), or the headlight aim will have to be re-adjusted.

3. On the 1977-90 vehicles, disengage the spring from the retaining ring, using a cotter pin removal tool, then remove the retaining ring screws.

4. On 1991–92 vehicles, open hood and reach in behind headlamp assembly twist the retainer counterclockwise to remove the bulb.

CAUTION: *Handle halogen bulbs with care! Do not touch bulb, to avoid damage and or personal injury always handle halogen bulbs by their base*

4. Pull the bulb and ring forward and then separate them. Unplug the electrical connector from the rear of the bulb.

5. Plug the new bulb into the electrical connector. Install the bulb into the retaining ring and then install the ring and the bulb. Install the headlight bezel.

HEADLIGHT AIMING

The headlights must be properly aimed to provide the best, safest road illumination. The lights should be checked for proper aim, and adjusted if necessary, after installing a new sealed

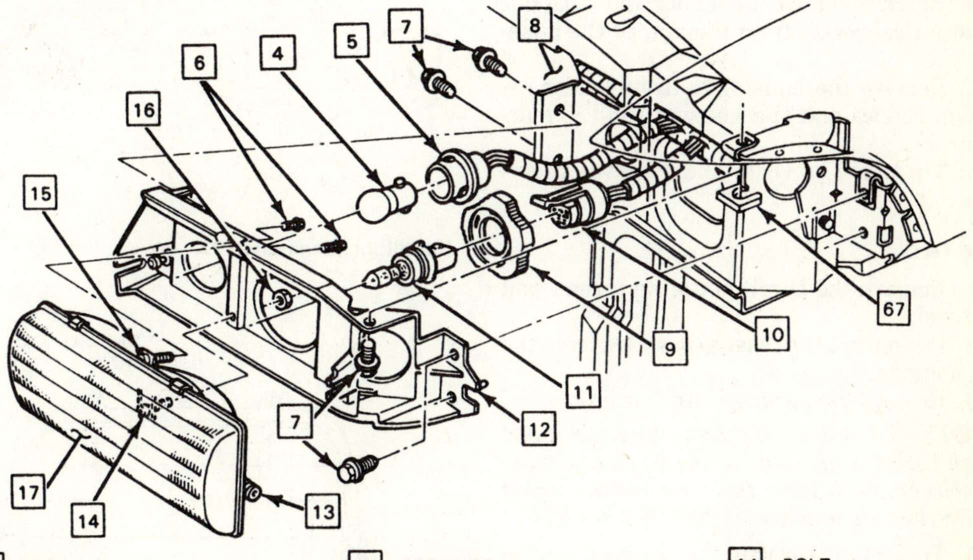

4 BULB	9 RETAINER	14 BOLT
5 SOCKET	10 CONNECTOR	15 NUT
6 SCREW	11 BULB/SOCKET	16 NUT
7 BOLT	12 BRACKET	17 HEADLAMP CAPSULE
8 RADIATOR SUPPORT	13 SLIDE PIVOT	67 SHIM

Headlight assembly, 1991–92

336 CHASSIS ELECTRICAL

beam unit if the front end sheet metal has been replaced. Certain state and local authorities have requirements for headlight aiming; these should be checked before adjustment is made.

NOTE: *The fuel tank should be about half full when adjusting the headlights. Tires should be properly inflated, and if a heavy load is carried in the trunk or in the cargo area of station wagons, it should remain there.*

Horizontal and vertical aiming of each sealed beam unit is provided by two adjusting screws, which move the mounting ring in the body against the body of the coil spring. There is no adjustment for focus; this is done during headlight manufacturing.

Signal and Marker Lights

NOTE: *Since the light housing capsules (on the late model vehicles) are constructed by sonic welding, the only service which can be performed are the replacement of the bulbs or the light housing.*

REMOVAL AND INSTALLATION

Front Turn Signal and Parking Lights

TYPE ONE

1. Reach up under the fender and twist out the electrical socket from the rear of the housing.
2. Remove the housing-to-front fender extension screws and the housing (pull it rearward).
3. To install, reverse the removal procedure.

TYPE TWO

1. Remove the bezel mounting screws and the bezel.
2. Disconnect the twist lock socket from the lens housing.
3. Remove the parking light housing.

NOTE: *To remove the bulb, turn the twist lock socket at the rear of the housing) counterclockwise $1/4$ turn, then remove the socket with the bulb; replace the bulb if defective.*

4. To install, reverse the removal procedures.

Side Marker Lights

NOTE: *Depending upon the year of the vehicle the side marker light assembly may be accessible by either reaching up from under the vehicle or opening the hood and reaching down into the engine compartment.*

1. Disconnect the negative battery cable.
2. Remove the electrical socket and the bulb.

NOTE: *To remove the bulb, turn the twist lock socket (at the rear of the housing) counterclockwise $1/4$ turn, then remove the socket with the bulb; replace the bulb if defective.*

3. Replace the bulb and reinstall the electrical socket assembly.

Rear Turn Signal, Brake and Parking Lights

NOTE: *Depending upon the year of the vehicle the light assembly may be accessible by either reaching up from under the vehicle or opening the trunk.*

1. Remove the tail light panel screws and the panel.
2. Disconnect the twist lock socket from the lens housing.

NOTE: *To remove the bulb, turn the twist lock socket (at the rear of the housing) counterclockwise $1/4$ turn, then remove the socket with the bulb; replace the bulb if defective.*

3. To install, reverse the removal procedures.

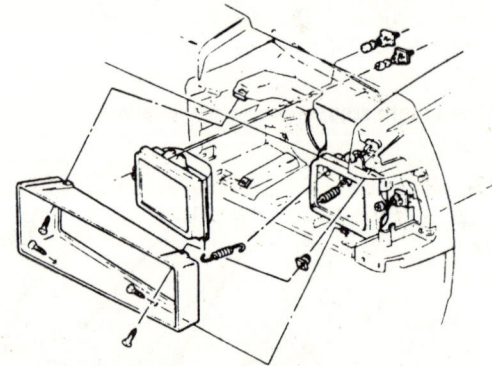

Headlight assembly, 1977–90

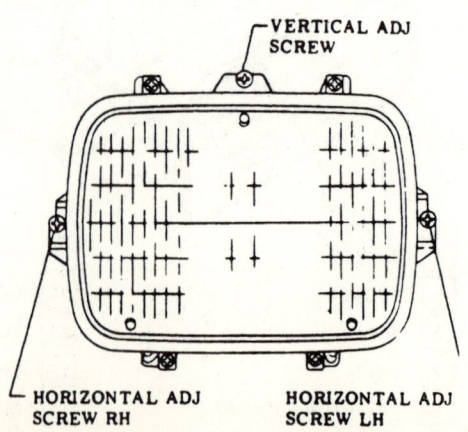

Headlight adjustment screw locations, round headlights similar

CHASSIS ELECTRICAL

TRAILER WIRING

Wiring the vehicle for towing is fairly easy. There are a number of good wiring kits available and these should be used, rather than trying to design your own. All trailers will need brake lights and turn signals as well as tail lights and side marker lights. Most states require extra marker lights for overly wide trailers. Also, most states have recently required back-up lights for trailers, and most trailer manufacturers have been building trailers with back-up lights for several years.

Additionally, some Class I, most Class II and just about all Class III trailers will have electric brakes.

Add to this number an accessories wire, to operate trailer internal equipment or to charge the trailer's battery, and you can have as many as seven wires in the harness.

Determine the equipment on your trailer and buy the wiring kit necessary. The kit will contain all the wires needed, plus a plug adapter set which included the female plug, mounted on the bumper or hitch, and the male plug, wired into, or plugged into the trailer harness.

When installing the kit, follow the manufacturer's instructions. The color coding of the wires is standard throughout the industry.

One point to note, some domestic vehicles, and most imported vehicles, have separate turn signals. On most domestic vehicles, the brake lights and rear turn signals operate with the same bulb. For those vehicles with separate turn signals, you can purchase an isolation unit so that the brake lights won't blink whenever the turn signals are operated, or, you can go to your local electronics supply house and buy four diodes to wire in series with the brake and turn signal bulbs. Diodes will isolate the brake and turn signals. The choice is yours. The isolation units are simple and quick to install, but far more expensive than the diodes. The diodes, however, require more work to install properly, since they require the cutting of each bulb's wire and soldering in place of the diode.

One final point, the best kits are those with a spring loaded cover on the vehicle mounted socket. This cover prevents dirt and moisture from corroding the terminals. Never let the vehicle socket hang loosely. Always mount it securely to the bumper or hitch.

NOTE: *For more information on towing a trailer please refer to Chapter 1.*

CIRCUIT PROTECTION

Fusible Links

A fusible link is a protective device used in an electrical circuit. When the current increases beyond a certain amperage, the fusible metal of the wire link will melt, thus breaking the electrical circuit and preventing further damage to any other components or wiring. Whenever a fusible link is melted because of a short circuit, correct the cause before installing a new one. Most models have four fusible links.

REPLACING FUSIBLE LINKS

1. Disconnect both battery cables. If the link is connected to the junction block or starter solenoid, disconnect it.
2. Cut the wiring harness right behind the link connector(s) and remove.
3. Strip the insulation off the harness wire back 1/2".
4. Position the clip around the new link and wiring harness or new connector and crimp it securely. Then, solder the connection, using rosin core solder and sufficient heat to guarantee a good connection. Repeat for the remaining connection.
5. Tape all exposed wiring with electrical tape. Where necessary, connect the link to the

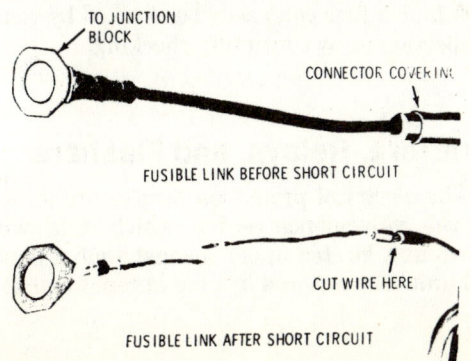

Fusible link

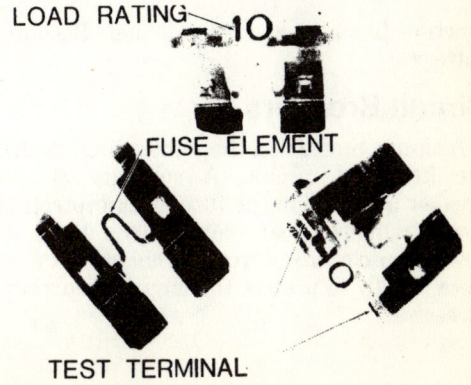

New style fuses

CHASSIS ELECTRICAL

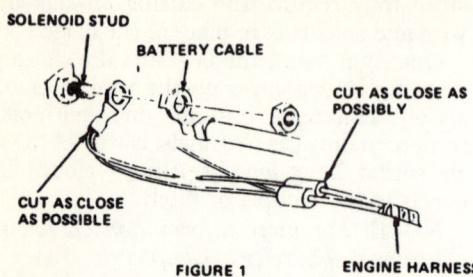

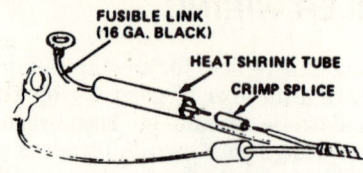

FIGURE 1
REMOVE BATTERY CABLE & FUSIBLE LINK FROM STARTER SOLENOID AND CUT OFF DEFECTIVE WIRE AS SHOWN TWO PLACES.

FIGURE 2
STRIP INSULATION FROM WIRE ENDS. PLACE HEAT SHRINK TUBE OVER REPLACEMENT LINK. INSERT WIRE ENDS INTO CRIMP SPLICE AS SHOWN. NOTE: PUSH WIRES IN FAR ENOUGH TO ENGAGE WIRE ENDS.

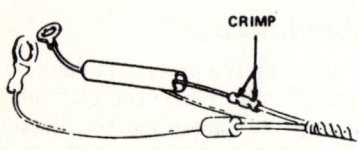

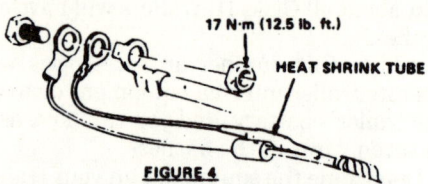

FIGURE 3
CRIMP SPLICE WITH CRIMPING TOOL TWO PLACES TO BIND BOTH WIRES.

FIGURE 4
SLIDE TUBE OVER SPLICE WITH SPLICE CENTERED IN TUBE. APPLY LOW TEMPERATURE HEAT TO SHRINK TUBE AROUND WIRES & SPLICE. REASSEMBLE LINKS & BATTERY CABLE.

Fusible link repair

Fuse Block

The fuse block on some models is located under the instrument panel next to the steering wheel and is a swing down unit. Other models have the fuse block located on the right side of the dash and access is gained through the glove box.

Each fuse block uses miniature fuses which are designed for increased circuit protection and greater reliability. The compact fuse is a blade terminal design which allows fingertip removal and replacement.

Although the fuses are interchangeable, the amperage values are molded in bold, color coded, easy to read numbers on the fuse body. Use only fuses of equal replacement value.

A blown fuse can easily be checked by visual inspection or by continuity checking.

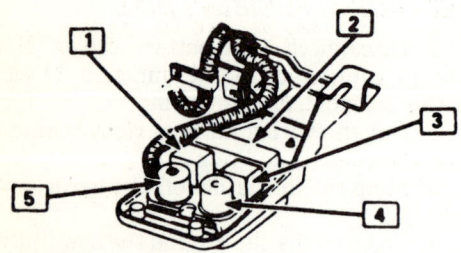

1. Horn relay
2. Seat belt–ignition key–headlight buzzer
3. Choke relay (vacant w/EFI)
4. Hazard flasher
5. Signal flasher

View of convenience center and components

junction block or started solenoid. Reconnect battery.

Circuit Breakers

A circuit breaker in the light switch protects the headlight circuit. A separate 30 amp breaker mounted in the fuse block protects the power window, seat, power door locks and power top circuits. Circuit breakers open and close rapidly to protect the circuit if current is excessive.

Buzzers, Relays, and Flashers

The electrical protection devices are located in the convenience center, which is a swing down unit located under the instrument panel. All units are serviced by plug-in replacements.

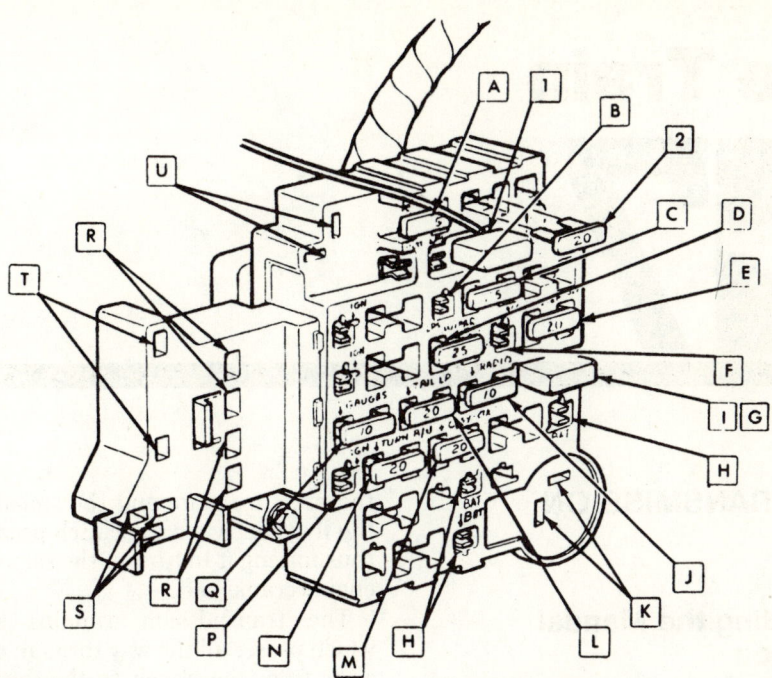

1. Circuit breaker
2. Fuse assembly and L22
A. Fuse-inst panel lamps
B. Receptacle-lamps
C. Fuse-heater and A/C (C60)
D. Fuse-windshield wiper
E. Fuse-stop and traffic hazard lamps
F. Receptacle-accessory
G. Power-accessory
H. Receptacle-battery
J. Fuse-radio (U63)
K. Receptacle-directional signal lamp flasher
L. Fuse-tail lamps
M. Fuse-clock, cigarette lighter and courtesy lamps
N. Fuse-Directional signal and back-up lamps
P. Receptacle-cruise control (K30) rear window defogger (C50) (all ignition control)
Q. Fuse-gauges
R. Receptacle-seat belt buzzer
S. Receptacle-horn relay
T. Receptacle-ignition key lock warning buzzer
U. Receptacle-traffic hazard flasher

New style fuse panel

Drive Train

7

MANUAL TRANSMISSION

Understanding the Manual Transmission

Because of the way the gasoline engine breathes, it can produce torque, or twisting force, only within a narrow speed range. Most modern engines must turn at about 2,500 rpm to produce their peak torque. By 4,500 rpm they are producing so little torque that continued increases in engine speed produce no power increases.

The transmission and clutch are employed to vary the relationship between engine speed and the speed of the wheels so that adequate engine power can be produced under all circumstances. The clutch allows engine torque to be applied to the transmission input shaft gradually, due to mechanical slippage. The vehicle can, consequently, be started smoothly from a full stop.

The transmission changes the ratio between the rotating speeds of the engine and the wheels by the use of gears. 3-speed or 4-speed transmissions are most common. The lower gears allow full engine power to be applied to the rear wheels during acceleration at low speeds.

The clutch driven plate is a thin disc, the center of which is splined to the transmission input shaft. Both sides of the disc are covered with a layer of material which is similar to brake lining and which is capable of allowing slippage without roughness or excessive noise.

The clutch cover is bolted to the engine flywheel and incorporates a diaphragm spring which provides the pressure to engage the clutch. The cover also houses the pressure plate. The driven disc is sandwiched between the pressure plate and the smooth surface of the flywheel when the clutch pedal is released, thus forcing it to turn at the same speed as the engine crankshaft.

The transmission contains a mainshaft which passes all the way through the transmission, from the clutch to the driveshaft. This shaft is separated at one point, so that front and rear portions can turn at different speeds.

Power is transmitted by a countershaft in the lower gears and reverse. The gears of the countershaft mesh with gears on the mainshaft, allowing power to be carried from one to the other. All the countershaft gears are integral with that shaft, while several of the mainshaft gears can either rotate independently of the shaft or be locked to it. Shifting from one gear to the next causes one of the gears to be freed from rotating with the shaft, and locks another to it. Gears are locked and unlocked by internal dog clutches which slide between the center of the gear and the shaft. The forward gears usually employ synchronizers: friction members which smoothly bring gear and shaft to the same speed before the toothed dog clutches are engaged.

The clutch is operating properly if:

1. It will stall the engine when released with the vehicle held stationary.

2. The shift lever can be moved freely between 1st and reverse gears when the vehicle is stationary and the clutch disengaged.

A clutch pedal free-play adjustment is incorporated in the linkage. If there is about 1-2" of motion before the pedal begins to release the clutch, it is adjusted properly. Inadequate free-play wears all parts of the clutch releasing mechanisms and may cause slippage. Excessive free-play may cause inadequate release and hard shifting of gears.

DRIVE TRAIN 341

Identification

See Chapter 1, which lists the basic types of manual transmissions and the various locations of their serial numbers. By finding the serial number on your transmission and comparing its location with the information there, you can readily determine the type of gearbox used.

Linkage Adjustment

Column Shift

1968

1. Located on the left side of the transmission case are two levers. Manipulate these levers until the transmission is in neutral. Depress the clutch pedal, start the engine, and release the pedal slowly. If the vehicle fails to move and the engine is still running with the pedal fully released, the transmission is in neutral. If the vehicle moves, the transmission is in gear and the levers should be repositioned until neutral is found. Loosen the swivel nuts on both shift rods.
2. Move the shift lever (on the column) to the neutral position. Raise the hood and locate the shifter tube levers on the steering column. Align the 1st and reverse lever with the 2nd and 3rd lever. Using a large L-shaped Allen wrench, hold these levers in alignment (most vehicles have alignment holes in the levers and an alignment plate) until the linkage is connected.
3. Make the final adjustments to align the shift rods and levers at the transmission into the neutral position. Road test the vehicle and check the shifting operation. If the adjustment is correct, the alignment pin should pass freely through all alignment holes. If not, readjustment is necessary.

1969

1. Turn the ignition switch to the OFF position.
2. Loosen the swivel nuts on both shift rods.
3. Place the column mounted shift lever in the reverse position. There are two levers located on the side of the transmission case. The lever to the front of the transmission controls 2nd and 3rd gears while the other lever controls 1st and reverse. Place this 1st and reverse lever into the reverse position. Push up on the 1st/reverse shift rod until the column lever is in the reverse detent position. Tighten the swivel nut.
4. Place the column lever and the transmission levers (located on the side of the case) in neutral. The shift tube levers are located on the steering column mast jacket. Make sure the column lever is in neutral and hold it in this position by inserting a pin through the alignment holes in the shift tube levers (a $1/16''$ Allen wrench is perfect).
5. Hold the 2nd/3rd shift rod steady, to prevent a change in adjustment and tighten the swivel locknut.
6. Remove the alignment pin from the shift tube levers and shift the column shift lever to the reverse position. Turn the ignition key to LOCK and check the ignition interlock control. If it binds, leave the control in LOCK and readjust the 1st/reverse rod at the swivel.
7. Move the column lever through the gear positions and return it to neutral. The alignment pin should pass freely through the alignment holes of the shift tube levers. If it doesn't loosen the swivel nuts and readjust.

1970-73

1. Place the shift lever (on the column) in reverse and the ignition switch in **OFF** position.
2. Raise the vehicle and support it with safety stands.
3. Loosen the locknuts on the shift rod swivels. Pull down slight on the 1st/reverse control rod on the lower steering column to remove any slack. Tighten the locknut at the transmission lever.
4. Unlock the ignition switch and shift the column lever into neutral. Position the shift tube levers, which are located on the lower steering column in neutral by aligning the lever alignment holes. Hold them in this position by inserting a $1/16''$ Allen wrench through the alignment holes.

1968 column shift adjustment

342 DRIVE TRAIN

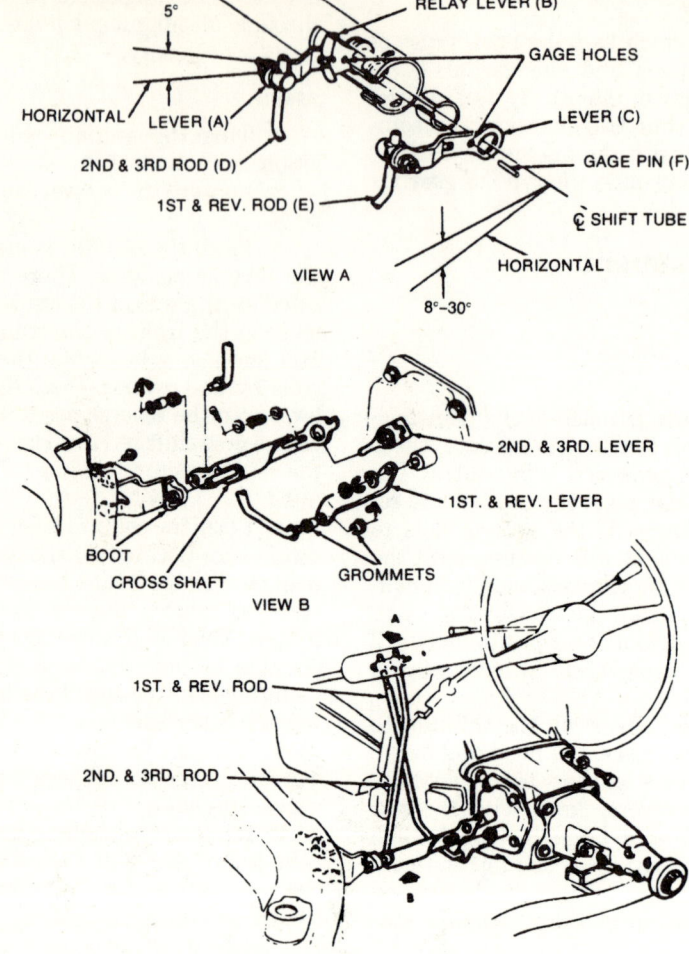

1969–73 column shift adjustment

5. Hold the 2nd/3rd shift rod steady and tighten the rod locknut.

6. Remove the alignment tool from the shift tube levers and check the shifting operation.

7. Place the column lever in reverse and check the movement of the ignition key. In reverse and only reverse, the key must turn freely in and out of LOCK.

Floor Shift

3-SPEED 1968-73

1. Loosen the locknuts on both shift rod swivels. The shift rods should pass freely through the swivels.

2. Move the floor shift to Neutral and install the locating gauge into the shifter bracket assembly.

NOTE: *The locating gauge is a piece of $1/8''$ thick flat stock $41/64''$ wide and $3''$ long.*

3. Position the levers on the transmission in Neutral. Turn the 1st/reverse shift rod nut down against the swivel, and then tighten the locknut against the swivel.

4. Turn the 2nd/3rd shift rod nut down against the swivel, and then tighten the locknut against the swivel.

NOTE: *On 1969-73 vehicles, skip Step 4 and perform Steps 5-8. Step 4 is the final step for 1968 vehicles.*

5. Remove the locating gauge, and shift into Reverse. Turn the ignition switch to the **Lock** position.

6. Loosen the swivel locknut on the reverse control rod. Pull down slightly on the control rod to take up any slack in the column mechanism, and then tighten the clevis jam nut.

7. The ignition switch should move easily in and out of lock. If there is any binding present, keep the switch in Lock and readjust the reverse control rod.

8. Check shift pattern for correct operation.

DRIVE TRAIN

4-SPEED 1968

1. Remove the control rods from the transmission levers, and position the levers in Neutral.
2. Move the floor shift lever into Neutral and insert a locating gauge into the bracket assembly (use the same gauge as the 3-speed floor shift).
3. Adjust the length of the control rods, and then secure the swivels with the jam nuts and install the clevis pins.
4. Remove the locating gauge, and check the shift pattern.

NOTE: *Muncie transmission levers have two control rod holes. Attaching the control rods in the lower holes will result in reduced shift lever travel and allow faster shifting, with increased shifting effort as a minor drawback.*

4-SPEED 1969-70

1. Turn the ignition switch to the **Off** position, except on 1970 models which should be turned to **Lock** position.
2. Loosen the swivel locknuts on the shift rods and reverse control rod. Position the transmission side cover levers to their Neutral detent position.
3. Place the floor shift in Neutral and insert a locating gauge (same gauge as 3-speed floor shift) into the lever bracket assembly.
4. Adjust all control rods swivels for easy entry into their respective levers.
5. Tighten the shift rod locknuts and remove the gauge.
6. Shift the lever into Reverse and pull down slightly on the reverse rod to remove the slack. Tighten the locknut.
7. The ignition switch should now be above to be moved easily into Lock, and it must not be possible to turn the key to Lock, when in any other position other than Reverse. Readjust the reverse rod, if necessary.
8. Check the shift pattern for correctness.

CLUTCH SWITCH ADJUSTMENT AND REPLACEMENT

A clutch operated neutral safety switch was used beginning in 1970. The ignition switch must be in the START position and the clutch must be fully depressed before the vehicle will start. The switch mounts to the clutch pedal arm. Removal of this switch is obvious and simple. This switch cannot be adjusted.

Back-up Light Switch

REMOVAL AND INSTALLATION

Column Shift

NOTE: *The back-up light switch is mounted on the mast jacket of the column assembly.*

1. Disconnect the wiring connectors at the switch terminals.
2. Remove the screws attaching the switch to the mast jacket and remove the switch.
3. Position the column shift tube in the Reverse detent using the 1st/Reverse lower shift lever.
4. Position the switch on the column with the drive tang in the slot in the shift tube and the right side of the tang contacting the right surface of the shift tube slot.
5. Holding the switch in position, install the two attaching screws.

NOTE: *Moving the gear selector lever out of Reverse position will break the shear pin.*

6. Plug the wiring connector on the switch.
7. Check the operation of the switch.

Floor Shift

NOTE: *The back-up light switch is mounted on the transmission.*

1. Raise the vehicle and support it safely.
2. Disconnect the switch wiring from the harness wiring at the inline connector.
3. Remove the bolt retaining the wiring attaching clip to the transmission.
4. Remove the wire clip retaining the Reverse lever rod to the switch.
5. Remove the screws retaining the switch and shield assembly to the transmission and remove the switch.

NOTE: *Do not remove the transmission-to-bracket retaining bolts.*

6. Reverse the above steps to install.

Transmission

REMOVAL AND INSTALLATION

1968-69 3-Speed and 4-Speed

1. Raise the vehicle and remove the driveshaft. On floor shift models, also remove the trim plate and shifter boot.
2. It may be necessary to disconnect the exhaust pipe at the manifold.
3. Disconnect the speedometer cable and, on floor shift models, disconnect the back-up light switch also.
4. Remove the crossmember-to-frame bolts. On the floor shift models, remove the bolts holding the control lever support to the crossmember.
5. Remove the transmission mount bolts.

DRIVE TRAIN

6. Using a suitable jack and a block of wood (to be placed between the jack and the engine), raise the engine slightly and remove or relocate the crossmember.

7. Remove the shift levers from the transmission side cover.

8. On floor shift models, remove the shifter assembly and stabilizer rod (if so equipped) situated between the shift lever assembly and the transmission.

9. Remove the transmission-to-bellhousing bolts. Remove the top bolts first and insert guide pins into the holes, then remove the bottom bolts.

10. Remove the transmission.

To Install:

11. Perform the following:

   a. Lift the transmission into position and insert the mainshaft into the bellhousing.

   b. Install the transmission-to-bellhousing bolts and lockwashers, and torque them to 50 ft. lbs.

   c. Install the transmission shift levers to the side cover. On floor shift models, install the stabilizer rod if equipped.

   d. Raise the engine slightly, position the crossmember, and install the bolts.

   e. Install the transmission mount bolts. On floor shift models, install the bolts holding the shift lever support to the crossmember.

NOTE: *Lubricate the tailshaft bushing before the driveshafts are installed.*

   f. Install the driveshaft and, if removed, install the exhaust pipe to manifold.

   g. Connect the speedometer cable and, on floor shift equipped vehicles, connect the backup light.

   h. Fill the transmission with lubricant.

1970-73 3-Speed and All 4-Speed

1. On floor shift models, remove the shift knob and console trim plate.

2. Raise and support the front end on jackstands.

3. Disconnect the speedometer cable and the TCS switch wiring.

4. Remove the driveshaft. As required, disconnect and move the exhaust system out of the way.

5. Remove the bolts securing the transmission mounts to the crossmember and also those bolts securing the crossmember to the frame. Remove the crossmember.

6. Remove the shift levers from the side of the transmission.

7. Disconnect the back drive rod from the bellcrank.

8. Remove the bolts from the shift control assembly and carefully lower the assembly until the shift lever clears the rubber shift boot. Remove the assembly from the vehicle.

9. Remove the transmission-to-bellhousing bolts and lift the transmission from the vehicle.

To Install:

10. Perform the following:

   a. Lift the transmission and insert the mainshaft into the bellhousing.

   b. Install and torque the transmission-to-clutch housing bolts and lockwashers.

   c. Install the shift lever.

   d. Install the shift levers to the transmission side cover.

   e. Connect the back drive rod to the bellcrank.

   f. Raise the engine high enough to position the crossmember. Install and tighten the crossmember-to-frame bolts and transmission mount to crossmember bolts.

   g. Install the driveshaft. As required, connect the exhaust system.

   h. Connect the speedometer cable and TCS wiring.

   i. Fill the transmission with the specified lubricant. If applicable, install the console trim plate and shift knob. Adjust the linkage.

CLUTCH

Understanding the Clutch

The purpose of the clutch is to disconnect and connect engine power from the transmission. A vehicle at rest requires a lot of engine torque to get all that weight moving. An internal combustion engine does not develop a high starting torque (unlike steam engines), so it must be allowed to operate without any load until it builds up enough torque to move the vehicle. Torque increases with engine rpm. The clutch allows the engine to build up torque by physically disconnecting the engine from the transmission, relieving the engine of any load or resistance. The transfer of engine power to the transmission (the load) must be smooth and gradual; if it weren't, drive line components would wear out or break quickly. This gradual power transfer is made possible by gradually releasing the clutch pedal. The clutch disc and pressure plate are the connecting link between the engine and transmission. When the clutch pedal is released, the disc and plate contact each other (clutch engagement), physically joining the engine and transmission. When the pedal is pushed in, the disc and plate separate (the clutch is disengaged), disconnecting the engine from the transmission.

The clutch assembly consists of the flywheel,

DRIVE TRAIN 345

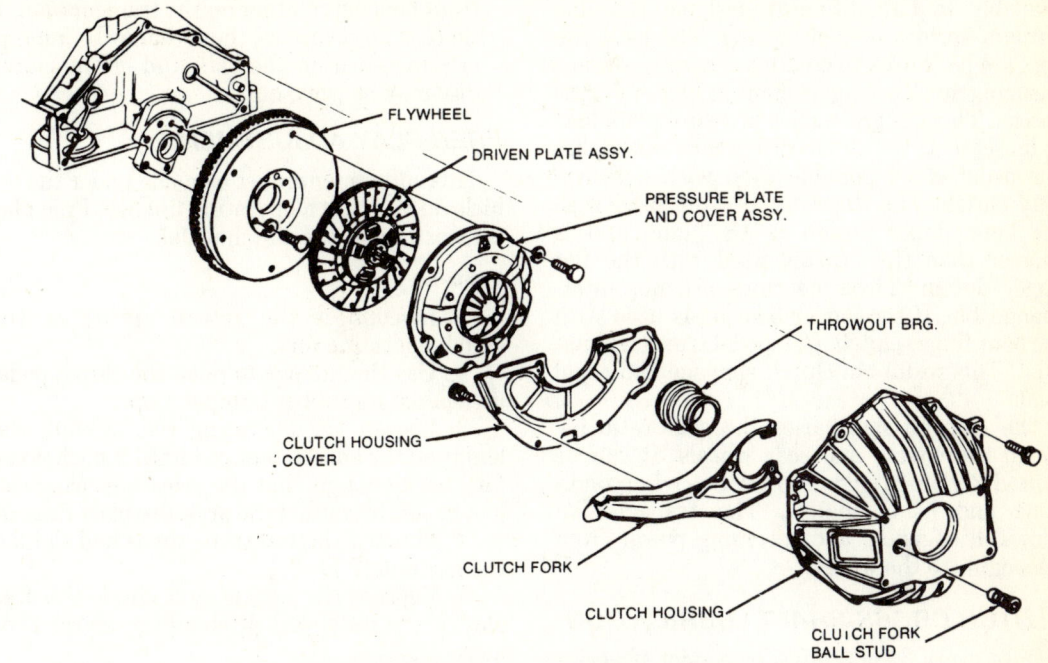

Exploded view of the clutch assembly

the clutch disc, the clutch pressure plate, the throwout bearing and fork, the actuating linkage and the pedal. The flywheel and clutch pressure plate (driving members) are connected to the engine crankshaft and rotate with it. The clutch disc is located between the flywheel and pressure plate, and splined to the transmission shaft. A driving member is one that is attached to the engine and transfers engine power to a driven member (clutch disc) on the transmission shaft. A driving member (pressure plate) rotates (drives) a driven member (clutch disc) on contact and, in so doing, turns the transmission shaft. There is a circular diaphragm spring within the pressure plate cover (transmission side). In a relaxed state (when the clutch pedal is fully released), this spring is convex; that it, it is dished outward toward the transmission. Pushing in the clutch pedal actuates an attached linkage rod. Connected to the other end of this rod is the throwout bearing fork. The throwout bearing is attached to the fork. When the clutch pedal is depressed, the clutch linkage pushes the fork and bearing forward to contact the diaphragm spring of the pressure plate. The outer edges of the spring are secured to the pressure plate and are pivoted on rings so that when the center of the spring is compressed by the throwout bearing, the outer edges bow outward and, by so doing, pull the pressure plate in the same direction - away from the clutch disc. This action separates the disc from the plate, disengaging the clutch and allowing the transmission to be shifted into another gear. A coil type clutch return spring attached to the clutch pedal arm permits full release of the pedal. Releasing the pedal pulls the throwout bearing away from the diaphragm spring resulting in a reversal of spring position. As bearing pressure is gradually released from the spring center, the outer edges of the spring bow outward, pushing the pressure plate into closer contact with the clutch disc. As the disc and plate move closer together, friction between the two increases and slippage is reduced until, when full spring pressure is applied (by fully releasing the pedal), The speed of the disc and plate are the same. This stops all slipping, creating a direct connection between the plate and disc which results in the transfer of power from the engine to the transmission. The clutch disc is now rotating with the pressure plate at engine speed and, because it is splined to the transmission shaft, the shaft now turns at the same engine speed. Understanding clutch operation can be rather difficult at first; As mentioned earlier, the clutch pedal return spring permits full release of the pedal and reduces linkage slack due to wear. As the linkage wears, clutch free-pedal travel will increase and free-travel will decrease as the clutch wears. Free-travel is actually throwout (release) bearing lash.

DRIVE TRAIN

The diaphragm spring type clutches used are available in two different designs: flat diaphragm springs or bent spring. The bent fingers are bent back to create a centrifugal boost ensuring quick re-engagement at higher engine speeds. This design enables pressure plate load to increase as the clutch disc wears and makes low pedal effort possible even with a heavy-duty clutch. The throwout bearing used with the bent finger design is $1 \frac{1}{4}''$ long and is shorter than the bearing used with the flat finger design. These bearings are not interchangeable. If the longer bearing is used with the bent finger clutch, free-pedal travel will not exist. This results in clutch slippage and rapid wear.

The transmission varies the gear ratio between the engine and rear wheels. It can be shifted to change engine speed as driving conditions and loads change. The transmission allows disengaging and reversing power from the engine to the wheels.

CLUTCH CROSS-SHAFT LUBRICATION

Once every 36,000 miles or sooner if necessary, remove the plug, install a fitting and lubricate with water resistant EP (Extreme Pressure) chassis lubricant.

LINKAGE INSPECTION

A clutch may have all the symptoms of going bad when the real trouble lies in the linkage. To avoid the unnecessary replacement of a clutch, make the following linkage checks:

a. Start the engine and depress the clutch pedal until it is about $1/2''$ from the floor mat and move the shift lever between 1st and reverse (1st and 2nd on a 4-speed) several times. If this can be done smoothly without any grinding, the clutch is releasing fully. If the shifting is not smooth, the clutch is not releasing fully and adjustment is necessary.

b. Check the condition of the clutch pedal bushings for signs of sticking or excessive wear.

c. Check the throwout bearing fork for proper installation on the ball stud. The fork could possibly be pulled off the ball if not properly lubricated.

d. Check the crossshaft levers for distortion or damage.

e. Check the vehicle for loose or damaged motor mounts. Bad motor mounts can cause the engine to shift under acceleration and bind the clutch linkage at the crossshaft. There must be some clearance between the crossshaft and the motor mount.

f. Check the throwout bearing clearance between the clutch spring fingers and the front bearing retainer on the transmission. If there is no clearance, the fork may be improperly installed on the ball stud or the clutch disc may be worn out.

FREE-PLAY ADJUSTMENT

This adjustment must be made under the vehicle on the clutch operating linkage. Free play is measured at the clutch pedal.

1968-70

1. Disconnect the return spring at the clutch operating fork.
2. Use the linkage to push the clutch pedal up against its rubber bumper stop.
3. Loosen the operating rod locknut and lengthen the adjustment rod until it pushes the fork back enough that the release bearing can just be left to contact the pressure plate fingers.
4. Shorten the rod three turns and tighten the locknut.
5. Replace the spring and check the free play at the pedal pad. It should be about 1" or more.

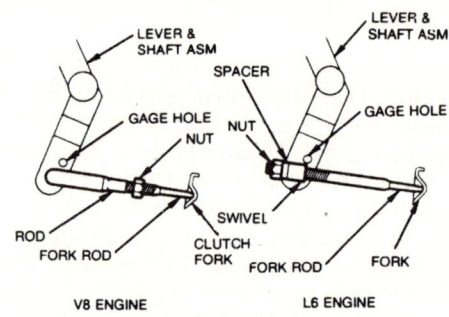

Clutch pedal freeplay adjustment, 1968–72

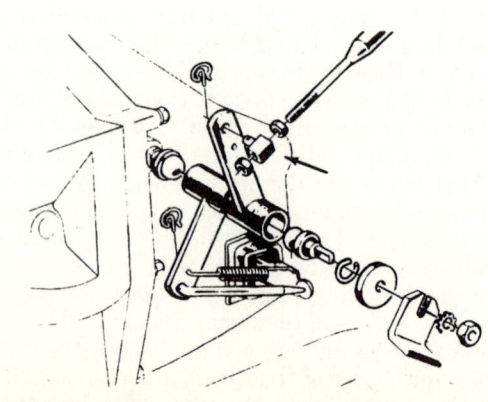

Clutch linkage adjustment is made at arrow

DRIVE TRAIN

1971-73

You can use this procedure on any earlier models that have a gauge hole in the clutch pivot shaft arm.

1. Disconnect the return spring at the clutch operating fork.
2. Use the linkage to push the clutch pedal up against its rubber bumper stop. More clearance can be obtained by loosening the rubber bumper bracket and moving the bracket.
3. Push the end if the clutch operating fork to the rear until the release bearing can just be felt to contact the pressure plate fingers.
4. Detach the front end of the operating rod from the clutch pivot shaft arm and place it in the gauge hole on the arm.
5. Loosen the locknut and lengthen the rod just enough to take all the play out of the linkage. Tighten the locknut.
6. Replace the operating rod in its original location.
7. Replace the return spring and check the free play at the pedal pad. It should be about 1" or more.

Driven Disc and Pressure Plate

REMOVAL

CAUTION: *The clutch driven disc contains asbestos, which has been determined to be a cancer causing agent. NEVER clean the clutch surfaces with compressed air! Avoid inhaling dust from the clutch surface! When cleaning the clutch surfaces use a commercially available brake cleaning fluid.*

1. Support the engine and remove the transmission.
2. Disconnect the clutch fork push rod and spring.
3. Remove the flywheel housing.
4. Slide the clutch fork from the ball stud and remove the fork from the dust boot. The ball stud is threaded into the clutch housing and may be replace, if necessary.
5. Install an alignment tool to support the clutch assembly during removal. mark the flywheel and clutch cover for reinstallation, if they do not already have X marks.
6. Loosen the clutch-to-flywheel attaching bolts evenly, one turn at a time, until spring pressure is released. Remove the bolts and clutch assembly.

INSTALLATION

1. Clean the pressure plate and flywheel face.
2. Support the clutch disc and pressure plate with an alignment tool. The driven disc is installed with the damper springs on the transmission side. On some 6-cylinder engines, the clutch disc is installed in the reverse manner with the damper springs to the flywheel side.
3. Turn the clutch assembly until the mark on the cover lines up with the mark on the flywheel, then install the bolts. Tighten down evenly and gradually to avoid distortion.
4. Remove the alignment tool.
5. Lubricate the ball socket and fork fingers at the release bearing end with high melting point grease. Lubricate the recess on the inside of the throwout bearing and throwout fork groove with a light coat of graphite grease.
6. Install the clutch fork and dust boot into the housing. Install the flywheel housing. Install the transmission.
7. Connect the fork push rod and spring. Lubricate the spring and pushrod ends.
8. Adjust the shift linkage and clutch pedal free-play.

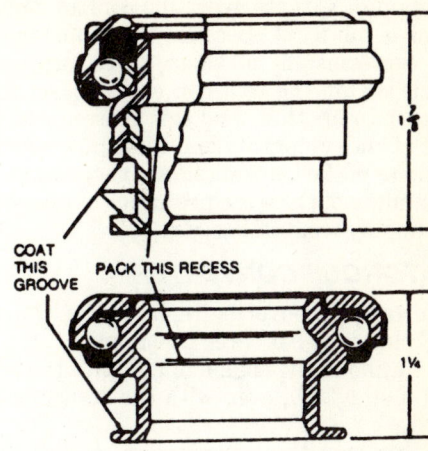

Clutch release bearing lubrication, flat finger type (top), bent finger type (bottom)

AUTOMATIC TRANSMISSION

Understanding Automatic Transmissions

The automatic transmission allows engine torque and power to be transmitted to the rear wheels within a narrow range of engine operating speeds. The transmission will allow the engine to turn fast enough to produce plenty of power and torque at very low speeds, while keeping it at a sensible rpm at high vehicle speeds. The transmission performs this job entirely without driver assistance. The transmission uses a light fluid as the medium for the transmission of power. This fluid also works in the operation of various hydraulic control circuits and as a lubricant. Because the transmission fluid performs all of these three functions, trouble within the unit can easily travel from one part to another. For this reason, and because of the complexity and unusual operating principles of the transmission, a very sound understanding of the basic principles of operation will simplify troubleshooting.

THE TORQUE CONVERTER

The torque converter replaces the conventional clutch. It has three functions:

1. It allows the engine to idle with the vehicle at a standstill, even with the transmission in gear.
2. It allows the transmission to shift from range to range smoothly, without requiring that the driver close the throttle during the shift.
3. It multiplies engine torque to an increasing extent as vehicle speed drops and throttle opening is increased. This has the effect of making the transmission more responsive and reduces the amount of shifting required.

The torque converter is a metal case which is shaped like a sphere that has been flattened on opposite sides. It is bolted to the rear end of the engine's crankshaft. Generally, the entire metal case rotates at engine speed and serves as the engine's flywheel.

The case contains three sets of blades. One set is attached directly to the case. This set forms the torus or pump. Another set is directly connected to the output shaft, and forms the turbine. The third set is mounted on a hub which, in turn, is mounted on a stationary shaft through a one-way clutch. This third set is known as the stator.

A pump, which is driven by the converter hub at engine speed, keeps the torque converter full of transmission fluid at all times. Fluid flows continuously through the unit to provide cooling.

Under low speed acceleration, the torque converter functions as follows:

The torus is turning faster than the turbine. It picks up fluid at the center of the converter and, through centrifugal force, slings it outward. Since the outer edge of the converter moves faster than the portions at the center, the fluid picks up speed.

The fluid then enters the outer edge of the turbine blades. It then travels back toward the center of the converter case along the turbine blades. In impinging upon the turbine blades, the fluid loses the energy picked up in the torus.

If the fluid were now to immediately be returned directly into the torus, both halves of the converter would have to turn at approximately the same speed at all times, and torque input and output would both be the same.

In flowing through the torus and turbine, the fluid picks up two types of flow, or flow in two separate directions. It flows through the turbine blades, and it spins with the engine. The stator, whose blades are stationary when the vehicle is being accelerated at low speeds, converts one type of flow into another. Instead of allowing the fluid to flow straight back into the torus, the stator's curved blades turn the fluid almost 90° toward the direction of rotation of the engine. Thus the fluid does not flow as fast toward the torus, but is already spinning when the torus picks it up. This has the effect of allowing the torus to turn much faster than the turbine. This difference in speed may be compared to the difference in speed between the smaller and larger gears in any gear train. The result is that engine power output is higher, and engine torque is multiplied.

As the speed of the turbine increases, the fluid spins faster and faster in the direction of engine rotation. As a result, the ability of the stator to redirect the fluid flow is reduced. Under cruising conditions, the stator is eventually forced to rotate on its one-way clutch in the direction of engine rotation. Under these conditions, the torque converter begins to behave almost like a solid shaft, with the torus and turbine speeds being almost equal.

THE PLANETARY GEARBOX

The ability of the torque converter to multiply engine torque is limited. Also, the unit tends to be more efficient when the turbine is rotating at relatively high speeds. Therefore, a planetary gearbox is used to carry the power output of the turbine to the driveshaft.

Planetary gears function very similarly to conventional transmission gears. However, their construction is different in that three elements make up one gear system, and, in that all three elements are different from one another. The three elements are: an outer gear that is shaped like a hoop, with teeth cut into the inner surface; a sun gear, mounted on a shaft and located at the very center of the outer gear; and a set of three planet gears, held by pins in a ring-like planet carrier, meshing with both the sun gear and the outer gear. Either the outer gear or the sun gear may be held stationary, providing more than one possible torque multiplication factor for each set of gears. Also, if all three gears are forced to rotate at the same speed, the gearset forms, in effect, a solid shaft.

Most modern automatics use the planetary gears to provide either a single reduction ratio of about 1.8:1, or two reduction gears: a low of about 2.5:1, and an intermediate of about 1.5:1. Bands and clutches are used to hold various portions of the gearsets to the transmission case or to the shaft on which they are mounted. Shifting is accomplished, then, by changing the portion of each planetary gearset which is held to the transmission case or to the shaft.

THE SERVOS AND ACCUMULATORS

The servos are hydraulic pistons and cylinders. They resemble the hydraulic actuators used on many familiar machines, such as bulldozers. Hydraulic fluid enters the cylinder, under pressure, and forces the piston to move to engage the band or clutches.

The accumulators are used to cushion the engagement of the servos. The transmission fluid must pass through the accumulator on the way to the servo. The accumulator housing contains a thin piston which is sprung away from the discharge passage of the accumulator. When fluid passes through the accumulator on the way to the servo, it must move the piston against spring pressure, and this action smooths out the action of the servo.

THE HYDRAULIC CONTROL SYSTEM

The hydraulic pressure used to operate the servos comes from the main transmission oil pump. This fluid is channeled to the various servos through the shift valves. There is generally a manual shift valve which is operated by the transmission selector lever and an automatic shift valve for each automatic upshift the transmission provides: i.e., 2-speed automatics have a low/high shift valve, while 3-speeds have a 1-2 valve, and a 2-3 valve.

There are two pressures which effect the operation of these valves. One is the governor pressure which is affected by vehicle speed. The other is the modulator pressure which is affected by intake manifold vacuum or throttle position. Governor pressure rises with an increase in vehicle speed, and modulator pressure rises as the throttle is opened wider. By responding to these two pressures, the shift valves cause the upshift points to be delayed with increased throttle opening to make the best use of the engine's power output.

Most transmissions also make use of an auxiliary circuit for downshifting. This circuit may be actuated by the throttle linkage or the vacuum line which actuates the modulator, or by a cable or solenoid. It applies pressure to a special downshift surface on the shift valve or valves.

The transmission modulator also governs the line pressure, used to actuate the servos. In this way, the clutches and bands will be actuated with a force matching the torque output of the engine.

Identification

The different types of pan gaskets used on the automatic transmissions are pictured below for ready identification.

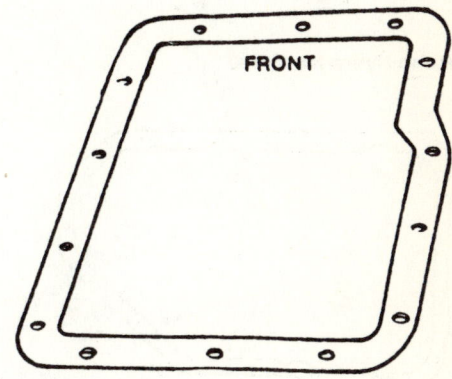

GM Powerglide

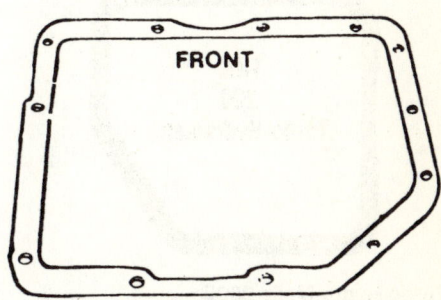

GM Turbo Hydra-Matic 250, 350, 375B

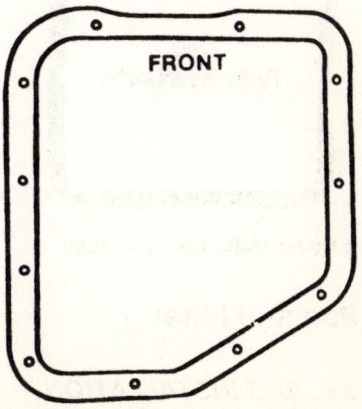

GM Turbo Hydra-Matic 200

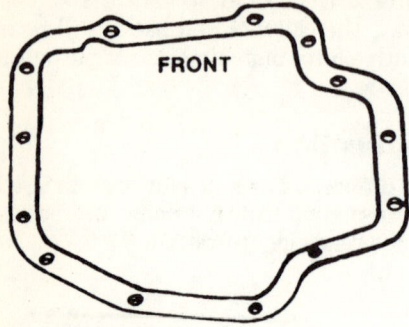

GM Turbo Hydra-Matic 400

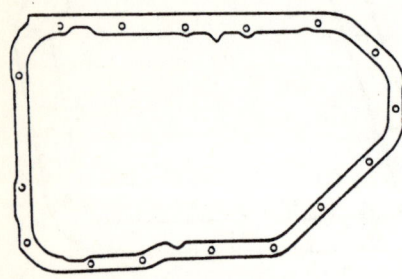

GM Turbo Hydra-Matic 200-4R

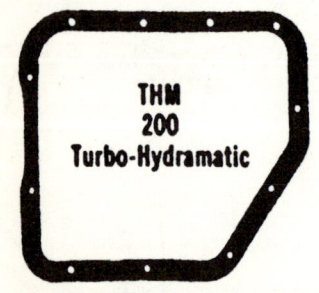

GM Turbo Hydra-Matic 200C

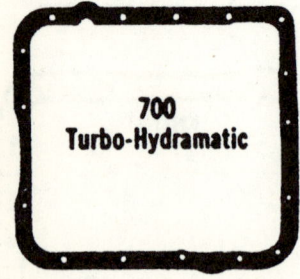

GM Turbo Hydra-Matic 700/THM-4L60

Fluid Pan and Filter

REMOVAL AND INSTALLATION

The fluid should be changed with the transmission warm. A 20 minute drive at highway

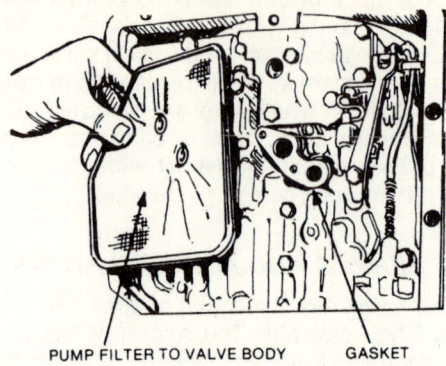

Removing the Turbo Hydra-Matic 350 transmission filter

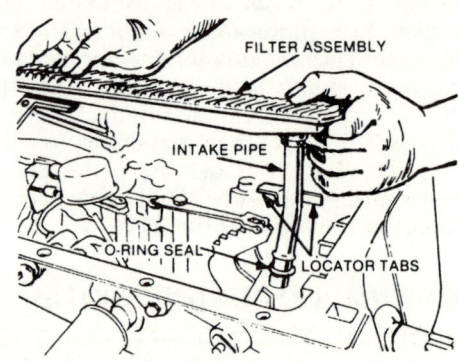

Removing the filter, intake pipe and O-ring on the Turbo Hydra-Matic 400

speeds should accomplish this.

1. Raise and support the vehicle on safety stands.
2. Depending on the type of transmission the crossmember may have to be removed from the vehicle, in order to remove the fluid pan bolts. If so, be sure to properly support the transmission before removing the crossmember.
3. Place a large pan under the transmission pan. Remove all the front and side pan bolts. Loosen the rear bolts about four turns.
4. Pry the pan loose and let it drain.
5. Remove the pan and gasket. Clean the pan thoroughly with solvent and air dry it. Be very careful not to get any lint from rags in the pan.
6. Remove the strainer to valve body screws, the strainer, and the gasket. Most transmissions will have a throw-away filter instead of a strainer. On the 400 transmission, remove the filter retaining bolt, filter, and intake pipe O-ring.
7. If there is a strainer, clean it in solvent and air dry.

DRIVE TRAIN 351

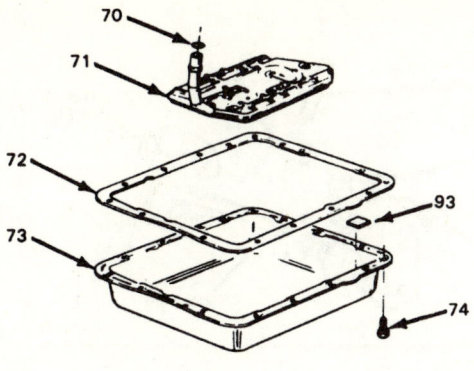

70 FILTER SEAL
71 TRANSMISSION OIL FILTER ASSEMBLY
72 TRANSMISSION OIL PAN GASKET
73 TRANSMISSION OIL PAN
74 OIL PAN SCREW
93 CHIP COLLECTOR MAGNET

Removing the filter, O-ring and gasket on the Turbo Hydra-Matic 700R/4L60

8. Install the new filter or cleaned strainer with a new gasket. Tighten the screws to 12 ft. lbs. On the 400, install a new intake pipe O-ring and a new filter, tightening the retaining bolt to 10 ft. lbs.
9. Install the pan with a new gasket. Tighten the bolts evenly to 12 ft. lbs. (8 for Powerglide).
10. Lower the vehicle and add the proper amount of DEXRON® or DEXRON®II automatic transmission fluid through the dipstick tube.

NOTE: *Refer to Chapter 1 for the recommended lubricant.*

11. Start the engine in Park and let it idle. Do not race the engine. Shift into each lever position, shift back into Park, and check the fluid level on the dipstick. The level should be $1/4''$ below ADD. Be very careful not to overfill. Recheck the level after the vehicle has been driven long enough to thoroughly warm up the transmission. Add fluid as necessary. The level should then be at FULL.

Adjustments

BAND ADJUSTMENT

There are no band adjustments possible or required for the Turbo Hydra-Matic 200, 200-4R, 700-4R, 350 or 400.

Low Band

POWERGLIDE

The low band must be adjusted at the first required fluid change or whenever there is slippage.
1. Position the shift lever in Neutral.
2. Remove the protective cap from the adjusting screw on the left side of the transmission.
3. Loosen the locknut $1/4$ turn and hold it with a wrench during the entire adjusting procedure.
4. Tighten the adjusting nut to 70 inch lbs., using a $7/32''$ allen wrench.
5. Back off the adjusting nut exactly three turns for a band used less than 6,000 miles. Back off exactly four turns for a band used 6,000 miles or more.
6. Torque the locknut to 15 ft. lbs. and replace the cap.

Intermediate Band

TURBO HYDRA-MATIC 250

The intermediate band must be adjusted with every required fluid change or whenever there is slippage.
1. Position the shift lever in Neutral.
2. Loosen the locknut on the right side of the transmission and tighten the adjusting screw to 30 inch lbs.
3. Back the screw out three turns and then tighten the locknut to 15 ft. lbs.

GASOLINE ENGINE SHIFT LINKAGE ADJUSTMENT

Powerglide Column Shift

1. The shift tube and lever assembly must be free in the mast jacket.
2. Lift the selector lever toward the steering wheel and allow the selector lever to be positioned in Drive by the transmission detent.
3. Release the selector lever. The lever should be prevented from engaging low, unless the lever is lifted.
4. Lift the selector lever toward the steering wheel and allow the lever to be positioned in Neutral by the transmission detent.
5. Release the selector lever. The selector lever should not be kept from engaging Reverse unless the lever is lifted. If the linkage is adjusted correctly, the selector lever should be prevented from moving beyond both the neutral detent and the Drive detent unless the lever is lifted to pass over the mechanical stop in the steering column.

If adjustment is necessary, perform the following steps:

352 DRIVE TRAIN

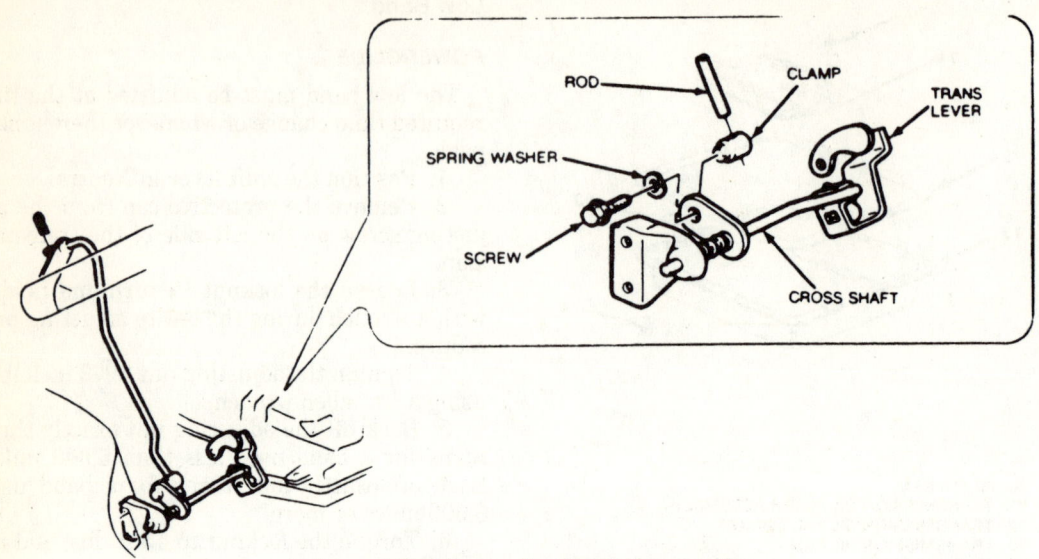

Column shift linkage adjustment, Powerglide

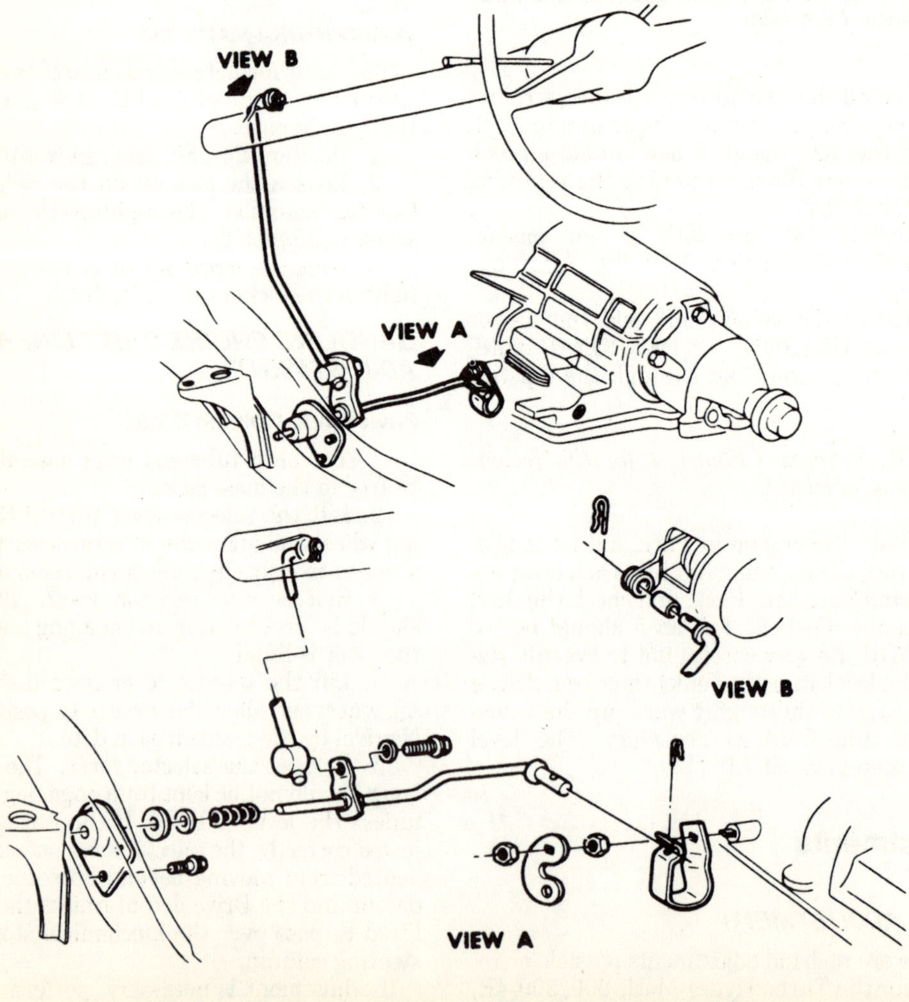

Column shift linkage adjustment, 1970 and later Powerglide

pointer and neutral start switch.

5. Check that the key cannot be removed fro the RUN position if the transmission selector is in Reverse and that the key can be removed with the selector in Park. Make sure the lever will not move from Park position with key out of ignition.

1975-81 Vehicles with Cable Linkage

1. Loosen the swivel at the lower end of the rod that comes from the steering column.
2. Loosen the pin at the transmission end of the cable.
3. Set the floorshift lever in the Drive detent.
4. Set the transmission lever in the Drive detent by moving it counterclockwise to the L1 detent, then clockwise three detent positions.
5. Tighten the nut on the pin at the transmission end of the cable.
6. Push the floorshift lever in Park and the ignition switch in LOCK.
7. Pull down lightly on the rod from the column and tighten its clamp nut.

1982 and Later Vehicles with Cable Linkage

1. Loosen the clamping screw at the shifting rod-to-equalizer lever.
2. Position the steering column shifting lever into the Neutral position.
3. Set the transmission lever into the Neutral position.
4. Finger tighten the equalizer lever clamping screw to the sifting rod.

NOTE: *While performing this operation, DO NOT exert force in any direction.*

5. Tighten the equalizer clamping screw.

THROTTLE VALVE LINKAGE ADJUSTMENT

1964-73 6-Cylinder Engine

1. Depress the accelerator pedal.
2. The bellcrank must be at the wide open throttle position.
3. The dash lever at the firewall must be $1/6$-$1/16''$ off its lever stop.
4. The transmission lever must be against the transmission internal stop.
5. Adjust the linkage to simultaneously obtain the conditions in Step 1-4.

1968-73 Powerglide with V8 Engine

1. Remove the air cleaner.
2. Disconnect the accelerator linkage at the carburetor.
3. Disconnect both return springs.

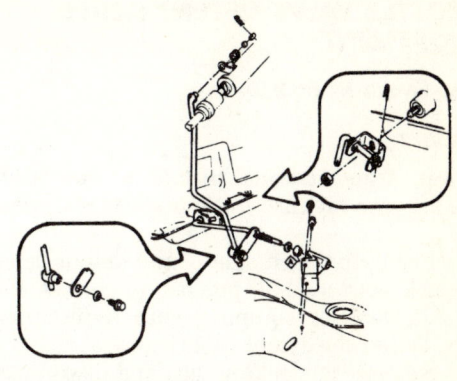

Column shift linkage, 1980 and later

4. Pull the throttle valve upper rod forward until the transmission is through the detent.
5. Open the carburetor to the wide open throttle position. Adjust the swivel on the end of the upper throttle valve rod so that the carburetor reaches wide open throttle position at the same time that the ball stud contact the end of the slot in the upper throttle valve rod. A tolerance of $1/32''$ is allowable.

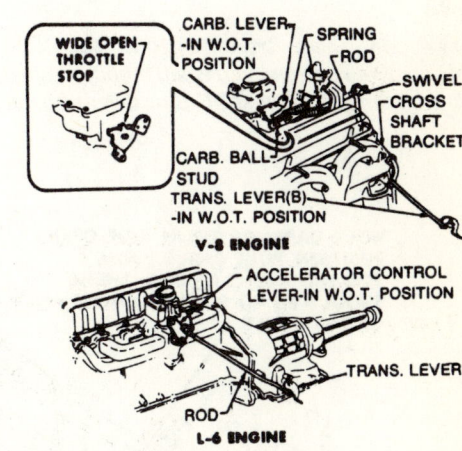

Adjusting the throttle valve linkage

Throttle valve adjustment, THM 200

DRIVE TRAIN

THROTTLE VALVE/DETENT CABLE ADJUSTMENT

Turbo Hydra-Matic 250, 350

1968-1977

These transmissions utilize a downshift cable between the carburetor and the transmission.

1. Pry up on each side of the detent cable snap-lock with a small pry bar to release the lock. On vehicles equipped with a retaining screw, loosen the detent cable screw.
2. Squeeze the locking tabs and disconnect the snap-lock assembly from the throttle bracket.
3. Place the carburetor lever in the wide open throttle position. Make sure that the lever is against the wide open stop. On vehicles equipped with Quadrajet carburetors, disengage the secondary lock-out before placing the lever in the wide open position.

NOTE: *The detent cable must be pulled through the detent position.*

4. With the carburetor lever in the wide open position, push the snap-lock on the cable or else tighten the retaining screw.

NOTE: *Do not lubricate the detent cable.*

1978-92

NOTE: *On these transmissions the T.V. cable controls line pressure, shift points, shift feel, part throttle downshifts and detent downshifts. The T.V. cable operates the throttle valve lever and bracket assembly in the control valve.*

1. Stop the engine.
2. Locate the TV cable adjuster near the car-

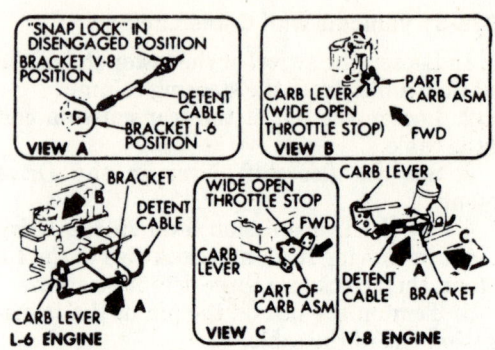

Detent cable adjustment—Turbo Hydra-matic 350 (250 similar)

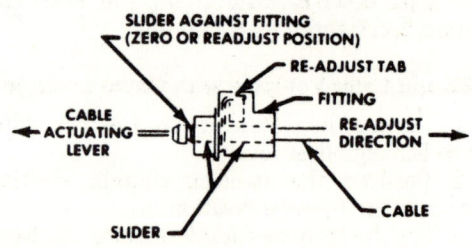

View of the TV cable adjuster, 1973

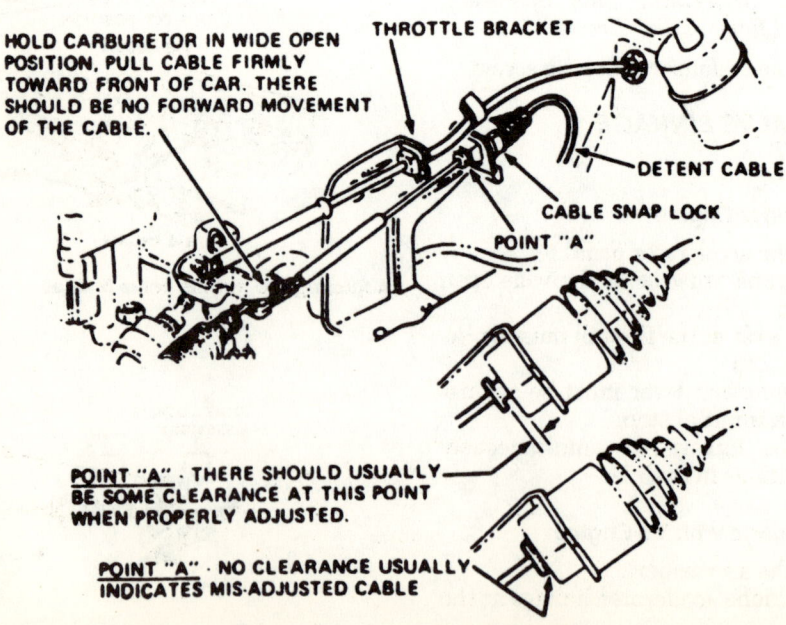

THM 200 downshift cable adjustment

buretor (or throttle body).

3. Depress and hold down the metal tab of the TV cable adjuster.

4. Move the slider until it stops against the fitting.

5. Release the adjuster tab.

6. Turn the carburetor (or throttle body) lever to the full throttle stop position and release it.

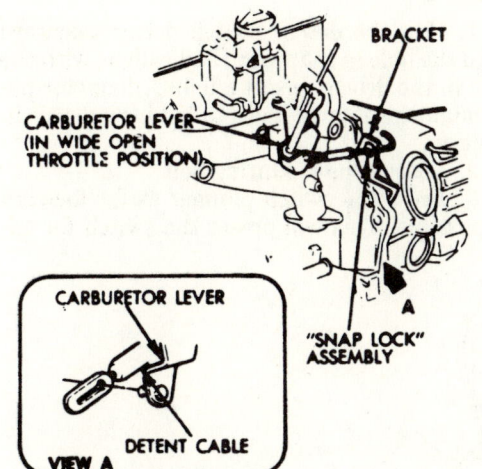

THM 200, 250, 350 detent cable adjustment

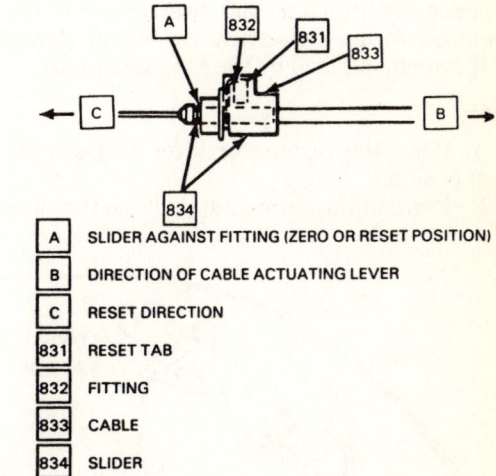

A	SLIDER AGAINST FITTING (ZERO OR RESET POSITION)
B	DIRECTION OF CABLE ACTUATING LEVER
C	RESET DIRECTION
831	RESET TAB
832	FITTING
833	CABLE
834	SLIDER

TV cable adjustment setting, THM 4L60

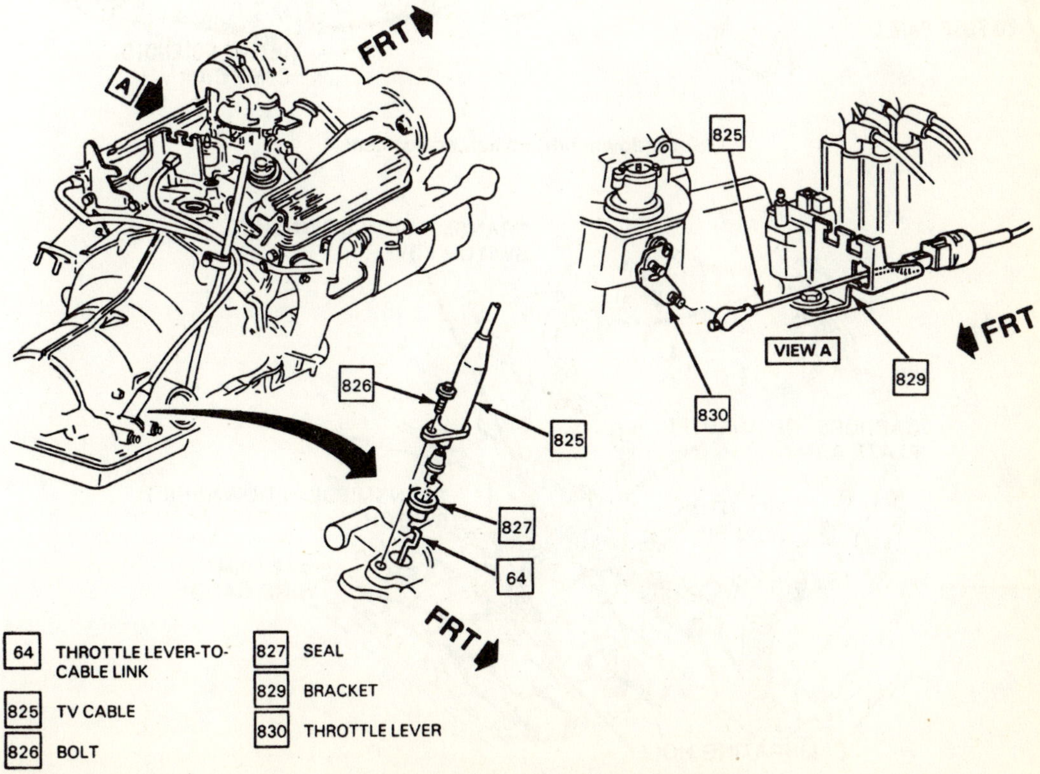

64	THROTTLE LEVER-TO-CABLE LINK
825	TV CABLE
826	BOLT
827	SEAL
829	BRACKET
830	THROTTLE LEVER

TV cable adjustment, 1992 THM 4L60

358 DRIVE TRAIN

NOTE: *By turning the carburetor (or throttle body) lever to the full throttle stop position, the TV cable will automatically adjust itself.*

DETENT SWITCH ADJUSTMENT

Turbo Hydra-matic 400 transmissions are equipped with an electrical detent, or downshift switch operated by the throttle linkage.

1968

1. Place the carburetor lever in the wide open position.
2. Position the automatic choke so that it is off.
3. Fully depress the switch plunger.
4. Adjust the switch mounting to obtain a distance of 0.05″ between the switch plunger and the throttle lever paddle.

1969-79

1. Pull the detent switch driver rearward until the hole in the switch body aligns with the hole in the driver. Insert a 0.092″ diameter pin through the aligned holes to hold the driver in position.
2. Loosen the mounting bolt.
3. Press the switch plunger as far forward as possible. This will preset the switch for ad-

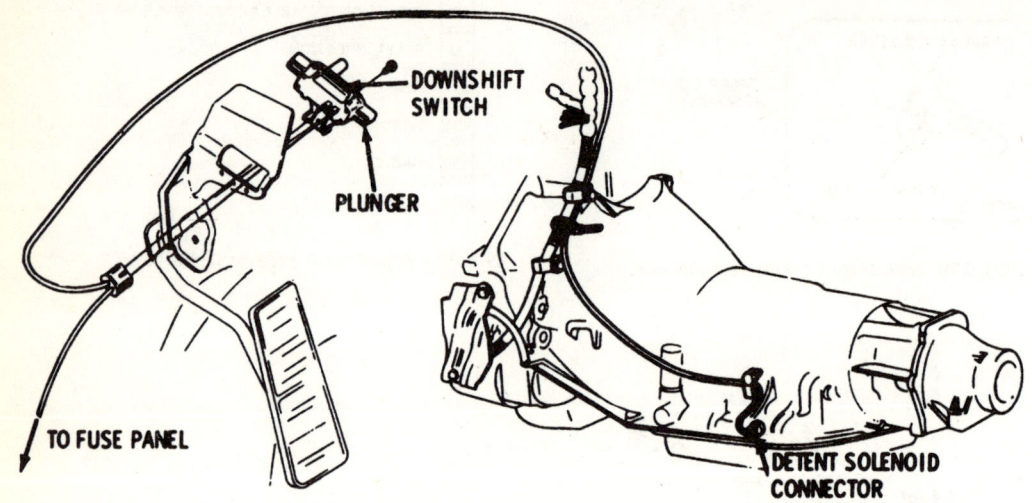

Detent (downshift) switch adjustment

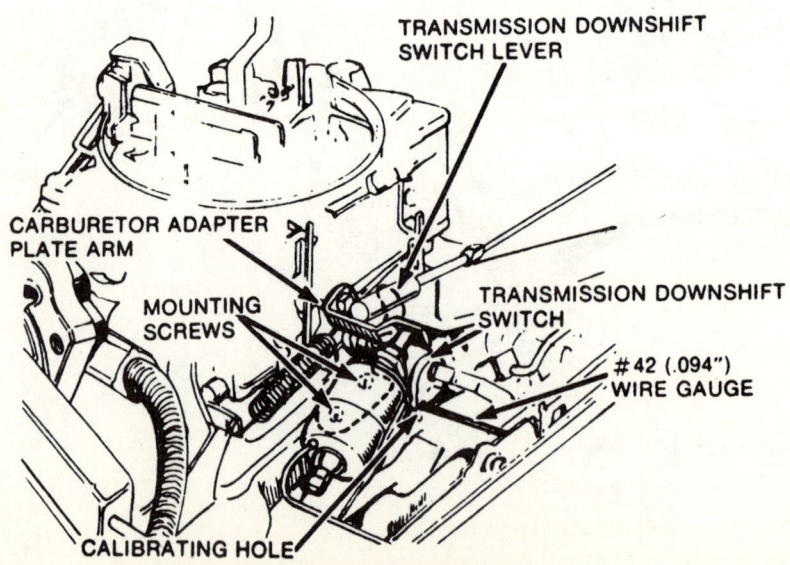

THM 400 downshift switch adjustment

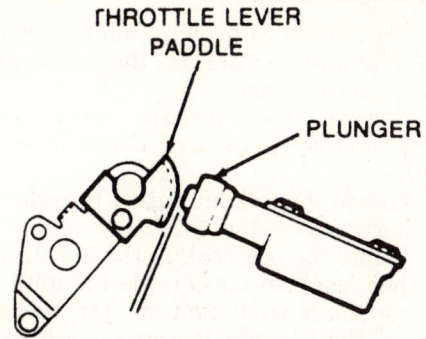

Adjusting the detent switch, 1968–72 THM 400

justment, which will occur on the first application of wide open throttle.

4. Tighten the mounting bolt and remove the pin.

NEUTRAL SAFETY/BACK-UP SWITCH REPLACEMENT

The neutral safety/back-up switch on automatic transmissions is combined in a single unit. The switch prevents the engine from being started in any transmission position except Neutral or Park. The switch is located on the upper side of the steering column under the instrument panel on column shift vehicles, and inside the shift console on floor shift models.

1968-81

1. Remove the console for access on floor shift models.
2. Disconnect the electrical connectors.
3. Remove the neutral switch.
4. Place 1968-70 column shift lever models in Drive, and 1971-81 models in neutral. Locate the lever tang against the transmission selector plate on column shift models. Place 1968-72 floor shift models in Drive, and mid-1972-88 models in Park.
5. Align the slot in the contact support with the hole in the switch. Insert a $1/32''$ pin in place. The switch is now aligned in Drive position.

NOTE: *1973-81 neutral safety switches have a shear-pin installed to aid in proper switch alignment so that insertion of a pin is unnecessary. Moving the shift lever from Neutral shears the pin.*

6. Place the contact support drive slot over the drive tang. Install the switch mounting screws.
7. Remove the aligning pin. Connect the elec-

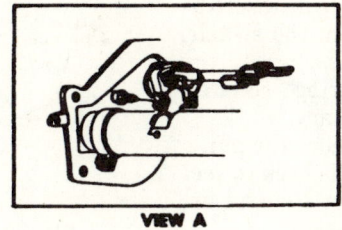
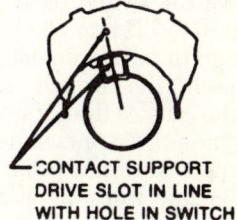
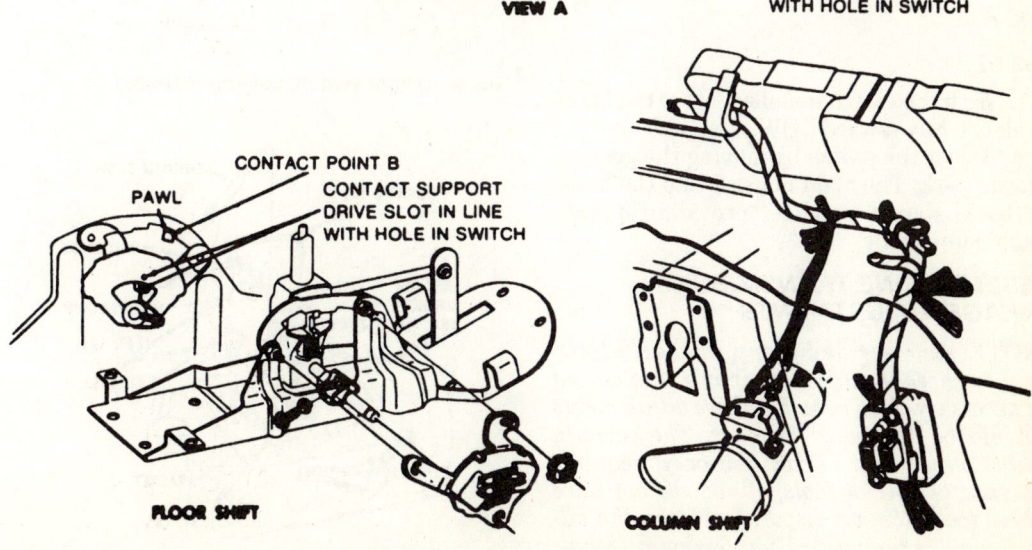

Installation of the neutral safety switch

DRIVE TRAIN

trical wiring, and replace the console.

8. Set the parking brake and hold your foot on the service brake pedal. Check to see that the engine will start only in Park or Neutral.

1982-92

1. Disconnect the wiring harness from the switch and remove the retaining screws and the switch.
2. To install, place the gear selector in NEUTRAL.
3. Align the actuator on the switch with a hole in the switch tube.
4. Position the rearward portion of the switch (connector side) to fit into the cutout in the lower jacket.
5. Push down on the front of the switch. The two tangs on the housing back will snap into place in rectangular holes in the jacket.
6. Adjust the switch by moving the gear selector to park. The main housing and the housing back should ratchet, providing proper switch adjustment.

NEUTRAL SAFETY SWITCH ADJUSTMENT

1968-81

1. Place shift lever in Neutral.
2. Move the switch until you can insert a gauge pin, 0.0972" into the hole in the switch and through to the alignment hole.
3. Loosen the switch securing screws. Remove the console first, if necessary.
4. Tighten the screws and remove the pin.
5. Step on the brake pedal and check to see that the engine will only start in Neutral or Park.

1982-92

1. With the switch installed, move the housing all the way toward LOW.
2. Adjust the switch by moving the gear selector to park. The main housing and the housing back should ratchet, providing proper switch adjustment.

DIESEL ENGINE TRANSMISSION LINKAGE ADJUSTMENTS

NOTE: *Before making any linkage adjustments, check the injection timing, and adjust if necessary. Also note that these adjustments should be performed together. The vacuum valve adjustment (THM 350 only) requires the use of special tools. If you do not have these tools at your disposal, refer to the adjustment to a qualified, professional technician.*

Throttle Rod Adjustment

1. If equipped with cruise control, remove the clip from the control rod, then remove the rod from the bellcrank.
2. Remove the throttle valve cable (THM200) or detent cable (THM350) from the bellcrank.
3. Loosen the locknut on the throttle rod, then shorten the rod several turns.
4. Rotate the bellcrank to the full throttle stop, then lengthen the throttle rod until the injection pump lever contacts the injection pump full throttle stop. Release the bell crank.
5. Tighten the throttle rod locknut.
6. Connect the throttle valve or detent cable and cruise control rod to the bellcrank. Adjust if necessary.

Throttle Valve (TV) Or Detent Cable Adjustment

Please refer to the illustration.

Transmission Vacuum Valve Adjustment

1. Remove the air cleaner assembly.
2. Remove the air intake crossover from the intake manifold. Cover the intake manifold pas-

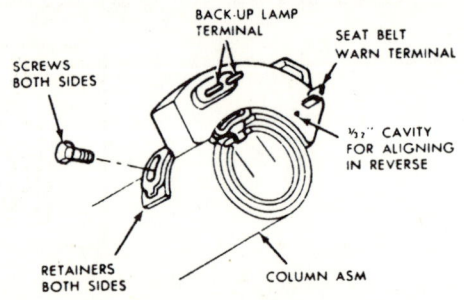

Back-up light switch, column mounted

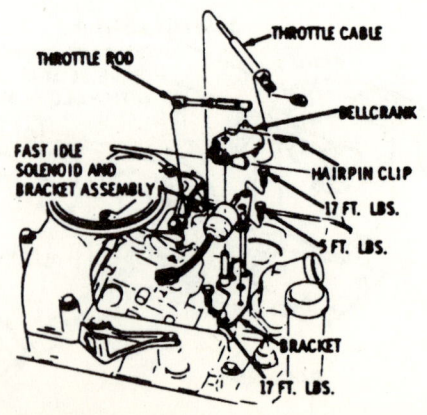

Diesel throttle linkage

DRIVE TRAIN 361

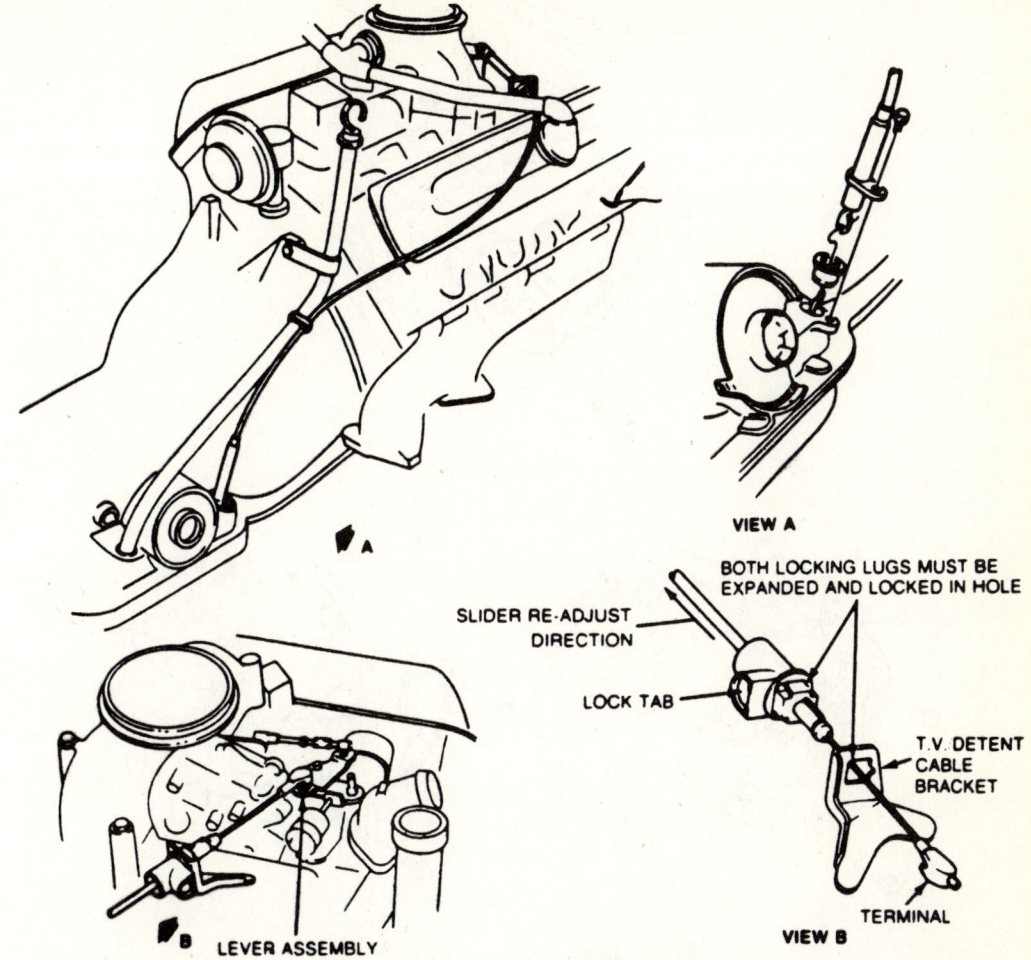

CABLE ADJUSTMENT PROCEDURE

1. Remove pump rod from lever assembly.
2. After installation to transmission, install cable fitting into cable bracket. **CAUTION** Slider must not be adjusted before or during assembly to bracket
3. Install cable terminal to lever assembly
4. Rotate the lever assembly to its full throttle stop position to automatically adjust slider on cable to correct setting
5. Release lever assembly & reconnect the pump rod to lever assembly

CABLE RE-ADJUSTMENT PROCEDURE

In case re-adjustment is necessary because of inadvertent adjustment before or during assembly, perform the following.

1. Remove pump rod from lever assembly, depress and hold metal lock tab.
2. Move slider through fitting in direction away from lever assembly until slider stops against fitting
3. Release metal lock tab
4. Repeat steps 4 & 5 of adjustment procedure

Diesel transmission, vacuum detent cable adjustment

sages to prevent foreign material from entering the engine.

3. Disconnect the throttle rod from the injection pump throttle lever.

4. Loosen the transmission vacuum valve-to-injection pump bolts.

5. Mark and disconnect the vacuum lines from the vacuum valve.

6. Attach a carburetor angle gauge adapter (Kent-Moore tool J-26701-15 or its equivalent) to the injection pump throttle lever. Attach an angle gauge (J-26701 or its equivalent) to the gauge adapter.

7. Turn the throttle lever to the wide open throttle position. Set the angle gauge to zero degrees.

8. Center the bubble in the gauge level.

9. Set the angle gauge to one of the follow-

362 DRIVE TRAIN

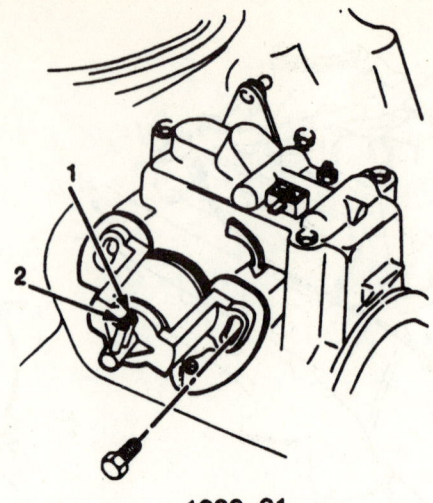

1980–81

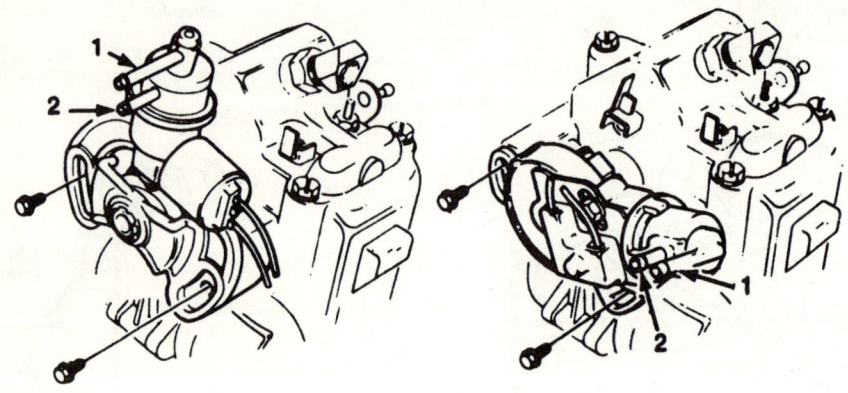

1982 AND LATER

Diesel transmission vacuum valve adjustment

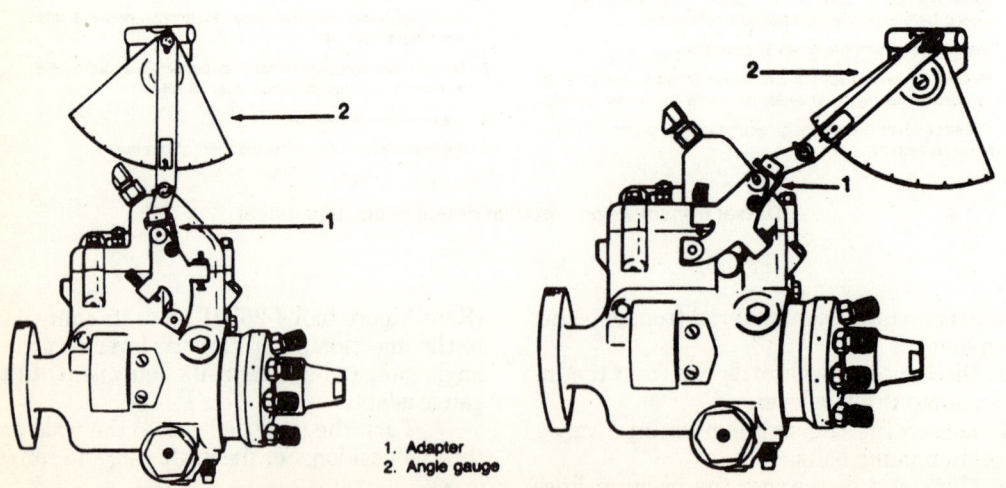

1. Adapter
2. Angle gauge

Angle gauge installation, with adaptor for diesel vacuum valve adjustment. The gauge is positioned differently, depending on the type of throttle lever used

DRIVE TRAIN

ing settings, according to the year and type of engine:

10. Attach a vacuum gauge to port 2 and a vacuum source (e.g. hand-held vacuum pump) to port 1 of the vacuum valve (as illustrated).
11. Apply 18-22 in.Hg of vacuum to the valve. Slowly rotate the valve until the vacuum reading drops to one of the following values:
12. Tighten the vacuum valve retaining bolts.
13. Reconnect the original vacuum lines to the vacuum valve.
14. Remove the angle gauge and adapter.
15. Connect the throttle rod to the throttle lever.
16. Install the air intake crossover, using new gaskets.
17. Install the air cleaner assembly.

Transmission

REMOVAL AND INSTALLATION

1. Open the hood and place protectors on the fenders. Remove the air cleaner assembly and cover the air intake.
2. Disconnect the negative battery cable. Disconnect the detent cable at its upper end.
3. Remove the transmission oil dipstick, and the bolt holding the dipstick tube if it is accessible.
4. Raise and support the vehicle safely.

NOTE: *If a floor pan reinforcement is used, remove it if it interferes with driveshaft removal and installation.*

5. Disconnect the speedometer cable at the transmission.
6. Disconnect the shift linkage at the transmission. Remove the driveshaft from the transmission.

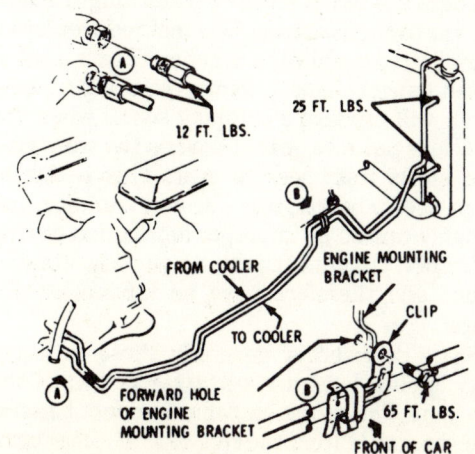

Transmission oil cooler lines

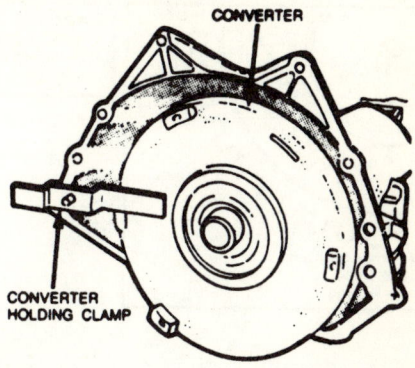

Torque converter holding tool, a C-clamp can also be used

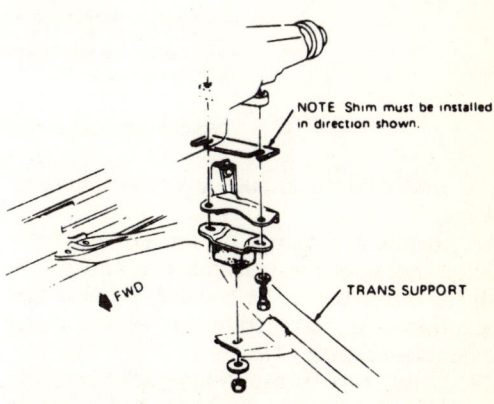

Some transmission rear mounts are shimmed

7. Disconnect all electrical leads at the transmission and any clips that hold these leads to the transmission case.
8. Remove the flywheel cover and matchmark the flywheel and torque converter for later assembly.
9. Remove the torque converter-to-flywheel bolts and/or nuts.
10. On gasoline engine vehicles, disconnect the catalytic converter support bracket.
11. Remove transmission support-to-transmission mount bolt and transmission support-to-frame bolts, and any insulators (if used).
12. Position the transmission jack under the transmission and raise it slightly.
13. Slide the transmission support rearward.
14. Loosen the transmission enough to gain access to the oil cooler lines and detent cable attachments.
15. Disconnect the oil cooler lines and detent cable. Plug all openings.
16. Support the engine and remove the engine-to-transmission bolts.
17. Disconnect the transmission assembly,

DRIVE TRAIN

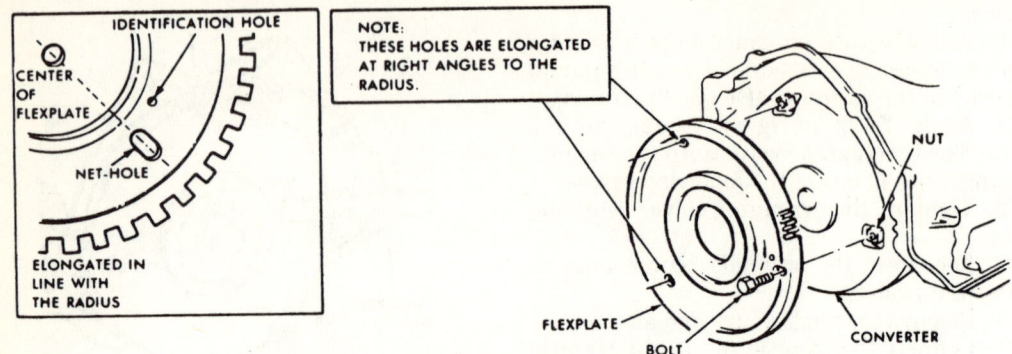

1. MOVE CONVERTER FORWARD TO CONTACT ATTACHING SURFACE ON FLEX-PLATE PRIOR TO TIGHTENING BOLTS.
2. ALIGN SLOT IN FLEXPLATE THAT HAS AN IDENTIFICATION HOLE NEAR IT, WITH AN ATTACHMENT HOLE IN CONVERTER. INSTALL BOLT AND NUT AND TIGHTEN TO SPECIFIED TORQUE. TIGHTEN ALL REMAINING BOLTS TO SPECIFIED TORQUE AS THEY ARE INSTALLED.

Typical transmission attachment, net hole design

being careful not to damage any cables, lines or linkage.

18. Install a C-clamp or torque converter holding tool onto the transmission housing to hold the converter to the housing. Remove the transmission assembly from the vehicle, using a transmission jack.

19. Using the transmission jack carefully move the transmission into position. Replace or connect all components removed above. When installing the flex pate-to-converter bolts, make sure that the weld nuts on the converter are flush with the flex plate and that the converter rotates freely by hand. Hand start the three bolts and tighten them finger tight, then torque them evenly.

20. Install a new oil seal on the oil filler tube before installing the tube. Torque the flywheel-to-converter bolts to 35 ft. lbs. Torque the transmission-to-engine bolts to 35 to 40 ft. lbs. Adjust the shift linkage and detent cable following the procedures in this chapter, and check the transmission fluid if the transmission was not drained previously.

DRIVELINE

Driveshaft and U-Joints

Chevrolet driveshafts are of the conventional, open type. Located at either end of the driveshaft is a U-joint or universal joint, which allows the driveshaft to move up and down to match the motion of the rear axle. The front U-joint connect the driveshaft to a slip jointed yoke. This yoke is internally splined, and allows the driveshaft to move in and out on the transmission splines. The rear U-joint is clamped or bolted to a companion flange fastened to the rear axle drive pinion. The rear U-joint is secured in the yoke in one of two ways. Dana and Cleveland design driveshafts use a conventional type snapring to hold each bearing cup in the yoke. The snapring fits into a groove located in each yoke end, just on top of the bearing cup. A Saginaw design driveshaft secures the U-joints differently. Nylon material is injected through a small hole in the yoke during manufacture, and flows along a circular groove between the U-joint and the yoke creating a non-metallic snapring.

There are two methods of attaching the rear U-joint to the rear axle. One method employs a pair of straps, while the other method is a set of bolted flanges. Band U-joints, requiring replacement, will produce a clunking sound when the vehicle is put into gear and when the transmission shifts from gear to gear. This is due to worn needle bearings or a scored trunnion end possibly caused by improper lubrication during assembly. U-joints require no periodic maintenance and therefore have no lubrication fittings.

Some driveshafts, generally those in heavy duty applications, use a damper as part of the slip joint. This vibration damper cannot be serviced separately from the slip joint. If either component goes bad, the two must be replaced as a unit.

DRIVE TRAIN

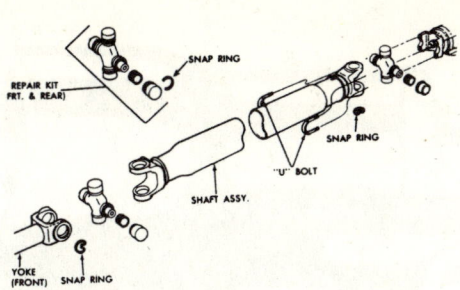

Exploded view of the driveshaft assembly

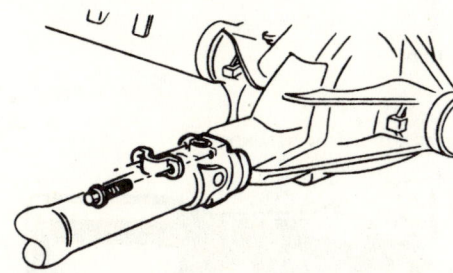

Strap type driveshaft mounting

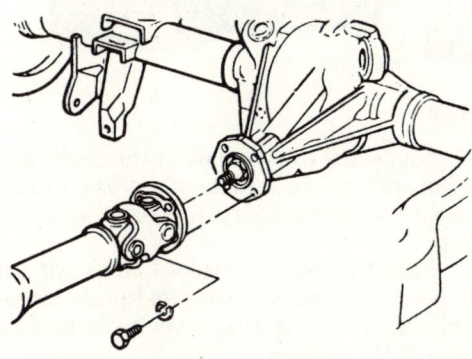

Flange type driveshaft mounting

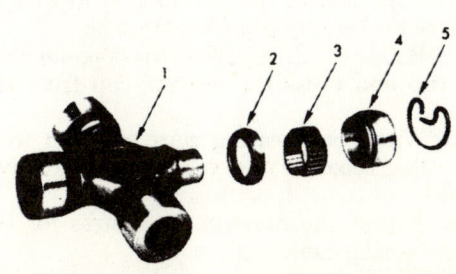

Cleveland universal joint

REMOVAL AND INSTALLATION

1. Raise the vehicle and support it safely.
2. Mark the relationship of the driveshaft to the differential flange so that they can be reassembled in the same position.
3. Disconnect the rear U-joint by removing the U-bolts or retaining straps.
4. To prevent the loss of the needle bearings, tape the bearing caps in place. If you are replacing the U-joint, this is not necessary.
5. Remove the driveshaft from the transmission by sliding it rearward. There will be some oil leakage from the rear of the transmission. It can be contained by placing a small plastic bag over the rear of the transmission and holding it in place with a rubber band.

To Install:

6. Insert the front yoke into the transmission so that the driveshaft splines mesh with the transmission splines.
7. Using the reference marks made earlier, align the driveshaft with the differential flange and secure it with the U-bolts or retaining straps.
8. Check fluid on automatic transmissions, gear lubricant on manual transmissions and add as necessary.

U-JOINT OVERHAUL

1968-81 Cleveland Type

1. Remove the driveshaft from the vehicle.

WARNING: *NEVER clamp the driveshaft tube in a vise, for this may dent the tube. Support the driveshaft horizontally and clamp on the yokes of the universal joints.*

2. Remove the lock rings from the ends of the trunion yoke.
3. Support the driveshaft in the horizontal position with the base plate of a press, so that the lower ear of the yoke is supported on a piece of $1\frac{1}{4}''$ I.D. pipe.
4. Place a socket on the upper bearing cup and press the lower bearing cup out of the yoke ear.

NOTE: *Since the bearing cup cannot be fully pressed from the yoke ear, grasp the cup in the jaws of a vice and work it form the yoke.*

5. Rotate the driveshaft to the opposite bearing cup and press the bearing cup from the yoke, using the same removal procedure.
6. With both bearing cups removed from the yoke, separate the yoke from the driveshaft.
7. Repeat the removal procedures for the other bearing cups.
8. Clean and inspect all of the parts.

NOTE: *If the used universal joints are going to be reinstalled, repack with new grease.*

366 DRIVE TRAIN

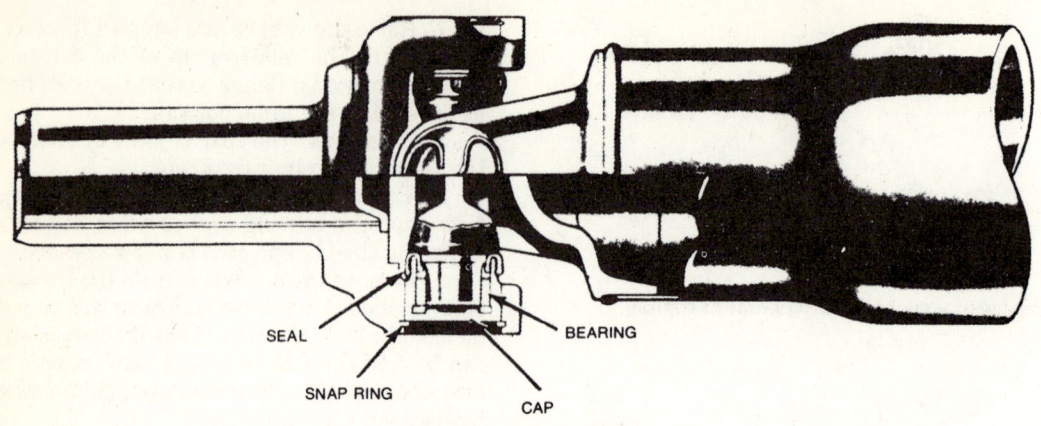

Cleveland type U-joint

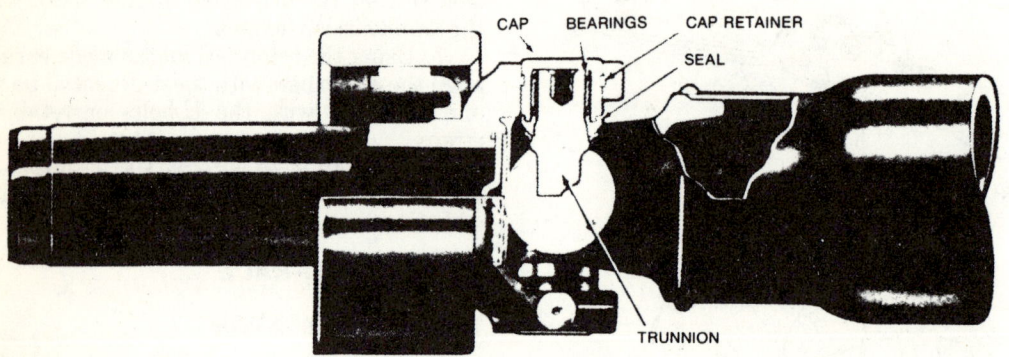

Saginaw type U-joint

To Install:

9. Use new universal joints or repack the old ones. Place a bearing cup part way into 1 side of the yoke (place the yoke ear to the bottom).

10. Insert the cross into the yoke so that the trunnion seats freely into the bearing cup.

11. Insert the opposite bearing cup part way into the yoke ear. Install the cross into the cup, making sure that both trunnions are straight and true with the bearing cups.

12. Using an arbor press, press the bearing cups into the yoke, making sure that the cross trunnions are free to turn. Install the bearing retainers.

13. Assemble the other side of the yoke in the same manner.

14. To complete the installation, reverse the removal procedures.

1979 and Later Saginaw Type

1. Remove the driveshaft from the vehicle.
NOTE: *NEVER clamp the driveshaft tube in a vise, for this may dent the tube. Support the driveshaft horizontally and clamp on the yokes of the universal joints.*

2. Support the driveshaft in the horizontal position with the base plate of a press, so that the lower ear of the yoke is supported on a $1^{1}/_{8}''$ socket.

3. Place the cross press tool J-9522-3 on the open horizontal bearing cups and press (shear the plastic retaining ring) the lower bearing cup out of the yoke ear.

NOTE: *If the bearing cup is not completely removed, lift the cross tool and place a spacer tool J-9522-5 between the seal and the bearing cup. Repeat the pressing procedure to drive the bearing cup from the yoke.*

4. Rotate the driveshaft to the opposite bearing cup and press the bearing cup from the yoke.

5. With both bearing cups removed from the yoke, separate the yoke from the driveshaft.

6. Repeat the removal procedures for the other bearing cups.

NOTE: *Since there are no bearing retainer grooves in the production bearing cups, the universal cannot be reused.*

7. Remove the remains of the sheared bearing cups and check for nicks in the yoke ears.

DRIVE TRAIN

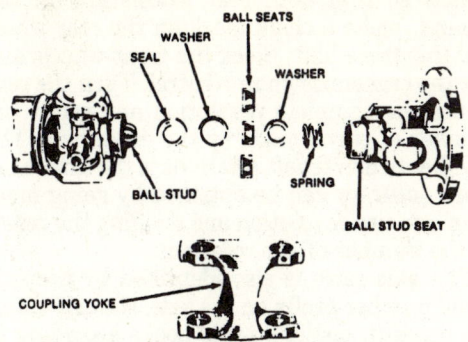

Exploded view of the constant velocity joint

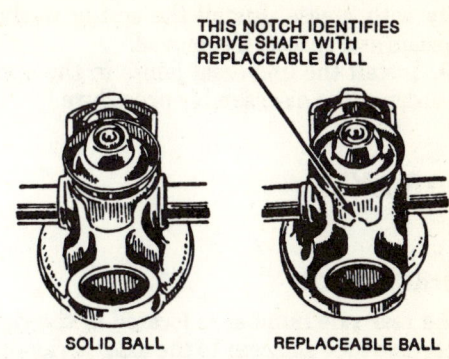

Solid and replaceable U-joint balls in the constant velocity joint

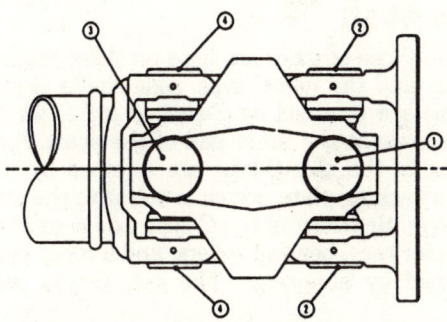

Constant velocity joint disassembly sequence

To Install:

8. Use new universal joints and place a bearing cup part way into one side of the yoke. Place the yoke ear to the bottom).

9. Insert the cross into the yoke so that the trunnion seat freely into the bearing cup.

10. Insert the opposite bearing cup part way into the yoke ear. Install the cross into the cup, making sure that both trunnions are straight and true with the bearing cups.

11. Using the press, press the bearing cups into the yoke, making sure that the cross trunnions are free to turn.

12. As soon as the bearing retainer groove(s) clears the yoke, stop pressing and install the bearing retainers onto the grooves.

NOTE: *It may be necessary to strike the yoke with a hammer to align the seating of the bearing retainers.*

13. Assemble the other side of the yoke in the same manner.

14. To complete the installation, reverse the removal procedures.

Constant Velocity Joint Overhaul

1. Using a punch, mark the link yoke and adjoining yokes before disassembly to ensure proper reassembly and driveshaft balance.

NOTE: *It is easier to remove the universal joint bearings from the flange yoke first. The first pair of flange yoke universal joint bearings to be removed is the pair in the link yoke.*

2. With the driveshaft in a horizontal position, solidly support the link yoke.

3. Apply force to the bearing cup on the opposite side using a $1^{1}/_{8}$" pipe or a socket the size of the bearing cup. Use a vise or a press to apply the force needed for removal. Force the cup inward as far as possible.

4. Remove the pieces of pipe and complete the removal of the protruding bearing cup by tapping around the circumference of the exposed portion of the bearing using a small hammer.

5. Reverse the position of the pipe and apply force to the exposed journal end. This will force the other bearing cup out of its bore and allow the removal of the flange.

NOTE: *There is a ball joint located between the two universals. The ball joint portion of this assembly is on the inner end of the flange yoke. Prior to 1973, the ball was not replaceable. Beginning 1973 the ball as well as the ball seat parts, are replaceable. Care must be taken not to damage the ball. The ball portion of this joint is on the driveshaft. To remove the seat, pry the seal out using a small pry bar.*

6. To remove the journal from the flange, use Steps 2-5.

7. Remove the universal joint bearings from the driveshaft using Steps 2-5. The first pair of bearing caps that should be removed is the pair in the link yoke.

8. Examine all parts for defects. Worn seats can be replaced with a kit. A worn ball will require the replacement of the entire shaft yoke and flange assembly.

9. Prior to installation fill the ball seat

cavity with grease. Install the spring washer, ball seats and spacer, if removed.

10. Install the universal joints in the opposite order of the disassembly procedure.

REAR AXLE

Identification

The rear axle number is located in the right or left axle tube adjacent to the axle carrier (differential). Anti-slip (positive traction) differentials are identified by a tab attached to the lower right section of the axle cover.

Determining Axle Ratio

An axle ratio is obtained by dividing the number of teeth on the drive pinion gear into the number of teeth on the ring gear. For instance, on a 4.11 ratio, the driveshaft will turn 4.11 times for every turn of the rear wheel.

The most accurate way to determine the axle ratio is to drain the differential, remove the cover and count the number of teeth on the ring and pinion.

An easier method is to jack and support the vehicle so that both rear wheels are off the ground. make a chalk mark on the rear wheel and the driveshaft. Block the front wheels and put the transmission in Neutral. Turn the rear wheel one complete revolution and count the number of turns made by the driveshaft. The number of driveshaft rotations is the axle ratio. More accuracy can be obtained by going more than one tire revolution and dividing the result by the number of tire rotations.

The axle ratio is also identified by the axle serial number prefix on the axle; the axle ratios are listed in dealer's parts books according to prefix number. Some axles have a tag on the cover, refer to chapter 1 for more identification information.

Axle Shaft

Two types of axles are used on these models the C- and the non-C type. Axle shafts in the C-type are retained by C-shaped locks, which fit grooves at the inner end of the shaft. Axle shafts in the non C-type are retained by the brake backing plate, which is bolted to the axle housing. Bearings in the C-type axle consist of an outer race, bearing rollers, and a roller cage retained by snaprings. The non C-type axle

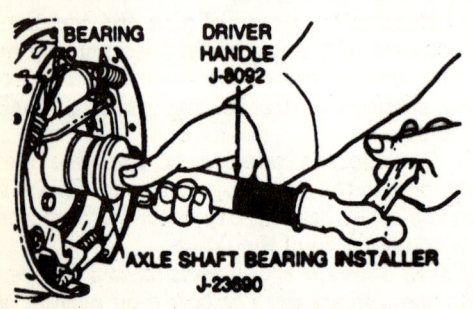

Installing axle bearing

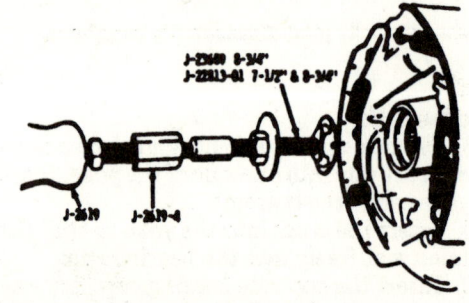

Removing axle bearing with bearing puller

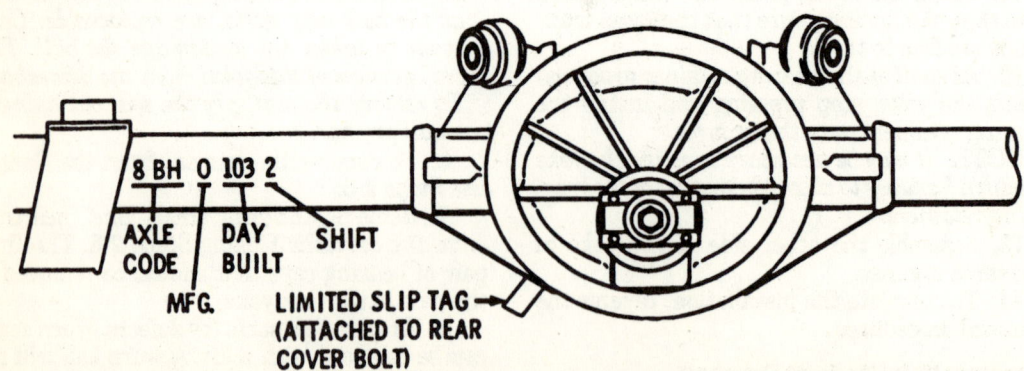

Rear axle identification. Manufacturing codes are C—Chevrolet Buffalo; G Chevrolet Gear and Axle div.

DRIVE TRAIN

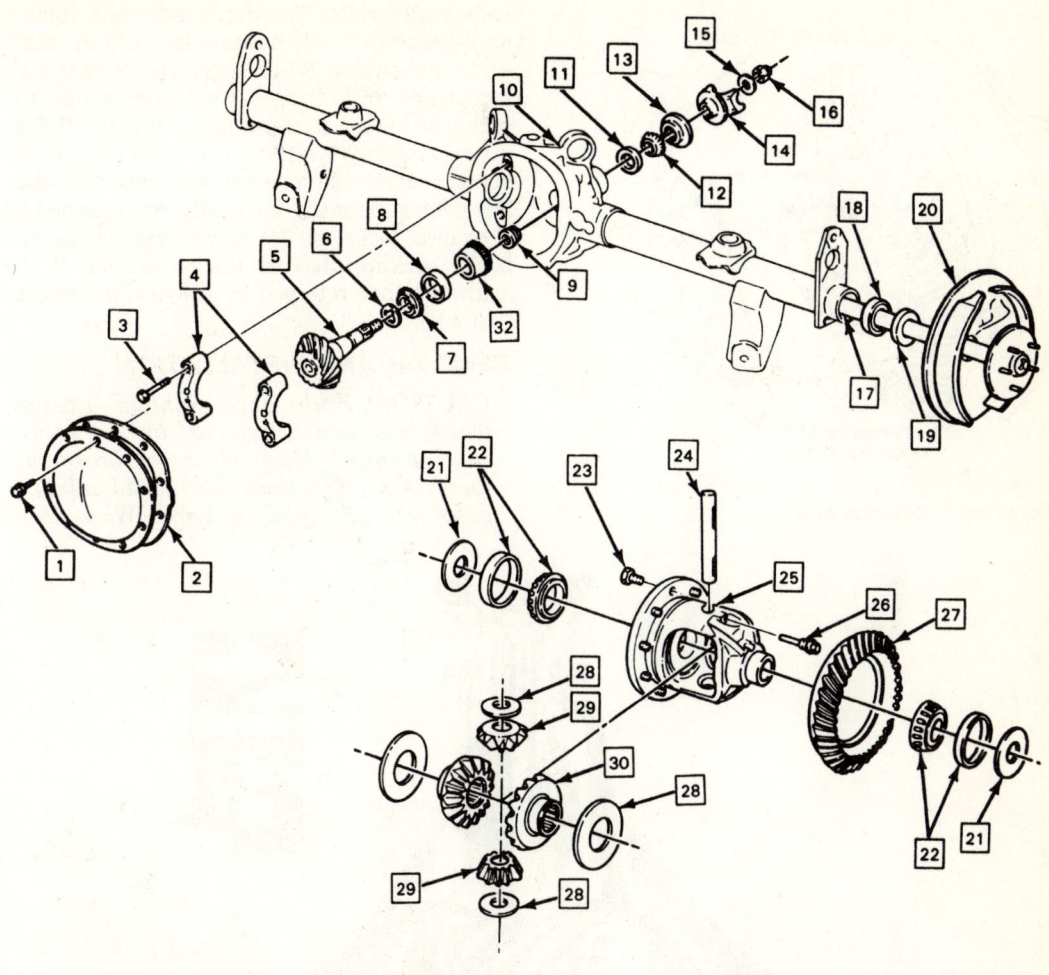

1	COVER BOLT	12	FRONT PINION BEARING	23	BOLT
2	COVER GASKET	13	PINION YOKE OIL SEAL	24	PINION GEAR SHAFT
3	DIFF. BEARING CAP BOLT	14	PINION YOKE	25	DIFFERENTIAL CASE
4	DIFF. BEARING CAP	15	WASHER	26	LOCK BOLT
5	DRIVE PINION	16	PINION YOKE NUT	27	RING GEAR
6	SHIM	17	AXLE SHAFT	28	THRUST WASHER
7	REAR PINION BEARING	18	BEARING ASM.	29	PINION GEAR
8	INNER RACE	19	OIL SEAL	30	SIDE GEAR
9	SPACER	20	BACKING PLATE	32	ABS SENSOR RING
10	REAR AXLE HOUSING	21	SHIM		
11	OUTER RACE	22	SIDE BEARING		

Exploded view of the standard rear axle

DRIVE TRAIN

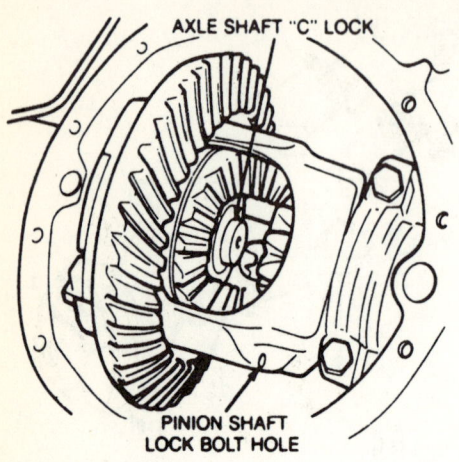

View of the C-lock rear axle

uses a unit roller bearing (inner race, rollers, and outer race), which is pressed onto the shaft up to a shoulder. When servicing C- or non C-type axles, it is imperative to determine the axle type before attempting any service. Before attempting any service to the drive axle or axle shaft, remove the axle carrier cover and visually determine if the axle shafts are retained by C-shaped locks at the inner end, or by the brake backing plate at the outer end. If the shafts are not retained by C-locks, proceed as follows.

REMOVAL AND INSTALLATION

CAUTION: *Brake shoes contain asbestos, which has been determined to be a cancer causing agent. Never clean the brake surfaces with compressed air! Avoid inhaling any dust from any brake surface! When clean-*

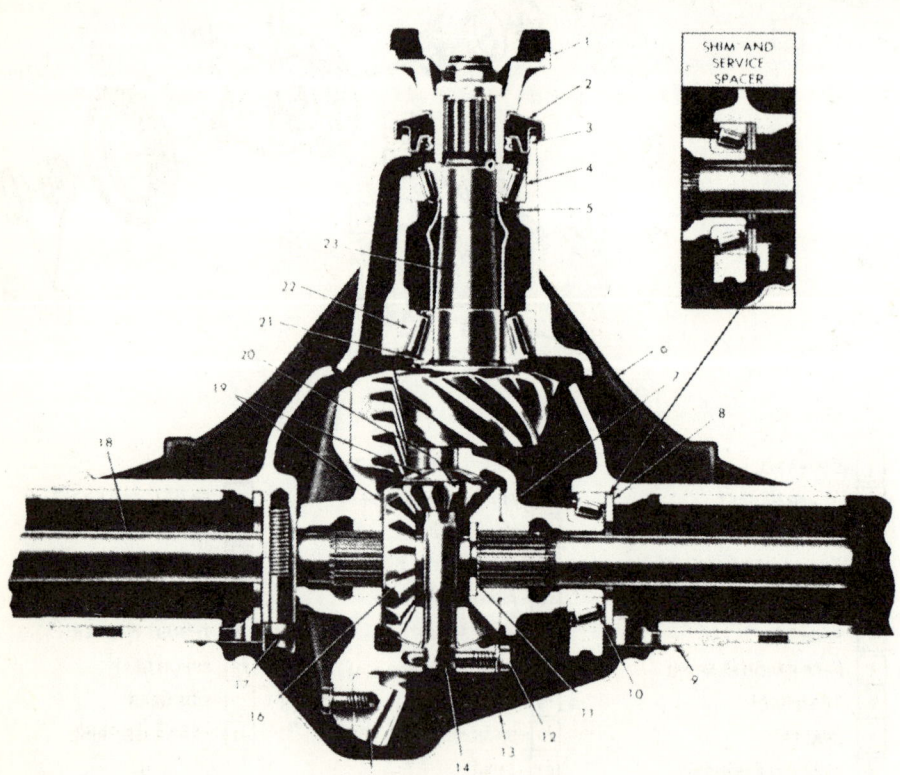

1. Companion flange
2. Deflector
3. Pinion oil seal
4. Pinion front bearing
5. Pinion bearing spacer
6. Differential carrier
7. Differential case
8. Shim
9. Gasket
10. Differential bearing
11. C-lock
12. Pinion shaft lockbolt
13. Cover
14. Pinion shaft
15. Ring gear
16. Side gear
17. Bearing cap
18. Axle shaft
19. Thrust washer
20. Differential pinion
21. Shim
22. Pinion rear bearing
23. Drive pinion

Cross section of the rear axle

DRIVE TRAIN 371

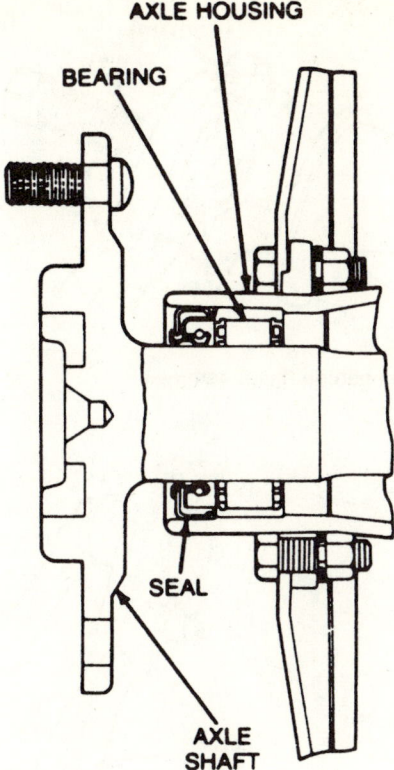

Axle shaft, cross section

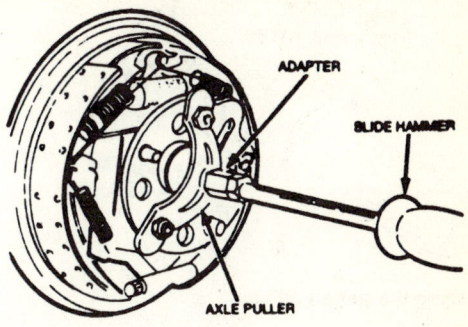

Removing the axle shaft

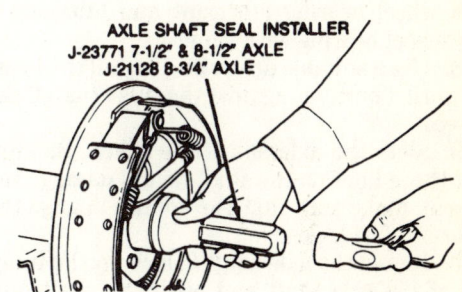

Use a seal installation tool to install axle seal

ing brake surfaces, use a commercially available brake cleaning fluid.

1968-79 Non C-Type

This design allows for a maximum axle shaft endplay of 0.022", which can be measured with a dial indicator. If endplay is found to be excessive, the bearing should be replaced. Shimming the bearing is not recommended as this ignores endplay of the bearing itself and could result in improper seating of the bearing.

1. Raise the vehicle and support it safely. Remove the wheel and brake drum.
2. Remove the nuts holding the retainer plate to the backing plate. Disconnect the brake line.
3. Remove the retainer and install nuts, fingertight, to prevent the brake backing plate from being dislodged.
4. Pull out the axle shaft and bearing assembly, using a slide hammer.
5. Using a chisel, nick the bearing retainer in three or four places. The retainer does not have to be cut, merely collapsed sufficiently, to allow the bearing retainer to be slid from the shaft.
6. Press off the bearing and install the new one by pressing it into position.
7. Press on the new retainer.

NOTE: *Do not attempt to press the bearing and the retainer on at the same time.*

8. Assemble the shaft and bearing in the housing being sure that the bearing is seated properly in the housing.
9. Install the retainer, drum, wheel and tire. Bleed the brakes.

1979 and Later C-type

If they are retained by C-shaped locks, proceed as follows:

1. Raise and support the vehicle safely. Remove the wheel and the brake drum.
2. The differential cover has already been removed (see Caution note). Remove the differential pinion shaft lock-screw and the differential pinion shaft.
3. Push the flanged end of the axle shaft toward the center of the vehicle and remove the C-lock from the end of the shaft.
4. Remove the axle shaft from the housing, being careful not to damage the oil seal.
5. Remove the oil seal by inserting the bottom end of the axle shaft behind the steel case of the oil seal. Pry the seal loose from the bore.
6. Seat the legs of the bearing puller behind the bearing. Seat a washer against the bearing

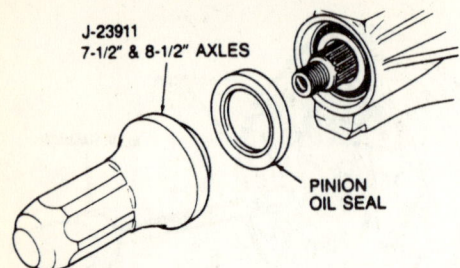

Installing the pinion oil seal

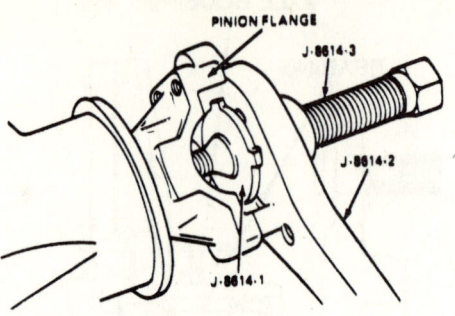

Pinion companion flange removal

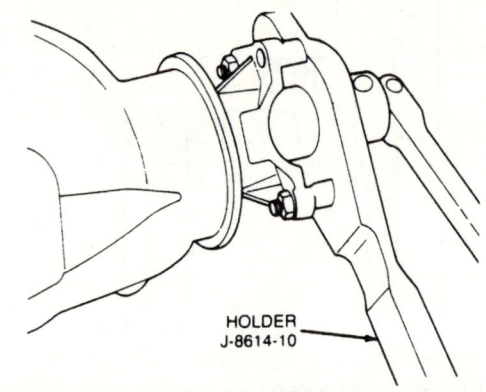

Pinion nut removal and installation

and hold it in place with a nut. Use a slide hammer to pull the bearing.

7. Pack the cavity between the seal lips with wheel bearing lubricant and lubricate a new wheel bearing with same.
8. Use a suitable driver and install the bearing until it bottoms against the tube. Install the oil seal.
9. Slide the axle shaft into place. Be sure that the splines on the shaft do not damage the oil seal. Make sure that the splines engage the differential side gear.
10. Install the axle shaft C-lock on the inner end of the axle shaft and push the shaft outward so that the C-lock seats in the differential side gear counterbore.
11. Position the differential pinion shaft through the case and pinions, aligning the hole in the case with the hole for the lockscrew.
12. Use a new gasket and install the carrier cover. Be sure that the gasket surfaces are clean before installing the gasket and cover.
13. Fill the axle with lubricant to the bottom of the filler hole.
14. Install the brake drum and wheels and lower the vehicle. Check for leaks and road test the vehicle.

Pinion Seal

REMOVAL AND INSTALLATION

1. Raise the vehicle and support it safely. Mark the driveshaft and the pinion companion flange so that they can be reassembled in the same position.
2. Disconnect the driveshaft from the pinion companion flange and support the propeller shaft up in the body tunnel by wiring it to the exhaust pipe.

NOTE: *If the joint bearing caps are not retained by a retainer strap, use a piece of tape to hold the bearing caps on their trunnions.*

3. Mark the position of the companion flange, pinion shaft and nut, so that the proper pinion bearing preload can be maintained upon reassembly.
4. Using Tool J-8614-10 to hold the pinion companion flange, remove the companion flange nut and washer.
5. With a suitable container in place to hold any fluid that may drain from thee rear axle, remove the pinion companion flange with Tool J-8614-10.
6. Remove the seal by driving it out with a blunt chisel.

WARNING: *Be careful not to damage the carrier.*

7. Examine the pinion companion flange for any nicks or damage. If so, replace it.
8. Examine the pinion seal bore in the carrier and remove any burrs.
9. Using Tool J-23911, install a new seal.
10. Apply special seal lubricant, No. 1050169 or equivalent to the O.D. of the pinion flange and sealing lip of the new seal.
11. Install the pinion companion flange and nut and tighten the nut $1/16''$ beyond the alignment marks.

Axle Assembly

REMOVAL AND INSTALLATION

1. Raise and support the vehicle safely.
CAUTION: *Make sure the rear axle assembly is supported safely.*

DRIVE TRAIN

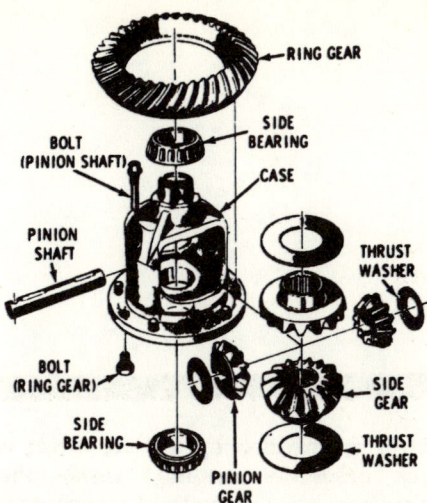

Conventional differential case assembly, all models similar

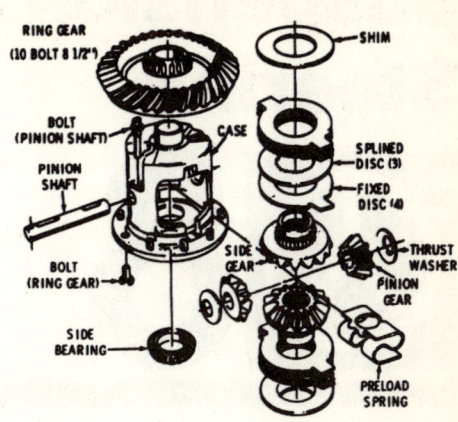

Disc type limited slip rear axle

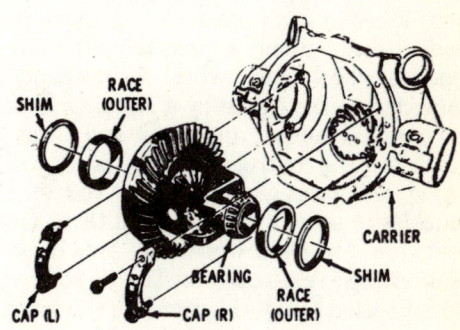

Conventional differential case and bearings

2. Disconnect the shock absorbers from the axle.
3. Mark the driveshaft and pinion flange, then disconnect the driveshaft and support out of the way.
4. Remove the brake line junction block bolt at the axle housing. If necessary disconnect the brake lines at the junction block.
5. Disconnect the upper control arms from the axle housing.
6. Lower the rear axle assembly and remove the springs.
7. Remove the rear wheels and drums.
8. Continue lowering the rear axle assembly and remove it from the vehicle.

To Install:
9. Raise the axle assembly on the hoist and install the springs.
10. Reconnect the upper and lower control arms.
11. Connect the brake lines.
12. Connect the driveshaft.
13. Connect the shock absorbers to the axle carrier.
14. Install the brake drums and wheels.
15. Bleed the brakes.

Suspension and Steering

8

FRONT SUSPENSION

The front suspension is designed to allow each wheel to compensate for changes in the road surface without appreciably affecting the opposite wheel. Each wheel is independently connected to the frame by a steering knuckle, ball joint assemblies, and upper and lower control arms. The control arms are specifically designed and positioned to allow the steering knuckles to move in a prescribed three dimensional arc. The front wheels are held in proper relationship to each other by two tie rods which are connected to steering arms on the knuckles and to an intermediate rod.

Coil chassis springs are mounted between the spring housings on the frame or front end sheet metal and the lower control arms. Ride control is provided by double, direct acting, shock absorbers mounted inside the coil springs and attached to the lower control arms by bolts and nuts. The upper portion of each shock absorber extends through the upper control arms frame bracket and is secured with two grommets, two grommet retainers, and a nut.

Side role of the front suspension is controlled by a spring steel stabilizer shaft. It is mounted in rubber bushings which are held to the frame side rails by brackets. The ends of the stabilizer are connected to the lower control arms by link bolts isolated by rubber grommets.

The upper control arm is attached to a cross shaft through isolating rubber bushing. The cross shaft, in turn, is bolted to frame brackets.

A ball joint assembly is riveted to the outer

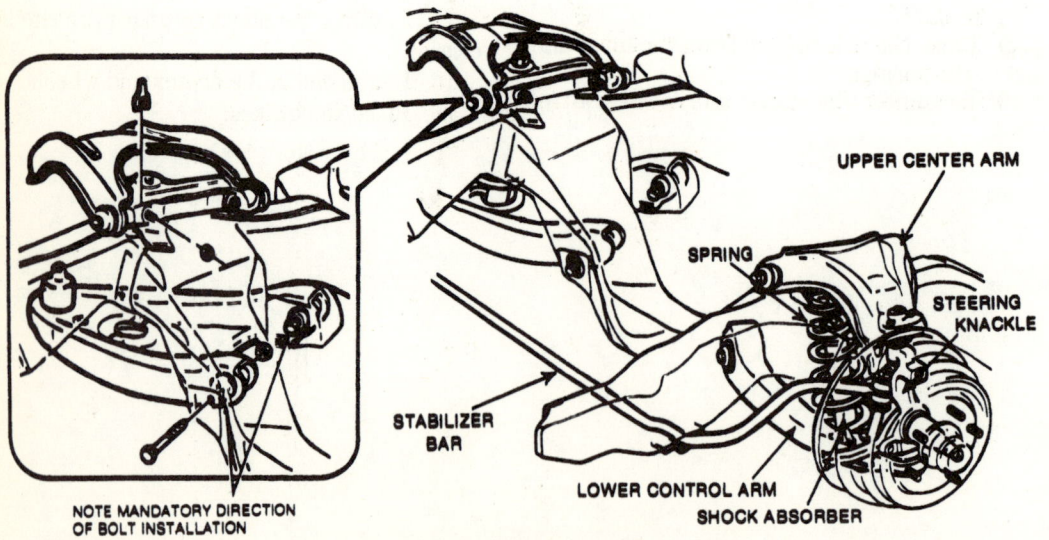

Front Suspension Assembly

SUSPENSION AND STEERING 375

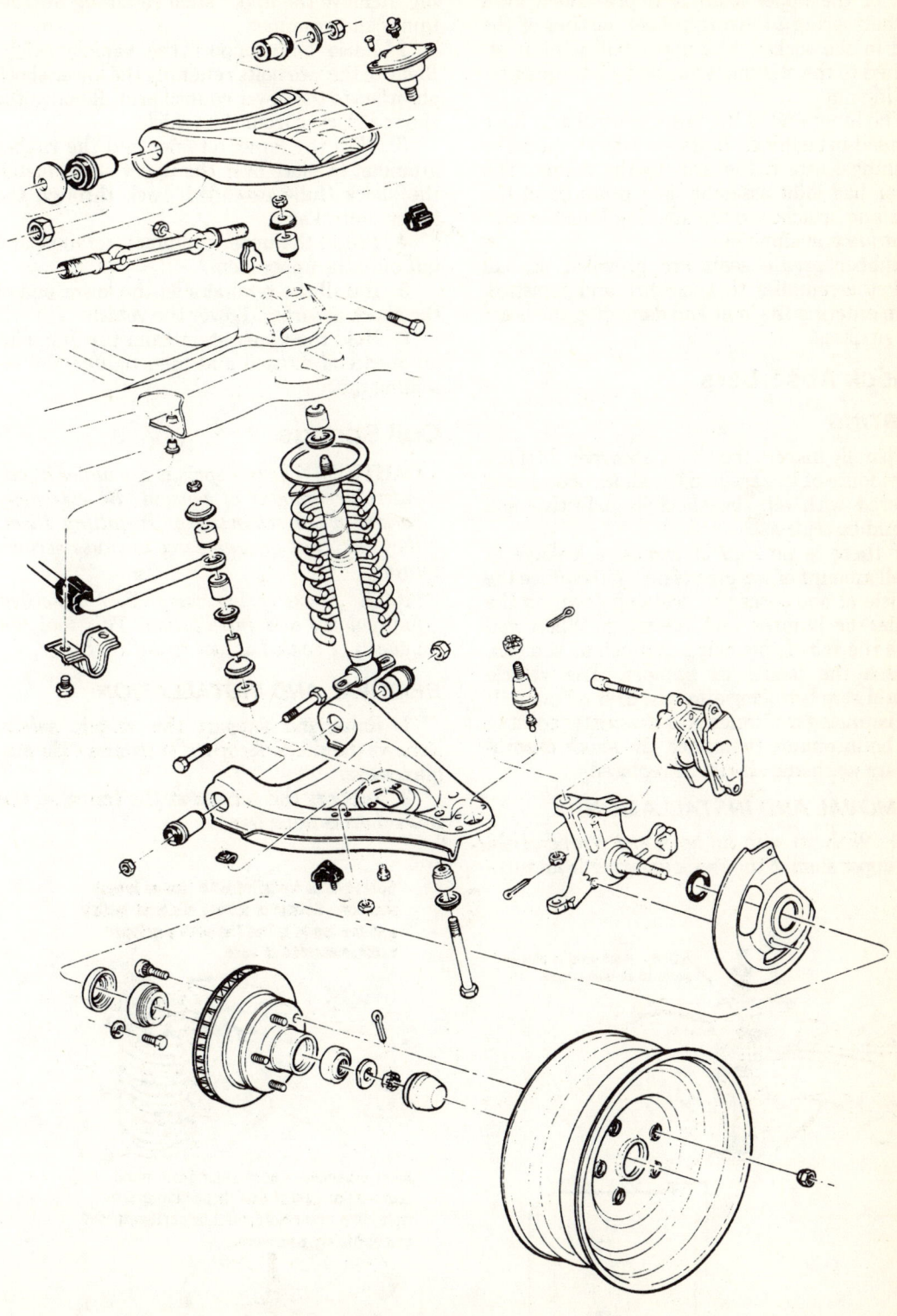

Exploded view of the front suspension

SUSPENSION AND STEERING

end of the upper arm. It is pre-loaded by a rubber spring to insure proper seating of the ball in the socket. The upper ball joint is attached to the steering knuckle by a torque prevailing nut.

The inner end of the lower control arm have pressed-in bushings. Bolts, passing through the bushings, attach the arm to the frame. The lower ball joint assembly is a press fit in the arm and attaches to the steering knuckle with a torque prevailing nut.

Rubber grease seals are provided at ball socket assemblies to keep dirt and moisture from entering the joint and damaging the bearing surfaces.

Shock Absorbers

TESTING

Visually inspect the shock absorber. If there is evidence of leakage and the shock absorber is covered with oil, the shock is defective and should be replaced.

If there is no sign of excessive leakage (a small amount of weeping is normal) bounce the vehicle at one corner by pressing down on the fender or bumper and releasing. When you have the vehicle bouncing as much as you can, release the fender or bumper. The vehicle should stop bouncing after the first rebound. If the bouncing continues past the center point of the bounce more than once, the shock absorbers are worn and should be replaced.

REMOVAL AND INSTALLATION

1. With an with an open end wrench hold the upper stem of the shock absorber from turning. Remove the upper stem retaining nut, retainer and grommet.
2. Raise and support the vehicle safely. Remove the two bolts retaining the lower shock absorber to the lower control arm. Remove the shock absorber from the vehicle.
3. With the lower retainer and the rubber grommet in place over the upper stem, install the shock (fully extended) back through the lower control arm.
4. Install the upper grommet, retainer and nut onto the upper stem.
5. Install the retainers on the lower end of the shock absorber. Lower the vehicle.
6. Hold the upper stem from turning with an open end wrench and then tighten the retaining nut.

Coil Springs

CAUTION: *The coil springs are under a considerable amount of tension. Be extremely careful when removing or installing them; they can exert enough force to cause serious injury.*

NOTE: *A coil spring compressor is needed for removal and installation. This tool can usually be rented at tool rental shops.*

REMOVAL AND INSTALLATION

1. Raise and support the vehicle safely. Remove the shock absorber. Disconnect the stabilizer bar.
2. Support the vehicle at the frame so the control arms hang free.

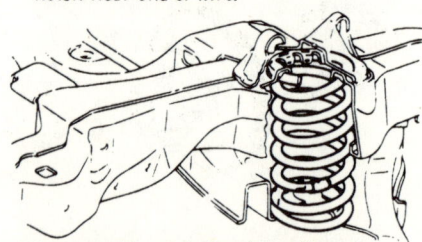

Spring to be installed with tape at lowest position. Bottom of spring is coiled helical, and the top is coiled flat with a gripper notch near end of wire.

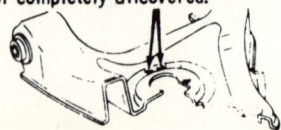

After assembly, end of spring coil must cover all or part of one inspection drain hole. The other hole must be partly exposed or completely uncovered.

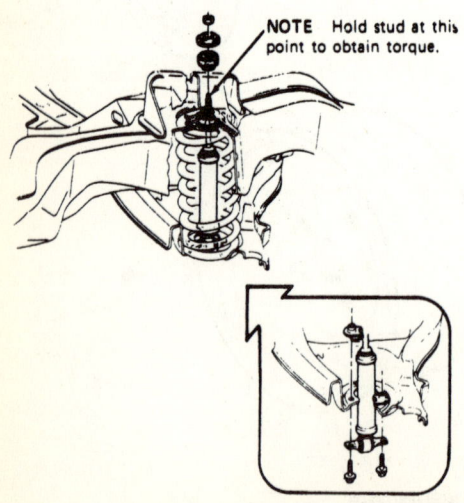

Shock absorber mounting locations

Coil spring positioning, 1983 model shown. All models similar

SUSPENSION AND STEERING 377

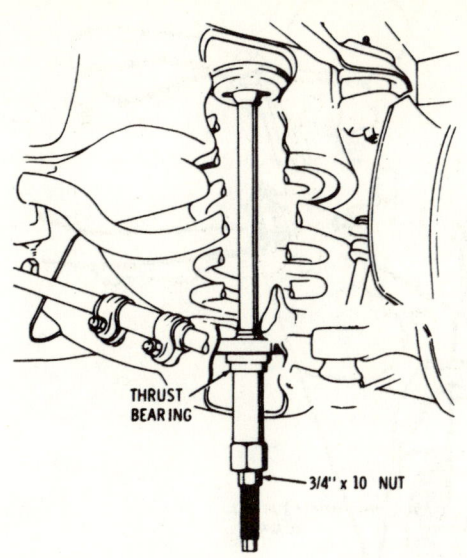

Front coil spring removal. Make sure the lock in the top of the spring compressor is in position when ever the tool is used

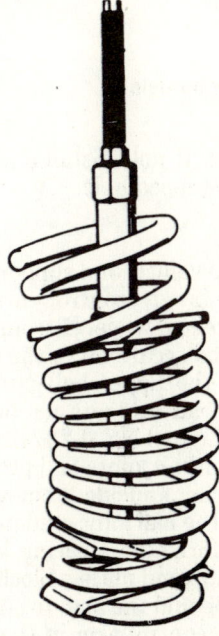

Spring compressed and ready to install

3. Disconnect the Anti-lock Brake Sensor and secure off to the side.
4. Support the inner end of the control arm using tool No. J-23028 and a floor jack.
5. Raise the jack enough to take the tension off the lower control arm pivot bolts.
6. Chain the spring to the lower control arm.

7. Remove first the rear, then the front pivot bolt.
8. Cautiously lower the jack until all spring tension is released.
9. Note the way in which the spring is installed in relation to the drain holes in the control arm and remove it.
10. On installation, position the spring to the control arm and raise into place.
11. Install the pivot bolts and reverse the rest of the removal procedures.

Ball Joints

INSPECTION

NOTE: *Before performing this inspection, make sure that the wheel bearings are adjusted correctly and that the control arm bushing are in good condition.*

1. Raise the vehicle by placing the jack under the lower control arm at the spring seat.
2. Raise the vehicle until there is a 1-2" clearance under the wheel.
3. Insert a bar under the wheel and pry upward. If the wheel raises more than $1/8"$, the ball joints are worn. Determine whether the upper or lower ball joint is worn by visual inspection while prying on the wheel.

NOTE: *Due to the distribution of forces in the suspension, the lower ball joint is usually the defective joint. Because of this, 1974-92 vehicles are equipped with wear indicators on the lower ball joint. As long as the indicator extends below the ball stud seat, replacement is unnecessary.*

UPPER BALL JOINT REPLACEMENT

1968-70

1. Support the vehicle by placing a jack under the outer end of the lower control arm.
2. Remove the wheel and tire assembly.
3. Remove the cotter pin and nut from the stud.
4. Remove the stud from the steering knuckle.
5. Cut off the ball joint rivets with a chisel.
6. It may be necessary to enlarge the stud attaching holes in the control arm to accept the larger $5/16"$ bolts. Inspect and clean the tapered hole in the steering knuckle. If the hole is damaged or deformed, the knuckle must be replaced.
7. Install the new joint and connect the stud to the steering knuckle. When installing the stud nut, never back off on the nut to align the cotter pin holes; always tighten the nut to the next hole.
8. Replacement ball joints may not include

SUSPENSION AND STEERING

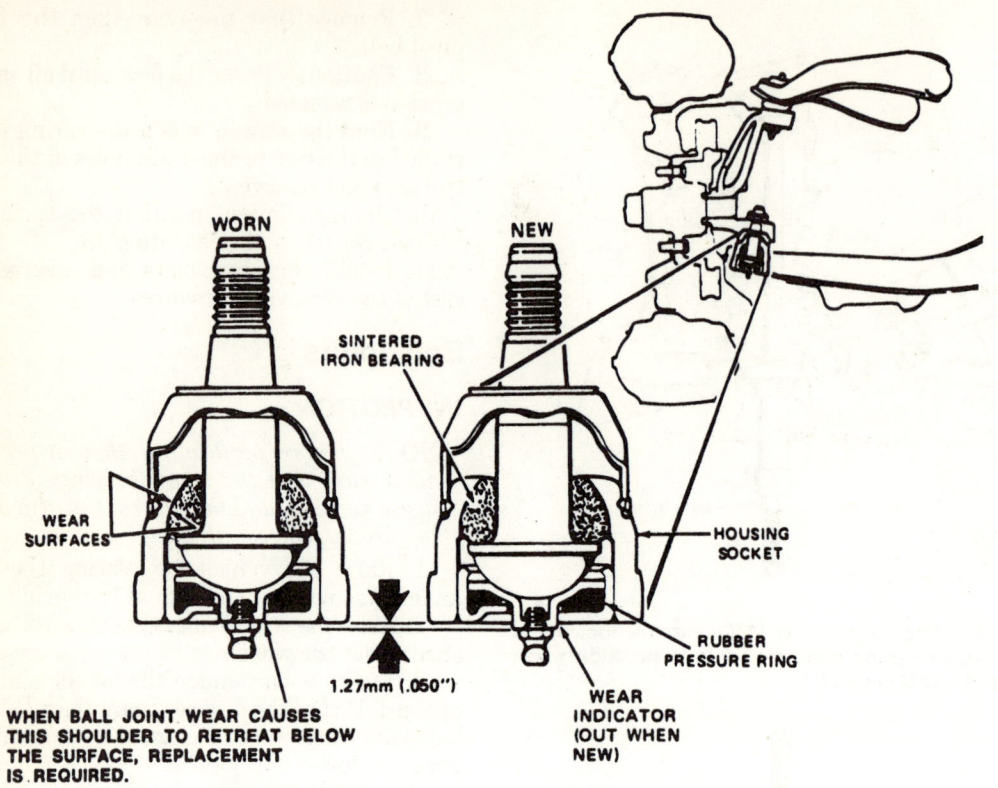

Lower ball joint wear indicator, 1974 and later models

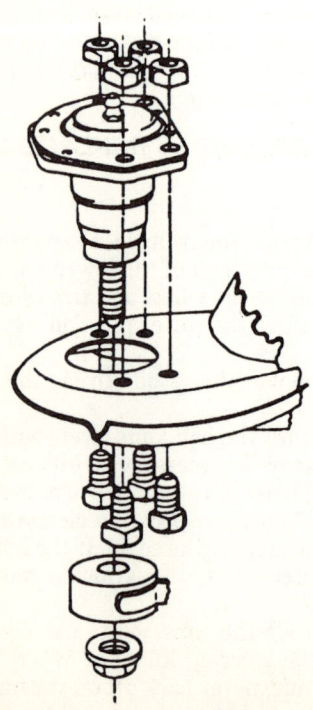

When installing the new upper ball joint, make sure that the nuts are on top, 1971 models and later

the lube fitting. If not, install a self-threading fitting into the tapped hole.

1971-92

1. Raise the vehicle and support it securely. Support the lower control arm securely. Remove the tire and wheel. Disconnect the ABS speed sensor and secure off to the side.

2. Remove the upper ball stud cotter pin and loosen the ball stud nut just one turn.

3. Locate the tool No. J-23742 between the upper and lower ball joints and press the joints out of the steering knuckle. Remove the tool.

4. Remove the ball joint stud nut, and separate the joint form the steering knuckle. Lift the upper arm up and place a block of wood between the frame and the arm to support it.

5. With the control arm in the raised position, drill a hole 1/4" deep into each rivet. Use a 1/8" drill bit.

6. Use a 1/2" drill bit and drill off the heads of each rivet.

7. Punch out the rivet using a small punch and then remove the ball joint.

8. Install the new ball joint using fasteners that meet the same grade specification as the original bolts. Bolts should come in from the bottom with the nuts going on top. Torque to 10 ft. lbs.

SUSPENSION AND STEERING

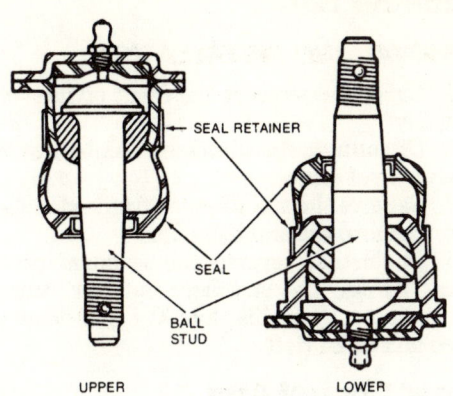

Cross section of the ball joints

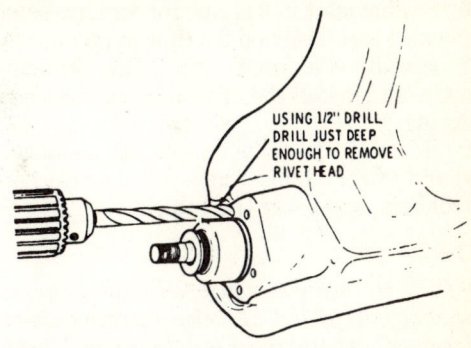

Use a 1/2 in. drill to drill the upper ball joint rivet heads, 1971 and later

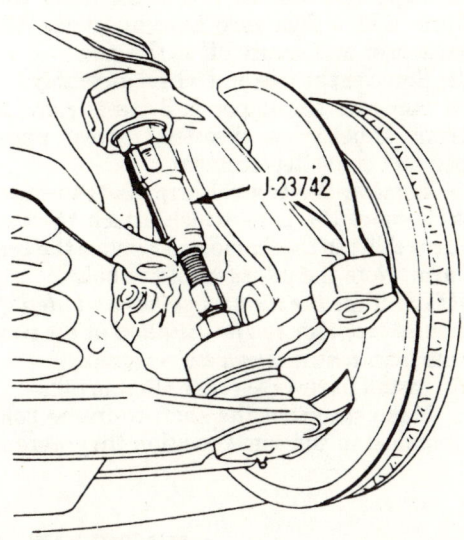

Disconnecting the lower ball joint, 1980 shown others similar

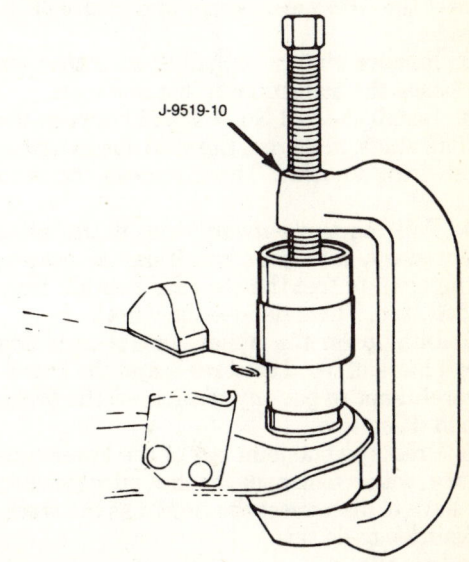

Pressing out the lower ball joint using a ball joint tool

9. Turn the ball stud cotter pin hole to the fore and aft position. Remove the block of wood from between the upper control arm and frame.

10. Clean and inspect the steering knuckle hole. Replace the steering knuckle if any out of roundness is noted.

11. Insert the ball joint stud into the steering knuckle then install and torque the stud nut to 60 ft. lbs. Install a new cotter pin. If nut must be turned to align cotter pin holes, tighten it further. Do not back off!

12. Install a lube fitting, and fill the joint with fresh grease.

13. Remove the lower control arm support and lower the vehicle.

LOWER BALL JOINT REPLACEMENT

1968-1970

1. Raise the vehicle and support it securely. Support the lower control arm with a floor jack.

2. Remove the wheel. If the vehicle has disc brakes, remove the caliper assembly.

3. Remove the lower ball stud. Then, using a tool designed for such work, press the ball stud out of the steering knuckle. Wire the steering knuckle out of the way so you'll have more room.

4. Press the joint out of the control arm with a tool designed for that purpose.

5. Start the replacement joint into the control arm with the air vent in the rubber boot facing inboard.

SUSPENSION AND STEERING

6. Set the joint in the control arm, pressing it in with a tool designed for that purpose.
7. Install the stud into the steering knuckle, and install the attaching nut and new cotter pin.
8. Reinstall the caliper assembly (as necessary) and wheel remove the jack supporting the control arm, and lower the vehicle.

1971-92

NOTE: *On vehicles equipped with wear indicating ball joint, Chevrolet recommends replacement of both upper and lower ball joints if only the lower ball joint is bad.*

1. Raise the vehicle and support it securely. Support the lower control arm with a jack. Disconnect the ABS speed sensor and secure off to the side.
2. Remove the lower ball stud cotter pin and loosen the ball stud nut just one turn.
3. Install the tool No. J-23742 between the two ball studs, and press the stud downward in the steering knuckle. Then, remove the stud nut.
4. Pull the tire outward and at the same time upward, with your hands on the bottom (of the tire), to free the steering knuckle from the ball stud. Then, remove the wheel.
5. Lift up on the upper control arm and place a block of wood between it and the frame. Be careful not to put any tension on the brake hose in doing this.
6. Press the ball joint out of the lower control arm with a tool made for that purpose. You may have to disconnect the tie rod at the steering knuckle to do this.

To Install:

7. Position the new ball joint, with the vent in the rubber boot facing inward, onto the lower control arm. Press the joint fully into the control arm with the tool No. J-9519-10 and J-9519-9.
8. Turn the ball stud cotter pin hole so it is fore and aft.
9. Remove the block of wood holding the upper control arm out of the way, and inspect the tapered hole in the steering knuckle. Remove any dirt from the hole. If the hole is out of round or there is other noticeable damage, replace the entire steering knuckle.
10. Insert the ball joint stud into the steering knuckle, install the stud nut, and torque it to 83 ft. lbs. Install a new cotter pin, aligning cotter pin holes in the nut and stud only through further tightening. Do not loosen the nut from the torque position.
11. Install a lube fitting and lube the joint. Reconnect the tie rod (as necessary), install the wheel, remove the jack supporting the lower control arm and lower the vehicle.

Stabilizer Bar

REMOVAL AND INSTALLATION

1. Raise and support the front of the vehicle safely.
2. Disconnect the stabilizer link bolt at the lower control arms.
3. Remove the stabilizer-to-frame clamps.
4. Remove the stabilizer bar.
5. To install, reverse the removal procedures. Torque the stabilizer-to-lower control arm bolts to 13 ft. lbs. and the stabilizer-to-frame bolt to 24 ft. lbs.

Upper Control Arm

REMOVAL AND INSTALLATION

1. Raise the vehicle and support it safely.
2. Support the outer end of the lower control arm with a floor jack. Disconnect the ABS speed sensor and secure off to the side.
3. Remove the tire and wheel assembly.
4. Separate the upper ball joint from the steering knuckle as described above under Upper Ball Joint Replacement.
5. Remove the control arm shaft-to-frame nuts. Remove the bolts which attach the control arm shaft to the frame and remove the control arm. Note the positions of the bolts.

NOTE: *Tape the shims together and identify them so that they can be installed in the positions from which they were removed.*

6. Install in reverse order of removal.
7. Make sure that the shaft-to-frame bolts are installed in the same position they were in

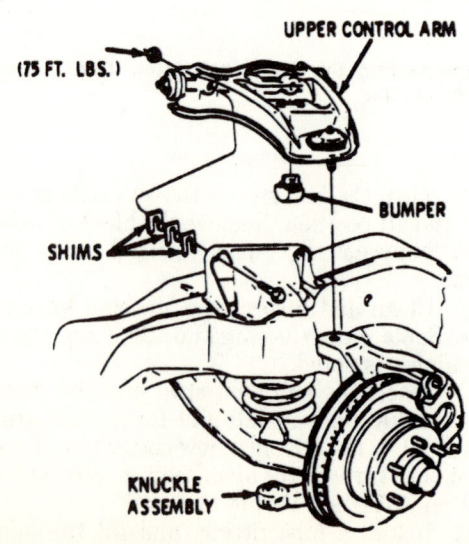

Upper control arm installation

SUSPENSION AND STEERING

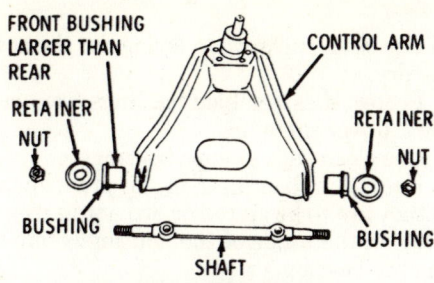

Upper control arm components, 1983 shown

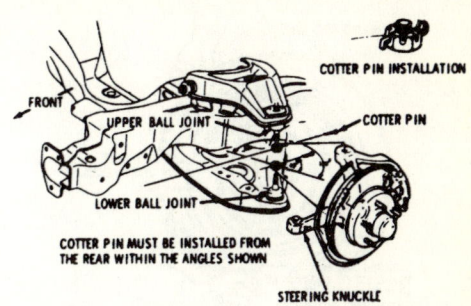

Steering knuckle cotter pin installation

before removal and that the shims are in their original positions.

8. Use free running nuts (not locknuts) to pull serrated bolts through the frame. Then install the locknuts. Tighten the thinner shim pack first.

9. After the vehicle has been lowered to the ground, bounce the front end to center the bushings and then tighten the bushing collar bolts to 45 ft. lbs.

10. Tighten the shaft-to-frame bolts to 80 ft. lbs. on 1968-73 vehicles and 90 ft. lb. on 1974-92 vehicles.

Lower Control Arm

REMOVAL AND INSTALLATION

1. Remove the spring as described earlier.
2. Remove the ball stud from the steering knuckle.
3. Remove the control arm pivot bolts and the control arm.
4. To install, reverse the above procedures. If any bolts are to be replaced, do so with bolts of equal strength and quality.

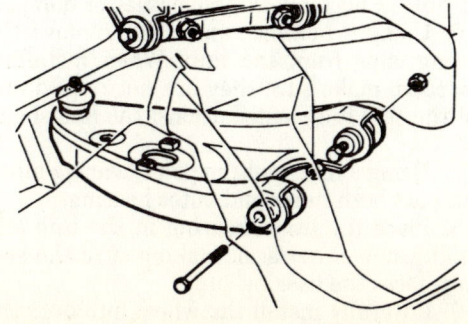

View of the lower control arm attachment bolts

Steering Knuckle

REMOVAL AND INSTALLATION

1. Siphon some fluid from the brake master cylinder. Raise the support the vehicle on jackstands. Remove the wheel and tire assembly.

2. Disconnect the ABS speed sensor and secure off to the side. Remove the caliper from the steering knuckle and support on a wire. Remove the grease cup, the cotter pin, the castle nut and the hub assembly. Remove the 3 bolts holding the shield to the steering knuckle.

3. Using the ball joint removal tool J-6627, disconnect the tie rod from the steering knuckle. Using ball joint removal tool J-23742, disconnect the ball joints from the steering knuckle.

4. Place a floor jack under the lower control arm (near the spring seat) and disconnect the ball joint form the steering knuckle.

5. Raise the upper control arm and disconnect the ball joint form the steering knuckle. Remove the steering knuckle from the vehicle.

6. To install, reverse the removal procedures. Torque the upper ball joint-to-steering knuckle nut to 65 ft. lbs. the lower ball joint-to-steering knuckle nut to 90 ft. lbs. and the tie rod-to-steering knuckle nut to 40 ft. lbs. Adjust the wheel bearings and refill the master cylinder.

Wheel Bearings

Properly adjusted bearings have a slightly loose feeling. Wheel bearings must never be preloaded. Preloading will damage the bearings and eventually the spindles. If the bearings are tool loose, they should be cleaned, inspected and then adjusted.

Hold the tire at the top and bottom and move the wheel in and out of the spindle. If the movement is greater than 0.008" for 1968-71 vehicles and 0.005" for 1974-92 vehicles, the bearings are too loose.

SUSPENSION AND STEERING

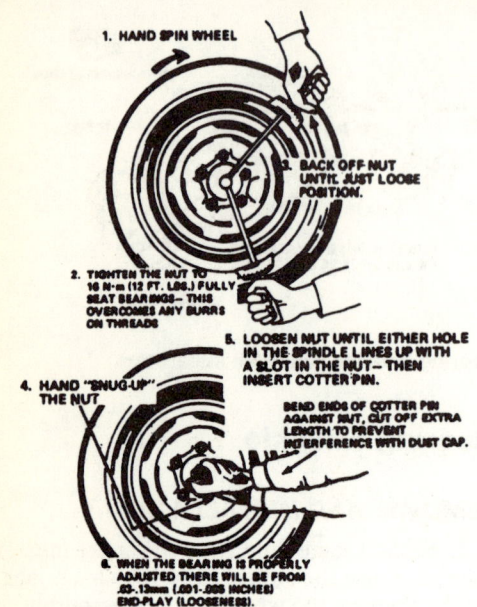

Wheel bearing adjustment

ADJUSTMENT

1. Raise and support the vehicle by the lower control arm.
2. Remove the hub cap, then remove the dust cap from the hub.
3. Remove the cotter pin and spindle nut.
4. Spin the wheel forward by hand. Tighten the nut until snug to fully seat the bearings.
5. Back off the nut $1/4$-$1/2$ turn until it is juCst loose, then tighten it finger-tight.
6. Loosen the nut until either hole in the spindle lines up with a slot in the nut and then insert the cotter pin. This may appear to be too loose, but it is the correct adjustment. The spindle nut should not be even fingertight.
7. Proper adjustment creates 0.001-0.008" endplay for 1968-73 vehicles and 0.001-0.005" endplay for 1974-88 vehicles.

REMOVAL AND INSTALLATION

Before handling the bearings, there are a few things that you should remember to do and not to do.

Remember to DO the following:
• Remove all outside dirt from the housing before exposing the bearing.
• Treat a used bearing as gently as you would a new one.
• Work with clean tools in clean surroundings.
• Use clean, dry canvas gloves, or at least clean, dry hands.
• Clean solvents and flushing fluids are a must.
• Use clean paper when laying out the bearings to dry.
• Protect disassembled bearings from rust and dirt. Cover them up.
• Use clean rags to wipe bearings.
• Keep the bearings in oil-proof paper when they are to be stored or are not in use.
• Clean the inside of the housing before replacing the bearing.

Do NOT do the following:
• Don't work in dirty surroundings.
• Don't use dirty, chipped or damaged tools.
• Try not to work on wooden work benches or use wooden mallets.
• Don't handle bearings with dirty or moist hands.
• Do not use gasoline for cleaning; use a safe solvent.
• Do not spin-dry bearings with compressed air. They will be damaged.
• Do not spin dirty bearings.
• Avoid using cotton waste or dirty cloths to wipe bearings.
• Try not to scratch or nick bearing surfaces.
• Do not allow the bearing to come in contact with dirt or rust at any time.

1968-69

1. Remove the wheel and tire assembly, and the brake drum or brake caliper.
2. On vehicles with disc brakes, remove the hub and disc as an assembly. Remove the caliper mounting bolts and insert a block between the brake pads as the caliper is removed. Remove the caliper and wire it out of the way.
3. Pry out the grease cap, cotter pin, spindle nut, and washer, then remove the hub. Do not drop the wheel bearings.
4. Remove the outer roller bearing assembly from the hub. The inner bearing assembly will remain in the hub and may be removed after prying out the inner seal. Discard the seal.
5. Clean all parts in solvent and allow them to air dry. Check for excessive wear or damage.
6. Using a hammer and drift, remove the bearing cups from the hub. When installing new cups, make sure they are not cocked and that they are fully seated against the hub shoulder.
7. Using a high melting point bearing lubricant, pact both inner and outer bearings.
8. Place the inner bearing in the hub and install a new inner seal, making sure the seal flange faces the bearing cup.
9. Carefully install the wheel hub over the spindle.
10. Using your hands, firmly press the outer

Wheel Alignment Specifications

Year	Caster Range (deg.)	Caster Pref Setting (deg.)	Camber Range (deg.)	Camber Pref Setting (deg.)	Toe-in (in.)	Steering Axis Inclin. (deg.)
1968–69	1/4P to 1 1/4P	1/4P	1/4N to 3/4P	1/4P	1/8 to 1/4	7 to 8
1970	1/4P to 1 1/4P	1/4P	1/4N to 3/4P	1/4P	1/8 to 1/4	7 to 8
1971	1 1/2N to 1/2N	1N	0 to 1P	1/2P	3/4 to 1/4	9 1/2 to 10 1/2
1972	1/2P to 1 1/2P	1P	0 to 1P	1/2P	3/16 to 5/16	9 1/2 to 10 1/2
1973	0 to 2P	1P	1/4P to 1 3/4P ①	1P	1/16N to 3/16P	10 1/2
1974	1/2P–1 1/2P	1P	1/2P–1 1/2P ①	1P ②	1/16 to 3/16	9 1/2
1975–76	1/2P–2 1/2P	1 1/2P	1/2P–1 1/2P ①	1P ②	1/16 to 3/16	9 7/64
1977–79	2P–4P	3P	0–1.6P	0.8P	1/16–5/16P	9.785
1980–81	2P–4P	—	0–1.6P	—	1/16–1/4P	—
1982–90	2P–4P	3P	0–1.6P	0.8P	1/16–1/4P	—
1991–92	2.5P–4.5P	3.5P	0–1.6P	0.8P	1/8–1/4P	—

N—Negative
P—Positive
① Left wheel given, right wheel is 1/4N to 1 1/4P, preferred 1/2P
② Left wheel given, right wheel is 1/2P

bearing into the hub. Install the spindle washer and nut, and adjust as instructed above.

1970-92

1. Raise and support the vehicle safely. Remove the hub and disc assembly.
2. Remove the outer roller bearing assembly from the hub. The inner bearing assembly can be removed after prying out the inner seal. Discard the seal.
3. Wash all parts in solvent and check for excessive wear or damage.
4. To replace the outer or inner race, knock out the old race with a hammer and brass drift. New races must be installed squarely and evenly to avoid damage.
5. Pack the bearings with a high melting-point bearing lubricant.
6. Light grease the spindle and inside of the hub.
7. Place the inner bearing in the hub race and install a new grease seal.
8. Carefully install the hub and disc assembly.
9. Install the outer wheel bearing.
10. Install the washer and nut and adjust the bearings according to the procedures outlined above.
11. Install the caliper and torque the mounting bolts to 35 ft. lbs.
12. Install the dust cap and the wheel, then lower the vehicle to the ground.

PACKING

Clean the wheel bearings thoroughly with solvent and check their condition before installation.

WARNING: *Do not blow the bearing dry with compressed air as this would allow the bearing to turn without lubrication.*

Apply a sizable daub of lubricant to the palm of one hand. Using your other hand, work the bearing into the lubricant so that the grease is pushed through the rollers and out the other side. Keep rotating the bearing while continuing to push the lubricant through it.

Front End Alignment

CAMBER

Camber is the inward or outward tilting of the front wheels from the vertical. When the wheels tilt outward at the top, the camber is said to be positive (+). When the wheels tilt inward at the top, the camber is said to be negative (-). The amount of tilt is measured in degrees from the vertical and this measurement is called the camber angle.

CASTER

Caster is the tilting of the front steering axis either forward or backward from the vertical. A backward tilt is said to be positive (+) and a forward tilt is said to be negative (-).

384 SUSPENSION AND STEERING

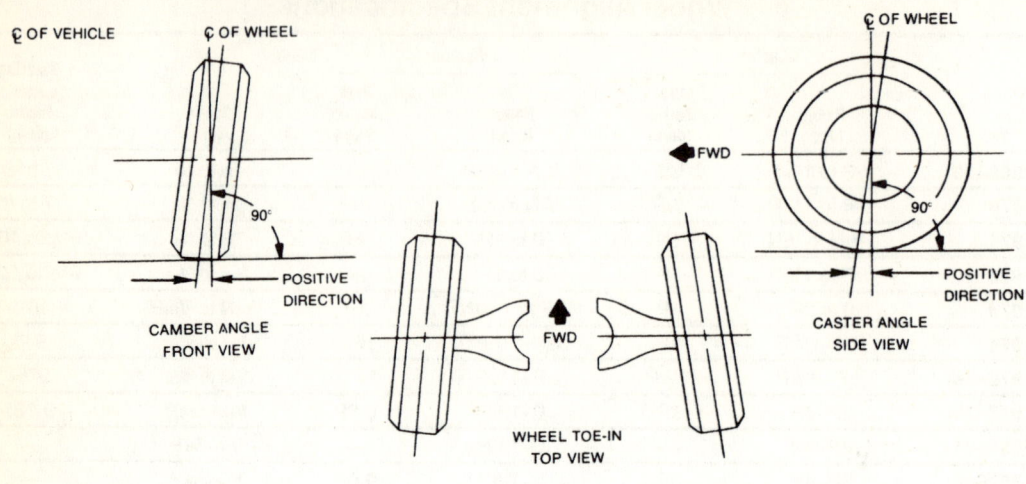

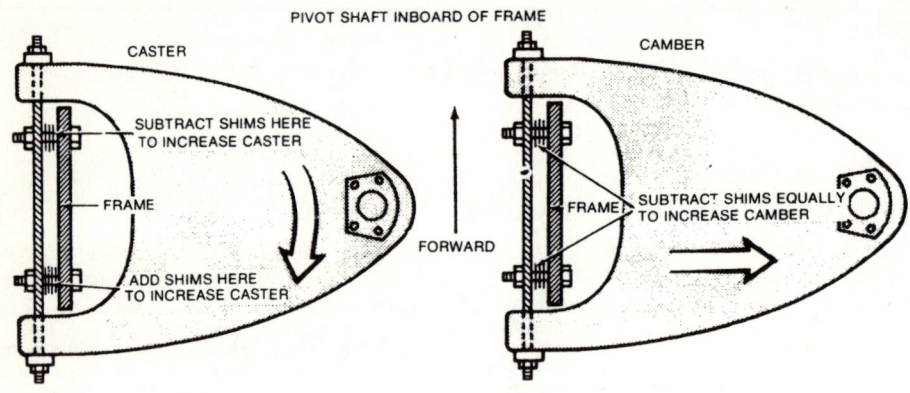

Caster and Camber adjustment

TOE-IN

Toe-in is the turning in of the front wheels. The actual amount of toe-in is normally only a fraction of a degree. The purpose of toe-in is to ensure parallel rolling of the front wheels. (Excessive toe-in or toe-out will cause tire wear).

CASTER/CAMBER ADJUSTMENT

Caster and camber can be adjusted by moving the position of the upper strut mount assembly. Moving the mount forward/rearward adjusts caster. Movement inboard/outboard adjusts camber.

TOE-IN ADJUSTMENT

1. Loosen the clamp bolts at each end of the steering tie rod adjustable sleeves.
2. With the steering wheel set straight ahead, turn the adjusting sleeves to obtain the proper adjustment.
3. When the adjustment has been completed, check to see that the number the threads showing on each end of the sleeve are equal. Also check that tie rod end housings are at the right angles to the steering arm.

REAR SUSPENSION

The rear axle assembly is attached to the frame through a link-type suspension system. Two rubber bushed lower control arms mounted between the axle assembly and the frame maintain fore and aft relationship of the axle assembly to the chassis. Two rubber bushed upper control arms, angularly mounted with respect to the centerline of the vehicle, control driving and braking torque and sideways movement of the axle assembly. The rigid axle holds the rear wheels in proper alignment.

The upper control arms are sorter than the lower arms, causing the differential housing to

SUSPENSION AND STEERING 385

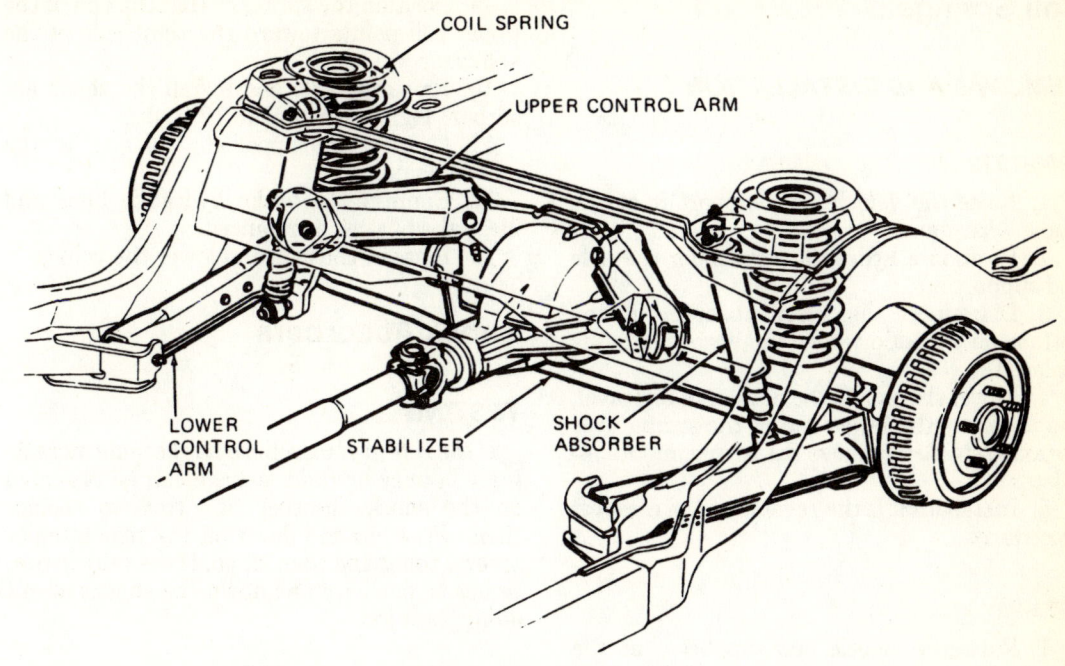

Rear suspension, all models

rock or tilt forward on compression. This rocking or titling lowers the rear propeller shaft to make possible the use of a lower tunnel in the rear floor pan area.

The rear upper control arms control drive forces, side sway and pinion nose angle. Pinion angle adjustment can greatly affect vehicle smoothness and must be maintained as specified.

The rear chassis springs are located between brackets on the axle tube and spring seats in the frame. The springs are held in the seat pilots by the weight of the vehicle and by the shock absorbers which limit axle movement during rebound.

Ride control is provided by two identical direct double acting shock absorbers angle-mounted between brackets attached to the axle housing and the rear spring seats.

Leaf Spring

REMOVAL AND INSTALLATION

1. Raise the vehicle on a hoist and place an adjustable jack under the axle.
2. Raise the axle until all tension is relieved from the spring.
3. Disconnect the shock absorber from the spring retainer plate.
4. Remove the upper shackle retaining bolt, then the front spring eye bolt.
5. Remove the spring/axle U-bolts, lower plate, spring pads, and spring.
6. Remove the shackle from the spring.
7. Before installing the spring, install the shackle on the rear ward end.
8. Replace the upper cushion on the spring, then insert the front of the spring into the frame and attach the rear shackle, leaving the bolt loose.
9. Install the lower spring pad and retainer plate, tighten the U-bolt nuts to 40 ft. lbs.
10. Tighten the rear shackle bolts to 80 ft. lbs.
11. Tighten the front eye bolt to 115 ft. lbs.
12. Attach the shock absorber to spring retainer plate, tighten to 65 ft. lbs.
13. Remove the jack and lower the vehicle.

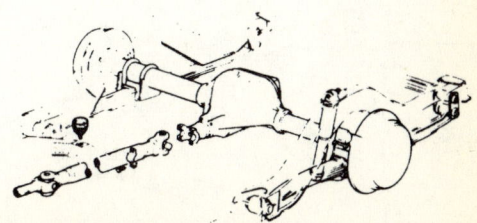

Rear leaf spring suspension, 1968–76 station wagon

SUSPENSION AND STEERING

Coil Springs

REMOVAL AND INSTALLATION

1968-1972

1. Raise the vehicle and support it at the frame with stands.
2. Position a hydraulic jack under the axle and support it.
3. Disconnect the shock absorber at the bottom on the side where the spring is being replaced.
4. Lower the axle to the bottom of its travel, and then pry the lower end of the spring over the axle retainer. Remove the spring and its insulator.
5. Installation is the reverse of the removal procedure.

1973-92

1. Raise the vehicle and support it at the frame with stands.
2. Position a hydraulic jack under the axle and support it.
3. Disconnect the shock absorber at the bottom on the side where the spring is being replaced. Disconnect the ABS speed sensor and secure off to the side.
4. Disconnect the brake hydraulic line at the junction block located on the axle housing. It may be possible to remove the junction block retaining bolt and position the junction block to the side.
5. Disconnect the upper control arm at the axle.
6. Lower the axle to the bottom of its travel, then pry the lower end of the spring over the axle retainer and remove the spring and its insulator.
   To Install:
7. Install the spring in its frame seat with its rubber insulator.
8. Pry the lower end of the spring over the vertical flange of the axle bracket spring seat.
9. Position the spring so that the end of the upper coil points toward the right side of the vehicle.
10. Raise the axle and install the shock absorber. Tighten the nut to 12 ft. lbs.
11. Connect the upper control arm to the axle housing.
12. Connect the brake hydraulic lines and bleed the brakes, as required.
13. Remove the jack and lower the vehicle.

Shock Absorbers

TESTING

If the ride of your vehicle has become increasingly bouncy or fluid leakage can be observed on the shock absorber, it's time to replace them. Push up and down on the rear bumper several times and then let go. If the vehicle continues to move up and down the shocks aren't doing their job.

REMOVAL AND INSTALLATION

1. Jack the rear of the vehicle up and support the axle.
2. Remove the two upper mounting bolts. Disconnect the air line if equipped with superlift shock absorbers.
3. While holding the stud hex with a wrench, remove the lower mounting nut.
4. Remove the shock absorber from the vehicle.
5. Position the new shock absorber in place. Install the upper mounting bolts handtight.
6. Install the lower stud into the housing bracket and loosely install the nut.
7. Tighten the two upper bolts to 12 ft. lbs.
8. While holding the stud hex, tighten the lower nut to 60 ft. lbs.
9. On superlift equipped vehicle, install the air hose and add approximately 10 psi of air.
10. Lower the vehicle. If equipped with superlift shock absorbers, check for air leaks.

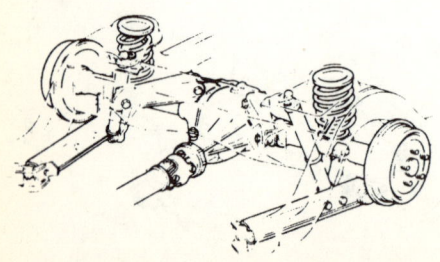

Typical coil spring rear suspension

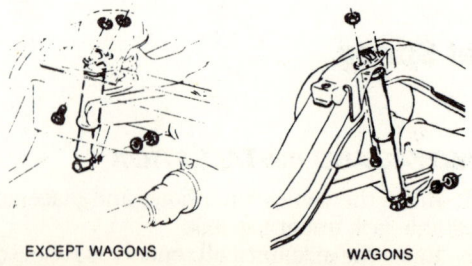

Rear shock absorber mounting

SUSPENSION AND STEERING

ELECTRONIC LEVEL CONTROL SYSTEM (ELC)

REMOVAL AND INSTALLATION

The electronic level control system, was used on 1988 and later vehicles and, keeps the rear of the vehicle level by automatically adjusting the rear trim height with varying vehicle loads. The system is activated when the ignition is on and excess weight is added to the vehicle. When the excess weight is removed from the vehicle, an exhaust solenoid connected to the battery positive, allows air to be released from the system even with the ignition off. This system is used in place of the standard rear shock absorbers.

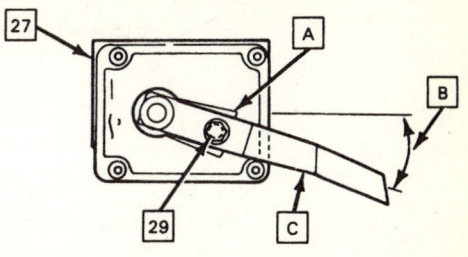

A	PLASTIC ARM	27	HEIGHT SENSOR ASSEMBLY
B	ARM ANGLE	29	LOCK BOLT
C	METAL ARM		

ELC height sensor adjustment

Air Lines and Fittings

The air lines include spring clip connections with molded sealing shoulders in the retainer and on the end of the air line with double O-ring seals. Before making any air line disconnection, clean the connector and surrounding area. Turn the spring clip to release the connector. To reassemble, lubricate the O-rings with petroleum jelly and push the air line fully into the fitting.

Height Sensor Assembly

1. Disconnect the battery ground cable.
2. Raise and safely support vehicle on jack stands.
3. Disconnect the electrical connector.
4. Remove the nut attaching link to upper control arm.
5. Remove the two bolts attaching height sensor assembly to rear crossmember.
6. Remove the height sensor from the vehicle.
7. Installation is the reverse of removal procedure.

Height Sensor Adjustment

NOTE: *The attaching link should be securely connected to the height sensor arm when any adjustments are made. For every 1 degree change in arm angle the trim height will change approximately $1/4$ inch. The arm angle may be changed a total of 5 degrees, resulting in a trim height change of approximately $1^1/4$ inches.*

1. Loosen the locknut securing the metal arm to the plastic arm.
2. To decrease the amount of suspension travel required to turn on compressor (To raise vehicle height), move the plastic arm to the top of the slot and tighten locknut.
3. To increase the amount of suspension travel required to turn on compressor (To lower vehicle height), move the plastic arm to the bottom of the slot and tighten locknut.

Compressor Assembly and Bracket

1. Disconnect the battery ground cable.
2. Raise and safely support vehicle on jack stands.
3. Disconnect the air hoses and tubing assembly at the compressor assembly.
4. Remove the bolts retaining the mounting bracket with compressor assembly to rear crossmembers.
5. Remove the bolt attaching the ABS wheel speed sensor wire clip from the compressor mounting bracket, and remove the ABS wire clip from the mounting bracket.
6. Disconnect the compressor electrical connector.
7. Remove the mounting bracket with compressor assembly from the rear crossmember.
8. Remove the screws attaching the compressor assembly to mounting bracket, and remove the compressor assembly.
9. Installation is the reverse of removal procedure.

Air Shocks

With the exception of the air lines, the air shocks are removed in the same manner as the standard rear shocks. Refer to that procedure

CAUTION: *When replacement of an air shock is necessary, always replace it with an exact matching shock. Failure to do so could result in, an ill handling vehicle, physical damage and or personal injury.*

388 SUSPENSION AND STEERING

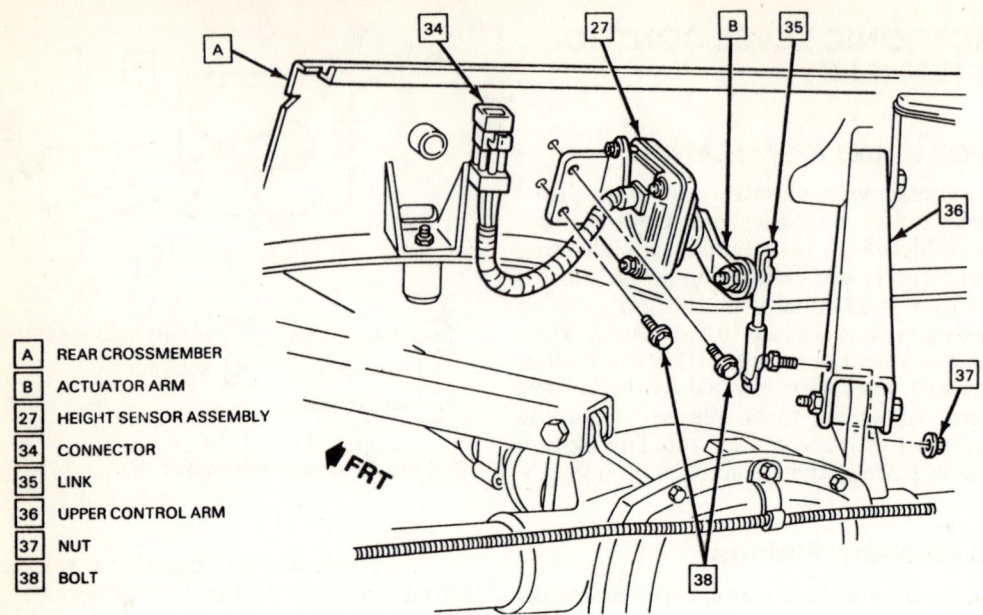

A	REAR CROSSMEMBER
B	ACTUATOR ARM
27	HEIGHT SENSOR ASSEMBLY
34	CONNECTOR
35	LINK
36	UPPER CONTROL ARM
37	NUT
38	BOLT

ELC height sensor mounting location

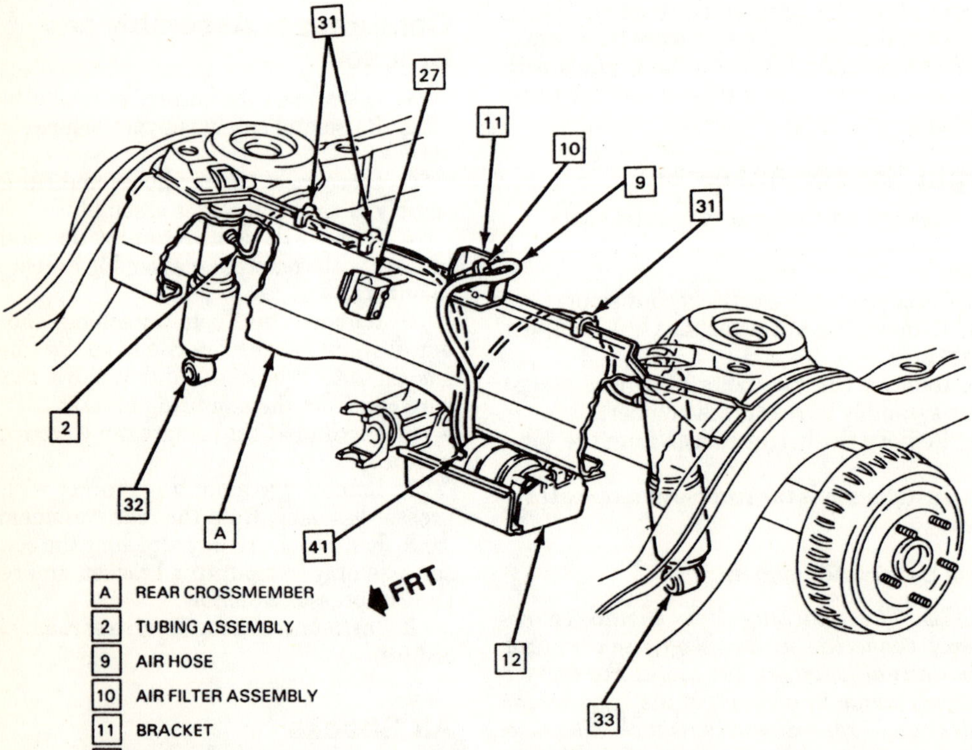

A	REAR CROSSMEMBER
2	TUBING ASSEMBLY
9	AIR HOSE
10	AIR FILTER ASSEMBLY
11	BRACKET
12	BRACKET
27	HEIGHT SENSOR ASSEMBLY
31	CLIP
32	RIGHT HAND AIR ADJUSTABLE SHOCK
33	LEFT HAND AIR ADJUSTABLE SHOCK
41	COMPRESSOR ASSEMBLY

Electronic Level Control system components

SUSPENSION AND STEERING

Rear Lower Control Arm

REMOVAL AND INSTALLATION

NOTE: *Remove and install ONLY one lower control arm at a time. If both arms are removed at the same time, the axle could roll or slip sideways, making installation of the arms very difficult.*

1. Raise and support the rear of the vehicle on jackstands under the rear axle. Remove the stabilizer bar, if equipped.
2. Remove the control arm attaching fasteners and the control arm.
3. To install, reverse the removal procedures. Torque the control arm-to-frame nut to 70 ft. lbs. and the control arm-to-axle bolt to 79 ft. lbs. If equipped with a stabilizer bar, torque the mounting fasteners to 35 ft.lbs.

NOTE: *Before torquing the fasteners, the weight of the vehicle must be resting on its wheels.*

Rear Upper Control Arm

REMOVAL AND INSTALLATION

NOTE: *Remove and install ONLY one upper control arm at a time. If both arms are removed at the same time, the axle could roll or slip sideways, making installation of the arms very difficult.*

1. Raise and support the rear of the vehicle on jackstands under the axle. support the nose of the axle with a jackstand.
2. Remove the upper control arm nut at the axle. Remove the ABS speed sensor and the height sensor link, if so equipped.

NOTE: *To remove the mounting bolt from the axle, it may be necessary to rock the axle. On some vehicles, it may be necessary to remove the lower shock absorber stud to provide clearance for the upper control arm removal.*

3. Remove the upper control arm-to-frame nut and bolt, then the control arm.
4. To install, reverse the removal procedures. Torque the upper control arm-to-axle nut to 70 ft. lbs., the upper control arm-to-axle bolt to 79 ft. lbs. and the upper control arm-to-frame bolt to 70 ft. lbs.

Stabilizer Bar

REMOVAL AND INSTALLATION

1. Raise and support the rear of the vehicle on jackstands under the frame.
2. Support the axle assembly with a floor jack.
3. Remove the stabilizer bar-to-lower control arm bolts and the stabilizer bar.
4. To install, reverse the removal procedures. Torque the stabilizer bar-to-lower control arm to 35 ft. lbs.

STEERING

All models have recirculating ball type steering. Forces are transmitted from a worm to a sector gear through ball bearings. Relay type steering linkage is used with a pitman arm connected to one end of the relay rod. The other end of the relay rod is connected to an idle arm which is attached to the frame. The relay rod is connected to the steering arms by two adjustable tie rods. most models are equipped with a collapsible steering column designed to collapse on impact, thereby reducing possible chest injuries during accidents. When making any repairs to the steering column or steering wheel, excessive pressure or force capable of collapsing the column must be avoided. Beginning 1969, the ignition lock, ignition switch, and an antitheft system were built into each column. The key cannot be removed unless the transmission is in Park (automatic) or Reverse (manual) with the switch in the Lock position. Placing

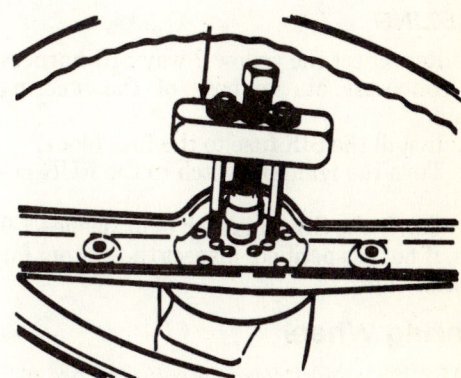

Steering wheel puller in position

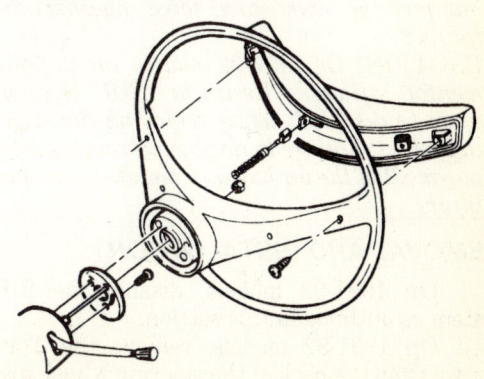

1969 deluxe steering wheel, 1970–89 standard steering wheel

SUSPENSION AND STEERING

the lock in the Lock position activates a rod within the column which locks the steering wheel and shift lever. On floorshift models, a back drive linkage between the floorshift and the column produces the same effect.

CAUTION: *On 1991-92 models, equipped with the Supplemental Inflatable Restraint System (SIR), the system must be disabled, before working on many parts of the steering and dash areas. Failure to do so may result in deployment of the air bag and possible personal injury.*

Supplemental Inflatable Restraint (SIR) System

DISABLING

1. Align the steering wheel so the vehicle wheels are pointing in the staright ahead position.
2. Turn the ignition switch to the LOCK position.
3. Remove the SIR fuse from the fuse block.
4. Disconnect the yellow 2-way SIR harness wire connector at the base of the steering column.

ENABLING

5. Reconnect the yellow 2-way SIR harness wire connector at the base of the steering column.
6. Install the SIR fuse to the fuse block.
7. Turn the ignition switch to the RUN position.
8. Verify the SIR indicator light flashes 7-9 times, if not as specified, inspect the system for a malfunction.

Steering Wheel

WARNING: *Most steering columns are collapsible. When replacing the wheel, do not hammer or exert any force against the column.*

CAUTION: *On 1991-92 models, the Supplemental Inflatable Restraint (SIR) System, must be disabled, before removing the steering wheel. Failure to do so may result in deployment of the air bag and possible personal injury.*

REMOVAL AND INSTALLATION

1. On 1991-92 models, disable the SIR system as outlined in this section.
2. On 1991-92 models, remove the Torx screws from the back of the steering wheel, disconnect the connector and remove the inflator module.
3. Disconnect the negative battery cable.

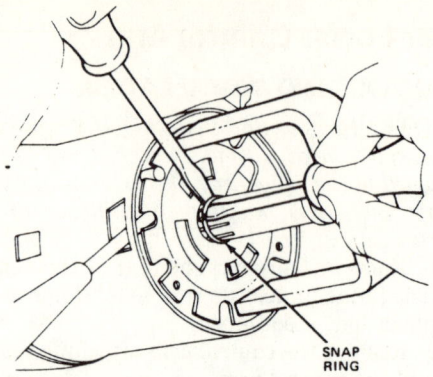

Removing the lockplate retaining ring

4. Remove the horn pad assembly or the horn button cap. Remove the snapring.
5. Remove the steering wheel nut.
6. Remove the upper horn insulator, receiver, and belleville spring.
7. Install the steering wheel puller, turn the puller bolt clockwise, and remove the wheel.
8. Installation is the reverse of the removal procedure. On 1991-92 models, connct the coil assembly, install the inflator module to the wheel and tighten the bolts to 25 inch lbs. Enable the SIR system.

Turn Signal Switch

REMOVAL AND INSTALLATION

1968

1. Disconnect the negative battery cable.
2. Disconnect the signal switch wiring from the wiring harness under the instrument panel.
3. Remove the steering wheel.
4. If applicable, remove the shift lever.
5. Remove the four-way flasher lever arm.
6. If equipped with an automatic transmission, remove the dial indicator housing and lamp assembly from the column.
7. Remove the mast jacket lower trim cover.
C8. Remove the C-ring and washers from the upper steering shaft.
9. Loosen the signal switch screws, move the switch counterclockwise, and remove it from the mast jacket.
10. Remove the upper support bracket assembly.

WARNING: *Support the column; do not allow it to be suspended by the lower reinforcement only.*

11. Remove the wiring harness protector and clip, and then reinstall the support bracket and finger-tighten the bolts.
12. Remove the shift lever bowl from the

SUSPENSION AND STEERING 391

mast jacket and disconnect it form the wiring harness.

13. Remove the three lockplate screws, being careful not to lose the three springs.

14. Disassemble the switch and upper bearing housing from the switch cover.

To Install:

15. Insert the upper bearing housing assembly and switch assembly into the switch cover.

16. Align the switch and bearing housing with the mounting holes in the cover and install the three mounting screws.

17. Slide the springs onto the screws and install the lockplate over the springs. Tighten the screws three turns into the lockplate.

18. Position the switch wire through the shift lever bowl and place the upper end assembly on top of the bowl.

19. Place the shift lever down and signal switch assembly on top of the jacket, insert the lockplate tangs into the slots.

20. Push down on the cover assembly and turn clockwise to lock the assembly into position.

21. Tighten the signal mounting screws.

22. Remove the mast jacket support bracket,

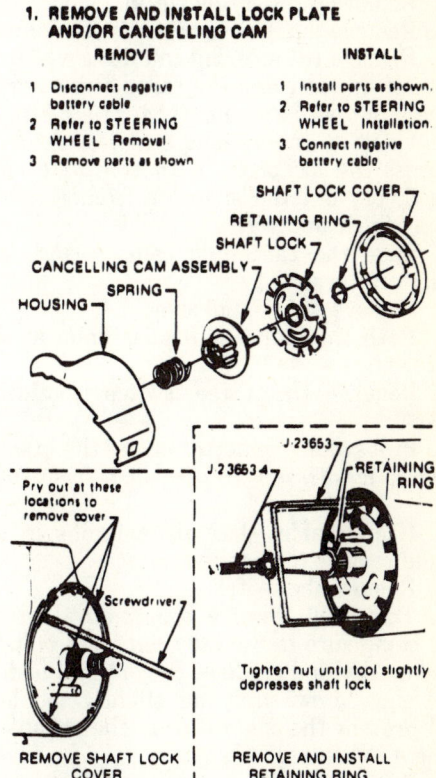

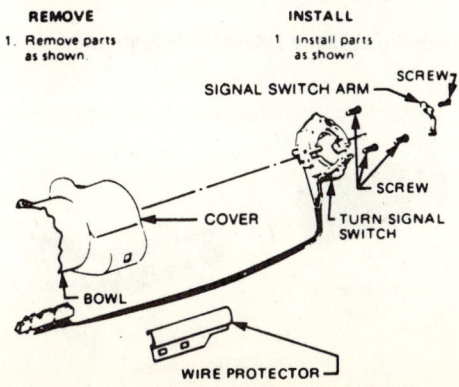

Turn signal switch removal, adjustable column

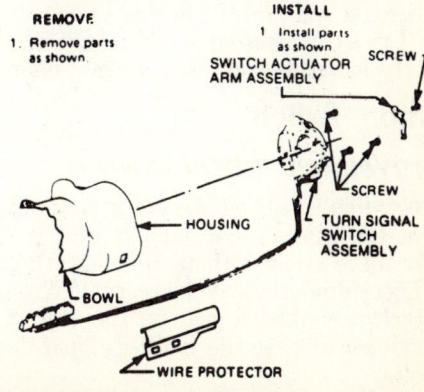

Turn signal switch removal, standard column 1980 model shown

SUSPENSION AND STEERING

then install the wiring, wiring cover and clip, and install and tighten the support brackets.

23. Install a C-ring onto the shaft.
24. Install the dial indicator and lamp assembly on the column if so equipped.
25. Install the mast jacket lower trim cover if so equipped.
26. Install the four-way flasher knob and the turn signal lever.
27. Install the shift lever.
28. Install the steering wheel.
29. Connect the wiring and battery cable.

1969-92

1. Remove the steering wheel.
2. Remove the trim cover from the column.
3. Remove the steering column cover from the shaft by removing the three screws or by prying it out with a small prybar (1976-88).
4. Using a compressing tool No. J-23653, compress the lockplate. With the plate compressed, pry out the snapring from its shaft grrove and throw away.
5. Slide the cancelling cam, spring, and washer off the shaft.
6. Remove turn signal lever.
7. Push the four-way flasher knob in and unscrew it.
8. Remove the three switch mounting screws.
9. Pull switch connector out of the bracket and wrap it with tape to prevent it from snagging.
10. If applicable, place tilt columns in the low and remove the harness cover.
11. Remove the switch.
12. To install, reverse the removal procedure, being sure to use only nuts and bolts of the same size and grade as the original fasteners. Using screws that are slightly too long could prevent the column from the collapsing during a collision.
13. When installing the canceling cam, spring, and washer, make sure that the switch is in neutral and that the flasher knob is out.
14. Use a compressing tool No. J-23653, compress the lockplate and install a new snapring.

Ignition Switch

REMOVAL AND INSTALLATION

The switch is located inside the channel section of the brake pedal support and is completely inaccessible without first lowering the steering column. the switch is actuated by a rod and rack assembly. A gear on the end of the lock cylinder engage the toothed upper end of the rod.

1. Disconnect the negative battery cable. Support and lower the steering column.
2. Place the ignition switch in the OFF-UNLOCKED position and move the actuating rod two detent from the top.
3. Remove the two mounting screws and the ignition switch assembly.
4. Before installing, place the new switch in the OFF-UNLOCKED position and make sure the ignition lock cylinder and the actuating rod are in the OFF-UNLOCKED (2nd detent from the top) position.
5. Install the actuating rod into the switch, mount the switch to the column and torque the mounting screws to 3 ft. lbs.

WARNING: *Use only the specified screws since over length screws could impair the collapsibility of the column.*

6. Install the steering column and torque the steering column-to-bracket nuts to 25 ft. lbs.

Lock Cylinder

REMOVAL AND INSTALLATION

1968

1. Disconnect the negative battery cable.
2. Reach under the dash and remove the electrical connectors from the ignition lock switch.
3. Remove the ignition lock switch retain-

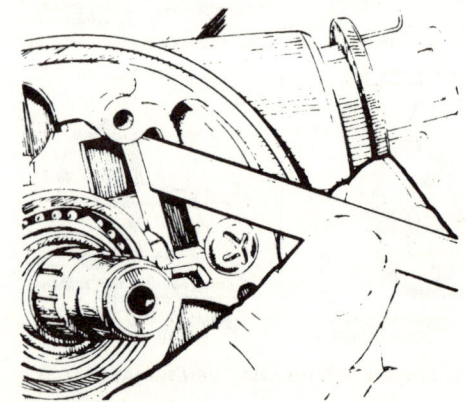

Removing the lock cylinder through 1977

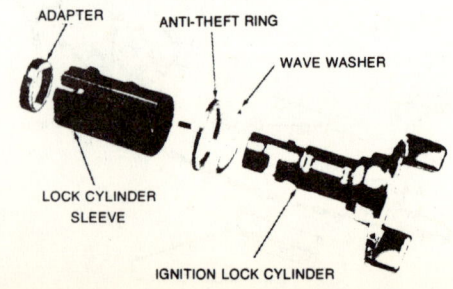

Exploded view of the lock cylinder

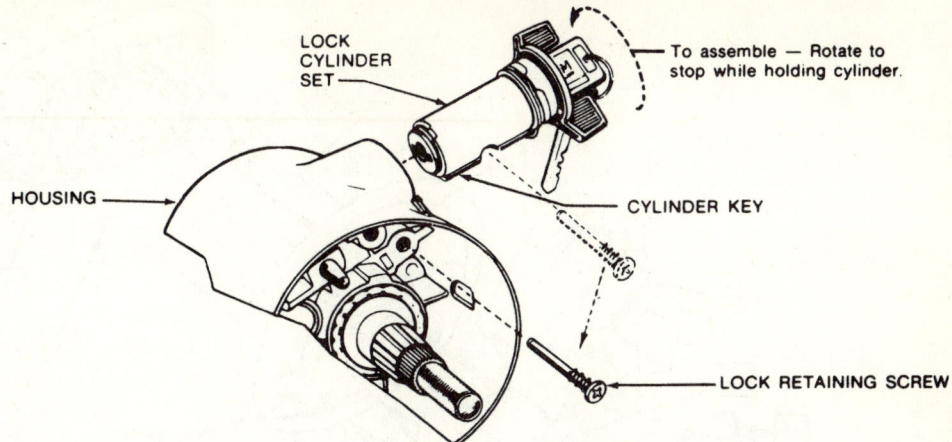
Removal and installation of the 1979 and later lock cylinder

ing clip from either the front or the rear of the assembly.

4. Remove the ignition lock assembly from the vehicle.

5. Installation is the reverse of the removal procedure.

1969-92

1. Disconnect the negative battery cable. Remove the steering wheel and the directional signal switch. It is not necessary to pull the wire harness out of the column. See the applicable procedures above.

2. Place the lock cylinder in Lock (up to 1970), or Run (1971-92).

3. Insert a small prybar into the turn signal housing slot. Keeping the screwdriver to the right side of the slot, break the housing flash loose and depress the spring latch at the lower end of the lock cylinder. Remove the lock cylinder.

NOTE: *Considerable force may be necessary to break this casting flash, but be careful not to damage any other parts. When ordering a new lock cylinder, specify a cylinder assembly. This will save assembling the cylinder washer, sleeve, and adaptor.*

4. To install, hold the lock cylinder sleeve and rotate the knob clockwise against the stop. Insert the cylinder into the housing, aligning the key and keyway. On 1969-76 vehicles hold a 0.070" drill between the lock bezel and the housing. On 1977-88 vehicles, push the cylinder into abutment of cylinder and sector. Rotate the cylinder counterclockwise, maintaining a light pressure until the drive section of the cylinder mates with the sector. Push in until the snapring pops into the grooves. Remove the drill. Check the operation of the cylinder. Install the direction signal switch and steering wheel.

Steering Column

REMOVAL AND INSTALLATION

b34 1968-90

1. **Disconnect the negative battery cable.**

NOTE: *If necessary to remove the steering wheel, be sure to use a steering wheel puller.*

2. Remove the nut/bolt from the upper intermediate shaft coupling, then separate the coupling from the lower end of the steering column.

3. If equipped with a column mounted shifter, disconnect the transmission control linkage from the column shift tube levers. If equipped with a floor shifter, disconnect the backdrive linkage.

4. At the steering column assembly, disconnect all of the electrical connectors.

5. Remove the floor pan cover-to-floor screws, the floor seal, and the cover.

6. Remove the steering column bracket-to-instrument panel nuts. If equipped with an automatic transmission, disconnect the shift position indicator pointer.

NOTE: *Once the steering column has been removed from the vehicle, be careful not to drop it (especially on it's end), lean on it or damage it in any way; the column is very susceptible to damage.*

7. Carefully remove the steering column from the vehicle.

To Install:

8. Position the steering column in place. Install the steering column bracket-to-instrument panel nuts. If equipped with an automatic transmission, disconnect the shift position indicator pointer.

9. Install the floor pan cover-to-floor screws, the floor seal, and the cover.

10. At the steering column assembly, connect all of the electrical connectors.

394 SUSPENSION AND STEERING

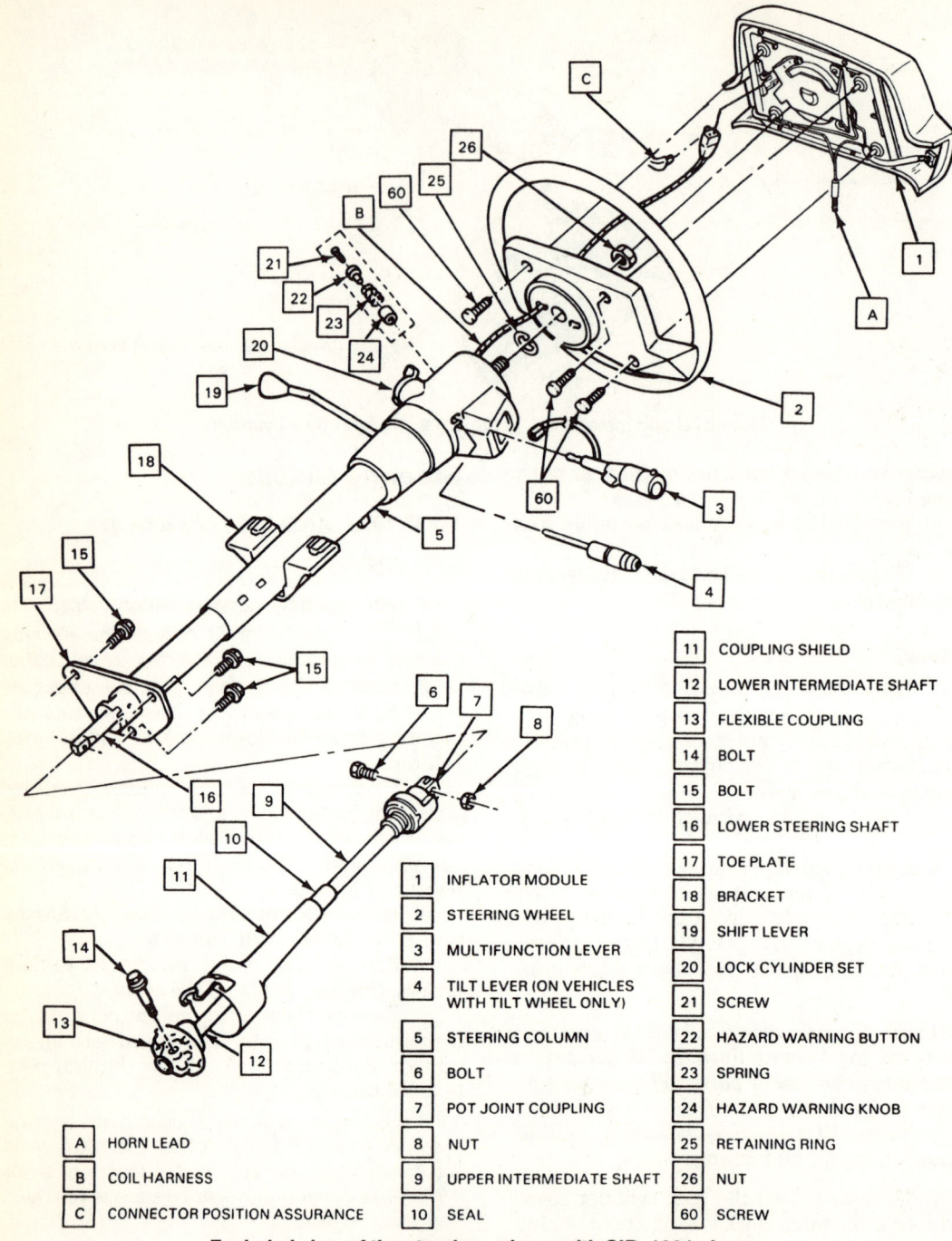

Exploded view of the steering column with SIR, 1991 shown

11. If equipped with a column mounted shifter, connect the transmission control linkage from the column shift tube levers. If equipped with a floor shifter, connect the backdrive linkage.

12. Install the nut/bolt from the upper intermediate shaft coupling, then separate the coupling from the lower end of the steering column.

13. If necessary to install the steering wheel.
14. Connect the negative battery cable.

1991-92

WARNING: *The wheels of the vehicle must be in the straight ahead position and the key must be in the LOCK position when removing or installing the steering column. Failure to do so will cause the coil assembly in* the steering column to become off center and possibly damage the coil or deploy the SIR module.

1. Disable the SIR system as outlined in this section.
2. Remove the stoplight switch.
3. Remove the bolt and nut from the joint coupler attaching the intermediate shaft to the steering column.
4 Disconnect the shift linkage from the steering column.
5. If necessary, remove the steering wheel.
6. Remove the steering column opening filler and the knee bolster and deflector.
7. Remove the bolts attaching the toe plate to the cowl and remove the shift indicator cable from the steering column.
8. Disconnect all the steering column electrical connections.
9. Remove the capsule nuts attaching the column support bracket to the instrument panel and remove the steering column from the vehicle.

To Install:
10. Tighten the capsule nuts attaching the column support bracket to the instrument panel to 20 ft. lbs.
11. Tighten the bolt and nut at the joint coupling to 40 ft. lbs.
12. The remainder of the installation is the reverse of removal.
13. Enable the SIR system as outlined in this section.

Manual Steering Gear

REMOVAL AND INSTALLATION

1. Raise and support the front of the vehicle safely.
2. Disconnect the steering shaft coupling.
3. Remove the pitman arm with a puller after marking the arm-to-shaft relationship.
4. Remove the steering gear-to-frame mounting bolts and remove the steering gear.
5. Reverse the removal steps to install the steering gear. Tighten the frame mounting bolts to 70 ft. lbs. Tighten the pitman shaft nut to 180 ft. lbs. and the steering coupling nuts to 20 ft. lbs.

ADJUSTMENT

1. Disconnect the negative battery cable.
2. At the steering wheel, remove the horn ring or button.
3. Remove the pitman arm-to-steering gear nut, then using the puller tool No. J-6632, pull the pitman arm from the steering gear.
   NOTE: *Be sure to mark the relationship of the pitman arm to the steering gear before it is removed.*

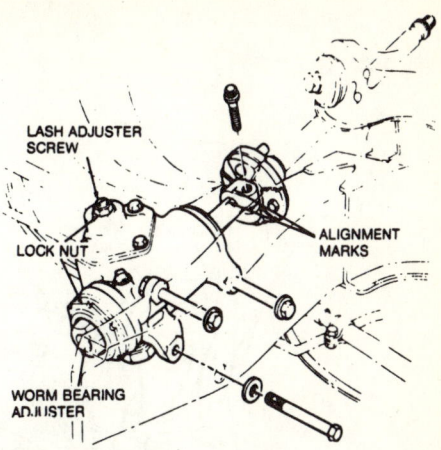

Typical steering gear mounting

4. Turn the steering wheel 1/2 turn from either stop, then loosen the sector shaft adjusting screw to eliminate the sector load.
   WARNING: *Turning the steering wheel too hard against the stops will damage the ball return guides.*
5. Place a 50 in. lbs. torque wrench on the steering wheel nut, then turn the wrench 90 degrees and observe the reading. If the force is less than 5-8 in. lbs. loosen the steering gear adjuster locknut and turn the worm thrust bearing adjuster to increase the preload.
6. Tighten the adjuster locknut and recheck the preload.
7. Reassemble the pitman arm to the pitman shaft, lining up the marks made during disassembly. Torque the retaining bolts to 70 ft. lbs.
8. Install the horn cap or ring and connect the battery cable.

Power Steering Gear

REMOVAL AND INSTALLATION

1. Raise and support the front of the vehicle safely.
2. Disconnect and plug the power steering lines. Disconnect the steering shaft coupling.
3. Remove the pitman arm with a puller after marking the arm-to-shaft relationship.
4. Remove the steering gear-to-frame mounting bolts and remove the steering gear.
5. Reverse the removal steps to install the steering gear. Tighten the frame mounting bolts to 70 ft. lbs. Tighten the pitman shaft nut to 180 ft. lbs. and the steering coupling nuts to 20 ft. lbs.

ADJUSTMENT

NOTE: *The steering gear must be removed from the vehicle in order to adjust the preload.*

SUSPENSION AND STEERING

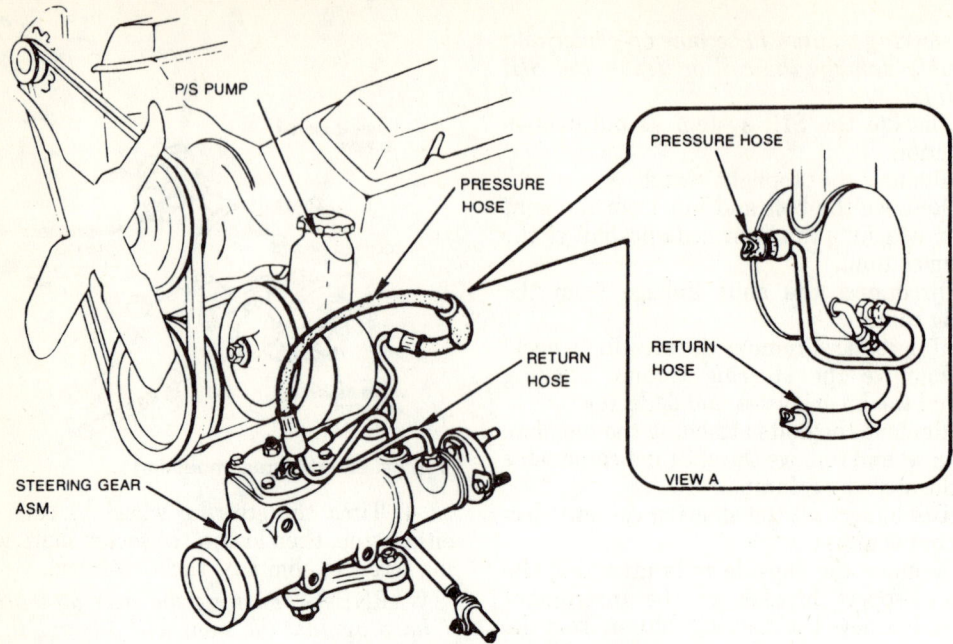

Power steering pump mounting

1. Rotate the stub shaft several times (from stop to stop) to drain the fluid from the steering gear.
2. Mount the steering gear in a vise and remove the adjuster plug locknut.
3. Using a spanner wrench, turn the adjusting plug clockwise until the plug and the thrust bearing are firmly bottomed (about 20 ft. lbs.).
4. Using the scribe mark on the housing (next to the hole in the adjuster plug), measure counterclockwise ($1/6$-$1/4''$) and mark the housing.
5. Turn the adjuster plug counterclockwise until the hole in the plug aligned with the second mark.
6. While holding the adjuster plug (to maintain position), tighten the locknut.
7. Using a inch lbs.. torque wrench and a $3/4''$ socket, turn the stub shaft to the right stop, then back $1/4$ turn and measure the drag. The reading must be within 4-10 in. lbs. (record the reading).
8. Rotate the stub shaft from stop to stop, the back to the center. Using the torque wrench, turn the stub shaft 45 degree to each side of center and check the reading.
9. Loosen the locknut and turn the pre-load adjusting screw clockwise until the over center (additional torque) reading of 4-8 in. lbs. (new gear, not to exceed 18 in. lbs.) or 4-5 in. lbs. (used gear, not to exceed 14 in. lbs.).
10. While holding the adjuster plug, tighten the locknut.
11. Reinstall the steering gear. Bleed the power steering system, as required.

Power Steering Pump

REMOVAL AND INSTALLATION

1. Disconnect the negative battery cable. Remove the hoses at the pump and tape the openings shut to prevent contamination.
2. Remove the power steering pump belt.
3. Loosen the retaining bolts and any braces, and remove the pump.
4. Install the pump on the engine with the retaining bolts hand-tight.
5. Connect and tighten the hose fittings. Install the power steering pump belt.
6. Fill the power steering pump with fluid and bleed by turning the pulley counterclockwise (viewed from the front). Stop the bleeding when air bubbles no longer appear.

SYSTEM BLEEDING

1. Fill the fluid reservoir.
2. Let the fluid stand undisturbed for two minutes, then crank the engine for about two seconds. Refill reservoir is necessary.
3. Repeat Steps 1 and 2 above until the fluid level remains constant after cranking the engine.
4. Raise the front of the vehicle until the wheels are off the ground, then start the engine. Increase the engine speed to about 1,500 rpm.
5. Turn the wheels to the left and right, checking the fluid level and refilling it necessary. If the oil is extremely foamy, allow the vehicle to stand a few minutes with the engine off, then repeat the process.

SUSPENSION AND STEERING

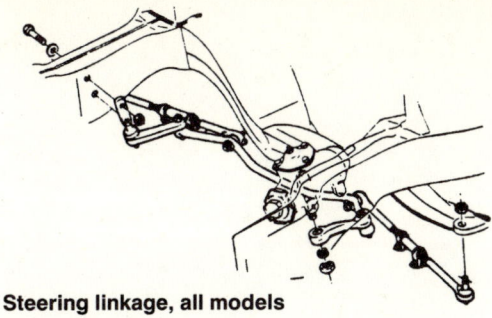

Steering linkage, all models

Steering Linkage

REMOVAL AND INSTALLATION

Tie Rod

1. Raise and support the vehicle safely.
2. Remove the cotter pins and nuts from the tie rod end studs.
3. Tap on the steering arm near the tie rod end (use another hammer as backing) or Ball Stud Puller tool No. J-6627 and pull down on the tie rod, if necessary, to free it.
4. Remove the inner ball stud in the same manner as the outer.
   WARNING: *DO NOT disengage the joint by driving a wedge between the joint and the knuckle, for damage to the seal may result.*
5. Loosen the clamp bolt and unscrew the ends if they are being replaced.
6. Lubricate the tie rod end threads with chassis grease if they were removed. Install each end assembly an equal distance from the sleeve.
7. Ensure that the tie rod end threads with chassis grease if they were removed. Install each end assembly an equal distance from the sleeve.
8. Install the stud nuts. Tighten the inner and outer end nuts to 35 ft. lbs.
9. Adjust the toe-in to specifications.
   NOTE: *Before tightening the sleeve clamps, ensure that the clamps are positioned so that adjusting sleeve sot is covered by the clamp.*

Idler Arm

1. Raise and support the vehicle safely.
2. Remove the idler arm-to-frame nut, washer, and bolt.
3. Remove the cotter pin and nut from the idler arm-to-relay rod ball end stud.
4. Tap the relay rod with a hammer, using another hammer as backing, or using the puller tool No. J-24319-01 to remove the relay rod from the idler arm.
5. Remove the idler arm.
6. Place the idler arm on the frame and install the retaining bolts and nuts. Tighten the nuts to 35 ft. lbs.

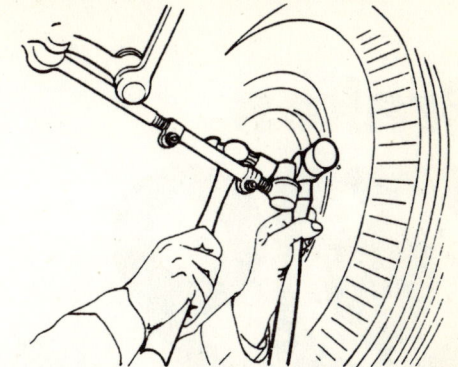

Freeing the tie rod end, use another hammer as backup

7. Position the relay rod on the idler arm. Ensure that the seal is on the stud. Install the nut and tighten to 35 ft. lbs. Install a cotter pin.

Relay Rod (Intermediate Rod)

1. Raise and support the vehicle safely.
2. Remove the inner tie rod ends from the relay rod.
3. If equipped, remove the damper from the relay rod.
4. Remove the relay rod stud nut and cotter pin from the pitman arm. Free the relay rod from the pitman arm using the puller tool No. J-24319-01, moving the steering linkage if necessary. Repeat this operation to remove the relay rod from the idler arm and remove the relay rod from the vehicle.
5. Install the relay rod on the idler arm. Tighten the nut to 35 ft. lbs.
6. Raise the relay and install it on the pitman arm. Tighten the nut to 45 ft. lbs.
7. Install the tie rod ends to the relay rod. Install the damper, as required. Adjust the toe-in.

Pitman Arm

1. Raise and support the vehicle safely.
2. Remove the pitman arm stud nut and cotter pin.
3. Tap the relay rod off the pitman arm, using another hammer as backing or the puller tool No. J-24319-01. Pull the relay rod off the pitman arm stud.
4. Remove the pitman arm nut and mark the arm-to-shaft relationship.
5. Remove the pitman arm using a puller tool No. J-6632.
6. Install the pitman arm on the shaft, aligning the previously made marks. Install the pitman shaft nut and tighten it to 180 ft. lbs.
7. Install the relay rod on the pitman. Tighten the nut to 45 ft. lbs. and install the cotter pin.

Brakes

BASIC OPERATING PRINCIPLES

Hydraulic systems are used to actuate the brakes of all automobiles. The system transports the power required to force the frictional surfaces of the braking system together from the pedal to the individual brake units at each wheel. A hydraulic system is used for two reasons.

First, fluid under pressure can be carried to all parts of an automobile by small pipes and flexible hoses without taking up a significant amount of room or posing routing problems.

Second, a great mechanical advantage can be given to the brake pedal end of the system, and the foot pressure required to actuate the brakes can be reduced by making the surface area of the master cylinder pistons smaller than that of any of the pistons in the wheel cylinders or calipers.

The master cylinder consists of a fluid reservoir and a double cylinder and piston assembly. Double type master cylinders are designed to separate the front and rear braking systems hydraulically in case of a leak.

Steel lines carry the brake fluid to a point on the vehicle's frame near each of the vehicle's wheels. The fluid is then carried to the calipers and wheel cylinders by flexible tubes in order to allow for suspension and steering movements.

In drum brake systems, each wheel cylinder contains two pistons, one at either end, which push outward in opposite directions.

In disc brake systems, the cylinders are part of the calipers. One cylinder in each caliper is used to force the brake pads against the disc.

All pistons employ some type of seal, usually made of rubber, to minimize fluid leakage. A rubber dust boot seals the outer end of the cylinder against dust and dirt. The boot fits around the outer end of the piston on disc brake calipers, and around the brake actuating rod on wheel cylinders.

The hydraulic system operates as follows: When at rest, the entire system, from the piston(s) in the master cylinder to those in the wheel cylinders or calipers, is full of brake fluid. Upon application of the brake pedal, fluid trapped in front of the master cylinder piston(s) is forced through the lines to the wheel cylinders. Here, it forces the pistons outward, in the case of drum brakes, and inward toward the disc, in the case of disc brakes. The motion of the pistons is opposed by return springs mounted outside the cylinders in drum brakes, and by spring seals, in disc brakes.

Upon release of the brake pedal, a spring located inside the master cylinder immediately returns the master cylinder pistons to the normal position. The pistons contain check valves and the master cylinder has compensating ports drilled in it. These are uncovered as the pistons reach their normal position. The piston check valves allow fluid to flow toward the wheel cylinders or calipers as the pistons withdraw. Then, as the return springs force the brake pads or shoes into the released position, the excess fluid reservoir through the compensating ports. It is during the time the pedal is in the released position that any fluid that has leaked out of the system will be replaced through the compensating ports.

Dual circuit master cylinders employ two pistons, located one behind the other, in the same cylinder. The primary piston is actuated directly by mechanical linkage from the brake pedal through the power booster. The secondary piston is actuated by fluid trapped between the two pistons. If a leak develops in front of the secondary piston, it moves forward until it bottoms against the front of the master cylinder, and the fluid trapped between the pistons will operate the rear brakes. If the rear brakes develop a leak, the primary piston will move forward until direct contact with the secondary piston takes place, and it will force the secondary piston to actuate the front brakes. In either case, the brake pedal moves farther when the brakes are applied, and less braking power is available.

All dual circuit systems use a switch to warn the driver when only half of the brake system is operational. This switch is located in a valve body which is mounted on the firewall or the frame below the master cylinder. A hydraulic piston receives pressure from both circuits, each circuit's pressure being applied to one end of the piston. When the pressures are in balance, the piston remains stationary. When one circuit has a leak, however, the greater pressure in that circuit during application of the brakes will push the piston to one side, closing the switch and activating the brake warning light.

In disc brake systems, this valve body also contains a metering valve and, in some cases, a proportioning valve. The metering valve keeps pressure from traveling to the disc brakes on the front wheels until the brake shoes on the rear wheels have contacted the drums, ensuring that the front brakes will never be used alone. The proportioning valve controls the pressure to the rear brakes to lessen the chance of rear wheel lock-up during very hard braking.

Warning lights may be tested by depressing the brake pedal and holding it while opening one of the wheel cylinder bleeder screws. If this does not cause the light to go on, substitute a new lamp, make continuity checks, and, finally, replace the switch as necessary.

The hydraulic system may be checked for leaks by applying pressure to the pedal gradually and steadily. If the pedal sinks very slowly to the floor, the system has a leak. This is not to be confused with a springy or spongy feel due to the compression of air within the lines. If the system leaks, there will be a gradual change in the position of the pedal with a constant pressure.

Check for leaks along all lines and at wheel cylinders. If no external leaks are apparent, the problem is inside the master cylinder.

Disc Brakes

BASIC OPERATING PRINCIPLES

Instead of the traditional expanding brakes that press outward against a circular drum, disc brake systems utilize a disc (rotor) with brake pads positioned on either side of it. Braking effect is achieved in a manner similar to the way you would squeeze a spinning phonograph

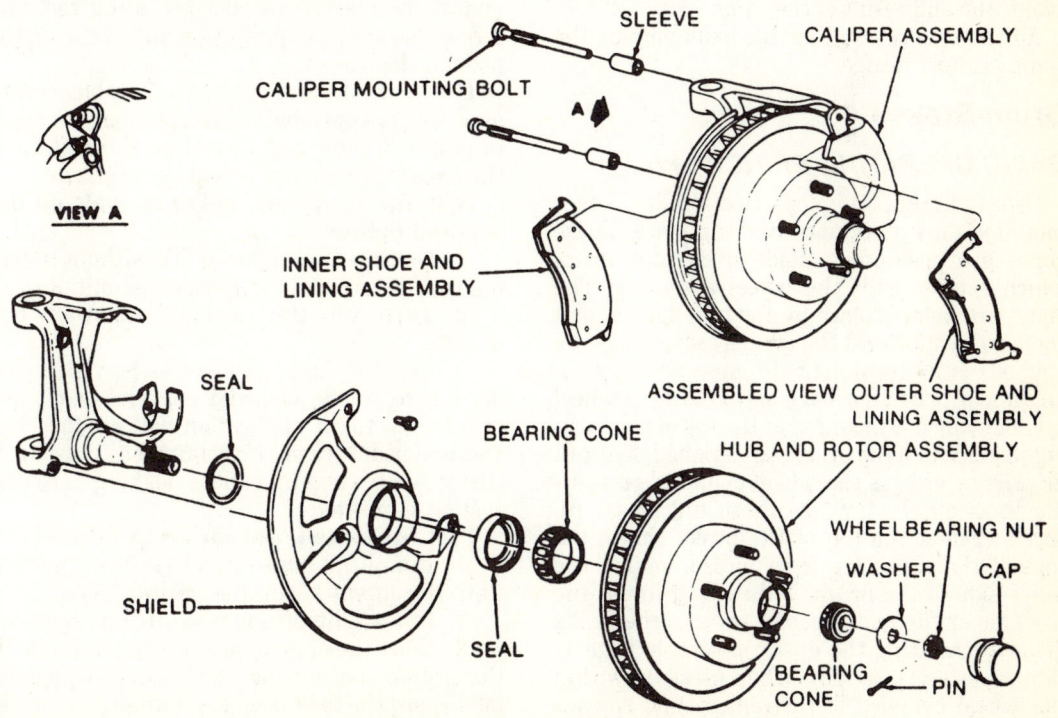

Exploded view of the disc brake assembly

record between your fingers. The disc (rotor) is a casting with cooling fins between the two braking surfaces. This enables air to circulate between the braking surfaces making them less sensitive to heat buildup and more resistant to fade. Dirt and water do not affect braking action since contaminants are thrown off by the centrifugal action of the rotor or scraped off the by the pads. Also, the equal clamping action of the two brake pads tends to ensure uniform, straight line stops. Disc brakes are inherently self-adjusting.

There are three general types of disc brake:
1. A fixed caliper.
2. A floating caliper.
3. A sliding caliper.

The fixed caliper design uses two pistons mounted on either side of the rotor (in each side of the caliper). The caliper is mounted rigidly and does not move.

The sliding and floating designs are quite similar. In fact, these two types are often lumped together. In both designs, the pad on the inside of the rotor is moved into contact with the rotor by hydraulic force. The caliper, which is not held in a fixed position, moves slightly, bringing the outside pad into contact with the rotor. There are various methods of attaching floating calipers. Some pivot at the bottom or top, and some slide on mounting bolts. In any event, the end result is the same.

All the cars covered in this book employ the sliding caliper design.

Drum Brakes

BASIC OPERATING PRINCIPLES

Drum brakes employ two brake shoes mounted on a stationary backing plate. These shoes are positioned inside a circular drum which rotates with the wheel assembly. The shoes are held in place by springs. This allows them to slide toward the drums (when they are applied) while keeping the linings and drums in alignment. The shoes are actuated by a wheel cylinder which is mounted at the top of the backing plate. When the brakes are applied, hydraulic pressure forces the wheel cylinder's actuating links outward. Since these links bear directly against the top of the brake shoes, the tops of the shoes are then forced against the inner side of the drum. This action forces the bottoms of the two shoes to contact the brake drum by rotating the entire assembly slightly (known as servo action). When pressure within the wheel cylinder is relaxed, return springs pull the shoes back away from the drum.

Most modern drum brakes are designed to self-adjust themselves during application when the vehicle is moving in reverse. This motion causes both shoes to rotate very slightly with the drum, rocking an adjusting lever, thereby causing rotation of the adjusting screw.

Power Boosters

Power brakes operate just as non-power brake systems except in the actuation of the master cylinder pistons. A vacuum diaphragm is located on the front of the master cylinder and assists the driver in applying the brakes, reducing both the effort and travel he must put into moving the brake pedal.

The vacuum diaphragm housing is connected to the intake manifold by a vacuum hose. A check valve is placed at the point where the hose enters the diaphragm housing, so that during periods of low manifold vacuum brake assist vacuum will not be lost.

Depressing the brake pedal closes off the vacuum source and allows atmospheric pressure to enter on one side of the diaphragm. This causes the master cylinder pistons to move and apply the brakes. When the brake pedal is released, vacuum is applied to both sides of the diaphragm, and return springs return the diaphragm and master cylinder pistons to the released position. If the vacuum fails, the brake pedal rod will butt against the end of the master cylinder actuating rod, and direct mechanical application will occur as the pedal is depressed.

The hydraulic and mechanical problems that apply to conventional brake systems also apply to power brakes, and should be checked for if the tests below do not reveal the problem.

Test for a system vacuum leak as described below:

1. Operate the engine at idle without touching the brake pedal for at least one minute.
2. Turn off the engine, and wait one minute.
3. Test for the presence of assist vacuum by depressing the brake pedal and releasing it several times. Light application will produce less and less pedal travel, if vacuum was present. If there is no vacuum, air is leaking into the system somewhere.

Test for system operation as follows:

1. Pump the brake pedal (with engine off) until the supply vacuum is entirely gone.
2. Put a light, steady pressure on the pedal.
3. Start the engine, and operate it at idle. If the system is operating, the brake pedal should fall toward the floor if constant pressure is maintained on the pedal.

Power brake systems may be tested for hydraulic leaks just as ordinary systems are tested.

BRAKES

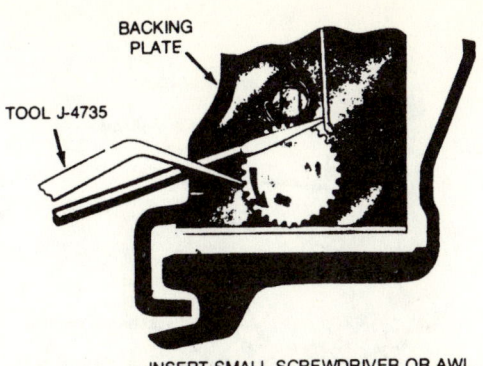

INSERT SMALL SCREWDRIVER OR AWL THROUGH BACKING PLATE SLOT AND HOLD ADJUSTER LEVER AWAY FROM SPROCKET BEFORE BACKING OFF BRAKE SHOE ADJUSTMENT

Brake adjustment

ADJUSTMENTS

Drum Brakes

1. Raise the vehicle and support it with safety stands.
2. Remove the rubber plug from the adjusting slot on the backing plate.
3. Insert a brake adjusting spoon into the slot and engage the lowest possible tooth on the starwheel. Move the end of the brake spoon downward to move the starwheel upward and expand the adjusting screw. Repeat this operation until the brakes lock the wheels.
4. Insert the proper brake tool or a piece of firm wire (coat hanger wire) into the adjusting slot and push the automatic adjuster lever out and free of the starwheel on the adjusting screw.
5. Holding the adjusting lever out of the way, engage the top most tooth possible on the starwheel with a brake adjusting spoon. Move the end of the adjusting spoon upward to move the adjusting screw starwheel downward and contact the adjusting screw. Back off the adjusting screw starwheel until the wheel spins freely with a minimum of drag. Keep track of the number of turns the starwheel is backed off.
6. Repeat the operation for the other side. When backing off the brakes on the other side, the adjusting lever must be backed off the same number of turns to prevent side-to-side brake pull.
7. Repeat this operation on all service brake assemblies.
8. When all brakes (front if so equipped and rear) are adjusted properly, lower the vehicle.
9. Road test the vehicle and make several stops, while backing the vehicle in reverse, to equalize all the wheels.

Brake Pedal

TRAVEL

The pedal travel is the distance which the pedal moves toward the floor from the fully released position. Inspection should be made with the brake pedal firmly depressed and when the brake system is cold. The brake pedal travel should be $2^{1}/_{4}''$ (1968-84), $2^{3}/_{4}''$ (1985 and later) or $3^{1}/_{3}''$ for vehicles equipped with hydro-boost system.

NOTE: *If equipped with power brakes, be sure to pump the brakes 3 times with the engine Off, before making the travel check.*

1. Under the dash, remove the pushrod-to-pedal clevis pin and separate the pushrod from the brake pedal.
2. Loosen the pushrod adjuster lock nut, then adjust the pushrod.
3. After the correct travel is established, tighten the pushrod adjuster locknut.

Brake Light Switch

REMOVAL AND INSTALLATION

When the brake pedal is in the fully released position, the stop light switch plunger should be fully depressed against the pedal arm. The switch is adjusted by moving it in or out.

1. Disconnect the negative battery cable. Disconnect the stop light switch electrical connectors.
2. Remove the switch from the bracket.
3. Make sure that the tubular clip is in the brake pedal mounting bracket.
4. Depress the brake pedal and insert the switch into the tubular clip until it seats on the clip, at this point a click will be heard.
5. Pull the brake pedal fully rearward against the pedal stop, until the clicking sounds can no longer be heard. At this point the switch is adjusting itself in the bracket.
6. Release the brake pedal, then pull the pedal rearward again to assure that the adjustment is complete.

Master Cylinder

REMOVAL AND INSTALLATION

NOTE: *Vehicles with disc brakes do not have a check valve in the front outlet port of the master cylinder. If one is installed, the front discs will quickly wear out due to residual hydraulic pressure holding the pads against the rotor.*

1. Disconnect the negative battery cable. Disconnect and plug the hydraulic lines from the master cylinder.
2. Remove the retaining nuts and the lock-

402 BRAKES

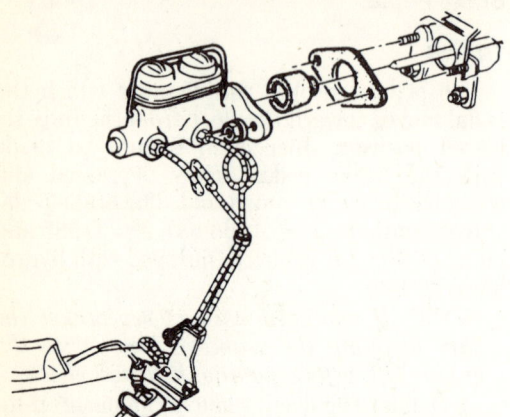

Master cylinder mounting-vehicles without power brakes

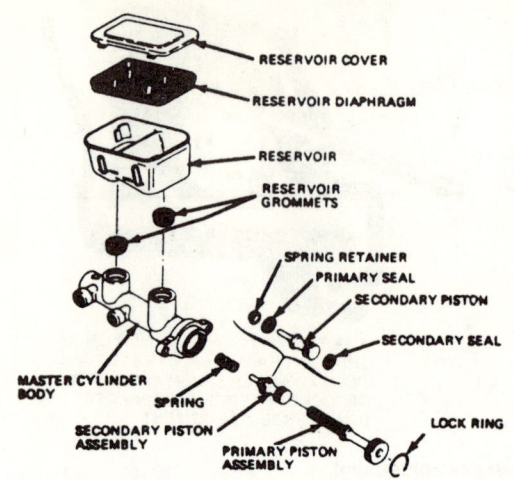

Master cylinder with removable reservoir

washers holding the cylinder to the cowl or the brake booster.

NOTE: *If equipped with non-power brakes, disconnect the pushrod at the brake pedal.*

3. Remove the master cylinder, the gasket and the rubber boot.

To install:

4. Position the master cylinder and torque the master cylinder mounting nuts to 22 ft. lbs. and the hydraulic lines to 18 ft. lbs. Refill (bench bleed master cylinder before installation—refer to the necessary procedure) the master cylinder, bleed the brake system and check the brake pedal free-play.

NOTE: *On non-powered brakes, position the master cylinder on the cowl, making sure that the pushrod goes through the rubber boot into the piston. Reconnect the pushrod clevis to the brake pedal. If equipped with power brakes, install the master cylinder on the power booster.*

MASTER CYLINDER BENCH BLEEDING

1. Place the master cylinder in a vise.
2. Connect two lines to the fluid outlet orifices, and into the reservoir.
3. Fill the reservoir with brake fluid.
4. Using a wooden dowel, depress the pushrod slowly, allowing the pistons to return. Do this several times until the air bubbles are all expelled—do not let master cylinder assembly run dry.

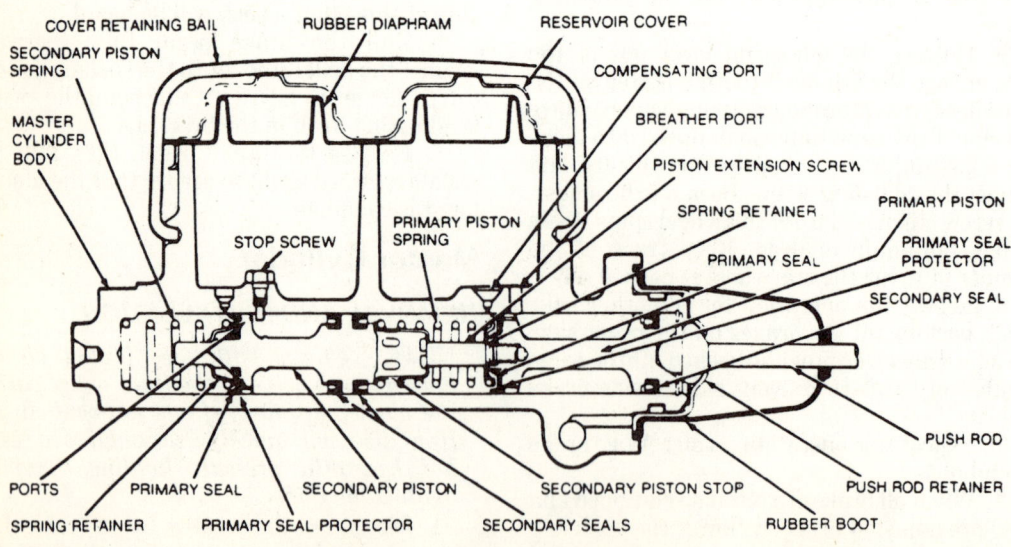

Cross section of a master cylinder

BRAKES 403

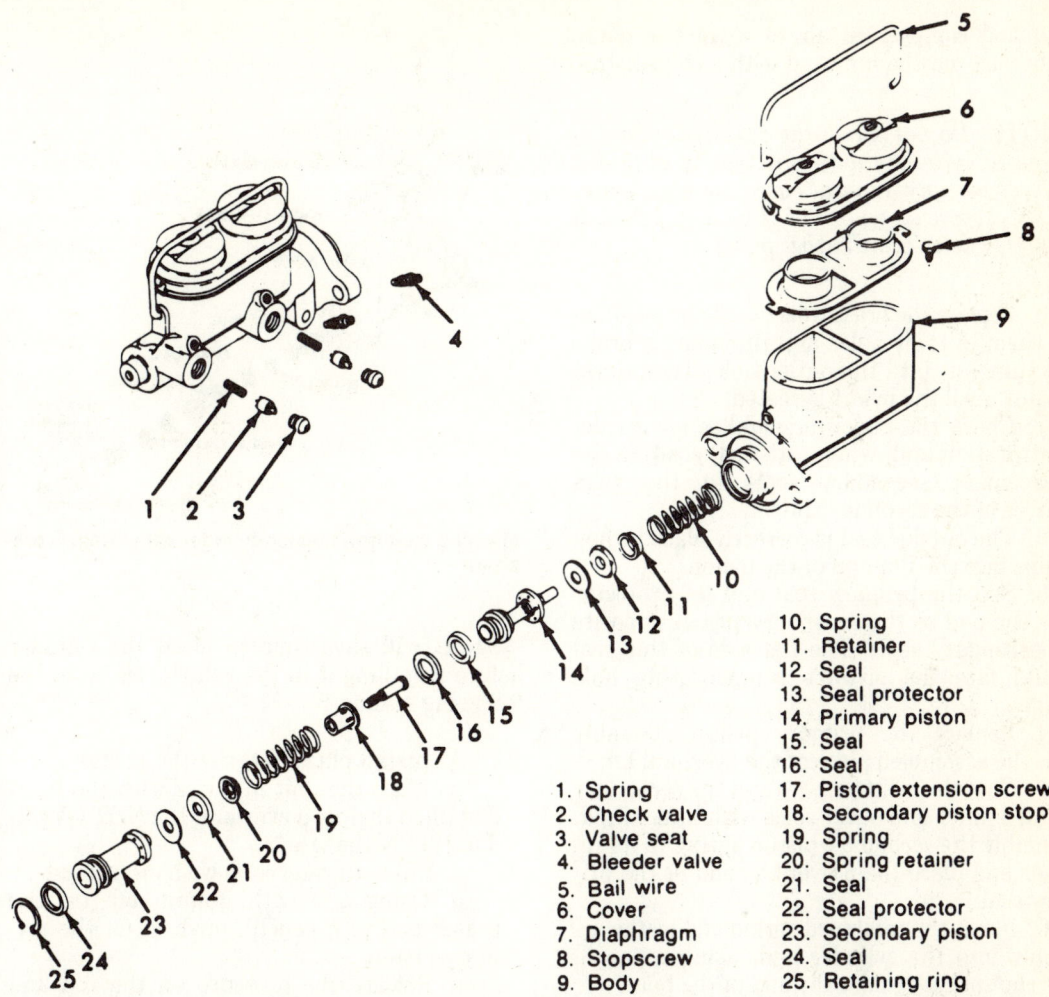

Exploded view of a master cylinder. Some 1980 and all 1981 models no longer use the bails to retain the reservoir cap

1. Spring
2. Check valve
3. Valve seat
4. Bleeder valve
5. Bail wire
6. Cover
7. Diaphragm
8. Stopscrew
9. Body
10. Spring
11. Retainer
12. Seal
13. Seal protector
14. Primary piston
15. Seal
16. Seal
17. Piston extension screw
18. Secondary piston stop
19. Spring
20. Spring retainer
21. Seal
22. Seal protector
23. Secondary piston
24. Seal
25. Retaining ring

5. Remove the bleeding tubes from the master cylinder, plug the outlets and install the caps.

OVERHAUL

NOTE: *Use this service procedure and exploded view illustrations as a guide for overhaul of the master cylinder assembly. If in doubt about overhaul condition or service procedure REPLACE the complete assembly with a new master cylinder assembly.*

1. Remove the master cylinder from the vehicle.
2. Remove the mounting gasket and boot, and the main cover. Empty the cylinder of all fluid.
3. Place the cylinder in a vise and remove the pushrod retainer and the secondary piston stop bolt that are found inside the front reservoir.
4. Remove the retaining ring and primary piston assembly.

5. Direct compressed air into the piston stop screw hole to force the secondary piston, spring, and retainer from the cylinder bore. If compressed air isn't available, use a hooked wire to pull out the secondary piston.

6. Check the brass tube fitting inserts and, if damaged, remove them; if not, leave them in place.

7. If insert replacement is necessary, thread a No. 6-32 x $5/8''$ self-tapping screw into the insert. Hook the end of the screw with a claw hammer and pull out the insert.

8. An alternative (but more troublesome) way to remove the inserts is to drill out the outlet holes with a $11/64''$ drill and then thread them with a $1/4''$-20 tap. Position a thick washer over the hole to serve as a spacer and then thread a $1/4''$-20 x $3/4''$ hex-head bolt into the insert and tighten the bolt until the insert is free.

9. Use only denatured alcohol or brake

fluid and compressed air to clean the parts. slight rust may be removed with crocus cloth.

NOTE: *Do not polish the aluminum bore of type A: cylinders with any type of abrasive. Never use any mineral based solvents (gasoline, kerosene, etc.) for cleaning. It will quickly deteriorate rubber parts.*

10. Replace the brass tube inserts by positioning them in their holes and threading a brake line tube nut into the outlet hole. Turn down the nut until the insert is seated.
11. Check the piston assemblies for correct identification and, when satisfied, position the replacement secondary seals in the twin grooves of the secondary piston.
12. The outside seal is correctly placed when its lips face the flat end of the piston.
13. Slip the primary seal and its protector over the end of the secondary piston opposite the secondary seals. The flat side of this seal should face the piston's compensating hole flange.
14. Replace the primary piston assembly with the assembled piece of the overhaul kit.
15. Coat the cylinder bore and the secondary piston's inner and outer seals with brake fluid. Assembly the secondary piston spring to its retainer and place them over the end of the primary seal.
16. Insert the combined spring and piston assembly into the cylinder and, using a pencil, seat the spring against the end of the bore.
17. Coat the primary piston seals with brake fluid and push it (pushrod receptacle end out) into the cylinder.
18. Hold the piston in and snap the retaining ring into place.
19. Continue to hold the piston down to make sure that all components are seated and insert the secondary piston stop screw in its hole in the bottom of the front reservoir. Torque the screw to 25-40 in. lbs.
20. Install the reservoir diaphragm and cover.

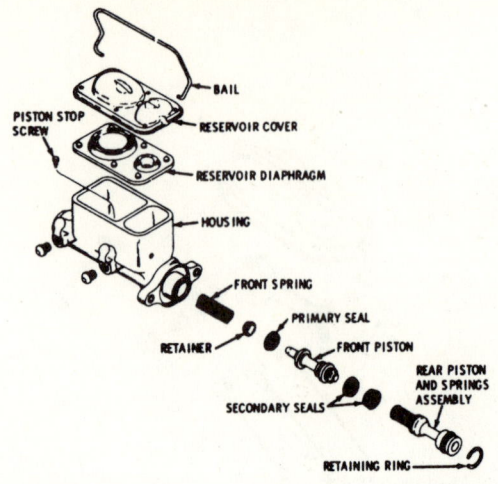

Moraine cast iron master cylinder with integral reservoir

21. It will save time to bleed the cylinder before installing it in the vehicle. Do so in the following manner:

  a. Install plugs in the outlet ports.
  b. Place the unit in a vise with the front end tilted slightly downward. DO NOT OVERTIGHTEN the vise.
  c. Fill both reservoirs with clean fluid.
  d. Using a smooth, round rod (try the eraser end of a pencil), push in on the primary piston.
  e. Release the pressure on the rod and watch for air bubbles in the fluid. Keep repeating this until the bubbles disappear.
  f. Loosen the vise and position the cylinder so the front end is tilted slightly upward. Repeat Steps D and E.
  g. Place the diaphragm cover on the reservoir.

NOTE: *Master cylinder overhaul on vehicles with power brakes is the same as above.*

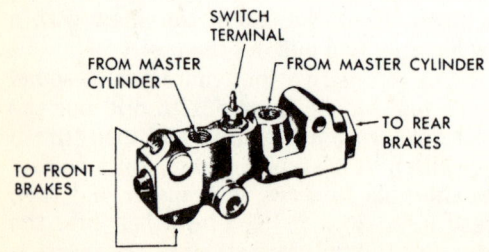

Combination valve assembly

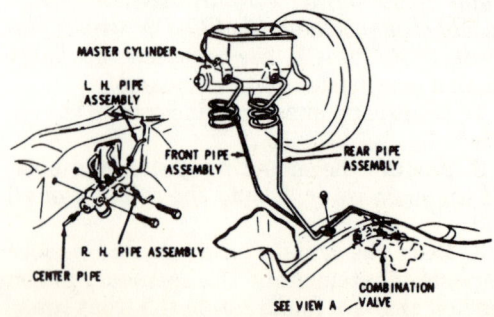

Combination valve mounting on front frame member

BRAKES 405

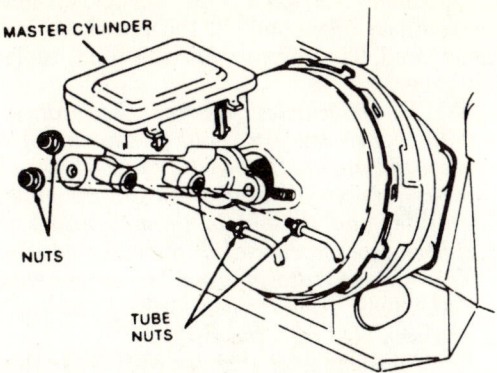

Master cylinder and brake booster assembly

Combination Valve

REMOVAL AND INSTALLATION

1. Disconnect the negative battery cable.
2. Disconnect the electrical lead from the switch.
3. Place rags under the unit to absorb any spilled brake fluid.
4. Clean any dirt from the hydraulic lines and the switch/valve assembly. Disconnect the hydraulic lines from the assembly. If necessary, loosen the line connections at the master cylinder. Tape the open line ends to prevent the entrance of dirt.
5. Remove the mounting screws and remove the switch/valve assembly.

To install:

6. Make sure that the new unit is clean and free of dust and lint. If in doubt, wash the new unit in clean brake fluid.
7. Place the new unit in position and install it to its mounting bracket with screws.
8. Remove the tape from the hydraulic lines and connect them to the unit. If necessary, tighten the line connections at the master cylinder.
9. Connect the electrical lead.
10. Connect the negative battery cable.
11. Bleed the brake system.

Power Brake Booster

REMOVAL AND INSTALLATION

Except Hydro-boost System

1. Disconnect the negative battery cable. Disconnect the vacuum hose from the vacuum check valve.
2. Unbolt the master cylinder and carefully move it aside without disconnecting the hydraulic lines.

NOTE: *If sufficient booster clearance cannot be obtained, it will be necessary to disconnect and cap the hydraulic lines from the master cylinder, then remove the master cylinder.*

3. Disconnect the pushrod at the brake pedal assembly.

NOTE: *Some brake boosters may be held on with sealant, this can be easily removed with tar remover.*

4. Remove the booster-to-cowl nuts and lockwashers and the booster from engine compartment.

To install:

5. Place the booster into position, connect the pushrod at the brake pedal assembly and torque the booster-to-cowl and the master cylinder-to-booster mounting nuts to 28 ft.lbs.

NOTE: *Make sure to check the operation of the stop lights. Allow the engine vacuum to build before applying the brakes. Bleed the hydraulic system if the lines were disconnected from the master cylinder.*

Hydro-Boost System

Hydro-Boost differs from conventional power brake systems, in that it operates from power steering pump fluid pressure rather than intake manifold vacuum.

The Hydro-Boost unit contains a spool valve with an open center which controls the strength of the pump pressure when braking occurs. A lever assembly controls the valve's position. A boost piston provides the force necessary to operate the conventional master cylinder on the front of the booster.

A reserve of at least two assisted brake applications is supplied by an accumulator which is spring loaded on earlier vehicles and pneumatic on later vehicles. The accumulator is an integral part of the Hydro-Boost II unit. The brakes can be applied manually if the reserve system is depleted.

All system checks, tests and troubleshooting procedure are the same for the two systems.

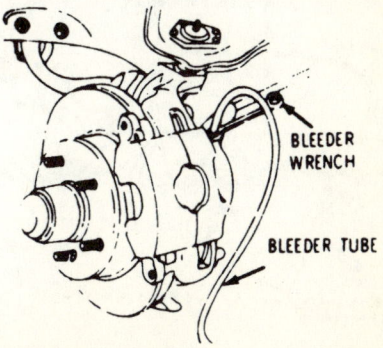

Wrench and plastic tubing fit on the bleeder nipple on each caliper

1. Disconnect the negative battery cable. Turn the engine off and pump the brake pedal 4 or 5 times to deplete the accumulator.
2. Remove the nuts from the master cylinder, then move the master cylinder away from the booster, with brake lines still attached.
3. Remove the fluid lines from the booster.
4. Remove the retainer and washer at the brake pedal.
5. Remove the attaching nuts retaining the booster fastened to the cowl and the booster.

To install:
6. Place the booster into position, reconnect the fluid lines, reconnect the to the brake pedal and torque the booster-to-cowl nuts to 15 ft. lbs. and the master cylinder-to-booster nuts to 20 ft. lbs. Bleed the power steering and hydro-booster system.

Bleeding

The hydraulic brake system must be bled any time one of the lines is disconnected or any time air enters the system. If the brake pedal feels spongy upon application, and goes almost to the floor but regains height when pumped, air has entered the system. It must be bled out. Check for leads that would have allowed the entry of air and repair them before bleeding the system. The correct bleeding sequence is: right rear wheel cylinder, left rear, right front and left front. If the master cylinder is equipped with bleeder valves, bleed them first then go to the wheel cylinder nearest the master cylinder (left front) followed by the right front, left rear, and right rear.

This method of bleeding requires two people, one to depress the brake pedal and the other to open the bleeder screws.

1. Clean the top of the master cylinder, remove the cover and fill the reservoirs with clean fluid. To prevent squirting fluid, replace the cover.
NOTE: *On vehicles with front disc brakes, it will be necessary to hold in the metering valve pin during the bleeding procedure. The metering valve is located beneath the master cylinder and the pin is situated under the rubber boot on the end of the valve housing. This may be taped in or held by an assistant.*
CAUTION: *Never reuse brake fluid which has been bled from the system.*
2. Fill the master cylinder with brake fluid.
3. Install a box-end wrench (or special tool brake bleeder wrench) on the bleeder screw on the right rear wheel.
4. Attach a length of small diameter, clear vinyl tubing to the bleeder screw. submerge the other end of the rubber tubing in a glass jar partially filled with clean brake fluid. Make sure the rubber tube fits on the bleeder screw snugly or you may be squirted with brake fluid when the bleeder screw is opened.
5. Have your friend slowly depress the brake pedal. As this is done, open the bleeder screw half a turn and allow the fluid to run through the tube. Close the bleeder screw, then return the brake pedal to its fully released position.
6. Repeat the procedure until no bubbles appear in the jar. Refill the master cylinder.
7. Repeat this procedure on the left rear right front, and left front wheels, in that order. Periodically refill the master cylinder so it does not run dry.
8. If the brake warning light is on, depress the brake pedal firmly. If there is no air in the system, the light will go out.

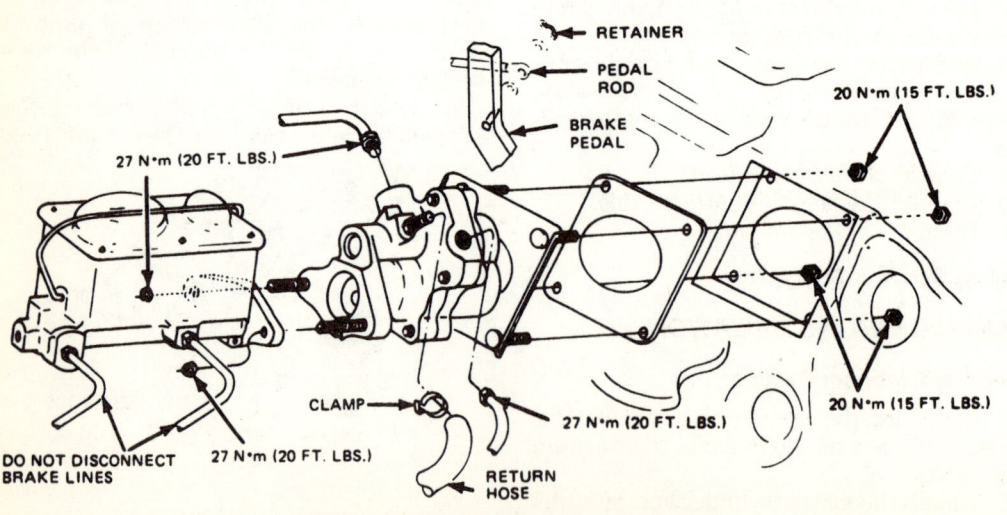

View of the Hydro-brake booster assembly

BRAKES

Brake Hoses and Lines

HYDRAULIC BRAKE LINE CHECK

The hydraulic brake lines and brake linings are to be inspected at the recommended intervals in the maintenance schedule. Follow the steel tubing from the master cylinder to the flexible hose fitting at each wheel. If a section of the tubing is found to be damaged, replace the entire section with tubing of the same type (steel, not copper), size, shape, and length. When installing a new section of brake tubing, flush clean brake fluid or denatured alcohol through to remove any dirt or foreign material from the line. When bending the tubing to fit the underbody contours, be careful not to kink or crack the line.

Check the flexible brake hoses that connect the steel tubing to each wheel cylinder. Replace the hose if it shows any signs of softening, cracking, or other damage. When installing a new front brake hose, position the hose to avoid contact with other chassis parts. Always place a NEW copper gasket over the hose fitting and thread the hose assembly into the front wheel cylinder. A new rear brake hose must be positioned clear of the exhaust pipe or shock absorber. Thread the hose into the rear brake tube connector. When installing either a new front or rear brake hose, engage the opposite end of the hose to the bracket on the frame. Install the horseshoe type retaining clip and connect the tube to the hose with the tube fitting nut.

Always bleed the system after hose or line replacement. Before bleeding, make sure that the master cylinder is topped up with high temperature, extra heavy duty brake fluid.

FRONT DRUM BRAKES

CAUTION: *Brake shoes contain asbestos, which has been determined to be a cancer causing agent. Never clean the brake surfaces with compressed air! Avoid inhaling any dust from any brake surface! When cleaning brake surfaces, use a commercially available brake cleaning fluid.*

Brake Drums and Shoes

REMOVAL AND INSTALLATION

NOTE: *For information on the wheel bearing, refer to Wheel Bearings in Chapter 1. Disassemble one side at a time so that you may use the unassembled side for reference during reassembly.*

1. Raise the vehicle and support it on safety stands.
2. Remove the front tire and wheel assembly. Remove the front drum by removing the spindle nut and cotter pin.
3. Free the brake shoe return springs, actuator pull-back spring, holddown pins and springs, and actuator assembly.

NOTE: *Special tools available from auto supply stores will ease removal of the spring and anchor pin, but the job may still be done with common hand tools.*

4. Disconnect the adjusting mechanism and spring, and remove the primary shoe. The primary shoe has a shorter lining than the secondary and is mounted at the front of the wheel.

To install:

5. Clean and inspect all brake parts.
6. Check the wheel cylinders for seal condition and leaking.
7. Repack wheel bearings and replace the seals.
8. Inspect the replacement shoes for nicks or burrs, lubricate the backing plate contact points, brake cable and levers, and adjusting screws and then assemble using the unassembled side for reference.

NOTE: *Make sure that the right and left hand adjusting screws are not mixed. You can prevent this by working on one side at a time. This will also provide you with a reference for reassembly. The star wheel should be nearest to the secondary shoe when correctly installed.*

When completed, make an initial adjustment as previously described.

NOTE: *Maintenance procedures for the metallic lining option are the same as those for standard linings. Do not substitute these linings in standard drums, unless they have been honed to a 20 micro-inch finish and equipped with special heat resistant springs.*

DRUM INSPECTION

1. Check the drums for any cracks, scores, grooves, or an out-of-round condition. Replace if cracked. Slight scores can be removed with fine emery cloth while extensive scoring requires turning the drum on a lathe.
2. Never have a drum turned more than 0.060 inch.

Wheel Cylinders

REMOVAL AND INSTALLATION

1. Clean away all dirt, crud and foreign material from around wheel cylinder. It is important that dirt be kept away from the brake line when the cylinder is disconnected.
2. Disconnect the inlet tube line.

408 BRAKES

3. Wheel cylinders are retained by two types of fasteners. One type uses a round retainer with locking clips, which attaches to the wheel cylinder on the back side of the brake backing plate. The other type simply uses two bolts, which screw into the wheel cylinder from the back side of the backing plate.

4. To remove the round retainer type cylinders, insert two pawls or pins into the access slots between the wheel cylinder pilot and the retainer locking tabs. Bend both tabs away simultaneously. The wheel cylinder can be removed, as the retainer is released.

5. To remove the cylinders, loosen and remove the bolts from the back side of the backing plate, and remove the cylinder.

To install:

6. To install the retainer type cylinder, position the wheel cylinder and hold it in place with a wooden block between the cylinder and the axle flange. Install the new retainer clip, using a $1^{1}/_{8}''$, 12-point socket extension. This tool will seat the retainer evenly.

7. On the bolt type cylinder, position the cylinder and install the bolts.

8. On both types of cylinder, torque the inlet type nuts.

9. Assemble the remaining brake components. Bleed the brakes.

FRONT DISC BRAKES

CAUTION: *Brake shoes contain asbestos, which has been determined to be a cancer causing agent. Never clean the brake surfaces with compressed air! Avoid inhaling any dust from any brake surface! When cleaning brake surfaces, use a commercially available brake cleaning fluid.*

Disc Brake Pads

INSPECTION

Brake pads should be inspected once a year or at 7,500 miles, whichever occurs first. Check both ends of the outboard shoe, looking in at each end of the caliper; then check the lining thickness of the inboard shoe, looking down through the inspection hole. Lining should be more than 0.020″ thick above the rivet (so that the lining is thicker than the metal backing). Keep in mind that any applicable state inspection standards that are more stringent take precedence. All 4 pads must be replaced if one shows excessive wear.

NOTE: *Most disc brake shoes have a wear indicator that makes a noise when the linings wear to a degree where replacement is neces-*

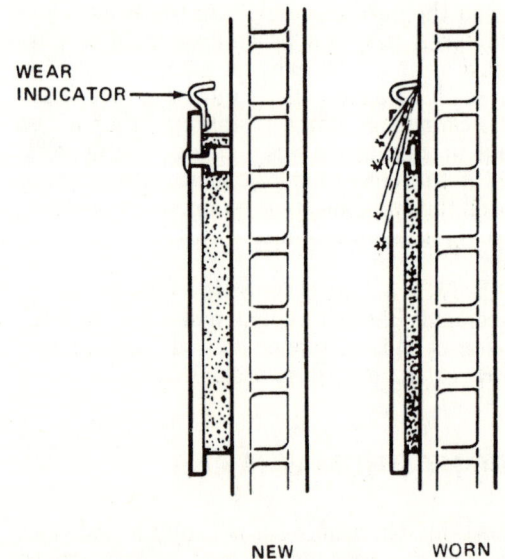

Front disc brake pad wear indicator

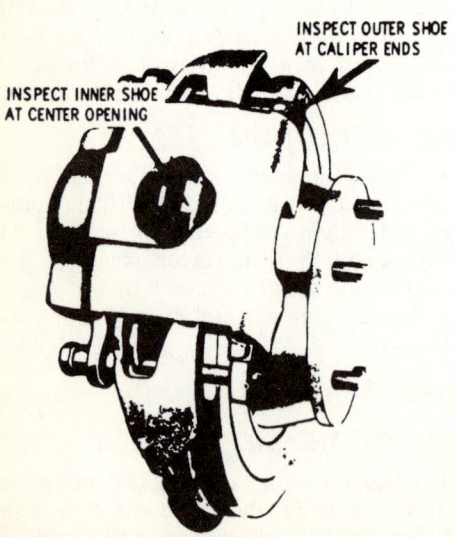

Disc brake pad inspection

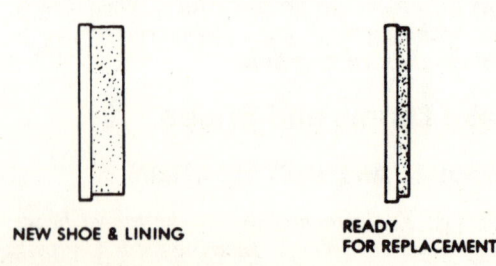

New and worn brake pads

BRAKES 409

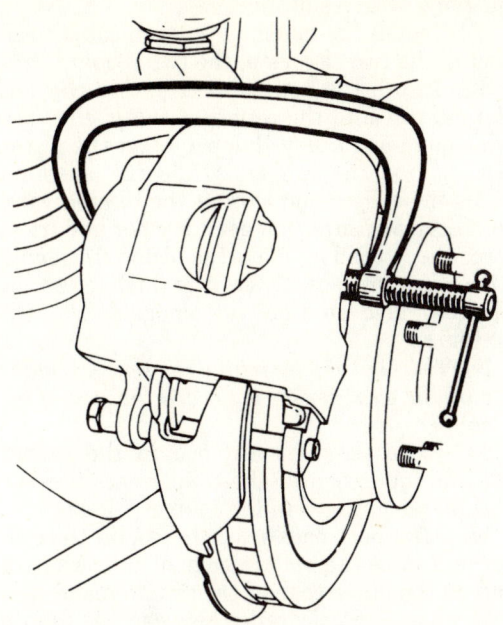

Use a C-clamp to seat the caliper piston

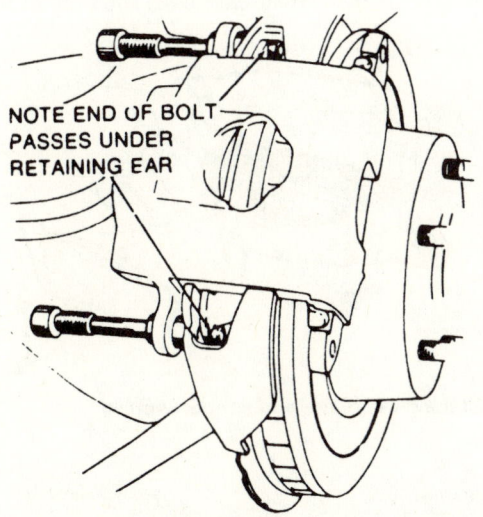

Caliper bolts must go under the pad retaining ears

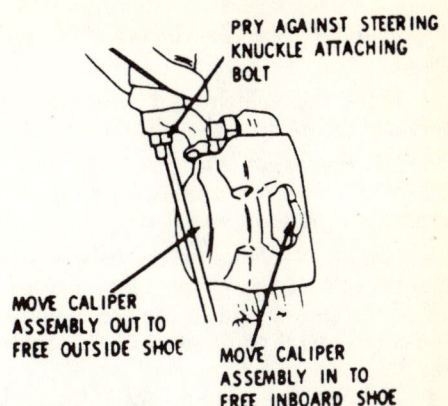

Compressing caliper with tool

NOTE: *The insertion of the thicker replacement pads will push the caliper pistons back into their bores and will cause a full master cylinder to overflow causing paint damage. In addition to siphoning fluid, it would be wise to keep the cylinder cover on during pad replacement.*

2. Raise the vehicle and support it with safety stands. Remove the wheels.

NOTE: *Replacing the pads on just one wheel will result in uneven braking. Always replace the pads on both wheels.*

3. Extract and discard the pad retaining pin cotter key.

4. Remove the retaining pin and, while removing one pad, insert its replacement before the piston has time to move outward. If you were too slow and the pistons were too fast, it will be necessary to use a wide-bladed putty knife to hold in the pistons while inserting the new pads. If this gives you difficulty open the bleed screw on the caliper and release some of the fluid, but do not allow the fluid to drain from the master cylinder. This may reduce the pressure and make it easier to push in on the pistons. After removing the outboard pad, inspect it and compare it with the inboard pad. They may be slightly different; if so, make sure that the replacement pads are installed correctly.

5. After installing the new pads, install the retaining pin and insert a new cotter pin.

6. Refill the master cylinder and bleed the system if necessary.

1969 And Later Models

1. Siphon off about $2/3$ of the brake fluid from a full master cylinder.

NOTE: *The insertion of the thicker replacement pads will push the piston back into its bore and will cause a full master cylinder to overflow causing paint damage. In addition to siphoning off fluid it would be wise to keep*

sary. The spring clip is an integral part of the inboard shoe and lining. When the brake pad reaches a certain degree of wear, the clip will contact the rotor and produce a warning noise.

REMOVAL AND INSTALLATION

1968 Model

1. Siphon off about $2/3$ of the brake fluid from the full master cylinder.

the cylinder cover on during pad replacement.

2. Raise the vehicle and support it with safety stands. Remove the wheels.

NOTE: *Replacing the pads on just one wheel will result in uneven braking. Always replace the pads on both wheels.*

3. Install a C-clamp on the caliper so that the solid side of the clamp rests against the back of the caliper and so the screw end rests against the metal part (shoe) of the outboard pad.

4. Tighten the clamp until the caliper moves enough to bottom the piston in its bore. Remove the clamp.

5. Remove the two allen-head caliper mounting bolts enough to allow the caliper to be pulled off the disc.

6. Remove the inboard pad and loosen the outboard pad. Place the caliper where it won't strain the brake hose. It would be best to wire it out of the way.

7. Remove the pad support spring clip from the piston.

8. Remove the two bolt ear sleeves and the four rubber bushings from the ears.

9. Brake pads should be replaced when they are worn to within $1/_{32}$" of the rivet heads.

10. Check the inside of the caliper for leakage and the condition of the piston dust boot.

To install:

11. Lubricate the two new sleeves and four bushings with a silicone spray.

12. Install the bushings in each caliper ear. Install the two sleeves in the two inboard ears.

13. Install the pad support spring clip and the old pad into the center of the piston. You will then push this pad down to get the piston flat against the caliper. While the assistant holds the caliper and loosens the bleeder valve to relieve pressure, you get a pry bar and try to force the old pad in, make the piston flush with the caliper surface. When it is flush, close the bleeder valve so that no air gets into the system.

NOTE: *On vehicles with wear sensors, make sure the wear sensor is toward the rear of the caliper.*

14. Place the outboard pad in the caliper with its top ears over the caliper ears and the bottom tab engaged in the caliper cutout.

15. After both pads are installed, lift the caliper and place the bottom edge of the outboard pad on the outer edge of the disc to make sure that there is no clearance between the tab on the bottom of the shoes and the caliper abutment.

16. Place the caliper over the disc, lining up the hole in the caliper ears with the hole in the

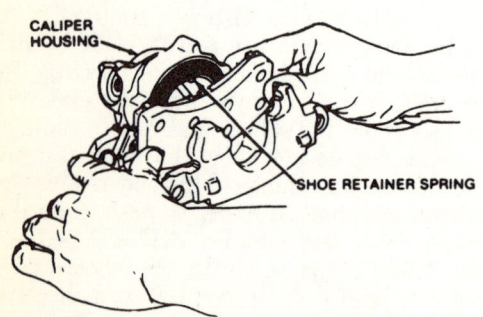

Install inboard brake shoe assembly

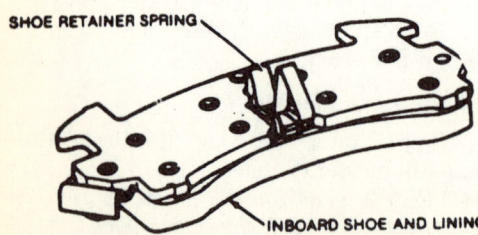

Proper retaining spring installation

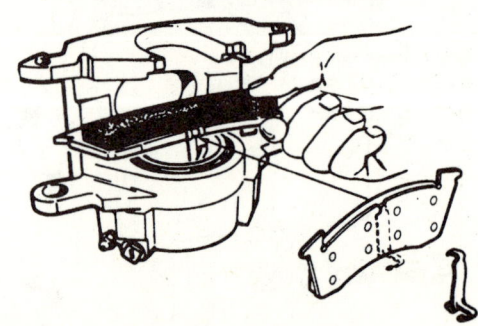

Installing the brake pad support spring

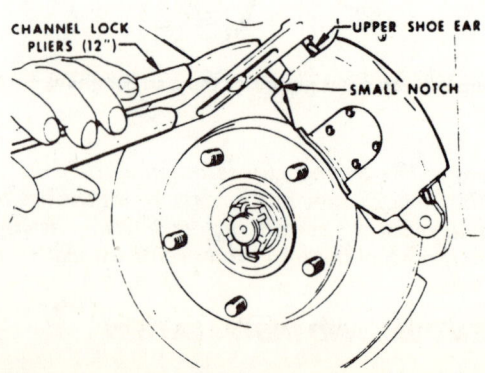

Use pliers to fit the brake pad to the caliper housing

BRAKES

mounting bracket. Make sure that the brake hose is not kinked.

17. Start the caliper-to-mounting bracket bolts through the sleeves in the inboard caliper ears and through the mounting bracket making sure the ends of the bolts pass under the retaining ears of the inboard shoe.

18. Push the mounting bolts through to engage the holes in the outboard shoes and the outboard caliper ears and then threading them into the mounting bracket.

19. Torque the mounting bolts to 35 ft. lbs. Pump the brake pedal to seat the linings against the rotors.

20. With a pair of channel lock pliers placed on the notch on the caliper housing, bend the caliper upper ears until no clearance exist between the shoe and the caliper housing.

21. Install the wheels, lower the car, and refill the master cylinder with fluid. Pump the brake pedal to make sure that it is firm. If it is not, bleed and adjust the brakes.

Caliper

REMOVAL AND INSTALLATION

1. Raise the vehicle and support it on safety stands.
2. Remove the tire and wheel assembly from the side on which the caliper is being removed.
3. Disconnect the brake hose at the support bracket. Tape the end of the line to prevent contamination.
4. Remove the cotter pin from the brake pad and retaining pin and remove the pin.
5. Remove the brake pads and identify them as inboard or outboard if they are being reused.
6. Remove the U-shaped retainer from the hose fitting and pull the hose from the bracket.
7. Remove the two caliper retaining bolts and also the caliper from its mounting bracket.

To install:

8. Install the brake pads. If the same pads are being reused, return them to their original places (outboard or inboard) as marked during removal. New pads will usually have an arrow on the back indicating the direction of disc rotation. See Brake Pad Replacement for details.
9. Install the brake hose into the caliper, passing the female end through the support bracket.
10. Make sure that the tube line is clean and connect the brake line nut to the caliper.
11. Install the hose fitting into the support bracket and install the U-shaped retainer. Turn the steering wheel from side to side to make sure that the hose doesn't interfere with the tire. If it does, turn the hose end one or two points in the bracket until the interference is eliminated.
12. After performing the above check, install the steel tube connector and tighten it.

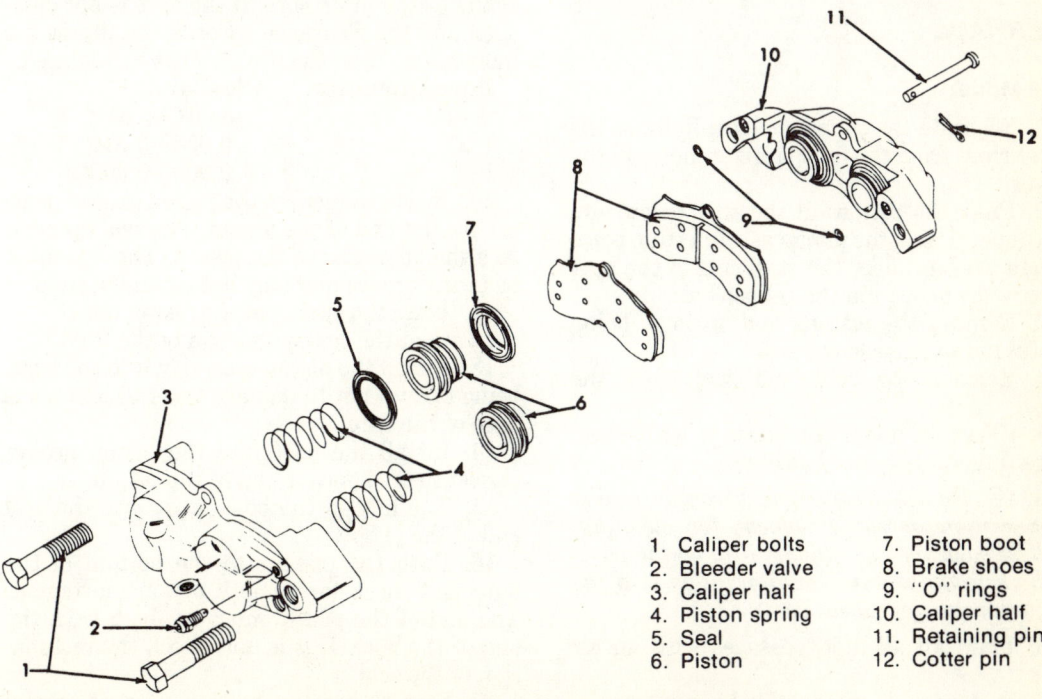

1. Caliper bolts
2. Bleeder valve
3. Caliper half
4. Piston spring
5. Seal
6. Piston
7. Piston boot
8. Brake shoes
9. "O" rings
10. Caliper half
11. Retaining pin
12. Cotter pin

Exploded view of the four piston brake caliper

412 BRAKES

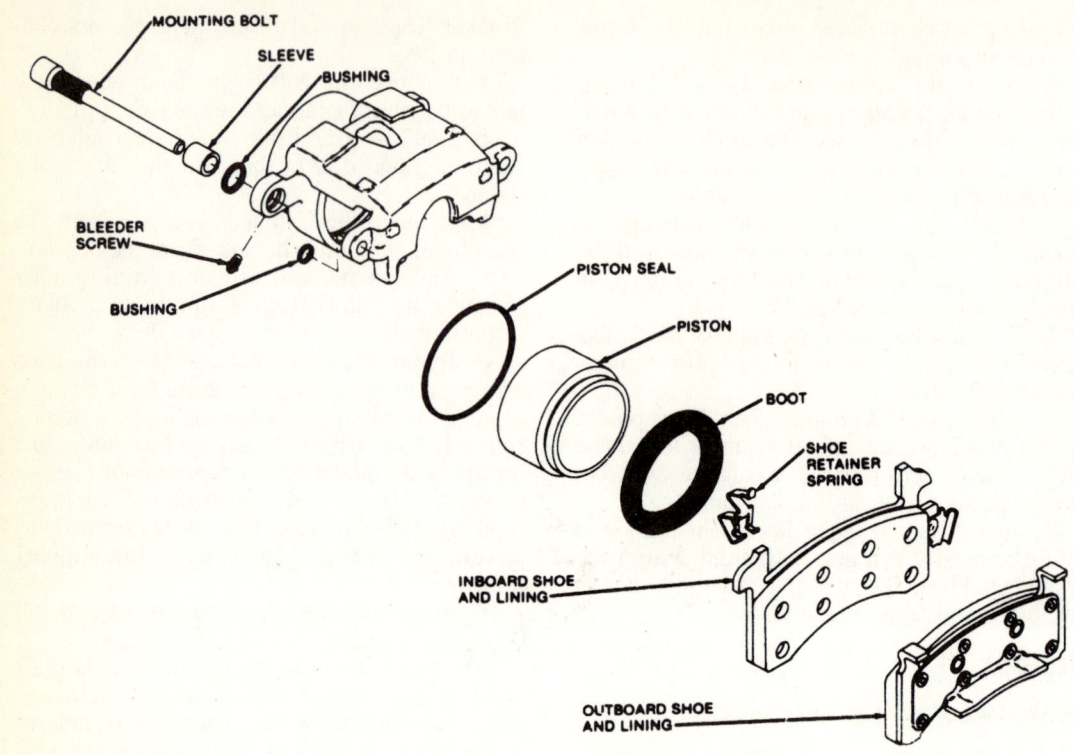

Exploded view of the single piston brake caliper

13. Bleed the brakes as instructed earlier in this chapter.
14. Install the wheels and lower the vehicle.

OVERHAUL

1968 Model

1. Separate the caliper halves. Remove the two O-rings from the fluid transfer holes in the caliper.
2. Push the piston all the way down into the caliper. Using the piston as a fulcrum, place a small pry bar under the steel ring in the boot and pry the boot from the caliper half.
3. Remove the pistons and springs, being careful not to damage the seal.
4. Remove the boot and seal from the piston.
5. Clean all metal components with clean brake fluid or denatured alcohol.

NOTE: *Do not use gasoline, kerosene, or any other mineral based solvent for cleaning. These solvents form an oily film on the parts which leads to fluid contamination and the deterioration of rubber parts.*

6. Blow out all fluid passages with an air hose.
7. Discard and replace all rubber parts.
8. Inspect all bores for scoring and pitting and replace is necessary. Minor flaws can be removed with very fine crocus cloth but do so with a circular motion.
9. Using a feeler gauge, check the clearance of the piston in its bore. If the bore is not damaged and the clearance exceeds the maximum limit below, then the piston must be replaced.

Bore Diameter	Clearance
$2^{1}/_{16}"$	0.0045-0.010"
$1^{7}/_{8}"$	0.0045-0.010"
$1^{3}/_{8}"$	0.0035-0.009"

10. Insert the seal in the piston groove nearest the flat end of the piston. The seal lip must face the large end of the piston. The lips must be in the groove and may not extend beyond.
11. Place the spring in the piston bore.
12. Coat the seal with clean brake fluid.
13. Install the piston assembly into the bore, being careful not to damage the seal lip on the edge of the bore.
14. Install the boot into the piston groove closest to the concave end of the piston.
15. The fold in the boot must face the seal end of the piston.
16. Push the pistons to the bottom of the bore and check for smooth piston movement. The end of the piston must be flush with the end of the bore. If it is not, check the installation of the seal.
17. Seat the piston boot so that its metal ring is even in the counterbore. The ring is even in the counterbore. The ring must be

flush or below the machined face of the caliper. If the ring is seated unevenly dirt and moisture could get into the bore.

18. Insert the O-rings around the fluid transfer holes at both ends of the caliper halves.

19. Lubricate the bolts with brake fluid, connect the caliper halves, and torque the bolts evenly.

20. While holding in the brake pistons with a putty knife, mount the caliper over the disc. Be careful not to damage the piston boots on the edge of the disc.

21. Install the two mounting bolts and torque evenly.

1969 And Later Models

1. Raise and support the vehicle on safety stands. Remove the tire and wheel assembly. Remove the disc brake pads.

2. Disconnect brake hose and plug the line.

3. Remove the U-shaped retainer from the fitting.

4. Pull the hose from the frame bracket and remove the caliper with the hose attached.

5. Clean the outside of the caliper with denatured alcohol.

6. Remove the brake hose and discard the copper gasket.

7. Remove the brake fluid from the caliper.

8. Place clean rags inside the caliper opening to catch the piston when it is released.

9. Apply compressed air to the caliper fluid inlet hole and force the piston out of its bore. Do not blow the piston out of its bore, use just enough pressure to each it out.

10. Use a small pry bar to pry the boot out of the caliper. Avoid scratching the bore.

11. Remove the piston seal from its groove in the caliper bore. Do not use a metal tool of any type for this operation.

NOTE: *Replace, do not reuse the boot, piston seal, rubber bushings, and sleeves.*

12. Blow out all passages in the caliper and bleeder valve. Clean the piston and piston bore with fresh brake fluid.

13. Examine the piston for scoring, scratches, or corrosion. If any of these conditions exist, the piston must be replaced because it is plated and cannot be refinished.

14. Examine the bore for the same defects. Light rough spots may be removed by rotating crocus cloth, using finer pressure, in the bore. Do not polish with an in-and-out motion or use any other abrasive.

15. Lubricate the piston bore and the new rubber pars with fresh brake fluid. Position the seal in the piston bore groove.

16. Lubricate the piston with brake fluid and assemble the boot into the piston groove so that the fold faces the open end of the piston.

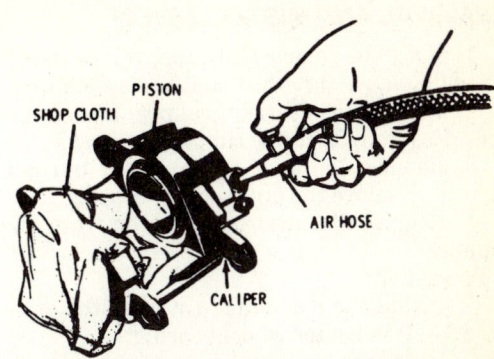

Piston removal using compressed air. Keep fingers out of the way of the piston when air is applied.

17. Insert the piston into the bore, taking care not to unseat the seal.

18. Force the piston to the bottom of the bore (this will require a force of 40-100 lbs.). seat the boot lip around the caliper counterbore. Proper seating of the boot is very important for sealing out contaminants.

19. Install the brake hose into the caliper with a NEW copper gasket.

20. Lubricate the new sleeves and rubber bushing. Install the bushings in the caliper ears. Install the sleeves so that the end toward the disc pad is flush with the machined surface.

NOTE: *Lubrication of the sleeves and bushings is essential to insure the proper operation of the sliding caliper design.*

21. Install the shoe support spring in the piston.

22. Install the disc pads in the caliper and remount the caliper.

23. Reconnect the brake hose to the steel brake line. Install the retainer clip. Bleed the brakes.

24. Replace the wheels, check the brake fluid level, check the brake pedal travel, and road test the vehicle.

Brake Disc

INSPECTION

1. Tighten the spindle nut to remove all wheel bearing play.

2. Install a dial indicator on the caliper so that its feeler will contact the disc about 1" below its outer edge.

3. Turn the disc and observe the runout reading. If the reading exceeds 0.002" (0.004" total reading), the disc should be replaced.

4. Measure the thickness of the rotor at 5 points around the circumference. Tolerance is 0.005".

5. Minimum thickness dimensions are cast into the caliper for reference.

414 BRAKES

REMOVAL AND INSTALLATION

1. Raise the vehicle and support it on safety stands. Remove the wheel and tire assembly.
2. Remove the brake caliper and brake pads. Position the brake caliper to the side.
3. Remove the dust cover, cotter pin and locknut. Remove the outer wheel bearing.
4. Remove the disc brake rotor from the spindle.

To install:

5. Install the disc brake rotor on the spindle. Install the outer wheel bearing.
6. Install the locknut. Adjust the wheel bearings.
7. Install the cotter pin and dust cover.
8. Install the brake caliper and brake pads.
9. Install the tire and wheel assembly. Lower the vehicle.

REAR DRUM BRAKES

CAUTION: *Brake shoes contain asbestos, which has been determined to be a cancer causing agent. Never clean the brake surfaces with compressed air! Avoid inhaling any dust from any brake surface! When cleaning brake surfaces, use a commercially available brake cleaning fluid.*

Brake Drum

REMOVAL AND INSTALLATION

1. Raise and support the vehicle on safety stands.
2. Remove the tire and wheel assembly.

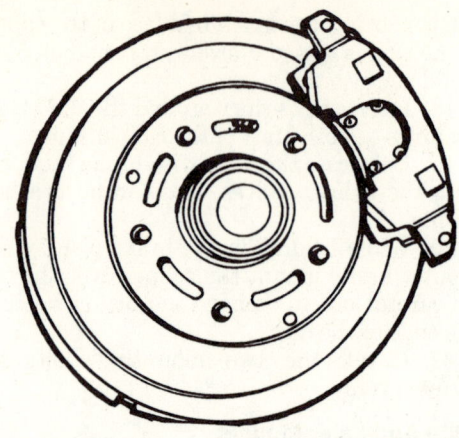

Discard dimension is stamped on the disc hub assembly

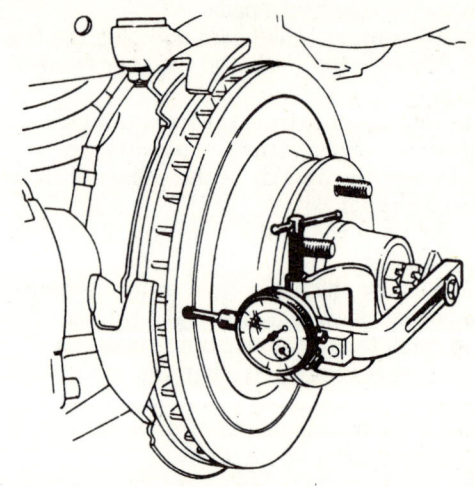

Use a dial indicator to determine brake disc runout

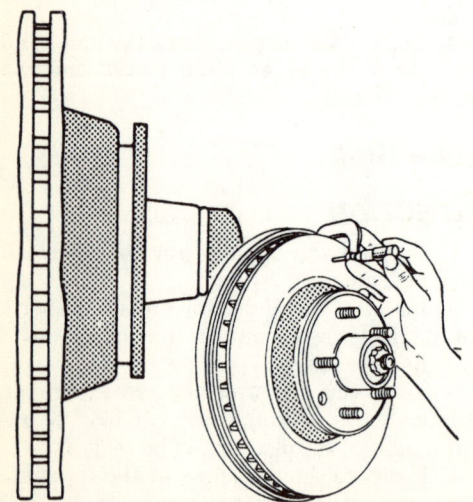

Parallelism

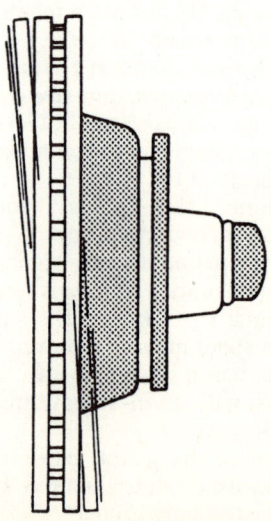

Excessive runout

BRAKES

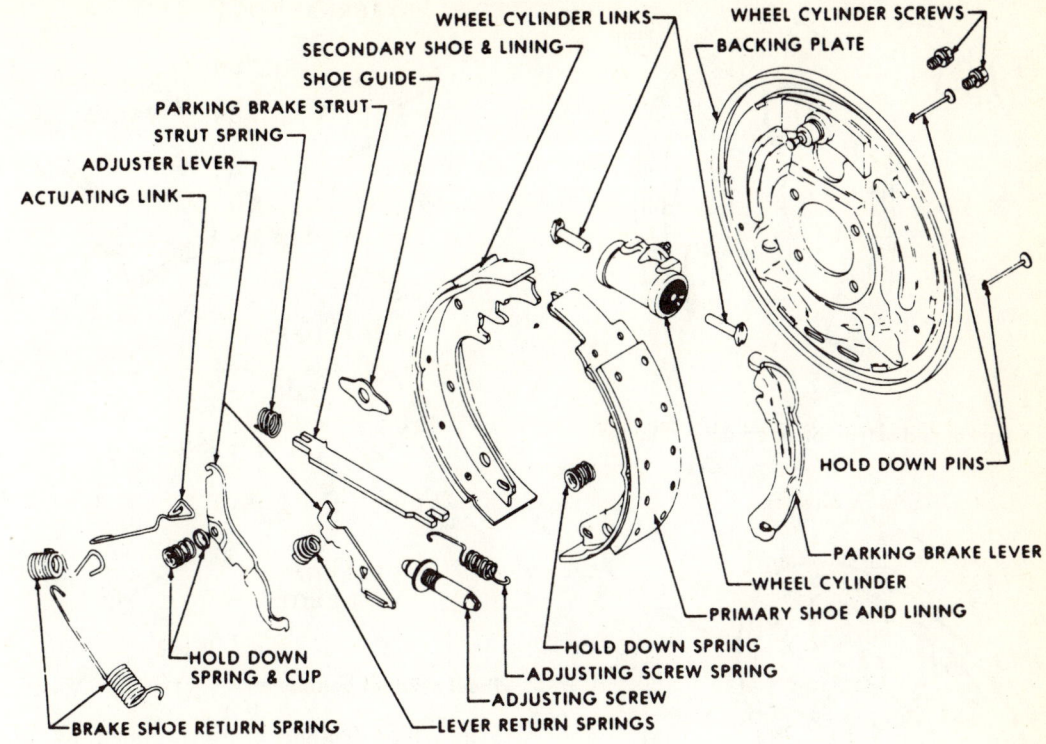

Exploded view of the brake drum brake assembly

3. Pull the brake drum off. It may be necessary to gently tap the rear edge of the drum to start it off the studs.

4. If extreme resistance to removal is encountered, it will be necessary to retract the adjusting screw. Knock out the access hole in the brake drum and turn the adjusting to retract the linings from the drum.

5. Install a replacement hole cover before re-installing drum.

6. Install the drums in the same position on the hub as removed.

DRUM INSPECTION

1. Check the drums for any cracks, scores, grooves, or an out-of-round condition. Replace if cracked. Slight scores can be removed with fine emery cloth while extensive scoring requires turning the drum on a lathe.

2. Never have a drum turned more than 0.060 inch.

Brake Shoes

REMOVAL AND INSTALLATION

NOTE: *Disassemble one side at a time so that you may use the unassembled side for reference during reassembly.*

1. Raise the vehicle and support it on safety stands.

2. Slacken the parking brake cable if necessary.

3. Remove the rear wheel and brake drum—refer to the necessary service procedure.

4. Free the brake shoe return springs, actuator pull-back spring, holddown pins and springs and actuator assembly.

NOTE: *Special tools available from auto supply stores will ease removal of the spring and anchor pin, but the job may still be done with common hand tools.*

5. Disconnect the adjusting mechanism and spring, and remove the primary shoe. The primary shoe has a shorter lining than the secondary and is mounted at the front of the wheel.

6. Disconnect the parking brake lever from the secondary shoe and remove the shoe.

To install:

7. Clean and inspect all brake parts.

8. Check the wheel cylinders for seal condition and leaking.

9. Inspect the replacement shoes for nicks or burrs, lubricate the backing plate contact points, brake cable and levers, and adjusting screws and then assemble using the unassembled side as a reference.

416 BRAKES

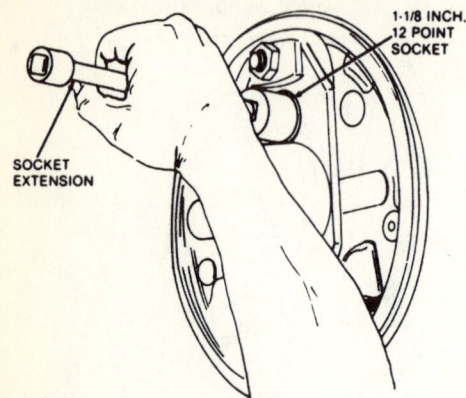

Use a socket and extension to seat the retainer

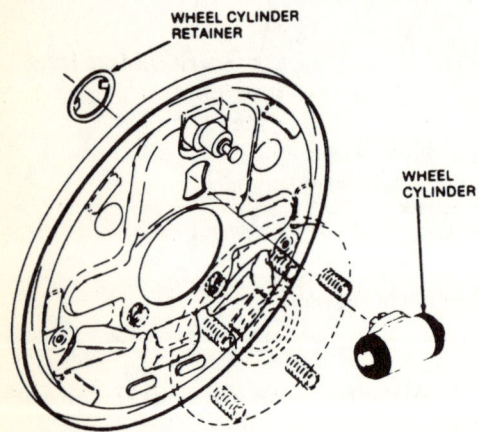

Wheel cylinder—retainer type

NOTE: *Make sure that the right and left hand adjusting screws are not mixed. You can prevent this by working on one side at a time. This will also provide you with a reference for reassembly. The star wheel should be nearest to the secondary shoe when correctly installed.*

When completed, make an initial adjustment as previously described.

NOTE: *Maintenance procedures for the metallic lining option are the same as those for standard linings. Do not substitute these linings in standard drums, unless they have been honed to a 20 micro-inch finish and equipped with special heat resistant springs.*

Wheel Cylinders

REMOVAL AND INSTALLATION

1. Clean away all dirt, crud and foreign material from around wheel cylinder. It is important that dirt be kept away from the brake line when the cylinder is disconnected.
2. Disconnect the inlet tube line.

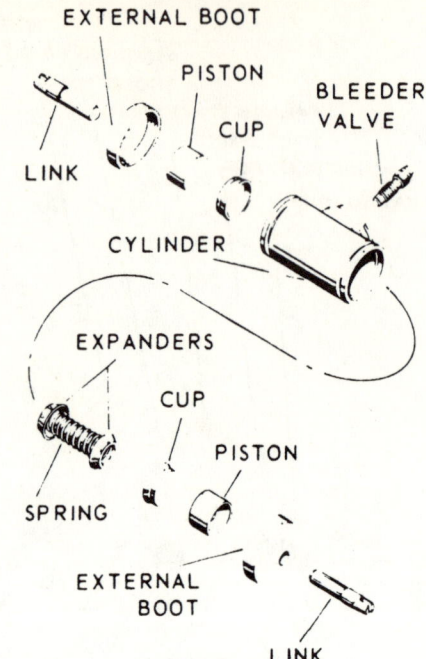

Wheel cylinder components

3. Wheel cylinders are retained by two types of fasteners. One type uses a round retainer with locking clips, which attaches to the wheel cylinder on the back side of the brake backing plate. The other type simply uses two bolts, which screw into the wheel cylinder from the back side of the backing plate.

4. To remove the round retainer type cylinders, insert two pawls or pins into the access slots between the wheel cylinder pilot and the retainer locking tabs. Bend both tabs away simultaneously. The wheel cylinder can be removed, as the retainer is released.

5. To remove the cylinders, loosen and remove the bolts from the back side of the backing plate, and remove the cylinder.

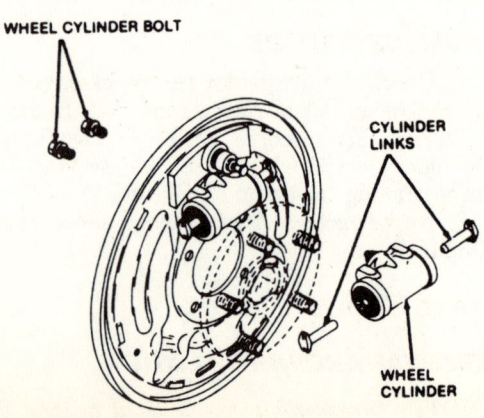

Wheel cylinder—bolt type

BRAKES 417

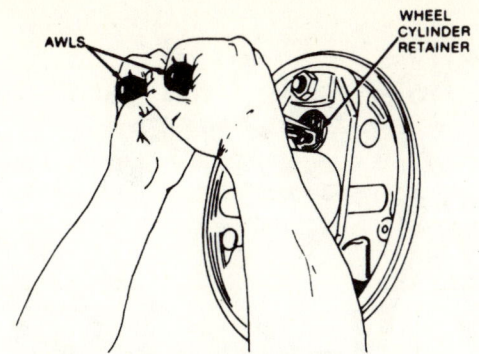

Bend retainer stubs with tools

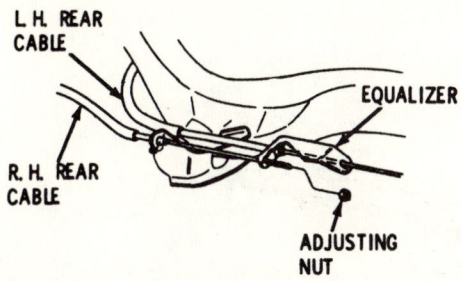

Parking brake cable adjustment

6. To install the retainer type cylinder, position the wheel cylinder and hold it in place with a wooden block between the cylinder and axle flange. Install a new retainer clip, using a $1^{1}/_{8}''$, 12-point socket and socket extension. This tool will seal the retainer evenly.

7. On the bolt type cylinder, position the cylinder and install the bolts.

8. On both types of cylinders, torque the inlet type nuts evenly.

9. Assemble the remaining brake components. Bleed the brakes.

Wheel Bearings

Refer to Chapter 1 for removal and installation, adjustment and packing procedures.

PARKING BRAKE

All vehicles are equipped with a foot operated ratchet type parking brake. A cable assembly connects this pedal to an intermediate cable by means of an equalizer. Adjustment is made at the equalizer. The intermediate cable connects with two rear cables and each of these cables enters a rear wheel.

Cable

REMOVAL AND INSTALLATION

Front Cable

1. Raise the vehicle and support it with safety stands.
2. Remove the adjusting nut at the equalizer.
3. Remove the spring retainer clip from the bracket.
4. Lower the vehicle. Disconnect the cable

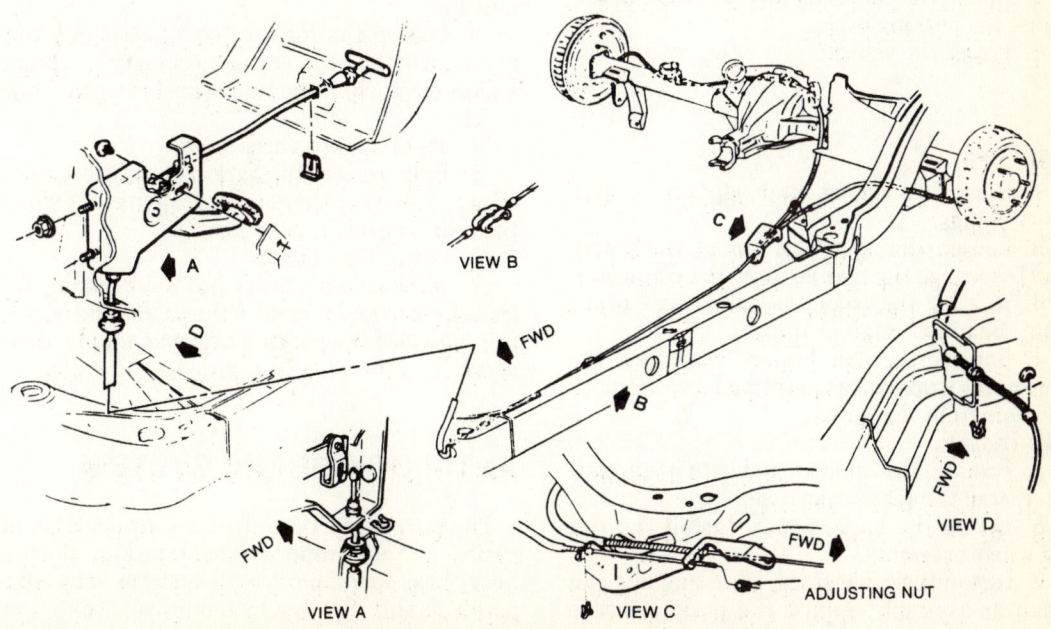

Parking brake cable routing 1977 and later models

418 BRAKES

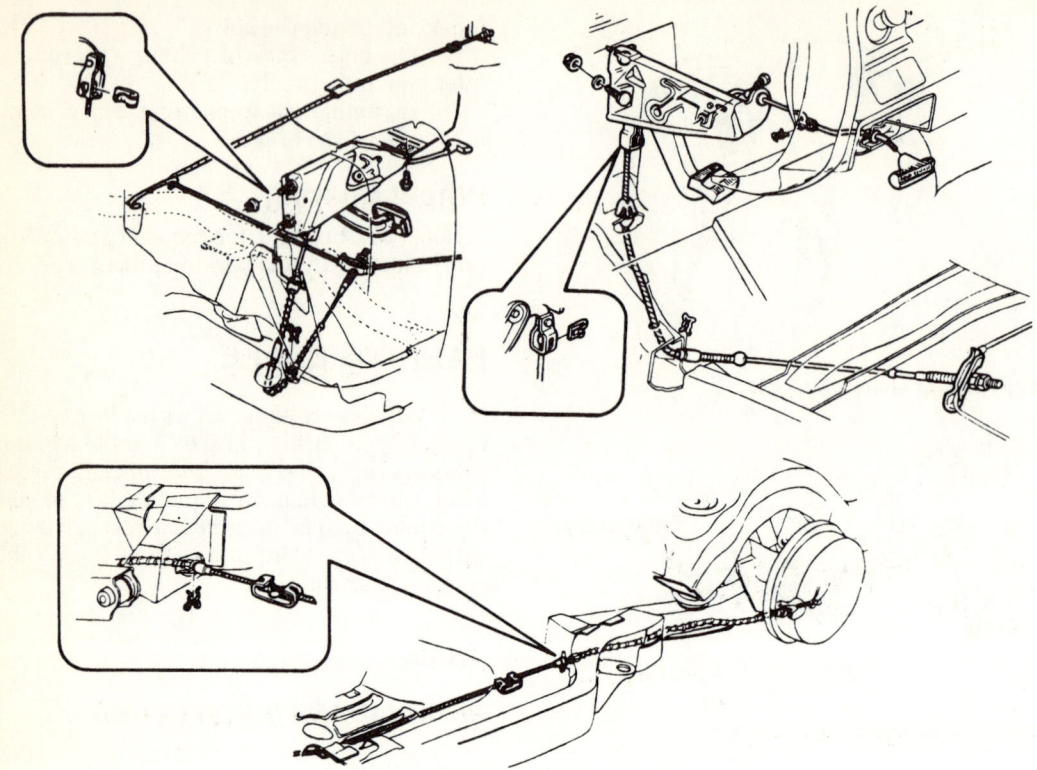

Parking brake cable routing 1968-76 models

from its mounting on the parking brake assembly. Remove the cable from the vehicle.

To install:

5. Position the cable assembly to its mounting. Install the spring clips and the retainer.
6. Install the adjusting nut at the equalizer. Adjust the parking brake.
7. Lower the vehicle.

Rear Cable

1. Raise the vehicle and support it with safety stands.
2. Loosen the adjusting nut at the equalizer. Disengage the rear cable at the connector.
3. Remove the wheel assembly and brake drum. Bend the retainer fingers.
4. Remove the rear brakes. Disengage the cable at the brake shoe operating lever. Remove the cable from the vehicle.

To install:

5. Position the cable assembly to its mounting. Install the rear brake assembly.
6. Install the brake drum. Install the tire and wheel assembly.
7. Install the cable at the adjusting nut and equalizer assembly. Adjust the parking brake cable.
8. Lower the vehicle.

ADJUSTMENT

1. Raise the vehicle and support it with safety stands.
2. Push down on the parking brake pedal so that it is two notches from the fully released position.
3. Loosen the forward equalizer check nut and adjust the rear nut as necessary to obtain a light drag when the rear wheel is turned forward.
4. Tighten both check nuts.
5. Fully release the parking brake lever and check to see that there is no drag present when the rear wheel is turned forward.
6. Lower the vehicle.
7. If the parking brake has to be forcibly released, clean and lubricate the cables and equalizer, and also the parking brake assembly, then check the cables for straightness and kinks.

ANTI-LOCK BRAKE SYSTEM

The purpose of the Anti-Lock Brake System (ABS) is to minimize wheel lockup during heavy braking on most road surfaces. The ABS performs this function by monitoring the speed of each wheel and controlling the brake fluid pressure to each front wheel and both rear

BRAKES 419

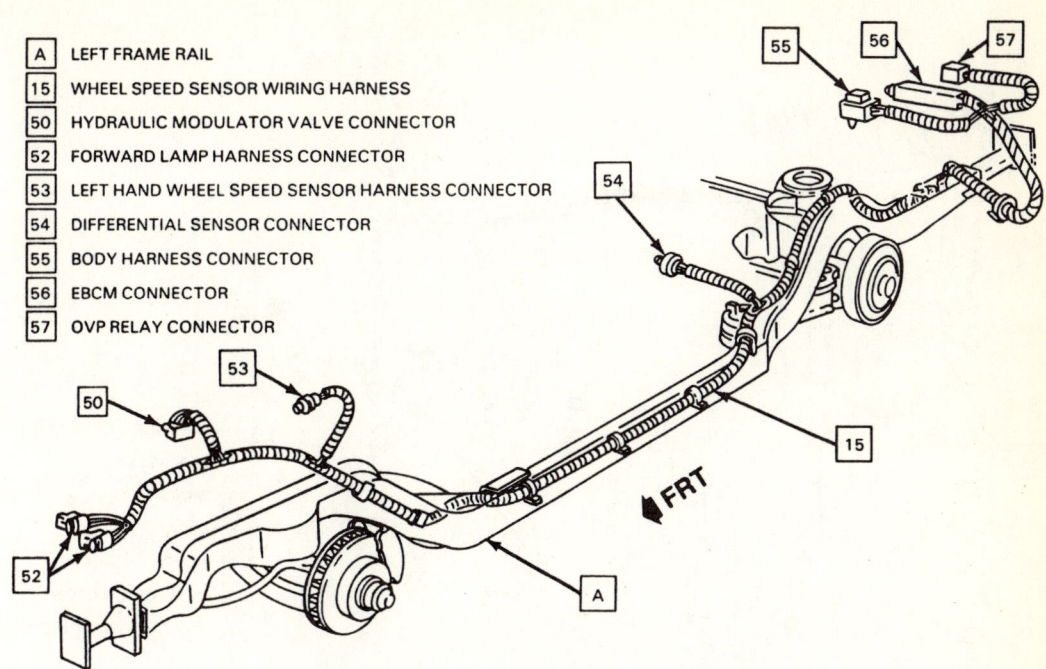

ABS wiring harness routing and connectors, Sedan

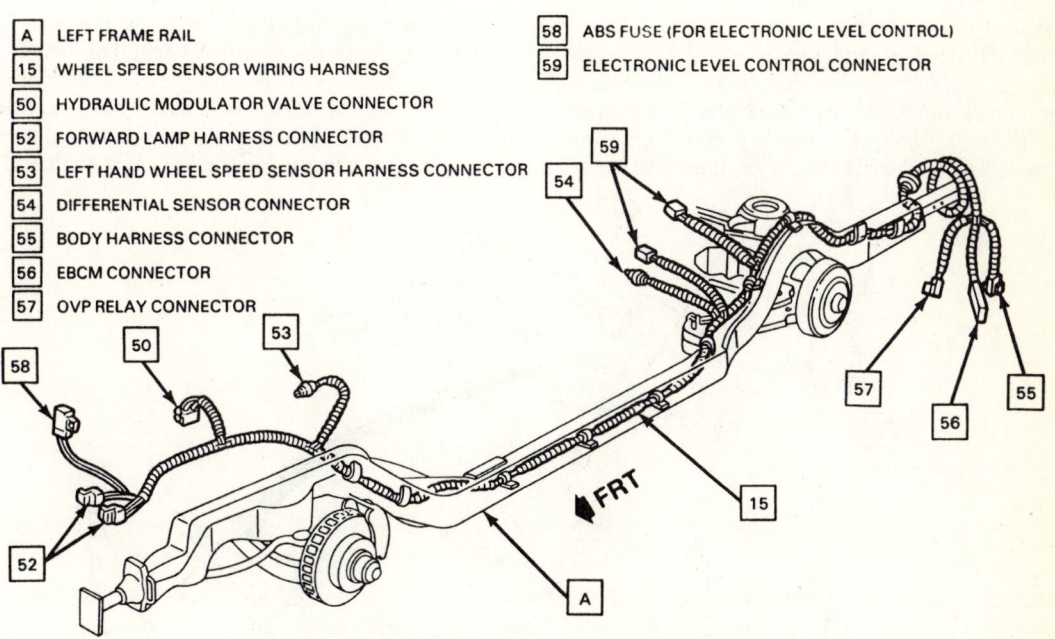

ABS wiring harness routing and connectors, Wagon

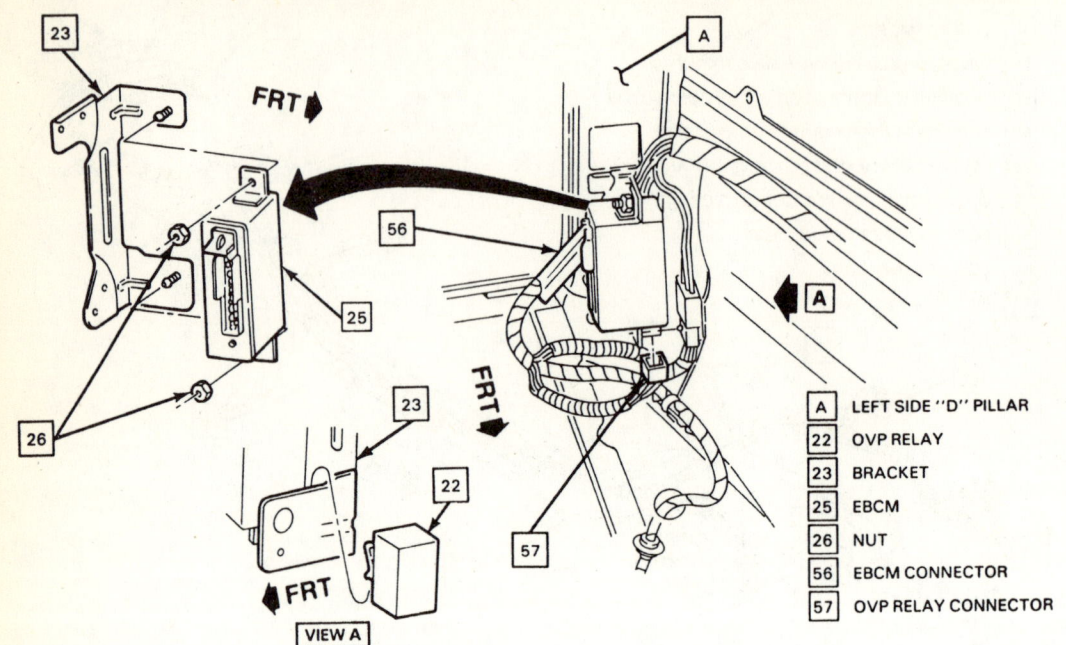

Electronic Brake Control Module and relay, Wagon

wheels during a braking maneuver. This allows you (as the driver) to retain directional stability and steering capability.

The ABS continuously monitors all of its components and uses several methods of determining a fault and notifying the driver of a system malfunction. When the vehicle is started, a functional check of the ABS electrical circuitry is performed. As the vehicle speed reaches 4 mph, a functional check of the hydraulic modulator takes place. During this check each valve is cycled and the pump motor is turned on briefly. You may hear or feel this check take place when the vehicle begins to move. This test will only occur once with each ignition startup and is considered normal operation. If a malfunc-

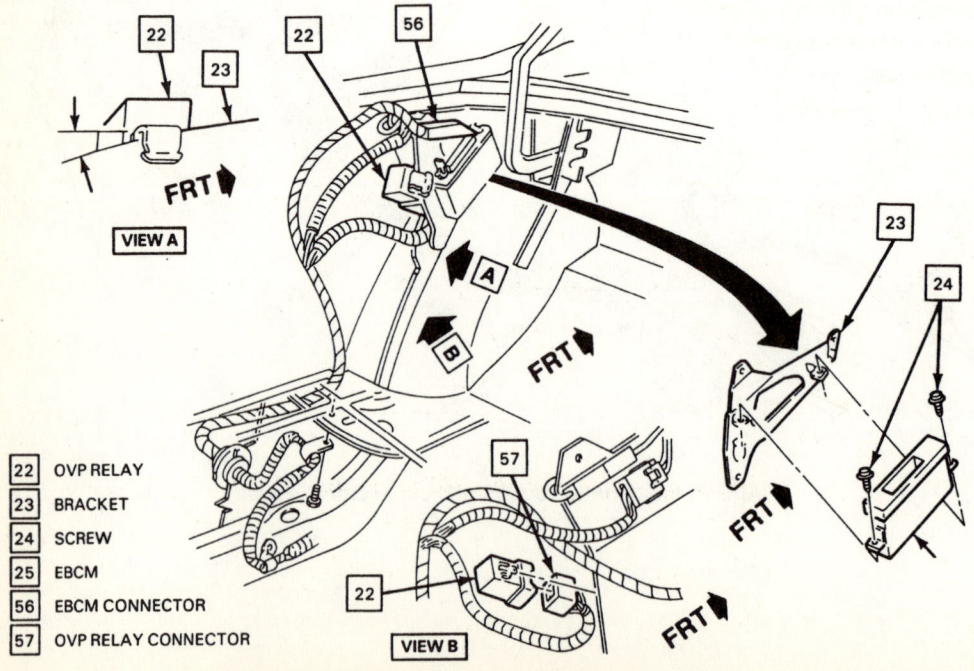

Electronic Brake Control Module and relay, Sedan

tion should occur the Brake warning lamp or the Anti-Lock warning lamp will either stay on or begin to blink.

NOTE: *A qualified technician should perform all diagnostic and repairs on the ABS, due to the complexity of this system. However some system components will have to be removed in order to obtain access to other parts.*

Left Front Wheel Speed Sensor

REMOVAL AND INSTALLATION

1. Disconnect the battery ground cable.
2. Raise and safely support vehicle on jack stands.
3. From under hood disconnect the wheel speed sensor wire harness connector, and remove from clip.
4. Remove tire and wheel assembly.
5. Remove the bolt attaching harness bracket to the frame rail.
6. Remove the wheel speed sensor assembly harness, with grommets, from brackets and combination valve brake pipe clip.

NOTE: *Note position of grommets and harness routing, to ease the installation process.*

7. Remove the sensor retaining bolt, and remove the wheel speed sensor from the steering knuckle assembly.
8. Clean the sensor, if signs of wear or damage replace it with a new one.

To Install:

The wheel speed sensors are a tight fit into the knuckle and should be pushed in by hand. DO NOT hammer the sensors into position. The knuckle must be given a coating of anti-corrosion sealer such as GM part # 1052856 or equivalent, DO NOT use grease!

9. Push sensor into coated steering knuckle assembly and install retaining bolt. Tighten to 71 in.lbs.
10. Install the wheel speed sensor assembly harness with grommets to brackets and combination valve brake pipe clip.

NOTE: *Proper installation of the wheel speed sensor assembly wire is critical to continued system operation. Failure to install the wire in brackets as shown, may result in contact with moving parts and/or over extension of the wire, resulting in circuit damage.*

11. Install the bracket and bolt to the frame rail. Tighten bolt to 89 in.lbs.
12. Connect the speed sensor connector to the wire harness connector and install harness into clip.
13. Install tire and wheel assembly and lower vehicle.
14. Connect the negative battery cable to battery.
15. Check operation of brake system.

Right Front Wheel Speed Sensor

REMOVAL AND INSTALLATION

1. Disconnect the battery ground cable.
2. Raise and safely support vehicle on jack stands.
3. From under hood disconnect the wheel speed sensor wire harness connector, and remove from clip.

NOTE: *The wheel speed sensor wires are labeled with either a white tag or white letter designating L (left) and R (right). It is of great importance that the speed sensors be installed in their proper sides.*

4. Remove tire and wheel assembly.
5. Remove the bolt attaching harness bracket to the frame rail.
6. Remove the wheel speed sensor assembly harness, with grommets, from brackets and combination valve brake pipe clip.

NOTE: *Note the position of grommets and harness routing, to ease the installation process.*

7. Remove the sensor retaining bolt, and remove the wheel speed sensor from the steering knuckle assembly.
8. Clean the sensor, if signs of wear or damage replace it with a new one.

To Install:

The wheel speed sensors are a tight fit into the knuckle and should be pushed in by hand. DO NOT hammer the sensors into position. The knuckle must be given a coating of anti-corrosion sealer such as GM part # 1052856 or equivalent, DO NOT use grease!

9. Push sensor into coated steering knuckle assembly and install retaining bolt. Tighten to 71 in.lbs.
10. Install the wheel speed sensor assembly harness with grommets to brackets and combination valve brake pipe clip.

NOTE: *Proper installation of the wheel speed sensor assembly wire is critical to continued system operation. Failure to install the wire in brackets as shown, may result in contact with moving parts and/or over extension of the wire, resulting in circuit damage.*

11. Install the bracket and bolt to the frame rail. Tighten bolt to 89 in.lbs.
12. Connect the speed sensor connector to the wire harness connector and install harness into clip.
13. Install tire and wheel assembly and lower vehicle.
14. Connect the negative battery cable to battery.
15. Check operation of brake system.

422 BRAKES

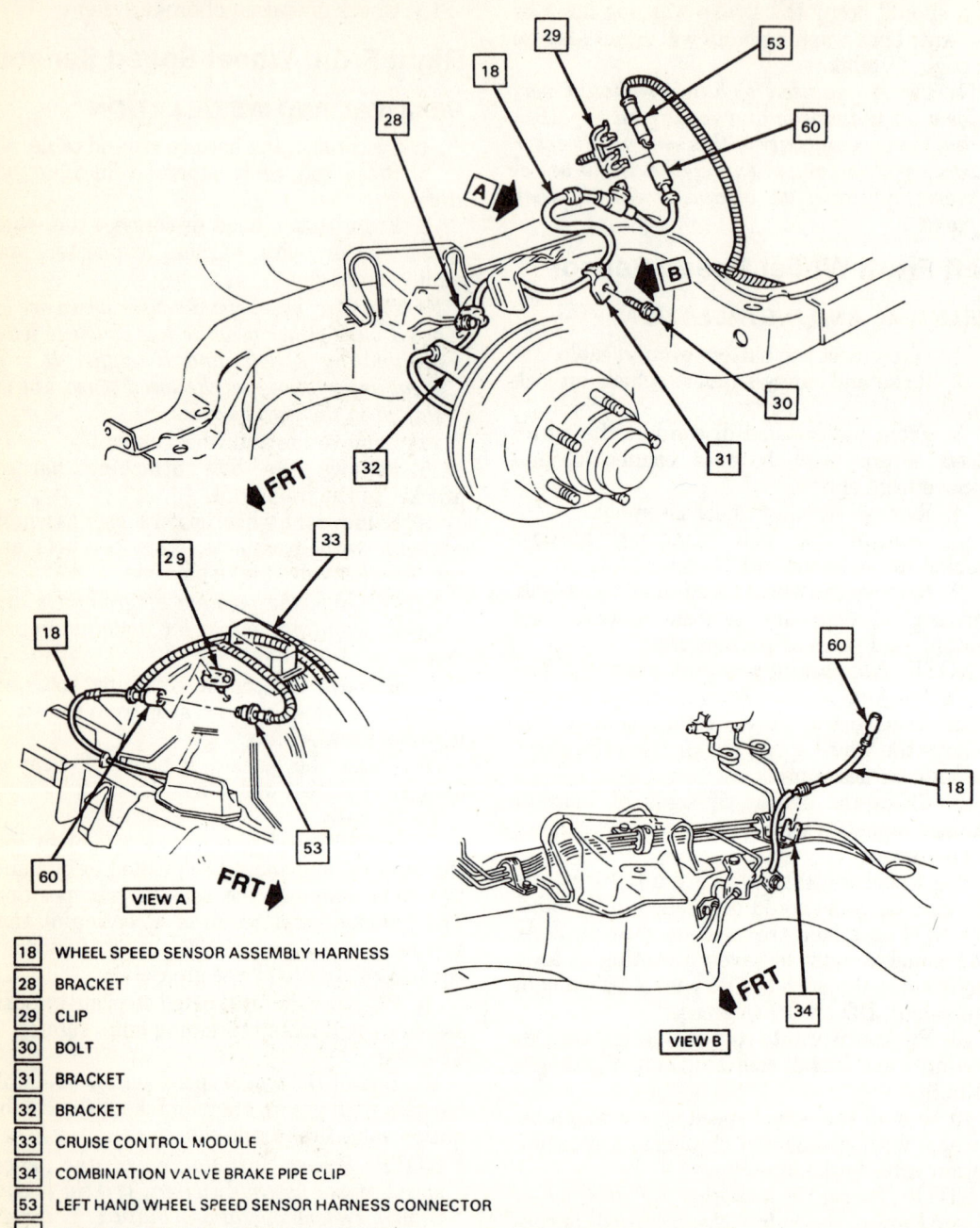

18	WHEEL SPEED SENSOR ASSEMBLY HARNESS
28	BRACKET
29	CLIP
30	BOLT
31	BRACKET
32	BRACKET
33	CRUISE CONTROL MODULE
34	COMBINATION VALVE BRAKE PIPE CLIP
53	LEFT HAND WHEEL SPEED SENSOR HARNESS CONNECTOR
60	WHEEL SPEED SENSOR ASSEMBLY CONNECTOR

Left front wheel speed sensor jumper harness

BRAKES 423

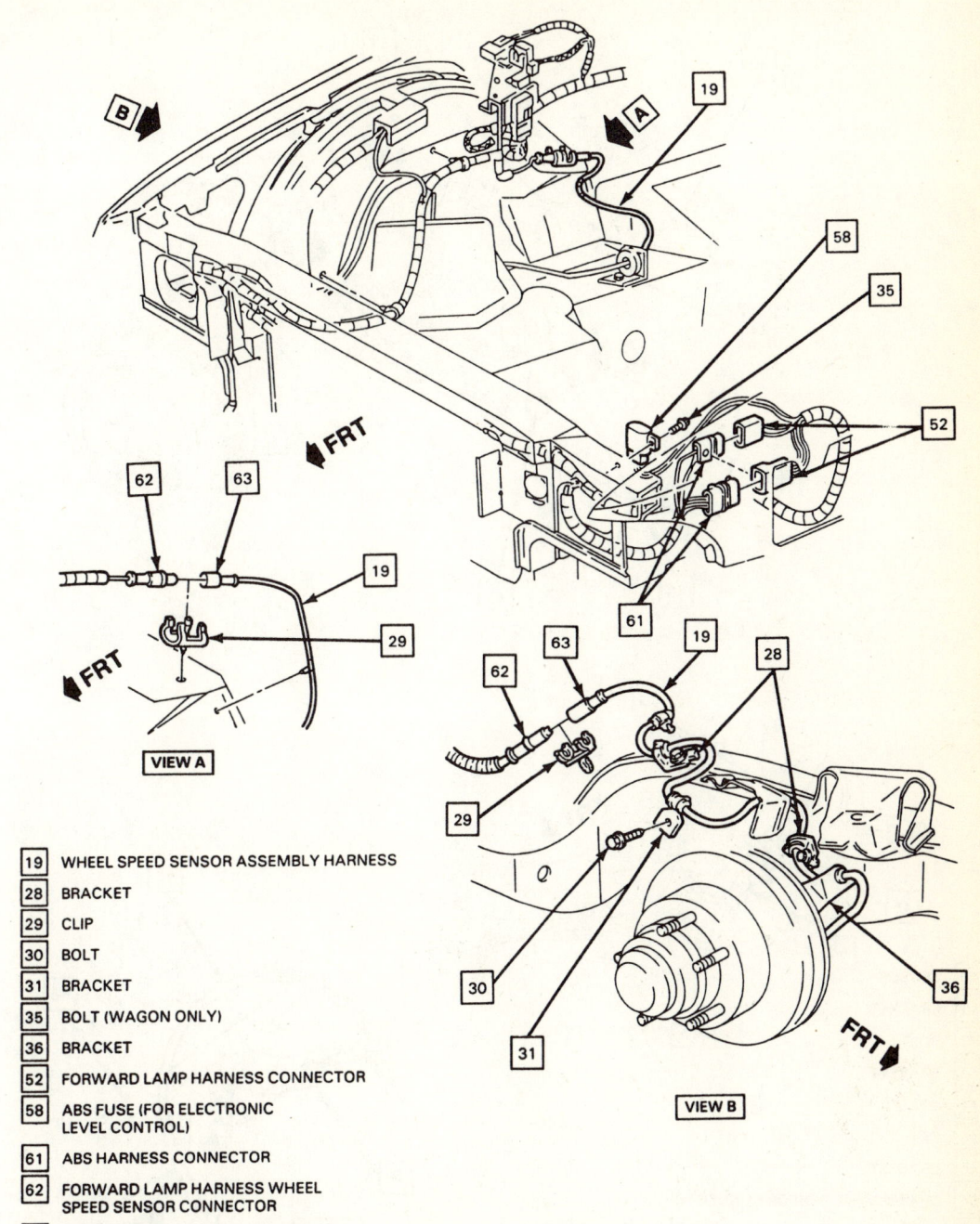

19	WHEEL SPEED SENSOR ASSEMBLY HARNESS
28	BRACKET
29	CLIP
30	BOLT
31	BRACKET
35	BOLT (WAGON ONLY)
36	BRACKET
52	FORWARD LAMP HARNESS CONNECTOR
58	ABS FUSE (FOR ELECTRONIC LEVEL CONTROL)
61	ABS HARNESS CONNECTOR
62	FORWARD LAMP HARNESS WHEEL SPEED SENSOR CONNECTOR
63	WHEEL SPEED SENSOR ASSEMBLY CONNECTOR

Right front wheel speed sensor jumper harness

424 BRAKES

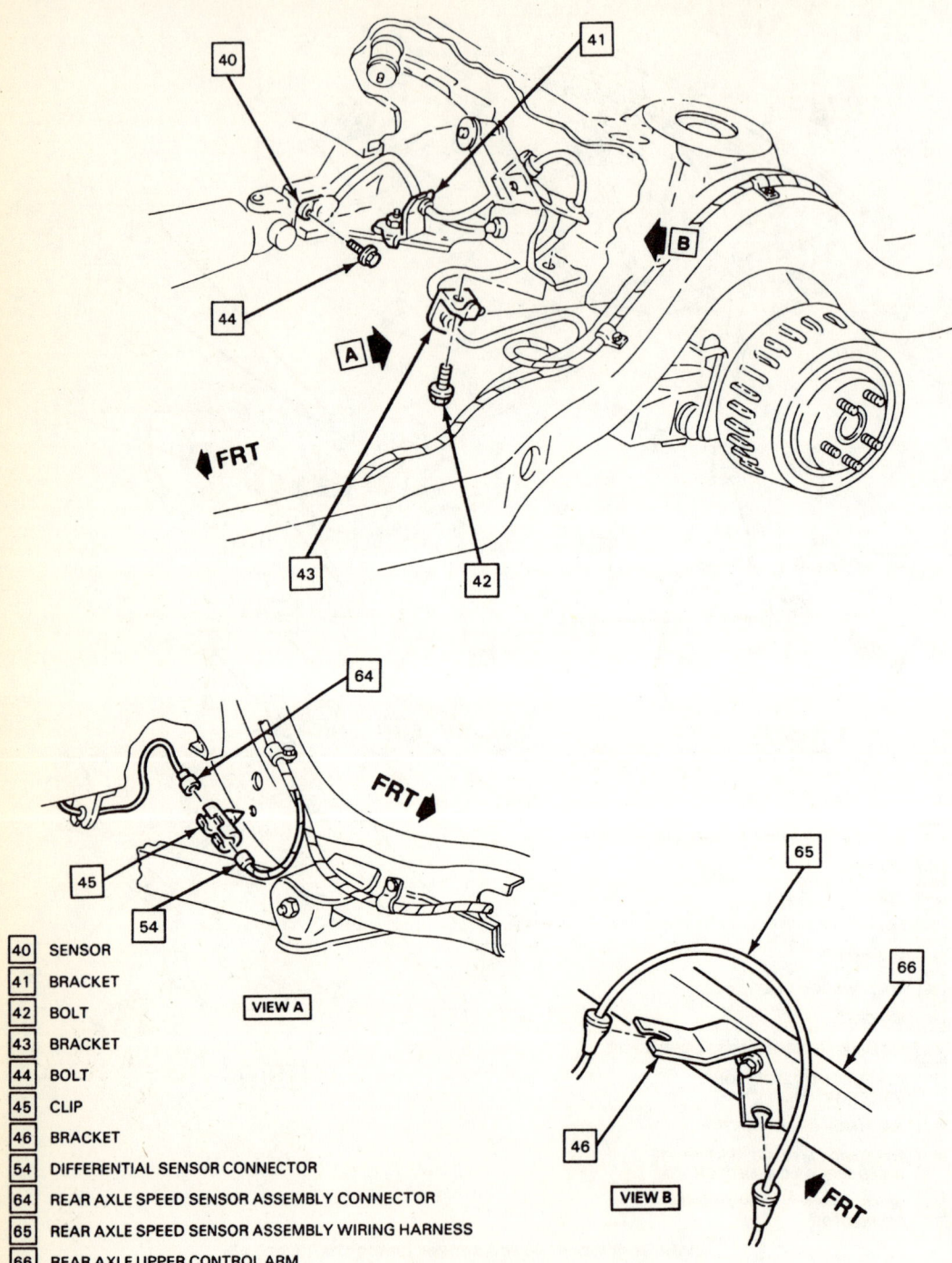

40	SENSOR
41	BRACKET
42	BOLT
43	BRACKET
44	BOLT
45	CLIP
46	BRACKET
54	DIFFERENTIAL SENSOR CONNECTOR
64	REAR AXLE SPEED SENSOR ASSEMBLY CONNECTOR
65	REAR AXLE SPEED SENSOR ASSEMBLY WIRING HARNESS
66	REAR AXLE UPPER CONTROL ARM

Rear axle speed sensor, 1991–92

Brake Specifications

All measurements given are (in.) unless noted

Year	Model	Master Cylinder Bore	Brake Disc Minimum Thickness	Brake Disc Maximum Run-Out	Brake Drum Diameter	Brake Drum Max. Machine O/S	Brake Drum Max. Wear Limit
1968	Exc. Sta. Wag.	1.0①	—	—	②	—	—
	Sta. Wag.	1.0①	—	—	②	—	—
1969–71	Exc. Sta. Wag.	1.0③	—	—	②	—	—
	Sta. Wag.	1.0③	—	—	②	—	—
1972–73	Exc. Sta. Wag.	—	1.215	.005	11.0	11.060	11.090
	Sta. Wag.	—	1.215	.005	12.0	12.060	12.090
1974–77	All	—	1.215	.002	9.50 11.0 12.0	④	⑤
1978–92	All	—	.965	.005	9.50 11.0	④	⑤

① Metallic brake lining 7/8 in.
② Disc 11.75 in.; Front and Rear drums 11.00 in.
③ With disc brake 1 1/8 in.
④ Maximum O/S 9.56—11.060—12.060 in.
⑤ Maximum wear limit 9.590—11.090—12.090 in.

Front Wheel Speed Sensor Ring

The front wheel speed sensor ring is an integral part of the brake rotor. The sensor ring is accessible for inspection by removing the brake rotor. If replacement of the wheel speed sensor ring is necessary, the brake rotor must be replaced.

Rear Axle Speed Sensor Ring

The rear axle speed sensor ring is an integral part of the rear axle differential pinion gear, and may be inspected by removing the rear speed sensor and using a flashlight and mirror to look through the mounting hole. If the sensor ring needs to be replaced, the entire pinion gear must be removed and replaced.

Rear Axle Speed Sensor

REMOVAL AND INSTALLATION

1. Disconnect the battery ground cable.
2. Raise and safely support vehicle on jack stands.
3. Disconnect the rear axle speed sensor assembly connector from the differential sensor connector. Unclip connectors from the clip and separate.
4. Note the position of the grommets and harness routing for reinstallation and remove the speed sensor assembly wiring harness with grommets from brackets.
5. Remove the speed sensor retaining bolt, and remove the rear axle speed sensor from axle housing.
6. Clean and inspect sensor for wear or damage replace as necessary.
7. Installation is the reverse of removal preceedure.

NOTE: *Proper installation of the wheel speed sensor assembly wire is critical to continued system operation. Failure to install the wire in brackets as shown, may result in contact with moving parts and/or over extension of the wire, resulting in circuit damage.*

CAUTION: *The wheel speed sensors are a tight fit into the knuckle and should be pushed in by hand. DO NOT hammer the sensors into position. The knuckle must be given a coating of anti-corrosion sealer such as GM part # 1052856 or equivalent, DO NOT use grease!*

Body 10

EXTERIOR

Doors

REMOVAL AND INSTALLATION

When removing the door, it is easier to remove the hinges with the door because the door side hinges are very accessible.

NOTE: *All factory installed door hardware attaching screws contain an epoxy thread-locking compound to ensure that the torque setting will be maintained. Service replacement screws may not contain a thread-locking compound. Such screws must be treated with No. 1052279 Loctite® 75 or equivalent. The adhesive is placed on the fastener prior to installation.*

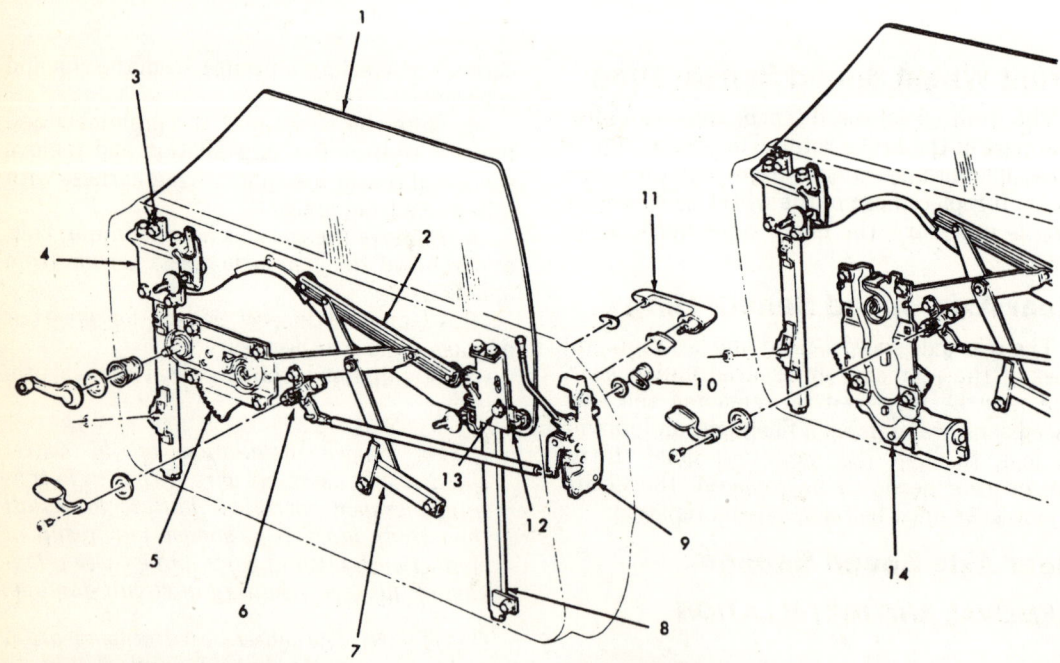

1. Front door window assembly
2. Lower sash channel cam
3. Window front upper stop
4. Front guide
5. Window regulator—manual
6. Door lock remote control
7. Inner panel cam
8. Rear guide
9. Door lock
10. Door lock cylinder
11. Door outside handle
12. Window rear upper stop (on window)
13. Window rear upper stop (on guide)
14. Window regulator—electric

Front door glass and related components without ventilator 1968-70

BODY

1. Disconnect the negative battery cable. Mark the position of the door hinges-to-body to make the installation easier.

2. If equipped with power operated components, remove the trim panel and detach the inner panel water deflector enough to disconnect the wiring harness from the components. Separate and remove the rubber conduit and the wiring harness from the door.

3. Using an assistant (to support the door), remove the upper and lower hinge-to-body bolts. Remove the door from the vehicle.

4. To install, reverse the removal procedures. Torque the hinge-to-body bolt to 15-21 ft. lbs.

ADJUSTMENT

The door adjustments are made possible through the use of floating anchor plates in the door and the body hinge pillars.

1. Remove the door lock striker from the body and allow the door to hang freely on it's hinges.

2. Using a door hinge alignment tool,

1. Support, front window bumper bolt
2. Support rear window bumper bolt
3. Stop, front up-travel bolt location ("B, C and E" styles)
4. Stop, rear up-travel bolt location ("B, C and E" styles)
5. Trim pad adjusting plates and stabilizer strips
6. Plate assembly, lower sash guide bolts
7. Guide assembly, lower sash upper bolts
8. Guide assembly, lower sash lower bolts
9. Glass bearing plate to inner panel bolts
10. Glass bearing plate adjusting stud
11. Guide assembly, lower sash upper adjustment access hole (rotated "cocked" glass adjustment)
12. Plate assembly, lower sash guide adjustment access hole (fore and aft adjustment)

Front door (open frame) hardware 1971-76

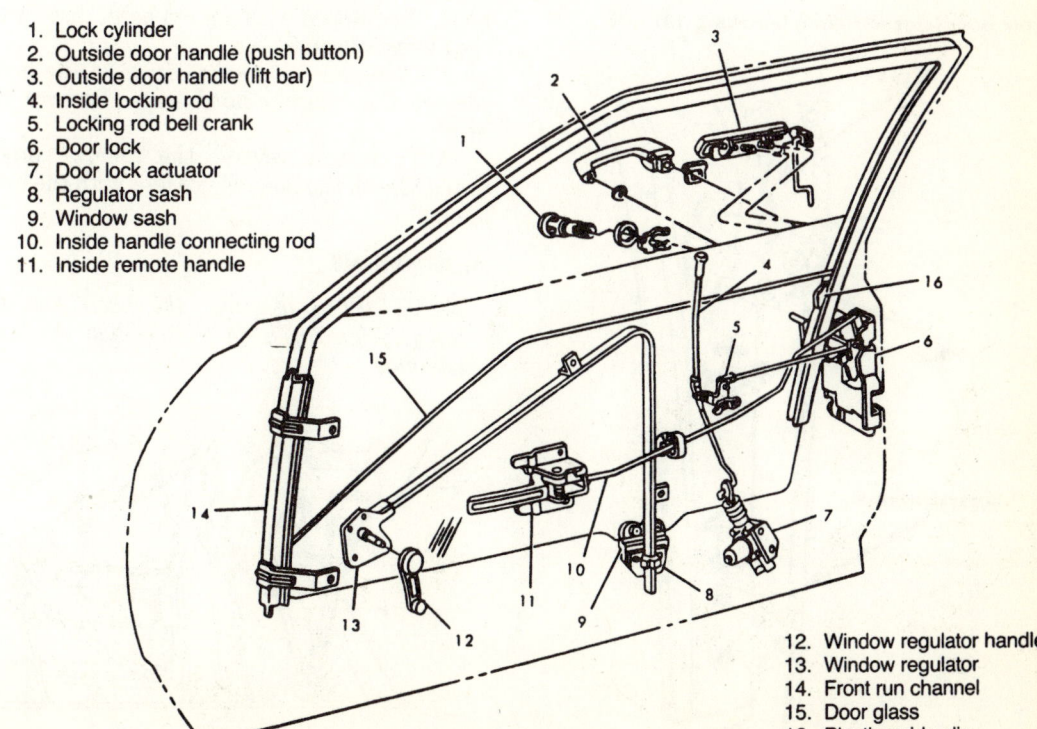

1. Lock cylinder
2. Outside door handle (push button)
3. Outside door handle (lift bar)
4. Inside locking rod
5. Locking rod bell crank
6. Door lock
7. Door lock actuator
8. Regulator sash
9. Window sash
10. Inside handle connecting rod
11. Inside remote handle
12. Window regulator handle
13. Window regulator
14. Front run channel
15. Door glass
16. Plastic guide clip

Front door glass and related components 1980 and later

BODY

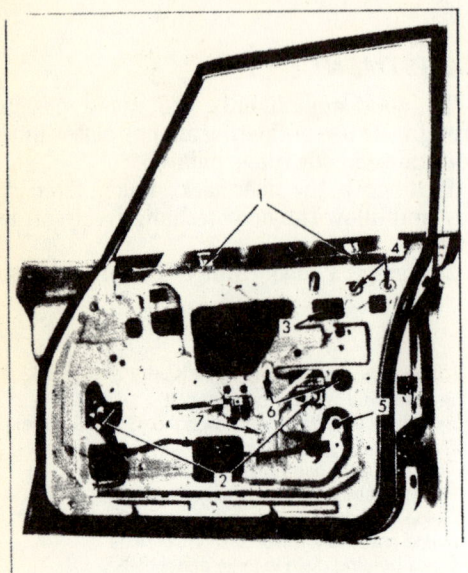

1. Window trim support hanger and stabilizer strip bolts
2. Window lower sash channel cam attaching stud nut access holes
3. Window rear guide to guide bracket bolt
4. Window rear guide bracket to inner panel bolts
5. Window rear guide lower bolt
6. Inner panel cam bolts
7. Window down-travel support bracket bolt

Front door (closed frame) hardware 1971-76

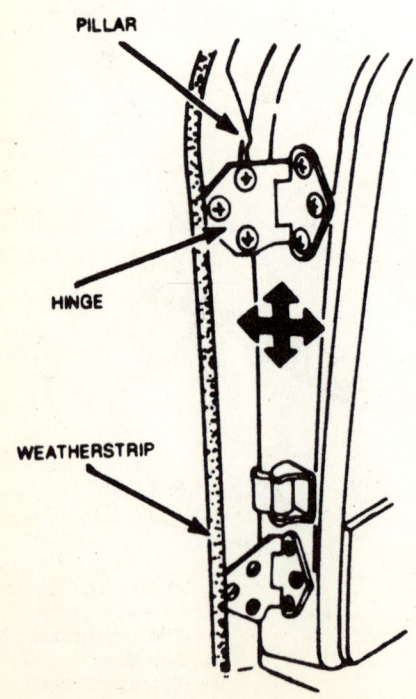

Adjusting the door hinge position

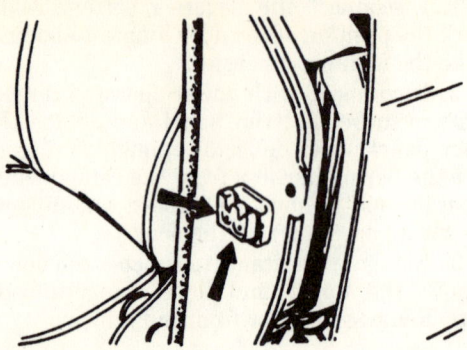

Adjusting the door striker

loosen the door hinge-to-body pillar bolts.

3. Using the tool attachments, adjust the door up/down and fore/aft.

NOTE: *If a rearward adjustment is made, it may be necessary to replace the jamb switch.*

4. At the door hinge pillar attachments, adjust the door in and out.

5. After adjusting the door, torque the door hinge-to-body pillar to 15-21 ft. lbs.

Hood

REMOVAL AND INSTALLATION

1. Disconnect the negative battery cable. Using a scratch awl, scribe the hinge onto the hood. If equipped with a hood light, disconnect the electrical connector.

2. Using an assistant to support the hood, remove the hinge-to-hood bolts. Remove the hood.

3. To install, reverse the removal procedures. Check the hood alignment with the hood latch.

ALIGNMENT

NOTE: *When aligning the hood and the latch, align the hood (first), then the latch (second).*

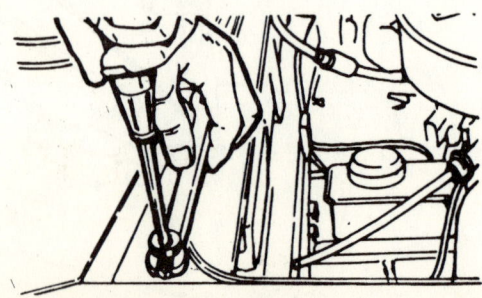

The hood is adjusted vertically by stop screws at the front and or rear position

CHILTON'S AUTO BODY REPAIR TIPS

EASY STEP-BY-STEP TIPS FROM PROS

Tools and Materials • Step-by-Step Illustrated Procedures
How To Repair Dents, Scratches and Rust Holes
Spray Painting and Refinishing Tips

With a little practice, basic body repair procedures can be mastered by any do-it-yourself mechanic. The step-by-step repairs shown here can be applied to almost any type of auto body repair.

TOOLS & MATERIALS

You may already have basic tools, such as hammers and electric drills. Other tools unique to body repair — body hammers, grinding attachments, sanding blocks, dent puller, half-round plastic file and plastic spreaders — are relatively inexpensive and can be obtained wherever auto parts or auto body repair parts are sold. Portable air compressors and paint spray guns can be purchased or rented.

Auto Body Repair Kits

The best and most often used products are available to the do-it-yourselfer in kit form, from major manufacturers of auto body repair products. The same manufacturers also merchandise the individual products for use by pros.

Kits are available to make a wide variety of repairs, including holes, dents and scratches and fiberglass, and offer the advantage of buying the materials you'll need for the job. There is little waste or chance of materials going bad from not being used. Many kits may also contain basic body-working tools such as body files, sanding blocks and spreaders. Check the contents of the kit before buying your tools.

BODY REPAIR TIPS

Safety

Many of the products associated with auto body repair and refinishing contain toxic chemicals. Read all labels before opening containers and store them in a safe place and manner.

• Wear eye protection (safety goggles) when using power tools or when performing any operation that involves the removal of any type of material.

• Wear lung protection (disposable mask or respirator) when grinding, sanding or painting.

Sanding

1 Sand off paint before using a dent puller. When using a non-adhesive sanding disc, cover the back of the disc with an overlapping layer or two of masking tape and trim the edges. The disc will last considerably longer.

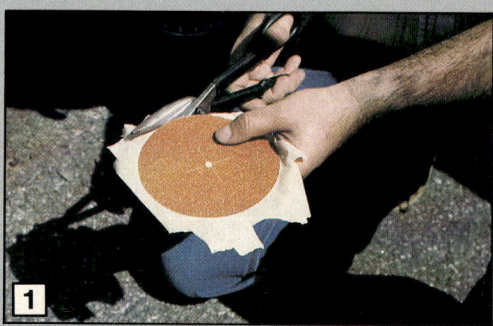

2 Use the circular motion of the sanding disc to grind *into* the edge of the repair. Grinding or sanding away from the jagged edge will only tear the sandpaper.

3 Use the palm of your hand flat on the panel to detect high and low spots. Do not use your fingertips. Slide your hand slowly back and forth.

WORKING WITH BODY FILLER

Mixing The Filler

Cleanliness and proper mixing and application are extremely important. Use a clean piece of plastic or glass or a disposable artist's palette to mix body filler.

1 Allow plenty of time and follow directions. No useful purpose will be served by adding more hardener to make it cure (set-up) faster. Less hardener means more curing time, but the mixture dries harder; more hardener means less curing time but a softer mixture.

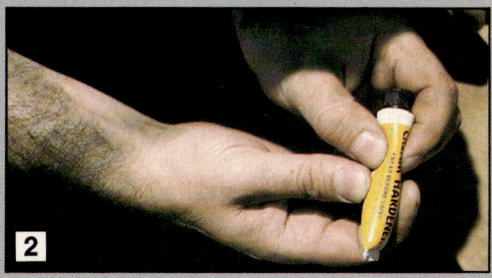

2 Both the hardener and the filler should be thoroughly kneaded or stirred before mixing. Hardener should be a solid paste and dispense like thin toothpaste. Body filler should be smooth, and free of lumps or thick spots.

Getting the proper amount of hardener in the filler is the trickiest part of preparing the filler. Use the same amount of hardener in cold or warm weather. For contour filler (thick coats), a bead of hardener twice the diameter of the filler is about right. There's about a 15% margin on either side, but, if in doubt use less hardener.

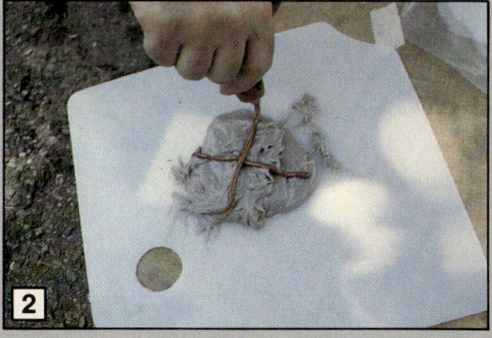

3 Mix the body filler and hardener by wiping across the mixing surface, picking the mixture up and wiping it again. Colder weather requires longer mixing times. Do not mix in a circular motion; this will trap air bubbles which will become holes in the cured filler.

Applying The Filler

1 For best results, filler should not be applied over 1/4" thick.

Apply the filler in several coats. Build it up to above the level of the repair surface so that it can be sanded or grated down.

The first coat of filler must be pressed on with a firm wiping motion.

Apply the filler in one direction only. Working the filler back and forth will either pull it off the metal or trap air bubbles.

REPAIRING DENTS

Before you start, take a few minutes to study the damaged area. Try to visualize the shape of the panel before it was damaged. If the damage is on the left fender, look at the right fender and use it as a guide. If there is access to the panel from behind, you can reshape it with a body hammer. If not, you'll have to use a dent puller. Go slowly and work the metal a little at a time. Get the panel as straight as possible before applying filler.

1 This dent is typical of one that can be pulled out or hammered out from behind. Remove the headlight cover, headlight assembly and turn signal housing.

2 Drill a series of holes ½ the size of the end of the dent puller along the stress line. Make some trial pulls and assess the results. If necessary, drill more holes and try again. Do not hurry.

3 If possible, use a body hammer and block to shape the metal back to its original contours. Get the metal back as close to its original shape as possible. Don't depend on body filler to fill dents.

4 Using an 80-grit grinding disc on an electric drill, grind the paint from the surrounding area down to bare metal. Use a new grinding pad to prevent heat buildup that will warp metal.

5 The area should look like this when you're finished grinding. Knock the drill holes in and tape over small openings to keep plastic filler out.

6 Mix the body filler (see Body Repair Tips). Spread the body filler evenly over the entire area (see Body Repair Tips). Be sure to cover the area completely.

7 Let the body filler dry until the surface can just be scratched with your fingernail. Knock the high spots from the body filler with a body file ("Cheesegrater"). Check frequently with the palm of your hand for high and low spots.

8 Check to be sure that trim pieces that will be installed later will fit exactly. Sand the area with 40-grit paper.

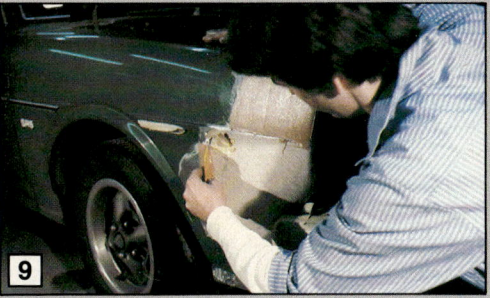

9 If you wind up with low spots, you may have to apply another layer of filler.

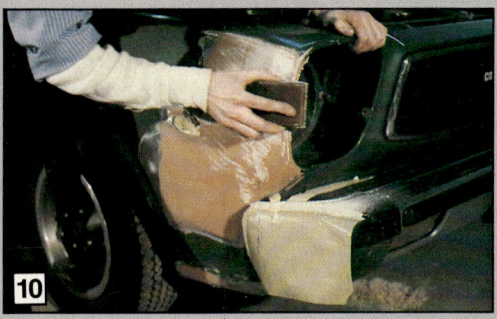

10 Knock the high spots off with 40-grit paper. When you are satisfied with the contours of the repair, apply a thin coat of filler to cover pin holes and scratches.

11 Block sand the area with 40-grit paper to a smooth finish. Pay particular attention to body lines and ridges that must be well-defined.

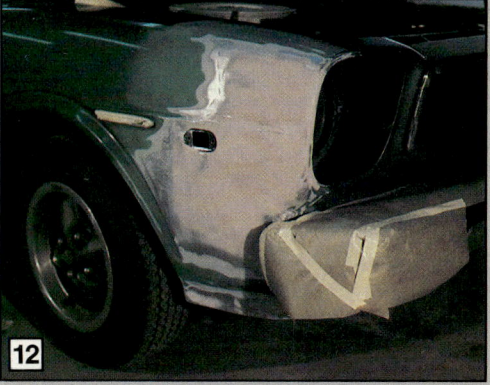

12 Sand the area with 400 paper and then finish with a scuff pad. The finished repair is ready for priming and painting (see Painting Tips).

Materials and photos courtesy of Ritt Jones Auto Body, Prospect Park, PA.

REPAIRING RUST HOLES

There are many ways to repair rust holes. The fiberglass cloth kit shown here is one of the most cost efficient for the owner because it provides a strong repair that resists cracking and moisture and is relatively easy to use. It can be used on large and small holes (with or without backing) and can be applied over contoured areas. Remember, however, that short of replacing an entire panel, no repair is a guarantee that the rust will not return.

1 Remove any trim that will be in the way. Clean away all loose debris. Cut away all the rusted metal. But be sure to leave enough metal to retain the contour or body shape.

2 Grind away all traces of rust with a 24-grit grinding disc. Be sure to grind back 3-4 inches from the edge of the hole down to bare metal and be sure all traces of paint, primer and rust are removed.

3 Block sand the area with 80 or 100 grit sandpaper to get a clear, shiny surface and feathered paint edge. Tap the edges of the hole inward with a ball peen hammer.

4 If you are going to use release film, cut a piece about 2-3" larger than the area you have sanded. Place the film over the repair and mark the sanded area on the film. Avoid any unnecessary wrinkling of the film.

5 Cut 2 pieces of fiberglass matte to match the shape of the repair. One piece should be about 1" smaller than the sanded area and the second piece should be 1" smaller than the first. Mix enough filler and hardener to saturate the fiberglass material (see Body Repair Tips).

6 Lay the release sheet on a flat surface and spread an even layer of filler, large enough to cover the repair. Lay the smaller piece of fiberglass cloth in the center of the sheet and spread another layer of filler over the fiberglass cloth. Repeat the operation for the larger piece of cloth.

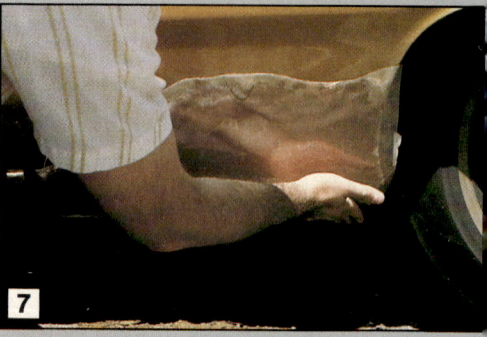

7 Place the repair material over the repair area, with the release film facing outward. Use a spreader and work from the center outward to smooth the material, following the body contours. Be sure to remove all air bubbles.

8 Wait until the repair has dried tack free and peel off the release sheet. The ideal working temperature is 60°-90° F. Cooler or warmer temperatures or high humidity may require additional curing time. Wait longer, if in doubt.

9 Sand and feather-edge the entire area. The initial sanding can be done with a sanding disc on an electric drill if care is used. Finish the sanding with a block sander. Low spots can be filled with body filler; this may require several applications.

10 When the filler can just be scratched with a fingernail, knock the high spots down with a body file and smooth the entire area with 80-grit. Feather the filled areas into the surrounding areas.

11 When the area is sanded smooth, mix some topcoat and hardener and apply it directly with a spreader. This will give a smooth finish and prevent the glass matte from showing through the paint.

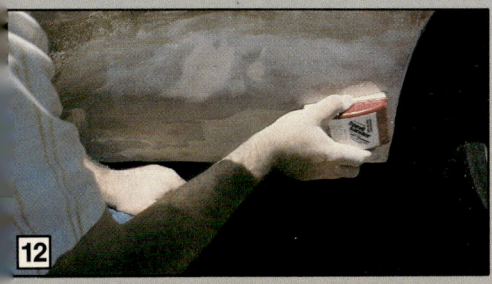

12 Block sand the topcoat smooth with finishing sandpaper (200 grit), and 400 grit. The repair is ready for masking, priming and painting (see Painting Tips).

Materials and photos courtesy Marson Corporation, Chelsea, Massachusetts

PAINTING TIPS

Preparation

1 SANDING — Use a 400 or 600 grit wet or dry sandpaper. Wet-sand the area with a 1/4 sheet of sandpaper soaked in clean water. Keep the paper wet while sanding. Sand the area until the repaired area tapers into the original finish.

2 CLEANING — Wash the area to be painted thoroughly with water and a clean rag. Rinse it thoroughly and wipe the surface dry until you're sure it's completely free of dirt, dust, fingerprints, wax, detergent or other foreign matter.

3 MASKING — Protect any areas you don't want to overspray by covering them with masking tape and newspaper. Be careful not get fingerprints on the area to be painted.

4 PRIMING — All exposed metal should be primed before painting. Primer protects the metal and provides an excellent surface for paint adhesion. When the primer is dry, wet-sand the area again with 600 grit wet-sandpaper. Clean the area again after sanding.

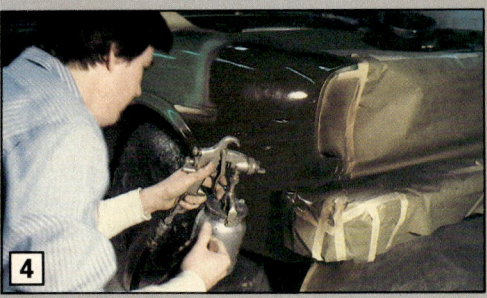

Painting Techniques

Paint applied from either a spray gun or a spray can (for small areas) will provide good results. Experiment on an old piece of metal to get the right combination before you begin painting.

SPRAYING VISCOSITY (SPRAY GUN ONLY) — Paint should be thinned to spraying viscosity according to the directions on the can. Use only the recommended thinner or reducer and the same amount of reduction regardless of temperature.

AIR PRESSURE (SPRAY GUN ONLY) — This is extremely important. Be sure you are using the proper recommended pressure.

TEMPERATURE — The surface to be painted should be approximately the same temperature as the surrounding air. Applying warm paint to a cold surface, or vice versa, will completely upset the paint characteristics.

THICKNESS — Spray with smooth strokes. In general, the thicker the coat of paint, the longer the drying time. Apply several thin coats about 30 seconds apart. The paint should remain wet long enough to flow out and no longer; heavier coats will only produce sags or wrinkles. Spray a light (fog) coat, followed by heavier color coats.

DISTANCE — The ideal spraying distance is 8"-12" from the gun or can to the surface. Shorter distances will produce ripples, while greater distances will result in orange peel, dry film and poor color match and loss of material due to overspray.

OVERLAPPING — The gun or can should be kept at right angles to the surface at all times. Work to a wet edge at an even speed, using a 50% overlap and direct the center of the spray at the lower or nearest edge of the previous stroke.

RUBBING OUT (BLENDING) FRESH PAINT — Let the paint dry thoroughly. Runs or imperfections can be sanded out, primed and repainted.

Don't be in too big a hurry to remove the masking. This only produces paint ridges. When the finish has dried for at least a week, apply a small amount of fine grade rubbing compound with a clean, wet cloth. Use lots of water and blend the new paint with the surrounding area.

WRONG

Thin coat. Stroke too fast, not enough overlap, gun too far away.

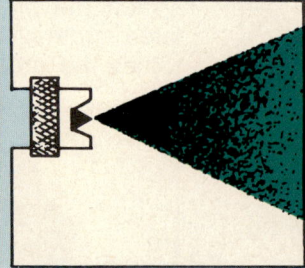

CORRECT

Medium coat. Proper distance, good stroke, proper overlap.

WRONG

Heavy coat. Stroke too slow, too much overlap, gun too close.

Hood

The hood hinge-to-body mount is slotted to provide forward and rearward movement. Adjust the hood so that it is flush with the body sheet metal.

1. Using a scratch awl, scribe the hinge outline onto the hood.
2. Loosen the appropriate screws and shift the hood into proper alignment with the vehicles sheet metal.

CAUTION: *Make sure that the rear of the hood is properly positioned at the cowl seal; proper sealing will restrict fumes from the*

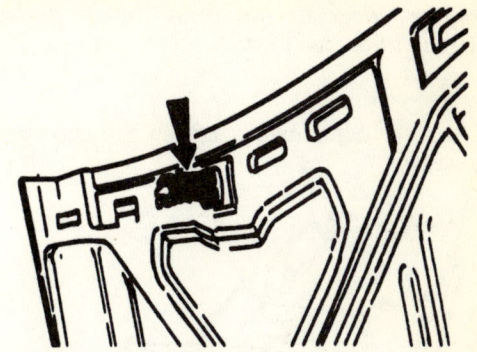

The hood pin can be adjusted for proper lock engagement

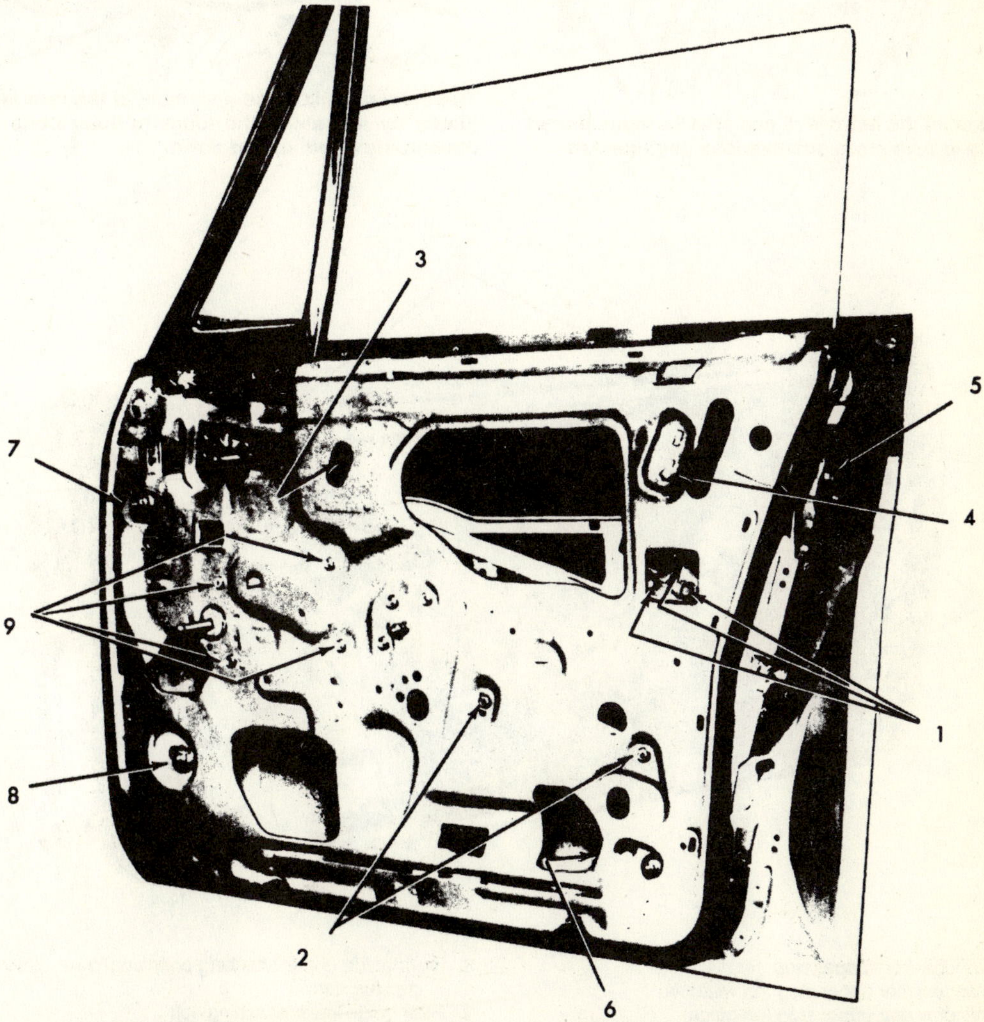

1. Window lower sash channel to rear guide plate bolts
2. Inner panel cam attaching bolts
3. Window front up-stop
4. Window rear up-stop
5. Rear guide upper attaching bolt
6. Rear guide to lower support bracket bolt
7. Ventilator lower frame adjusting stud and nut
8. Ventilator division channel lower adjusting stud and nut
9. Window regulator attaching bolts

Front door glass and related components with ventilator and open frame 1968-70

430 BODY

engine compartment from being pulled through the cowl vent.

3. After adjustment, tighten the appropriate screws.

Latch

The hood latch assembly is mounted on a plate with elongated holes, which allow vertical adjustment.

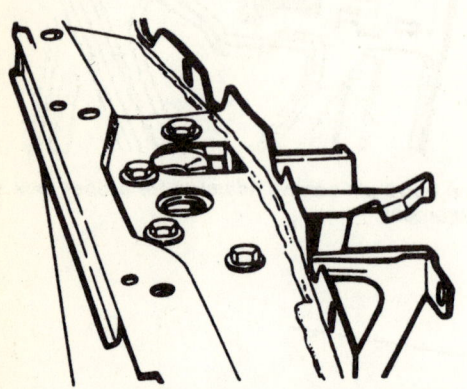

The base of the hood lock can also be repositioned slightly to give more positive lock engagement

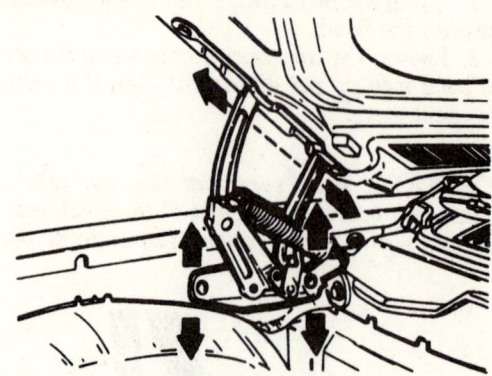

The height of the hood assembly at the rear is adjusted by loosening the hinge-to-body bolts and moving the hood up and down

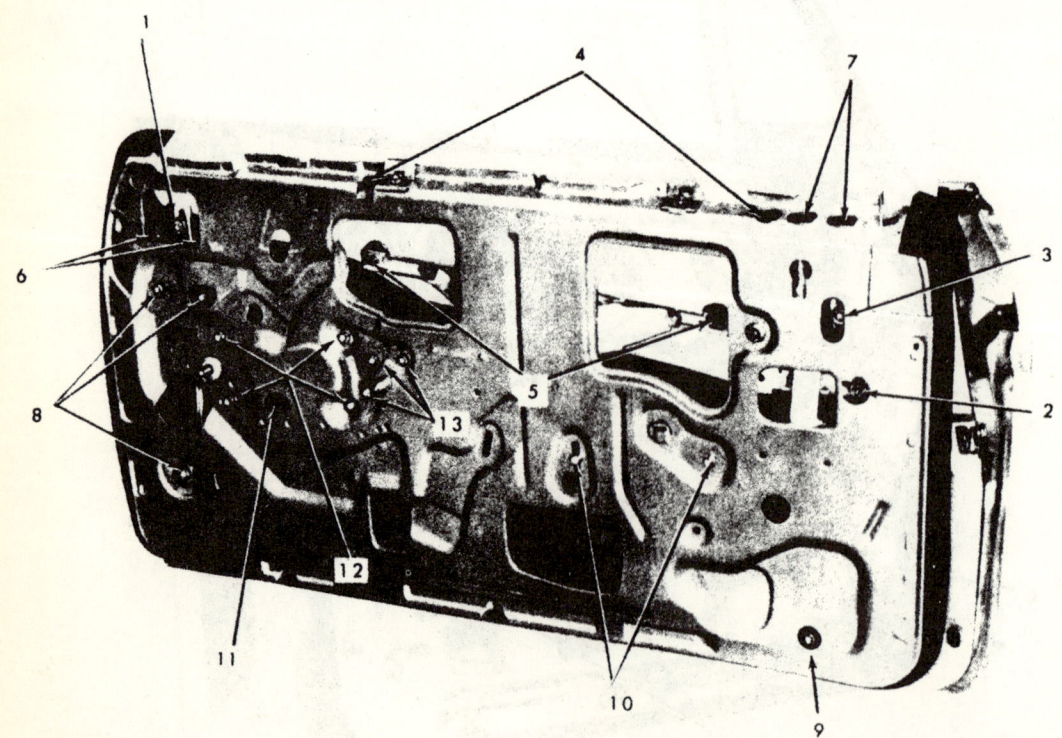

1. Window front upper stop
2. Window rear upper stop (on window)
3. Window rear upper stop (on guide)
4. Window stabilizer strip assemblies
5. Window lower sash channel cam stud nuts
6. Front guide to upper support bracket bolts
7. Rear guide upper attaching bolts
8. Front guide upper attaching bolts and lower attaching stud nut
9. Rear guide lower attaching bolt
10. Inner panel cam attaching bolts
11. Sector gear stop bolts
12. Window regulator attaching bolts
13. Door lock remote control attaching bolts

Front door glass and related components without ventilator and open frame 1968-70

BODY

Striker

The striker, on the hood, adjusts laterally to align with the hood latch assembly.

Trunk Lid

REMOVAL AND INSTALLATION

1. Disconnect the negative battery cable. Open the trunk lid and place protective coverings over the rear fenders to protect the paint from damage.
2. Mark the location of the hinge-to-trunk lid bolts and disconnect the electrical connections and wiring from the lid, if equipped.
3. Using an assistant to support the lid, remove the hinge-to-lid bolts and the lid from the vehicle.

NOTE: *Some later vehicles use gas cushioned shock type assemblies to aid in supporting the trunk lid assembly. Remove the retaining clips and remove the shock assemblies from the trunk lid before removing the hinge retaining bolts.*

4. To install, reverse the removal procedures. Adjust the position of the trunk lid to the body. Rear compartment torque rods are adjustable to increase or decrease operating effort. To increase the amount of effort needed to raise the rear compartment lid or to decrease the amount of effort to close the lid, reposition the end of the rod to a lower torque rod adjusting notch. To decrease the amount of effort needed to raise the rear compartment lid or to increase the amount of effort to close the lid, reposition the end of the rod to a higher torque rod adjusting notch. Refer to the illustration. Prop the trunk lid in full-open position to keep lid from falling when torque rods are disengaged from the torque rod bracket.

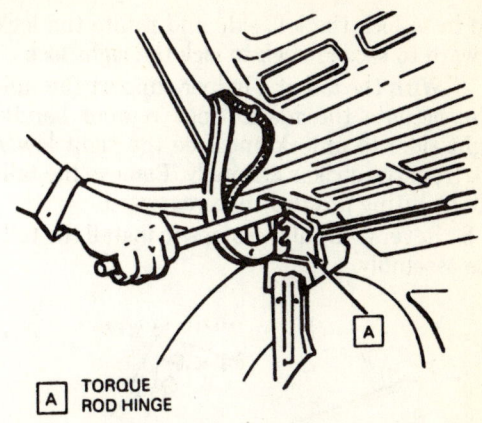

Removing or adjusting the torque rod

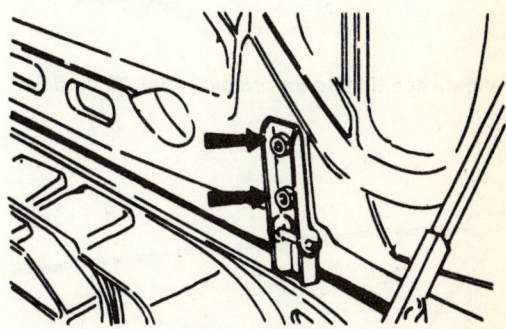

Loosen the hinge bolts to permit fore-and-aft and horizontal adjustment

Tailgate

REMOVAL AND INSTALLATION

1968-70 Models

1. Disconnect the negative battery cable. Partially open the tailgate to achieve a neutral torque rod position or until tension on the torque rod has been relieved.
2. With the tension relieved remove the torque rod assist link retainer to body attaching bolts.
3. Support the tailgate in the full open position and remove the support cable to left upper hinge and striker assembly attaching bolt.
4. On models equipped with electrical options in the tailgate, remove the inner panel water deflector and access hole cover. Disconnect the wiring harness connectors and pull the wiring from the tailgate.
5. With the aid of a helper, remove the left lower hinge to body attaching bolts.
6. Using a suitable tool, manually lock the right and left upper locks. Push the arm down

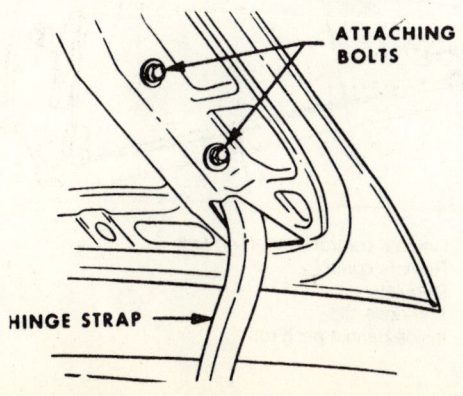

View of the trunk lid-to-hinge bolts

432 BODY

and in to lock the left side and rotate the lock forward to second click to lock the right lock.

7. With the aid of a helper support the tailgate, actuate the inside door remote handle (right side) to unlock and free the right lower lock from the striker assembly. Remove the tailgate by lifting upward then rearward.

8. Reverse the above steps to install the tailgate assembly.

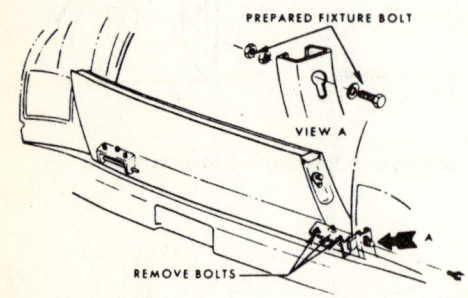

Tailgate positioning for removal 1968-70 models

1971-76 Models

NOTE: *Before removing the tailgate assembly fabricate a $1/4$-20×1". fixture bolt to serve as a roller stop for the removal of the tailgate assembly.*

1. Be sure an protect the surface of the tailgate assembly by covering the rear bumper upper surface with tape or equivalent.

2. Disconnect the negative battery cable.

3. Raise the tailgate to the latched position and remove the left side access cover.

NOTE: *Be sure to maintain the firm grip on the tailgate assembly when performing the following removal steps.*

4. Remove the regulator hinge arm to tailgate retaining screws. On manual tailgates actuate the control knob to unlock the gate.

5. Lower the gate to clear the lock assembly and allow the gate to pivot rearward about 45 degrees.

6. Insert the homemade fixture bolt into the key slot at the upper right end of the channel guide assembly below the roller. This will

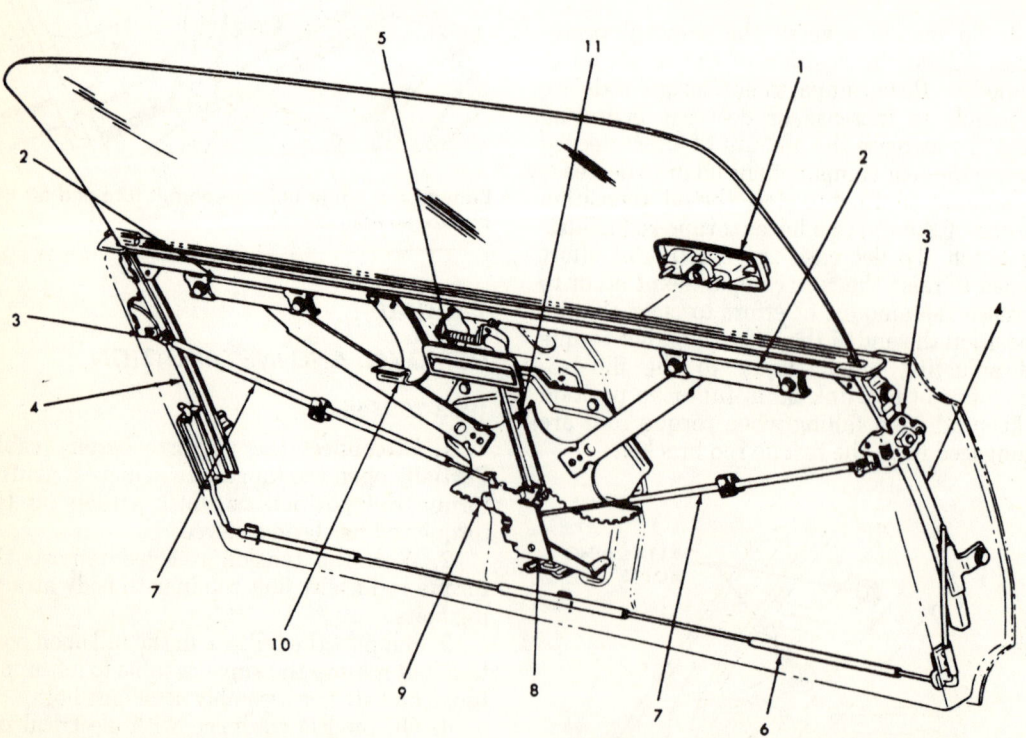

1. Outside handle or key switch
2. Sash channel cams
3. Locks
4. Lower glass run channels
5. Inside handle
6. Torque rod
7. Remote control connecting rods
8. Remote control
9. Regulator
10. Anti-rattle clip
11. Inside handle push rod

Tailgate assembly 1968-70 models

hold the synchronizing tube assembly in position during the tailgate removal process.

7. Scribe the location of the right lower roller support on the tailgate. Remove the retaining bolts.

8. Move the right side of the tailgate rearward to clear the body opening.

9. Slide the tailgate to the right and out of engagement with the left torque roller shaft assembly.

10. Remove the tailgate assembly from the vehicle.

To install:

11. Position the tailgate in place on the vehicle. Slide the tailgate to the right and engage it with the left torque roller shaft assembly.

12. Using the scribe marks align the components. Install the retaining bolts.

13. Remove the homemade fixture bolt.

14. Install the regulator hinge arm to tailgate retaining screws.

15. Raise the tailgate to the latched position and install the left side access cover.

16. Connect the negative battery cable. Adjust the tailgate as required. Remove the protective tape from the bumper.

1977-92 Models

1. Disconnect the negative battery cable.

2. Open the tailgate until the tension is relieved from the torque rod. Remove the torque rod assist link retainer to body bolts.

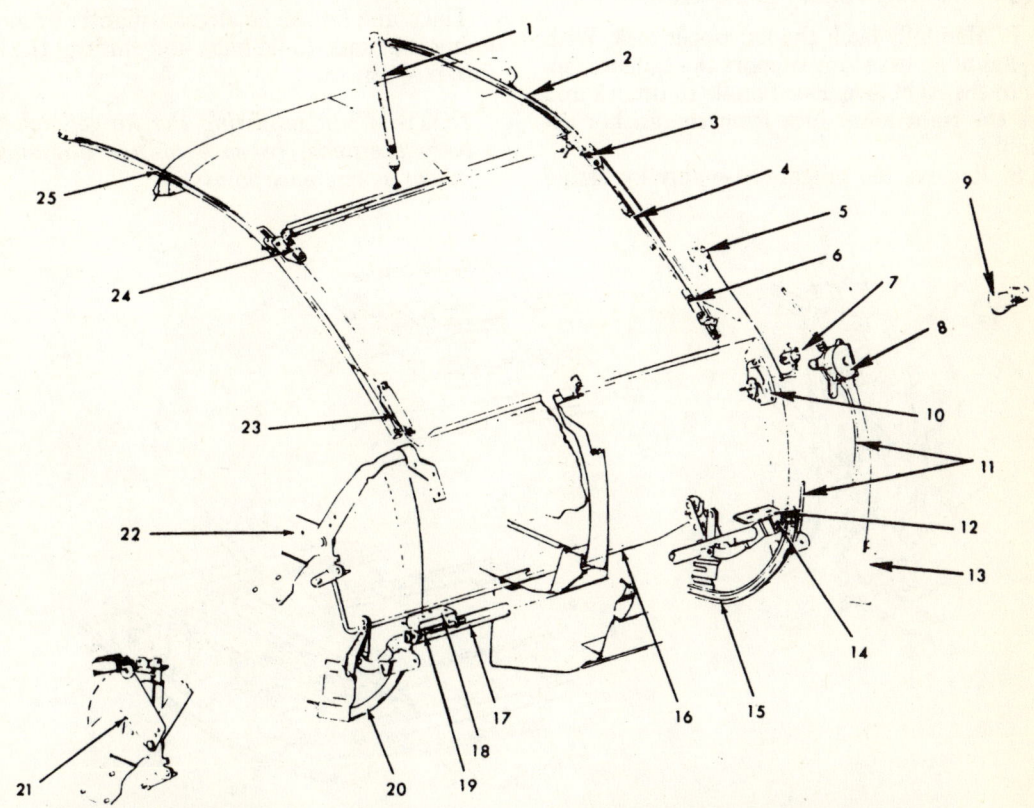

1. Window lift spring and silencer
2. Window guide cam assembly—right
3. Window guide roller assembly—upper right
4. Window drive cable assembly
5. Window stop cable and clip assembly—(manual gate)
6. Window guide roller assembly—lower right
7. Lock and release switch assembly—(manual gate)
8. Window regulator assembly
9. Window and gate switch assembly (power gate)
10. Tailgate lock assembly
11. Window regulator storage conduit
12. Tailgate roller support—right
13. Window motor assembly
14. Torque roller and shaft assembly—right
15. Tailgate guide channel assembly—right
16. Tailgate hinge torque rod
17. Tailgate synchronizing torque
18. Tailgate roller support—left
19. Torque roller and shaft assembly—left
20. Tailgate guide channel assembly—left
21. Lift arm hinge and regulator assembly—(electric)
22. Lift arm hinge assembly—(manual)
23. Window guide roller assembly—lower left
24. Window guide roller assembly—upper left
25. Window guide cam assembly—left

Tailgate assembly 1971-76 models

BODY

3. Close the tailgate and open it as a gate.

4. Support the tailgate in the full open position. Remove the support cable to the left upper hinge striker assembly. Remove the left side quarter trim.

5. With the aid of an assistant, remove the left lower hinge to body retaining bolts.

6. Disconnect the electrical harness connector. Remove the screws retaining the electrical conduit to the tailgate assembly.

NOTE: *With the tailgate open in the gate position and the right and left upper locks manually engaged, the tailgate is in a vulnerable position and could drop from the right lower lock if the locking rod and the outside handle are activated. Do not pull on the rod or activate the outside handle, as damage to the tailgate or personal injury could result.*

7. Manually latch the left upper lock. With the aid of an assistant support the tailgate, activate the right side door handle to unlock and free the right lower lock from the striker assembly.

8. Remove the tailgate assembly by lifting upward and than rearward.

To install:

9. Position the tailgate assembly in place on the vehicle.

10. Install the retaining bolts.

11. Install the electrical conduit to the tailgate assembly.

12. Connect the electrical harness connector.

13. Install the left quarter trim.

14. Install the left upper hinge striker assembly. Adjust the tailgate assembly, as required.

15. Connect the negative battery cable.

ALIGNMENT

Trunk Lid

The trunk lid can be aligned slightly by loosening the hinge-to-lid bolts and shifting the lid into position.

NOTE: *When adjusting the hinge/latch-to-body positions, be sure to use alignment marks as reference points.*

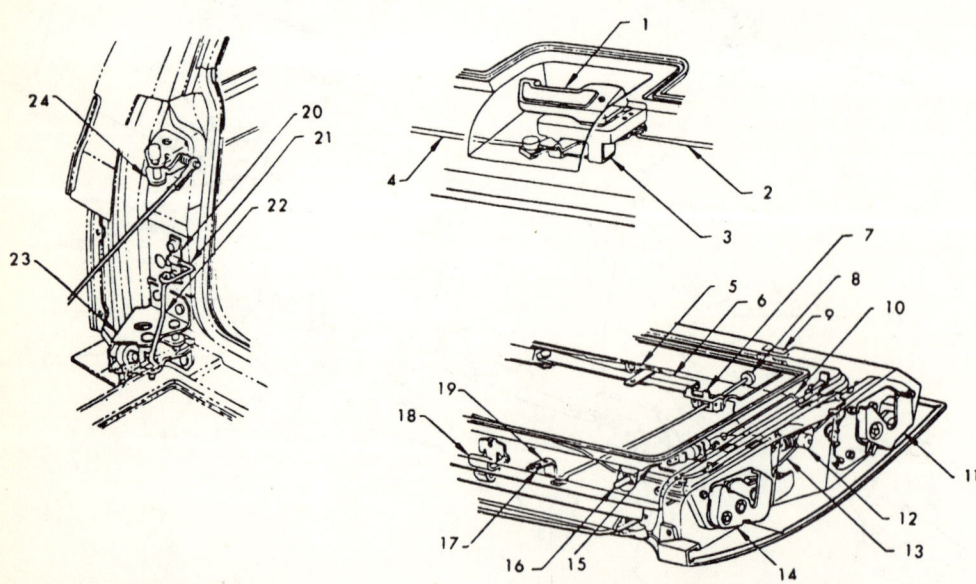

1. Handle—tailgate remote control
2. Rod—tailgate lock remote control
3. Control asm.—tailgate lock remote
4. Rod asm.—tailgate lock remote control to lock left side
5. Cam—tailgate glass regulator
6. Cam—tailgate inner panel
7. Plate—tailgate window guide right side
8. Bumper—tailgate window down stop
9. Retainer asm.—torque rod tailgate hinge
10. Rod—tailgate window blockout
11. Lock asm.—tailgate lower right side
12. Actuator asm.—electric lock
13. Rod asm.—tailgate upper to lower lock connection
14. Lock asm.—tailgate upper right side
15. Rod—tailgate inside locking to lock
16. Knob—door inside locking rod
17. Stop—tailgate window up
18. Retainer asm.—tailgate belt trim support
19. Stop—tailgate window up
20. Retainer—tailgate hinge torque rod link
21. Link—tailgate hinge torque rod
22. Rod—torque tailgate hinge
23. Hinge asm.—tailgate lower lt. side
24. Striker asm.—tailgate upper hinge lt. side

Tailgate assembly 1977 and later models

BODY

Tailgate

The tailgate can be aligned slightly by loosening the hinge and/or the latch bolts, then adjusting them. The bottom of the tailgate on 1977-92 vehicles can be adjusted in or out by adding or removing shims between the hinge and body.

NOTE: *When adjusting the hinge/latch-to-body positions, be sure to use alignment marks as reference points.*

Bumpers

REMOVAL AND INSTALLATION

Front

1. Disconnect the negative battery cable. Raise and support the vehicle safely.
2. Properly support the bumper. As required, disconnect all the necessary electrical connections at the turn signal assemblies and the cornering light housings.
3. On the vehicles without the energy absorber type bumper, remove the bolts from the frame and remove the bumper.
4. On vehicles equipped with the energy absorber type bumper, remove the bolts from the reinforcement to the energy absorber (each side) and remove the bumper assembly.
5. Installation is the reverse of the removal procedure.

Rear

1. Disconnect the negative battery cable. Raise and support the vehicle safely.
2. Properly support the bumper. As required, disconnect all the necessary electrical connections at the tail light assembly housings.
3. On the vehicles without the energy absorber type bumper, remove the bolts from the frame and remove the bumper.
4. On vehicles equipped with the energy absorber type bumper, remove the bolts from the reinforcement to the energy absorber (each side) and remove the bumper assembly.
5. Installation is the reverse of the removal procedure.

Grille

REMOVAL AND INSTALLATION

1. Disconnect the negative battery cable. As required, open the hood.
2. Remove the sheet metal screws that retain the grille assembly to its mounting.
3. On some vehicles, the headlight trim rings may have to be removed before the grille assembly can be removed from the vehicle.
4. Remove the grille assembly from the vehicle.
5. Installation is the reverse of the removal procedure.

Outside Mirrors

NOTE: *The mirror glass face may be replaced by placing a piece of tape over the glass then breaking the mirror face. Adhesive back mirror faces are available.*

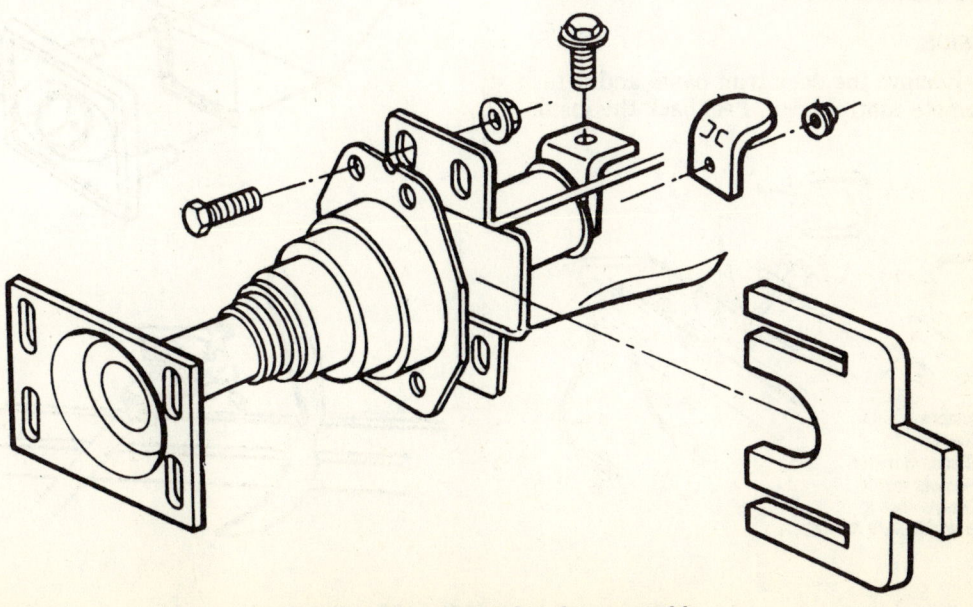

Front bumper energy absorber assembly

436 BODY

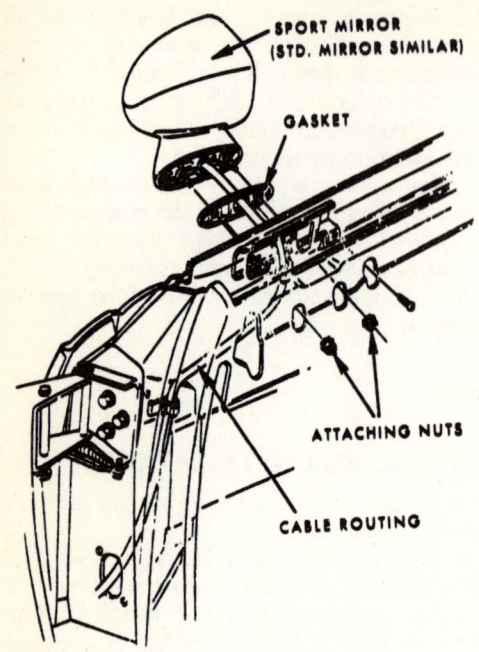

Installation of a right door outside mirror

REMOVAL AND INSTALLATION

Standard Mirror

1. Remove the door trim panel.
2. Remove the mirror base to door outer panel stud nuts and remove the mirror from the door.
3. Install the base gasket and reverse the above to install.

Manual Remote Mirror

LEFT SIDE

1. Remove the door trim panel and detach the remote control lever. Peel back the insulator and water deflector to gain access to the mirror cable.
2. Detach the cable from any retaining tabs in the door.
3. Remove the attaching nuts and remove the mirror and cable assembly from the door.
4. Install the base gasket reverse the above to install.

RIGHT SIDE

1. Remove the door trim panel and detach the remote control lever. Peel back the insulator and water deflector to gain access to the mirror cable.
2. On models, with the instrument panel mounted control, remove the set screw from the control knob.
3. Remove the shroud side finishing panel as follows:
    a. Remove the sill plate screws and sill plate.
    b. Remove the litter container if so equipped.
    c. Remove the screw retaining the hinge pillar pinch-weld upper finishing lace.
    d. Grasp the shroud finishing panel at the forward edge toward the dash panel and pull inward to disengage the plastic retaining clip, and slide the panel rearward.
4. Feed the remote cable through the shroud and rubber conduit between the door and pillar

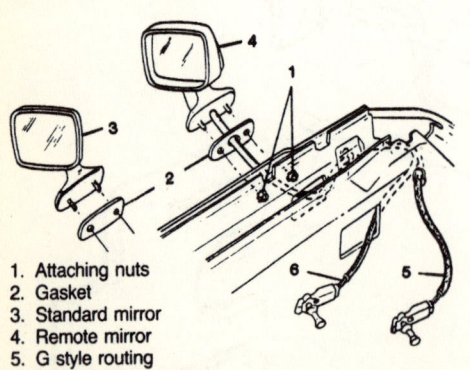

1. Attaching nuts
2. Gasket
3. Standard mirror
4. Remote mirror
5. G style routing
6. B and D style routing

Installation of a left door outside mirror

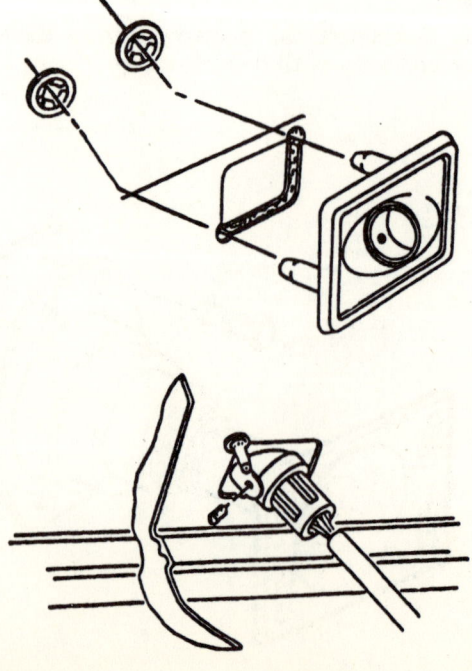

View of the remote mirror cable and escutcheon

BODY

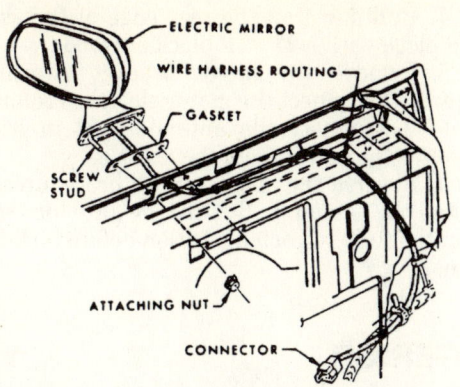

Installation of a power operated outside mirror-left side

and detach the cable from any retaining tabs in the door.

5. Remove the attaching nuts and remove the mirror and cable assembly from the door.

6. Reverse the above to install. Make sure the remote mirror operates properly before installing the trim.

Power Operated Mirror

RIGHT AND LEFT

1. Remove the door trim panel and disconnect the wire harness at the connector. Peel back the insulator pad and water deflector enough to gain access to the wire harness.

2. Detach the harness from any retaining tabs in the door.

3. Remove the attaching nuts and remove the mirror and harness assembly from the door.

4. Reverse the above to install. Make sure mirror operates correctly before installing the door panel.

Antenna

REMOVAL AND INSTALLATION

Manual Type

1. Unscrew the mast from the top of the fender.

2. Unscrew the nut and the bezel from the top of the fender.

3. On later vehicles it may be necessary to remove a bolt/screw which retains the base of the antenna under the fender. This bolt/screw is accessible under the hood.

4. Disconnect the antenna lead from the antenna. On some later vehicles the antenna lead may even plug into another lead under the hood.

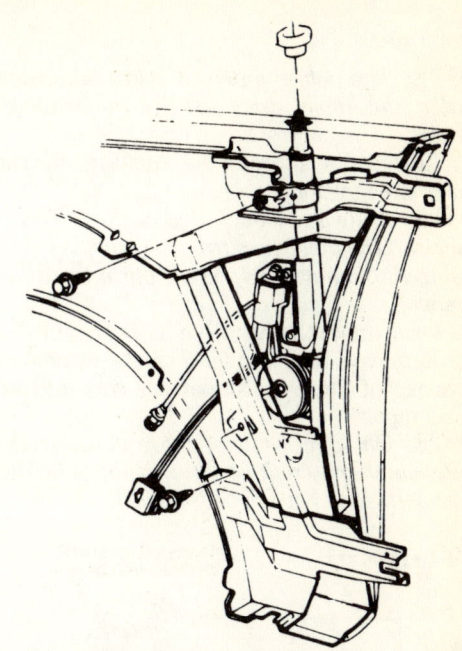

Power antenna assembly installation

5. Reach under the fender and remove the antenna base.

6. When installing the antenna to the fender make sure the retaining nut is tight. A loose antenna or one that does not make good contact at the fender can cause radio interference.

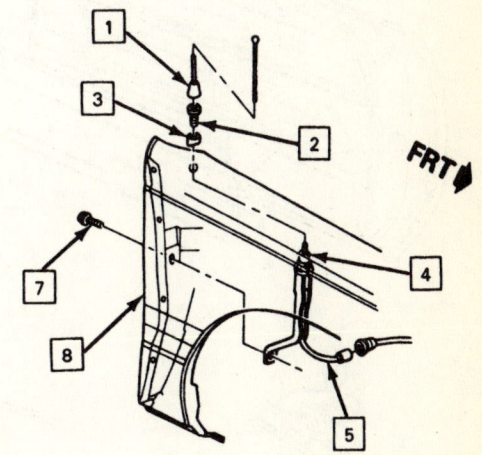

1	ANTENNA MAST	5	ANTENNA LEAD
2	NUT	7	SCREW
3	BEZEL	8	FENDER
4	BASE ASSEMBLY		

Fixed type radio antenna mounting

BODY

Power Type

NOTE: *The power antenna relay is located under the dash area in the convenience center.*

1. Lower the antenna by turning off the radio or the ignition.

NOTE: *If the mast has failed in the UP position, and the mast or entire assembly is being replaced, the mast may be cut off to facilitate removal.*

2. Disconnect the negative battery cable.
3. Remove the fender skirt attaching screws except those to the battery tray and radiator support.

NOTE: *On some vehicles there is an access plate which may reduce the amount of fender skirt bolts that have to be removed.*

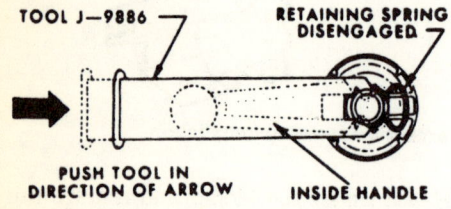

Removing the clip from the inside the door handle

4. Pull down on the rear edge of the skirt and block with a 2" × 4" block of wood.
5. Remove the motor bracket attaching screws. Disconnect the motor electrical connections. Disconnect the antenna lead in wire. Remove the motor from the vehicle.
6. Reverse the above service procedures to install the assembly. Be sure the antenna mast is in the fully retracted position before installation.

INTERIOR

Front Door Panels

REMOVAL AND INSTALLATION

1. Disconnect the negative battery cable. Remove the door handles and the locking knob from the inside of the doors. If equipped with remote control mirrors, remove the remote mirror escutcheon, then disengage the end of the mirror control cable from the escutcheon.

NOTE: *If equipped with door pull handles, remove the screws through the handle into*

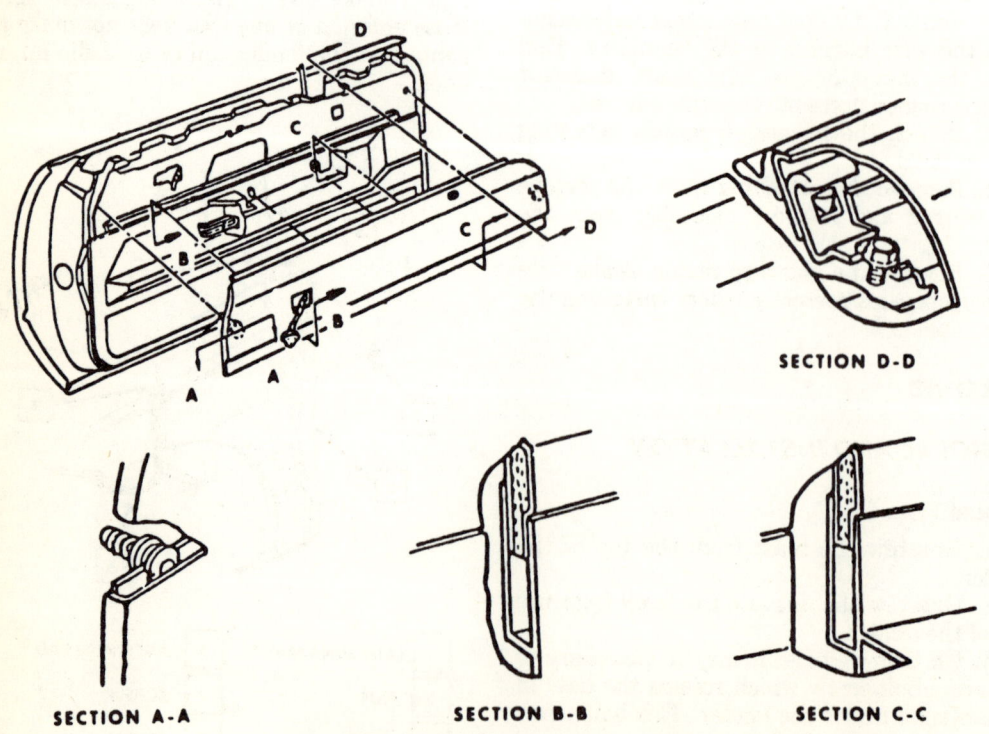

Exploded view of the two-piece door trim

the door inner panel.

2. If equipped with a switch cover plate in the door armrest, remove the cover plate screws, then disconnect the switches and the cigar lighter, if equipped from the electrical harness.

3. If equipped with an integral armrest, remove the screws inserted through the pull cup into the armrest hanger support. If equipped with an armrest applied after the door trim installation, remove the armrest-to-inner panel screws.

4. If equipped with two-piece trim panels, disengage the retainer clips from the front and the rear of the upper trim panel, using tool No. BT-7323A, then lift the upper door trim and slide it slightly rearward to disengage it from the door inner panel at the beltline.

NOTE: *If equipped with electric switches in the door trim panel, disconnect the electrical connectors from the switch assembly.*

5. Along the upper edge of the lower trim panel, remove the mounting screws. At the lower edge of the panel, insert tool No. BT-7323A between the inner panel and the trim panel, then disengage the retaining clips from around the outer perimeter. To remove the lower panel, push the panel down and outward to disengage it from the door.

NOTE: *If equipped with courtesy lights, disconnect the wiring harness.*

6. If equipped with an insulator pad glued to the door inner panel, remove the pad with a putty knife by separating it from the inner panel.

7. To install, reverse the removal service removal procedures.

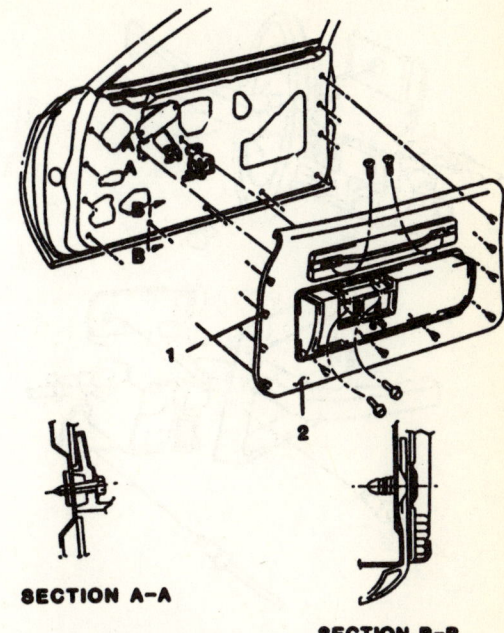

1. Trim fastener locations
2. Door trim assembly

Exploded view of the door trim panel

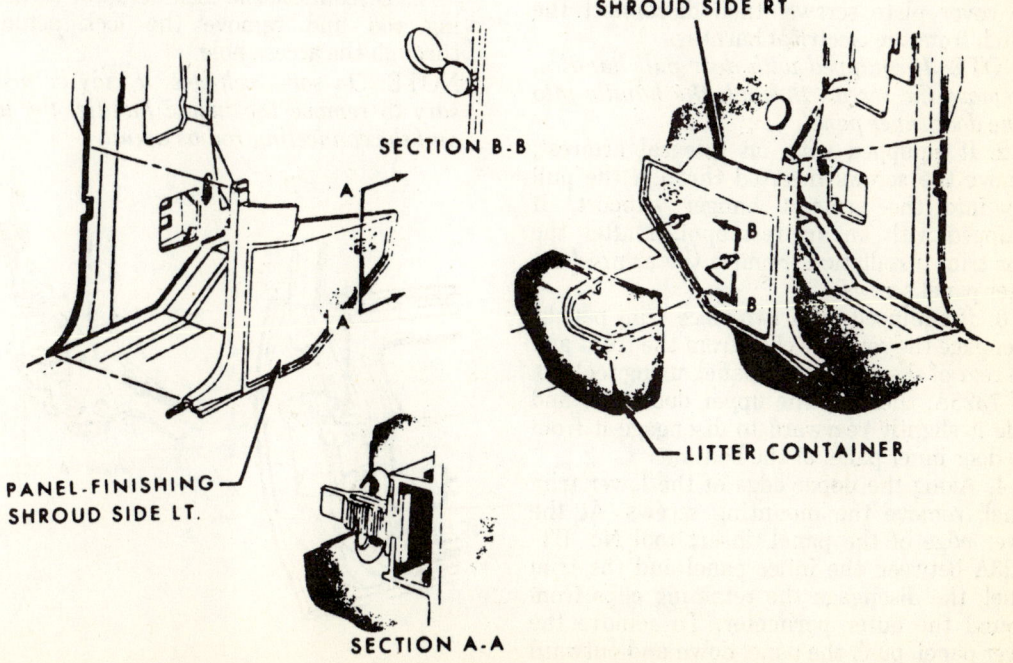

Shroud side finishing panel

BODY

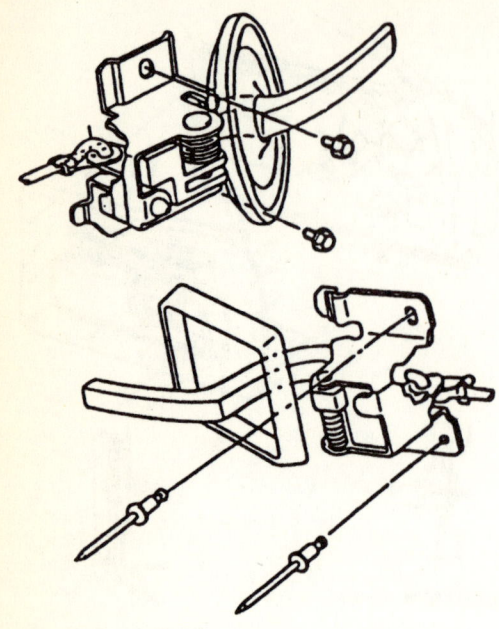

View of the door lock remote control handle

Rear Door Panels

REMOVAL AND INSTALLATION

1. Disconnect the negative battery cable. Remove the door handles and the locking knob from the inside of the doors. If equipped with a switch cover plate in the door armrest, remove the cover plate screws, then disconnect the switch from the electrical harness.

NOTE: *If equipped with door pull handles, remove the screws through the handle into the door inner panel.*

2. If equipped with an integral armrest, remove the screws inserted through the pull cup into the armrest hanger support. If equipped with an armrest applied after the door trim installation, remove the armrest-to-inner panel screws.

3. If equipped with two-piece trim panels, disengage the retainer clips from the front and the rear of the upper trim panel, using tool No. BT-7323A, then lift the upper door trim and slide it slightly rearward to disengage it from the door inner panel at the beltline.

4. Along the upper edge of the lower trim panel, remove the mounting screws. At the lower edge of the panel, insert tool No. BT-7323A between the inner panel and the trim panel, the disengage the retaining clips from around the outer perimeter. To remove the lower panel, push the panel down and outward to disengage it from the door.

NOTE: *If equipped with courtesy lights, disconnect the wiring harness.*

5. If equipped with an insulator pad glued to the door inner panel, remove the pad (with a putty knife) by separating it from the inner panel.

6. To install, reverse the service removal procedures.

Door Locks

The door locks uses a fork bolt lock design which includes a safety interlock feature. The door is securely closed when the door lock fork bolt engages the striker bolt.

REMOVAL AND INSTALLATION

NOTE: *Never attempt to make repairs to the lock actuator assembly, replace the assembly.*

1. Disconnect the negative battery cable. Remove the door trim, then detach the insulator pad, if equipped and the inner panel water deflector enough to access the door lock.

2. Disconnect the electrical connector from the actuator assembly. If equipped with vacuum operated power locks, disconnect the vacuum line from the actuator assembly.

3. Remove the electric lock actuator by performing the following procedures;

   a. Using a center punch, drive the center pins out of pop rivets.

   b. Using a $1/4''$ drill bit, drill the heads off of the pop rivets.

   c. Disconnect the lock actuator connecting rod and remove the lock actuator through the access hole.

NOTE: *On some vehicles, it may be necessary to remove the inside handle, the lock and the connecting rod as a unit.*

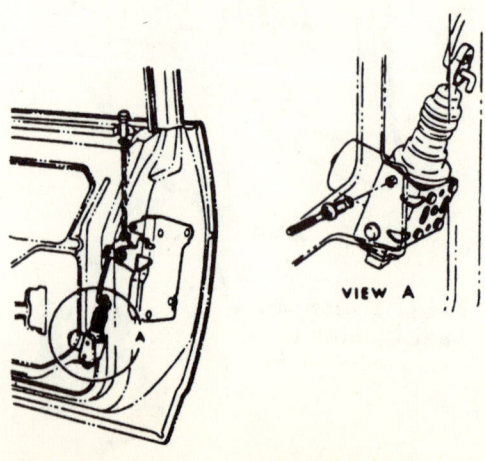

View of the power door lock actuator assembly

BODY 441

4. Installation is the reverse of the removal procedure. When attaching the power door lock actuator to the door, use $1/4''$ x $1/2''$ pop rivets or nuts and bolts.

Front Door Glass

REMOVAL AND INSTALLATION

Closed Style Window Frame

1968-70 Models

1. Disconnect the negative battery cable. Remove the door panel.
2. If equipped, remove the front door ventilator. Loosen the window glass run channel lower retaining bolt. Remove the inner panel cam.
3. Slide the window lower sash channel cam off the window regulator lift arm and balance the arm rollers at the same time.
4. Remove the window outboard of the door upper frame and the window assembly.
5. Installation is the reverse of the service removal procedure.

1971-76 Models

1. Disconnect the negative battery cable. Remove the door panel.
2. Loosen the window stabilizer strips. Place the window in the $3/4$ down position.
3. Remove the window lower sash channel cam to glass retaining stud nuts. Tilt the front edge of the glass downward. Remove the inboard of the door upper frame and the window.
4. Installation is the reverse of the removal procedure. Adjust the window, as required.

1977-92 Models

1. Close the window and tape the glass to the door frame. Disconnect the negative battery cable. Remove the door panel.
2. Remove the bolts retaining the lower sash channel the regulator sash. On coupe doors, remove the rubber down stop at the bottom of the door.
3. Attach the window regulator and lower

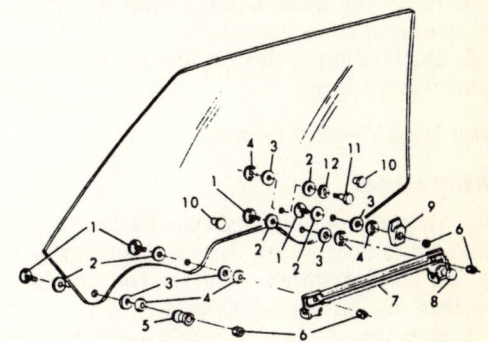

1. Spanner bolt
2. Bushing
3. Washer (plastic)
4. Spanner nut
5. Front up stop
6. Attaching nut
7. Lower sash channel cam
8. Rear guide roller—on lower sash
9. Rear up stop
10. Stabilizer button
11. Front guide roller—on glass
12. Washer (metal)—front guide roller

Front door glass and related components 1977 and later models (bolt type)

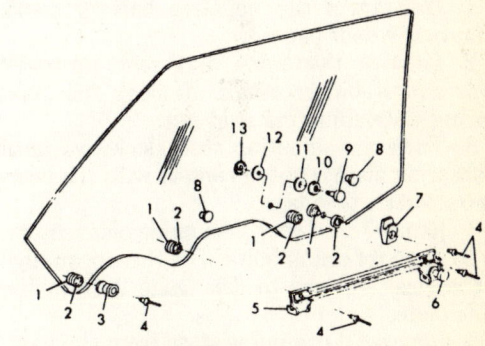

1. Rivet bushing
2. Rivet retainer
3. Front up stop
4. Rivet
5. Lower sash channel cam
6. Rear guide roller—on sash channel
7. Rear up stop
8. Stabilizer button
9. Front guide roller—on glass
10. Washer (metal)—front guide roller
11. Bushing
12. Washer (plastic)
13. Spanner nut

Front door glass and related components 1977 and later models (rivet type)

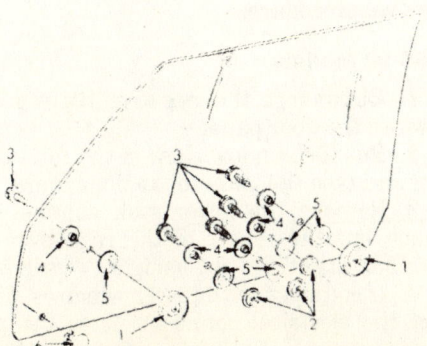

Front door glass and related components 1971-76 models

the window regulator to the full down position. Remove the regulator sash.

4. While supporting the glass, remove the tape and lower the window to the full down position. Slide the glass forward and remove the guide clip from the window run channel.

5. Raise the glass while tilting it forward and remove it from the vehicle.

6. Installation is the reverse of the service removal procedure.

Open Style Window Frame

1968-70 Models

1. Disconnect the negative battery cable. Remove the door panel. Lower the window to the full down position. Remove the up-travel stop from the lower sash channel.

2. Roll the window 1/2 way up and remove the bolts that secure the lower window sash channel to the rear run channel guide plate. Disengage the guide plate from the sash channel.

3. Roll the window completely up. Remove the inner panel cam bolts. Tilt the upper corner of the glass inboard and rotate the glass counterclockwise until the lower sash channel cam is close to being parallel with the beltline.

4. Slide the glass rearward and disengage it from the regulator lift arm. Remove the glass from the vehicle.

5. Installation is the reverse of the removal procedure. Adjust the glass, as required.

1971-76 Models

1. Disconnect the negative battery cable. Remove the door panel.

2. Remove the front and rear up-travel stops and stabilizer strips. Remove the glass bearing plate adjusting stud nut.

3. Turn the adjusting stud clockwise until the bearing plate is out of contact with the bearing button on the glass.

4. Remove the lower sash guide plate assembly to glass retaining bolts. Tilt the upper edge of the glass inboard to disengage it from the guide plate.

5. Remove the window glass from the vehicle by lifting straight up.

6. Installation is the reverse of the removal procedure. Adjust the glass, as required.

ADJUSTMENT PROCEDURE

1. Disconnect the negative battery cable. Remove the door panel.

2. To rotate the window, loosen the front and rear up-stops, adjust the inner panel cam and the up-stops, then tighten the screws.

3. To adjust the window's upper inboard and outboard edge, perform the following procedures:

 a. Position the window in the partially down position.

 b. Loosen the vertical guide upper support (lower) screws, which are accessible through the inner panel access holes.

 c. Loosen the pin assembly screws, the rear up-stop screw and the front belt stabilizer screw.

 d. Adjust the vertical guide upper support and pin assembly (in or out as required, then tighten the screws, adjust and tighten the other components.

NOTE: *When adjusting glass, make sure that it remains inboard of the blow-out clip, when cycled.*

4. If the window is too far forward or rearward, position the window partially down , loosen the vertical guide (upper and lower) screws, then adjust as required.

5. If the window is too high or low in it's Up position, adjust the front and rear up-travel stops.

6. If the window is too high or low in the Down position, adjust the down-travel stop.

7. If the window binds during the up and down operation, adjust the front and/or rear belt stabilizer pin assemblies.

Front Door Regulator

REMOVAL AND INSTALLATION

1968-79 Models

1. Disconnect the negative battery cable. Remove the door panel. Remove the window glass.

2. If equipped with power windows, disconnect the electrical connector from the power window motor.

3. Remove the window regulator retaining bolts. Remove the window regulator from the vehicle.

4. Installation is the reverse of the service removal procedure.

1980-92 Models

1. Disconnect the negative battery cable. Remove the door panel.

2. Put the window glass in the full up position and tape the glass to the door frame.

3. Remove the lower sash channel bolts. Punch out the center of the regulator rivets and than drill them out using a 1/4" drill bit.

4. If equipped with power windows, disconnect the electrical connector from the power motor assembly. Remove the regulator from its mounting.

5. Installation is the reverse of the removal procedure. Be sure to use new rivets or nuts

and bolts to retain the regulator to its mount on the door.

Front Door Electric Window Motor

REMOVAL AND INSTALLATION

1. Disconnect the negative battery cable. To properly remove the power window motor from the vehicle the window regulator must first be removed from the vehicle.
2. Refer to the Window Regulator Removal and Installation procedures for the proper information.

NOTE: *The regulator arm is under tension, due to the mounting of the electric window motor, and if the following operation is not performed, serious damage could result.*

4. Drill a $1/4''$ hole through the regulator backing plate and sector gear. Install a $ú/_{16}''$ bolt through the hole, but do not tighten the nut.
5. Remove the motor to regulator retaining bolts or rivets. Remove the motor from the regulator.
6. Installation is the reverse of the service removal procedure.

Rear Door Glass

REMOVAL AND INSTALLATION

1. Disconnect the negative battery cable. Remove the trim panel. Raise the window to the full up position and tape the glass to the door frame.
2. Remove the lower sash channel retaining bolts. Lower the window to the full down position. Remove the regulator sash.
3. Disengage the front edge of the glass from the glass channel retainer. Slide the glass forward and tilt it up slightly.
4. Using care, remove the glass from its mounting.
5. Installation is the reverse of the removal procedure. Before installing the trim panel, check for proper window operation.

Rear Door Regulator

REMOVAL AND INSTALLATION

1. Disconnect the negative battery cable. Remove the trim panel. Remove the window.
2. Remove the regulator assembly retaining bolts or rivets. Remove the regulator from the vehicle.
3. Installation is the reverse of the service removal procedure.

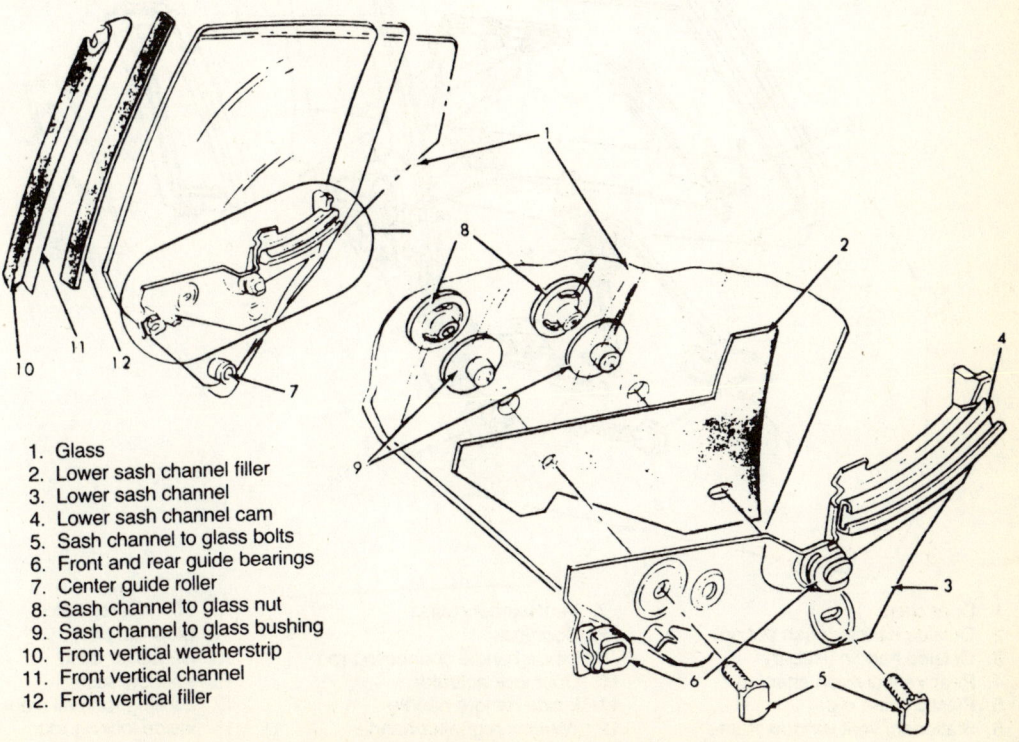

1. Glass
2. Lower sash channel filler
3. Lower sash channel
4. Lower sash channel cam
5. Sash channel to glass bolts
6. Front and rear guide bearings
7. Center guide roller
8. Sash channel to glass nut
9. Sash channel to glass bushing
10. Front vertical weatherstrip
11. Front vertical channel
12. Front vertical filler

Rear window assembly 1971-76 coupe style models

444   BODY

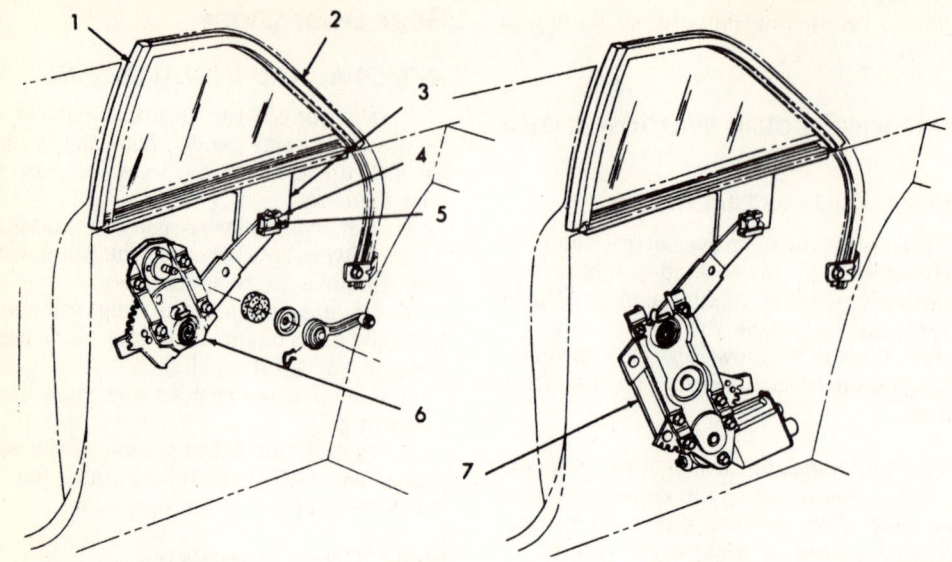

1. Front glass run channel
2. Rear glass run channel
3. Window lower sash
4. Clip retainer
5. Pivot pin
6. Regulator (manual)
7. Regulator (electric)

Rear window assembly 1968-70 coupe style models

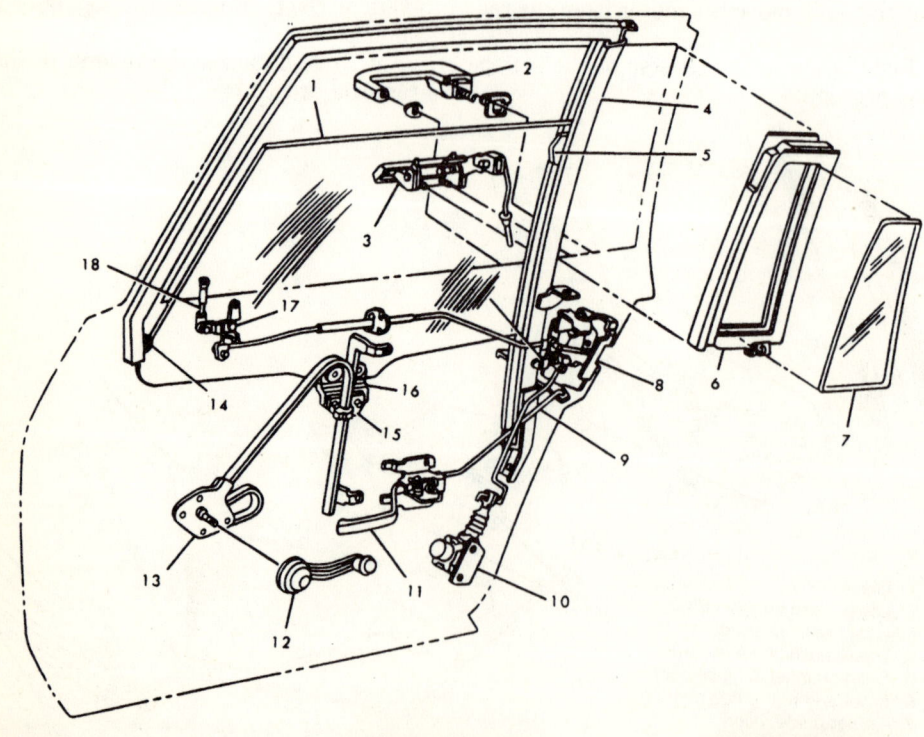

1. Door glass
2. Outside handle (push button)
3. Outside handle (lift bar)
4. Rear vent division channel
5. Plastic guide clip
6. Stationary vent window frame
7. Vent window glass
8. Door lock
9. Inside handle connecting rod
10. Door lock actuator
11. Inside remote handle
12. Window regualtor handle
13. Window regulator
14. Blow out clip
15. Regualtor sash
16. Window sash
17. Locking rod bell crank
18. Inside locking rod

Rear door hardware and related components 1980 and later models

BODY

Rear Door Electric Window Motor

REMOVAL AND INSTALLATION

1. Disconnect the negative battery cable. Remove the trim panel. Remove the window.
2. Disconnect the electrical connection from the electrical motor assembly.
3. Remove the regulator retaining bolts or rivets and remove the regulator and window motor as an assembly.

NOTE: *The regulator arm is under tension, due to the mounting of the electric window motor, and if the following operation is not performed, serious damage could result.*

4. Drill a $1/4''$ hole through the regulator backing plate and sector gear. Install a $ú/_{16}''$ bolt through the hole, but do not tighten the nut.
5. Separate the motor assembly from the window regulator.
6. Installation is the reverse of the service removal procedure.

Inside Rear View Mirror

REPLACEMENT

1. Disconnect the negative battery cable.
2. If the mirror is mounted on the upper windshield moulding, remove the retaining screws. If equipped with a map light, pull the mirror downward and disconnect the light electrical connector. Remove the mirror.
3. If the mirror is attached to the windshield, remove the mounting screw and remove the mirror from the vehicle.
4. Installation is the reverse of the service removal procedure.

Seats

REMOVAL AND INSTALLATION

1. Move the seat to the full up position. Remove the seat retaining bolts.
2. Move the seat to the full back position. Remove the seat retaining bolts.
3. Disconnect the negative battery cable.
4. If equipped with electric seats disconnect the seat motor electrical connector.
5. Disconnect the seat belts from their mountings. With the aid of an assistant, remove the seat from the vehicle.
6. Installation is the reverse of the service removal procedure.

Power Seat Motor

REMOVAL AND INSTALLATION

1. Disconnect the negative battery cable. Disconnect the electrical harness connector.
2. Remove the seat retaining bolts. Remove the seat from the vehicle. Place it upside down on a protected workbench.
3. Disconnect the motor feed wires from the motor control relay.
4. Remove the motor mounting screws and the transmission-to-motor screws, then move the motor away to disengage it from the rubber coupling.
5. Installation is the reverse of the removal procedure.

NOTE: *When installing the motor, make sure that the rubber coupling is properly engaged at the motor and the transmission.*

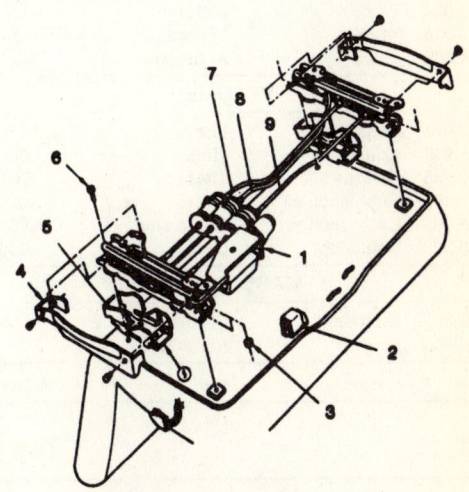

1. Transmission assembly
2. Seat relay
3. Nut
4. Adjuster track lower cover
5. Adjuster track upper cover
6. Adjuster-to-seat frame attaching bolts
7. Horizontal drive cable
8. Rear vertical drive cable
9. Front vertical drive cable

View of the six-way power seat adjusters

Mechanic's Data

11

General Conversion Table

Multiply By	To Convert	To	
LENGTH			
2.54	Inches	Centimeters	.3937
25.4	Inches	Millimeters	.03937
30.48	Feet	Centimeters	.0328
.304	Feet	Meters	3.28
.914	Yards	Meters	1.094
1.609	Miles	Kilometers	.621
VOLUME			
.473	Pints	Liters	2.11
.946	Quarts	Liters	1.06
3.785	Gallons	Liters	.264
.164	Cubic inches	Liters	61.02
16.39	Cubic inches	Cubic cms.	.061
28.32	Cubic feet	Liters	.0353
MASS (Weight)			
28.35	Ounces	Grams	.035
.4536	Pounds	Kilograms	2.20
—	To obtain	From	Multiply by

Multiply By	To Convert	To	
AREA			
6.45	Square inches	Square cms.	.155
.836	Square yds.	Square meters	1.196
FORCE			
4.448	Pounds	Newtons	.225
.138	Ft. lbs.	Kilogram/meters	7.23
1.356	Ft. lbs.	Newton-meters	.737
.113	In. lbs.	Newton-meters	8.844
PRESSURE			
.068	Psi	Atmospheres	14.7
6.89	Psi	Kilopascals	.145
OTHER			
1.104	Horsepower (DIN)	Horsepower (SAE)	.9861
.746	Horsepower (SAE)	Kilowatts (KW)	1.34
1.609	Mph	Km/h	.621
.425	Mpg	Km/L	2.35
—	To obtain	From	Multiply by

Tap Drill Sizes

National Coarse or U.S.S.

Screw & Tap Size	Threads Per Inch	Use Drill Number
No. 5	40	39
No. 6	32	36
No. 8	32	29
No. 10	24	25
No. 12	24	17
1/4	20	8
5/16	18	F
3/8	16	5/16
7/16	14	U
1/2	13	27/64
9/16	12	31/64
5/8	11	17/32
3/4	10	21/32
7/8	9	49/64

National Coarse or U.S.S.

Screw & Tap Size	Threads Per Inch	Use Drill Number
1	8	7/8
1 1/8	7	63/64
1 1/4	7	1 7/64
1 1/2	6	1 11/32

National Fine or S.A.E.

Screw & Tap Size	Threads Per Inch	Use Drill Number
No. 5	44	37
No. 6	40	33
No. 8	36	29
No. 10	32	21

National Fine or S.A.E.

Screw & Tap Size	Threads Per Inch	Use Drill Number
No. 12	28	15
1/4	28	3
6/16	24	1
3/8	28	Q
7/16	20	W
1/2	20	29/64
9/16	18	33/64
5/8	18	37/64
3/4	16	11/16
7/8	14	13/16
1 1/8	12	1 3/64
1 1/4	12	1 11/64
1 1/2	12	1 27/64

MECHANICS DATA

Drill Sizes In Decimal Equivalents

Inch	Decimal	Wire	mm	Inch	Decimal	Wire	mm	Inch	Decimal	Wire & Letter	mm	Inch	Decimal	Letter	mm	Inch	Decimal	mm
1/64	.0156		.39		.0730	49			.1614		4.1		.2717		6.9		.4331	11.0
	.0157		.4		.0748		1.9		.1654		4.2		.2720	I		7/16	.4375	11.11
	.0160	78			.0760	48			.1660	19			.2756		7.0		.4528	11.5
	.0165		.42		.0768		1.95		.1673		4.25		.2770	J		29/64	.4531	11.51
	.0173		.44	5/64	.0781		1.98		.1693		4.3		.2795		7.1	15/32	.4688	11.90
	.0177		.45		.0785	47			.1695	18			.2810	K			.4724	12.0
	.0180	77			.0787		2.0	11/64	.1719		4.36	9/32	.2812		7.14	31/64	.4844	12.30
	.0181		.46		.0807		2.05		.1730	17			.2835		7.2		.4921	12.5
	.0189		.48		.0810	46			.1732		4.4		.2854		7.25	1/2	.5000	12.70
	.0197		.5		.0820	45			.1770	16			.2874		7.3		.5118	13.0
	.0200	76			.0827		2.1		.1772		4.5		.2900	L		33/64	.5156	13.09
	.0210	75			.0846		2.15		.1800	15			.2913		7.4	17/32	.5312	13.49
	.0217		.55		.0860	44			.1811		4.6		.2950	M			.5315	13.5
	.0225	74			.0866		2.2		.1820	14			.2953		7.5	35/64	.5469	13.89
	.0236		.6		.0886		2.25		.1850	13		19/64	.2969		7.54		.5512	14.0
	.0240	73			.0890	43			.1850		4.7		.2992		7.6	9/16	.5625	14.28
	.0250	72			.0906		2.3		.1870		4.75		.3020	N			.5709	14.5
	.0256		.65		.0925		2.35	3/16	.1875		4.76		.3031		7.7	37/64	.5781	14.68
	.0260	71			.0935	42			.1890		4.8		.3051		7.75		.5906	15.0
	.0276		.7	3/32	.0938		2.38		.1890	12			.3071		7.8	19/32	.5938	15.08
	.0280	70			.0945		2.4		.1910	11			.3110		7.9	39/64	.6094	15.47
	.0292	69			.0960	41			.1929		4.9	5/16	.3125		7.93		.6102	15.5
	.0295		.75		.0965		2.45		.1935	10			.3150		8.0	5/8	.6250	15.87
	.0310	68			.0980	40			.1960	9			.3160	O			.6299	16.0
1/32	.0312		.79		.0981		2.5		.1969		5.0		.3189		8.1	41/64	.6406	16.27
	.0315		.8		.0995	39			.1990	8			.3228		8.2		.6496	16.5
	.0320	67			.1015	38			.2008		5.1		.3230	P		21/32	.6562	16.66
	.0330	66			.1024		2.6		.2010	7			.3248		8.25		.6693	17.0
	.0335		.85		.1040	37		13/64	.2031		5.16		.3268		8.3	43/64	.6719	17.06
	.0350	65			.1063		2.7		.2040	6		21/64	.3281		8.33	11/16	.6875	17.46
	.0354		.9		.1065	36			.2047		5.2		.3307		8.4		.6890	17.5
	.0360	64			.1083		2.75		.2055	5			.3320	Q		45/64	.7031	17.85
	.0370	63		7/64	.1094		2.77		.2067		5.25		.3346		8.5		.7087	18.0
	.0374		.95		.1100	35			.2087		5.3		.3386		8.6	23/32	.7188	18.25
	.0380	62			.1102		2.8		.2090	4			.3390	R			.7283	18.5
	.0390	61			.1110	34			.2126		5.4		.3425		8.7	47/64	.7344	18.65
	.0394		1.0		.1130	33			.2130	3		11/32	.3438		8.73		.7480	19.0
	.0400	60			.1142		2.9		.2165		5.5		.3445		8.75	3/4	.7500	19.05
	.0410	59			.1160	32		7/32	.2188		5.55		.3465		8.8	49/64	.7656	19.44
	.0413		1.05		.1181		3.0		.2205		5.6		.3480	S			.7677	19.5
	.0420	58			.1200	31			.2210	2			.3504		8.9	25/32	.7812	19.84
	.0430	57			.1220		3.1		.2244		5.7		.3543		9.0		.7874	20.0
	.0433		1.1	1/8	.1250		3.17		.2264		5.75		.3580	T		51/64	.7969	20.24
	.0453		1.15		.1260		3.2		.2280	1			.3583		9.1		.8071	20.5
	.0465	56			.1280		3.25		.2283		5.8	23/64	.3594		9.12	13/16	.8125	20.63
3/64	.0469		1.19		.1285	30			.2323		5.9		.3622		9.2		.8268	21.0
	.0472		1.2		.1299		3.3		.2340	A			.3642		9.25	53/64	.8281	21.03
	.0492		1.25		.1339		3.4	15/64	.2344		5.95		.3661		9.3	27/32	.8438	21.43
	.0512		1.3		.1360	29			.2362		6.0		.3680	U			.8465	21.5
	.0520	55			.1378		3.5		.2380	B			.3701		9.4	55/64	.8594	21.82
	.0531		1.35		.1405	28			.2402		6.1		.3740		9.5		.8661	22.0
	.0550	54		9/64	.1406		3.57		.2420	C		3/8	.3750		9.52	7/8	.8750	22.22
	.0551		1.4		.1417		3.6		.2441		6.2		.3770	V			.8858	22.5
	.0571		1.45		.1440	27			.2460	D			.3780		9.6	57/64	.8906	22.62
	.0591		1.5		.1457		3.7		.2461		6.25		.3819		9.7		.9055	23.0
	.0595	53			.1470	26			.2480		6.3		.3839		9.75	29/32	.9062	23.01
	.0610		1.55		.1476		3.75	1/4	.2500	E	6.35		.3858		9.8	59/64	.9219	23.41
1/16	.0625		1.59		.1495	25			.2520		6.		.3860	W			.9252	23.5
	.0630		1.6		.1496		3.8		.2559		6.5		.3898		9.9	15/16	.9375	23.81
	.0635	52			.1520	24			.2570	F		25/64	.3906		9.92		.9449	24.0
	.0650		1.65		.1535		3.9		.2598		6.6		.3937		10.0	61/64	.9531	24.2
	.0669		1.7		.1540	23			.2610	G			.3970	X			.9646	24.5
	.0670	51		5/32	.1562		3.96		.2638		6.7		.4040	Y		31/64	.9688	24.6
	.0689		1.75		.1570	22		17/64	.2656		6.74	13/32	.4062		10.31		.9843	25.0
	.0700	50			.1575		4.0		.2657		6.75		.4130	Z		63/64	.9844	25.0
	.0709		1.8		.1590	21			.2660	H			.4134		10.5	1	1.0000	25.4
	.0728		1.85		.1610	20			.2677		6.8	27/64	.4219		10.71			

GLOSSARY

AIR/FUEL RATIO: The ratio of air to gasoline by weight in the fuel mixture drawn into the engine.

AIR INJECTION: One method of reducing harmful exhaust emissions by injecting air into each of the exhaust ports of an engine. The fresh air entering the hot exhaust manifold causes any remaining fuel to be burned before it can exit the tailpipe.

ALTERNATOR: A device used for converting mechanical energy into electrical energy.

AMMETER: An instrument, calibrated in amperes, used to measure the flow of an electrical current in a circuit. Ammeters are always connected in series with the circuit being tested.

AMPERE: The rate of flow of electrical current present when one volt of electrical pressure is applied against one ohm of electrical resistance.

ANALOG COMPUTER: Any microprocessor that uses similar (analogous) electrical signals to make its calculations.

ARMATURE: A laminated, soft iron core wrapped by a wire that converts electrical energy to mechanical energy as in a motor or relay. When rotated in a magnetic field, it changes mechanical energy into electrical energy as in a generator.

ATMOSPHERIC PRESSURE: The pressure on the Earth's surface caused by the weight of the air in the atmosphere. At sea level, this pressure is 14.7 psi at 32°F (101 kPa at 0°C).

ATOMIZATION: The breaking down of a liquid into a fine mist that can be suspended in air.

AXIAL PLAY: Movement parallel to a shaft or bearing bore.

BACKFIRE: The sudden combustion of gases in the intake or exhaust system that results in a loud explosion.

BACKLASH: The clearance or play between two parts, such as meshed gears.

BACKPRESSURE: Restrictions in the exhaust system that slow the exit of exhaust gases from the combustion chamber.

BAKELITE: A heat resistant, plastic insulator material commonly used in printed circuit boards and transistorized components.

BALL BEARING: A bearing made up of hardened inner and outer races between which hardened steel balls roll.

BALLAST RESISTOR: A resistor in the primary ignition circuit that lowers voltage after the engine is started to reduce wear on ignition components.

BEARING: A friction reducing, supportive device usually located between a stationary part and a moving part.

BIMETAL TEMPERATURE SENSOR: Any sensor or switch made of two dissimilar types of metal that bend when heated or cooled due to the different expansion rates of the alloys. These types of sensors usually function as an on/off switch.

BLOWBY: Combustion gases, composed of water vapor and unburned fuel, that leak past the piston rings into the crankcase during normal engine operation. These gases are removed by the PCV system to prevent the buildup of harmful acids in the crankcase.

BRAKE PAD: A brake shoe and lining assembly used with disc brakes.

BRAKE SHOE: The backing for the brake lining. The term is, however, usually applied to the assembly of the brake backing and lining.

BUSHING: A liner, usually removable, for a bearing; an anti-friction liner used in place of a bearing.

BYPASS: System used to bypass ballast resistor during engine cranking to increase voltage supplied to the coil.

CALIPER: A hydraulically activated device in a disc brake system, which is mounted straddling the brake rotor (disc). The caliper contains at least one piston and two brake pads. Hydraulic pressure on the piston(s) forces the pads against the rotor.

CAMSHAFT: A shaft in the engine on which are the lobes (cams) which operate the valves. The camshaft is driven by the crankshaft, via

GLOSSARY

a belt, chain or gears, at one half the crankshaft speed.

CAPACITOR: A device which stores an electrical charge.

CARBON MONOXIDE (CO): A colorless, odorless gas given off as a normal byproduct of combustion. It is poisonous and extremely dangerous in confined areas, building up slowly to toxic levels without warning if adequate ventilation is not available.

CARBURETOR: A device, usually mounted on the intake manifold of an engine, which mixes the air and fuel in the proper proportion to allow even combustion.

CATALYTIC CONVERTER: A device installed in the exhaust system, like a muffler, that converts harmful byproducts of combustion into carbon dioxide and water vapor by means of a heat-producing chemical reaction.

CENTRIFUGAL ADVANCE: A mechanical method of advancing the spark timing by using fly weights in the distributor that react to centrifugal force generated by the distributor shaft rotation.

CHECK VALVE: Any one-way valve installed to permit the flow of air, fuel or vacuum in one direction only.

CHOKE: A device, usually a movable valve, placed in the intake path of a carburetor to restrict the flow of air.

CIRCUIT: Any unbroken path through which an electrical current can flow. Also used to describe fuel flow in some instances.

CIRCUIT BREAKER: A switch which protects an electrical circuit from overload by opening the circuit when the current flow exceeds a predetermined level. Some circuit breakers must be reset manually, while most reset automatically

COIL (IGNITION): A transformer in the ignition circuit which steps up the voltage provided to the spark plugs.

COMBINATION MANIFOLD: An assembly which includes both the intake and exhaust manifolds in one casting.

COMBINATION VALVE: A device used in some fuel systems that routes fuel vapors to a charcoal storage canister instead of venting them into the atmosphere. The valve relieves fuel tank pressure and allows fresh air into the tank as the fuel level drops to prevent a vapor lock situation.

COMPRESSION RATIO: The comparison of the total volume of the cylinder and combustion chamber with the piston at BDC and the piston at TDC.

CONDENSER: 1. An electrical device which acts to store an electrical charge, preventing voltage surges.
2. A radiator-like device in the air conditioning system in which refrigerant gas condenses into a liquid, giving off heat.

CONDUCTOR: Any material through which an electrical current can be transmitted easily.

CONTINUITY: Continuous or complete circuit. Can be checked with an ohmmeter.

COUNTERSHAFT: An intermediate shaft which is rotated by a mainshaft and transmits, in turn, that rotation to a working part.

CRANKCASE: The lower part of an engine in which the crankshaft and related parts operate.

CRANKSHAFT: The main driving shaft of an engine which receives reciprocating motion from the pistons and converts it to rotary motion.

CYLINDER: In an engine, the round hole in the engine block in which the piston(s) ride.

CYLINDER BLOCK: The main structural member of an engine in which is found the cylinders, crankshaft and other principal parts.

CYLINDER HEAD: The detachable portion of the engine, fastened, usually, to the top of the cylinder block, containing all or most of the combustion chambers. On overhead valve engines, it contains the valves and their operating parts. On overhead cam engines, it contains the camshaft as well.

DEAD CENTER: The extreme top or bottom of the piston stroke.

DETONATION: An unwanted explosion of the air/fuel mixture in the combustion chamber caused by excess heat and compression, advanced timing, or an overly lean mixture. Also referred to as "ping".

DIAPHRAGM: A thin, flexible wall separating two cavities, such as in a vacuum advance unit.

DIESELING: A condition in which hot spots in the combustion chamber cause the engine to run on after the key is turned off.

DIFFERENTIAL: A geared assembly which allows the transmission of motion between drive axles, giving one axle the ability to turn faster than the other.

DIODE: An electrical device that will allow current to flow in one direction only.

DISC BRAKE: A hydraulic braking assembly consisting of a brake disc, or rotor, mounted on an axle, and a caliper assembly containing, usually two brake pads which are activated by hydraulic pressure. The pads are forced against the sides of the disc, creating friction which slows the vehicle.

DISTRIBUTOR: A mechanically driven device on an engine which is responsible for electrically firing the spark plug at a predetermined point of the piston stroke.

DOWEL PIN: A pin, inserted in mating holes in two different parts allowing those parts to maintain a fixed relationship.

DRUM BRAKE: A braking system which consists of two brake shoes and one or two wheel cylinders, mounted on a fixed backing plate, and a brake drum, mounted on an axle, which revolves around the assembly. Hydraulic action applied to the wheel cylinders forces the shoes outward against the drum, creating friction, slowing the vehicle.

DWELL: The rate, measured in degrees of shaft rotation, at which an electrical circuit cycles on and off.

ELECTRONIC CONTROL UNIT (ECU): Ignition module, amplifier or igniter. See Module for definition.

ELECTRONIC IGNITION: A system in which the timing and firing of the spark plugs is controlled by an electronic control unit, usually called a module. These systems have no points or condenser.

ENDPLAY: The measured amount of axial movement in a shaft.

ENGINE: A device that converts heat into mechanical energy.

EXHAUST MANIFOLD: A set of cast passages or pipes which conduct exhaust gases from the engine.

FEELER GAUGE: A blade, usually metal, of precisely predetermined thickness, used to measure the clearance between two parts. These blades usually are available in sets of assorted thicknesses.

F-HEAD: An engine configuration in which the intake valves are in the cylinder head, while the camshaft and exhaust valves are located in the cylinder block. The camshaft operates the intake valves via lifters and pushrods, while it operates the exhaust valves directly.

FIRING ORDER: The order in which combustion occurs in the cylinders of an engine. Also the order in which spark is distributed to the plugs by the distributor.

FLATHEAD: An engine configuration in which the camshaft and all the valves are located in the cylinder block.

FLOODING: The presence of too much fuel in the intake manifold and combustion chamber which prevents the air/fuel mixture from firing, thereby causing a no-start situation.

FLYWHEEL: A disc shaped part bolted to the rear end of the crankshaft. Around the outer perimeter is affixed the ring gear. The starter drive engages the ring gear, turning the flywheel, which rotates the crankshaft, imparting the initial starting motion to the engine.

FOOT POUND (ft.lb. or sometimes, ft. lbs.): The amount of energy or work needed to raise an item weighing one pound, a distance of one foot.

FUSE: A protective device in a circuit which prevents circuit overload by breaking the circuit when a specific amperage is present. The device is constructed around a strip or wire of a lower amperage rating than the circuit it is designed to protect. When an amperage higher than that stamped on the fuse is present in the circuit, the strip or wire melts, opening the circuit.

GEAR RATIO: The ratio between the number of teeth on meshing gears.

GENERATOR: A device which converts mechanical energy into electrical energy.

HEAT RANGE: The measure of a spark plug's ability to dissipate heat from its firing end. The higher the heat range, the hotter the plug fires. **HUB:** The center part of a wheel or gear.

HYDROCARBON (HC): Any chemical compound made up of hydrogen and carbon. A major pollutant formed by the engine as a byproduct of combustion.

HYDROMETER: An instrument used to measure the specific gravity of a solution.

INCH POUND (in.lb. or sometimes, in. lbs.): One twelfth of a foot pound.

INDUCTION: A means of transferring electrical energy in the form of a magnetic field. Principle used in the ignition coil to increase voltage.

INJECTION PUMP: A device, usually mechanically operated, which meters and delivers fuel under pressure to the fuel injector.

INJECTOR: A device which receives metered fuel under relatively low pressure and is activated to inject the fuel into the engine under relatively high pressure at a predetermined time.

INPUT SHAFT: The shaft to which torque is applied, usually carrying the driving gear or gears.

INTAKE MANIFOLD: A casting of passages or pipes used to conduct air or a fuel/air mixture to the cylinders.

JOURNAL: The bearing surface within which a shaft operates.

KEY: A small block usually fitted in a notch between a shaft and a hub to prevent slippage of the two parts.

MANIFOLD: A casting of passages or set of pipes which connect the cylinders to an inlet or outlet source.

MANIFOLD VACUUM: Low pressure in an engine intake manifold formed just below the throttle plates. Manifold vacuum is highest at idle and drops under acceleration.

MASTER CYLINDER: The primary fluid pressurizing device in a hydraulic system. In automotive use, it is found in brake and hydraulic clutch systems and is pedal activated, either directly or, in a power brake system, through the power booster.

MODULE: Electronic control unit, amplifier or igniter of solid state or integrated design which controls the current flow in the ignition primary circuit based on input from the pick-up coil. When the module opens the primary circuit, the high secondary voltage is induced in the coil.

NEEDLE BEARING: A bearing which consists of a number (usually a large number) of long, thin rollers.

OHM:(Ω) The unit used to measure the resistance of conductor to electrical flow. One ohm is the amount of resistance that limits current flow to one ampere in a circuit with one volt of pressure.

OHMMETER: An instrument used for measuring the resistance, in ohms, in an electrical circuit.

OUTPUT SHAFT: The shaft which transmits torque from a device, such as a transmission.

OVERDRIVE: A gear assembly which produces more shaft revolutions than that transmitted to it.

OVERHEAD CAMSHAFT (OHC): An engine configuration in which the camshaft is mounted on top of the cylinder head and operates the valves either directly or by means of rocker arms.

OVERHEAD VALVE (OHV): An engine configuration in which all of the valves are located in the cylinder head and the camshaft is located in the cylinder block. The camshaft operates the valves via lifters and pushrods.

OXIDES OF NITROGEN (NOx): Chemical compounds of nitrogen produced as a byproduct of combustion. They combine with hydrocarbons to produce smog.

OXYGEN SENSOR: Used with the feedback system to sense the presence of oxygen in the exhaust gas and signal the computer which can reference the voltage signal to an air/fuel ratio.

PINION: The smaller of two meshing gears.

GLOSSARY

PISTON RING: An open ended ring which fits into a groove on the outer diameter of the piston. Its chief function is to form a seal between the piston and cylinder wall. Most automotive pistons have three rings: two for compression sealing; one for oil sealing.

PRELOAD: A predetermined load placed on a bearing during assembly or by adjustment.

PRIMARY CIRCUIT: Is the low voltage side of the ignition system which consists of the ignition switch, ballast resistor or resistance wire, bypass, coil, electronic control unit and pick-up coil as well as the connecting wires and harnesses.

PRESS FIT: The mating of two parts under pressure, due to the inner diameter of one being smaller than the outer diameter of the other, or vice versa; an interference fit.

RACE: The surface on the inner or outer ring of a bearing on which the balls, needles or rollers move.

REGULATOR: A device which maintains the amperage and/or voltage levels of a circuit at predetermined values.

RELAY: A switch which automatically opens and/or closes a circuit.

RESISTANCE: The opposition to the flow of current through a circuit or electrical device, and is measured in ohms. Resistance is equal to the voltage divided by the amperage.

RESISTOR: A device, usually made of wire, which offers a preset amount of resistance in an electrical circuit.

RING GEAR: The name given to a ring-shaped gear attached to a differential case, or affixed to a flywheel or as part a planetary gear set.

ROLLER BEARING: A bearing made up of hardened inner and outer races between which hardened steel rollers move.

ROTOR: 1. The disc-shaped part of a disc brake assembly, upon which the brake pads bear; also called, brake disc.
2. The device mounted atop the distributor shaft, which passes current to the distributor cap tower contacts.

SECONDARY CIRCUIT: The high voltage side of the ignition system, usually above 20,000 volts. The secondary includes the ignition coil, coil wire, distributor cap and rotor, spark plug wires and spark plugs.

SENDING UNIT: A mechanical, electrical, hydraulic or electromagnetic device which transmits information to a gauge.

SENSOR: Any device designed to measure engine operating conditions or ambient pressures and temperatures. Usually electronic in nature and designed to send a voltage signal to an on-board computer, some sensors may operate as a simple on/off switch or they may provide a variable voltage signal (like a potentiometer) as conditions or measured parameters change.

SHIM: Spacers of precise, predetermined thickness used between parts to establish a proper working relationship.

SLAVE CYLINDER: In automotive use, a device in the hydraulic clutch system which is activated by hydraulic force, disengaging the clutch.

SOLENOID: A coil used to produce a magnetic field, the effect of which is to produce work.

SPARK PLUG: A device screwed into the combustion chamber of a spark ignition engine. The basic construction is a conductive core inside of a ceramic insulator, mounted in an outer conductive base. An electrical charge from the spark plug wire travels along the conductive core and jumps a preset air gap to a grounding point or points at the end of the conductive base. The resultant spark ignites the fuel/air mixture in the combustion chamber.

SPLINES: Ridges machined or cast onto the outer diameter of a shaft or inner diameter of a bore to enable parts to mate without rotation.

TACHOMETER: A device used to measure the rotary speed of an engine, shaft, gear, etc., usually in rotations per minute.

THERMOSTAT: A valve, located in the cooling system of an engine, which is closed when cold and opens gradually in response to engine heating, controlling the temperature of the coolant and rate of coolant flow.

TOP DEAD CENTER (TDC): The point at which the piston reaches the top of its travel on the compression stroke.

GLOSSARY

TORQUE: The twisting force applied to an object.

TORQUE CONVERTER: A turbine used to transmit power from a driving member to a driven member via hydraulic action, providing changes in drive ratio and torque. In automotive use, it links the driveplate at the rear of the engine to the automatic transmission.

TRANSDUCER: A device used to change a force into an electrical signal.

TRANSISTOR: A semi-conductor component which can be actuated by a small voltage to perform an electrical switching function.

TUNE-UP: A regular maintenance function, usually associated with the replacement and adjustment of parts and components in the electrical and fuel systems of a vehicle for the purpose of attaining optimum performance.

TURBOCHARGER: An exhaust driven pump which compresses intake air and forces it into the combustion chambers at higher than atmospheric pressures. The increased air pressure allows more fuel to be burned and results in increased horsepower being produced.

VACUUM ADVANCE: A device which advances the ignition timing in response to increased engine vacuum.

VACUUM GAUGE: An instrument used to measure the presence of vacuum in a chamber.

VALVE: A device which control the pressure, direction of flow or rate of flow of a liquid or gas.

VALVE CLEARANCE: The measured gap between the end of the valve stem and the rocker arm, cam lobe or follower that activates the valve.

VISCOSITY: The rating of a liquid's internal resistance to flow.

VOLTMETER: An instrument used for measuring electrical force in units called volts. Voltmeters are always connected parallel with the circuit being tested.

WHEEL CYLINDER: Found in the automotive drum brake assembly, it is a device, actuated by hydraulic pressure, which, through internal pistons, pushes the brake shoes outward against the drums.

454 INDEX

A

Air cleaner 15
Air conditioning
  Blower 312
  Charging 34
  Compressor 146
  Condenser 147
  Control head 313
  Evaporator 320
  Gauge sets 32
  General service 30
  Leak testing 37
  Preventive maintenance 31
  Safety precautions 32
  Sight glass check 36
  System checks
Air pump 202
Alternator
  Alternator precautions 103
  Operation 100
  Removal and installation 104
  Specifications 107
Alignment, wheel
  Camber 383
  Caster 383
  Toe 384
Antenna 437
Antifreeze 30
Anti-lock brake system 418
Automatic transmission
  Adjustments 351
  Application chart 13
  Back-up light switch 359
  Filter change 350
  Fluid change 350
  Identification 349
  Linkage adjustments 351
  Neutral safety switch 359
  Operation 347
  Pan removal 350
  Removal and installation 363
Axle
  Rear 368

B

Back-up light switch
  Automatic transmission 359
  Manual transmission 343
Ball joints
  Inspection 377
  Removal and installation 377-379
Battery
  Fluid level and maintenance 20
  Jump starting 50, 52
  Removal and installation 109
Bearings
  Axle 368
  Engine 174, 179, 185
  Wheel 381
Bellcrank
Belts 24
Brakes
  Anti-lock brake system 418
  Bleeding 406
  Brake light switch 401
  Combination valve 405
  Disc brakes (Front)
    Caliper 411
    Operating principles 399
    Pads 408
    Rotor (Disc) 413
  Drum brakes (Front)
    Adjustment 401
    Drum 407
    Operating principals 400
    Shoes 407
    Wheel cylinder 407
  Drum brakes (Rear)
    Adjustment 401
    Drum 414
    Operating principals 400
    Shoes 415
    Wheel cylinder 416
  Fluid level 46
  Hydra-Boost system 405
  Hoses and lines 407
  Master cylinder 401
  Operation 398
  Parking brake
    Adjustment 418
    Removal and installation 417
  Power booster
    Operating principals 400
    Removal and installation 405
  Specifications 425
Breaker points 65
Bumpers 435

C

Calipers
  Overhaul 412
  Removal and installation 411
Camber 383
Camshaft and bearings
  Service 172
  Specifications 251
Capacities Chart 54
Carburetor
  Adjustments 81, 251, 286
  Overhaul 249
  Removal and Installation 249
  Specifications 251
Caster 383
Catalytic converter 190

Charging system 100
Chassis electrical system
  Circuit protection 387
  Heater and air conditioning 312
  Lighting 334
  Windshield wipers 38, 322
Chassis lubrication 47
Circuit breakers 338
Circuit protection 337
Clutch
  Adjustment 346
  Operation 344
  Removal and installation 347
Coil (ignition) 76
Combination manifold 139
Combination valve 405
Combination switch 390
Compression testing 116
Compressor (A/C) 146
Condenser
  Air conditioning 147
  Ignition 65
Connecting rods and bearings
  Service 179
  Specifications 127-129
Control arm
  Lower 381, 389
  Upper 380, 389
Cooling system
Crankcase ventilation valve 18, 192
Crankshaft
  Service 185
  Specifications 127-129
Crankshaft damper 164
Cylinder head 148
Cylinders 175

D

Diesel fuel system 17
Disc brakes 408
Distributor
  Breaker points 65
  Condenser 65
  HEI system 71
  Removal and installation 97
Door glass 441, 443
Door locks 440
Doors
  Glass 441, 443
  Locks 440
  Regulator 442, 443
  Removal and installation 426
Door trim panel 438, 440
Drive axle (rear)
  Axle shaft 368-370
  Axle shaft bearing 368-370
  Fluid recommendations 45

Identification 13
Lubricant level 45
Pinion oil seal 372
Ratios 368
Removal and installation 373
Driveshaft
  Front 364
Drive Train 340
Drum brakes 407, 414
Dwell angle 70

E

EGR valve 196
Electrical
  Chassis
    Battery 20
    Circuit breakers 338
    Fuses 338
    Fusible links 337
    Heater and air conditioning 312
    Jump starting 50, 52
    Spark plug wires 65
  Engine
    Alternator 100
    Coil 76
    Distributor 71
    Ignition module 77, 97
    Starter 109
Electronic Ignition 71
Electronic Level Control System 387
Emission controls
  Air Management 204
  Air pump 202
  Catalytic Converter 190
  Computer Command Control
  System 213
  Early Fuel Evaporation 209
  Evaporative canister 194-196
  Exhaust Gas Recirculation (EGR)
  system 196
  Oxygen (O_2) sensor 228
  PCV valve 18, 192
  Thermostatically controlled air
  cleaner 199
  Transmission controlled spark
  system 207
Engine
  Camshaft 172
  Combination manifold 139
  Compression testing 116
  Connecting rods and bearings 179
  Crankshaft 185
  Crankshaft damper 164
  Cylinder head 148
  Cylinders 175
  Design 113
  Exhaust manifold 143

Fluids and lubricants 39
Flywheel 188
Front (timing) cover 164
Front seal 168
Identification 6-10
Intake manifold 139
Lifters 157
Main bearings 185
Oil pan 158
Oil pump 161
Overhaul tips 113
Piston pin 177
Pistons 175
Rear main seal 182
Removal and installation 117
Rings 177
Rocker cover 133
Spark plug wires 65
Specifications 119
Thermostat 138
Timing chain and gears 169, 170
Tools 114
Valve guides 154
Valve lash adjustment 136
Valve lifters 157
Valves 153
Valve springs 156
Valve stem oil seals 156
Water pump 147
Evaporative canister 194-197
Evaporator 320
Exhaust Manifold 143
Exhaust pipe 189
Exhaust system 188

F

Filters
   Air 15
   Crankcase 193
   Fuel 16, 17, 292
   Oil 43
Firing orders 65
Flashers 378
Fluids and lubricants
   Automatic transmission 45
   Battery 20
   Chassis greasing 47
   Coolant 46
   Drive axle 45
   Engine oil 39
   Fuel 42
   Manual transmission 44
   Master cylinder
      Brake 46
   Power steering pump 47
Flywheel and ring gear 188
Front bumper 435

Front brakes 407, 408
Front suspension
   Ball joints 377-379
   Description 374
   Knuckles 381
   Lower control arm 381
   Shock absorbers 376
   Springs 376
   Stabilizer bar 380
   Upper control arm 380
   Wheel alignment 383
Front wheel bearings 381
Fuel injection
   Fuel meter cover 290
   Fuel pump 289
   Idle air control valve 291
   Injectors 290
   Operation 286
   Relieving fuel system pressure 17
   Throttle body 290
Fuel filter 16
Fuel pump
   Electric 289
   Mechanical 248, 292
Fuel system
   Carbureted 249
   Diesel 17, 292
   Gasoline Fuel injection 286
Fuel tank 297
Fuses and circuit breakers 338
Fusible links 337

G

Gearshift linkage
   Adjustment
      Automatic 351
      Manual 341
Generator (See alternator)
Glass
   Door 440, 443
Glossary 448
Grille 435

H

Hazard flasher 378
Headlights 334
Headlight switch 332
Heater
   Blower 312
   Control head 318
   Core 313
Hoisting 51
Hood 428
Hoses

INDEX

Brake 407
Coolant 24
How to Use This Book 1

I

Identification
  Axle 13
  Engine 6-10
  Serial number 6
  Transmission
    Automatic 11-13
    Manual 11-13
  Vehicle 6
Idle speed and mixture adjustment 81, 91
Idler arm 397
Ignition
  Coil 76, 96
  Electronic 71, 91
  Lock cylinder 392
  Module 77, 97
  Switch 334
  Timing 78
Injectors, fuel 290
Instrument cluster 326
  Cluster 326
  Radio 320
  Speedometer cable 332
Intake manifold 139

J

Jacking points 51
Jump starting 50, 52

K

Knuckles 381

L

Lighting
  Headlights 334
  Signal and marker lights 336
Lower ball joint 379
Lower control arm 381, 389
Lubrication
  Automatic transmission 45
  Body 47
  Chassis 47
  Differential 45
  Engine 39

Manual transmission 44

M

Main bearings 185
Maintenance intervals 53-57
Manifolds
  Combination 139
  Intake 139
  Exhaust 143
Manual steering gear
  Adjustments 395
  Removal and installation 395
Manual transmission
  Application chart 13
  Identification 341
  Linkage adjustment 341
  Operation 340
  Removal and installation 343
Marker lights 336
Master cylinder
  Brake 401
Mechanic's data 446
Mirrors 435, 445
Module (ignition) 77, 97
Muffler 189
Multi-function switch 390

N

Neutral safety switch 359

O

Oil and fuel recommendations 39
Oil and filter change (engine) 43
Oil level check
  Differential 45
  Engine 42
  Transmission
    Automatic 45
    Manual 44
Oil pan 158
Oil pump 161
Oxygen (O_2) sensor 228

P

Parking brake 417
Piston pin 177
Pistons 175
Pitman arm 397

Pivot pins 177
PCV valve 18, 192
Points 65
Power brake booster 405
Power seat motor 445
Power steering gear
   Adjustments 395
   Removal and installation 395
Power steering pump
   Fluid level 47
   Removal and installation 396
Power windows 443, 445
Preventive Maintenance Charts 53-57
Pushing 50

R

Radiator 147
Radiator cap 29
Radio 320
Rear axle
   Axle shaft 368-370
   Axle shaft bearing 368-370
   Fluid recommendations 45
   Identification 13
   Lubricant level 45
   Pinion oil seal 372
   Ratios 368
   Removal and installation 373
Rear brakes 414
Rear bumper 435
Rear main oil seal 182
Rear suspension
   Control arms 389
   Electronic Level Control 387
   Shock absorbers 386
   Springs 385-386
Rear wheel bearings 368
Regulator
   Operation 105
   Removal and installation 105
   Testing and adjustment 106
Rings 177
Rocker arms 135
Rotor (Brake disc) 413
Routine maintenance 15

S

Safety notice 1
Seats 445
Serial number location 6
Shock absorbers 376, 386
Spark plugs 58
Spark plug wires 65
Special tools 4

Specifications Charts
   Alternator and regulator 107
   Brakes 417
   Camshaft 125-126
   Capacities 54
   Carburetor 251
   Crankshaft and connecting rod 127-129
   Fastener markings and torque standards 115
   General engine 118-120
   Piston and ring 130
   Preventive Maintenance 53-57
   Starter 111
   Torque 131
   Tune-up 59-61
   Valves 123
   Wheel alignment 383
Speedometer cable 332
Springs 376, 385, 386
Stabilizer bar 380
Starter
   Removal and installation 109
   Specifications 111-112
Steering column 393
Steering gear
   Manual 395
   Power 395
Steering knuckles 381
Steering linkage
   Idler arm 397
   Pitman arm 397
   Relay rod 397
   Tie rod ends 396
Steering wheel 390
Stripped threads 114
Suspension 374, 384
Switches
   Back-up light 343, 359
   Brake light 401
   Headlight 332
   Ignition switch 334
   Multi-function switch 390
   Windshield wiper 332

T

Tailgate 431
Tailpipe 189
Thermostat 138
Throttle body 290
Tie rod ends 396
Timing (ignition) 78
Timing chain and gears 169, 170
Timing gear cover 164
Tires
   Design 38
   Inflation 38
   Rotation 39

Storage 39
Toe-in 334
Tools
Torque specifications 131
Towing 50
Trailer towing 49
Transmission
   Application charts 13
   Automatic 347
   Manual 340
   Routine maintenance 44-45
Trunk lid 431
Tune-up
   Condenser 65
   Distributor 71
   Dwell angle 70
   Idle speed 81, 91
   Ignition timing 78
   Points 65
   Procedures 58
   Spark plugs and wires 58-65
   Specifications 59-61
Turn signal flasher 338
Turn signal switch 390

U

U-joints
   Identification 364
   Overhaul 365
   Replacement 365
Understanding the manual transmission 340
Upper ball joint 377
Upper control arm 380, 389

V

Vacuum diagrams 231
Valve guides 154
Valve lash adjustment 79
Valve lifters 157
Valve service 153
Valve specifications 123
Valve springs 156
Vehicle identification 6

W

Water pump 147
Wheel alignment
   Adjustment 383
   Specifications 383
Wheel bearings
   Front wheel 381
   Rear wheel 368
Wheel cylinders 407, 416
Window glass 441, 443
Window regulator 441, 443
Windshield wipers
   Arm 322
   Blade 322
   Linkage 324
   Motor 322
   Rear window wiper 326
   Refills 38
   Windshield wiper switch 332
Wiring
   Spark plug 65
   Trailer 337